JOURNAL OF REPRODUCTION AND FERTILITY

SUPPLEMENT 54

REPRODUCTION IN DOMESTIC RUMINANTS IV

Proceedings of the Fifth International Symposium
on
Reproduction in Domestic Ruminants

Colorado Springs, Colorado, USA
1–5 August 1998

Edited by W. W. Thatcher, E. K. Inskeep, G. D. Niswender
and C. Doberska

COVER ILLUSTRATION

Cover figures were provided by authors of papers in the Neuroendocrine Relationships, Reproductive Technology and The Corpus Luteum sessions. Top: schematic diagram of the hormonal interactions in the gonadotrope; Centre: the first calves produced after artificial insemination of semen that had been sorted to contain the X or Y chromosome (i.e. 'sexed semen'); Bottom: representation of the interactions of the protein kinase A and protein kinase C second messenger pathways on the synthesis and biological activity of steroidogenic acute regulatory (StAR) protein in luteal cells.

Journal of **REPRODUCTION** *and fertility*
1999

First published 1999

ISSN 0449 3087
ISBN 0 906545 34 X

JOURNALS OF REPRODUCTION AND FERTILITY LTD

The Journals of Reproduction and Fertility Ltd produce the *Journal of Reproduction and Fertility* and *Reviews of Reproduction.*

The *Journal of Reproduction and Fertility* publishes novel and significant papers on the molecular biology, biochemistry and physiology of reproduction, including lactation, and early embryogenesis in man and other animals. The *Journal* is the official organ of the Society for the Study of Fertility.

The *Journal* also publishes Supplements which are distinct from the regular issues, are not associated with any particular volume and are known by their serial number and date.

Reviews of Reproduction publishes topical, concise, well-illustrated reviews, contributed by leading research workers, on basic mechanisms in reproductive biology and comments on recent developments in the field, new hypotheses and highlights of scientific meetings. Articles are commissioned by the Editorial Board in consultation with the International Advisory Panel, and are peer-reviewed by experts in the field.

Published by **The Journals of Reproduction and Fertility Ltd.**

Agents for distribution: **Portland Press, PO Box 32, Commerce Way, Whitehall Industrial Estate, Colchester, CO2 8HP, Essex, UK; Tel: 44(0)1206 796351; FAX: 44(0)1206 799331; e-mail: sales@portlandpress.co.uk**

Printed by **Cambridge University Press**

CONTENTS

FOLLICULAR DEVELOPMENT

NEUROENDOCRINE RELATIONSHIPS

COMPARATIVE REPRODUCTIVE FUNCTION: IMPLICATIONS FOR MANAGEMENT

THE CORPUS LUTEUM

MALE FUNCTION AND FERTILITY

EMBRYONIC SURVIVAL

LOCAL CELLULAR AND TISSUE COMMUNICATION

NUTRITION AND METABOLIC SIGNALLING

REPRODUCTIVE TECHNOLOGY

ABSTRACTS

PREFACE

The Fifth International Symposium on Reproduction in Domestic Ruminants was held in Colorado Springs, Colorado, USA, on August 1–5, 1998. This meeting followed the general format of previous, highly successful symposia which had been convened in the Blue Mountains in Australia (1980), Ithaca, New York, USA (1986), Nice, France (1990), and Townsville, Australia (1994). This supplement contains the proceedings of the Fifth Symposium where 38 scientific presentations were made in nine scientific sessions which included: Follicular Development; Neuroendocrine Relationships; Comparative Reproductive Function; Implications for Management; The Corpus Luteum; Male Function and Fertility; Embryonic Survival; Local Cellular and Tissue Communication; Nutrition and Metabolic Signalling; and Reproductive Technology. Forty abstracts were also presented in two poster sessions and these abstracts are included in these proceedings.

A highlight of the Fifth International Symposium was the presentation of the first Pharmacia & Upjohn Award to Dr William Hansel for his many contributions to our understanding of reproduction in domestic ruminants. Seven financial sponsors made important contributions towards a very pleasant, thought-provoking meeting.

ORGANIZING COMMITTEE

Local Arrangements:
G. D. Niswender
R. V. Anthony
T. M. Nett
K. P. McNatty
H. R. Sawyer
K. A. Thomas

Program Co-Chairs:
E. K. Inskeep
W. W. Thatcher

Members:
I. Clarke, Australia
M-A. Driancourt, France
D. L. Hamernik, USA
G. King, Canada
A. Lopez-Sebastian, Spain
G. B. Martin, Australia
Y. Mori, Japan
L. Zarco-Quintero, Mexico
R. J. Scaramuzzi, UK
D. Schams, Germany
M. Shemesh, Israel
M-A. Sirard, Canada
D. C. Wathes, UK

The organisers gratefully acknowledge financial sponsorship from the following companies:

InterAg, Hamilton, New Zealand
Intervet International BV, Boxmeer, The Netherlands
Monsanto Company, St Louis, Missouri, USA
Pharmacia & Upjohn Company, Kalamazoo, Michigan, USA
Select Sires, Plain City, Ohio, USA
United States Department of Agriculture, Washington, DC, USA
XY, Inc., Fort Collins, Colorado, USA

DELEGATES

Brenda Alexander
Mohammad Ali
Divakar Ambrose
Russell V. Anthony
David Armstrong
Daniel R. Arnold
Alejandro Arreguin-Arevalo
Edward Austin
Ciro M. Barros
Ross Bathgate
Sylvie Bilodeau-Goeseels
Mario Binelli
Karin Bodensteiner
Olga U. Bolden
Ruth Braw-Tal
Chris Burke
W. Ronald Butler
Bruce K. Campbell
Fernando J. Cavazos
Daniel Cavestany
Young G. Chung
Robert Collier
Robert Cushman
Robert A. Dailey
Michael L. Day
Mel DeJarnette
Francisco J. Diaz
Steph M. Dieleman
Michael G. Diskin
Christine Doberska
Hilary Dobson
William J. Enright
Alex Evans
Lee A. Fitzpatrick
Anthony Flint
Marcel Frajblat
Carlos Galina
H. Allen Garverick
Thomas Geary
Jin Gong
Robert L. Goodman
Hari O. Goyal
Carlos G. Gutierrez
Harold D. Hafs
William Hansel
Dale F. Hentges
Ali Honaramooz
Wilfredo Huanca
Keith Inskeep
James Ireland
Hector Jimenez-Severiano
Maureen Keller-Wood
David Kiesling
James E. Kinder
Jeffrey W. Koch
Freddie Kojima
Lisa J. Kulick
Carlos Lamothe
Elizabeth A. Lane
John W. Lemaster
Paul E. Lewis
Sharon K. Lewis
Hugo Leyva-Ocariz
Antonio Lopez-Sebastian
Donald D. Lunstra
Jacques G. Lussier
Chris McGrath
Eric W. McIntush
Roni Mamluk
Ricardo C. Mattos
Monika Mihm
Robert A. Milvae
Akio Miyamoto
Frederico P. Moreira
Gary Moss
Martin L. Mussard
Terry Nett
Gordon D. Niswender
Bruce J. Nosky
Troy Ott
Andrew Padula
Joy Pate
David J. Patterson
George A. Perry
Susan Powell
James K. Pru
Richard Pursley
Ronald D. Randel
Carl C. Rasor
Norman Rawlings
William Ricke
Andy J. Roberts
Lindsay Robertson
Robert S. Robinson
Helen Rodgers
M. Keith Rollyson
Miriam Rosenberg
Janice E. Rowell
Siwat Sangsritavong
Roberto Sartori Filho
Heywood R. Sawyer
Brian L. Sayre
Rex J. Scaramuzzi
Roy W. Silcox
Patrick Silva
Inderjeet Singh
Marc-Andre Sirard
A. Lowell Slyter
George W. Smith
Michael M. Smith
Marsha C. Sousa
Randy L. Stanko
John N. Stellflug
Grant Stone
Fredrick Stormshak
Trista A. Strauch
Rob Taft
William Thatcher
Milton G. Thomas
Jose Luiz M. Vasconlelos
William Vivanco
Alvin C. Warnick
Jonathan E. Wheaton
Scott Whisnant
Jack Whittier
Anne Wiley
Gary Williams
Suzannah A. Williams
Milo C. Wiltbank
Katie Woad
Stacey L. Wood
Joel V. Yelich

FOLLICULAR DEVELOPMENT

Chair
M. A. Driancourt

Journal of Reproduction and Fertility Supplement **54**, 3–16

Control of early ovarian follicular development

K. P. McNatty[1], D. A. Heath[1], T. Lundy[1], A. E. Fidler[1], L. Quirke[1], A. O'Connell[1], P. Smith[1], N. Groome[2] and D. J. Tisdall[3]

[1]Wallaceville Animal Research Centre, PO Box 40063, Upper Hutt, New Zealand; [2]Oxford Brookes University, Gipsy Lane, Headington, Oxford, OX3 0BP, UK; [3]Virus Research Unit, Department of Microbiology, University of Otago, PO Box 56, Dunedin, New Zealand

Early follicular growth refers to the development of an ovarian follicle from the primordial to early antral phase. In sheep and cows these phases of growth can be classified by the configuration of granulosal cells in the largest cross–section of the follicle as types 1 (primordial), 1a (transitory) 2 (primary), 3 and 4 (preantral) and 5 (early antral). Follicles classified as type 1 may be highly variable within each species with respect to number of granulosal cells and diameter of oocyte. Much of the variation in granulosal cell composition of type 1 follicles may occur at formation and this may account for the variability in granulosal cell composition throughout subsequent stages of growth. There appear to be important differences among species (for example sheep and cattle) in the number and function of granulosal cells relative to the diameter of the oocyte during the initiation of follicular growth. There is evidence that most, if not all, of the growth phases from types 1 to 5 are gonadotrophin-independent and that follicles develop in a hierarchical manner. In sheep, cows and pigs, numerous growth factor, growth factor receptor and gonadotrophin receptor mRNAs and peptides (for example c-kit, stem cell factor, GDF-9, β_B and β_A activin/inhibin subunit, α inhibin subunit, follistatin, FGF-2, EGF, EGF-R, TGF$\beta_{1, 2 \text{ and } 3}$ FSH-R and LH-R) are expressed in a phase of growth (for example types 1–5)-specific and cell-specific manner. However, the roles of many of these factors remain to be determined.

Introduction

The aim of this review is to describe the localization of growth factors and receptors for both growth factors and gonadotrophins during the development of an ovarian follicle from the primordial to early antral phase of growth. Particular emphasis is focused on current understanding of early follicular growth in domestic ruminants. Where appropriate some references to the extensive literature in other species is included.

Classification of Early Follicular Growth Stages

A classification system for small bovine follicles was described by Braw-Tal and Yossefi (1997). In this classification system preantral follicles and the smallest antral follicles are classified as types 1–5: type 1 refers to primordial follicles (one layer of 'flattened' granulosal cells, granulosal cell), type 1a to transitory follicles (one layer of cells that are a mixture of 'flattened' and cuboidal granulosal cell), type 2 to primary follicles (one or two layers of cuboidal granulosal cells), type 3 to small preantral (two to four layers of granulosal cells), type 4 to large preantral (four to six layers of granulosal cells) and type 5 to small antral follicles (more than five layers of granulosal cells). The inclusion of type 1a follicles acknowledges the finding that in cows 82.5% of the follicle population with one layer of

Table 1. Classification and characterization of small ovine follicles

Follicle (type)	n	Layers of granulosal cells (shape)	Number of granulosal cells in largest cross-section	Total number of granulosal cells in follicle	Oocyte diameter (μm)	Follicle diameter (μm)
Primordial	195	1	4^{a}	16^{a}	34.6^{a}	40.8^{a}
(1)		(all flattened)	(1,11)	(3,52)	(22.8,52.3)	(28.1,60.5)
Transitory	53	1	9^{b}	39^{b}	40.6^{b}	50.8^{b}
(1a)		(one or more cuboidal)	(4,18)	(9,136)	(27.3,53.0)	(37.3,64.0)
Primary	109	1–2	19^{c}	128^{c}	52.1^{c}	75.2^{c}
(2)		(all cuboidal)	(3,49)	(30,520)	(31.0,80.0)	(49.7,118.8)
Small preantral	38	> 2–4	64^{d}	637^{d}	72.9^{d}	128.5^{d}
(3)			(19,152)	(127,2174)	(40.6,92.0)	(63.5,191.0)
Large preantral	18	> 4–6	187^{e}	2104^{e}	87.8^{e}	194.1^{e}
(4)			(116,308)	(1090,3464)	(76.9,100.2)	(164.2,256.3)
Small antral	23	> 5	475^{f}	11649^{f}	118.8^{f}	326.9^{f}
(5)			(199,1128)	(3425,51447)	(90.9,141.6)	(191.5,450.0)

Values are geometric means (and ranges). Values in columns not sharing a common alphabetical superscript are significantly different ($P < 0.05$, ANOVA on $\log_e$ transformed data). Data from Lundy *et al.* (1999).

Table 2. Expression of growth factors and receptors during early follicular growth in sheep

Cell type		C-kit	SCF	β_B activin/ inhibin	FSHR	α-Inhibin	Follistatin	β_B activin/ inhibin	TGFβ 1	TGFβ 2/3	FGF_2[+]	LHR
Granulosal												
Follicle type	1		● ⊙					⊙				
	1a		● ⊙	● ⊙				⊙				
	2		● ⊙	● ○	● ⊙		⊙	⊙	⊙	⊙	⊙	
	3		● ⊙	● ○	● ⊙	● ⊙	● ⊙	⊙	⊙	⊙	⊙	
	4	● ⊙	● ⊙	● ○	● ⊙	● ⊙	● ⊙	⊙	⊙	⊙	⊙	
	5	● ⊙	● ⊙	● ○	● ⊙	● ⊙	● ⊙	● ⊙	⊙	⊙	⊙	
Oocyte		● ⊙	⊙	○			⊙	⊙	⊙	⊙	⊙	
		(1–5)	(1–5)								(1–2)	
Theca interna									● ⊙	● ⊙	⊙	● ⊙
												(4–5)

●, mRNA; ⊙, protein; ○, localized increasingly in oocyte plasma membrane and zona pellucida
( ) = follicle type in which expression is observed, [+]Data from studies in cows, SCF = stem cell factor

granulosal cells has a least one cuboidal granulosal cell (317 follicles examined; van Wezel and Rogers, 1996). In ewes a similar analysis revealed that 24.2% of follicles with one layer of granulosal cells (n = 215 follicles) contained one or more cuboidal cells (Lundy *et al.*, 1998). A classification system for small ovarian follicles in the ewe is shown in Table 1.

The results for primordial follicles in ewes with respect to oocyte diameter and the number of granulosal cells in the largest cross-section are similar to those for cows (Braw-Tal and Yossefi, 1997). The total number of granulosal cells around the oocyte in the ewe was determined using the nucleator and fractionator techniques of Gundersen (1988). The total population of granulosal cells in primordial follicles together with oocyte diameter are highly variable (Table 1). The variability in granulosal cells in type 1 follicles may arise at the time when follicles are first forming (Hirshfield and De Santi, 1995). There appear to be important differences among species (for example between sheep and cattle) in both the number and activity of granulosal cells relative to the diameter of the oocyte during the initiation of follicular growth (Braw-Tal and Yossefi, 1997; Lundy *et al.*, 1998). In

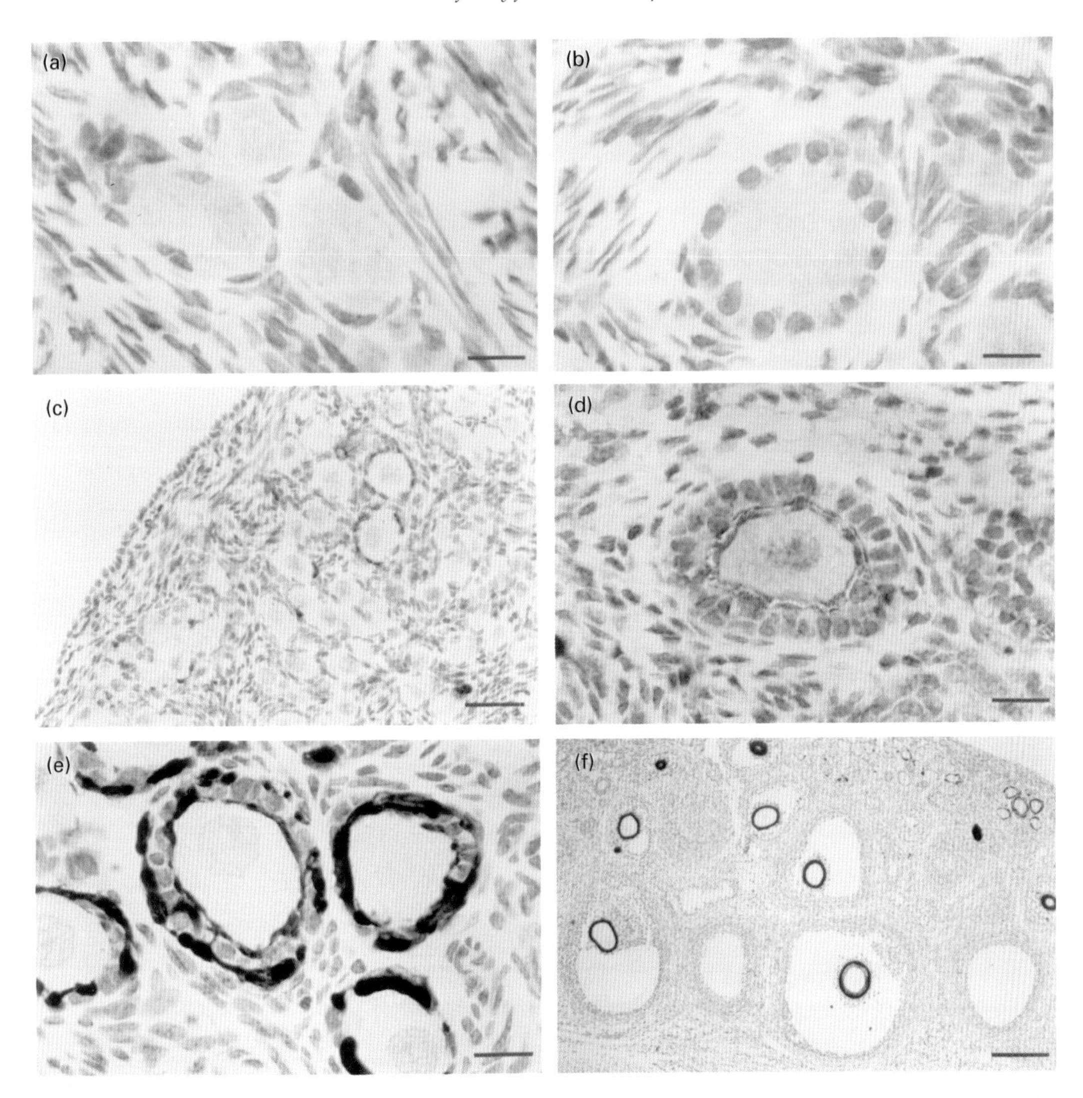

Fig. 1. Immunostaining for growth factor receptors or growth factors in ovarian follicles of the sheep ovary: (a) c-kit localized in oocytes of type 1 and 1a follicles and (b) in an oocyte of a type 2 follicle; (c) stem cell factor (SCF) localized to granulosal cells of type 1 follicles and (d) the oocyte and granulosal cells of a type 2 follicle; (e) inhibin β_B localized to granulosal cells of type 1a and 2 follicles and (f) zona pellucida of larger preantral and antral follicles. All immunostaining was performed with paraformaldehyde fixed paraffin wax embedded 5 µm sections. All immunostaining was blocked when the antibodies were either preincubated with their respective antigens or replaced by non-immune serum. The immunohistochemical procedures used were those described by Hsu *et al.* (1981). The c-kit antibody (C19) was supplied by Santa Cruz Biotechnology Inc, Santa Cruz (CA). The stem cell factor antibody (RGAS005) and recombinant ovine stem cell factor (+1 to +206 amino acid sequence RGAS006) were generated at the Wallaceville Animal Research Centre. Inhibin β_B antibody was clone C5 described in Groome *et al.* (1996). Scale bars represent (a) and (b) 13.3 µm, (c) 40 µm, (d) 16 µm, (e) 16 µm and (f) 160 µm.

cows and rats, very few granulosal cells in type 1 follicles immunostain for proliferating cell nuclear antigen (PCNA), an essential nuclear protein involved in cell proliferation (Oktay *et al*, 1995; Wandji *et al.*, 1996a). By contrast in ewes, about 30% or more of the granulosal cells in type 1 follicles immunostain for PCNA (Lundy *et al.*, 1998). Therefore a key question that remains unanswered is

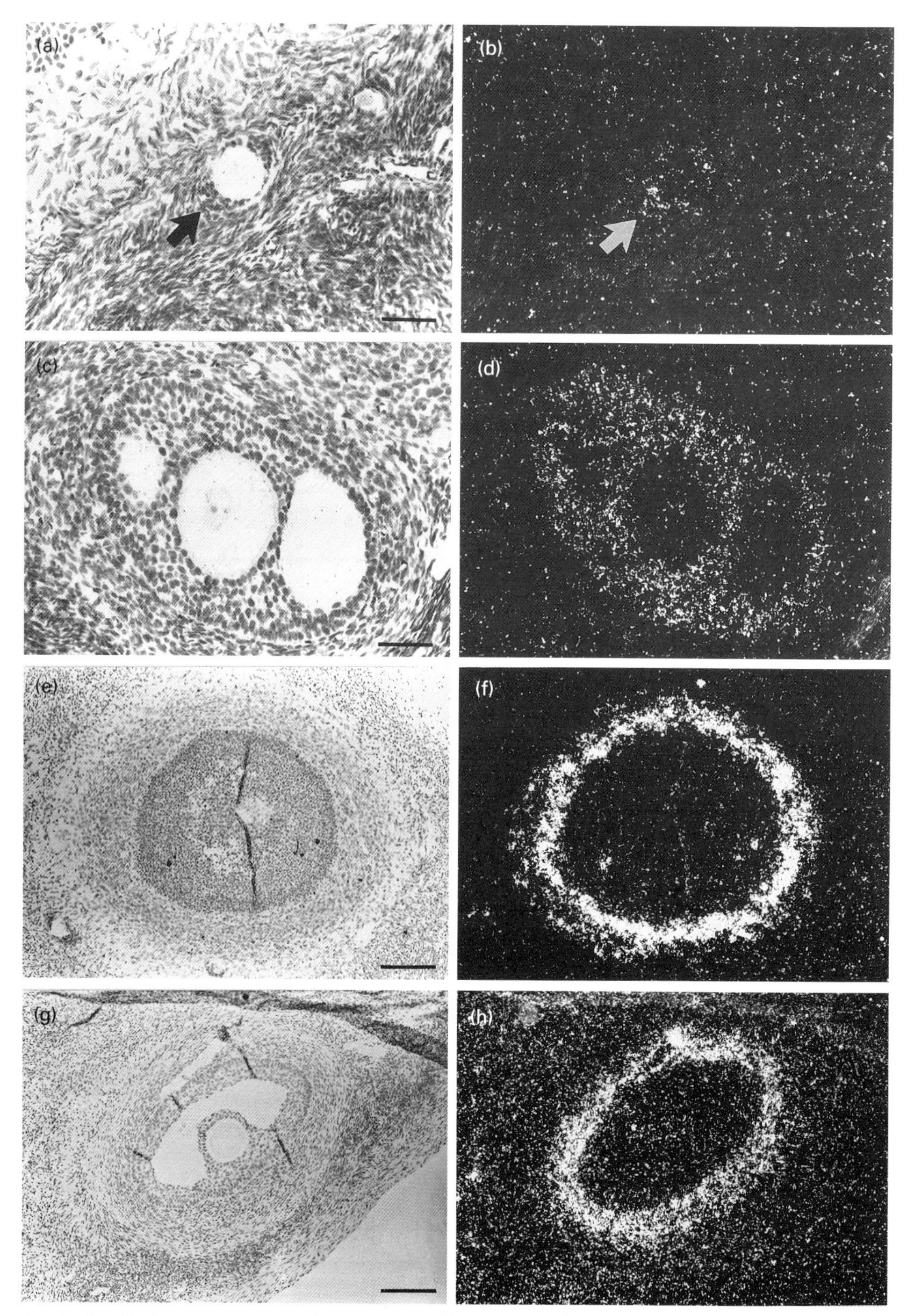

Fig. 2. Localization of gonadotrophin receptor mRNAs in small follicles of the sheep ovary: (a) light field and (b) dark field views of a type 1 and type 3 follicle with evidence of hybridization of ^{33}P-labelled FSH-receptor antisense riboprobe to granulosal cells of the type 3 (arrowed) but not type 1 follicle; (c) light field and (d) dark field views of a type 5 follicle showing hybridization of the FSH-receptor antisense riboprobe to granulosal cells; (e) light field and (f) dark field views of ^{33}P-labelled LH receptor antisense riboprobe hybridized to theca interna of a type 4–5 follicle; (g) light field and (h)

whether type 1 follicles are a pool of quiescent follicles. Compared with the uncertainty about the growth status of type 1 follicles, the collective evidence from several species is that type 1a follicles have entered the growth phase. In cows, ewes and rats, cuboidal granulosal cells in type 1a follicles may immunostain for PCNA or incorporate tritiated thymidine (Oktay *et al.*, 1995; Wandji *et al.*, 1996a; Braw-Tal and Yossefi, 1997; Lundy *et al.*, 1998). Moreover in ewes, type 1a follicles contain significantly more granulosal cells and the mean oocyte diameters are larger compared with type 1 follicles (Table 1). As ovine follicles transit through the types 1 to 3 growth phases, the mean total number of granulosal cells undergo at least six doublings before entering the type 4 phase which in turn may include two doubling cycles. Thus when considering factors that might regulate early follicular growth it may be important to consider: (i) the classification system being used; (ii) the animal model being studied; (iii) whether type 1 follicles are all quiescent follicles and; (iv) the likelihood that types 2, 3 and 4 follicles each represent more than one cell-doubling cycle with the possibility that each cycle is under some form of regulatory control. It is suggested that future studies need to focus more closely on the characteristics of the types of follicle being studied. Some of the growth regulatory factors associated with these early phases of growth are discussed below (see also Table 2).

Growth Factors and Receptors

C-kit/Stem cell factor

The tyrosine kinase receptor c-kit and its ligand, stem cell factor (SCF), have been localized to oocytes and granulosal cells, respectively (Motro and Bernstein, 1993). In the mouse, inhibition of the interaction between SCF and c-kit prevents the transformation of primordial follicles to primary follicles without blocking the formation of primordial follicles (Huang *et al.*, 1993; Yoshida *et al.*, 1997). In the sheep ovary, SCF mRNA has been detected in granulosal cell and c-kit mRNA in the oocyte at all stages of follicular growth from the primordial phase (Clark *et al.*, 1996; Tisdall *et al.*, 1997). Furthermore, in sheep, c-kit protein can be localized to oocytes of primordial and growing follicles and SCF protein to granulosal cells and oocytes of both primordial and primary follicles (Fig. 1). These findings are consistent with the view that activation of the c-kit tyrosine kinase system by SCF is an important factor in the growth of primordial follicles. One interesting finding from the study of Yoshida *et al.* (1997) was that the granulosal cell ceased to proliferate after the administration of c-kit antibody to mice. Thus a very early signal from oocytes might be necessary to promote proliferation of granulosal cells. Potential candidates might be growth differentiating factor 9 (GDF-9) (Dong *et al.*, 1996), epidermal growth factor or its receptor (Singh *et al.*, 1995) or a product from the X chromosome, since sheep that are homozygous for an X-linked mutation (FecX^1) contain primordial and primary follicles with enlarging oocytes (expressing c-kit) but with no corresponding increase in the number of granulosal cells (P. Smith, T. Lundy and K. P. McNatty, unpublished).

Gonadotrophin receptors

Follicle stimulating hormone (FSH) is unlikely to be a critical factor for initiating the growth of primordial follicles. There is convincing evidence from sheep, humans, cows and pigs that the gene for the FSH receptor (FSH-R) is not expressed until the follicle has reached the type 2–3 stage of development (Tisdall *et al.*, 1995; Xu *et al.*, 1995; Yuan *et al.*, 1996; Oktay *et al.*, 1997). At this and all

---

dark field views of ^{33}P-labelled p450scc antisense riboprobe hybridized to theca interna of a type 5 follicle. All sense probes showed no evidence of specific hybridization (see Tisdall *et al.*, 1995 for methodology). The ovine LH-R cDNA was obtained from G. Niswender, Colorado State University, Fort Collins, CO and the bovine p450scc from M. Waterman, Vanderbilt University School of Medicine, Nashville, TN. Scale bars represent (a,b,c,d) 80 μm; (e,f,g,h) 160 μm.

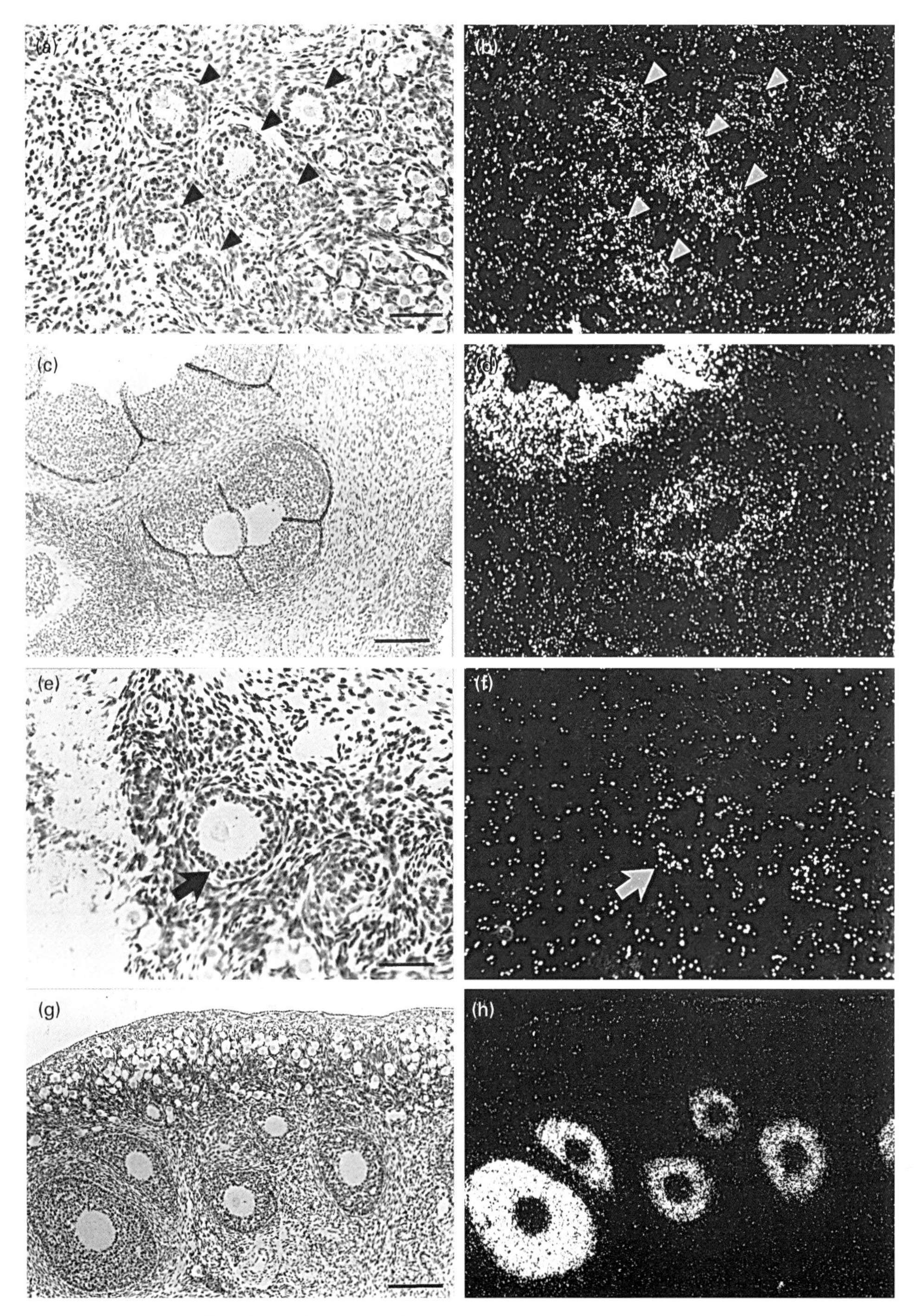

Fig. 3. Localization of inhibin and activin subunit mRNAs in small follicles of the sheep ovary: (a) light field and (b) dark field views of type 2 and 3 follicles (arrowheads) with evidence of hybridization of inhibin/activin β_B subunit antisense riboprobe to granulosal cells. Note there is no increase above background in type 1 follicles (b); (c) light field and (d) dark field views of type 5 and larger follicles with hybridization of inhibin/activin β_A subunit antisense riboprobe to granulosal cells; (e) light field and (f) dark field views of a type 3 follicle (arrows) with hybridization of inhibin α subunit antisense

subsequent stages of development FSH-R mRNA is localized exclusively to granulosal cells (Fig. 2). Autoradiographic analysis of ^{125}I-labelled FSH binding to preantral follicles is consistent with the presence of FSH-R in granulosal cells of type 2–4 but not type 1 or 1a follicles (Wandji *et al.*, 1992a). It is likely that the FSH-R is functionally active during preantral development because culture of preantral bovine, hamster and human follicles in serum-free media supplemented with FSH leads to significant increases in the number of granulosal cells or uptake of BrdU or thymidine by granulosal cells (Roy and Greenwald, 1989; Hulshof *et al.*, 1995; Wandji *et al.*, 1996a). Moreover, granulosal cells in large preantral–early antral follicles can synthesize cAMP or lactate in response to FSH *in vitro* (McNatty *et al.*, 1992; Boland *et al.*, 1993).

Although it can be demonstrated that FSH has stimulatory effects on granulosal cell proliferation and function in preantral follicles it is not an essential factor for proliferation of granulosal cells (Hirshfield, 1985) or the formation of a theca interna (Magarelli *et al.*, 1996). FSH-deficient mice produced by gene targeting of embryo stem cells (Kumar *et al.*, 1997) or hypohysectomized hamsters, rats or sheep all contain normally developing small follicles (Hirshfield, 1985; McNatty *et al.*, 1990; Wang and Greenwald, 1993). Furthermore, slices of ovarian cortex from the cow cultured in media devoid of gonadotrophins led to an increase in the number of primary and small preantral follicles and a corresponding decrease in the number of primordial follicles (Wandji *et al.*, 1996a; Braw-Tal and Yossefi, 1997) indicating that the initiation of follicular growth is likely to involve paracrine or autocrine rather than endocrine factors.

In sheep, the theca interna develops in type 3 follicles and in cattle in type 4 follicles (Braw-Tal and Yossefi, 1997; Lundy *et al.*, 1998). Luteinizing hormone receptor (LH-R) mRNA is demonstrable in theca interna of type 4–5 preantral follicles of pigs, cows and sheep (Xu *et al.*, 1995; Yuan *et al.*, 1996; Fig. 2). In rodents, the mRNAs encoding the intra- and extracellular domains of the LH-R appear concomitantly with the appearance of differentiated thecal cells but increased expression of LH-R during follicular growth is a gonadotrophin-dependent event (Sokka *et al.*, 1996; O'Shaughnessy *et al.*, 1997). In cattle, pigs and sheep the mRNAs for $P450_{scc}$ and $P450_{17\text{-hydroxylase}}$ enzymes are present in theca interna of type 4–5 follicles, which also synthesize androgens *in vitro* (Fig. 2; McNatty *et al.*, 1986; Yuan *et al.*, 1996). In domestic ruminants there is evidence that type 4–5 follicles have functional FSH-R and LH-R in granulosal cells and thecal cells, respectively, and that in ewes these follicles are capable of synthesizing progestins, androgens and oestrogens *in vitro* (McNatty *et al.*, 1986; Yuan *et al.*, 1996; Wandji *et al.*, 1996b).

Inhibin/activin subunits

Inhibins and activins are dimeric growth factors of the transforming growth factor β superfamily. The subunits inhibin α, inhibin/activin β_A and inhibin/activin β_B together with the activin receptors type I, IIA and IIB are expressed in ovarian cells during follicular development (Roberts *et al.*, 1993; Cameron *et al.*, 1994). In sheep ovaries the mRNA and peptide for β_B inhibin/activin subunit is first detected in granulosal cells of type 1a–3 follicles (Figs 1 and 3). The smallest follicles containing the β_B inhibin/activin peptide (using antisera clone C5; Groome *et al.*, 1996) were those containing one layer of granulosal cells but with at least one cuboidal cell (Fig. 1). At later stages of growth the β_B peptide was found increasingly in the zona pellucida although gene expression continued in granulosal cells (Fig. 1). However, no hybridization of the antisense β_B probe was noted in the oocyte, theca interna or interstitial cells. Unlike the early appearance of β_B mRNA in granulosal cells that for inhibin/activin β_A was not found until follicles reached the type 5 stage of growth (Torney *et al.*, 1989; Braw-Tal, 1994; Tisdall *et al.*, 1994). At this stage and thereafter β_A inhibin/activin mRNA was localized exclusively to granulosal cells (Fig. 3). In contrast to the β_A

---

riboprobe to granulosal cells; (g) light field and (h) dark field views of types 3–5 but not type 1–2 follicles with hybridization of follistatin subunit antisense riboprobe to granulosal cells. For details on α-inhibin and β_A inhibin/activin *in situ* procedures and ^{33}P-labelled riboprobes see Braw-Tal *et al.* (1994) and Tisdall *et al.* (1994). The β_B cDNA was supplied by R. Rodgers (Flinders University of South Australia, SA, Australia). Scale bars represent (a, b, e, f) 80 μm; (c, d, g, h) 160 μm.

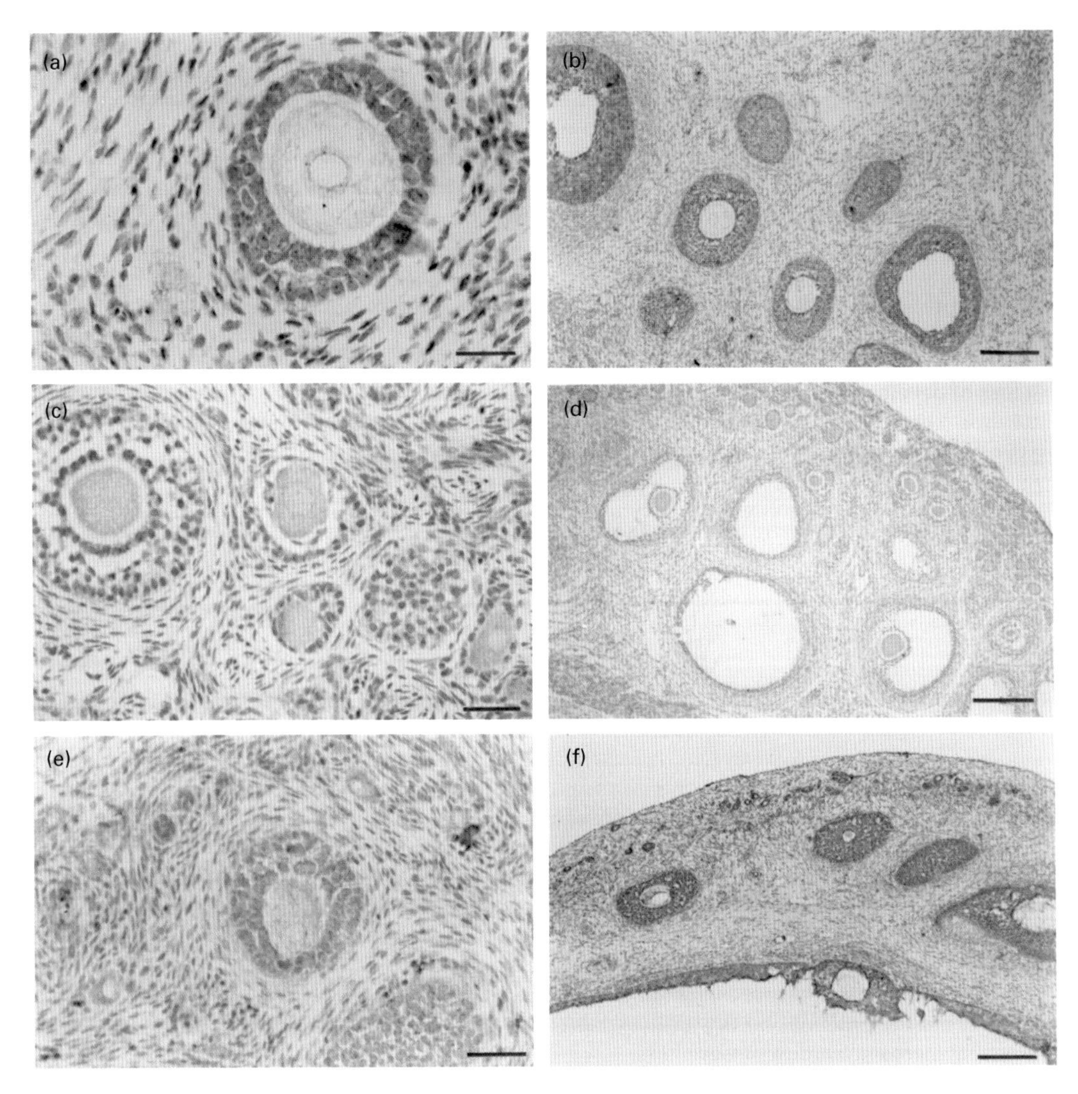

Fig. 4. Immunostaining for inhibin and/or activin subunits and follistatin in ovarian follicles of the sheep ovary: inhibin-α subunit immunostaining in type 3 follicle (a) and in type 4–5 follicles (b). Note intense immunostaining in granulosal cells and light immunostaining in some oocytes. (c,d) Inhibin/activin β_A subunit immunostaining in granulosal cells and oocytes of small (type 2–5) and both small and large antral follicles; (e,f) follistatin immunostaining in oocytes and granulosal cells of both preantral and antral follicles. For details on histological and immunohistochemical procedures see legend to Fig. 1. For details on the inhibin-α subunit antisera (clone R1) and β_A inhibin/activin antisera (clone E4) see Groome *et al.* (1990) and Groome and Lawrence (1991). The follistatin antisera (RGAS005) was generated against a 315 amino acid recombinant ovine follistatin (RGAS006) and both products were produced at the Wallaceville Animal Research Centre. Scale bars represent (a) 26.7 µm; (b, d, f) 160 µm; (c, e) 40 µm.

inhibin/activin mRNA results, immunohistochemical studies (using antisera Clone E4; Groome and Lawrence, 1991) localized the β_A inhibin/activin peptide to oocytes as well as to granulosal cell at all stages of follicular growth (for example primordial, primary, preantral and antral growth) (Fig. 4). One interpretation of this finding is that the monoclonal antibody to β_A peptide binds β_A inhibin/activin subunit which originated from an extragonadal source via the blood supply.

Collectively these results provide evidence for the presence of β inhibin/activin subunit peptides in oocytes or granulosal cells of preantral follicles.

Inhibin α subunit mRNA was first observed in granulosal cells of developing ovine follicles with two or three layers of granulosal cells (Fig. 1). Thereafter the hybridization signal increased exclusively in granulosal cells of both preantral and antral follicles (Braw-Tal, 1994). Similarly immunohistochemical studies with an inhibin α-subunit antibody (Clone R1; Groome *et al.*, 1990) demonstrated specific binding to granulosal cells of follicles with two or three layers of granulosal cells with no binding to theca interna, although a low level of binding was associated with the oocyte (Fig. 4).

The presence of β activin/inhibin subunits at several sites may be indicative of the presence of activin receptors in both oocytes and granulosal cell or other factors that bind the β activin/inhibin subunits (for example follistatin). Of interest is the finding that type 1a–2 follicles represent the only stages of development in which β_B activin/inhibin expression occurs in the absence of the α inhibin subunit. Mice homozygous for a deletion of the inhibin α subunit gene secrete large amounts of activin A and B and develop sex cord–stromal tumours (Matzuk *et al.*, 1996); these tumours in the female may arise from uncontrolled proliferation of granulosal cells. In normal animals the rate of proliferation of granulosal cells may be influenced in part by the endogenous production of inhibin in type 3 and larger follicles thereby blocking the action of activin via the type II activin receptor (Martens *et al.*, 1997). In domestic ruminants it remains to be determined when the activin receptors first appear and whether activin B has an important role in either upregulating FSH receptors (see Findlay, 1993 for review) or in promoting the development of a type 3 follicle.

Follistatin

Overexpression of follistatin in transgenic mice inhibits follicular growth at the primary and subsequent stages of development (Guo *et al.*, 1998). Thus follistatin may regulate the actions of activin or other members of the transforming growth factor β (TGFβ) family (for example GDF-9).

In sheep, follistatin gene expression was first observed in granulosal cells of type 3 follicles (Braw-Tal, 1994; Braw-Tal *et al.*, 1994) (Fig. 3). Thereafter it was observed in almost all preantral, and non-atretic antral follicles throughout their different growth phases. Follistatin protein was first observed in granulosal cells of type 2 follicles and at subsequent phases of growth. The earlier detection of protein over mRNA for follistatin may be the result of differences in sensitivities in the methods used or to the protein originating from an extracellular source. In addition, follistatin peptide can be localized to oocytes in types 1–5 follicles as well as large preantral follicles (Dr R. Braw-Tal, personal communication; Fig. 4).

Collectively the immunohistochemical data infer that follistatin and activin and perhaps to a lesser extent inhibin are associated with the oocyte throughout follicular growth. In this context, exposure of oocytes to excess activin A stimulates meiotic maturation in follicle-enclosed oocytes and oocyte degeneration in immature follicles (Woodruff *et al.*, 1990; Erickson *et al.*, 1995). During preantral and early antral follicular growth, follistatin may act as a binding protein to prevent premature oocyte maturation. It is worth noting for most follicular growth phases that the β_A activin/inhibin peptide and follistatin are consistently co-localized where the two peptides have been studied together. In contrast the β_B activin/inhibin peptide does not co-localize with follistatin. Instead the β_B subunit is strongly associated with both the oocyte plasma membrane and zona pellucida inferring a different mechanism of action.

TGFβ and FGF

Transforming growth factor β is known to be produced by bovine thecal cells and to influence granulosal cell proliferation (Skinner *et al.*, 1987; Wandji *et al.*, 1996b). In the sheep ovary $TGF\beta_1$ mRNA was observed in stromal/interstitial tissues and first observed in theca interna in type 4 or 5 follicles, whereas $TGF\beta_3$ was noted mainly in smooth muscle cells around blood vessels in the theca

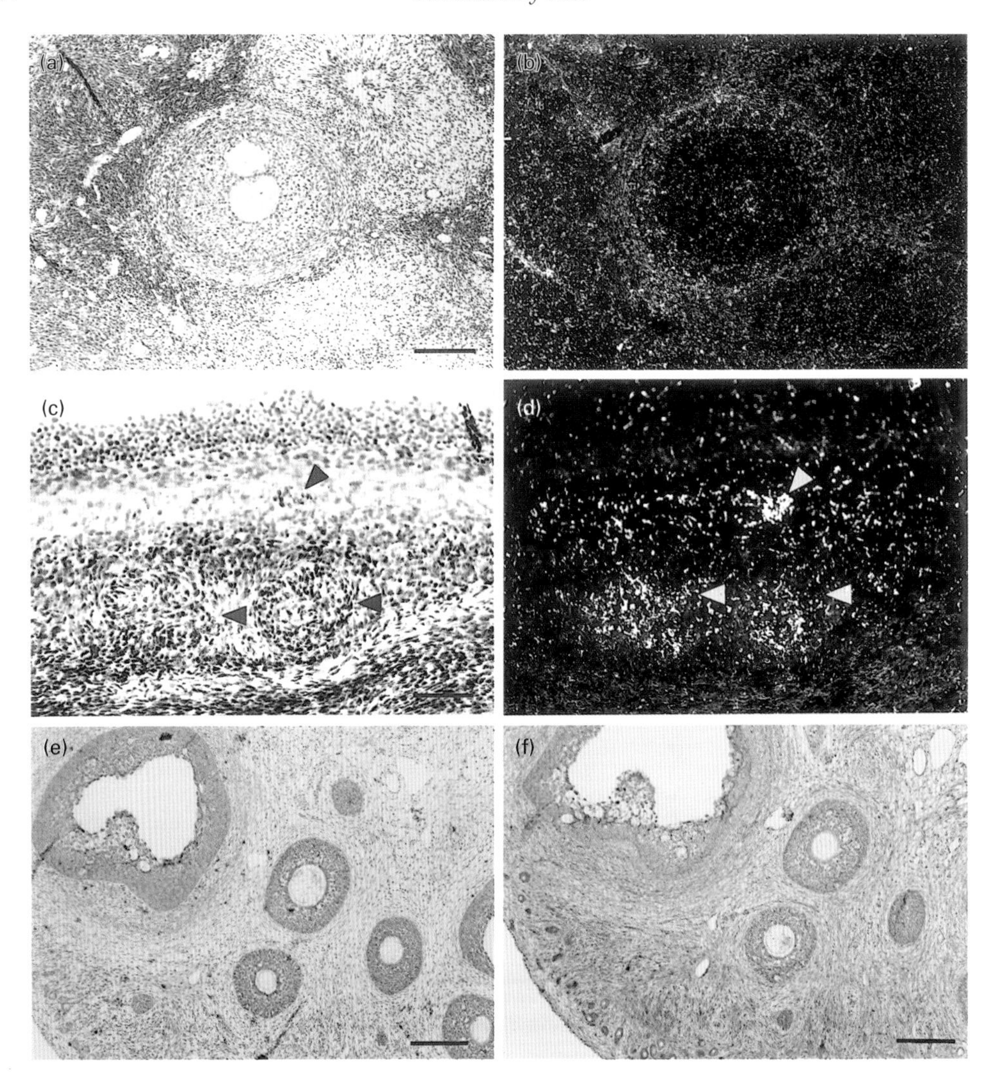

Fig. 5. Localisation of transforming growth factor β in the sheep ovary: (a) light field and (b) dark field views of a type 5 follicle with evidence of hybridization of the ^{33}P-labelled TGFβ_1 antisense riboprobe to theca interna and areas of the adjacent stroma/ interstitium but not granulosal cells; (c) light field and (d) dark field views of large antral follicle with evidence of hybridization of the ^{33}P-labelled TGFβ_3 antisense riboprobe to smooth muscle cells of blood vessels (arrowheads) and thecal interna but not granulosal cells; (e) immunostaining for TGFβ_1 in oocytes, granulosal cells of preantral and antral follicles and to a lesser extent in the theca interna and interstitium; (f) immunostaining for TGF$\beta_{2,3}$ indicating widespread localization in the ovary. The TGFβ_1 riboprobe was from an RT-PCR derived 1172 bp ovine cDNA for the complete coding region. The TGFβ_3 riboprobe was from an RT–PCR derived 960 bp ovine cDNA. All sense probes showed no evidence of specific hybridization. The TGFβ_1 antibody (clone TB-21) was obtained from Anogen Inc, Mississauga, Ontario and the anti-TGF$\beta_{2,3}$ (clone AB-1) from Oncogene Science Inc, Uniondale, NY. For further details on immunohistochemical procedures see Fig. 1. Scale bars represent (a, b) 80 μm; (c, d) 73 μm; (e, f) 160 μm.

layer (Fig. 5). The pattern of TGFβ_2 expression was similar to that of TGFβ_1 (D. Tisdall, unpublished). By contrast TGFβ_1 peptide was widespread throughout the ovary (Fig. 5): it was observed in oocytes, especially in primordial follicles and it was prominent in granulosal cells at the type 2 stage of development and to a lesser extent in the theca interna. Immunostaining with antibody that

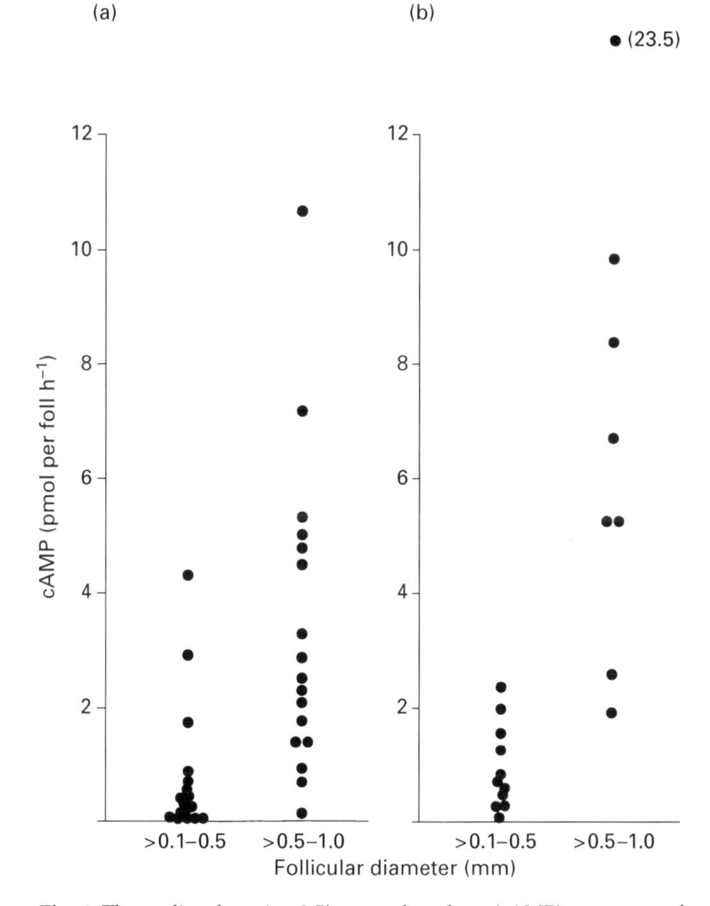

Fig. 6. The cyclic adenosine 3,5′-monophosphate (cAMP) responses of individual large preantral–early antral ovine follicles (follicle types 4 and 5) dissected from an individual (a) intact or (b) long-term (i.e. 60 days) hypophysectomized ewe. Each follicle from each ewe was incubated separately for 1 h in 1 ml Dulbecco's phosphate-buffered saline + 0.1% w v^{-1} bovine serum albumin + ovine LH (NIH-LH-S23, 1 μg ml^{-1}) + and FSH (NIH-FSH-S11, 1 μg ml^{-1}) (see McNatty *et al.*, 1986 for experimental details).

recognizes TGFβ$_2$ and TGFβ$_3$ but not TGFβ$_1$ indicated widespread distribution of these peptides throughout the ovary (see Fig. 5).

Granulosal cells in culture express the gene encoding fibroblast growth factor 2 (FGF-2) (Neufield *et al.*, 1987). However, the ontogeny of FGF-2 gene expression in developing follicles and other ovarian cellular sources is not well understood. Immunocytochemical studies reveal that FGF-2 peptide is widespread throughout the bovine ovary being present in oocytes of primordial and primary follicles, granulosal cells of large preantral and antral follicles, theca interna, ovarian surface epithelium and smooth muscle cells surrounding blood vessels (Van Wezel *et al.*, 1995). Moreover FGF receptors are found in granulosal cells of type 2 follicles (Wandji *et al.*, 1992b).

In addition to the localization of TGFβs and FGF-2 to oocytes of primordial and primary follicles, epidermal growth factor (EGF) has been localized to oocytes of primordial follicles of hamsters and pigs and TGFα to oocytes of primary follicles (Roy and Greenwald, 1990; Singh *et al.*, 1995). Collectively, therefore, oocytes in primordial follicles appear to be bathed in growth factors, some of which are likely to have stimulatory effects on granulosal cells (i.e. stem cell factor, activin, FGF-2 and EGF), whereas others are likely to be inhibitory (follistatin, TGFβs) (Gospodarowicz *et al.*, 1986; Li *et al.*, 1995; Wandji *et al.*, 1996b; Guo *et al.*, 1998). The relative importance of these growth factors or their receptors in the initiation of follicular growth remains to be determined.

Hierarchical Development

In domestic ruminants, initiation of follicular growth begins in late fetal life and continues without interruption during infancy, pregnancy, lactation and the oestrous cycle. From studies in rodents it has been proposed that follicles grow sequentially and continue to grow until they either become atretic or ovulate (Peters *et al.*, 1975). In studies of granulosal cell populations and steroid concentrations in follicular fluid of antral follicles from sheep, cows and humans it is evident that, at any time, no two follicles share an identical cell composition or steroid microenvironment (for example McNatty, 1978). These data are consistent with the notion that follicles develop in a hierarchical manner. Further evidence to support this notion was obtained from examining the gonadotrophin-induced cyclic AMP responses of late preantral–early antral follicles isolated from either hypophysectomized or control ewes (Fig. 6). These data also infer that preantral follicles may develop in a hierarchical manner and that this pattern of development is, at least in part, independent of gonadotrophins.

Conclusions

The growth of a primordial follicle to the early antral phase involves eight doublings of the population of granulosal cells and a 3–4-fold enlargement of the oocyte. In domestic ruminants most, if not all, of these growth phases can occur in the absence of gonadotrophins. However, based upon evidence from animal mutants, many of these growth phases are dependent on locally produced growth factors or receptors including c-kit, stem cell factor, members of the transforming growth factor superfamily (for example β_A inhibin/activin, α-inhibin, GDF-9) and follistatin. The evidence from studies in domestic ruminants shows that these growth factors together with receptors for both growth factors and gonadotrophins are expressed during early follicular growth in a stage- and cell-specific manner.

References

Boland NI, Humpherson PG, Leese HJ and Gosden RG (1993) Pattern of lactate production and steroidogenesis during growth and maturation of mouse ovarian follicles *in vitro. Biology of Reproduction* **48** 798–806

Braw-Tal R (1994) Expression of mRNA for follistatin and inhibin/activin subunits during follicular growth and atresia *Journal of Molecular Endocrinology* **13** 253–264

Braw-Tal R and Yossefi S (1997) Studies *in vivo* and *in vitro* on the initiation of follicle growth in the bovine ovary *Journal of Reproduction and Fertility* **109** 165–171

Braw-Tal R, Tisdall DJ, Hudson NL, Smith P and McNatty KP (1994) Follistatin but not α or β_A inhibin subunit mRNA is expressed in ovine fetal ovaries in late gestation *Journal of Molecular Endocrinology* **13** 1–9

Cameron VA, Nishimura E, Mathews LS, Lewis KA, Sawchenkko PE and Vale WW (1994) Hybridization histochemical localization of activin receptor subtypes in rat brain, pituitary, ovary and testis *Endocrinology* **134** 799–808

Clark DE, Tisdall DJ, Fidler AE and McNatty KP (1996) Localization of mRNA encoding c-kit during the initiation of folliculogenesis in ovine fetal ovaries *Journal of Reproduction and Fertility* **106** 329–335

Dong J, Albertini DF, Nishimori K, Kumar TR, Lu N and Matzuk M (1996) Growth differentiation factor-9 is required during early ovarian folliculogenesis *Nature* **383** 531–535

Erickson GF, Kokka S and Rivier C (1995) Activin causes premature superovulation *Endocrinology* **136** 4804–4813

Findlay JR (1993) An update on the roles of inhibin, activin and follistatin as local regulators of folliculogenesis *Biology of Reproduction* **48** 15–23

Gospodarowicz D, Neufield G and Schweigener L (1986) Fibroblast growth factor *Molecular Cellular Endocrinology* **46** 187–204

Groome N and Lawrence M (1991) Preparation of monoclonal antibodies to the beta A subunit of ovarian inhibin using a synthetic peptide immunogen *Hybridoma* **10** 309–316

Groome N, Hancock J, Betteridge A, Lawrence M and Craven R (1990) Monoclonal and polyclonal antibodies reactive with the 1–32 amino terminal sequence of the alpha subunit of human 32K inhibin *Hybridoma* **9** 31–42

Groome NP, Illingworth PJ, O'Brien M, Pai R, Rodger FE, Mather JP and McNeilly AS (1996) Measurement of dimeric inhibin B throughout the human menstrual cycle *Journal of Clinical Endocrinology and Metabolism* **81** 1401–1405

Gundersen HJG (1988) The nucleator *Journal of Microscopy* **151** 3–21

Guo Q, Kumar T, Woodruff T, Hadswell L, De Mayo F and Matsuk M (1998) Overexpression of mouse follistatin causes reproductive defects in transgenic mice *Molecular Endocrinology* **12** 96–106

Hirshfield AN (1985) Comparison of granulosa cell proliferation in small follicles of hypophysectomized, prepubertal and mature rats *Biology of Reproduction* **32** 979–987

Hirshfield AN and De Santi AM (1995) Patterns of ovarian cell proliferation in rats during the embryonic period and the first three weeks post partum *Biology of Reproduction* **53** 1208–1221

Hsu SM, Raine L and Fanger H (1981) Use of a activin–biotin–peroxidase complex (ABC) in immunoperoxidase techniques: a comparison between ABC and unlabelled antibody (MAP) procedures *Journal of Histochemistry and Cytochemistry* **29** 577–580

Huang E, Manova K, Packer AI, Sanchez S, Bachvarova RF and Besmer P (1993) The murine steel panda mutation affects kit ligand expression and growth of early ovarian follicles *Developmental Biology* **157** 100–109

Hulshof SCJ, Figueiredo JR, Beckers JF, Bevers MM, van der Donk JA and van den Hurk R (1995) Effects of fetal bovine serum, FSH and 17β-estradiol on the culture of bovine preantral follicles *Theriogenology* **44** 217–226

Kumar T, Wang Y, Lu N and Matzuk M (1997) Follicle stimulating hormone is required for ovarian follicle maturation but not male fertility *Nature Genetics* **15** 201–204

Li R, Phillips DM and Mather JP (1995) Activin promotes ovarian follicular development *in vitro. Endocrinology* **136** 849–856

Lundy T, Smith P, O'Connell A, Hudson NL and McNatty KP (1999) Populations of granulosa cells in small follicles of the sheep ovary *Journal of Reproduction and Fertility* **115** 251–262

McNatty KP (1978) Cyclic changes in antral fluid hormone concentrations in humans. In *Clinics in Endocrinology and Metabolism* Vol. 7 no 3 pp 577–600 Eds GT Ross and MB Lipsett. WB Saunders Co. Ltd, London

McNatty KP, Kieboom LE, McDiarmid J, Heath DA and Lun S (1986) Adenosine cyclic 3′,5′-monophosphate and steroid production by small ovarian follicles from Booroola ewes with and without a fecundity gene *Journal of Reproduction and Fertility* **76** 471–480

McNatty KP, Heath DA, Hudson N and Clarke IJ (1990) Effect of long-term hypophysectomy on ovarian follicle populations and gonadotrophin-induced adenosine cyclic 3′,5′-monophosphate output by follicles from Booroola ewes with or without the F gene *Journal of Reproduction and Fertility* **90** 515–522

McNatty KP, Smith P, Hudson NL, Heath DA, Lun S and O W-S (1992) Follicular development and steroidogenesis. In *Local Regulation of Ovarian Function* pp 21–38 Eds N-O Sjoberg, L Hamberger, PO Jansen, Ch Owman and HJT Coelingh-Bennink. Parthenon Publishing, Park Ridge, New Jersey

Magarelli P, Zachow R and Magoffin D (1996) Developmental and hormonal regulation of rat theca-cell differentiation factor secretion in ovarian follicles *Biology of Reproduction* **55** 416–420

Martens JW, de Winter JP, Timmerman MA, McLuskey A, van Schauk RH, Themmen AP and de Jong FH (1997) Inhibin interferes with activin signalling at the level of the activin receptor complex in Chinese hamster ovary cells *Endocrinology* **138** 2928–2936

Matzuk M, Kumar T, Shou W, Coerver K, Lau A, Behringer R and Finegold M (1996) Transgenic models to study the roles of inhibins and activins in reproduction, oncogenesis and development *Recent Progress in Hormone Research* **51** 123–137

Motro B and Bernstein A (1993) Dynamic changes in ovarian c-kit and steel expression during the estrous reproductive cycle *Developmental Dynamics* **197** 69–79

Neufield G, Ferrara N, Schweigerer L, Mitchell R and Gospodarowicz D (1987) Bovine granulosa cells produce basic fibroblast growth factor *Endocrinology* **121** 597–603

Oktay K, Schenken R and Nelson JT (1995) Proliferating cell nuclear antigen marks the initiation of follicular growth in the rat *Biology of Reproduction* **53** 295–301

Oktay K, Briggs D and Gosden RG (1997) Ontogeny of follicle-stimulating hormone receptor gene expression in isolated human follicles *Journal of Clinical Endocrinology and Metabolism* **82** 3748–3751

O'Shaughnessy PJ, McLelland D and McBride MW (1997) Regulation of luteinizing hormone-receptor and follicle-stimulating hormone-receptor messenger ribonucleic acid levels during development in the neonatal mouse ovary *Biology of Reproduction* **57** 602–608

Peters H, Byskov AG, Himelstein-Braw R and Faber M (1975) Follicle growth: the basic event in the mouse and human ovary *Journal of Reproduction and Fertility* **45** 559–566

Roberts VJ, Barth S, el-Roeiy and Yen SS (1993) Expression of inhibin/activin subunits and follistatin messenger ribonucleic acids and proteins in ovarian follicles and the corpus luteum during the human menstrual cycle *Journal of Clinical Endocrinology and Metabolism* **77** 1402–1410

Roy SK and Greenwald GS (1989) Hormonal requirements for the growth and differentiation of hamster preantral follicles in long-term culture *Journal of Reproduction and Fertility* **87** 103–114

Roy SK and Greenwald G (1990) Immunohistochemical localization of epidermal growth factor-like activity in the hamster ovary with a polyclonal antibody *Endocrinology* **126** 1309–1317

Singh B, Rutledge JM and Armstrong DT (1995) Epidermal growth factor and its receptor gene expression and peptide localisation in porcine ovarian follicles *Molecular Reproduction and Development* **40** 391–399

Skinner MK, Keshi-Oja J, Osteen K and Moses HL (1987) Ovarian theca cells produce transforming growth factor-beta which can regulate granulosa cell growth and differentiation *Endocrinology* **121** 786–792

Smith P, O W-S, Corrigan KA, Smith T, Lundy T, Davis GH and McNatty KP (1997) Ovarian morphology and endocrine characteristics of female sheep fetuses that are heterozygous or homozygous for the Inverdale prolificacy gene (FecX[1]) *Biology of Reproduction* **57** 1183–1192

Sokka TA, Hamalainen TM, Kaipa A, Warren DW and Huhtaniemi IT (1996) Development of luteinizing hormone action in the perinatal rat ovary *Biology of Reproduction* **55** 663–670

Tisdall DJ, Hudson N, Smith P and McNatty KP (1994) Localization of ovine follistatin and α and β_A inhibin mRNA in the sheep ovary during the oestrous cycle *Journal of Molecular Endocrinology* **12** 181–193

Tisdall DJ, Watanabe K, Hudson NL, Smith P and McNatty KP (1995) FSH receptor gene expression during ovarian follicle development in sheep *Journal of Molecular Endocrinology* **15** 273–281

Tisdall DJ, Quirke LD, Smith P and McNatty KP (1997) Expression of the ovine stem cell factor during folliculogenesis in late fetal and adult ovaries *Journal of Molecular Endocrinology* **18** 127–135

Torney AH, Hodgson YM, Forage R and de Kretser DM (1989) Cellular localization of inhibin mRNA in the bovine ovary by *in situ* hybridization *Journal of Reproduction and Fertility* **86** 391–399

van Wezel IL and Rodgers RJ (1996) Morphological characterization of bovine primordial follicles and their environment *in vivo. Biology of Reproduction* **55** 1003–1011

van Wezel IL, Umapathysivam K, Tilley WD and Rodgers RJ (1995) Immunohistochemical localization of basic fibroblast growth factor in bovine ovarian follicles *Molecular and Cellular Endocrinology* **115** 133–140

Wandji S-A, Pelletier G and Sirard M-A (1992a) Ontogeny and cellular localization of ^{125}I-labelled insulin-like growth factor-I, ^{125}I-labelled follicle-stimulating hormone and ^{125}I-labelled human chorionic gonadotropin binding sites in ovaries from bovine fetuses and neonatal calves *Biology of Reproduction* **47** 814–822

Wandji S-A, Pelletier G and Sirad M-A (1992b) Ontogeny and cellular localization of ^{125}I-labelled basic fibroblast growth factor and ^{125}I-labelled epidermal growth factor binding sites in ovaries from bovine fetuses and neonatal calves *Biology of Reproduction* **47** 807–813

Wandji S-A, Sosen V, Voss AK, Eppig JJ and Fortune JE (1996a) Initiation *in vitro* of growth of bovine primordial follicles *Biology of Reproduction* **55** 942–948

Wandji S-A, Eppig JJ and Fortune JE (1996b) FSH and growth factors affect the growth and endocrine functions *in vitro* of granulosa cells of bovine preantral follicles *Theriogenology* **45** 817–832

Wang X-N and Greenwald GS (1993) Hypophysectomy of the cyclic mouse. 1. Effects on folliculogenesis, oocyte growth and follicle-stimulating hormone and human chorionic gonadotrophin receptors *Biology of Reproduction* **48** 585–594

Woodruff TK, Lyan RJ, Hansen SE, Rice GC and Mather JP (1990) Inhibin and activin locally regulate rat ovarian folliculogenesis *Endocrinology* **127** 3196–3205

Xu ZZ, Garverick HA, Smith GW, Smith MF, Hamilton SA and Young-Quist RS (1995) Expression of follicle-stimulating hormone and luteinizing hormone receptor messenger ribonucleic acids in bovine follicles during the first follicular wave *Biology of Reproduction* **53** 951–957

Yuan W, Lucy MC and Smith MF (1996) Messenger ribonucleic acid for insulin-like growth factors-1 and -11, insulin-like growth factor binding protein-2, gonadotropin receptors, and steroidogenic enzymes in porcine follicles *Biology of Reproduction* **55** 1045–1054

Yoshida H, Takakura N, Kataoka H, Kunisada T, Okamura H and Nishikawa S-I (1997) Stepwise requirement of c-kit tyrosine kinase in mouse ovarian follicle development *Developmental Biology* **184** 122–137

Journal of Reproduction and Fertility Supplement **54**, 17–32

Comparative patterns of follicle development and selection in ruminants

G. P. Adams

Veterinary Anatomy, Western College of Veterinary Medicine, University of Saskatchewan, Saskatoon, Saskatchewan, S7N 5B4, Canada

Expanding technological capabilities, particularly in ultrasonography and molecular endocrinology, have bridged the gap between form and function of the ovary, and have been a catalyst for intense research activity in this area during the last decade. However, the study of follicular dynamics is still in its infancy in ruminant species other than cattle, and controversy persists regarding the pattern of follicular growth and the existence of follicular dominance. The bovine model of ovarian function is presented as a foundation for concepts surrounding the control of follicular development in ruminants, and to place in context the results of recent studies in sheep, goats, muskoxen, cervids and camelids. This comparative approach is used to determine important generalities that appear to be applicable, as fundamental physiological phenomena, to all ruminant species. Although clear differences in follicular dynamics are evident, differences appear to be specific rather than general, and the following conclusions are consistent with results reported in ruminant species to date: (1) follicles grow in a wave-like fashion; (2) periodic surges in circulating FSH are associated with follicular wave emergence; (3) selection of a dominant follicle involves a decline in FSH and acquisition of LH responsiveness; (4) periodic anovulatory follicular waves continue to emerge until occurrence of an LH surge (that is, at the time of luteolysis during the ovulatory season or during transition from the anovulatory season); (5) within species, there is a positive relationship between the duration of the oestrous cycle and the number of follicular waves; (6) progesterone suppresses LH secretion and growth of the dominant follicle; (7) the duration of the interwave interval is a function of follicular dominance, and is negatively correlated with circulating FSH; (8) follicular dominance in all species is more pronounced during the first and last follicular waves of the oestrous cycle; and (9) pregnancy, the prepubertal period and seasonal anoestrus are characterized by regular, periodic surges in FSH and emergence of anovulatory follicular waves.

Introduction

The kinetics of follicular development are best characterized in cattle. The bovine model is presented first to introduce the concept of follicular wave dynamics and the fundamental mechanisms involved. Historically, there has been a lack of scientific consensus regarding the pattern of ovarian follicular development within and among species. However, results of recent studies, using a number of different methods, are increasingly consistent and it appears that the wave phenomenon is common to cattle of different breeds, to several domestic and wild ruminant species, and to different reproductive and lactational states. The intent of this review is to provide a comparative overview of ovarian follicular development in ruminants.

Bovine Model of Ovarian Function

Follicular dynamics during the oestrous cycle

Greater than 95% of oestrous cycles are composed of either two or three follicular waves (Fig. 1; Ginther *et al.*, 1989a; Savio *et al.*, 1988; Sirois and Fortune, 1988; Adams, 1994). Single-wave cycles have been reported in heifers at the time of puberty (Evans *et al.*, 1994a) and in mature cows during the first interovulatory interval after calving (Murphy *et al.*, 1990; Savio *et al.*, 1990). Four-wave cycles are observed occasionally in *Bos indicus* (Rhodes *et al.*, 1995; Zeitoun *et al.*, 1996), but most oestrous cycles composed of four or more follicular waves are accompanied by a prolonged interovulatory interval as a result of delayed luteolysis or failure to ovulate (Adams *et al.*, 1992a; Roche and Boland, 1991). The proportion of animals with two- versus three-wave cycles differs between studies; some report a majority of two-wave cycles (> 80%, Ginther *et al.*, 1989a; Rajamahendran and Taylor, 1990; Ahmad *et al.*, 1997) and others report a majority of three-wave cycles (> 80%, Sirois and Fortune 1988), while others have observed a more even distribution (Evans *et al.*, 1994a; Savio *et al.*, 1990). Although the subject has not been studied extensively, there appears to be no clear breed- or age-specific preference for a particular follicular wave pattern, and no difference in pregnancy rate was detected between two- and three-wave animals (Ahmad *et al.*, 1997). In a study of the effects of nutrition on follicular dynamics (Murphy *et al.*, 1991), cattle fed a low energy ration had a greater proportion of three-wave cycles than those fed higher energy rations. Whether the pattern (that is, either two-wave or three-wave) within individuals is reproducible has not been investigated. The evolutionary reason for a two- or a three-wave cycle, or indeed for the wave-like pattern itself, is unclear, but differences between wave patterns are distinct and have important implications regarding schemes for ovarian synchronization and superstimulation (Adams, 1994).

The wave pattern of follicular development refers to periodic, synchronous growth of a group of antral follicles. In cattle, follicular wave emergence is characterized by the sudden (within 2–3 days) growth of 8–41 (average = 24) small follicles that are initially detected by ultrasonography at a diameter of 3–4 mm (Pierson and Ginther, 1987a; Savio *et al.*, 1988; Sirois and Fortune, 1988; Ginther *et al.*, 1989a, 1996b). For about 2 days, growth rate is similar among follicles of the wave, then one follicle is selected to continue growth (dominant follicle) while the remainder become atretic and regress (subordinate follicles). In both two- and three-wave oestrous cycles, emergence of the first follicular wave occurs consistently on the day of ovulation (day 0). Emergence of the second wave occurs on day 9 or 10 for two-wave cycles, and on day 8 or 9 for three-wave cycles. In three-wave cycles, a third wave emerges on day 15 or 16. Under the influence of progesterone (for example, dioestrus), dominant follicles of successive waves undergo atresia (Bergfelt *et al.*, 1991). The dominant follicle present at the onset of luteolysis becomes the ovulatory follicle, and emergence of the next wave is delayed until the day of the ensuing ovulation. The corpus luteum begins to regress earlier in two-wave cycles (day 16) than in three-wave cycles (day 19) resulting in a correspondingly shorter oestrous cycle (20 days versus 23 days, respectively). Hence, the mean interval of 21 days occurs only as an average between two- and three-wave cycles.

FSH and the wave phenomenon

Each follicular wave is preceded by a surge in circulating FSH; hence, cows with two-wave cycles have two FSH surges and three-wave cycles have three surges (Fig. 1; Adams *et al.*, 1992b). This observation has been confirmed in several subsequent studies (Ginther *et al.*, 1996a). Initial detection of follicles (3 mm) emerging within a wave occurs during the incline in the FSH surge and continues until FSH decreases to pre-surge values after about 48–72 h (Ginther *et al.*, 1996b). The decline in circulating FSH is a result of negative feedback from products of the emerging follicles (Adams *et al.*, 1992b, 1993b; Gibbons *et al.*, 1997) and the following nadir in FSH effectively prevents new wave emergence (Bergfelt *et al.*, 1994). Oestradiol and inhibin suppress FSH secretion *in vivo* and *in vitro*, and are likely the most important follicular products responsible for the suppressive effects involved in the wave phenomenon; however, follicular factors mediating the effects on FSH are not yet defined clearly.

Although it has been assumed that the primary source of such factors is the dominant follicle, recent results suggest that all follicles that are ≥ 5 mm in diameter in a wave help to suppress FSH secretion (Gibbons *et al.*, 1997). Studies of follicular dynamics in pregnant cows (Ginther *et al.*, 1996a) and prepubertal calves (Adams *et al.*, 1994; Evans *et al.*, 1994b) have offered a good opportunity for examining the relationships between FSH and follicular wave emergence. Observations on pregnant and prepubertal animals support the notion of a positive relationship between the magnitude of the FSH surge and the number of follicles in a wave.

The selection mechanism

The mechanism of selection of the dominant follicle is based on differential responsiveness of follicles within a wave to FSH and LH (Ginther *et al.*, 1996b). The mechanism involves, in the first instance, the post-surge decline in FSH. The time of selection (defined as divergence in growth profiles of dominant versus subordinate follicles) coincided with the first significant decrease in FSH concentrations (Adams *et al.*, 1992b), and selection could be delayed with exogenous FSH (Adams *et al.*, 1993a). The second important aspect in selection of the dominant follicle is a change to LH responsiveness. The transient rise in FSH permits sufficient follicular growth so that some (not all) follicles acquire LH responsiveness. This responsiveness imbues the follicle with the ability to survive without FSH. At the time that the growth profiles of the dominant and subordinate follicles begin to diverge, about 2 days after wave emergence, the follicle destined to become dominant apparently has more LH receptors and the competitive advantage over incipient subordinate follicles. However, subordinate follicles can achieve dominance if the original dominant follicle is removed (Adams *et al.*, 1993b; Gibbons *et al.*, 1997) or if exogenous FSH is supplied (Adams *et al.*, 1993a). Furthermore, the competition for LH among multiple dominant follicles (that is, superstimulated with FSH) is apparent by the smaller maximum diameter attained compared with single dominant follicles (Adams *et al.*, 1993a). In this regard, comparative observations in sheep (see section below) provide a rationale for the hypothesis that there is a relative difference in LH responsiveness between monovular and polyovular species such that more follicles attain the ability to use LH in the latter.

During the growing and static phases, continued secretion of follicular products from the dominant follicle causes FSH to be suppressed to its nadir, and together with continued suppression of LH as a consequence of luteal phase progesterone secretion, the dominant follicle ceases its metabolic functions and begins to die. In this way, the dominant follicle may play a role in its own demise as well as that of its subordinates. Upon cessation of follicle-product secretion, FSH is again allowed to surge. This surge has no effect on the dying dominant follicle, but stimulates emergence of the next wave. The ovarian cycle then continues. Relief from progestational suppression (that is, luteolysis) allows LH pulse frequency to increase, permitting further growth of the dominant follicle and markedly higher circulating concentrations of oestradiol, which results in the LH surge and ovulation.

Ovarian asymmetry

Asymmetry in follicular dynamics in the left and right ovaries of an individual has been the subject of much study, and has been used to elucidate local versus systemic mechanisms of control of ovarian function. Some workers have reported greater follicular activity in the right ovary in cattle and a higher incidence of right-side ovulation (approximately 60%; reviewed in Pierson and Ginther, 1987b), whereas others report that there are no differences (Ginther *et al.*, 1989b; Sirois and Fortune, 1988). Ginther (1989b) concluded that the dominant follicle effects follicle suppression by systemic rather than local channels. A positive intra-ovarian effect of the corpus luteum on the development of small antral follicles (≤ 3 mm) has been documented in sheep (Dufour *et al.*, 1972) and cattle (Pierson and Ginther, 1987b), but this effect did not extend to the large dominant follicles (Pierson and Ginther, 1987b; Ginther *et al.*, 1989b). On the contrary, the corpus luteum of pregnancy has been

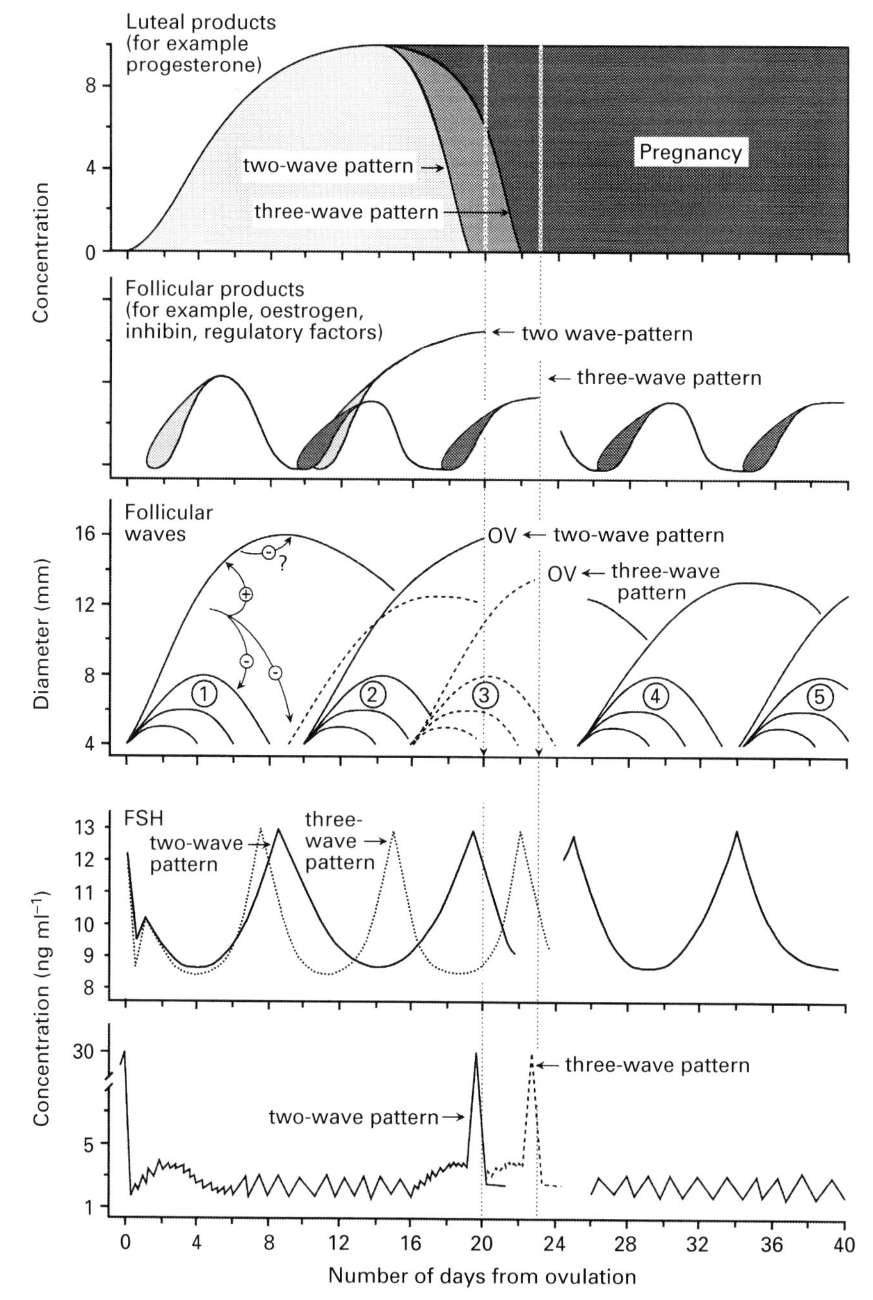

Fig. 1. Proposed model of bovine ovarian follicular wave dynamics during two-wave (——) and three-wave (------) interovulatory intervals (OV, ovulation), and early pregnancy (anovulatory; for example waves 4, 5). Follicle diameter profiles are represented in the middle panel, luteal and follicular products are represented in the upper panels, and gonadotrophin profiles are depicted in the lower panels. Shapes drawn for follicular products represent the relative number of follicles contributing to the pool at a given time. Shapes taper as subordinate follicles of each wave regress, leaving only the dominant follicle as the main producer. Episodic pulses of LH are schematic and do not represent actual pulse frequency and amplitude. Hypotheses indicated by arrows in the middle panel are (1) the dominant follicle suppresses its subordinates and emergence of the next follicular wave, and (2) the dominant follicle contributes to self-growth and self-demise. (Modified from Adams and Pierson, 1995.)

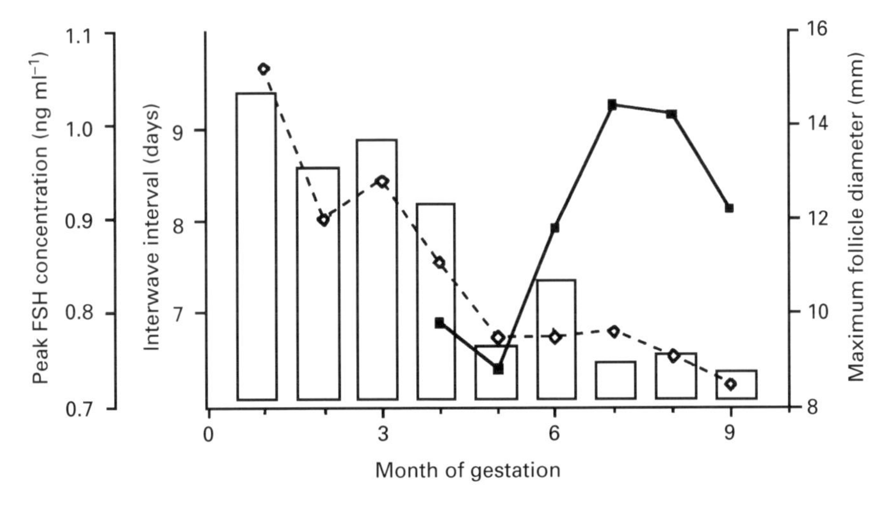

Fig. 2. Interrelationship between circulating concentrations of FSH (■), maximum diameter of the dominant follicle (◇), and the interwave interval (bars) in pregnant cattle. Values are means. (Data from Bergfelt *et al.*, 1991; Ginther *et al.*, 1996a.)

associated with a negative intra-ovarian effect on the dominant follicle (Ginther *et al.*, 1989b). Although the luteal–follicular relationships during the first two waves were similar in pregnant and nonpregnant cows (that is, no differential effects between ovaries), dominant follicles of successive waves during pregnancy were more frequently (75–80%) found in the ovary contralateral to the corpus luteum. The cause of the negative local association between the corpus luteum and the follicles is unknown but it may be directly related to the conceptus rather than to the corpus luteum (Bergfelt *et al.*, 1991; Thatcher *et al.*, 1991). These findings have important implications regarding mechanisms controlling ovarian function and the roles of locally produced ovarian peptides. It appears that the two ovaries act primarily as a single unit, and follicular counterparts between ovaries influence each other through systemic rather than local routes.

Pregnancy and postpartum period

Regular periodic emergence of anovulatory follicular waves has been detected throughout pregnancy, except for the last 21 days when follicles ≥ 6 mm in diameter were not detected (Bergfelt *et al.*, 1991; Ginther *et al.*, 1996a). In the presence of progesterone (endogenous or exogenous; Bergfelt *et al.*, 1991), anovulatory follicular waves emerged at regular intervals. However, the maximum diameter of the dominant follicle of successive waves decreased, and was associated with a successive decrease in the interwave interval (that is, the period of dominance became shorter; Fig. 2). An initial decrease occurred immediately after the first follicular wave of pregnancy, and a subsequent sharp decrease occurred after the fourth month of pregnancy. Progesterone suppresses growth of the dominant follicle in a dose-dependent manner (Adams *et al.*, 1992a), and the progressive decrease in follicular dominance seen in pregnant cattle was attributed to rising progesterone during mid-pregnancy, and rising oestrogen near term (Ginther *et al.*, 1996a). The mean interval between successive FSH surges (6.8 days) and follicular waves (6.9 days) was nearly identical and, of 118 waves examined, 83% emerged within 1 day of the peak in the FSH surge. In addition to the temporal relationships, the magnitude and frequency of FSH surges were influenced by the size of the dominant follicle. Waves in which the dominant follicle reached a diameter of ≥ 10 mm were associated with longer intervals between successive FSH surges and lower concentrations of FSH (Fig. 2). Conversely, waves with smaller dominant follicles (that is, 6–9 mm) were associated

Table 1. Characteristics of post-partum ovarian events in dairy and beef cattle

End point	Dairy cattle	Beef cattle
Emergence of first follicular wave (days postpartum)	4 (2–7)	≤ 10
Percentage ovulating from first postpartum wave	74	11
First ovulation (days postpartum)	21 (10–55)	31
First oestrus (days postpartum)	59 (17–139)	—
Percentage with a short interovulatory interval (≤ 14 days)	25	78

with shorter intervals and higher concentrations of FSH. A similar inter-relationship between FSH and follicular wave dynamics was demonstrated in prepubertal calves.

Recrudescence of follicular wave development occurs early in the postpartum period in both beef and dairy cattle (Table 1, Rajamahendran and Taylor, 1990; Savio *et al.*, 1990; Ginther *et al.*, 1996a). First ovulation was not accompanied by oestrous behaviour in 17 of 18 (94%) postpartum dairy cows, and the duration of the first postpartum interovulatory interval varied depending on when the first follicle destined to ovulate emerged. Short postpartum anovulatory periods (about 14 days) were followed by cycles of normal duration (18–21 days), whereas longer postpartum anovulatory periods (21–25 days) were followed by short cycles (< 14 days). Short cycles were associated with shorter luteal phases, a smaller corpus luteum, and lower circulating progesterone concentrations.

There were small differences in postpartum ovarian function between dairy and beef cattle and these were differences in magnitude rather than in basic nature (Table 1). The first ovulation occurred later in beef than in dairy cattle and rarely from the dominant follicle of the first postpartum wave (Murphy *et al.*, 1990). In the majority of beef cows (78%), ovulation occurred from the second, third or fourth postpartum follicular wave, and as in dairy cattle, first ovulations occurring after 20 days (16 of 18 cows) were followed by a short cycle (14 of 16 cows). Slightly earlier resumption of ovulatory cyclicity in dairy cows may be attributed to the effects of calf suckling in beef cattle, and greater selection pressure for this characteristic in dairy cattle.

Prepubertal period

Ovarian follicular dynamics in prepubertal heifers have only recently been investigated (Adams *et al.*, 1994; Evans *et al.*, 1994a, b). Transrectal ultrasonography was used to monitor daily changes in follicular development in calves from 2 weeks of age; calves were monitored at regular intervals during the first year of life. Puberty was defined as the time of the first ovulation, and was determined to be 52–56 weeks (12–13 months) in the Hereford-cross heifers used in the study. Follicular development occurred in a wave-like fashion, similar to that in adults, at all ages examined. Individual follicle development was characterized by growing, static and regressing phases, and periodic surges in serum concentrations of FSH were associated with follicular wave emergence (Fig. 3). FSH surges lasted for a mean of 3 days and were maximal 1 day before wave emergence.

It appears that adult-like interplay between the ovaries and the hypothalamo–pituitary axis begins to emerge at about 14 weeks of age. Periodic surges in FSH elicited periodic emergence of follicular waves, which in turn periodically suppressed FSH. The gradual increase in LH during the first year of life (mean and pulse frequency; Evans *et al.*, 1994a) appears to be responsible for the gradual increase in dominant follicle diameter, which in turn causes a longer period of dominance and progressively longer interwave intervals. Finally, at puberty, sufficient LH is released to induce ovulation, and follicular wave development similar to that of the prepubertal period is punctuated by regular ovulation thereafter. Unlike the prepubertal period, follicular wave development in sexually mature cattle is influenced by progesterone during regular luteal phases.

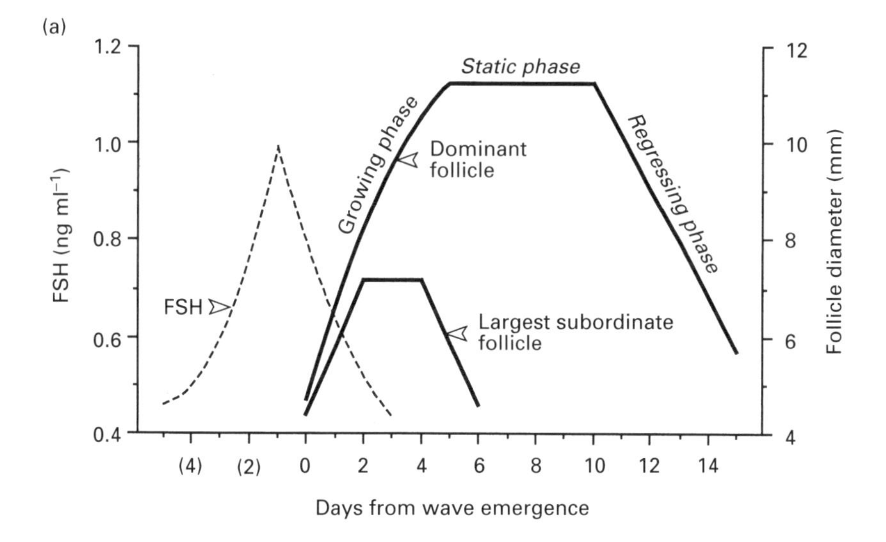

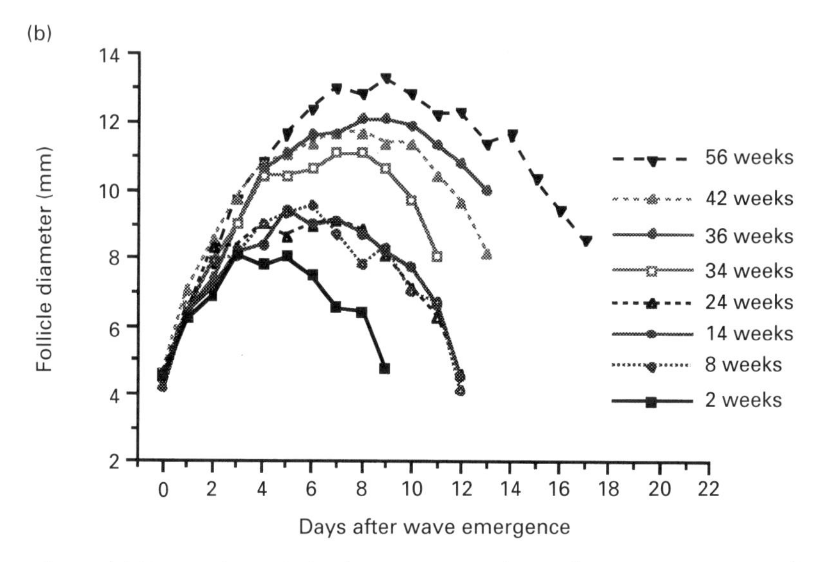

Fig. 3. (a) Temporal association between surges in circulating concentrations of FSH and emergence of anovulatory follicular waves in prepubertal heifers (regression analysis; Adams *et al.*, 1994), and (b) increasing magnitude of mean diameter profiles of the dominant follicle of successive anovulatory follicular waves in calves between 2 and 56 weeks of age. (Data from Evans *et al.*, 1994a,b.)

The diameter profile of the dominant as well as the largest subordinate follicle in prepubertal calves increased with age (Figs. 3 and 4). The increase was greatest from 2–8 weeks of age and again between 24 and 40 weeks of age, in temporal association with increases in mean concentrations of LH (Evans *et al.*, 1994b). The number of follicles detected (≥ 3 mm in diameter) also increased markedly between 8 and 14 weeks of age and then decreased thereafter, in temporal association with the early rise and subsequent decrease in FSH concentrations. The early rise in circulating concentrations of LH and FSH between 4 and 14 weeks of age may reflect initial maturation of the hypothalamo–pituitary axis and subsequent sensitivity to negative feedback by ovarian steroids.

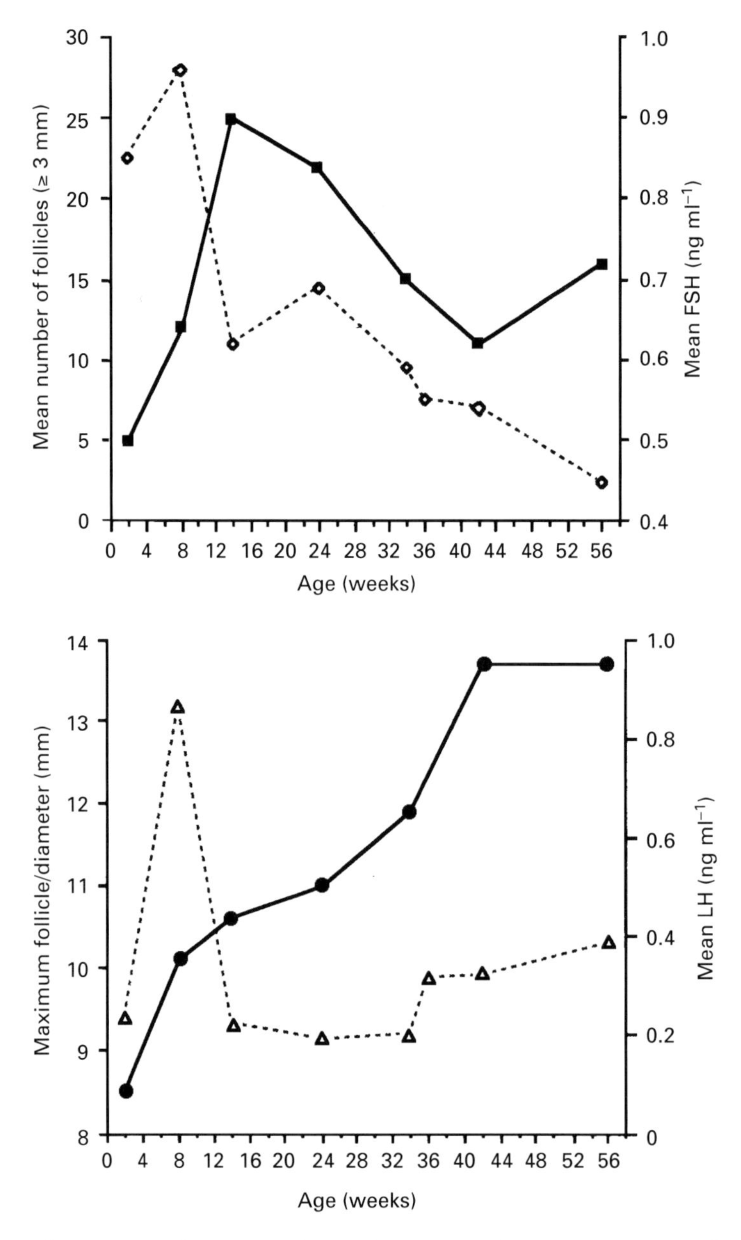

Fig. 4. Temporal association (means) between circulating concentrations of gonadotrophins (◇, FSH; △, LH) and ovarian follicle development (■, number of follicles; ●, maximum diameter of follicles) in calves from birth to puberty. Mean onset of puberty (age at first ovulation) was 56 weeks (data from Evans *et al.*, 1994a,b).

This notion is supported by data illustrated in Fig. 4, and in observed changes in the interval between the emergence of dominant follicles of successive waves (interwave interval). The interwave interval was relatively long (8–9 days) at 2 and 8 weeks of age, but was significantly shorter (6.8 days) at 14 weeks. The interwave interval gradually increased (7–9 days) thereafter until puberty.

At the onset of puberty, the first ovulatory cycle was short (7.7 ± 0.2 days) and the first ovulation occurred after the dominant follicle had entered the static phase: that is, the dominant follicle was older at the time of ovulation than its counterpart of later cycles. The first corpus luteum (from the aged ovulatory follicle) was smaller and shorter-lived than corpora lutea of subsequent cycles, thus resulting in a short interovulatory interval. The second interovulatory interval was of normal duration (20.3 ± 0.5 days) and was composed of two ($n = 3$) or three ($n = 7$) follicular waves. These results demonstrate a striking similarity between the onset of puberty in young heifers and recrudescence of ovulatory cyclicity in postpartum cows.

Other Ruminant Species

Sheep and goats

For over three decades understanding of the pattern of ovarian follicular development in sheep has been clouded by a lack of consensus among reports. Follicular growth has been described as continuous and independent of the stage of the oestrous cycle (Hutchinson and Robertson, 1966; Yenikoye *et al.*, 1989; Schrick *et al.*, 1993; Lopez-Sebastian *et al.*, 1997). Conversely, results from a number of studies involving serial hormone (oestradiol and FSH) measurement, histomorphology, repeated laparoscopy and daily ultrasonography support the notion that follicular growth is wave-like in sheep (reviewed in Noel *et al.*, 1993; Ravindra *et al.*, 1994; Ginther *et al.*, 1995; Bartelewski *et al.*, 1998a,b). Similar controversy has surrounded folliculogenesis in goats (Ginther and Kot, 1994) and the confusion is not unlike that experienced with respect to follicular dynamics in cattle during the 1980s (Adams and Pierson, 1995). Apart from inconsistencies in the number of follicular waves reported, temporal relationships between follicular and endocrine data from recent studies provide compelling support for the concept that follicular growth in sheep and goats occurs in a wave-like pattern.

In an initial study involving daily transrectal ultrasonography of Western White Face ewes (Rambouillet × Columbia), the emergence of follicles from a pool of follicles ≤ 2 mm in diameter was detected on most days of the oestrous cycle, but there was a significant increase on days 2 and 11 (day 0 = ovulation; Ravindra *et al.*, 1994). Similarly, in another study involving serial ultrasonography of Polypay ewes (Ginther *et al.*, 1995), an organized pattern of development was not detected in follicles that reached only 3 or 4 mm, but follicles that grew to ≥ 5 mm in diameter emerged at regular intervals during the oestrous cycle, leading authors to conclude that the majority of oestrous cycles consisted of four or more follicular waves (Table 2). There was a tendency for FSH to increase in concentration 2–3 days before wave emergence and there was a close correlation between the number of waves and the number of FSH peaks during the oestrous cycle (4.1 ± 0.3 and 4.5 ± 0.3, respectively) and between the duration of the interwave interval and the interval between FSH peaks (4.0 ± 0.3 days and 3.6 ± 0.2 days, respectively).

Most recent ultrasound studies (Bartelewski *et al.*, 1998b) confirm a distinct wave-like pattern of follicle development during the oestrous cycle in both non-prolific (Western White Face) and prolific (Finn) breeds of sheep (Table 2). No differences between breeds were found in the pattern of follicular development except that in the Finn sheep, the diameter of the dominant follicle was slightly smaller (5.6 ± 0.2 mm versus 6.7 ± 0.2 mm), FSH concentrations were higher around the day of ovulation, the dominant follicle from the penultimate wave of the cycle often ovulated along with that of the ultimate wave, and the ovulation rate was greater (2.7 ± 0.2 versus 1.8 ± 0.2). Consistent with the previous study, the number of follicular waves and the number of FSH peaks per cycle did not differ (3.7 ± 0.2 and 3.8 ± 0.1), and the interwave interval was highly correlated with the interpeak interval in FSH. In addition, the number of peaks in circulating oestradiol concentration did not differ from the number of follicular waves per cycle (3.5 ± 0.2 and 3.8 ± 0.1), and the interwave interval was highly correlated with the interval between oestradiol peaks. Similar findings have been reported for goats (Ginther and Kot, 1994): the predominant pattern (75%) consisted of four follicular waves emerging at intervals of 3–4 days during a 23 day oestrous cycle

Table 2. Comparative aspects of ovarian follicular waves during the oestrous cycle among domestic and wild ruminants

Species	Number of waves/cycle (% of cycles)	IWI (days)	IOI (days)	Maximum follicle diameter (mm)			
				Wave 1[a]	Wave 2	Wave 3	Wave 4
Cattle	2 (> 95%)[b]	10	20	15	15	—	—
	3	8	23	15	12	15	—
Sheep	3 (8–29%)		9–16				
	4 (60–80%)	3.5–4.5	16–17	5–7	4–6	4–6	5–7
	> 4 (0–34%)		22–24				
Goat	4	3–4	23	9	7	7	10
Musk oxen	3–4	5	23	10	7	6 (minor) 10 (major)	10

IWI = interwave interval; IOI = interovulatory interval.
[a]In all species, the first follicular wave emerges on the day of ovulation ± 1 day.
[b]Some report a majority (>80%) of either 2 waves or 3 waves.

Table 3. Characteristics of ovarian follicular waves in musk oxen ($n = 4$) during an annual cycle

Characteristic	Anovulatory season	Transition[a]	Ovulatory season[b]
Interovulatory interval (days)	—	6.5 ± 0.5	22.8 ± 1.1
Number of waves per cycle	—	1	3–4
Interwave interval (days)	6.3 ± 0.3	—	5.4 ± 0.2
Maximum follicle diameter (mm)	9.5 ± 0.3	10.5 ± 0.7	9.1 ± 0.9

[a] The first interovulatory interval of the ovulatory season.
[b]The second interovulatory interval of the ovulatory season.

(Table 2). Few changes were noted in the characteristics of follicular waves from the beginning to the end of the anovulatory season in ewes (March–July; Bartelewski *et al.*, 1998a). Periodic fluctuations in FSH were associated with regular wave emergence, and circulating LH concentrations were suppressed throughout the anovulatory season. No differences were detected in maximum follicle diameter, interwave interval or circulating concentrations of FSH during successive follicular waves of the anovulatory period. The number of 2–4 mm follicles increased transiently early in the anovulatory period (Ravindra and Rawlings, 1997; Bartelewski *et al.*, 1998a).

A confounding aspect of studying follicular dynamics in sheep and goats is the apparent difference in the nature or magnitude of follicle dominance compared with that of cattle. In sheep, the wave pattern has been detected only in follicles destined to grow to ≥ 5 mm in diameter; consequently, very few follicles (that is, one to three per wave) are detectable for characterizing the wave pattern, complete with follicle selection and dominance. Indeed, there is controversy about whether follicular dominance exists in sheep (Driancourt *et al.*, 1991; Lopez-Sebastian *et al.*, 1997), and if it does, it is certainly less distinct than in cattle. However, the following observations support the notion of dominance in sheep, particularly during the first and last waves of the cycle: (1) emergence of follicular waves associated with a follicle larger than all others was detected during metoestrus and pro-oestrus in sheep (Ravindra *et al.*, 1994; Ginther *et al.*, 1995; Bartelewski *et al.*, 1998b) and goats (Ginther and Kot, 1994), (2) following prostaglandin-induced luteolysis on various days of the oestrous cycle, the proportion of ewes that ovulated the largest follicle at the time of

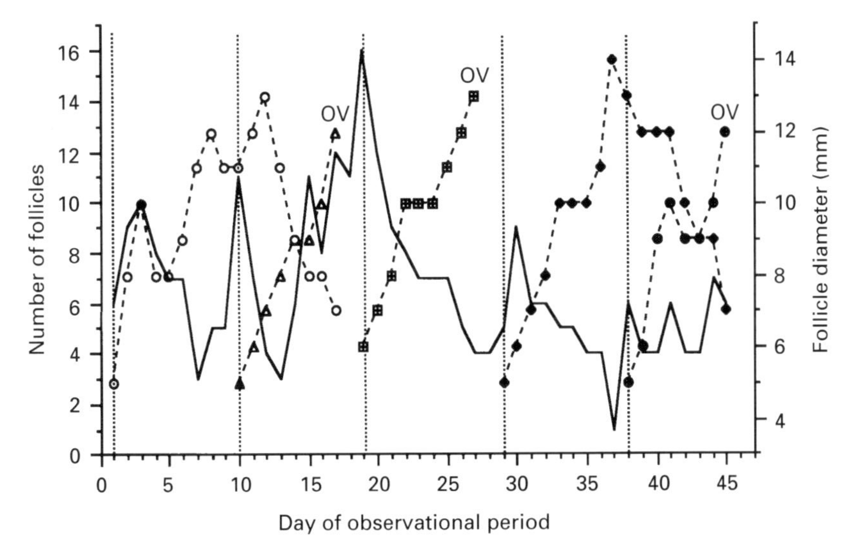

Fig. 5. Changes in the number of follicles ≥ 4 mm (——) relative to the diameter of successive dominant follicles (------) in a Wapiti hind during transition from the anovulatory period to the first 2 ovulatory cycles of the season (September to November). Vertical dotted lines indicate wave emergence. OV, ovulation (R. B. McCorkell and G. P. Adams, unpublished).

treatment and the interval to oestrus varied relative to the day of treatment (Houghton *et al.*, 1995), and (3) follicular and ovulatory responses to superstimulatory gonadotrophin treatment were influenced by the status of the follicular wave at the time of treatment, and the presence of a large growing follicle at the time treatment was initiated was associated with lower follicle recruitment, fewer ovulations, and fewer embryos (Rubianes *et al.*, 1997). These observations are consistent with those made in cattle in which variation in the ovulatory response to prostaglandin treatment and ovarian superstimulation have been attributed directly to the status of follicular dominance at the time treatment was initiated (reviewed in Adams, 1994).

Follicular waves in sheep and goats, as in horses (Ginther, 1993), may be better characterized as major waves, with a clearly discernible dominant follicle (occurring at the beginning and end of the oestrous cycle), and minor waves, with no clear dominant follicle (occurring during dioestrus). Results of independent studies suggest that the dominant follicle of waves emerging early or late in the oestrous cycle grow to a larger diameter (Table 2) and are associated with a longer lifespan than that of mid-cycle waves (Ravindra *et al.*, 1994; Ginther and Kot, 1994; Ginther *et al.*, 1995; Bartelewski *et al.*, 1998b). The suppressive effects of progesterone on the growth of the dominant follicle have been clearly documented (Adams *et al.*, 1992a; Johnson *et al.*, 1996) and provide a rationale for the hypothesis that the absence of dominance in minor waves is a result of progesterone-induced suppression of LH during dioestrus, whereas the presence of dominance in major waves is a result of a relative lack of progesterone suppression during metoestrus and pro-oestrus.

Musk oxen

In the first detailed study of ovarian follicular dynamics in a wild species, daily transrectal ultrasonography was conducted on a group of four captive musk oxen. Follicular waves were apparent during both the ovulatory (Hoare *et al.*, 1997) and anovulatory seasons (S. Parker and G. P. Adams, unpublished; Tables 2 and 3). Only one wave was detected during the first (short) cycle of the ovulatory season. During the second (long) cycle of the ovulatory season, one musk ox had three waves and the remaining three animals had four waves. Only the dominant follicle of the last wave of the oestrous cycle ovulated; the dominant follicle of other waves regressed slowly over a period of a few days. Dominance was manifested in the first and last follicular waves of the oestrous cycle in each of the four animals (major waves), whereas the other waves in all but one instance appeared to be minor waves. Wave characteristics and indistinct follicular dominance during dioestrus are remarkably similar to that observed in sheep and goats (cited above), species to which the musk ox is most closely related. Insight of this kind is important in the design of appropriate artificial breeding systems and, in this respect, the musk ox may provide a useful model for the endangered takin (*Budorcas taxicolor*). A detailed knowledge of ovarian events may also be critical to the interpretation of the response of wild populations to environmental stress.

Cervids

A paucity of information is available on ovarian function in deer. Two recent papers demonstrate different approaches to the study of follicular development, and the approaches have led to different conclusions. In one study (McLeod *et al.*, 1996), ovaries excised from hinds on different days after oestrous synchronization were dissected to determine the number and size of follicles, and to distinguish between healthy versus atretic follicles. A wide variation among hinds was observed in the total number of follicles present on a given day and in the percentage of follicles that were healthy. Results led the authors to conclude that there was a lack of an obvious pattern of follicle development and that a large oestrogenic follicle was present in all animals at all stages of the oestrous cycle. The second study (Asher *et al.*, 1997) involved transvaginal ultrasonographic monitoring of surgically modified hinds to study follicular changes on a daily basis. The ovaries were sutured to the peritoneal surface of the vaginal fornix to permit consistent access during daily examinations. Results were equivocal in that they observed a highly variable and generally non-synchronous pattern of growth and regression of the largest follicle, but emergence of new follicles 3 mm in diameter was more pronounced on days 1 and 14 of the oestrous cycle. These authors also stated that most oestrous cycles were associated with either two or three consecutive large follicles and that the emergence of one or two new follicles ≥ 3 mm in diameter occurred only during the early growth phase of the presumptive preovulatory follicle or after the demise of the preceding large follicle. In the only other study of follicular dynamics in cervids that the author is aware of, the ovaries of ten Wapiti (North American elk) were examined daily by transrectal ultrasonography during the transition from the anovulatory to the ovulatory seasons (September–November; R. B. McCorkell and G. P. Adams, unpublished). Preliminary inspection of data suggests a distinct wave-like pattern during the first (short; 10.7 days) and second (long, 19.0 days) oestrous cycles of the ovulatory season (Fig. 5).

Camelids

Unlike other ruminant species, camelids (llama, alpaca, guanaco, vicuna, and dromedary and bactrian camels) are induced or reflex ovulators. Hence, camelids possess three naturally occurring reproductive statuses which may be expected to influence follicular and luteal dynamics: (1) the unstimulated anovulatory condition, (2) ovulatory but nonpregnant (for example, non-fertile mating), and (3) ovulatory and pregnant. An initial ultrasound study of the effects of lactational and reproductive status on patterns of follicle growth and regression (Adams *et al.*, 1990) documented a distinct wave-like pattern of follicular development in llamas. Llamas were examined daily by transrectal ultrasonography for ≥ 30 days and results demonstrated that follicular activity occurred in waves for all reproductive statuses and that lactation and the presence of a corpus luteum were associated with depressed follicular development. Waves of follicular activity were indicated by periodic increases in the number of follicles detected and an associated emergence of a dominant follicle that grew to ≥ 7 mm. The dominant follicle of a wave was first identified at a diameter of 3–4 mm, and subordinate follicles did not exceed 7 mm in diameter. The emergence of the first anovulatory dominant follicle was detected, on average, 3 days after ovulation in females mated to a

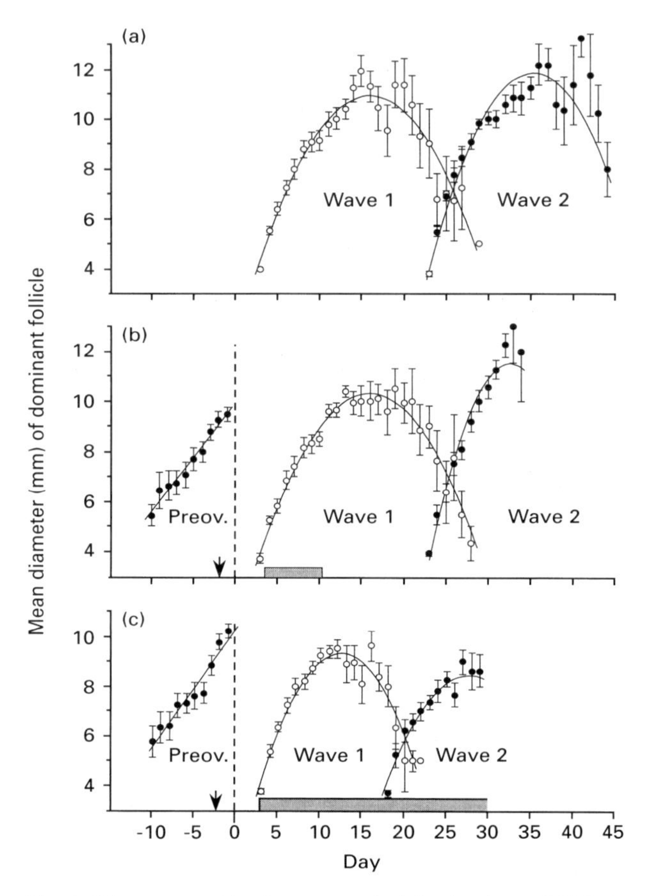

Fig. 6. Diameter profiles (mean ± SEM) of the dominant follicle in (a) anovulatory llamas (non-mated), (b) ovulatory non-pregnant llamas (mated to vasectomized males), and (c) ovulatory pregnant llamas (mated to intact males). Arrows indicate mean day of mating (ovulation = day 0) and shaded bars indicate the days of detection of the corpus luteum for the ovulatory groups (Adams *et al.*, 1990). Preov.; preovulatory.

vasectomized or intact male. The interwave interval was 19.8 ± 0.7 days in unmated and vasectomy–mated llamas and 14.8 ± 0.6 days in pregnant llamas (Fig. 6). Lactation was associated with an interwave interval that was shortened, on average, by 2.5 ± 0.5 days. Maximum diameter of anovulatory dominant follicles ranged from 9 to 16 mm and was greater for non-pregnant llamas than for pregnant llamas (anovulatory group, 12.1 ± 0.4 mm; ovulatory non-pregnant group, 11.5 ± 0.2 mm; pregnant group, 9.7 ± 0.2 mm). In addition, lactation was associated with a smaller maximum diameter of dominant follicles averaged over all reproductive statuses (10.4 ± 0.2 versus 11.7 ± 0.3 mm). The presence (ovulatory non-pregnant group) and persistence (pregnant group) of a corpus luteum was associated with a depression in the number of follicles detected and a smaller diameter profile of dominant follicles (Fig. 6).

Reports in alpacas using a laparoscopic technique (cited in Bravo *et al.*, 1990), and in llamas using transrectal palpation or ultrasonography (Bravo *et al.*, 1990), are consistent with the presence of a wave-like pattern of follicular growth in camelids. However, notable differences in the character of follicular growth were found between the studies. The interval between emergence of successive follicles was not stated, but examination of follicular profiles from individual alpacas shows an approximate interwave interval of 15 days. Similar results were obtained in the llama study; the apparent dominant follicle spanned approximately 14 days and the apparent mean interwave interval was 11.1 days. In studies with both alpaca and llama, the dominant follicle of successive waves alternated regularly between left and right ovaries in over 80% of the intervals. In contrast, results of others (Adams *et al.*, 1990) indicated that the growth and regression profile of the dominant anovulatory follicle of each wave lasted 20–25 days and the interwave interval was 20 days for non-pregnant llamas. Moreover, the incidence of alternation of successive dominant follicles between left and right ovaries did not differ from the incidence of non-alternation. The differences between studies seem too great to reconcile, but lactational status may have contributed to some of the disparity in results. However, the issue has become more confusing by a recent report in which the interwave interval for postpartum llamas was only 8 days (Bravo *et al.*, 1995).

The wave-like pattern of follicular development has also been documented in dromedary camels (Skidmore *et al.*, 1995), and wave characteristics are similar to those of llamas. As in llamas, distinct follicular dominance was manifest by a strong inverse relationship between the number of follicles detected and the diameter of the largest follicle. The interwave interval for unmated camels was 18.2 days.

Conclusion

The process of follicular development permits some (one in monovular species) follicles to continue to grow and have the potential to ovulate (appropriately a competitive process with much physiological reserve or excess), while at the same time minimizes attrition from the reserve pool by suppressing recruitment between waves. Periodic follicular suppression of FSH preserves the resources of the ovary by preventing continuous recruitment of large antral follicles, 99% of which are lost to atresia. To date, the wave pattern of follicular development has been demonstrated in every species in which it has been examined including cattle, sheep, goats, horses, camelids, and some wild ungulates. Concepts emerging from results of recent studies of ovarian function in these species are consistent with mechanisms implied in the bovine model (Fig. 1). Differences in follicular dynamics between ruminant species appear to be more in detail rather than in essence. Data from many studies involving serial hormone measurement, gross and histological examination of excised ovaries, repeated laparoscopy, and daily ultrasonography support the notion that (1) follicles grow in a wave-like fashion; (2) periodic surges in circulating FSH are associated with follicular wave emergence; (3) selection of a dominant follicle involves a decline in FSH and acquisition of LH responsiveness; (4) periodic anovulatory follicular waves continue to emerge until occurrence of an LH surge (that is, at the time of luteolysis during the ovulatory season or during transition from the anovulatory season); (5) within species, there is a positive relationship between the duration of the oestrous cycle and the number of follicular waves; (6) progesterone is suppressive to LH secretion and to the growth of the dominant follicle; (7) the duration of the interwave interval is a function of follicular dominance, and is negatively correlated with circulating FSH; (8) follicular dominance in all species is more pronounced during the first and last follicular waves of the oestrous cycle; and (9) pregnancy, the prepubertal period, and seasonal anoestrus are characterized by regular, periodic surges in FSH and emergence of anovulatory follicular waves.

References

Adams GP (1994) Control of ovarian follicular wave dynamics in cattle: implication for synchronization and superstimulation *Theriogenology* **41** 25–30

Adams GP and Pierson RA (1995) Bovine model for study of ovarian follicular dynamics in humans *Theriogenology* **43** 113–120

Adams GP, Sumar J and Ginther OJ (1990) Effects of lactational status and reproductive status on ovarian follicular waves in llamas (*Lama glama*) *Journal of Reproduction and Fertility* **90** 535–545

Adams GP, Matteri RL and Ginther OJ (1992a) The effect of progesterone on growth of ovarian follicles, emergence of follicular waves and circulating FSH in heifers *Journal of Reproduction and Fertility* **95** 627–640

Adams GP, Matteri RL, Kastelic JP, Ko JCH and Ginther OJ (1992b) Association between surges of follicle stimulating hormone and the emergence of follicular waves in heifers *Journal of Reproduction and Fertility* **94** 177–188

Adams GP, Kot K, Smith CA and Ginther OJ (1993a) Selection of a dominant follicle and suppression of follicular growth in heifers *Animal Reproduction Science* **30** 259–271

Adams GP, Kot K, Smith CA and Ginther OJ (1993b) Effect of the dominant follicle on regression of its subordinates in heifers *Canadian Journal of Animal Science* **73** 267–275

Adams GP, Evans ACO and Rawlings NC (1994) Follicular waves and circulating gonadotrophins in 8-month-old prepubertal heifers *Journal of Reproduction and Fertility* **100** 27–33

Ahmad N, Townsend EC, Dailey RA and Inskeep EK (1997) Relationship of hormonal patterns and fertility to occurrence of two or three waves of ovarian follicles, before and after breeding, in beef cows and heifers *Animal Reproduction Science* **49** 13–28

Asher GW, Scott IC, O'Neill KT, Smith JF, Inskeep EK and Townsen EC (1997) Ultrasonographic monitoring of antral follicle development in red deer (*Cervus elaphus*) *Journal of Reproduction and Fertility* **111** 91–99

Bartelewski PM, Beard AP, Cook SJ and Rawlings NC (1998a) Ovarian follicular dynamics during anoestrus in ewes *Journal of Reproduction and Fertility* **113** 275–285

Bartelewski PM, Beard AP, Cook SJ, Chandolia RK, Honoramooz A and Rawlings NC (1998b) Ovarian antral follicular dynamics and their relationships with endocrine variables throughout the oestrous cycle in breeds of sheep differing in prolificacy *Journal of Reproduction and Fertility* **115** 111–124

Bergfelt DR, Kastelic JP and Ginther OJ (1991) Continued periodic emergence of follicular waves in nonbred progesterone-treated heifers *Animal Reproduction Science* **24** 193–204

Bergfelt DR, Plata-Madrid H and Ginther OJ (1994) Counteraction of the follicular fluid inhibitory effect of follicular fluid by administration of FSH in heifers *Canadian Journal of Animal Science* **74** 633–639

Bravo PW, Fowler ME, Stabenfeldt GH and Lasley BL (1990) Ovarian follicular dynamics in the llama *Biology of Reproduction* **43** 579–585

Bravo PW, Lasley BL and Fowler ME (1995) Resumption of ovarian follicular activity and uterine involution in the postpartum llama *Theriogenology* **44** 783–791

Driancourt MA, Webb R and Fry RC (1991) Does follicular dominance occur in ewes? *Journal of Reproduction and Fertility* **93** 63–70

Dufour J, Ginther OJ and Casida LE (1972) Intraovarian relationship between corpora lutea and ovarian follicles in ewes *American Journal of Veterinary Research* **33** 1445–1446

Evans ACO, Adams GP and Rawlings NC (1994a) Endocrine and ovarian follicular changes leading up to the first ovulation in prepubertal heifers *Journal of Reproduction and Fertility* **100** 187–194

Evans ACO, Adams GP and Rawlings NC (1994b) Follicular and hormonal development in prepubertal heifers from 2 to 36 weeks of age *Journal of Reproduction and Fertility* **102** 463–470

Gibbons JR, Wiltbank MC and Ginther OJ (1997) Functional interrelationships between follicles greater than 4 mm and the follicle-stimulating hormone surge in heifers *Biology of Reproduction* **57** 1066–1073

Ginther OJ (1993) Major and minor follicular waves during the equine estrous cycle *Journal of Equine Veterinary Science* **13** 18–25

Ginther OJ and Kot K (1994) Follicular dynamics during the ovulatory season in goats *Theriogenology* **42** 987–1001

Ginther OJ, Knopf L and Kastelic JP (1989a) Temporal associations among ovarian events in cattle during oestrous cycles with two and three follicular waves *Journal of Reproduction and Fertility* **87** 223–230

Ginther OJ, Kastelic JP and Knopf L (1989b) Intraovarian relationships among dominant and subordinate follicles and the corpus luteum in heifers *Theriogenology* **32** 787–795

Ginther OJ, Kot K and Wiltbank MC (1995) Associations between emergence of follicular waves and fluctuations in FSH concentrations during the estrous cycle in ewes *Theriogenology* **43** 689–703

Ginther OJ, Kot K, Kulick LJ, Martin S and Wiltbank MC (1996a) Relationships between FSH and ovarian follicular waves during the last six months of pregnancy in cattle *Journal of Reproduction and Fertility* **108** 271–279

Ginther OJ, Wiltbank MC, Fricke PM, Gibbons JR and Kot K (1996b) Selection of the dominant follicle in cattle *Biology of Reproduction* **55** 1187–1194

Hoare EK, Parker SE, Flood PF and Adams GP (1997) Ultrasonic imaging of reproductive events in muskoxen *Rangifer* **17** 119–123

Houghton JAS, Liberati N, Schrick FN, Townsend EC, Dailey RA and Inskeep EK (1995) Day of estrous cycle affects follicular dynamics after induced luteolysis in ewes *Journal of Animal Science* **73** 2094–2101

Hutchinson JSM and Robertson HA (1966) The growth of the follicle and corpus luteum in the ovary of the sheep *Research in Veterinary Science* **7** 17–24

Johnson SK, Dailey RA, Inskeep EK and Lewis PE (1996) Effect of peripheral concentrations of progesterone on follicular growth and fertility in ewes *Domestic Animal Endocrinology* **13** 69–79

Lopez-Sebastian A, Gonzalez de Bulnes A, Santiago Moreno J, Gomez-Brunet A, Townsend EC and Inskeep EK (1997) Patterns of follicular development during the estrous cycle in monovular Merino del Pais ewes *Animal Reproduction Science* **48** 279–291

McLeod BJ, Meikle LM, Heath DA, McNatty KP, Fisher MW and Whaanga AJ (1996) Ovarian follicle development in the red deer hind *New Zealand Society of Animal Production* **56** 370–372

Murphy MG, Boland MP and Roche JF (1990) Pattern of follicular growth and resumption of ovarian activity in post-partum beef suckler cows *Journal of Reproduction and Fertility* **90** 523–533

Murphy MG, Enright WJ, Crowe MA, McConnell K, Spicer LJ, Boland MP and Roche JF (1991) Effect of dietary intake on pattern of growth of dominant follicles during the estrous cycle in beef heifers *Journal of Reproduction and Fertility* **92** 333–338

Noel B, Bister JL and Paquay R (1993) Ovarian follicular dynamics in Suffolk ewes at different periods of the year *Journal of Reproduction and Fertility* **99** 695–700

Pierson RA and Ginther OJ (1987a) Follicular populations during the estrous cycle in heifers: I. Influence of day *Animal Reproduction Science* **14** 165–176

Pierson RA and Ginther OJ (1987b) Follicle populations during oestrous cycle in heifers: II. Influence of right and left sides and intraovarian effect of the corpus luteum *Animal Reproduction Science* **14** 177–186

Rajamahendran R and Taylor C (1990) Characterization of ovarian activity in postpartum dairy cows using ultrasound imaging and progesterone profiles *Animal Reproduction Science* **22** 171–180

Ravindra JP and Rawlings NC (1997) Ovarian follicular dynamics in ewes during the transition from anoestrus to the breeding season *Journal of Reproduction and Fertility* **110** 279–289

Ravindra JP, Rawlings NC, Evans ACO and Adams GP (1994) Ultrasonographic study of ovarian follicular dynamics in ewes during the oestrous cycle *Journal of Reproduction and Fertility* **101** 501–509

Rhodes JM, De'ath G and Entwistle KW (1995) Animal and temporal effects on ovarian follicular dynamics in Brahman heifers *Animal Reproduction Science* **38** 265–277

Roche JF and Boland MP (1991) Turnover of dominant follicles in cattle of different reproductive states *Theriogenology* **35** 81–90

Rubianes E, Ungerfeld R, Vinoles C, Rivero A and Adams GP (1997) Ovarian response to gonadotrophin treatment initiated relative to wave emergence in ultrasonographically monitored ewes *Theriogenology* **47** 1479–1488

Savio JD, Keenan L, Boland MP and Roche JF (1988) Pattern of growth of dominant follicles during the oestrous cycle of heifers *Journal of Reproduction and Fertility* **83** 663–671

Savio JD, Boland MP, Hynes N and Roche JF (1990) Resumption of follicular activity in the early postpartum period of dairy cows *Journal of Reproduction and Fertility* **88** 569–579

Schrick FN, Surface RA, Pritchard JY, Dailey RA, Townsend EC and Inskeep EK (1993) Ovarian structures during the estrous cycle and early pregnancy in ewes *Biology of Reproduction* **49** 1133–1140

Sirois J and Fortune JE (1988) Ovarian follicular dynamics during the estrous cycle in heifers monitored by real-time ultrasonography *Biology of Reproduction* **39** 308–317

Skidmore JA, Billah M and Allen WR (1995) The ovarian follicular wave pattern in the mated and non-mated dromedary camel (*Camelus dromedarius*) *Journal of Reproduction and Fertility* **49** 545–548

Thatcher WW, Driancourt MA, Terqui M and Badinga L (1991) Dynamics of ovarian follicular development in cattle following hysterectomy and during early pregnancy *Domestic Animal Endocrinology* **8** 223–234

Yenikoye A, Mariana JC and Celeux G (1989) Follicular growth during the oestrous cycle in Peul sheep *Animal Reproduction Science* **21** 201–211

Zeitoun MM, Rodriguez HF and Randel RD (1996) Effect of season on ovarian follicular dynamics in Brahman cows *Theriogenology* **45** 1577–1581

Journal of Reproduction and Fertility Supplement **54**, 33–48

Molecular mechanisms regulating follicular recruitment and selection

R. Webb[1], B. K. Campbell[2†], H. A. Garverick[3], J. G. Gong[4], C. G. Gutierrez[4*] and D. G. Armstrong[4]

[1]*Division of Agriculture and Horticulture, School of Biological Sciences, University of Nottingham, Sutton Bonington Campus, Loughborough, Leicestershire, LE12 5RD, UK;* [2]*Department of Obstetrics and Gynaecology, University of Edinburgh, 37 Chalmers Street, Edinburgh, UK;* [3]*Department of Animal Sciences, University of Missouri-Columbia, Columbia, MO 65211, USA; and* [4]*Division of Development and Reproduction, Roslin Institute (Edinburgh), Roslin, Midlothian EH25 9PS, UK*

Ovarian follicular growth and development is an integrated process encompassing both extraovarian signals, such as gonadotrophins and metabolic hormones, and intraovarian factors. Follicular development has been classified into gonadotrophin-independent and -dependent phases. In the latter, FSH provides the primary drive for follicular recruitment and LH is required for continued development of follicles to the preovulatory stage. A transient increase in circulating FSH precedes the recruitment of a group of follicles, and these recruited follicles are characterized by expression of mRNAs encoding P450scc and P450arom in granulosal cells. As follicles mature, there is a transfer of dependency from FSH to LH, which may be part of the mechanism(s) involved in selection of follicles for continued growth. Indeed, changes in the pattern of expression of mRNA for gonadotrophin receptors and steroid enzymes within follicular cells appear to be closely linked to changes in peripheral concentrations of gonadotrophins. The mechanism of selection of dominant follicles still requires clarification, but seems to be linked to the timing of mRNA expression encoding LHr and 3β-hydroxysteroid dehydrogenase (3βHSD) in granulosal cells. Additional intraovarian systems, including the ovarian IGF and activin/inhibin systems, also exert a role. For example, it appears that the development of follicular dominance in cows is associated with the FSH-dependent inhibition of the expression of mRNA encoding insulin-like growth factor binding protein 2 (IGFBP-2) in granulosal cells. In conclusion, the integration of these endocrine signals and intraovarian factors within follicles determines whether follicles continue to develop and become dominant or are diverted into apoptotic pathways leading to atresia.

Introduction

Ovarian follicular growth is a developmental process during which follicles sequentially acquire a number of characteristics, each of which is an essential prerequisite for further development. The number of follicles that reach the ovulatory stage is regulated in a species- and breed-specific manner. However, many key mechanisms involved in this developmental process are still not understood, including (i) factors regulating the initiation of primordial follicle growth, (ii) control of antrum formation, (iii) mechanisms controlling follicle recruitment, selection and dominance, and (iv) the process of follicular atresia, the end-point of > 99% of follicles. However, molecular techniques have been used in domestic species over recent years to elucidate the patterns of expression of local follicular factors involved in this differentiative process.

*Present Address: Fac. Med. Vet. Zool., Universidad Nacional Autonome de Mexico, Mexico, IDF 04510.
†Present Address: School of Human Development, Queens Medical Centre, University of Nottingham, NG7 2UH, UK

This review will concentrate on some primary factors involved in follicular recruitment, selection and dominance. Recent work investigating patterns in gene expression will be reviewed, as this has provided new insights into possible key mechanisms. A brief outline of the patterns of follicular growth and the relationship with gonadotrophins and metabolic hormones will be described, since the control mechanisms involve the interaction of systemic hormones and intrafollicular factors.

Definitions of Stages of Follicular Development

For this review the following generally accepted definitions have been used:
Recruitment – gonadotrophins stimulation of a pool of rapidly growing follicles.
Selection – a process whereby one or more of these recruited follicles are selected to continue to develop further.
Dominance – the mechanism(s) by which the dominant follicle(s), the number of which is species- and breed-specific, undergoes rapid development in an environment where growth and development of other follicles are suppressed.

Early Follicular Growth

Growth of antral follicles to 2–4 mm in cattle and 1–2 mm in sheep is thought to be independent of gonadotrophins. Follicles can grow to this size in either the absence of gonadotrophins or the presence of very low concentrations of gonadotrophins (Campbell *et al.*, 1995; Webb and Armstrong, 1998). However, recent studies investigating follicular growth patterns using transrectal ultrasonography show that small follicles (approximately 2 mm) do exhibit waves of growth, as discussed later, suggesting that they are responsive to gonadotrophins.

Gene expression in preantral and early antral follicles

Gonadotrophin receptors. FSH receptor (FSHr) mRNA is localized specifically to both mural and cumulus granulosal cells and can be detected in follicles with only one or two layers of granulosal cells (Xu *et al.*, 1995a). The role of the FSHr in preantral and early antral follicular growth is unknown. However, gonadotrophins do not seem to be required for the activation of bovine primordial follicles during culture of bovine ovarian cortical slices (Wandji *et al.*, 1996). In contrast, late preantral follicles in culture show increased growth in response to FSH (Ralph *et al.*, 1995). We have recently developed a system in which bovine preantral follicles can be sustained in long-term culture and proceed to develop an antrum (Gutierrez *et al.*, 1997a). In this system follicles were responsive to the stimulatory effects of both FSH and a number of growth factors including IGF-I and EGF.

Expression of LH receptor (LHr) mRNA is localized to thecal cells during the preantral and early antral stages of growth, and expression is detected when the theca interna forms around the granulosal cells (Xu *et al.*, 1995a,b; Bao *et al.*, 1997a).

Steroidogenic enzymes. mRNA for steroidogenic enzymes, cytochrome P450 side-chain cleavage (P450scc), cytochrome P450 17α-hydroxylase (P450c17), and 3β-hydroxysteroid dehydrogenase (3β-HSD) are expressed soon after formation of the theca interna. Expression of steroid acute regulatory protein (StAR) mRNA has also been detected in thecal cells (Bao *et al.*, 1997b). Expression of mRNA for these enzymes tends to increase with growth of these early antral follicles (Xu *et al.*, 1995a,b; Bao *et al.*, 1997a,c). Cytochrome P450 aromatase (P450arom) is localized solely to granulosal cells, but expression cannot be found in non-recruited follicles < 4 mm in diameter. Apparently, the main steroid hormones produced by preantral and early antral follicles are pregnenolone, progesterone and androgen from thecal cells. This finding is in agreement with earlier work that measured *in vitro* steroid production and steroid concentrations in follicular fluid (see Webb and Gauld, 1987; Skyer *et al.*, 1987).

The role of gonadotrophins in the induction of mRNA encoding gonadotrophin receptor and steroidogenic enzymes in preantral and early antral follicles is unknown. However, mRNA expression of FSHr in granulosal cells and expression of LHr, P450scc and P450c17 in thecal cells was not different in follicles < 4 mm in diameter in heifers with normal oestrous cycles compared with heifers in which follicles were arrested at approximately 4 mm in diameter due to GnRH agonist inhibition of FSH and LH secretion (H. A. Garverick, J. G. Gong, B. Baxter, D. G. Armstrong, B. K. Campbell and R. Webb, unpublished).

Waves of Follicular Growth in Cattle and Sheep

Two or three major phases of growth of large follicles occur during the bovine oestrous cycle, and the ovulatory follicle is selected at about 3 days before ovulation (see Ginther *et al.*, 1996; Webb and Armstrong, 1998). Each wave of follicular development is characterized by the simultaneous emergence of medium-sized (> 4–8 mm in diameter) follicles from a pool of smaller follicles. A dominant follicle emerges and continues to develop, while the others undergo atresia. The dominant follicle remains dominant for a few days, until it too becomes atretic and regresses, to be replaced within approximately 5 days by the next dominant follicle from the next follicular wave. If luteal regression takes place, the dominant follicle, free from the restrictive hormonal milieu imposed by the corpus luteum upon the hypothalamus–pituitary gland, will continue to develop (up to 20 mm in diameter) and will trigger the hormonal cascade leading to ovulation. Follicular waves appear to be constitutive, because they are present before puberty, throughout most of pregnancy and during the post-partum period, as well as during the oestrous cycle (Ginther *et al.*, 1996; Webb and Armstrong, 1998).

In sheep, transrectal ultrasonography has proved more difficult to perform and interpret than in cattle because of problems of anatomical access and the smaller size and greater number of ovulatory follicles. Although evidence from both histological and ink marking studies support the occurrence of follicular waves (see Campbell *et al.*, 1995), a number of studies using ultrasonography have reported random emergence of ovulatory-sized follicles throughout the sheep oestrous cycle (Schrick *et al.*, 1993; Ravindra and Rawlings, 1997). Recent results, using the ovarian autotransplant model during seasonal anoestrus (Souza *et al.*, 1996) and during the follicular (Souza *et al.*, 1997a,b) and luteal (Souza *et al.*, 1998) phases of the oestrous cycle, indicate that large antral follicles in sheep exhibit wave-like cycles (Fig. 1). All of these studies indicate that there is a period of functional dominance characterized by high oestradiol and inhibin A secretion, shorter than the period of morphological dominance, although dominant follicles are not the only source of inhibin A. Therefore, in sheep, follicular size alone is not an adequate parameter to assign dominance.

The overall pattern of follicle turnover in sheep during the luteal phase appears to be similar to that in cows, but there are clear species differences. In sheep, the wave interval is just 4–5 days, perhaps reflecting the smaller diameter of the dominant follicle, so that the ovulatory wave in sheep is likely to be either the third or the fourth, rather than either the second or the third, as in cattle. Furthermore, sheep can have more than one dominant follicle per wave, depending on the ovulation rate of the breed (Fig. 1), compared to usually one dominant follicle per wave in cows, indicating that follicular dominance is not so intense in sheep (Driancourt *et al.*, 1991).

Follicular Waves Associated with Patterns of Hormone Secretion

Although it is clear that waves of dominant follicle development occur in both sheep and cattle, the endocrine and local mechanisms associated with this pattern of development have not been fully elucidated. It is well established, in both cattle (Adams *et al.*, 1992) and sheep (Figs 1 and 2), that the emergence of a follicular wave is preceded by a transient increase in FSH and that the secretion of FSH is regulated by oestradiol and inhibin (Campbell *et al.*, 1995). During the first follicular wave

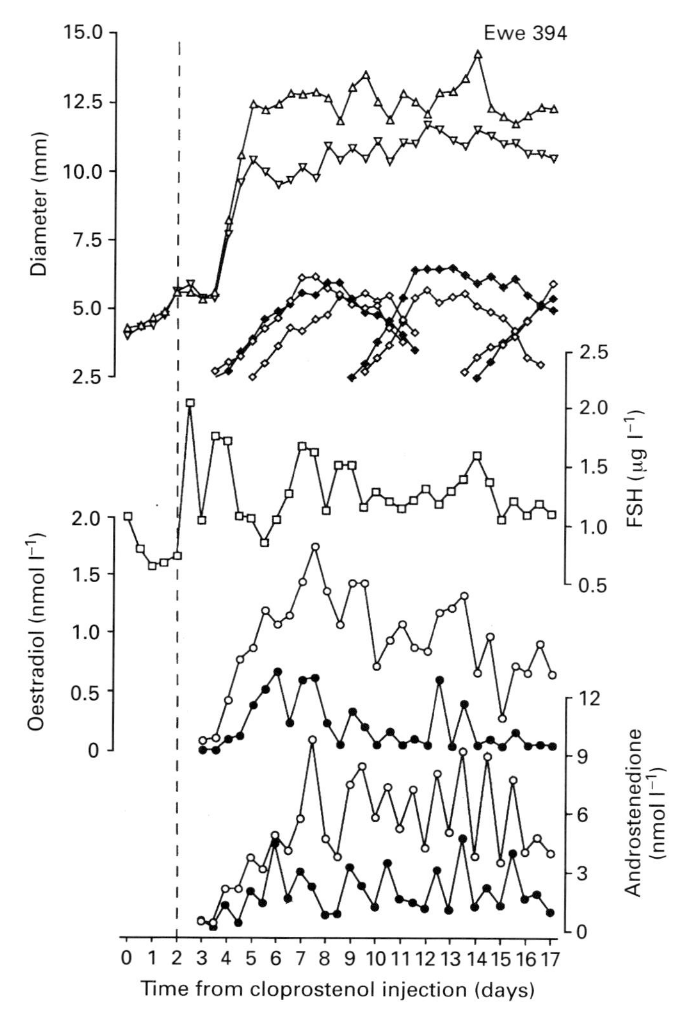

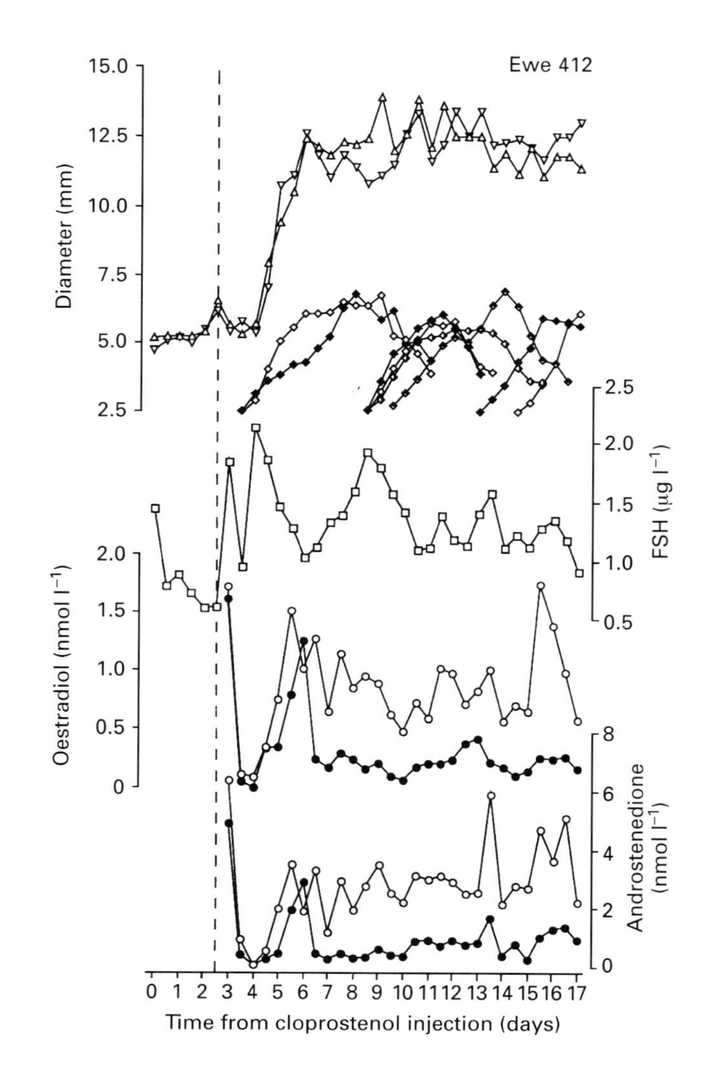

Fig. 1. For legend see facing page.

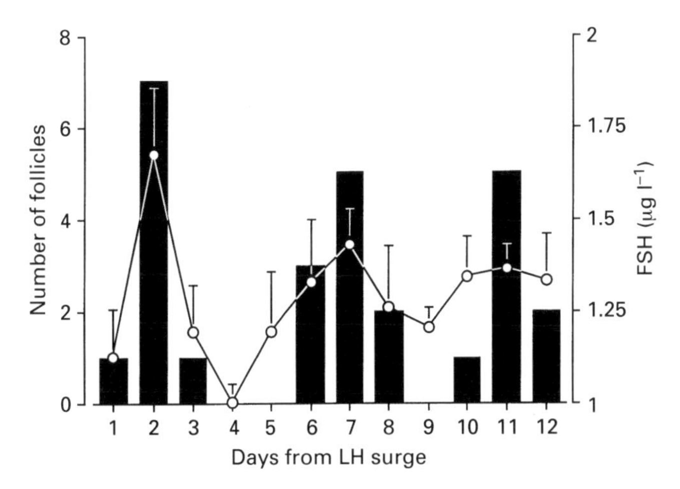

Fig. 2. Relationship between emergence of dominant follicles (solid bars) and mean concentration of FSH (open circles) in jugular venous blood (± SEM, $n = 5$) during the luteal phase of the ovine oestrous cycle. Adapted from Souza *et al.* (1998).

there is compelling evidence to support this explanation: secretion of both oestradiol (Knight, 1996; Souza *et al.*, 1997a, 1997b, 1998) and inhibin A (Souza *et al.*, 1997b, 1998) are positively related to follicular growth and inversely related to FSH. However, over subsequent waves, these relationships are less evident. Ovarian oestradiol secretion is dependent on both the presence of a large oestrogenic follicle in the ovary and appropriate LH stimulation (Campbell *et al.*, 1995).

During the first follicular wave, LH pulse frequency is high and ovarian oestradiol secretion reflects this. However, during subsequent follicular waves LH pulse frequency declines as a result of luteal progesterone. Hence secretion of oestradiol no longer predicts the secretory capacity of the dominant follicle accurately (Souza *et al.*, 1998). Indeed, wave 2 and 3 dominant follicles in sheep secrete the same amount of androstenedione, but less oestradiol, when challenged with LH. Therefore, it seems likely that a period of exposure to high frequency LH pulses is required before follicles acquire the ability to secrete normal amounts of oestradiol (Souza *et al.*, 1998). As indicated, in contrast to oestradiol, inhibin A, measured in the sheep autotransplant model, is not solely derived from the dominant follicle(s) and is secreted in relatively high and constant amounts throughout the luteal phase. Thus the contribution of multiple follicles to ovarian secretion of inhibin A explains the lack of association between inhibin A and development of the dominant follicle during the mid–late luteal phase. However, as inhibin A secretion remains high during this phase of the ovarian cycle, when oestradiol secretion is low, it appears that inhibin A is the major regulator of FSH during the mid–late luteal phase. Thus, although it is clear that oestradiol and inhibin A control FSH secretion during the first follicular wave, the precise regulation of the FSH fluctuations associated with waves 2 and 3 awaits studies on the pattern of secretion of dimeric inhibin B.

We have recently developed an experimental model to investigate the relative importance of FSH and LH during different stages of follicular development. In this model, heifers are

---

Fig. 1. Dynamics of ovulatory follicles and/or corpus luteum (triangle) and dominant follicles (diamonds) from the three waves of follicular development during the luteal phase (top panel) and concentration of FSH (squares) in jugular venous plasma and concentration of oestradiol and androstenedione in ovarian venous plasma during the luteal phase from two representative ewes. The results show basal steroid concentrations (filled circles) and the steroid concentrations following a GnRH-challenge (250 ng i.v.) (open circles). The dotted line indicates the time of the onset of the LH surge. Adapted from Souza *et al.* (1998).

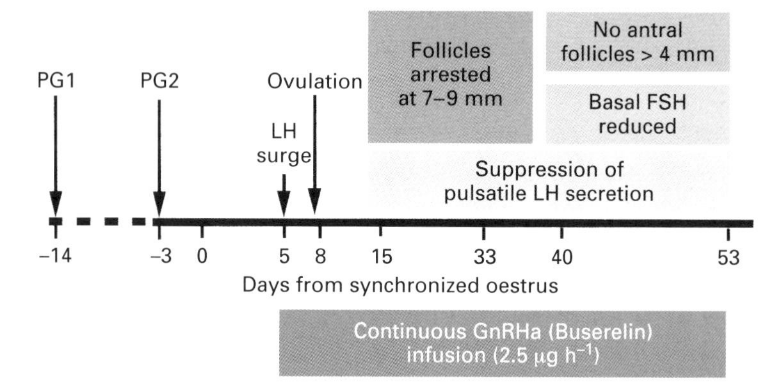

Fig. 3. Schematic summary of ovarian follicular dynamics, as measured by ultrasound per rectum, and association with patterns of gonadotrophin secretion in heifers treated with a GnRH agonist. Two injections of prostaglandin $F_{2\alpha}$ (PG1 and PG2) were given to synchronize oestrous cycles. Note that growth was arrested initially at 7–9 mm diameter and then at < 4 mm in diameter.

continuously infused with GnRH agonist (Fig. 3). We found that the growth of follicles is arrested at 7–9 mm in diameter when pulsatile LH secretion is suppressed (Gong *et al.*, 1995, 1996). When the secretion of basal FSH was also reduced, no antral follicles > 4 mm in diameter were observed (Gong *et al.*, 1996). Using this model in conjunction with FSH, LH, or LH and FSH infusion, we have shown that FSH is initially required for the development of gonadotrophin responsive follicles (Campbell *et al.*, 1995, 1998a; Gong *et al.*, 1997). However, large antral follicles can transfer their dependence on gonadotrophins from FSH to LH (Campbell *et al.*, 1995; Gong *et al.*, 1996b). Furthermore, adequate LH pulsatile support appears to be required to maintain the ovulatory competence of the preovulatory follicles under decreased FSH concentrations (Campbell *et al.*, 1995a). Indeed increased pulsatile secretion of LH has also been associated with the extended lifespan of dominant follicles (Sirois and Fortune, 1990). How do these changes in patterns of hormone release and follicular growth correlate with changes within the follicles themselves?

Expression of mRNA for Gonadotrophin Receptors and Steroidogenic Enzymes during Follicular Growth

Gene expression during recruitment

In cattle, follicular recruitment is generally thought to be gonadotrophin dependent and is considered to occur when follicles are stimulated to grow beyond 4 mm in diameter. Recruitment is also characterized by initiation of growth of a cohort of up to seven follicles that continue to grow to 8–9 mm in diameter. Thereafter usually one follicle in the cohort diverges rapidly from the others and continues to mature.

Gonadotrophin receptors. FSHr and LHr remain localized to granulosal and thecal cells, respectively, during recruitment and growth of the cohort (Fig. 4; Xu *et al.*, 1995a,b; Bao *et al.*, 1997a,c). During this time there is little change in expression of mRNA encoding FSHr and LHr (Table 1).

Steroidogenic enzymes. The initiation of simultaneous mRNA expression of P450scc and P450arom in granulosal cells of most follicles of 4–6 mm in diameter is associated with follicular recruitment in cattle (Xu *et al.*, 1995a,b; Bao *et al.*, 1997a). During later stages when the recruited follicles reach

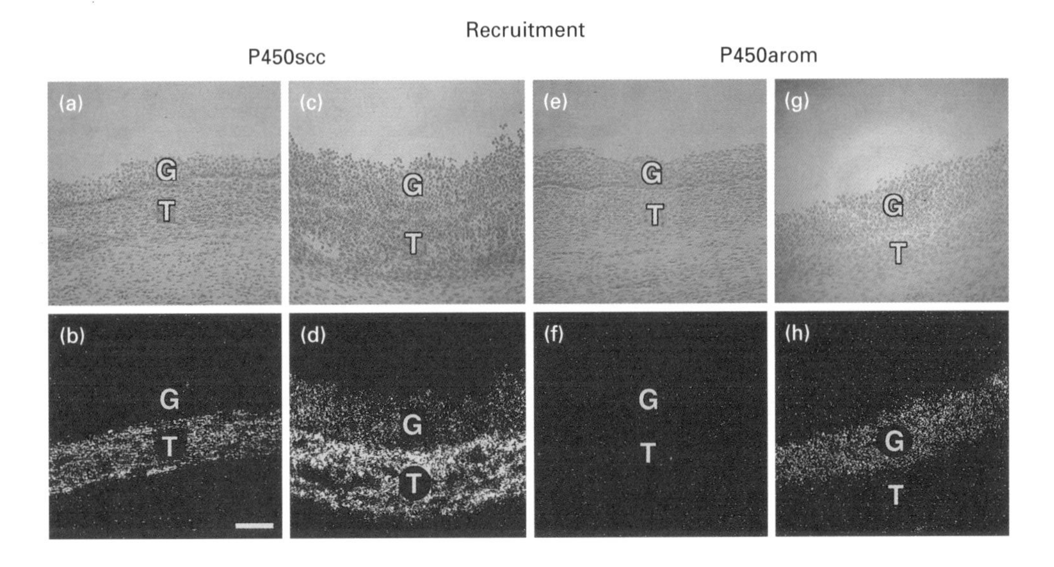

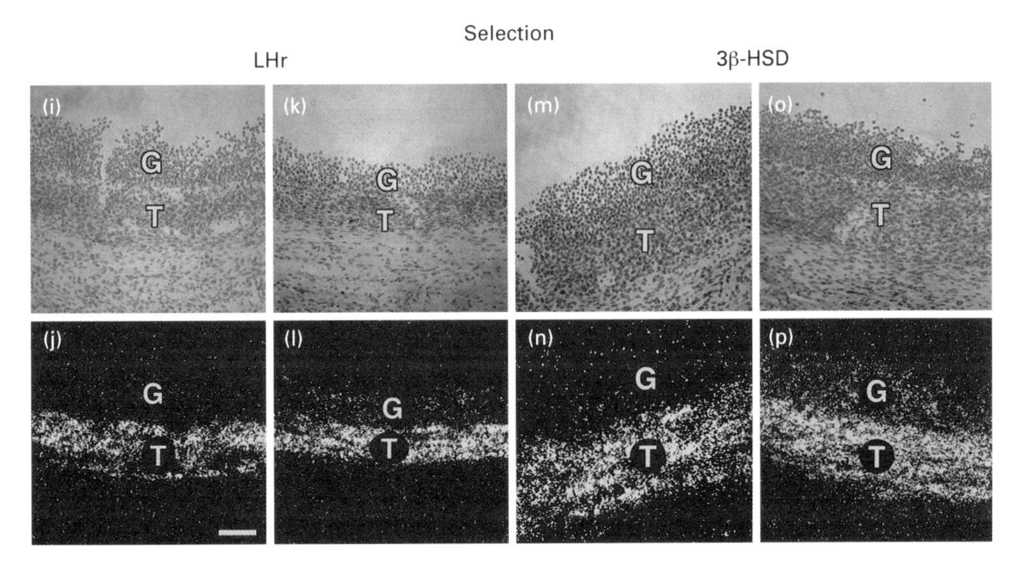

Fig. 4. *In situ* hybridization of luteinizing hormone receptor (LHr) and steroidogenic enzymes, cytochrome P450 side-chain cleavage (P450scc), cytochrome P450 aromatase (P450arom) and 3β-hydroxysteroid dehydrogenase (3β-HSD) mRNAs in cryosections of bovine ovarian follicles collected around the times of recruitment and selection of follicles during the first follicular wave. (a,b) Bright- and dark-field views of a 5 mm healthy follicle with no specific hybridization of P450scc in granulosal cells (not recruited). (c,d) Bright- and dark-field views of a healthy 6 mm recruited follicle collected at 24 h with specific hybridization for P450scc in both thecal and granulosal cells. (e,f) Bright- and dark-field views of a 5 mm healthy follicle with no specific hybridization of P450arom in granulosal cells (not recruited). (g,h) Bright- and dark-field views of a healthy 6 mm recruited follicle collected at 24 h with specific hybridization of P450arom in both thecal and granulosal cells. (i,j) Bright- and dark-field views of a healthy 5 mm follicle with hybridization of LHr localized to thecal cells. (k,l) Bright- and dark-field views of a 9 mm healthy follicle with specific hybridization of LHr to both thecal and granulosal cells. (m,n) Bright- and dark-field views of a 4 mm healthy follicle with specific hybridization of 3β-HSD localized to thecal cells. (o,p) Bright- and dark-field views of a 9 mm healthy follicle with hybridization of 3β-HSD to both thecal and granulosal cells. G: granulosal cells; T: thecal cells. Scale bars represent 50 μm.

Table 1. Expression of messenger RNA encoding gonadotrophin receptors and steroidogenic enzymes during recruitment and selection of bovine ovarian follicles

	Not recruited[a]		Recruited[a]		Selected	
	Granulosa	Theca	Granulosa	Theca	Granulosa	Theca
Gonadotrophin receptors						
FSH	+	–	+	–	++	–
LH	–	+	–	+	+*	++
Steroidogenic enzymes						
P450scc	–	+	+*	+	++	++
P450c17	–	+	–	+	–	++
P450arom	–	–	+*	–	++	–
3β-HSD	–	+	–	+	+*	++
Steroidogenic acute regulatory protein (StAR)	–	+	–	+	–	++

[a] Differences in the intensity of expression have not been tested between not recruited and recruited follicles.
* Indicates when mRNA expression is first detected.
+ or ++ denotes amount of expression.

6–9 mm in diameter, all follicles in the cohort express P450scc and P450arom mRNA in the granulosal cells. During this time some of the 4–5 mm follicles that were apparently recruited do not express P450scc and P450arom. The number of follicles that express P450scc and P450arom during the early stages of recruitment is similar to the number of follicles that continue growth during later stages of recruitment. Therefore, recruitment of follicles that continue growth beyond 4–6 mm in diameter may be associated with expression of mRNA P450scc and P450arom in granulosal cells (Table 1; Fig. 4).

Expression of P450scc and P450arom mRNA is likely to be important for continued growth, since all follicles that continue to grow beyond 4–6 mm in diameter, after the initial stages of recruitment, expressed mRNA for P450scc and P450arom (Bao *et al.*, 1997c). It is at this stage of growth that follicles develop the capability to produce significant quantities of oestradiol. This is consistent with previous reports that follicles less than 5 mm in diameter do not produce oestradiol (Skyer *et al.*, 1987). Induction of P450scc and P450arom mRNA is probably due to a transient increase in circulating FSH that precedes initiation of each wave of follicular growth (Adams *et al.*, 1992; Fig. 4). LH may not be involved in follicular recruitment or mRNA P450scc and P450arom expression, since LHr mRNA is not detected in granulosal cells during recruitment. If LH is involved, the effect is likely to be an indirect one through stimulation of thecal androgen synthesis. In addition, follicles grow to 7–9 mm in diameter when LH, but not FSH, is inhibited (Fig 3; Gong *et al.*, 1996).

Gene Expression during Selection

When the cohort of follicles in cattle reach 8–9 mm in diameter, there is rapid divergence whereby one follicle increases rapidly in size, becomes larger than the other follicles and becomes the dominant follicle (Ginther *et al.*, 1996). Divergence of the selected follicle seems to occur about 36 to 48 h after initiation of a follicular wave. In cattle, divergence of the selected follicle appears to be associated with initiation of mRNA expression of LHr and 3β-HSD in granulosal cells (Xu *et al.*, 1995b; Bao *et al.*, 1997c; Fig. 4). Whether selection of the dominant follicle or granulosal cell mRNA LHr or 3β-HSD expression occurs first, or whether they occur simultaneously, is unclear. Evans and Fortune (1997) reported an increase in both size and oestradiol secretion in one follicle of the cohort before detection of LHr and 3β-HSD mRNA expression in granulosal cells. Similarly, Bodensteiner *et al.* (1996) reported an increase in oestradiol concentration in the selected dominant follicle before an increase in the numbers of gonadotrophin receptors. Regardless, all dominant follicles express mRNA for LHr, 3β-HSD and P450arom (Xu et al, 1995a; Bao *et al.*, 1997a,c). However, in this study, the number of receptors for LH included those in the thecal cells as well as those in the granulosal cells. In addition, divergence (identification) of the selected follicle occurs when circulating concentration of FSH, which has been decreasing from shortly after recruitment, reaches its nadir. Thus, the follicle that is the most functionally developed can survive in an environment of decreasing FSH concentration. Hence the first follicle to develop LHr in granulosal cells would be able to respond to LH, as well as FSH, and to survive in an environment unable to support the other follicles (Gong *et al.*, 1996; Figs 3 and 4). In rodents, induction of the LHr in granulosal cells is dependent on the action of FSH and oestradiol (Segaloff *et al.*, 1990). Thus, the follicle with the highest concentration of oestradiol would be the first follicle to develop LHr in the granulosal cells and hence, allow granulosal cells of the selected follicle to become responsive to LH, as well as FSH, and survive in the face of declining serum FSH.

Continued growth of selected follicles is generally accompanied by increased expression of gonadotrophin receptors, steroidogenic enzymes and StAR, and selected and dominant follicles have higher mRNA expression than subordinate and atretic follicles. Despite the increases in mRNA expression of follicles during development, selection probably cannot be determined by the differential mRNA expression of either LHr, P450scc, P450c17, 3β-HSD or StAR in thecal cells, FSHr and P450arom in granulosal cells, or P450scc in granulosal and thecal cells (Table 1). This is because more than one follicle, of approximately the same size, expresses similar amounts of these mRNAs indicating that either the current techniques are too insensitive or selection depends on the expression of other upstream factors that remain to be determined.

Gene Expression during Dominance and Atresia

Follicular dominance

Dominant follicles continue to grow for a few days after selection. Expression of mRNA for the gonadotrophin receptors, steroidogenic enzymes and StAR generally increase in thecal and granulosal cells during the growing phase, and follicles produce greater amounts of oestradiol (Xu *et al.*, 1995a,b; Bao *et al.*, 1997a,c). Thus, dominant follicles acquire increased capability to produce steroids during their development, supporting previously published work investigating follicular steroid production (see Campbell *et al.*, 1995; Webb and Armstrong, 1998). In addition, the patterns of mRNA expression are in agreement with previous work in sheep and cattle that has found that LH can support dominant follicle development. However, if the LH pulse frequency is too low, for example during the middle of the luteal phase, dominant follicle growth will not continue. After luteolysis, mRNA expression for P450scc, P450c17 and 3βHSD, but not P450arom, increases (Tian *et al.*, 1995). Concurrently, follicular fluid concentrations of androstenedione and oestradiol increase, under the influence of increased pulsatile LH release. The significant increase in expression of mRNAs encoding the steroidogenic enzymes and follicular fluid steroid concentrations are likely due to increased pulse frequency of LH secretion during the preovulatory period (Campbell *et al.*, 1995). The increase in LH pulse frequency may increase mRNA expression for the steroidogenic enzymes necessary for synthesis of androgen precursors for oestradiol production. Thecal cell production of androgens in cattle may be the rate-limiting step for follicular oestradiol production (Badinga *et al.*, 1992), but the fact that the oestradiol:androgen ratio remains approximately one throughout the follicular phase in sheep (Campbell *et al.*, 1990) would appear to make this unlikely unless there are key species differences.

Follicular atresia

If luteolysis does not occur during the growing phase of the dominant follicle, the fate is atresia. Expression of gonadotrophin receptor mRNAs, steroidogenic enzymes and StAR decrease rapidly with atresia, and a decline in expression occurs earlier than morphological signs of atresia are observed (Xu *et al.*, 1995a,b). Atresia of dominant follicles appears to be initiated between days 4 and 6 in the non-ovulatory follicular wave. Expression of mRNAs for FSHr in granulosal cells, LHr in thecal cells, P450scc in granulosal and thecal cells, and P450c17 in thecal cells decreases markedly between days 4 and 6 of the follicular wave. Interestingly, expression of mRNAs for LHr and P450arom in granulosal cells is still high on day 6, but declines by day 8 of the follicular wave (Xu *et al.*, 1995a,b). Atresia of unselected follicles from the cohort appears to be similar to atresia of dominant follicles.

Additional Extraovarian Regulators of Follicular Growth

Although follicular development is primarily regulated by gonadotrophins, other systemic factors have been shown to alter follicular growth patterns. Pharmacological administration of recombinant GH increased the number of antral follicles without altering gonadotrophin concentrations (Gong *et al.*, 1991; de la Sota *et al.*, 1993; Gong *et al.*, 1993). Moreover, a reduction in GH and IGF-I concentrations after immunization of prepubertal heifers against GHRH inhibited the development of follicles > 7 mm in diameter (Cohick *et al.*, 1996). However, pharmacological manipulation of GH concentration may not reflect its physiological action. Recently we demonstrated that flushing heifers (200% maintenance of a low fibre diet) stimulated an increase in the number of small (< 4 mm) follicles, despite reduced GH concentrations. However, there was high insulin concentration compared with controls (Gutierrez *et al.*, 1997b) indicating that GH may not act directly to alter follicular development. Indeed, direct administration of GH into the ovarian artery did not stimulate ovarian steroid secretion in the autotransplanted sheep (Campbell *et al.*, 1995) and the identification of follicular GH receptors has proved difficult (Lucy *et al.*, 1993). GH may act through differential responses of IGF-I and insulin. IGF-I is a potent stimulator of steroidogenesis and proliferation of both granulosal and thecal cells *in vitro* (Campbell *et al.*, 1996, 1998b; Gutierrez *et al.*, 1997c). Insulin also stimulates proliferation and steroidogenesis of granulosal and thecal cells *in vitro* (Campbell *et al.*, 1998b; Gutierrez *et al.*, 1997d). The bioavailability of IGFs is regulated by their association with a family of specific IGFBPs which in turn are affected by nutrition (Webb and Armstrong, 1998). Hence, systemic metabolic factors can influence follicular recruitment and selection. However, in addition to this extraovarian IGF system there is also an intraovarian system that may function in concert to alter the response of follicles to gonadotrophins.

Intraovarian Regulation of Follicular Growth

A range of follicular growth factors are now known to be involved in the regulation of follicular growth, including the TGF-β superfamily, FGFs, EGF and TGFα and cytokines as well as the IGFs. Many of the intraovarian growth factor signalling systems act through tyrosine kinase receptors that regulate granulosal and thecal cell differentiation in a coordinated manner through interaction with gonadotrophin–cAMP-stimulated mechanisms. The integration of the endocrine and intraovarian mechanisms provide the necessary signals that either stimulate follicular growth or divert the follicle into apoptotic pathways resulting in follicular atresia. This section will concentrate on components of two of the most intensively studied families, namely the IGF and activin–inhibin systems.

Ovarian IGF system

The IGF system, occupying a central position within the 'network' of intraovarian signals, includes the IGF ligands (IGF-I and -II), at least six IGF-binding proteins (IGFBP-1 to -6), type 1 and type 2 IGF receptors and specific IGFBP proteases. To date, expression of mRNAs encoding IGFBP-2 to -5 have been found in bovine follicles and expression of IGFBP-2, -4 and -5 in ovine follicles (Armstrong and Webb, 1997; Webb and Armstrong, 1998).

Insulin-like growth factors I and II. IGFs stimulate granulosal and thecal cell proliferation and differentiation and have been identified as follicular survival factors. There is considerable species variation in the patterns of mRNA expression of IGF ligands and BPs during folliculogenesis. We have detected the expression of mRNA encoding IGF-II in thecal tissue of bovine ovarian follicles (Armstrong and Webb, 1997) and a similar spatial distribution has been described in sheep (Perks *et al.*, 1995). The expression of mRNA encoding IGF-I in ruminants remains controversial. Leeuwenberg *et al.* (1995) detected IGF-I mRNA in ovine granulosal and thecal tissue, and Yuan *et al.* (1998) detected IGF-I mRNA in bovine granulosal cells. In contrast, Perks *et al.* (1995) failed to detect the expression of mRNA encoding IGF-I in ovine follicles. Similarly, we were unable to detect expression of IGF-I mRNA in bovine follicles by *in situ* hybridization (Fig. 5). In support of this last observation we recently demonstrated that non-luteinized bovine granulosal cells do not produce IGF-I in serum-free cultures (Gutierrez *et al.*, 1997c).

IGF-binding proteins. The bioactivity of IGFs are controlled by their association with IGFBPs. As with the IGFs, the spatial expression of these binding proteins within ovarian follicles is species specific (Armstrong and Webb, 1997). In cows (Armstrong *et al.*, 1998) and sheep (Besnard *et al.*, 1996), expression of mRNA encoding IGFBP-4 and -2 is restricted to thecal and granulosal tissue, respectively. The spatial and temporal patterns of expression of mRNA encoding components of the IGF system in the bovine follicle are summarized in Figs 5 and 6.

In ovarian cell culture systems examined so far, IGFBPs attenuate the actions of IGFs (Mason *et al.*, 1992; Monget *et al.*, 1993; Spicer *et al.*, 1997). A decrease in follicular IGFBP production would therefore be expected to enhance the biological activity of locally produced IGFs, resulting in increased follicular response to gonadotrophins. The observed decrease in the concentration of IGFBP-2, -4 and -5 in follicular fluid during the development of dominance supports this hypothesis (Armstrong *et al.*, 1996). In cows (Armstrong *et al.*, 1998) and sheep (Besnard *et al.*, 1996) the decrease in IGFBP-2 concentration in follicular fluid during follicular growth was shown to be due to a loss of expression of mRNA encoding IGFBP-2 in granulosal cells in dominant follicles. Using serum-free bovine granulosal cell cultures, we have shown that FSH, at physiological concentrations, inhibits expression of mRNA encoding IGFBP-2 (Armstrong *et al.*, 1998). These results indicate that a key feature in the development of follicular dominance in cattle is the FSH-dependent inhibition of the expression of mRNA encoding IGFBP-2 in granulosal cells (Fig. 6). The resultant increase in IGF bioactivity in these follicles would increase FSH responsiveness of their granulosal cells.

Activin–inhibin system

The TGF-β superfamily comprises a range of proteins, including members of the activin–inhibin system, with the potential to act as intraovarian regulators. mRNAs encoding TGF-β are expressed in thecal cells from both mammalian and non-mammalian species (Armstrong and Webb, 1997). In cows, TGFβs inhibit granulosal and thecal cell proliferation while enhancing gonadotrophin-stimulated steroidogenesis (Roberts and Skinner, 1991).

Inhibin–activin family. Expression of mRNA for members of the inhibin–activin family appears to be initiated in a sequential and co-ordinated way during early follicle development. In sheep, mRNA expression of B-inhibin–activin subunit appears concomitant with expression of FSHr at 1–2 layers of cuboidal granulosal cells followed by inhibin α-subunit and follistatin at more than 2–4 layers of granulosa cells and finally βA-inhibin–activin subunit during early antral development (Eckery *et al.*, 1996). As ovine follicles progress from small (< 2 mm), to medium-sized (2–4 mm) and to large (> 4 mm) there is a progressive increase in P450c17 expression in the theca and inhibin α and βA-subunit, LH-receptor and P450arom expression in the granulosal cells. In contrast expression for βB-inhibin–activin subunit in granulosal cells and LH receptor in thecal cells remains relatively constant (B. K. Campbell, L. M. Harkness, D. G. Armstrong, H. A. Garverick and R. Webb,

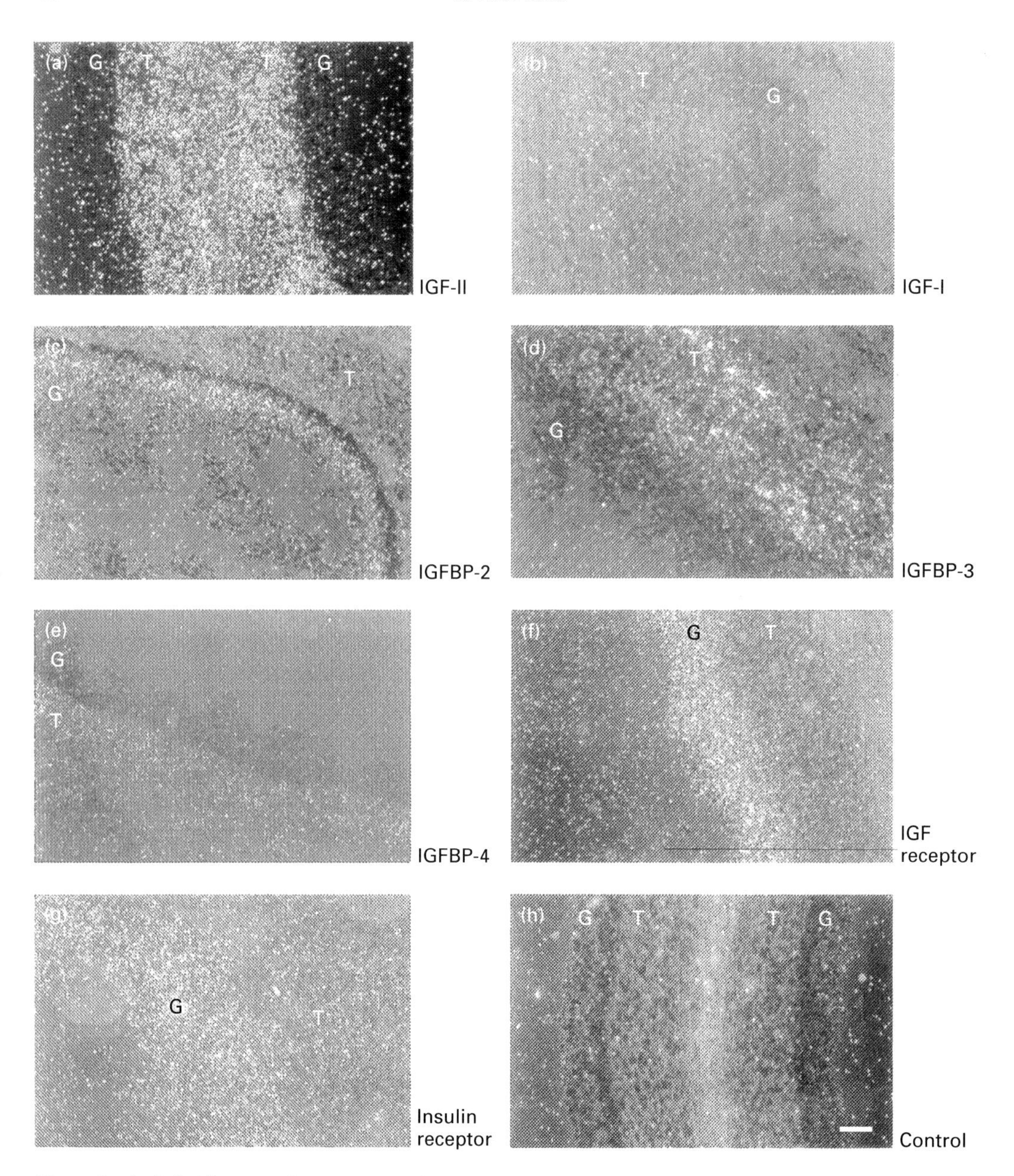

Fig. 5. *In situ* hybridization of mRNAs encoding (a) IGF-II, (b) IGF-I, (c) IGFBP-2, (d) IGFBP-3, (e) IGFBP-4, (f) type 1 IGF receptor, (g) insulin receptor and typical control section probed with sense IGF-II RNA in bovine follicles. G and T represent granulosal and thecal cells, respectively. Scale bar represents 100 µm.

unpublished). The association between P450arom and βA-subunit expression is particularly interesting as both these factors are expressed precociously in medium-sized follicles in sheep carrying the *FecB* gene (Fig. 7). This is a major gene that results in a marked increase in prolificacy associated with the ovulation of significantly more follicles at a smaller size. The correlation between

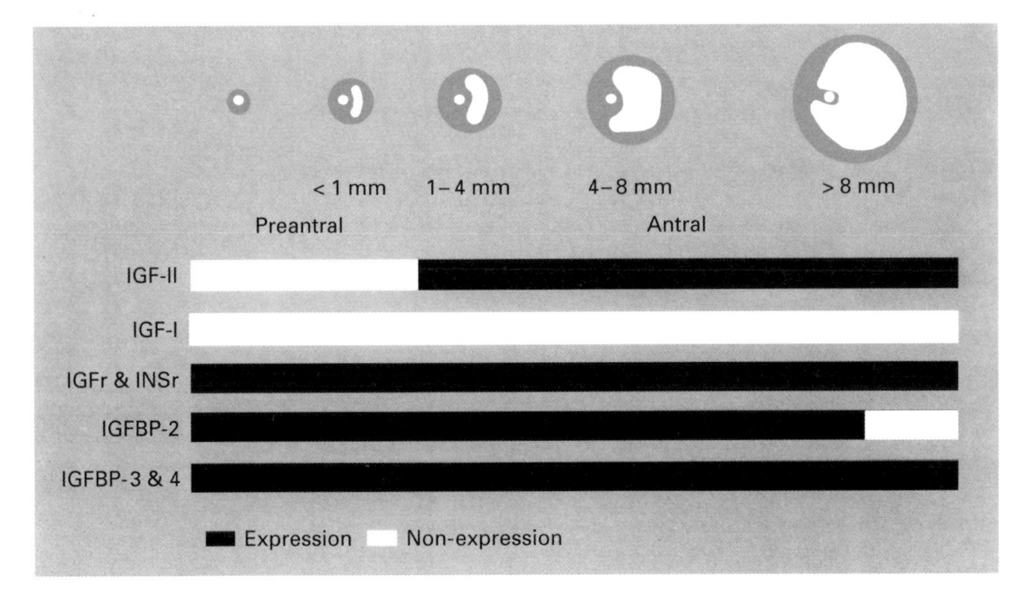

Fig. 6. The relationship between the temporal patterns of expression of mRNAs encoding components of the IGF system during bovine follicular development.

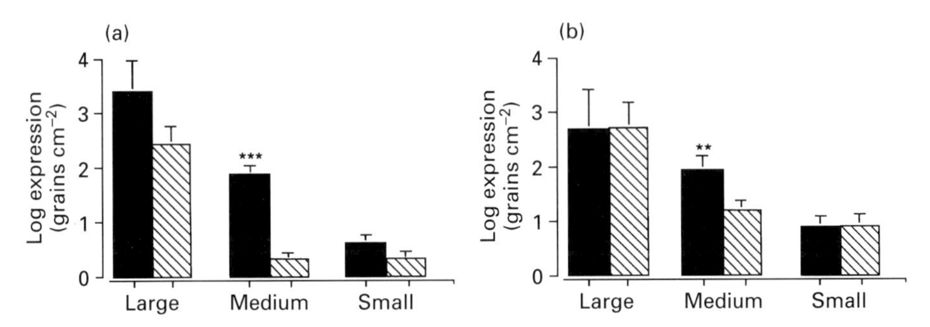

Fig. 7. Expression of mRNA, determined by *in situ* hybridization, for (a) cytochrome P450 aromatase and (b) the inhibin βA subunit in the granulosal cells of large (> 4 mm), medium-sized (2–4 mm) and small (< 2 mm) ovarian follicles of sheep with (black columns) and without (hatched columns) the *FecB* gene. Ovaries were recovered on day 4 of the oestrous cycle. Asterisks indicate significant differences (** $P < 0.01$; *** $P < 0.001$).

P450arom and βA-subunit expression and terminal follicle development is in agreement with results from experiments *in vitro*. These experiments demonstrated that FSH-stimulated differentiation of granulosal cells, from small follicles, results in a dose-responsive induction of inhibin A and oestradiol (Campbell *et al.*, 1997). Furthermore, inhibin A has been shown to modulate both oestradiol production by granulosal cells and androgen production by thecal cells *in vitro* (Knight, 1996; Campbell and Webb, 1995). Thus, inhibin βA-subunit expression would appear to be an essential component of the differentiative cascade.

Conclusions

Follicular recruitment, selection and the development of dominance involve the integration of systemic and intra-follicular mechanisms. Gonadotrophins provide the primary drive, particularly during the final stages of follicular development, although other extraovarian signals, including metabolic factors, can influence patterns of follicular growth. Indeed, it seems that, in cattle, FSH stimulates follicle growth up to 7 mm in diameter, followed by a requirement for LH in the final stages of follicular growth and maturation. Recent evidence on the pattern of gene expression in follicles has also demonstrated that there are a significant number of protein or peptide factors produced by follicles. The precise timing of mRNA expression, and hence production of these local factors, appears to be involved in the mechanisms of recruitment, selection and dominance. For example, it appears that a key feature in the development of follicular dominance in cows is the FSH-dependent inhibition of the expression of mRNA encoding IGFBP-2 in granulosal cells. The information from these molecular studies together with results using physiologically relevant *in vitro* granulosal and thecal cell culture systems demonstrate that these locally produced factors can amplify, attenuate or mediate the effects of circulating gonadotrophins on granulosal and thecal cell function. The optimum integration of these control systems determines both the response of individual follicles to gonadotrophins and whether it continues to develop to become dominant or is diverted into apoptotic pathways leading to eventual atresia.

Original research presented in this review from our laboratories was supported by the Ministry of Agriculture, Fisheries and Food (MAFF), the Office of Science and Technology (OST), the European Union (EU), the Biotechnology and Biological Sciences Research Council (BBSRC), the Sino-British Friendship Scholarship Scheme (SBFSS), the National Council of Science and Technology of Mexico (CONACYT) and the British Council. The authors thank Elanco Animal Health for generously providing rGH and rIGF-I and Hoechst Animal Health for the gift of GnRH agonist. They also thank K. J. Wood for synthesizing the IGF-II and IGF receptor probes (Fig. 5).

References

Adams GP, Matteri RL, Kastelic JP, Ko JCH and Ginther OJ (1992) Association between surges of follicle-stimulating hormone and the emergence of follicular waves in heifers *Journal of Reproduction and Fertility* **94** 177–188

Armstrong DG and Webb R (1997) Ovarian follicular dominance: novel mechanisms and protein factors *Reviews of Reproduction* **2** 139–146

Armstrong DG, Hogg CO, Campbell BK and Webb R (1996) Insulin-like growth factor (IGF) binding protein production by primary cultures of ovine granulosa and theca cells. The effect of IGF-I, gonadotropin and follicle size *Biology of Reproduction* **55** 1163–1171

Armstrong DG, Baxter G, Gutierrez CG, Hogg CO, Glazyrin AL, Campbell BK, Bramley TA and Webb R (1998) Insulin-like growth factor binding protein -2 and -4 mRNA expression in bovine ovarian follicles: effect of gonadotrophins and developmental status *Endocrinology* **139** 2146–2154

Badinga L, Driancourt MA, Savio JD, Wolfsen D, Drost M, de la Sota RL and Thatcher WW (1992) Endocrine and ovarian responses associated with the first-wave dominant follicle in cattle *Biology of Reproduction* **47** 871–883

Bao B, Garverick HA, Smith GW, Smith MF, Salfen BE and Youngquist RS (1997a) Expression of messenger RNA encoding 3β-hydroxysteroid dehydrogenase/Δ^5–Δ^4 isomerase during recruitment and selection of bovine ovarian follicles: identification of dominant follicles by expression of 3β-HSD mRNA within the granulosa cell layer *Biology of Reproduction* **56** 1466–1473

Bao B, Calder MD, Xie S, Smith MF, Youngquist RS and Garverick HA (1997b) Steroidogenic acute regulatory protein messenger ribonucleic acid (mRNA) expression is limited to theca of healthy bovine follicles *Biology of Reproduction Supplement* (Abstract 195)

Bao B, Garverick HA, Smith GW, Smith MF, Salfen BE and Youngquist RS (1997c) Changes in messenger RNA encoding LH receptor, cytochrome P450 side chain cleavage, and aromatase are associated with recruitment and selection of bovine ovarian follicles *Biology of Reproduction* **56** 1158–1168

Besnard N, Pisselet C, Monniaux D, Locatelli A, Benne F, Gasser F, Hatey F and Monget P (1996) Expression of messenger ribonucleic acids of insulin-like growth factor binding proteins-2, -4 and -5 in the ovine ovary: localization and changes during growth and atresia of antral follicles *Biology of Reproduction* **55** 1356–1367

Bodensteiner KJ, Wiltbank MC, Bergfelt DR and Ginther OJ (1996) Alterations in follicular estradiol and gonadotropin receptors during development of bovine antral follicles *Theriogenology* **45** 499–512

Campbell BK and Webb R (1995) Evidence that inhibin has paracrine and autocrine actions in controlling ovarian function in sheep *Journal of Reproduction and Fertility Abstract Series* **15** Abstract 140

Campbell BK, Baird DT, McNeilly AS and Scaramuzzi RJ (1990) Ovarian secretion rates and peripheral concentrations of inhibin in normal and androstenedione-immune ewes with an autotransplanted ovary *Journal of Endocrinology* **127** 285–296

Campbell BK, Scaramuzzi RJ and Webb R (1995) Control of antral follicle development and selection in sheep and cattle *Journal of Reproduction and Fertility Supplement* **49** 335–350

Campbell BK Scaramuzzi RJ and Webb R (1996) Induction and maintenance of oestradiol and immunoreactive inhibin production with FSH by ovine granulosa cells cultured in serum free media *Journal of Reproduction and Fertility* **106** 7–16

Campbell BK Groome N and Baird DT (1997) Effect of dose and time of exposure to FSH on oestradiol and dimeric inhibin A production by undifferentiated ovine granulosa cells in serum-free culture *Journal of Reproduction and Fertility Abstract Series* **19** Abstract 32

Campbell BK, Baird DT and Webb R (1998a) Effects of dose of LH on androgen production and luteinization of ovine theca cells cultured in a serum-free system *Journal of Reproduction and Fertility* **112** 69–77

Campbell BK Dobson H and Scaramuzzi RJ (1998b) Ovarian function in ewes made hypogonadal with GnRH-antagonist and stimulated with FSH in the presence or absence of low amplitude LH pulses *Journal of Endocrinology* **156** 213–222

Cohick WS, Armstrong JD, Withacre MD, Lucy MC, Harvey RW and Campbell RM (1996) Ovarian expression of insulin-like growth factor-I (IGF-I), IGF binding proteins and growth hormone (GH) receptor in heifers actively immunized against GH-releasing factor *Endocrinology* **137** 1670–1677

de la Sota RL, Lucy RL, Staples CR and Thatcher WW (1993) Effects of recombinant bovine somatotropin (Sometribove) on ovarian function in lactating and nonlactating dairy cows *Journal of Dairy Science* **76** 1002–1013

Driancourt MA, Webb R and Fry RC (1991) Does follicular dominance occur in ewes? *Journal of Reproduction and Fertility* **93** 63–70

Eckery DC, Tisdall, DJ, Heath DA and McNatty KP (1996) Morphology and function of the ovary during fetal and early neonatal life: comparison between the sheep and brushtail possum (*Trichosurus vulpecula*) *Animal Reproduction Science* **42** 1–4

Evans ACO and Fortune JE (1997) Selection of the dominant follicle in cattle occurs in the absence of differences in the expression of messenger ribonucleic acid for gonadotropin receptors *Endocrinology* **138** 2963–2971

Ginther OJ, Wiltbank MC, Fricke PM, Gibbons JR and Kot K (1996) Selection of the dominant follicle in cattle *Biology of Reproduction* **55** 1187–1194

Gong JG, Bramley TA and Webb R (1991) The effect of recombinant bovine somatotrophin on ovarian function in heifers: follicular populations and peripheral hormones *Biology of Reproduction* **45** 941–949

Gong JG, Bramley TA and Webb R (1993) The effect of recombinant bovine somatotrophin on ovarian follicular growth and development in heifers *Journal of Reproduction and Fertility* **97** 247–254

Gong JG, Bramley TA, Gutierrez CG, Peters AR and Webb R (1995) Effects of chronic treatment with a gonadotrophin-releasing hormone agonist on peripheral concentrations of FSH and LH, and ovarian function in heifers *Journal of Reproduction and Fertility* **105** 263–270

Gong JG, Campbell BK, Bramley TA, Gutierrez CG, Peters AR and Webb R (1996) Suppression in the secretion of follicle-stimulating hormone and luteinizing hormone, and ovarian follicle development in heifers continuously infused with a gonadotropin-releasing hormone agonist *Biology of Reproduction* **55** 68–74

Gong JG, Campbell BK and Webb R (1997) Stimulation of ovarian follicles to the preovulatory size by infusion with FSH alone in GnRH-agonist-treated heifers *Journal of Reproduction and Fertility Abstract Series* **19** Abstract 162

Gutierrez CG, Ralph JH, Wilmut I and Webb R (1997a) Follicle growth and antrum formation of bovine preantral follicles in long-term *in vitro* culture *Journal of Reproduction and Fertility Abstract Series* **19** Abstract 31

Gutierrez CG, Oldham J, Bramley TA, Gong JG, Campbell BK and Webb R (1997b) The recruitment of ovarian follicles is enhanced by increased dietary intake in heifers *Journal of Animal Science* **75** 1876–1884

Gutierrez CG, Campbell BK, Armstrong DG and Webb R (1997c) Insulin-like growth factor-I (IGF-I) production by bovine granulosa cells *in vitro* and peripheral IGF-I measurement in cattle serum: an evaluation of IGFBP extraction protocols *Journal of Endocrinology* **153** 231–240

Gutierrez CG, Campbell BK and Webb R (1997d) Development of a long-term bovine granulosa cell culture system: induction and maintenance of estradiol production, response to follicle-stimulating hormone, and morphological characteristics *Biology of Reproduction* **56** 608–616

Knight PG (1996) Roles of inhibins, activins and follistatin in the female reproductive system *Frontiers of Neuroendocrinology* **17** 476–509

Leeuwenberg BR, Hurst PR and McNatty KP (1995) Expression of IGF-I mRNA in the ovine ovary *Journal of Molecular Endocrinology* **15** 251–258

Lucy MC, Collier RJ, Kitchell ML, Dibner JJ, Hauser SD and Krivi GG (1993) Immunohistochemical and nucleic acid analysis of somatotropin receptor populations in the bovine ovary *Biology of Reproduction* **48** 1219–1227

Mason HD, Wills D, Holly JMP, Cwyfan-Hughs SC, Seppala M and Franks S (1992) Inhibitory effects of insulin-like growth factor-binding proteins on steroidogenesis by human granulosa cells in culture *Molecular and Cellular Endocrinology* **89** R1–R4

Monget P, Monniaux D, Pisselet C and Durand P (1993) Changes in insulin-like growth factor-I (IGF-I), IGF-II and their binding proteins during growth and atresia of ovine ovarian follicles *Endocrinology* **132** 1438–1446

Perks CM, Denning-Kendall PA, Gilmour RS and Wathes DC (1995) Localization of messenger ribonucleic acids for insulin-like growth factor I (IGF-I), IGF-II and the type 1 IGF receptor in the ovine ovary throughout the estrous cycle *Endocrinology* **136** 5266–5273

Ralph JH, Wilmut I and Telfer EE (1995) *In vitro* growth of bovine preantral follicles and the influence of FSH on follicular oocyte diameters *Journal of Reproduction and Fertility Abstract Series* **15** Abstract 12

Ravindra JP and Rawlings NC (1997) Ovarian follicular dynamics in ewes during the transition from anoestrus to the breeding season *Journal of Reproduction and Fertility* **110** 279–289

Roberts AJ and Skinner MK (1991) Transforming growth factor-α and -β differentially regulate growth and steroidogenesis of bovine thecal cells during antral follicle development *Endocrinology* **129** 2041–2048

Schrick FN, Surface RA, Pritchard JY, Dailey RA, Townsend EC and Inskeep EK (1993) Ovarian structures during the estrous cycle and early pregnancy in ewes *Biology of Reproduction* **49** 1133–1140

Segaloff DL, Wang H and Richards JS (1990) Hormonal regulation of luteinizing hormone/chorionic gonadotropin receptor mRNA in rat ovarian cells during follicular development and luteinization *Molecular Endocrinology* **4** 1856–1865

Sirois J and Fortune JE (1990) Lengthening the bovine estrous cycle with low levels of exogenous progesterone: a model for studying ovarian follicular dominance *Endocrinology* **127** 916–925

Skyer DM, Garverick HA, Youngquist RS and Krause GF (1987) Ovarian follicular populations and *in vitro* steroidogenesis on three different days of the bovine estrous cycle *Journal of Animal Science* **64** 1710–1716

Souza CJH, Campbell BK and Baird DR (1996) Follicular dynamics and ovarian steroid secretion in sheep during anoestrus *Journal of Reproduction and Fertility* **108** 101–106

Souza CJH, Campbell BK and Baird DT (1997a) Follicular dynamics and ovarian steroid secretion in sheep during the follicular and early luteal phases of the estrous cycle *Biology of Reproduction* **56** 483–488

Souza CJH, Campbell BK, Baird DT and Webb R (1997b) Secretion of inhibin A and follicular dynamics throughout the estrous cycle in sheep with and without the Booroola gene (*Fec*β) *Endocrinology* **138** 5333–5340

Souza CJH, Campbell BK and Baird DT (1998) Follicular development and concentrations of ovarian steroids and inhibin A in jugular venous plasma during the luteal phase of the oestrous cycle *Journal of Endocrinology* **156** 563–572

Spicer LJ, Stewart RC, Avarey P, Francisco CC and Keefer BE (1997) Insulin-like growth factor-binding protein-2 and -3: their biological effects in bovine thecal cells *Biology of Reproduction* **56** 1458–1465

Tian XC, Berndtson AK and Fortune JE (1995) Differentiation of bovine preovulatory follicles during the follicular phase is associated with increases in messenger ribonucleic acid for cytochrome P450 side-chain cleavage, 3β-hydroxysteroid dehydrogenase, and P450 17α-hydroxylase, but not P450 aromatase *Endocrinology* **136** 5102–5110

Wandji SA, Srsen V, Voss AK, Eppig JJ and Fortune JE (1996) Initiation *in vitro* of growth of bovine primordial follicles *Biology of Reproduction* **55** 942–948

Webb R and Armstrong DG (1998) Control of ovarian function; effect of local interactions and environmental influences on follicular turnover in cattle: a review *Livestock Production Science* **53** 95–112

Webb R and Gauld IK (1987) Endocrine control of follicular growth in the ewe. In *Follicular Growth and Ovulation Rate in Farm Animals* pp 107–118 Eds JF Roche and D O'Callaghan. Martinus Nijhoff, Dordrecht

Xu ZZ, Garverick HA, Smith GW, Smith MF, Hamilton SA and Youngquist RS (1995a) Expression of follicle-stimulating hormone and luteinizing hormone receptor messenger ribonucleic acids in bovine follicles during the first follicular wave *Biology of Reproduction* **53** 951–957

Xu ZZ, Garverick HA, Smith GW, Smith MF, Hamilton SA and Youngquist RS (1995b) Expression of messenger RNA encoding cytochrome P450 side-chain cleavage, cytochrome P450 17α-hydroxylase and cytochrome P450 aromatase in bovine follicles during the first follicular wave *Endocrinology* **136** 981–989

Yuan W, Bao B, Garverick HA, Youngquist RS and Lucy MC (1998) Follicular dominance in cattle is associated with divergent patterns of ovarian gene expression for insulin-like growth factor (IGF) -I, IGF-II and IGF binding protein-2 in dominant and subordinate follicles *Domestic Animal Endocrinology* **15** 55–63

Journal of Reproduction and Fertility Supplement **54**, 49–59

Role of growth hormone in development and maintenance of follicles and corpora lutea

M. C. Lucy, C. R. Bilby*, C. J. Kirby†, W. Yuan‡ and C. K. Boyd

164 Animal Sciences Research Center, University of Missouri, Columbia, MO 65211, USA

Growth hormone (GH) is a pituitary hormone that affects animal growth, metabolism, lactation, and reproduction. Many of the effects of GH are mediated by insulin-like growth factor I (IGF-I) which is synthesized in liver and ovary in response to GH. Insulin-like growth factor I synergizes with gonadotrophins (LH and FSH) to stimulate growth and differentiation of ovarian cells. There are species differences in the effects of GH in reproductive biology. In most species, ovarian follicles and corpora lutea are potential sites for GH action because the GH receptor is found within granulosal cells as well as corpora lutea. However, growth hormone does not control ovarian IGF-I in all species and, in ruminants, endocrine IGF-I from liver may be the principal mediator of GH action. In cattle, administration of GH increases the number of small antral ovarian follicles but does not increase the number of large antral (dominant) follicles. Growth hormone may antagonize some aspects of dominant follicular function because dominant follicles are shorter-lived in GH-treated cattle. The corpora lutea has increased growth and steroidogenesis in response to GH. Growth hormone-induced steroidogenesis in cultured granulosal and luteal cells depends on IGF-I release after GH treatment. Bovine and ovine granulosal cells do not release IGF-I in response to GH *in vitro* and, therefore, are less responsive to GH. These results demonstrate that GH is required for normal reproductive function in ruminant as well as nonruminant species.

Introduction

Growth hormone (GH) is a product of the pituitary somatotroph. As its name implies, GH is involved in animal growth. However, its actions are not confined to the growing animal and it is now clear that numerous metabolic and physiological processes (including reproduction) of the adult animal are controlled partially by GH. After its release from the pituitary, GH can act on a variety of tissues because cell-surface receptors for GH are widely distributed throughout the ruminant body (Lucy *et al.*, 1998; Fig. 1). Reproductive tissues that contain mRNA for GH receptor include hypothalamus, pituitary, corpus luteum, ovarian follicle, oviduct, endometrium, myometrium, and placenta (Kirby *et al.*, 1996; Lucy *et al.*, 1998). The highest amount of GH receptor is in the liver, where GH binding causes an increase in the synthesis and secretion of insulin-like growth factor I (IGF)-I. Insulin-like growth factor-I complexes with one of a series of IGF-binding proteins (IGFBP) and then travels as an endocrine hormone to stimulate several additional physiological and metabolic processes including those required for reproduction (Spicer and Echternkamp, 1995; Armstrong and Webb, 1997).

Ovarian IGF Physiology

The focus of this review is GH and ovarian function. A brief discussion of ovarian IGF-I physiology is necessary because many of the actions of GH are mediated by IGF-I. Insulin-like growth factor II

*Present address: Monsanto Company, St. Louis, MO 63198, USA.
†Present Address: College of Veterinary Medicine, North Carolina State University, Raleigh, NC 27606, USA.
‡Present Address: Department of Gynecology and Obstetrics, Stanford University School of Medicine, Stanford, CA 94305, USA.

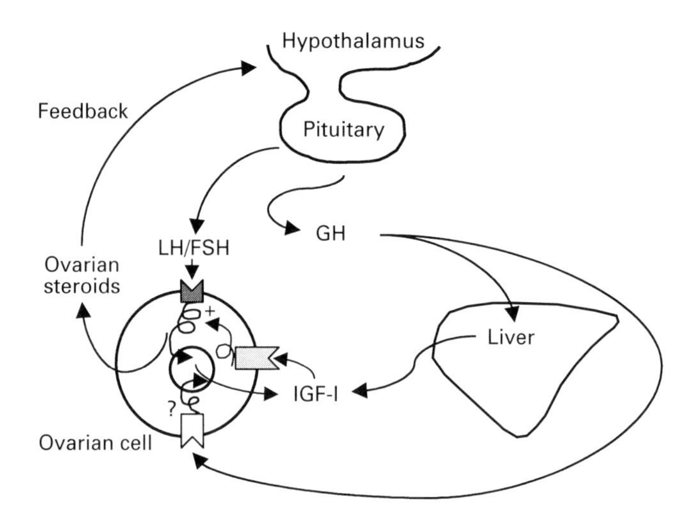

Fig. 1. Model for the actions of growth hormone (GH) and insulin-like growth factor I (IGF-I) on ovarian cells. The model represents general mechanisms that may or may not be proven in all ovarian cell types (thecal, granulosal and luteal) or in all species. Some of the effects of GH are mediated directly by GH at the ovary. The ovarian response may involve the synthesis and secretion of IGF-I by ovarian cells. An endocrine effect is also associated with GH because GH can bind to liver receptors and increase blood IGF-I that can affect ovarian function. Insulin-like growth factor I is synergistic with gonadotrophins for its effect on ovarian cells because gonadotrophins increase IGF-I and IGF receptor synthesis and IGF-I increases gonadotrophin receptor expression and second messenger systems. The synergy increases ovarian cell steroidogenesis. Ovarian function is linked to hypothalamic and pituitary function because oestradiol feeds back positively on GH secretion which can ultimately lead to increased concentrations of IGF-I reaching the ovary. In addition, nutritionally induced changes in liver function can alter ovarian function by modifying the amount of endocrine IGF-I.

has similar effects to IGF-I on ovarian cells but has lower potency than IGF-I and is not under GH control. Granulosal, thecal and luteal cells are sites for IGF action (Spicer and Echternkamp, 1995; Armstrong and Webb, 1997). Endocrine IGF-I (from liver) is not the only source of ovarian IGF-I because ovarian cells (granulosal and luteal cells) synthesize IGF-I and contribute to the total amount of IGF-I that reaches the ovary (endocrine plus local ovarian sources). Within the ruminant follicle, IGF-II produced by the theca may be the most important, locally produced IGF, because thecal IGF-II synthesis is greater than granulosal cell IGF-I synthesis (Yuan *et al.*, 1998). There is synergy between IGF-I and gonadotrophins (LH and FSH) that explains some of the actions of the IGF on ovarian cells. A maximal effect of IGF-I or IGF-II is only observed when cells are treated in combination with either FSH or LH (Spicer and Echternkamp, 1995). Gonadotrophins maintain IGF action by stimulating IGF-I and IGF receptor expression. Furthermore, IGF-I increases gonadotrophin responsiveness by stimulating the adenylate cyclase complex (Fig. 1).

Nutritional Regulation of GH and IGF-I

The release of IGF-I in response to GH within liver is regulated by nutritional status (energy and protein intake relative to requirements; McGuire *et al.*, 1992). The nutritional regulation of IGF-I

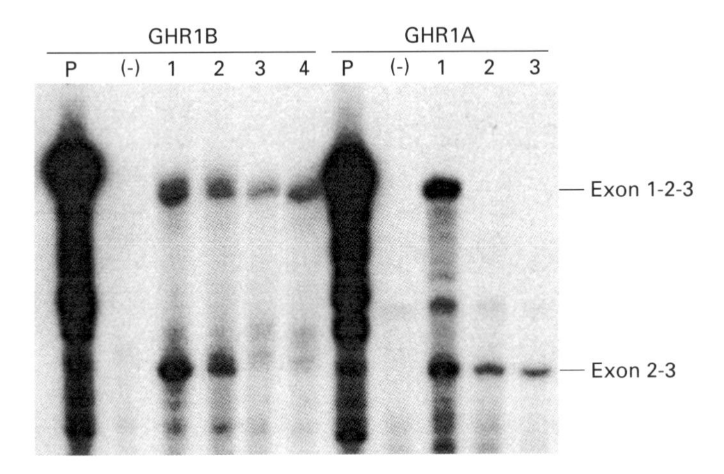

Fig. 2. Ribonuclease protection assay showing alternative splicing of GH receptor mRNA. The GH receptor mRNA is present in two forms (GHR1A and GHR1B) which are controlled by different promoters (1A and 1B). The GHR1B assay yields a full-length protected fragment (exon 1-2-3) for liver (lanes 1 and 2) and corpus luteum (lanes 3 and 4) showing that the 1B promoter is active in liver and corpora lutea. An exon 2-3 fragment is found in liver but not in corpora lutea. Therefore, other alternatively spliced forms of the GH receptor are found in liver but not in corpora lutea. The GHR1A assay yields a full-length protected fragment in liver (lane 1) but only the exon 2-3 fragment in corpora lutea (lanes 2 and 3). The 1A promoter, therefore, is active in liver but not corpora lutea. The exon 2-3 fragment represents GHR1B mRNA found in corpora lutea and liver. P = undigested probe; (-) = negative control.

should not be ignored when considering the effects of GH on reproduction because IGF-I is an important regulator of follicular and luteal function. Animals fed adequate nutrition have the highest concentration of blood IGF-I. Blood IGF-I is highly correlated with follicular fluid IGF-I because the majority of IGF-I in follicular fluid is derived from blood (Leeuwenberg *et al.*, 1996). Postpartum anoestrous cattle have lower blood IGF-I concentrations than cattle that have resumed oestrous cycles (Roberts *et al.*, 1997). Blood GH concentrations are correlated inversely with blood IGF-I because IGF-I is the primary negative feedback regulator of GH secretion. Therefore, animals with increased blood GH usually have low blood IGF-I. An exception to this relationship is found in animals treated with exogenous GH where both GH and IGF-I are increased. An example of the interplay between liver and ovary within the GH–IGF-I endocrine system is found in postpartum cattle. Insulin-like growth factor I is decreased in postpartum cattle because energy requirements exceed nutrient intake (McGuire *et al.*, 1992). Cattle in poor body condition or cows failing to improve body condition during lactation also have low blood IGF-I. However, blood GH concentrations are higher in postpartum cows with low concentrations of blood IGF-I. Later in the postpartum period, when nutritional deficiencies are corrected, blood IGF-I increases. Greater blood concentrations of IGF-I act on the hypothalamus and pituitary to decrease GH secretion. These changes in GH and IGF-I, which are a consequence of nutritionally induced changes in hepatic function, have a direct effect on the ovary by modifying the amounts GH and IGF-I in blood. Improved postpartum ovarian function is correlated with lower blood concentrations of GH and higher blood IGF-I (Roberts *et al.*, 1997).

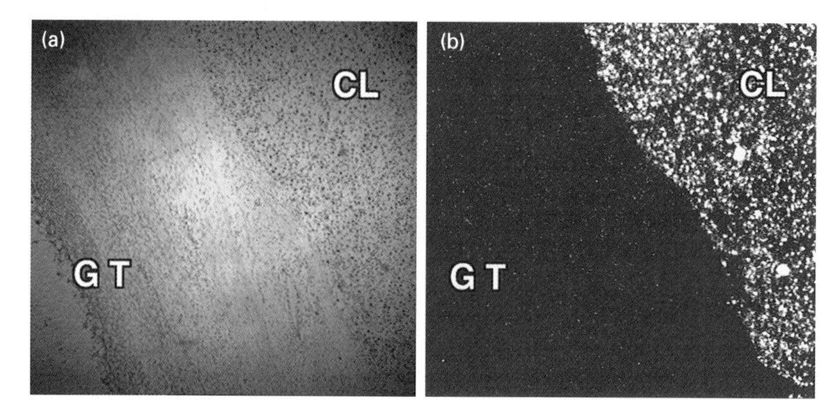

Fig. 3. Growth hormone receptor mRNA in bovine ovary measured by *in situ* hybridization. Photographs were taken with brightfield (a) and darkfield (b). The corpus luteum (CL) shows intense hybridization for GH receptor mRNA. Both granulosal cells (G) and thecal cells (T) of the follicle are negative for GH receptor mRNA. Original magnification × 42. Reprinted from Yuan and Lucy (1996) with permission from Elsevier Science.

Direct Actions of GH on the Ovary

Localization of GH receptors within the ovary

Growth hormone receptors are members of the cytokine–haematopoietic receptor superfamily that includes GH and prolactin receptors, as well as cytokine and haematopoietic hormone receptors. The GH receptor mediates the actions of GH by binding GH, dimerizing and transducing an intracellular signal via Janus Kinase (JAK) and signal transducers and activators of transcription (STAT). Other second messenger systems, including insulin receptor substrate (IRS-1), phosphatidylinositol-3-kinase, and mitogen-activated protein (MAP) kinase, may mediate the action of GH under different conditions or cell types. The GH receptor mRNA is almost identical in cattle and sheep and contains 4160 base pairs that encode 634 amino acids. Alternative exon 1 splicing of the GH receptor mRNA occurs in a variety of species including cattle and sheep. Two different promoters transcribe two GH receptor mRNAs with alternatively spliced exon 1 sequences. One promoter (bovine/ovine 1A) has liver-specific activity. A second promoter (bovine/ovine 1B) is active in adult liver but is also active within non-hepatic tissues including the reproductive tract (Lucy *et al.*, 1998; Fig. 2).

If GH has a direct effect on the ovary then the GH receptor should be present within ovarian cells. In cattle, the GH receptor was found within the large luteal cells (Lucy *et al.*, 1993). The presence of GH receptor was demonstrated by using several methods including immunohistochemistry (Lucy *et al.*, 1993), ribonuclease protection assay (Lucy *et al.*, 1993) and *in situ* hybridization (Yuan and Lucy, 1996). It is not known why the GH receptor was not localized specifically within large luteal cells in pigs (Yuan and Lucy, 1996).

There may be species differences for GH receptor expression in ovarian follicles. In humans (Sharara and Nieman, 1994) and rats (Carlsson *et al.*, 1993), GH receptor was detected in granulosal cells as well as corpora lutea. In cattle, GH receptor expression in follicles was approximately 20-fold lower when compared with corpora lutea (Lucy *et al.*, 1993). Furthermore, when histological sections containing both follicles and corpora lutea were examined, corpora lutea contained abundant GH receptor protein (Lucy *et al.*, 1993) or mRNA (Yuan and Lucy, 1996), whereas neighbouring follicles were negative for GH receptor (Fig. 3). However, when examined by reverse transcriptase PCR, the GH receptor was detected in bovine granulosal cells, cumulus cells and the oocyte (Izadyar *et al.*, 1997). In addition, the GH receptor was detected in the granulosal cells and oocytes of ovine small follicles by *in situ* hybridization (Eckery *et al.*, 1997). These results indicate that ruminant follicles contain the GH receptor but the concentration of GH receptor in follicles may be considerably lower than in corpora lutea.

Does GH control ovarian IGF-I?

Both follicles and corpora lutea have GH receptor and IGF-I mRNA. In liver, GH controls hepatic IGF-I synthesis. One important question, therefore, is whether GH controls ovarian IGF-I synthesis. Growth hormone-dependent, ovarian IGF-I synthesis has been shown in rats, pigs, and rabbits (Spicer and Echternkamp, 1995). In hypophysectomized ewes, LH increased luteal GH receptor mRNA and GH increased luteal IGF-I mRNA (Juengel *et al.*, 1997). However, in other studies GH failed to increase ovarian IGF-I synthesis either *in vitro* (Wathes *et al.*, 1995) or in intact cattle (Kirby *et al.*, 1996). In addition, immunization of heifers against GH releasing hormone (GRF) decreased blood GH and IGF-I but did not change the ovarian IGF-I mRNA concentration (Cohick *et al.*, 1996). Data showing GH-dependent follicular IGF-I synthesis have not been reported for sheep or cattle. Therefore, follicular IGF-I is probably not locally controlled by GH. Instead, endocrine IGF-I, under GH control, influences ovarian function through its contribution to follicular fluid IGF-I (Leeuwenberg *et al.*, 1996). However, the physiological importance of locally produced (ovarian) IGF-I and endocrine IGF-I (hepatic, GH dependent) is debated because the availability of IGF-I from different sources may depend on the interaction of IGF-I with locally produced and serum-derived IGFBP (Yuan *et al.*, 1998).

In Vivo Effects of GH on Reproduction

There is probably no absolute requirement for GH in reproduction, because women with inactivating GH receptor mutations (Laron dwarfs; Menashe *et al.*, 1991) and cattle with abnormal GH receptor expression (Chase *et al.*, 1998) are capable of reproduction. A knockout mouse for the GH receptor was also fertile (Zhou *et al.*, 1997). Although reproduction is possible in each of these conditions, the efficiency of reproduction is low. Hence, there is a facilitatory but not obligatory role for GH in reproductive processes.

Confounding effects of GH and IGF-I in vivo

A limitation of all *in vivo* studies using exogenous GH is the confounding effects of increased blood IGF-I after GH treatment. Insulin-like growth factor I is a potent ovarian growth factor (Spicer and Echternkamp, 1995; Armstrong and Webb, 1997). Therefore, the increase in IGF-I that occurs after GH treatment confounds the direct effects of GH on the ovary. Furthermore, the increase in blood GH followed by the increase in blood IGF-I leads to an unphysiological relationship between blood GH and blood IGF-I concentrations. In untreated animals, blood GH and blood IGF-I are inversely correlated because IGF-I feeds back negatively on GH secretion. In GH-treated animals, blood GH and blood IGF-I are positively correlated (that is, GH-treated animals have high blood GH concentrations and high IGF-I). When examined in heifers fed high-energy diets, increased follicular growth was negatively correlated with serum GH because greater nutrient intake increased insulin and IGF-I but suppressed GH (Gutierrez *et al.*, 1997). Therefore, the results of *in vivo* studies using exogenous GH should be interpreted with caution. Effects of exogenous GH *in vivo* are either direct effects of GH or IGF-I or a combined effect of both hormones.

Number of ovarian follicles

Administration of exogenous GH increased IGF-I as well as the number of recruited follicles (2–9 mm in diameter; Gong *et al.*, 1997; Kirby *et al.*, 1997). The increased number of recruited follicles can be maintained for at least 84 days and may persist for at least 21 days after GH treatment (Kirby *et al.*, 1997). The greater number of recruited follicles did not lead to increased numbers of selected follicles (> 10 mm diameter). Therefore, additional follicles in the recruited pool cannot proceed to larger size classes when stimulated with GH (selection process is unchanged). In controlled studies, ovulation rate was not changed in cattle treated with GH (Kirby *et al.*, 1997). Thus, ruminants are different from mice because exogenous GH (or a GH transgene) increases ovulation rate and litter size in mice (Cecim *et al.*, 1995). A higher proportion of twin births was reported in GH-treated cattle (Cole *et al.*, 1991). However, the twinning response varied across herds, and may reflect an interaction of GH with either genetic or environmental factors. Indeed, Bilby and Lucy (1997) found that other factors including parity and number of corpora lutea had a greater effect on follicular growth than did exogenous GH. Pigs are similar to cattle in this respect because GH increased the number of small follicles (Spicer *et al.*, 1992).

Hypophysectomized ewes did not develop preovulatory follicles unless GH and FSH were administered (Eckery *et al.*, 1997). This indicates that GH has a direct effect on small follicles. However, the study did not preclude an indirect effect of GH on the ovary through increased IGF-I after GH treatment. Most lines of evidence support an endocrine IGF-I effect rather than a direct GH effect on ruminant follicles. First, very little GH receptor mRNA or protein is found within ruminant follicles (Lucy *et al.*, 1993; Yuan and Lucy, 1996; Eckery *et al.*, 1997). Second, heifers treated with increasing doses of GH failed to have greater growth of antral follicles when the GH dose was below the threshold for increased IGF-I (Gong *et al.*, 1997). Third, cattle selected for multiple births have higher blood and follicular fluid IGF-I concentrations (Echternkamp *et al.*, 1990). Fourth, heifers immunized against GRF had low blood IGF-I, delayed puberty, fewer large antral follicles, but equivalent ovarian IGF-I mRNA compared with control heifers (Cohick *et al.*, 1996; Schoppee *et al.*, 1996). Finally, cattle with a liver GH receptor deficiency, causing high blood GH with low blood IGF-I, had one quarter of the number of small antral follicles compared with control cattle (Chase *et al.*, 1998).

The reason why the number of antral follicles is increased in animals supplemented with GH is not known. *In vitro*, IGF-I increases the number of gonadotrophin binding sites and the activity of gonadotrophin second messenger systems (Spicer and Echternkamp, 1995). Perhaps greater gonadotrophin action caused by GH or IGF-I can lead to an increase in follicular growth. *In vivo*, GH increased follicular fluid IGF-I but did not increase gonadotrophin binding sites in bovine follicles (Andrade *et al.*, 1996). Greater gonadotrophin receptor concentration, therefore, does not occur after GH treatment in cattle. The possibility that GH or IGF-I increases the activity of gonadotrophin second messenger pathways *in vivo* without changing the number of gonadotrophin receptors has not been addressed. An additional possibility is that GH supplementation decreases atresia of the growing pool and leads to a greater number of antral follicles. In nonruminant granulosal cells, GH and IGF-I decrease apoptosis (Kaipia and Hsueh, 1997). In cattle, a GH-mediated decrease in atresia (Cushman *et al.*, 1996) indicates that GH increases antral follicle populations by reducing atresia.

Dominant and subordinate follicles

Exogenous GH does not affect the growth rate or size of dominant follicles (Kirby *et al.*, 1997). However, the size of second largest follicles is increased in GH-treated cows. The increase in second largest follicle diameter is associated with greater development of the recruited pool of ovarian follicles (Gong *et al.*, 1997; Kirby *et al.*, 1997). Although the absolute size of the dominant follicle was not changed, the duration of the dominance phase in the first wave dominant follicle was shortened by about 2 days in GH-treated cattle. This led to an earlier emergence of the second wave dominant follicle (Kirby *et al.*, 1997). The shift towards a reduced period of dominance was also associated with a shift in the timing of the mid-cycle peak in blood concentration of FSH (Kirby *et al.*, 1997). The reason for the faster turnover in dominant follicles of GH-treated cattle is unknown. Bovine dominant follicles are dependent on LH for the maintenance of dominance and GH treatment decreases LH secretion (Schemm *et al.*, 1990). In pigs, GH decreases LH/hCG binding sites within follicles (Spicer *et al.*, 1992) but a similar response was not observed in cattle (Andrade *et al.*, 1996). *In vitro*, IGF-I antagonized insulin-induced oestradiol synthesis (Spicer *et al.*, 1993). Increased concentrations of

GH and IGF-I, therefore, do not necessarily prolong or improve the function of dominant follicles. Instead, greater concentrations of GH and IGF-I may accelerate the series of events that lead to dominant follicle atresia and cause premature turnover of mid-cycle dominant follicles.

Puberty

Heifers immunized against GRF have lower blood concentrations of GH and IGF-I and reach puberty at an older age than control heifers (Cohick *et al.*, 1996; Schoppee *et al.*, 1996). The effect of GRF immunization on puberty occurred despite normal patterns of LH secretion (Schoppee *et al.*, 1996). The delay in the timing of puberty in GRF-immunized heifers was caused by inadequate follicular oestradiol production that failed to trigger an LH surge. A synergistic relationship between GH, IGF-I and LH for puberty was demonstrated, therefore, because heifers with normal LH failed to reach puberty when GH and IGF-I were inadequate. Nutrient-restricted heifers also had delayed puberty and low IGF-I. However, unlike the GRF-immunized heifers, blood GH was increased and blood LH was decreased by nutrient restriction (Schoppee *et al.*, 1996). Undernutrition, therefore, delays puberty through a combined effect of decreased IGF-I and LH. Treating heifers with GH increased body growth but did not decrease age at puberty or increase the number of small follicles (Hall *et al.*, 1994). These results suggest that the initiation of LH secretion in peripubertal heifers is the most important factor that determines age at puberty. Growth hormone and IGF-I may play a role in puberty but their effects are permissive to LH.

Growth and steroidogenesis of corpora lutea

Growth hormone is required for growth and development of the corpus luteum in ruminants because decreased corpora lutea weight in hypophysectomized ewes could be restored to near normal size with exogenous GH and LH (Juengel *et al.*, 1997). In addition to a direct effect of GH on the corpora lutea, there may also be a requirement for endocrine IGF-I in corpora lutea growth. Cattle with a liver GH receptor deficiency (high blood GH concentrations but low blood IGF-I concentrations) had smaller corpora lutea and shorter luteal phases (Chase *et al.*, 1998). Nutrient-restricted heifers had smaller corpora lutea, greater blood GH, and lower blood IGF-I concentrations (Vandehaar *et al.*, 1995). In these heifers, the IGF-I mRNA in corpora lutea was not changed while the amount of IGF-I mRNA in liver was decreased by undernutrition. An endocrine mechanism involving low blood IGF-I, therefore, explained reduced corpus luteum size in nutrient-restricted heifers (Vandehaar *et al.*, 1995). However, in other studies of nutrient-restricted heifers, exogenous GH restored blood IGF-I to control concentrations but failed to increase corpora lutea weight (Yung *et al.*, 1996). The effect of exogenous GH on the corpora lutea of intact or normal-fed cattle is equally unclear. The corpora lutea of dairy cattle treated with GH from days 1–17 of an oestrous cycle were 60% heavier than those of control cattle (Lucy *et al.*, 1995). The increase in corpora lutea weight occurred without a change in IGF-I mRNA concentration within the corpora lutea (Kirby *et al.*, 1996). Greater plasma progesterone concentrations were reported in dairy cows treated with GH (Schemm *et al.*, 1990; Gallo and Block, 1991). However, in other studies, GH tended to decrease plasma progesterone concentrations (Kirby *et al.*, 1997). The inconsistencies in corpora lutea responses to GH suggest that other physiological factors may over-ride any stimulatory effect of exogenous GH on the corpora lutea. One concern for studies of GH in lactating animals is the confounding effect of increased milk production and loss of body condition on corpora lutea function in GH-treated animals. In one study of lactating cows, a period of anoestrus occurred after GH treatment (Waterman *et al.*, 1993). Therefore, GH-induced changes in milk production may compromise corpora lutea function and confound reproductive effects of GH.

***In Vitro* Effects of GH on Ovarian Cells**

One method to avoid the confounding effects of endocrine IGF-I in tests of GH action is to treat ovarian cells with GH *in vitro*. The *in vitro* treatments can be tested further with IGF-I neutralizing antibodies to determine whether a GH-dependent IGF-I release is responsible for the effects of GH on reproductive cells. One detraction for most *in vitro* studies of GH is the supraphysiological doses (10^2–10^5 ng ml^{-1}) used to show an effect of GH. This is of great concern when pituitary GH is used because of the possible contamination of the GH preparation with either LH or FSH. There are no data that show the direct activation of the JAK-STAT pathway by GH in ovarian cells. One focus of future studies should be the elucidation of GH signalling pathways in ovarian cells cultured *in vitro*. Furthermore, a physiological dose of recombinant GH should be used.

Granulosal cells

Porcine granulosal cells increase progesterone secretion in response to GH (Spicer and Echternkamp, 1995). An effect of GH on oestradiol synthesis in cultured human granulosal cells has also been demonstrated (Barreca *et al.*, 1993). The effect of GH on steroidogenesis was blocked by the addition of a neutralizing IGF-I antibody. Therefore, the responses to GH *in vitro* may be secondary to an increase in IGF-I that occurs after GH treatment. There is no consensus for the effects of GH on ruminant granulosal cells. In bovine granulosal cells isolated from small or large follicles, GH inhibited oestradiol synthesis and inhibited proliferation of cells from large follicles (Spicer and Stewart, 1996). Other workers also reported an inhibitory effect of GH on the proliferation of granulosal cells from large follicles (Gong *et al.*, 1993) but showed a stimulatory effect of GH on oestradiol secretion (Gong *et al.*, 1994). The inhibitory effect of GH on cell proliferation may be explained partially by the inhibitory effect of GH on IGF-I mRNA (and perhaps protein) in bovine granulosal cells (Spicer *et al.*, 1993). Growth hormone did not affect progesterone secretion from granulosal cells isolated from large follicles but progesterone secretion and proliferation were increased when granulosal cells were isolated from small follicles and co-treated with insulin (Spicer and Stewart, 1996). In ovine granulosal cells co-treated with insulin, GH also increased progesterone secretion in long-term culture (Wathes *et al.*, 1995). One conclusion from these studies is that GH will increase steroidogenesis in cultured granulosal cells when GH causes the release of IGF-I (pig and human). The IGF-I may increase steroidogenesis itself or may be permissive to the effect of GH. An increase in steroidogenesis may not occur in ruminants because GH does not increase IGF-I in cultured granulosal cells. It may be necessary to supplement cell culture media with either insulin or IGF-I to detect any effect of GH on steroidogenesis in cultured ruminant granulosal cells.

Cumulus cells

Growth hormone increased *in vitro* maturation of bovine oocytes as well as cumulus expansion by an IGF-I independent mechanism (Izadyar *et al.*, 1997). Further analyses showed that the stimulatory effects of GH were not mediated by tyrosine kinase activation. Instead, a cAMP second messenger pathway was suggested (Izadyar *et al.*, 1997). This result was unexpected because cAMP is not a traditional GH receptor second messenger. Nevertheless, the data implicate GH in follicular control of oocyte development.

Thecal cells

Rat thecal cells increased androgen synthesis in response to GH (Apa *et al.*, 1996a). However, the response to GH was different from that of granulosal cells, because an IGF-I antibody could not neutralize the effect of GH. In rats, therefore, GH may act on thecal cells through an IGF-I-independent mechanism. The response of bovine thecal cells to GH depended on LH-responsiveness of the cells. Growth hormone increased androstenedione secretion in thecal cells that responded well to LH. Thecal cells that responded poorly to LH did not have increased androstenedione secretion after GH treatment (Spicer and Stewart, 1996).

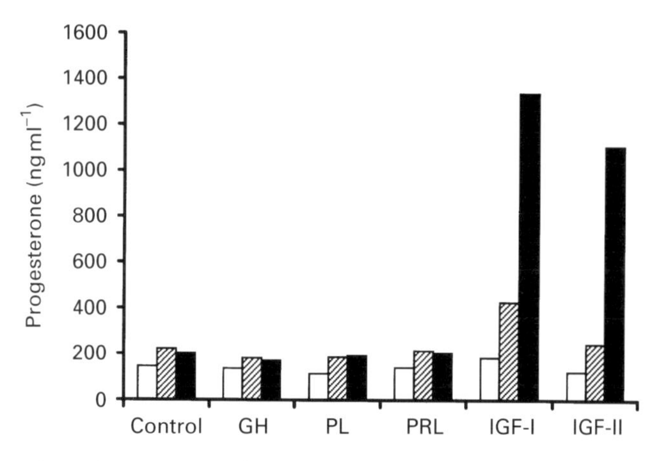

Fig. 4. Concentration of progesterone in tissue culture media from a mixed population of bovine luteal cells (oestrous cycle day 10) treated with 1 ng ml^{-1} (□), 10 ng ml^{-1} (▨) or 100 ng ml^{-1} (■) bovine LH as well as increasing dosages of control (no hormone) or recombinant hormones [bovine growth hormone (GH), bovine placental lactogen (PL), bovine prolactin (PRL), insulin-like growth factor I (IGF-I) or IGF-II]. Data represent secretion between 48 and 72 h of cell culture. Both IGF-I and IGF-II increase progesterone secretion with increasing dosage. Other hormones (GH, PL and PRL) have no effect on progesterone secretion. Data are means for duplicate wells from each of three heifers (pooled standard error = 89.5) (Lucy and Collier, unpublished).

Luteal cells

Bovine and ovine corpora lutea have increased secretion of progesterone in response to IGF-I in microdialysis systems (Sauerwein *et al.*, 1992; Khan-Dawood *et al.*, 1994). Furthermore, tyrosine kinase-mediated second messenger pathways for IGF-I and insulin were demonstrated for cultured bovine luteal cells (Chakravorty *et al.*, 1993). There is no equivalent demonstration of the GH receptor second messenger system in luteal cells from any species. Nevertheless, *in vitro* data suggest an effect of GH on luteal cell steroidogenesis and oxytocin secretion. Liebermann and Schamms (1994) increased progesterone secretion and caused oxytocin release by treating bovine corpora lutea with GH in a microdialysis system. In cultured cells, GH increased progesterone secretion in human luteal cells and the effect of GH could be blocked by an IGF-I neutralizing antibody (Apa *et al.*, 1996b). Therefore, it appears that the effects of GH on progesterone synthesis may depend on IGF-I synthesis. There is very little evidence that GH will increase steroidogenesis in cultured ruminant luteal cells. We were unable to show an effect of GH, prolactin, or placental lactogen on bovine luteal cells. At the same time, addition of IGF-I and IGF-II resulted in greater progesterone concentrations in luteal cell cultures (Fig. 4).

Conclusions

Growth hormone is involved in many aspects of ovarian physiology in ruminants. Most importantly, GH increases the growth and development of antral follicles and increases the growth and steroidogenesis of the corpora lutea. These actions of GH are usually synergistic with IGF-I and gonadotrophins (Fig. 1). An important issue that should be addressed is the relative importance of GH compared with IGF-I for ovarian function. Many of the perceived effects of GH on the ovary can be explained by changes in blood IGF-I that occur when GH causes hepatic IGF-I synthesis and secretion. The ruminant may be different from pigs, humans, and laboratory animals in which GH has a direct effect on the ovary through the control of ovarian IGF-I. The presence of GH receptors in ruminant corpora lutea and follicles presumes a direct action for GH. The direct actions of GH on the ovary, however, may be less important than the ovarian actions of endocrine IGF-I that are ultimately under nutritional and GH control.

References

Andrade LP, Rhind SM, Wright IA, McMillen SR, Goddard PJ and Bramley TA (1996) Effects of bovine somatotropin (bST) on ovarian function in postpartum beef cows *Reproduction, Fertility and Development* **8** 951–960

Apa R, Caruso A, Andreani CL, Miceli F, Lazzarin N, Mastrandrea M, Ronsisvalle E, Mancuos S and Lanaone A (1996a) Growth hormone stimulates androsterone synthesis by rat theca–interstitial cells *Molecular and Cellular Endocrinology* **118** 95–101

Apa R, Di Simone N, Ronsisvalle E, Miceli F, de Feo D, Caruso A, Lanzone A and Mancuso S (1996b) Insulin-like growth factor (IGF)-I and IGF-II stimulate progesterone production by human luteal cells: role of IGF-I as mediator of growth hormone action *Fertility and Sterility* **66** 235–239

Armstrong DG and Webb R (1997) Ovarian follicular dominance: the role of intraovarian growth factors and novel proteins *Reviews of Reproduction* **2** 139–146

Barreca A, Artini PG, Del Monte P, Ponzani P, Pasquini P, Cariola G, Volpe A, Genazzani AR, Giordano G and Minuto F (1993) *In vivo* and *in vitro* effect of growth hormone on estradiol secretion by human granulosa cells *Journal of Clinical Endocrinology and Metabolism* **77** 61–67

Bilby CR and Lucy MC (1997) Reproductive responses to bovine somatotropin (bST) in primiparous (P) and multiparous (M) cows with one or two corpora lutea (CL) *Journal of Dairy Science* **80** (Supplement 1) Abstact 151

Carlsson B, Nilsson A, Isaksson OGP and Billig H (1993) Growth hormone–receptor messenger RNA in the rat ovary: regulation and localization *Molecular and Cellular Endocrinology* **95** 59–66

Cecim M, Kerr J and Bartke A (1995) Effects of bovine growth hormone (bGH) transgene expression or bGH treatment on reproductive functions in female mice *Biology of Reproduction* **52** 1144–1148

Chakravorty A, Joslyn MI and Davis JS (1993) Characterization of insulin and insulin-like growth factor-I actions in the bovine luteal cell: regulation of receptor tyrosine kinase activity, phosphatidylinositol-3-kinase, and deoxyribonucleic acid synthesis *Endocrinology* **133** 1331–1340

Chase CC, Jr, Kirby CJ, Hammond AC, Olson TA and Lucy MC (1998) Patterns of ovarian growth and development in cattle with a growth hormone receptor deficiency *Journal of Animal Science* **76** 212–219

Cohick WS, Armstrong JD, Whitacre MD, Lucy MC, Harvey RW and Campbell RM (1996) Ovarian expression of insulin-like growth factor-I (IGF-I), IGF binding proteins, and growth hormone (GH) receptor in heifers actively immunized against GH-releasing factor *Endocrinology* **137** 1670–1677

Cole WJ, Madsen KS, Hintz RL and Collier RJ (1991) Effect of recombinantly-derived bovine somatotropin on reproductive performance of dairy cattle *Theriogenology* **36** 573–595

Cushman RA, DeSouza JC, Hedgpeth VS and Britt JH (1996) Alteration of bovine folliculogenesis by long-term treatment with estradiol (E2) and bovine somatotropin (bST) *Journal of Animal Science* **74** (Supplement 1) Abstract 220

Echternkamp SE, Spicer LJ, Gregory KE, Canning SF and Hammond JM (1990) Concentrations of insulin-like growth factor-I in blood and ovarian follicular fluid of cattle selected for twins *Biology of Reproduction* **43** 8–14

Eckery DC, Moeller CL, Nett TM and Sawyer HR (1997) Localization and quantification of binding sites for follicle-stimulating hormone, luteinizing hormone, growth hormone, and insulin-like growth factor I in sheep ovarian follicles *Biology of Reproduction* **57** 507–513

Gallo GF and Block E (1991) Effects of recombinant bovine somatotropin on hypophyseal and ovarian functions of lactating dairy cows *Canadian Journal of Animal Science* **71** 343–353

Gong JG, McBride D, Bramley TA and Webb R (1993) Effects of recombinant bovine somatotrophin, insulin-like growth factor-I and insulin on the proliferation of bovine granulosa cells *in vitro*. *Journal of Endocrinology* **139** 67–75

Gong JG, McBride D, Bramley TA and Webb R (1994) Effects of recombinant bovine somatotrophin, insulin-like growth factor-I and insulin on bovine granulosa cell steroidogenesis *in vitro*. *Journal of Endocrinology* **143** 157–164

Gong JG, Baxter G, Bramley TA and Webb R (1997) Enhancement of ovarian follicle development in heifers by treatment with recombinant bovine somatotropin: a dose–response study *Journal of Reproduction and Fertility* **110** 91–97

Gutierrez CG, Oldham J, Bramley TA, Gong JG, Campbell BK and Webb R (1997) The recruitment of ovarian follicles is enhanced by increased dietary intake in heifers *Journal of Animal Science* **75** 1876–1884

Hall JB, Schillo KK, Fitzgerald BP and Bradley NW (1994) Effects of recombinant bovine somatotropin and dietary energy intake on growth, secretion of luteinizing hormone, follicular development, and onset of puberty in beef heifers *Journal of Animal Science* **72** 709–718

Izadyar F, Colenbrander B and Bevers MM (1997) Stimulatory effect of growth hormone on *in vitro* maturation of bovine oocytes is exerted through the cyclic adenosine 3′,5′-monophosphate signaling pathway *Biology of Reproduction* **57** 1484–1489

Juengel JL, Nett TM, Anthony RV and Niswender GD (1997) Effects of luteotrophic and luteolytic hormones on expression of mRNA encoding insulin-like growth factor I and growth hormone receptor in the ovine corpus luteum *Journal of Reproduction and Fertility* **110** 291–298

Kaipai A and Hsueh AJW (1997) Regulation of follicular atresia *Annual Review of Physiology* **59** 349–363

Khan-Dawood FS, Gargiulo AR and Dawood MY (1994) *In vitro* microdialysis of the ovine corpus luteum of pregnancy:

effects of insulin-like growth factor on progesterone secretion *Biology of Reproduction* **51** 1299–1306

Kirby CJ, Thatcher WW, Collier RJ, Simmen FA and Lucy MC (1996) Effects of growth hormone and pregnancy on expression of growth hormone receptor, insulin-like growth factor-I and insulin-like growth factor binding protein-2 and -3 genes in bovine uterus, ovary and oviduct *Biology of Reproduction* **55** 996–1002

Kirby CJ, Smith MF, Keisler DH and Lucy MC (1997) Follicular function in lactating dairy cows treated with sustained release bovine somatotropin *Journal of Dairy Science* **80** 273–285

Leeuwenberg BR, Hudson NL, Moore LG, Hurst PR and McNatty KP .(1996) Peripheral and ovarian IGF-I concentrations during the ovine oestrous cycle *Journal of Endocrinology* **148** 281–289

Liebermann J and Schams D (1994) Actions of somatotropin on oxytocin and progesterone release from the microdialysed bovine corpus luteum *in vitro. Journal of Endocrinology* **143** 243–250

Lucy MC, Collier RJ, Kitchell MA, Dibner JJ, Hauser SD and Krivi GG (1993) Immunohistochemical and nucleic acid analysis of somatotropin receptor populations in the bovine ovary *Biology of Reproduction* **48** 1219–1227

Lucy MC, Thatcher WW, Collier RJ, Simmen FA, Ko Y, Savio JD and Badinga L (1995) Effects of somatotropin on the conceptus, uterus, and ovary during maternal recognition of pregnancy in cattle *Domestic Animal Endocrinology* **12** 73–82

Lucy MC, Boyd CK, Koenigsfeld AT and Okamura CS (1998) Expression of somatotropin receptor messenger ribonucleic acid in bovine tissues *Journal of Dairy Science* **81** 1889–1895

McGuire MA, Vicini JL, Bauman DE and Veenhuizen JJ (1992) Insulin-like growth factors and binding proteins in ruminants and their nutritional regulation *Journal of Animal Science* **70** 2901–2910

Menashe Y, Sack J and Mashinach S (1991) Spontaneous pregnancies in two women with Laron-type dwarfism: are growth hormone and circulating insulin-like growth factor mandatory for induction of ovulation? *Human Reproduction* **6** 670–671

Roberts AJ, Nugent RA, Klindt J and Jenkins TG (1997) Circulating insulin-like growth factor I, insulin-like growth factor binding proteins, growth hormone, and resumption of estrus in postpartum cows subjected to dietary energy restriction *Journal of Animal Science* **75** 1909–1917

Sauerwein H, Miyamoto A, Gunther J, Meyer HHD and Schams D (1992) Binding and action of insulin-like growth factors and insulin in bovine luteal tissue during the oestrous cycle *Journal of Reproduction and Fertility* **96** 103–115

Schemm SR, Deaver DR, Griel LC, Jr and Muller LD (1990) Effects of recombinant bovine somatotropin on luteinizing hormone and ovarian function in lactating dairy cows *Biology of Reproduction* **42** 815–821

Schoppee PD, Armstrong JD, Harvey RW, Whitacre MD, Felix A and Campbell RM (1996) Immunization against growth hormone releasing factor or chronic feed restriction initiated at 3.5 months of age reduces ovarian response to pulsatile administration of gonadotrophin-releasing hormone at 6 months of age and delays onset of puberty in heifers *Biology of Reproduction* **55** 87–98

Sharara FI and Nieman LK (1994) Identification and cellular localization of growth hormone receptor gene expression in the human ovary *Journal Clinical Endocrinology and Metabolism* **79** 670–672

Spicer LJ and Echternkamp SE (1995) The ovarian insulin and insulin-like growth factor system with an emphasis on domestic animals *Domestic Animal Endocrinology* **12** 223–245

Spicer LJ and Stewart RE (1996) Interaction among bovine somatotropin, insulin, and gonadotrophins on steroid production by bovine granulosa and theca cells *Journal of Dairy Science* **79** 813–821

Spicer LJ, Klindt J, Buonomo FC, Maurer R, Yen JT and Echternkamp SE (1992) Effect of porcine somatotropin on number of granulosa cell luteinizing hormone/human chorionic gonadotrophin receptors, oocyte viability, and concentrations of steroids and insulin-like growth factors I and II in follicular fluid of lean and obese gilts *Journal of Animal Science* **70** 3149–3157

Spicer LJ, Alpizar E and Echternkamp SE (1993) Effects of insulin, insulin-like growth factor I, and gonadotrophins on bovine granulosa cell proliferation, progesterone production, estradiol production, and(or) insulin-like growth factor I production *in vitro. Journal of Animal Science* **71** 1232–1241

Vandehaar MJ, Sharma BK and Fogwell RL (1995) Effect of dietary energy restriction on the expression of insulin-like growth factor-I in liver and corpus luteum of heifers *Journal of Dairy Science* **78** 832–841

Waterman DF, Silvia WJ, Hemken RW, Heersche G, Jr, Swenson TS and Eggert RG (1993) Effect of bovine somatotropin on reproductive function in lactating dairy cows *Theriogenology* **40** 1015–1028

Wathes DC, Perks CM and Davis AJ (1995) Regulation of insulin-like growth factor-I and progesterone synthesis by insulin and growth hormone in the ovine ovary *Biology of Reproduction* **53** 882–889

Yuan W and Lucy MC (1996) Messenger ribonucleic acid expression for growth hormone receptor, luteinizing hormone receptor, and steroidogenic enzymes during the estrous cycle and pregnancy in porcine and bovine corpora lutea *Domestic Animal Endocrinology* **13** 431–444

Yuan W, Bao B, Garverick HA, Youngquist RS and Lucy MC (1998) Follicular dominance in cattle is associated with divergent patterns of ovarian gene expression for insulin-like growth factor (IGF)-I, IGF-II, and IGF binding protein-2 in dominant and subordinate follicles *Domestic Animal Endocrinology* **15** 55–63

Yung MC, VandeHaar MJ, Fogwell RL and Sharma BK (1996) Effect of energy balance and somatotropin on insulin-like growth factor I in serum on weight and progesterone of corpus luteum in heifers *Journal of Animal Science* **74** 2239–2244

Zhou Y, Xu BC, Maheshwari HG, He L, Reed M, Lozykowski M, Okada S, Cataldo L, Coschigamo K, Wagner TE, Baumann G and Kopchick JJ (1997) A mammalian model for Laron syndrome produced by targeted disruption of the mouse growth hormone receptor/binding protein gene (the Laron mouse) *Proceedings of the National Academy of Sciences, USA* **94** 13215–13220

Journal of Reproduction and Fertility Supplement **54**, 61–71

Regulation of follicle waves to maximize fertility in cattle

J. F. Roche[1], E. J. Austin[1], M. Ryan[1], M. O'Rourke[2], M. Mihm[3] and M. G. Diskin[2]

[1]*Faculty of Veterinary Medicine, University College Dublin, Ireland;* [2]*Teagasc Research Centre, Athenry, Co. Galway, Ireland; and* [3]*University of Glasgow Veterinary School, Bearsden Road, Glasgow G61 1QH, UK*

Cattle have recurrent follicular waves every 7–10 days in most physiological situations; an FSH increase is associated with emergence of the wave and LH pulse frequency determines the fate of the dominant follicle. To control oestrus with hormones it is necessary to ensure that either induced corpus luteum regression or the termination of a progestogen treatment coincides with the selection of the dominant follicle during the wave, to give a precise onset of oestrus and high fertility. The exogenous administration of progesterone or progestagen blocks the normal turnover of the dominant follicle once the corpus luteum regresses. Thus, the effects of duration of dominance of the preovulatory follicle on onset of oestrus and fertility were examined. The variation in onset of oestrus was reduced but occurred 5–9 h later after 4 versus 8 days of dominance; pregnancy rate was also affected with dominance periods of 2–4, 4–8 and > 10 days resulting in 0, 10–15% or 20–50% reduction in pregnancy rates, respectively. The necessity for short duration of dominance of the preovulatory follicle means that to ensure high fertility the follicular wave needs to be regulated when using hormones to control oestrus. Two approaches were examined, namely the use of GnRH or oestradiol at time of progesterone intravaginal releasing device insertion. The effect of 250 μg of synthetic GnRH on the fate of an existing follicle wave was to ovulate the dominant follicle (20/20 cows) and a new wave emerged 1.6 ± 0.3 days later; however, there was no effect of GnRH on the wave if administered before dominant follicle selection. The effect of oestradiol concentrations on suppression of FSH in ovariectomized heifers showed that increasing oestradiol to 10–15 pg ml^{-1} caused a 37 ± 6.9% decrease in FSH for 24 h, with a subsequent increase to pretreatment values by 57 ± 13 h. In cyclic heifers, increasing oestradiol to > 10 pg ml^{-1} in conjunction with progesterone treatment at emergence of the first wave of the cycle affected the current follicle wave by either preventing dominant follicle selection or decreasing diameter of the dominant follicle, without consistently affecting the interval to new wave emergence. Increase of oestradiol after dominance, however, delayed new wave emergence by 2–5 days. A better understanding of the hormonal control of follicle waves will lead to development of improved hormonal regimens to control oestrus sufficiently to give high pregnancy rates to a single AI without recourse to detection of oestrus.

Introduction

The critical requirements for effective synchronization of oestrus in cattle are a predictable high oestrous response during a specified 12–24 h period and high pregnancy rates to a single breeding after treatment. To date, these two requirements have not been met successfully, despite our increasing knowledge of follicular development (Ginther *et al.*, 1996). Research in the early 1960s showed that progesterone–progestagen treatments of > 14 days duration resulted in good synchrony of oestrus but low pregnancy rates. Subsequently, the treatment period was reduced to 9–12 days by administration of oestradiol as a luteolytic agent at the start of the progestagen treatment (Wiltbank and Kasson, 1968; Roche, 1974); this procedure resulted in normal pregnancy rates but greater variability in onset of oestrus. The discovery of prostaglandin $F_{2\alpha}$ ($PGF_{2\alpha}$) as the endogenous luteolysin led to its use at or near the end of progestagen treatments; its use increased the oestrous response in animals with a corpus luteum. However, it failed to give sufficient precision of onset of oestrus to allow high pregnancy rates to a single planned AI (Odde, 1990). With the advent of ovarian ultrasonography, it became clear that follicular status at the end of a progestagen treatment or at the time of induced luteolysis affected the interval to oestrus. Thus, effective hormonal regulation of the ovarian cycle is now known to require both the strategic induction of premature regression of the corpus luteum and the presence of a recently selected dominant follicle at the end of the treatment period.

Antral follicle growth in cattle occurs in distinct wave-like patterns during the ovarian cycle and the postpartum period (reviewed in Roche *et al.*, 1998). The emergence of each new wave is stimulated by a transient (1–2 day) increase in FSH (Adams *et al.*, 1992; Sunderland *et al.*, 1994). Selection of the dominant follicle occurs during declining FSH concentrations, and the dominant follicle then maintains FSH at nadir concentrations until it either ovulates or succumbs to atresia, depending on the pattern of LH secretion. During the final stage of selection of the dominant follicle, there appears to be a transition from mainly FSH to LH dependency, but the cause and specific time frame of this change are unknown. The consequences are that continued growth and oestradiol production by the dominant follicle are dependent on increased LH pulse frequency, which, if prolonged, can lead to persistence of the dominant follicle (Stock and Fortune, 1993; Savio *et al.*, 1993; Taft *et al.*, 1996). Therefore, hormonal treatments that modify both FSH and LH clearly affect the fate of the follicular wave. Manipulation of the follicular wave, in turn, may alter systemic hormone concentrations, the intrafollicular environment, and the competency of the oocyte to be ovulated. However, successful hormonal regulation of oestrus must not compromise oocyte competency, ovulation, fertilization, early embryonic development or subsequent luteal function.

Each follicular wave has an inherent life span of 7–10 days as it progresses through the different stages of development, namely emergence, selection, dominance and atresia or ovulation. Thus, a dominant follicle capable of ovulation is present only at specific times during each wave. Therefore, the interval from exogenous luteolysis or withdrawal of a progestagen treatment to oestrus depends on the stage of the follicular wave at the end of treatment: cattle with a selected dominant follicle will be in oestrus within 2–3 days, but those before dominant follicle selection will not be in oestrus for 3–7 days. Thus, induced luteolysis needs to be synchronized with selection of the preovulatory dominant follicle. Current progestagen strategies to synchronize oestrus in cattle are reviewed, with reference to regulation of follicular wave and duration of dominance of the preovulatory follicle, onset of oestrus and fertility.

Precision of Onset of Oestrus

The variability in onset of oestrus after the end of a synchronization treatment will determine whether animals can be inseminated at pre-arranged times or at a detected oestrus to obtain high pregnancy rates to AI. In order to achieve maximum precision in oestrous onset, it is necessary to have a recently selected dominant follicle present at the end of treatment. New wave emergence needs to be synchronized during the treatment period, because both the stage of the follicular wave and the duration of dominance cause variation of 24 h in the duration of the follicular phase. Oestradiol concentrations of heifers with a dominant follicle of > 4 days of dominance are higher at the end of a progestagen treatment; subsequent follicular phases are short and onset of oestrus is 6–12 h earlier than in heifers with a dominant follicle of short (2–3 days) duration of dominance (Austin *et al.*, 1999). Synchrony of oestrus is optimal when the duration of dominance is either consistently short (< 4 days) or very long (10–12 days).

Duration of Dominance of the Preovulatory Follicle

The duration of dominance of the preovulatory follicle can affect pregnancy rate. Thus, it is necessary that all animals have a recently selected dominant follicle at the end of a progestagen treatment or at the time of induced luteolysis using $PGF_{2\alpha}$. Recurrent follicular waves are regulated systemically by FSH and LH, and locally by specific growth factors namely inhibins, activins, insulin-like growth factor I (IGF-I) and their respective binding proteins.

Persistent dominant follicle

Turnover of the dominant follicle is regulated by LH pulse frequency. During the luteal phase (Cupp *et al.*, 1995) or early postpartum anoestrous period (Stagg *et al.*, 1998), an LH pulse frequency of 1 pulse every 3–6 h results in loss of dominance and atresia of the dominant follicle. During the follicular phase, the high frequency–low amplitude LH pulse pattern stimulates sufficient oestradiol synthesis to induce oestrus and the preovulatory gonadotrophin surge (Rahe *et al.*, 1980). However, an intermediate LH pulse frequency of 1 pulse every 1–2 h causes an extension of the period of dominance (persistence) (Savio *et al.*, 1993; Stock and Fortune, 1993; Taylor *et al.*, 1993; Mihm *et al.*, 1999), which maintains not only continued growth of the dominant follicle (Taft *et al.*, 1996), but also its biochemical health, based on its high oestrogen activity, reduced amounts of the 34 kDa form of dimeric inhibin, and decreased amounts of the lower molecular weight forms of IGF-binding proteins (de la Sota *et al.*, 1994, 1996). During an extended period of dominance, the persistent dominant follicle suppresses emergence of a new follicular wave (functional dominance; Sirois and Fortune, 1990; Savio *et al.*, 1993; Stock and Fortune, 1993). Any dominant follicle becomes persistent in the presence of either subluteal concentrations (40–50% of normal luteal phase concentrations) of progesterone (which occur after endogenous luteolysis and more than 3–4 days after insertion of an intravaginal progesterone releasing device) or after use of a norgestomet ear implant combined with luteolysis. Neither protocol suppresses LH pulse frequency to the same extent as that achieved during the luteal phase (Kojima *et al.*, 1992).

Fertility

There is an association between persistency of the ovulatory follicle and decreased fertility in progestagen synchronized heifers (Savio *et al.*, 1993; Sanchez *et al.*, 1993; Co-operative Regional Research Project, 1996). However, the duration of dominance necessary to initiate this fertility decline, and whether the decline is gradual or abrupt, requires elucidation. Hence, the association between the duration of dominance of the ovulatory follicle and subsequent fertility to AI was examined in our laboratory using progestagen-treated heifers (Mihm *et al.*, 1994; Austin *et al.*, 1999). The duration of dominance of the second dominant follicle was controlled by causing corpus luteum regression with $PGF_{2\alpha}$ in heifers and insertion of a 3 mg norgestomet implant (Intervet Ltd, Boxmeer, Netherlands) at emergence or first day of dominance for different durations to maintain the dominant follicle. Dominance periods of up to 8 days resulted in high pregnancy rates, which were reduced as the period of dominance was increased: 89, 68, 78, 71, 52 and 12% for periods of dominance of 2, 4, 6, 8, 10 and 12 days at oestrus, respectively (Fig. 1).

The factors that cause these low pregnancy rates are not known but neither the ovulatory ability of a persistent dominant follicle, nor subsequent luteal function seem impaired (Stock and Fortune, 1993; Mihm *et al.*, 1994; Co-operative Regional Research Project, 1996). Prolonged increase of circulating oestradiol concentrations before oestrus may compromise uterine function, and thus reduce subsequent implantation rates (Butcher and Pope, 1979; Bryner *et al.*, 1990). However, pregnancy rates of heifers after embryo transfer in the presence or absence of a persistent dominant follicle were similar (Wehrman *et al.*, 1997); likewise, fertility was normal after AI when the subsequent dominant follicle after the persistent follicle was allowed to ovulate (Fike *et al.*, 1997). Thus, the absence of detrimental effects of the increased oestradiol from a persistent dominant follicle on embryo survival indicates that intrafollicular biochemical alterations or oocyte

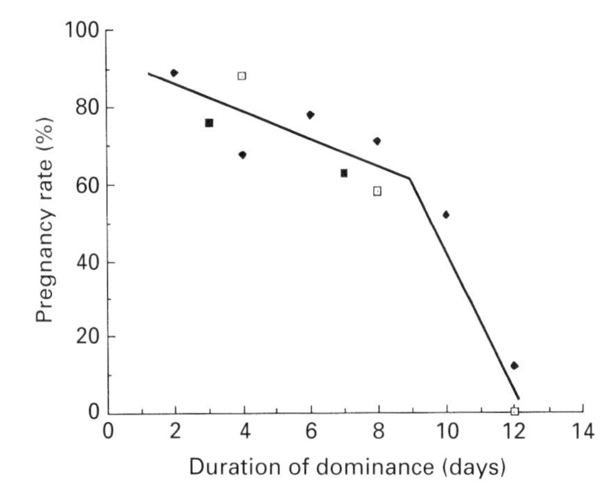

Fig. 1. Pregnancy rates in heifers after ovulation of follicles of different durations of dominance. The data are derived from Mihm *et al.* 1994 (□); and Austin *et al.*, 1998 (◆ and ■). Heifers were given a single injection of a $PGF_{2\alpha}$ analogue (PG) at second wave emergence and inseminated at a detected oestrus ($n = 9$) to achieve a duration of dominance of 1–2 days). Heifers were given PG as described above, but to achieve longer periods of dominance implants of norgestomet were also inserted at the time of PG injection for different periods as follows: 2 days ($n = 44$), 4 days ($n = 18$), 6 days ($n = 67$), 8 days ($n = 23$) and 12 days ($n = 31$), which resulted in durations of dominance of the preovulatory follicle of 4, 6, 8, 10 and 12 days, respectively. In all cases heifers detected in oestrus subsequent to implant removal were inseminated approximately 12 h later.

incompetency are potentially important factors in the decreased fertility after ovulation of the persistent dominant follicle (Ahmad *et al.*, 1995).

Changes may occur within the oocyte of a persistent dominant follicle after continued exposure to prolonged high LH pulse frequency. In rats, dissociation of the resumption of meiosis from the LH surge was induced by exogenous LH pulses, leading to the ovulation of excessively 'aged' oocytes (Mattheij *et al.*, 1994). In cattle, premature germinal vesicle breakdown and progression to metaphase I occurred in six of eight oocytes recovered from preovulatory follicles with a dominance period of 12 days (associated with severely reduced fertility) that were collected before the predicted onset of the gonadotrophin surge (Mihm *et al.*, 1999). This finding is in agreement with the findings of Revah and Butler (1996), who reported the dispersion of nuclear material in oocytes from single or multiple persistent follicles. Oocytes within persistent follicles may, therefore, experience alterations in both the timing and nature of nuclear and cytoplasmic maturation, leading to potential inhibition of normal polar body extrusion, sperm decondensation, or normal transgression to the activation of the embryonic genome. Such asynchrony between oocyte and follicular maturation may be one factor involved in the reduction of pregnancy rates after ovulation of follicles with an extended duration of dominance (Mihm *et al.*, 1999).

Synchronization of New Wave Emergence

It is necessary to have a recently selected dominant follicle of short duration of dominance at the end of a progestagen or $PGF_{2\alpha}$ treatment regimen to give a precise onset of oestrus and high pregnancy rates to a single timed AI. The changing dependency of follicles on gonadotrophins during the wave and the intricacies of local control mechanisms that play a key role in regulating the sequential progression of follicles through different physiological stages of the wave make it difficult to develop a simple exogenous hormonal treatment that gives predictable new wave emergence in > 95% of treated animals irrespective of stage of wave at time of treatment (Bo *et al.*, 1995; Roche *et al.*, 1998). Thus, the induction of new wave emergence using exogenous hormonal treatments imposed at different stages of the follicular wave requires the predictable induction of a transient increase in FSH to give new wave emergence and normal growth of dominant follicle after its selection.

GnRH

Hormones to control the follicular wave are usually administered at the start of synchrony treatments to cause demise of the existing wave, and subsequent predictable new wave emergence. Removal of the dominant follicle or antral cohort follicles before selection (either by cauterization or ultrasound-guided transvaginal follicle aspiration) leads to a transient FSH increase and consistent new wave emergence within 1–2 days (Bergfelt *et al.*, 1994). However, such approaches are impractical on the farm. Functional removal of the dominant follicle or cohort follicles by either inducing ovulation or their rapid luteinization with exogenously induced gonadotrophin release is an alternative approach. Thus, GnRH has been used to cause predictable new wave emergence in some cattle synchronization protocols, that is, either in combination with $PGF_{2\alpha}$ (Twagiramungu *et al.*, 1995; Schmidt *et al.*, 1996) or as part of a progesterone regimen (Ryan *et al.*, 1995). The GnRH-induced release of LH and FSH is acute, and its magnitude may be affected by the endocrine status of the animal or the stage of the follicular wave at the time of injection. Accordingly, Ryan *et al.* (1998) examined the effects of injection of dairy cows with 250 μg of a synthetic GnRH (gonadorelin, Sanofi, France) on the gonadotrophin release, fate of existing follicular wave and interval to new wave emergence (i) before or after selection of the dominant follicle and (ii) when progesterone concentrations were above or below 1 ng ml^{-1}. GnRH induced a coincident LH and FSH surge, the magnitude of which was independent of days postpartum, progesterone concentration or stage of follicular wave. However, the effect of GnRH on the existing follicular wave was dependent on the presence or absence of a dominant follicle. GnRH administered after dominant follicle selection caused its ovulation (20/20) and predictable new wave emergence 1.6 ± 0.3 days later; if GnRH was administered before selection, it had no effect on the progression of the existing follicular wave and a dominant follicle was formed 3.6 ± 0.5 days later. In all cows treated after dominant follicle selection, the induced gonadotrophin surge was followed by a transient increase in FSH but not LH, which was associated with new wave emergence. It was concluded that GnRH synchronizes new wave emergence only when administered in the presence of a functional dominant follicle. Furthermore, smaller doses of gonadorelin (25 or 100 μg) are only partially effective (100 μg) or incapable (25 μg) of ovulating a luteal phase dominant follicle, and do not affect the progression of the wave when administered before formation of the dominant follicle (Mihm *et al.*, 1998). The dichotomy of GnRH effects on the progression of a follicular wave is a limitation that needs to be considered when using it as a treatment to synchronize new wave emergence at the start of progesterone or $PGF_{2\alpha}$ synchronizing regimens. The subsequent use of $PGF_{2\alpha}$ to cause regression of induced corpus luteum is mandatory when GnRH is used to synchronize follicle waves.

The use of oestradiol and progesterone alone and in combination

The primary dependence of the follicular wave on gonadotrophin support has resulted in the use of steroids to suppress FSH and LH and thus, terminate the existing wave. However, FSH and LH are differentially regulated. In the case of FSH, the dominant follicle is the key regulator of the recurrent increases that take place, but the specific roles of oestradiol, different inhibin forms and other putative FSH regulators are not known. LH is regulated by GnRH; hence, administration of progesterone is important in the regulation of LH pulse frequency in the cyclic cow, although changes in LH pulse frequency occur during the luteal phase that are difficult to explain by changes

Table 1. Effect of oestradiol benzoate (ODB) administration at the time of progesterone-releasing intravaginal device (PRID) insertion on mean (± SEM) peak plasma concentrations of oestradiol (E_2) and FSH, duration of oestradiol increase and rate of decline of concentrations of oestradiol and FSH in ovariectomized heifers pretreated with PRID.

ODB treatment (mg)	Number of heifers (n)	Peak concentration of E_2 (pg ml^{-1})	Time (h) to reach maximum concentrations of E_2	Time (h) to reach concentration of E_2 of < 3 pg ml^{-1}	Maximum decline of FSH expressed as % decline from pretreatment values	Time (h) from ODB treatment to first FSH increase after nadir	Time (h) for FSH to return to pretreatment concentrations
Controls							
0.0	3	0.1 ± 0.05[a]			0.0[a]		
Injection (mg)							
1.0	4	18 ± 7.1[b]	15 ± 4.2	97 ± 19.4[a]	37 ± 6.9[b]	32 ± 14.9[a]	65 ± 16.9[a]
2.5	6	24 ± 5.8[b]	11 ± 3.4	125 ± 15.8[ab]	40 ± 5.7[b]	34 ± 12.2[a]	57 ± 13.8[a]
5.0	4	72 ± 7.1[c]	12 ± 4.2	163.5 ± 19.4[b]	70 ± 6.9[c]	45 ± 14.9[b]	183 ± 16.9[b]

Within column, means with different superscripts are significantly different ($P < 0.05$).

in progesterone or oestradiol concentrations (Cupp *et al.*, 1995). Oestradiol and progesterone administered together can affect the progression of an existing wave, and hence new wave emergence (Bo *et al.*, 1995; Caccia and Bo, 1998). However, the optimum dose of exogenous oestradiol to suppress FSH consistently for a specific time period is not clear in cattle. Thus, an experiment was carried out in ovariectomized heifers to determine the effect of different blood plasma concentrations of oestradiol at the time of progesterone administration on FSH suppression, to select effective doses for manipulation of follicular waves (O'Rourke et al., 1997). The results (Table 1) show that (i) blood plasma concentrations of oestradiol reached maximum concentrations within 11–15 h after intramuscular injection of oestradiol benzoate in oil, irrespective of dose administered, and (ii) administration of 5 mg of oestradiol benzoate resulted in higher ($P < 0.05$) blood plasma concentrations for longer periods than 1.0 or 2.5 mg doses (Table 1). The resulting increased concentrations of oestradiol in blood plasma caused a significant reduction in FSH concentrations (Fig. 2). However, the extent of the decline in FSH was greater and the time for FSH to reach pretreatment concentrations was longer ($P < 0.05$) for heifers administered 5.0 mg oestradiol benzoate compared with those administered 1.0 or 2.5 mg oestradiol benzoate (O'Rourke *et al.*, 1997). Blood plasma concentrations of oestradiol 2–3 times greater than pro-oestrous concentrations (1.0 mg ODB injection group) resulted in a 37 ± 6.9% decrease in FSH, which was maintained for 32 ± 14.9 h. FSH concentrations returned to pretreatment values by 57 ± 13.8 h. Thus, high concentrations of oestradiol have only a transitory suppressive effect on FSH, and concentrations begin to increase in the presence of high but declining blood plasma concentrations of oestradiol.

In cyclic heifers, oestradiol administered at the start of progesterone treatment will transiently affect FSH and consequently follicular wave dynamics. Suppression of the peri-ovulatory FSH increase delays new wave emergence (Turzillo and Fortune, 1993), but the effects of decreasing FSH at other stages of the follicular wave on current wave or new wave emergence are unknown. For selecting optimum doses of oestradiol benzoate to manipulate transient increases in FSH and hence, new wave emergence in cyclic cattle, it is essential to determine the optimum blood plasma concentrations of oestradiol required at different stages of the follicular wave to give predictable new wave emergence. Three different oestradiol benzoate treatments were administered to cyclic heifers treated with a progesterone intravaginal releasing device (PRID, Sanofi Ltd, France) for 12 days to achieve low (2–4 pg ml^{-1} using a 10 mg oestradiol benzoate capsule), medium (15–20 pg ml^{-1} using a 0.75 mg oestradiol benzoate injection) or high (40–60 pg ml^{-1} using a 5 mg oestradiol benzoate injection) blood plasma concentrations. Heifers were treated at specific stages of development of the first follicular wave of the cycle, that is, on the day of wave emergence, and the

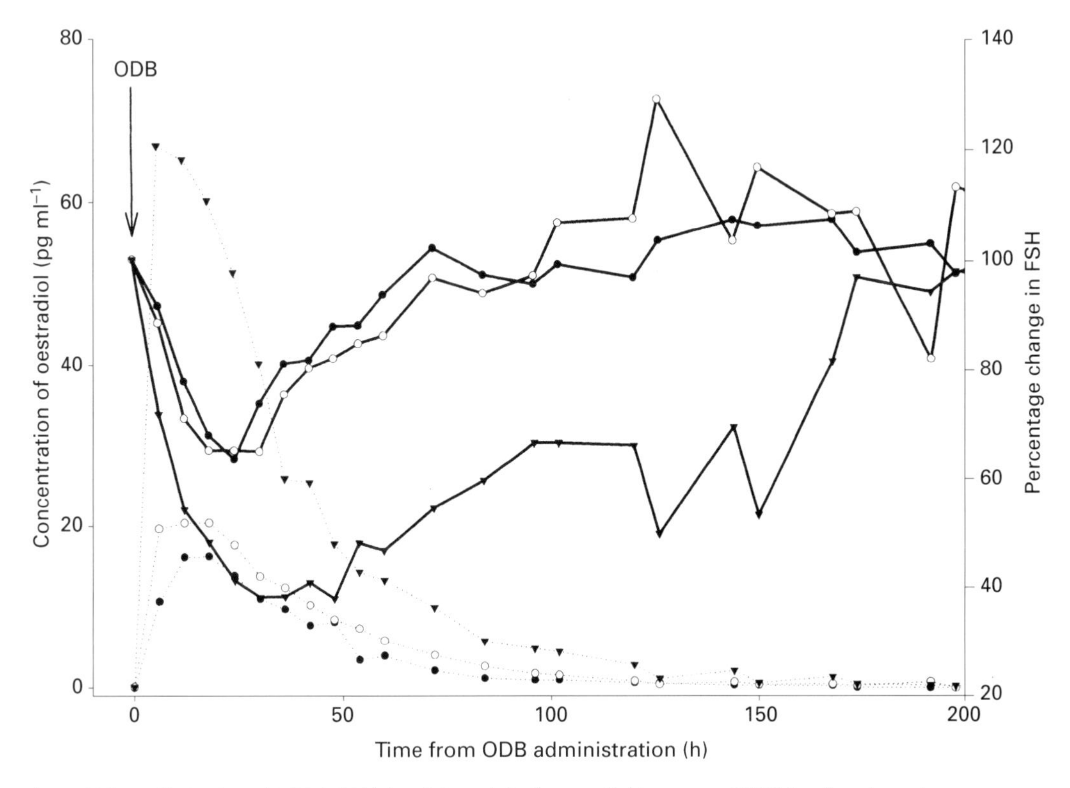

Fig. 2. Effect of injection of 1.0 (●), 2.5 (○) or 5.0 mg (▼) of oestradiol benzoate (ODB) in oil at time of progesterone releasing device (PRID) insertion on mean plasma concentrations of oestradiol (dotted lines) and percentage change in FSH (solid lines) in ovariectomized heifers pretreated with progesterone releasing devices ($n = 4$ heifers per dose). The range in SEM for oestradiol was 0.05–3.5 pg ml^{-1} and for FSH was 0.5–5.5 ng ml^{-1}.

first or fourth day of dominance before next new wave emergence. The results indicate that (i) progesterone with or without oestradiol benzoate (intravaginal capsule, or 0.75 or 5 mg intramuscular injections) administered in early metoestrus can either prevent selection of the first dominant follicle, or when selection has occurred, this treatment results in a decrease in dominant follicle diameter without consistently changing the timing of next new wave emergence, and (ii) all oestradiol benzoate treatments administered before the next predicted transient FSH increase delayed new wave emergence by 4–5 days (Table 2; O'Rourke *et al.*, 1998).

Thus, exogenous oestradiol administered with progesterone suppresses formation of or decreases the diameter of the dominant follicle when administered before or during emergence of the wave, presumably due to the suppression of FSH. However, once again the effects on the existing wave and interval to new wave emergence after administration of different oestradiol concentrations at the time of PRID insertion were variable, and dependent on stage of wave at time of treatment. In addition, oestradiol benzoate administered with progesterone in early metoestrus appeared to advance corpus luteum regression by 3–5 days as detected by ultrasound, thus confirming the luteolytic function of oestradiol when administered during early corpus luteum formation (Lemon, 1975), but the mechanism is not clear.

Synchronization of Pro-oestrous Oestradiol Rise after Treatment

The pattern and precision of onset of the synchronized oestrus are dependent on the time of luteolysis in relation to dominant follicle development, and the efficacy of hormones used to

Table 2. The effect of dose and route (intravaginal application in gelatin capsule versus intramuscular injection in oil) of oestradiol benzoate (ODB) given at the time of PRID insertion on wave dynamics when administered at different stages of development of the first dominant follicle (DF) in cyclic beef heifers

		Treatment				
Parameter	Follicle stage at treatment*	Control	PRID alone	PRID + ODB capsule 10 mg	PRID + ODB injection 0.75 mg	PRID + ODB injection 5.0 mg
Number forming first dominant follicle	Emergence	10/10[a]	4/7[bc]	4/6[ac]	1/6[bc]	2/7[bc]
Maximum size of first dominant follicle (mm)	Dominance	13.1 ± 0.5[a]	11.4 ± 0.6[b]	10.8 ± 0.60[b]		
Interval from treatment to new wave emergence (days)	Emergence	8.3 ± 0.7[a]	6.6 ± 0.8[ab]	6.5 ± 1.1[ab]	5.5 ± 0.5[b]	8.2 ± 0.8[a]
	Dominance	5.0 ± 0.6[a]	5.6 ± 0.8[ab]	5.3 ± 0.8[ab]	7.0 ± 0.9[ab]	7.3 ± 0.8[b]
	Dominance + 3	2.0 ± 0.5[a]	3.0 ± 0.7[a]	5.9 ± 0.7[b]	6.1 ± 0.7[b]	7.0 ± 0.7[b]
Interwave-interval (days)	Emergence	8.3 ± 0.7[a]	6.6 ± 0.8[ab]	6.5 ± 1.1[ab]	5.5 ± 0.9[b]	8.3 ± 0.8[a]
	Dominance	8.3 ± 0.7[a]	8.6 ± 0.8[a]	8.2 ± 0.9[a]	10.0 ± 0.9[b]	10.3 ± 0.9[b]
	Dominance + 3	8.3 ± 0.7[a]	8.6 ± 0.8[a]	11.7 ± 0.8[b]	12.3 ± 0.8[b]	13.1 ± 0.8[b]

[a,b,c] Within row, means with different superscripts are significantly different ($P < 0.05$).
*Determined by daily ultrasound examination.

synchronize follicular waves. Fertility depends on the duration of persistency of the dominant follicle during progestagen treatments. High fertility to AI after progesterone or progestagen administration requires a short duration of treatment (7–12 days) because it has been shown that treatment for 14 or more days results in decreased pregnancy rates to AI (Roche, 1978; Macmillan and Peterson, 1993). However, with the known relationship between duration of dominance of the ovulatory dominant follicle and fertility in heifers, it is important to re-examine the effect of duration of treatment on fertility, especially because recent results indicate that progesterone treatment for 8 days results in higher fertility (Ryan *et al.*, 1995) than treatment for 12 days. Shortening the duration of progesterone treatment to 7–8 days requires the use of $PGF_{2\alpha}$ as a luteolysin at or near the end of treatment, because oestradiol administered at the start of a progesterone treatment only shortens the duration of the ovarian cycle by 2–6 days when administered in early metoestrus. Thus, the time of occurrence and precision of the onset of oestrus are affected by the duration of progesterone treatment, the use and timing of $PGF_{2\alpha}$ administration and the specific hormone(s) used to synchronize follicular waves.

Owing to the variability in onset of oestrus caused by different developmental stages of the dominant follicle at the end of treatment (Austin *et al.*, 1998), another approach has been used to improve precision of onset of oestrus, that is to induce the synchronous occurrence of the proestrous oestradiol increase (McDougall *et al.*, 1992) by its exogenous administration at a specific time after the end of treatment. The use of 1.0 mg oestradiol benzoate administered 12–24 h after the end of a progesterone treatment increased the number of cows in oestrus the second day after treatment (Ryan *et al.*, 1995). This earlier onset of oestrus in some cows could result in ovulation of a small dominant follicle shortly after emergence of a new wave, which raises the question of the fertility consequences of induction of early ovulation of the dominant follicle. Thus, a trial was carried out (212 beef cows and 85 heifers) with the following aims: (1) to determine whether duration of PRID treatment for either 8 or 12 days affects oestrous and pregnancy response and (2) to determine whether 1 mg oestradiol benzoate administered 24 h after PRID removal will affect synchrony of onset of oestrus or pregnancy rates.

The overall oestrous response was similar for treatment for 12 days (94/112; 0.84) and 8 days (100/114; 0.88), but was higher ($P < 0.05$) after administration of 1.0 mg oestradiol benzoate 24 h after PRID removal (103/112; 0.92) compared with no oestradiol benzoate (91/114; 0.80) ($P < 0.01$). There was no interaction between the post-treatment use of oestradiol and duration of treatment on overall pregnancy rate ($P > 0.05$). The shorter duration of progesterone treatment tended to increase pregnancy rate (8 days 61/114 (0.54) and 12 days 47/112 (0.42); $P < 0.10$) but the effect was not significant. It appears that (i) pregnancy rate may be better ($P < 0.1$) after treatment for 8 days than after treatment for 12 days and (ii) oestradiol administered 24 h after PRID removal significantly increased the oestrous response without adversely affecting pregnancy rate. Thus, the use of oestradiol benzoate after treatment has the potential to increase the number of heifers or cows in oestrus and decrease the variation in onset of oestrus.

Conclusions

Major progress has been made in our understanding of the dynamics and regulation of follicular waves. It is essential now to use this knowledge to develop more effective hormonal regimens to regulate the time of predictable occurrence of oestrus in farm animals. It is a major scientific challenge to develop a hormonal treatment to cause predictable new wave emergence when administered at different physiological stages of a follicular wave in cattle. The most predictable and reliable induction of new wave emergence is achieved by dominant follicle ablation or ovulation. However, high GnRH doses that cause ovulation (250 µg) or lower doses (25 or 100 µg) that may not do not affect the morphological development of the wave when administered before dominant follicle selection. The alternative use of a combination of oestradiol and progesterone is equally problematic, and the interval to new wave emergence depends on the stage of the follicular wave at the time of treatment. As this could be due to changes in the physiological status of small antral follicles during the progression of a wave, the controlled increase in FSH needs to be synchronized with antral follicle function for new wave emergence to occur. It is necessary to have a short duration of dominance of the preovulatory follicle for maximum fertility and decreased variability in the onset of oestrus. Persistence of the dominant follicle for 8 days causes a slight reduction in pregnancy rates, but after 10 days, this decline is abrupt. Thus, the shorter the duration of progesterone treatment, the better the chance of achieving high pregnancy rates; however, progestagen treatments of < 12 days will also require the use of $PGF_{2\alpha}$ at or towards the end of treatment, because of the variable but slow luteolytic effects of exogenous oestradiol when administered at the start of a progestagen treatment. Variability in onset of oestrus even when a dominant follicle is present can be overcome by obtaining a synchronous increase in oestradiol by exogenous administration of oestradiol benzoate 24 h after the end of treatment; this procedure decreases the variability in onset of oestrus without compromising fertility. This approach has the potential to give sufficiently good synchrony of oestrus and high pregnancy rates after a single predetermined AI without recourse to detection of oestrus. Thus, it is important to understand the mechanisms regulating follicular waves and oocyte competency, to develop better hormonal methods to breed cattle to high genetic merit sires with significantly decreased labour input, but maintaining the goal of overall high pregnancy rates to AI.

Part of the research work in this paper was funded by Department of Agriculture EU Stimulus Fund and Sanofi Sante Nutrition Animale, Libourne, France.

References

Adams GP, Matteri RL, Kastelic JP, Ko JCH and Ginther OJ (1992) Association between surges of follicle-stimulating hormone and the emergence of follicle waves in heifers *Journal of Reproduction and Fertility* **94** 177–188

Ahmad N, Schrick FN, Butcher RL and Inskeep EK (1995) Effect of persistent follicles on early embryonic losses in beef cows *Biology of Reproduction* **52** 1129–1135

Austin EJ, Mihm M, Ryan MP, Williams DH and Roche JF (1999)

Effect of duration of dominance of the ovulatory follicle on onset of oestrus and fertility in heifers *Journal of Animal Science* **77** (in press)

Bergfelt DR, Plata-Madrid H and Ginther OJ (1994) Counteraction of the follicular inhibitory effect of follicular fluid by administration of FSH in heifers *Canadian Journal of Animal Science* **74** 633–639

Bo GA, Adams GP, Pierson RA and Mapletoft RJ (1995) Exogenous control of follicular wave emergence in cattle *Theriogenology* **43** 31–40

Bryner RW, Garcia-Winder M, Lewis PE, Inskeep EK and Butcher RL (1990) Changes in hormonal profiles during the estrous cycle in old lactating beef cows *Domestic Animal Endocrinology* **7** 181–190

Butcher RL and Pope RS (1979) Role of estrogen during prolonged estrous cycles of the rat on subsequent embryonic death or development *Biology of Reproduction* **21** 491–495

Caccia M and Bo GA (1998) Follicle wave emergence following treatment of CIDR-B implanted beef cows with estradiol benzoate and progesterone *Theriogenology* **49** 341 (Abstract)

Cooperative Regional Research Project, NE-161 (1996) Relationship of fertility to patterns of ovarian follicular development and associated hormonal profiles in dairy cows and heifers *Journal of Animal Science* **74** 1943–1952

Cupp AS, Stumpf TT, Kojima FN, Werth LA, Wolfe MW, Roberson MS, Kittok RJ and Kinder JE (1995) Secretion of gonadotrophins change during the luteal phase of the bovine oestrous cycle in the absence of corresponding changes in progesterone or 17β-oestradiol *Animal Reproduction Science* **37** 109–119

de la Sota RL, Good TEM, Ireland JLH, Ireland JJ, Simmen FA and Thatcher WW (1994) Analysis of different forms of intrafollicular inhibin and inhibin α-subunits during ovarian follicular dominance in cattle *Biology of Reproduction* **50** Abstract 162

de la Sota RL, Simmen FA, Diaz T and Thatcher WW (1996) Insulin-like growth factor system in bovine first-wave dominant and subordinate follicles *Biology of Reproduction* **55** 803–812

Fike KE, Wehrman ME, Bergfeld EGM, Kojima FN and Kinder JE (1997) Prolonged increased concentrations of 17β-estradiol associated with development of persistent ovarian follicles do not influence conception rates in beef cattle *Journal of Animal Science* **75** 1363–1367

Ginther OJ, Wiltbank MC, Fricke PM, Gibbons JR and Kot K (1996) Selection of the dominant follicle in cattle *Biology of Reproduction* **55** 1187–1194

Kojima N, Stumpf TT, Cupp AS, Werth LA, Roberson MS, Wolfe MW, Kittok RJ and Kinder JE (1992) Exogenous progesterone and progestins as used in estrous synchrony regimes do not mimic the corpus luteum in regulation of luteinizing hormone and 17β-estradiol in circulation of cows *Biology of Reproduction* **47** 1009–1017

Lemon M (1975) The effect of oestrogens alone or in association with progestagens on the formation and regression of the corpus luteum of the cyclic cow *Annales de Biologie Animale Biochimie Biophysique* **15** 243–253

McDougall S, Burke CR, Macmillan KL and Williamson NB (1992) The effect of pretreatment with progesterone on the oestrous response to oestradiol-17β benzoate in the post-partum dairy cow *Proceedings of the New Zealand Society of Animal Production* **52** 157–160

Macmillan KL and Peterson AJ (1993) A new intravaginal progesterone releasing device for cattle (CIDR-B) for oestrous synchronisation, increasing pregnancy rates and the treatment of post-partum anoestrus *Animal Reproduction Science* **33** 1–25

Mattheij JAM, Swarts JJM, Hurks HMH and Mulder K (1994) Advancement of meiotic resumption in Graafian follicles by LH in relation to preovulatory ageing of rat oocytes *Journal of Reproduction and Fertility* **100** 65–70

Mihm M, Baguisi A, Boland MP and Roche JF (1994) Association between the duration of dominance of the ovulatory follicle and pregnancy rate in beef heifers *Journal of Reproduction and Fertility* **102** 123–130

Mihm M, Curran N, Hyttel P, Knight PG, Boland MP and Roche JF (1999) Effect of dominant follicle persistance on follicular fluid oestradiol and inhibin and on oocyte maturation in heifers *Journal of Reproduction and Fertility* **117** (in press)

Mihm M, Deletang F and Roche JF (1998) The gonadotrophin and ovarian response to an intermediate or low dose of gonadorelin in beef heifers: influences of dose, follicle status and progesterone environment *Journal of Reproduction and Fertility Abstract Series* **21** Abstract 74

Odde KG (1990) A review of synchronization of estrus in postpartum cattle *Journal of Animal Science* **68** 817–830

O'Rourke M, Diskin MG, Sreenan JM and Roche JF (1997) The effect of dose and method of oestradiol administration on plasma concentrations of E2 and FSH in long-term ovariectomised beef heifers *Journal of Reproduction and Fertility Abstract Series* **19** Abstract 148

O'Rourke M, Diskin MG, Sreenan JM and Roche JF (1998) Effect of different concentrations of oestradiol administered during the first follicle wave in association with PRID insertion on follicle wave dynamics and oestrous response in beef heifers *Journal of Reproduction and Fertility Abstract Series* **21** Abstract 15

Rahe CH, Owens RE, Fleeger JL, Newton HJ and Harms PG (1980) Pattern of plasma luteinizing hormone in the cyclic cow: dependence upon the period of the cycle *Endocrinology* **107** 498–503

Revah I and Butler WR (1996) Prolonged dominance of follicles and reduced viability of bovine oocytes *Journal of Reproduction and Fertility* **106** 39–47

Roche JF, Mihm M, Diskin M and Ireland JJ (1998) A review of regulation of follicle growth in cattle *Journal of Animal Science* **76** (supplement 3; 16–23)

Roche JF (1974) Effect of short-term progesterone treatment on oestrous response and fertility in heifers *Journal of Reproduction and Fertility* **40** 433–440

Roche JF (1978) Control of oestrus in cattle using progesterone coils *Animal Reproduction Science* **1** 145–154

Ryan DP, Snijders S, Yakuub H and O'Farrell KJ (1995) An evaluation of estrus synchronization programs in reproductive management of dairy herds *Journal of Animal Science* **73** 3687–3695

Ryan M, Mihm M and Roche JF (1998) Effect of GnRH given before or after dominance on gonadotrophin response and the fate of that follicle wave in postpartum dairy cows *Journal of Reproduction and Fertility Abstract Series* **21** Abstract **61**

Schmidt EJP, Diaz T, Drost M and Thatcher WW (1996) Use of a gonadotropin-releasing hormone agonist or human chorionic gonadotropin for timed insemination in cattle *Journal of Animal Science* **74** 1084–1091

Sanchez T, Wehrman ME, Bergfeld EG, Peters KE, Kojima FN,

Cupp AS, Mariscal V, Kittok RJ, Rasby RJ and Kinder JE (1993) Pregnancy rate is greater when the corpus luteum is present during the period of progestin treatment to synchronize time of estrus in cows and heifers *Biology of Reproduction* **49** 1102–1107

Savio JD, Thatcher WW, Badinga L, de la Sota RL and Wolfenson D (1993) Regulation of dominant follicle turnover during the oestrous cycle in cows *Journal of Reproduction and Fertility* **97** 197–203

Sirois J and Fortune JE (1990) Lengthening of the bovine estrous cycle with two levels of exogenous progesterone: a model for studying ovarian follicular dominance *Endocrinology* **127** 916–925

Stagg K, Spicer LJ, Sreenan JM, Roche JF and Diskin MG (1998) Effect of calf isolation on follicular wave dynamics, gonadotrophin and metabolic hormone changes and interval to first ovulation in beef cows fed either of two levels of energy postpartum *Biology of Reproduction* **59** 777–783

Stock AE and Fortune JE (1993) Ovarian follicular dominance in cattle: relationship between prolonged growth of the ovulatory follicle and endocrine parameters *Endocrinology* **132** 1108–1114

Sunderland SJ, Crowe MA, Boland MP, Roche JF and Ireland JJ (1994) Selection, dominance and atresia of follicles during the oestrous cycle of heifers *Journal of Reproduction and Fertility* **101** 547–555

Taft R, Ahmad N and Inskeep EK (1996) Exogenous pulses of luteinizing hormone cause persistence of the largest bovine ovarian follicle *Journal of Animal Science* **74** 2985–2991

Taylor C, Rajamahendran R and Walton JS (1993) Ovarian follicular dynamics and plasma luteinizing hormone concentrations in norgestomet-treated heifers *Animal Reproduction Science* **32** 173–184

Turzillo AM and JE Fortune (1993) Effects of suppressing plasma FSH on ovarian follicular dominance in cattle *Journal of Reproduction and Fertility* **98** 113–119

Twagiramungu H, Guilbault LA and Dufour JJ (1995) Synchronization of ovarian follicular waves with gonadotropin–releasing hormone agonist to increase the precision of estrus in cattle: a review *Journal of Animal Science* **73** 3141–3151

Wehrman ME, Fike KE, Melvin EJ, Kojima FN and Kinder JE (1997) Development of a persistent ovarian follicle and associated elevated concentrations of 17β-estradiol preceding ovulation does not alter the pregnancy rate after embryo transfer in cattle *Theriogenology* **47** 1413–1421

Wiltbank JN and Kasson CW (1968) Synchronization of estrus in cattle with an oral progestational agent and an injection of an estrogen *Journal of Animal Science* **27** 113–116

NEUROENDOCRINE RELATIONSHIPS

Chair
D. L. Hamernik

Journal of Reproduction and Fertility Supplement **54**, 75–86

Regulation of GnRH receptor gene expression in sheep and cattle

A. M. Turzillo and T. M. Nett*

Animal Reproduction and Biotechnology Laboratory, Department of Physiology, Colorado State University, Fort Collins, CO 80523, USA

The GnRH receptor plays a pivotal role in reproduction. This review summarizes current knowledge of the regulation of GnRH receptor gene expression by endocrine factors in sheep and cattle. Expression of the GnRH receptor gene, measured by steady-state amounts of GnRH receptor messenger RNA (mRNA), is maximal during the preovulatory period. The molecular events leading to maximal GnRH receptor gene expression are probably triggered by decreased circulating concentrations of progesterone at luteolysis. Because GnRH is a positive homologous regulator of its own receptor, increased pulsatile GnRH after removal of negative feedback effects of progesterone stimulates expression of the GnRH receptor gene early in the preovulatory period. Oestradiol is also a positive regulator of GnRH receptor gene expression, and increased serum concentrations of oestradiol from developing follicles probably maintain high abundance of GnRH receptor mRNA later in the preovulatory period. Since increased amount of GnRH receptor mRNA precedes maximal numbers of GnRH receptors before the LH surge, increased expression of the GnRH receptor gene is an important mechanism by which maximal sensitivity of gonadotrophs to GnRH is achieved. Future efforts should be directed towards elucidating the molecular mechanisms underlying transcriptional regulation of the GnRH receptor gene in ruminants by endocrine factors.

Introduction

The pituitary receptor for GnRH is a critical component of the reproductive axis. Interaction of GnRH with its receptor stimulates synthesis and secretion of the gonadotrophins that are essential for gonadal function. Numbers of GnRH receptors throughout the oestrous cycle have been characterized (Nett *et al.*, 1987; Nett, 1990), and changes in numbers of GnRH receptors are believed to be important in regulating sensitivity of the anterior pituitary gland to GnRH (Wise *et al.*, 1984). Until 1992, the structural characteristics of the GnRH receptor were unknown and molecular probes for examining regulation of the GnRH receptor gene were not available. Cloning of complementary DNAs (cDNAs) encoding the GnRH receptor in several species including domestic ruminants (Brooks *et al.*, 1993; Kakar *et al.*, 1993) has revealed the basic structure of this receptor. In mice and rats, the GnRH receptor cDNA encodes a 327 amino acid protein while the ovine, bovine, human and pig cDNAs encode 328 amino acids (reviewed by Sealfon *et al.*, 1997). The GnRH receptor is believed to have seven transmembrane domains typical of G protein-coupled receptors. However, the receptor is unique among G protein-coupled receptors in that it lacks a cytoplasmic C-terminal tail (Sealfon *et al.*, 1997). Isolation of these GnRH receptor cDNAs was an essential first step towards increasing understanding of the gene encoding the GnRH receptor, and led to a wide range of studies on the regulation of GnRH receptor gene expression.

This review will focus on the regulation of GnRH receptor gene expression by endocrine factors in ewes. At least four hormones are known to affect the ovine GnRH receptor gene: oestradiol, GnRH, progesterone and inhibin. Evidence for regulation of GnRH receptor gene expression by each of these hormones during the oestrous cycle will be discussed. Although there is much less information available regarding GnRH receptor gene expression in the cow, findings from bovine studies are included where appropriate. Current efforts to elucidate molecular mechanisms underlying expression of the GnRH receptor gene in sheep are also discussed.

GnRH Receptor Gene Expression During the Oestrous Cycle

The most dynamic time during the oestrous cycle with respect to pituitary–ovarian interactions is the preovulatory period, when serum concentrations of progesterone decline as a result of luteolysis and serum concentrations of oestradiol and inhibin rise with development of the preovulatory follicle. This pattern of ovarian hormone secretion leads to important changes at the hypothalamus and a marked increase in release of GnRH (Moenter *et al.*, 1991). Throughout the oestrous cycle in ewes and cows, pituitary concentrations of GnRH receptors change four- to tenfold (Nett *et al.*, 1987; Nett, 1990). In ewes, numbers of GnRH receptors remain static during much of the luteal phase, but increase during the preovulatory period (Crowder and Nett, 1984; Brooks *et al.*, 1993; Turzillo *et al.*, 1994; Hamernik *et al.*, 1995). It seems likely that this increase contributes to maximal sensitivity of gonadotrophs to GnRH at this time, and thus is an important step in the series of events leading to the ovulatory LH surge. To determine whether the large number of GnRH receptors during the preovulatory period is the result of increased expression of the GnRH receptor gene, several investigators have measured steady-state concentrations of GnRH receptor mRNA in ovine pituitary glands after induction of luteolysis (Fig. 1). Concentrations of GnRH receptor mRNA are increased as early as 12 h after luteolysis (Turzillo *et al.*, 1994). Amounts of GnRH receptor mRNA remain high at 24 h and 48 h but return to pretreatment (luteal) values at 72 h and 96 h (Brooks *et al.*, 1993; Hamernik *et al.*, 1995). Amounts of GnRH receptor mRNA corresponded closely to numbers of GnRH receptors in each of these studies. In the early preovulatory period, increased GnRH receptor gene expression preceded an increase in the number of GnRH receptors (Fig. 1; Turzillo *et al.*, 1994) and maximal numbers of GnRH receptors were observed later in the preovulatory period near the onset of the LH surge (Crowder and Nett, 1984; Hamernik *et al.*, 1995). These temporal relationships support the hypothesis that increased amounts of ovine GnRH receptor mRNA lead to greater numbers of GnRH receptors which in turn maximize pituitary sensitivity to GnRH in preparation for the LH surge.

An increase in GnRH receptor gene expression has been observed during the preovulatory period in cows (A. M. Turzillo, T. M. Nett, A. Roberts and S. E. Echternkamp, unpublished). However, whether there are concomitant preovulatory increases in numbers of bovine GnRH receptors is unclear, since Leung *et al.* (1984) and Nett *et al.* (1987) did not observe higher concentrations of GnRH receptors during the preovulatory period compared with the luteal phase. However, in both studies, maximal concentrations of GnRH receptors were observed during pro-oestrus. A more detailed study in which pituitary glands are collected at several times after luteolysis is needed to clarify the relationship between GnRH receptor gene expression and numbers of GnRH receptors during the follicular phase in cows.

Collectively, these findings provide strong evidence that the increase in numbers of GnRH receptors during the preovulatory period in sheep and cattle are mediated by increased expression of the GnRH receptor gene. The next logical question is, what are the endocrine factors involved in stimulating GnRH receptor gene expression during the oestrous cycle? In the following sections, how changes in secretion of ovarian and hypothalamic hormones may affect expression of the GnRH receptor gene will be examined.

The Role of Oestradiol

The importance of oestradiol in regulating hypothalamic–pituitary physiology in ruminants is well documented. Treatment with oestradiol can induce an LH surge in ewes and cows, and one mechanism by which oestradiol exerts this effect is by stimulating hypothalamic secretion of GnRH.

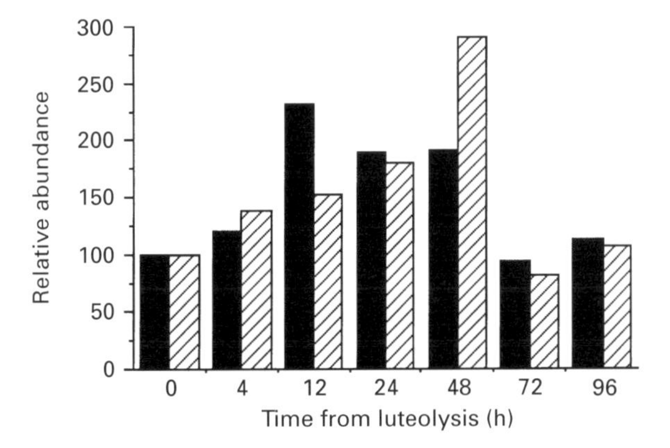

Fig. 1. Relative abundance of GnRH receptor mRNA (mean steady-state levels; ■) and concentrations of GnRH receptors (▨) in anterior pituitary tissue collected from ewes before and after induction of luteolysis. Data were combined from Brooks *et al.* (1993); Turzillo *et al.* (1994); and Hamernik *et al.* (1995).

Oestradiol also has important effects on pituitary function and increases pituitary sensitivity to GnRH in ewes and cows, apparently by increasing numbers of GnRH receptors (Moss *et al.*, 1981; Schoenemann *et al.*, 1985; Turzillo *et al.*, 1994; Hamernik *et al.*, 1995). The stimulatory effect of oestradiol on numbers of GnRH receptors is evident in the absence of hypothalamic input, thus demonstrating a direct pituitary site of action (Gregg and Nett, 1989). Similar effects are observed in cultured ovine pituitary cells, in which oestradiol induces a two- to threefold increase in GnRH-stimulated LH secretion (Huang and Miller, 1980; Moss and Nett, 1980) and increases numbers of GnRH receptors (Laws *et al.*, 1990a; Gregg *et al.*, 1990). More recently, it has been reported that oestradiol increases GnRH receptor gene expression *in vivo* (Turzillo *et al.*, 1994; Hamernik *et al.*, 1995), and this effect can occur in the absence of GnRH (Turzillo *et al.*, 1995b; Adams *et al.*, 1997) and in cultured ovine pituitary cells (Fig. 2; Wu *et al.*, 1994). Because serum concentrations of oestradiol increase markedly during the preovulatory period, these observations appear to indicate that oestradiol is the endocrine factor responsible for increasing GnRH receptor gene expression before the ovulatory LH surge, which in turn leads to greater numbers of GnRH receptors on gonadotrophs.

Although the importance of oestradiol during the preovulatory period is indisputable, there is evidence to indicate that oestradiol may not be the endocrine factor responsible for initiating the events leading to maximal numbers of GnRH receptors. To characterize the temporal relationships among GnRH receptor gene expression and endocrine changes during the early preovulatory period, we measured serum concentrations of oestradiol and progesterone and amounts of GnRH receptor mRNA in ewes during the 24 h after induction of luteolysis with prostaglandin $F_{2\alpha}$ ($PGF_{2\alpha}$; Turzillo *et al.*, 1994). Concentrations of GnRH receptor mRNA were increased at 12 h after treatment with $PGF_{2\alpha}$. This increase was associated with a 50% decrease in circulating concentrations of progesterone, but occurred before serum concentrations of oestradiol began to rise (Fig. 3). Thus it appears that the initial molecular events leading to increased expression of the GnRH receptor gene during the early preovulatory period in ewes do not require increased pituitary exposure to oestradiol, but instead are associated more closely with decreased concentrations of progesterone. Oestradiol probably maintains increased GnRH receptor gene expression later in the preovulatory period. Therefore, progesterone may exert negative effects on GnRH receptor gene expression during the oestrous cycle, which led us to consider the potential mechanisms underlying this effect of progesterone.

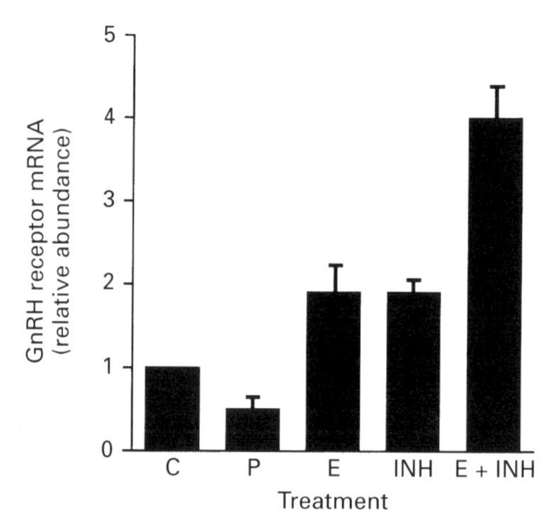

Fig. 2. Mean (± SEM) concentrations of GnRH receptor mRNA in cultured ovine pituitary cells treated for 48 h with control medium (C), progesterone (P; 100 nmol l^{-1}), oestradiol (E; 10 nmol l^{-1}), an enriched preparation of porcine follicular inhibin (INH; equivalent to 10 ng ml^{-1} pure 32 kDa porcine inhibin), or the combination of E + INH. Adapted from Wu *et al.* (1994).

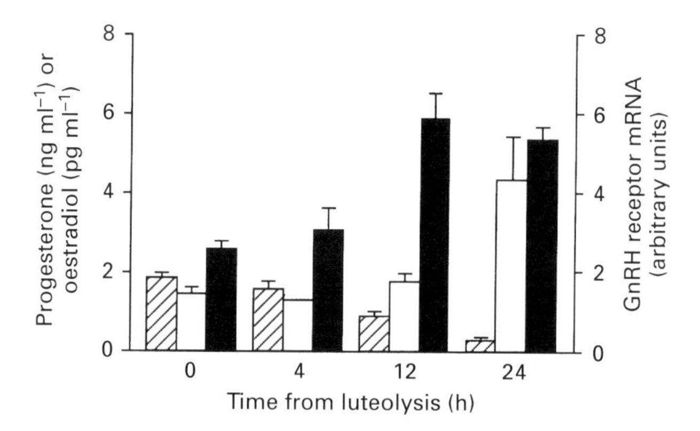

Fig. 3. Concentrations of progesterone (▨) and oestradiol (□) in serum and GnRH receptor mRNA in pituitary tissue (■) collected from ewes 0 h, 4 h, 12 h and 24 h after induction of luteolysis with $PGF_{2\alpha}$. Data are means ± SEM. Note that increased concentrations of GnRH receptor mRNA at 12 h occurred before serum concentrations of oestradiol were increased and were associated with a 50% decrease in serum concentrations of progesterone. Adapted from Turzillo *et al.* (1994).

Progesterone: A Negative Regulator of GnRH Receptor Gene Expression

Several lines of evidence implicate progesterone as a negative regulator of ovine GnRH receptor gene expression. First, numbers of GnRH receptors and expression of the GnRH receptor gene are relatively low during the luteal phase when concentrations of progesterone are maximal, and

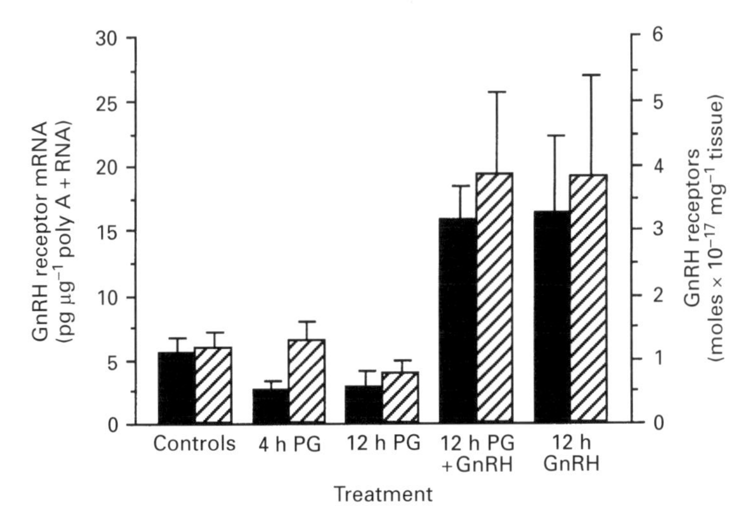

Fig. 4. Pituitary concentrations of GnRH receptor mRNA (■) and GnRH receptors (▨) in ewes during the anoestrous season. Follicular growth and ovulation were stimulated pharmacologically, and luteolysis was induced 11 or 12 days later with $PGF_{2\alpha}$. Pituitary tissues were collected 4 h or 12 h after treatment with $PGF_{2\alpha}$ (PG), 12 h after treatment with $PGF_{2\alpha}$ and GnRH (12 h PG + GnRH), or 12 h after treatment with GnRH only (12 h GnRH). GnRH was administered at intervals of 1 h. Control ewes were not treated with $PGF_{2\alpha}$ or GnRH. Data are means ± SEM. Adapted from Turzillo *et al.* (1995b).

increase when progesterone falls during the demise of the corpus luteum. Second, the stimulatory effects of oestradiol on GnRH receptor gene expression and numbers of GnRH receptors are prevented during the luteal phase when endogenous progesterone is high (Brooks and McNeilly, 1994; Turzillo *et al.*, 1998). The third line of evidence comes from a series of studies conducted *in vitro*. In cultured ovine pituitary cells, progesterone decreases responsiveness to GnRH (Batra and Miller, 1985), numbers of receptors for GnRH (Laws *et al.*, 1990b) and amounts of mRNA encoding the GnRH receptor (Fig. 2; Wu *et al.*, 1994). Furthermore, progesterone can attenuate the stimulatory effects of oestradiol on GnRH-stimulated LH secretion (Batra and Miller, 1985) and numbers of GnRH receptors (Sealfon *et al.*, 1990) *in vitro*. From these studies it seems logical to hypothesize that progesterone has inhibitory effects on GnRH receptors *in vivo*, and that the decrease in circulating concentrations of progesterone at luteolysis removes this inhibition, allowing increased expression of the GnRH receptor gene followed by increased numbers of GnRH receptors.

There are at least two mechanisms by which progesterone might exert its effects on the GnRH receptor gene. One mechanism involves direct effects of progesterone on the pituitary gland. This mechanism seems plausible in the light of evidence that progesterone affects GnRH receptor gene expression *in vitro*, in the absence of hypothalamic influences. Alternatively, progesterone might regulate the GnRH receptor gene indirectly via negative feedback effects on secretion of GnRH. Decreasing concentrations of progesterone during luteolysis lead to increased pulsatile secretion of GnRH. As discussed later, GnRH is an important regulator of its own receptor and pulsatile GnRH increases the number of GnRH receptors in ewes (Khalid *et al.*, 1987; Hamernik and Nett, 1988; Turzillo *et al.*, 1995b). Therefore, increased stimulation of the pituitary gland by GnRH during luteolysis could lead to increased expression of the GnRH receptor gene.

To determine whether progesterone affects GnRH receptor gene expression *in vivo* directly at the pituitary or indirectly via increased GnRH secretion, we designed an experiment in which decreased concentrations of progesterone could be achieved without a concurrent rise in pulsatile GnRH (Turzillo *et al.*, 1995b). Ovulation and subsequent luteal formation were induced in anoestrous ewes and luteolysis was initiated on day 11 or 12 of the induced oestrous cycle. Because of photoperiodic inhibition of the GnRH pulse generator during anoestrus, an increase in GnRH secretion following luteolysis did not occur in this experimental model. If effects of progesterone on GnRH receptor gene expression occur directly at the pituitary gland and are independent of GnRH, decreasing concentrations of progesterone in this model should result in increased amounts of GnRH receptor mRNA. This was not the case (Fig. 4). In contrast, when pulsatile GnRH was administered each hour for 12 h after induction of luteolysis, concentrations of GnRH receptor mRNA were increased. This effect of GnRH was also obvious in ewes in which luteolysis was not induced and serum concentrations of progesterone remained high (Fig. 4). From these results, we conclude that the increase in GnRH receptor mRNA during the early preovulatory period is mediated by an indirect mechanism involving decreased negative feedback of progesterone on pulsatile hypothalamic GnRH secretion. Increased stimulation of the pituitary gland by GnRH then causes heightened expression of the GnRH receptor gene, thus initiating the molecular events that lead to increased concentrations of GnRH receptor mRNA and synthesis of GnRH receptors before the LH surge.

Homologous Regulation by GnRH

GnRH is secreted from the ovine hypothalamus in a pulsatile fashion and the importance of this pattern of GnRH secretion to pituitary function is well established. A marked increase in GnRH release occurs during the follicular phase in ewes (Moenter *et al.*, 1991), and this increase is required for a normal LH surge. In addition to stimulating pulsatile LH secretion during the preovulatory period, GnRH is also a homologous regulator of its own receptor and probably serves to increase numbers of GnRH receptors before the ovulatory LH surge. Sheep have been used to develop several excellent experimental paradigms to study the regulation of GnRH receptors by GnRH. One of these paradigms takes advantage of the large size of sheep, which allows surgical disconnection of the hypothalamus from the pituitary gland (hypothalamic–pituitary disconnection, HPD). This procedure effectively deprives the pituitary gland of GnRH (and other hypothalamic hormones), but does not interfere with the hypophyseal blood supply and allows the pituitary gland to remain viable despite the absence of hypothalamic input. In ovariectomized ewes, HPD results in decreased numbers of GnRH receptors (Clarke *et al.*, 1987; Gregg and Nett, 1989; Turzillo *et al.*, 1995a) which can be restored by treatment with pulsatile GnRH (Clarke *et al.*, 1987; Hamernik and Nett, 1988). Significant reductions in numbers of GnRH receptors also occur when the pituitary gland is deprived of GnRH using non-surgical methods. Brooks and McNeilly (1994) observed decreased numbers of GnRH receptors after treating ewes during the oestrous cycle with an antagonist of GnRH, and Sakurai *et al.* (1997) reported similar results in wethers passively immunized against GnRH. From these studies, it is clear that continuous stimulation of the pituitary gland by GnRH is required to maintain normal numbers of GnRH receptors. However, there is some discrepancy regarding the effect of removing GnRH on GnRH receptor gene expression. Brooks and McNeilly (1994) and Sakurai *et al.* (1997) found that treatment of ovary-intact ewes with GnRH antagonist or passive immunization of wethers against GnRH resulted in decreased concentrations of GnRH receptor mRNA. This is in contrast to our observations in ovariectomized ewes in which amounts of GnRH receptor mRNA did not change significantly during the 3 days after HPD (Turzillo *et al.*, 1995a) or 6 days after passive immunization against GnRH (Turzillo and Nett, 1997). Although reasons for the lack of agreement among these studies are not obvious, the disparate results may be due to the use of different experimental models and the time after removal of GnRH when measurements were made.

Despite different effects of removing GnRH on GnRH receptor gene expression, administration of pulsatile GnRH to GnRH-deficient anoestrous ewes increased pituitary amounts of GnRH receptor mRNA (Turzillo *et al.*, 1995b) and numbers of GnRH receptors (Khalid *et al.*, 1987; Hamernik and Nett, 1988; Turzillo *et al.*, 1995b). These observations strengthen the claim that pulsatile GnRH is a positive regulator of GnRH receptor gene expression.

Information regarding effects of pulsatile GnRH on bovine GnRH receptors is limited. Numbers of GnRH receptors were increased by pulsatile GnRH in prepubertal bull calves (Rodriguez and

Wise, 1991) but in nutritionally anoestrous cows, treatment with pulsatile GnRH did not affect pituitary concentrations of GnRH receptors or GnRH receptor mRNA (Vizcarra *et al.*, 1997). Because secretion of LH is reduced in nutritionally anoestrous cows (Richards *et al.*, 1989), it is presumed that release of GnRH is also reduced. However, it is possible that the decrease in GnRH caused by nutritional restriction in the study of Vizcarra *et al.* (1997) was sufficient to affect LH secretion but not GnRH receptors. This result could explain why treatment with exogenous GnRH pulses was ineffective in stimulating further increases in pituitary concentrations of GnRH receptors or GnRH receptor mRNA.

Collectively, the bulk of evidence indicates that pulsatile stimulation by GnRH is required to maintain tissue concentrations of GnRH receptor and GnRH receptor mRNA in the ruminant pituitary gland. Therefore, marked increases in GnRH secretion probably play an important role in upregulation of GnRH receptors during the preovulatory period.

It is important to note that effects of GnRH on GnRH receptors differ markedly depending on the pattern of GnRH administration. After continuous exposure to GnRH, the anterior pituitary gland becomes refractory to further challenge with GnRH in ewes (Nett *et al.*, 1981) and cows (Lamming and McLeod, 1988). This desensitization is marked by decreased tissue concentrations of GnRH receptors (Nett *et al.*, 1981; Crowder *et al.*, 1986; Vizcarra *et al.*, 1997) and GnRH receptor mRNA (Vizcarra *et al.*, 1997; Turzillo *et al.*, 1998). Similarly, chronic treatment of ewes or wethers with GnRH agonists causes downregulation of GnRH receptors and reduces GnRH receptor gene expression (Brooks and McNeilly, 1994; Wu *et al.*, 1994). GnRH receptors are internalized after binding GnRH (Hazum *et al.*, 1980), and it is likely that the reduction in numbers of GnRH receptors induced by continuous GnRH treatment reflects this internalization. However, recent observations that decreased GnRH receptor gene expression occurs in conjunction with downregulation of GnRH receptors provide evidence that pituitary desensitization following continuous exposure to GnRH is mediated by reduced *de novo* synthesis of GnRH receptors as well as by internalization of existing GnRH receptors on gonadotrophs. These findings may have important physiological relevance. Secretion of GnRH increases 40-fold during the preovulatory period in ewes, and remains high for several hours after the LH surge (Moenter *et al.*, 1991). Since concentrations of GnRH receptor mRNA and numbers of GnRH receptors decrease after the LH surge (Crowder and Nett, 1984; Hamernik *et al.*, 1995), we speculate that termination of the LH surge may be due to pituitary desensitization mediated by decreased GnRH receptor gene expression and downregulation of GnRH receptors caused by extended exposure to high concentrations of GnRH.

Does Inhibin Regulate GnRH Receptor Gene Expression?

Inhibin is a gonadal peptide that selectively suppresses secretion of FSH. Inhibin is secreted by large antral ovine and bovine follicles, and increases in circulating concentrations of inhibin during the follicular phase in sheep and cattle have been reported (Padmanabhan *et al.*, 1984; Findlay *et al.*, 1990). Therefore, like oestradiol, ovarian inhibin may be an important endocrine regulator of pituitary function during the preovulatory period. Evidence that inhibin may affect GnRH receptors comes from studies conducted *in vitro*. In cultured ovine pituitary cells, inhibin enhanced GnRH-stimulated secretion of LH (Miller and Huang, 1985; Muttukrishna and Knight, 1990), increased numbers of GnRH receptors (Laws *et al.*, 1990b; Gregg *et al.*, 1991) and increased amounts of GnRH receptor mRNA (Fig. 2; Wu *et al.*, 1994). In the study by Wu *et al.* (1994), the combination of oestradiol and inhibin resulted in an additive effect on amounts of GnRH receptor mRNA. Positive effects of inhibin *in vitro* support the hypothesis that increasing concentrations of inhibin during the preovulatory period contribute to increased GnRH receptor gene expression and greater numbers of GnRH receptors observed during this time. However, there is no evidence to suggest that inhibin regulates GnRH receptors *in vivo*. In fact, treatment of ewes during the luteal phase for 4 or 9 days with bovine or ovine follicular fluid as a source of inhibin did not affect tissue concentrations of GnRH receptors or GnRH receptor mRNA (Brooks *et al.*, 1992; Brooks and McNeilly, 1994). Similarly, we observed no changes in numbers of GnRH receptors or GnRH receptor gene expression in

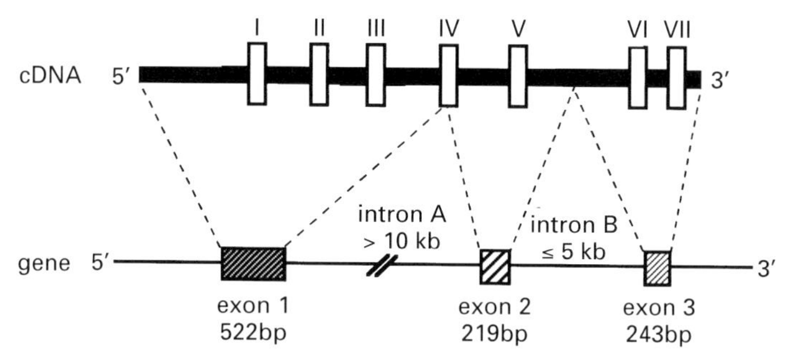

Fig. 5. Illustration of complementary DNA (cDNA) and gene encoding the ovine GnRH receptor. Transmembrane domains of the cDNA are represented by roman numerals I–VII. The coding sequence of the receptor is divided among three exons and two introns. Sizes of exons and introns are indicated in base pairs (bp) or kilobases (kb), respectively. (Reproduced with permission from Campion *et al.*, 1996).

ovariectomized ewes treated with bovine follicular fluid (Turzillo and Nett, 1997). The absence of an effect of treatment with bovine follicular fluid cannot be explained by a lack of inhibin bioactivity since serum concentrations of FSH and pituitary amounts of mRNA encoding FSHβ subunit were reduced ≥ 70% and 90%, respectively. Collectively, the results of studies conducted *in vivo* do not support the idea that inhibin is an endocrine regulator of GnRH receptors. Reasons for the lack of agreement between findings obtained *in vitro* versus *in vivo* are unclear, but may be related to the potential role of inhibin as an intra-pituitary regulatory factor. Inhibin subunits are produced in rat gonadotrophs (Roberts *et al.*, 1988), and local production of these subunits may be involved in transcriptional activation of the GnRH receptor gene (Fernandez-Vazquez *et al.*, 1996). Whether there is a functional system involving paracrine or autocrine actions or inhibin in the ruminant pituitary gland is not yet certain. However, it is possible that the effects of inhibin on numbers of GnRH receptors and GnRH receptor gene expression *in vitro* may reflect intra-pituitary regulation rather than endocrine effects of inhibin of ovarian origin.

The Gene Encoding Ovine GnRH Receptor

From the preceding paragraphs, it is easy to understand how isolation of cDNAs encoding GnRH receptors has expanded knowledge of GnRH receptor gene expression in cattle and sheep. Concurrent with studies on the endocrine regulation of GnRH receptor gene expression, we became interested in exploring the molecular mechanisms underlying transcriptional regulation of the GnRH receptor gene in ruminants. To begin to address this issue, we cloned the ovine GnRH receptor gene (Fig. 5; Campion *et al.*, 1996). This gene comprises three exons and two introns, and occurs as a single copy gene. Although there is considerable identity (≥ 60%) between the nucleotide sequences of the proximal 5′ flanking regions of the ovine and murine (Zhou *et al.*, 1994; Clay *et al.*, 1995) GnRH receptor genes, there are striking differences among species with respect to the DNA regulatory elements required for cell-specific expression. Transcriptional activity of the murine GnRH receptor gene is conferred by three cis-acting elements that lie within 500 bp of the proximal promoter (Duval *et al.*, 1997). These elements include an activating protein-1 (AP-1) binding site, presumably activated by the fos/jun family of transcription factors; a binding site for the orphan nuclear receptor, steroidogenic factor 1 (SF-1; reviewed by Parker and Schimmer, 1997); and GnRH receptor activating sequence, for which the trans-acting factor has yet to be identified. Of these three elements, only SF-1 appears to be a conserved mechanism for regulation of the ovine GnRH receptor promoter. Transcriptional activity of this promoter is increased following transient co-transfection of COS-7 cells with an expression vector for SF-1 (Quirk, 1997), indicating that SF-1 may regulate

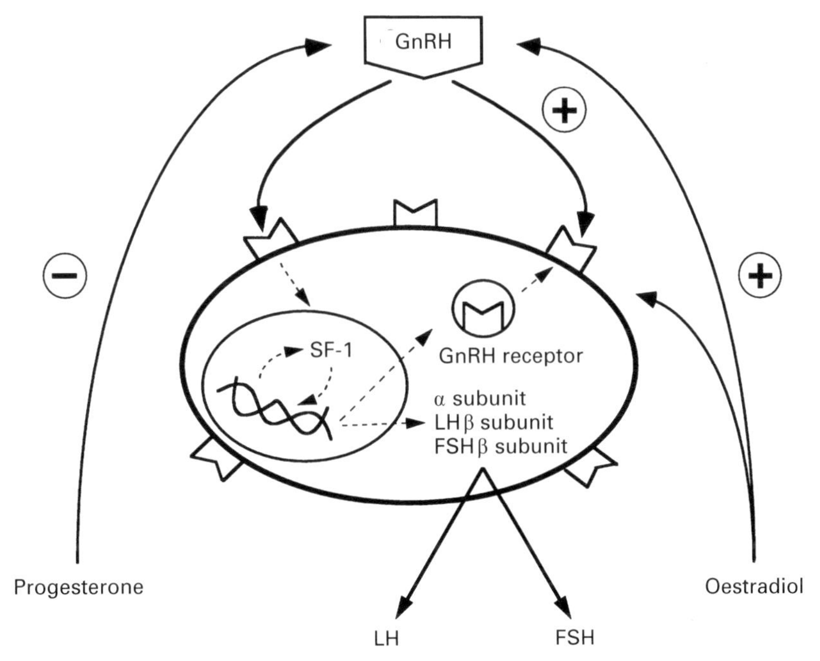

Fig. 6. Current working model of the endocrine and molecular regulation of GnRH receptor gene expression in gonadotrophs of domestic ruminants. Pulsatile GnRH from the hypothalamus stimulates GnRH receptor gene expression, and this effect appears to be modulated by negative feedback effects of progesterone from the corpus luteum. Oestradiol from the ovary also stimulates expression of the GnRH receptor gene. Although this effect of oestradiol is exerted directly on the pituitary gland, increased secretion of GnRH during positive feedback by oestradiol may also contribute to increased GnRH receptor gene expression. The transcription factor steroidogenic factor 1 (SF-1) may be an important regulator of several genes expressed by ruminant gonadotrophs, including the GnRH receptor gene. Increased expression of the GnRH receptor gene is believed to result in synthesis of GnRH receptors and their expression on the plasma membrane of gonadotrophs. Since interaction of GnRH with its receptor is absolutely necessary for synthesis of the gonadotrophin subunits (α, LHβ and FSHβ, regulated expression of the GnRH receptor gene and synthesis of GnRH receptors are critical steps in the production of LH and FSH.

transcription of the GnRH receptor gene in sheep. In the light of evidence that SF-1 also regulates transcription of the bovine LHβ subunit gene (Keri and Nilson., 1996) and the gene for SF-1 is expressed in ovine gonadotrophs (Turzillo *et al.*, 1997), SF-1 is emerging as a possible common transcriptional regulator of several genes that define the ruminant gonadotrope.

In addition to cell-specific expression, it is of interest to identify potential molecular mechanisms that mediate responsiveness of the GnRH receptor gene to endocrine factors. A DNA element that confers responsiveness of the ovine GnRH receptor gene to changes in concentrations of intracellular cAMP has been identified (C. M. Clay, personal communication). Transcriptional activity of the proximal promoter is increased in the presence of forskolin (a pharmacological activator of adenylyl cyclase), and this response appears to be mediated via a cAMP response element capable of binding cAMP response element binding protein. These findings implicate cAMP as a potential regulator of the GnRH receptor gene. In cultured ovine pituitary cells, binding of GnRH to ovine gonadotrophs increases cAMP and addition of a cAMP derivative can mimic the LH-releasing effect of GnRH (Adams *et al.*, 1979). Furthermore, treatment of rat pituitary cells with analogues of cAMP increased numbers of GnRH receptors (Young *et al.*, 1984). Thus it is intriguing to speculate that the stimulatory effects of GnRH on tissue concentrations of GnRH receptor mRNA and GnRH receptors in sheep and cattle may be mediated at the molecular level by cAMP. Additional studies are needed to explore further the role of cAMP in regulation of GnRH receptor gene expression by GnRH and other endocrine factors.

Conclusion

Coordinated changes in sensitivity of the anterior pituitary gland to GnRH are required for normal reproductive cyclicity in cattle and sheep. Therefore, knowledge of the mechanisms regulating GnRH receptors is not only valuable to our basic understanding of pituitary function, but is also relevant to the development of improved methods for controlling fertility in domestic ruminants. We and others have characterized patterns of GnRH receptor gene expression and the endocrine regulation of these patterns during the oestrous cycle in sheep and cattle. It is clear that both hypothalamic (GnRH) and ovarian hormones (progesterone and oestradiol) influence the expression of the GnRH receptor gene and numbers of GnRH receptors (Fig. 6). Because of the discrepancy regarding effects of inhibin on GnRH receptor gene expression *in vitro* versus *in vivo*, the role of ovarian inhibin as an endocrine regulator of the GnRH receptor gene in domestic ruminants remains uncertain. Recent evidence indicates that SF-1 may be a common transcriptional regulator of several genes expressed by gonadotrophs in domestic ruminants. Isolation of the gene encoding the ovine GnRH receptor will allow further study of the molecular mechanisms by which transcription of the GnRH receptor gene is influenced by changes in the endocrine milieu.

References

Adams BM, Sakurai H and Adams TE (1997) Effect of oestradiol on mRNA encoding GnRH receptor in pituitary tissue of orchidectomized sheep passively immunized against GnRH *Journal of Reproduction and Fertility* **111** 207–212

Adams TE, Wagner TOF, Sawyer HR and Nett TM (1979) GnRH interaction with anterior pituitary. II Cyclic AMP as an intracellular mediator in the GnRH activated gonadotroph *Biology of Reproduction* **21** 735–747

Batra SK and Miller WL (1985) Progesterone decreases the responsiveness of ovine pituitary cultures to luteinizing hormone-releasing hormone *Endocrinology* **117** 1436–1440

Brooks J and McNeilly AS (1994) Regulation of gonadotropin-releasing hormone receptor mRNA expression in the sheep *Journal of Endocrinology* **143** 175–182

Brooks J, Crow WJ, McNeilly JR and McNeilly AS (1992) Relationship between gonadotrophin subunit gene expression, gonadotrophin-releasing hormone receptor content and pituitary and plasma gonadotrophin concentrations during the rebound release of FSH after treatment of ewes with bovine follicular fluid during the luteal phase of the cycle *Journal of Molecular Endocrinology* **8** 109–118

Brooks J, Taylor PL, Saunders PTK, Eidne KA, Struthers WJ and McNeilly AS (1993) Cloning and sequencing of the sheep pituitary gonadotropin-releasing hormone receptor and changes in expression of its mRNA during the estrous cycle *Molecular and Cellular Endocrinology* **94** R23–R27

Campion CE, Turzillo AM and Clay CM (1996) The gene encoding the ovine gonadotropin-releasing hormone (GnRH) receptor: cloning and initial characterization *Gene* **170** 277–280

Clarke IJ, Cummins JT, Crowder ME and Nett TM (1987) Pituitary receptors for gonadotropin-releasing hormone in ovariectomized–hypothalamo pituitary disconnected ewes. I Effect of changing frequency of gonadotropin-releasing hormone pulses *Biology of Reproduction* **37** 749–754

Clay CM, Nelson SM, DiGregorio GB, Campion CE, Wiedemann AL and Nett RJ (1995) Cell-specific expression of the mouse gonadotropin-releasing hormone (GnRH) receptor is conferred by elements residing within 500 bp of proximal 5′ flanking region *Endocrine* **3** 615–622

Crowder ME and Nett TM (1984) Pituitary content of gonadotropins and receptors for gonadotropin-releasing hormone (GnRH) and hypothalamic content of GnRH during the periovulatory period of the ewe *Endocrinology* **114** 234–239

Duval DL, Nelson SE and Clay CM (1997) The tripartite basal enhancer of the gonadotropin-releasing hormone (GnRH) receptor gene promoter regulates cell-specific expression through a novel GnRH receptor activating sequence *Molecular Endocrinology* **11** 1814–1821

Fernandez-Vasquez G, Kaiser UB, Albarracin CT and Chin WW (1996) Transcriptional activation of the gonadotropin-releasing hormone receptor gene by activin A *Molecular Endocrinology* **10** 356–366

Findlay JK, Clarke IJ and Robertson DM (1990) Inhibin concentrations in ovarian and jugular venous plasma and the relationship of inhibin with follicle-stimulating hormone and luteinizing hormone during the ovine estrous cycle *Endocrinology* **126** 528–535

Gregg DW and Nett TM (1989) Direct effects of estradiol-17β on the number of gonadotropin-releasing hormone receptors in the ovine pituitary *Biology of Reproduction* **40** 288–293

Gregg DW, Allen MC and Nett TM (1990) Estradiol-induced increase in number of gonadotropin-releasing hormone receptors in cultured ovine pituitary cells *Biology of Reproduction* **43** 1032–1036

Gregg DW, Schwall RH and Nett TM (1991) Regulation of gonadotropin secretion and number of gonadotropin-releasing hormone receptors by inhibin, activin-A and estradiol *Biology of Reproduction* **44** 725–732

Hamernik DL and Nett TM (1988) Gonadotropin-releasing hormone increases the amount of messenger ribonucleic acid for gonadotropins in ovariectomized ewes after hypothalamic–pituitary disconnection *Endocrinology* **122** 959–966

Hamernik DL, Clay CM, Turzillo AM, VanKirk EA and Moss GE (1995) Estradiol increases amounts of messenger ribonucleic acid for gonadotropin-releasing hormone receptors in sheep *Biology of Reproduction* **53** 179–185

Hazum E, Cuatrecasas PP, Marian J and Conn PM (1980) Receptor-mediated internalization of fluorescent gonadotropin-releasing hormone by pituitary gonadotropes *Proceedings of the National Academy of Sciences USA* **77** 6692–6695

Huang ES and Miller WL (1980) Effects of estradiol-17β on basal and luteinizing hormone releasing hormone-induced secretion of luteinizing hormone and follicle stimulating hormone by ovine pituitary cell culture *Biology of Reproduction* **23** 124–134

Kakar SS, Rahe CH and Neill JD (1993) Molecular cloning, sequencing and characterizing the bovine receptor for gonadotropin releasing hormone (GnRH) *Domestic Animal Endocrinology* **10** 335–342

Keri RA and Nilson JH (1996) A steroidogenic factor-1 binding site is required for activity of the luteinizing hormone b subunit promoter in gonadotropes of transgenic mice *Journal of Biological Chemistry* **271** 10782–10785

Khalid M, Haresign W and Hunter MG (1987) Pulsatile GnRH administration stimulates the number of pituitary GnRH receptors in seasonally anoestrous ewes *Journal of Reproduction and Fertility* **79** 223–230

Lamming GE and McLeod BJ (1988) Continuous infusion of GnRH reduces the LH response to an intravenous GnRH injection but does not inhibit endogenous LH secretion in cows *Journal of Reproduction and Fertility* **82** 237–246

Laws SC, Webster JC and Miller WL (1990a) Estradiol alters the effectiveness of gonadotropin-releasing hormone (GnRH) in ovine pituitary cultures: GnRH receptors versus responsiveness to GnRH *Endocrinology* **127** 381–386

Laws SC, Beggs MJ, Webster JC and Miller WL (1990b) Inhibin increases and progesterone decreases receptors for gonadotropin-releasing hormone in ovine pituitary cell culture *Endocrinology* **127** 373–380

Leung K, Padmanabhan V, Convey EM, Short RE and Staigmiller RB (1984) Relationship between pituitary responsiveness to Gn-RH and number of Gn-RH-binding sites in pituitary glands of beef cows *Journal of Reproduction and Fertility* **71** 267–277

Miller WL and Huang ES (1985) Secretion of ovine luteinizing hormone *in vitro*: differential positive control by 17β-estradiol and a preparation of porcine ovarian inhibin *Endocrinology* **117** 907–911

Moenter SM, Caraty A, Locatelli A and Karsch FJ (1991) Pattern of gonadotropin-releasing hormone (GnRH) secretion leading up to ovulation in the ewe: existence of a preovulatory GnRH surge *Endocrinology* **129** 1175–1182

Moss GE and Nett TM (1980) GnRH interaction with anterior pituitary. IV. Effect of estradiol-17β on GnRH-mediated release of LH from ovine pituitary cells obtained during the breeding season, anestrous season, and period of transition into or out of the breeding season *Biology of Reproduction* **23** 398–403

Moss GE, Crowder ME and Nett TM (1981) GnRH–receptor interaction. VI. Effect of progesterone and estradiol on hypophyseal receptors for GnRH, and serum and hypophyseal concentrations of gonadotropins in ovariectomized ewes *Biology of Reproduction* **25** 938–944

Muttukrishna S and Knight PG (1990) Effects of crude and highly purified bovine inhibin (Mr 32,000 form) on gonadotrophin production by ovine pituitary cells *in vitro*: inhibin enhances gonadotrophin-releasing hormone-induced release of LH *Journal of Endocrinology* **127** 149–159

Nett TM (1990) Regulation of genes controlling gonadotropin secretion *Journal of Animal Science* **68** (Supplement 2) 3–17

Nett TM, Crowder ME, Moss GE and Duello TM (1981) GnRH receptor interaction. V. Down-regulation of pituitary receptors for GnRH in ovariectomized ewes by infusion of homologous hormone *Biology of Reproduction* **24** 1145–1155

Nett TM, Cermak D, Braden T, Manns J and Niswender G (1987) Pituitary receptors for GnRH and estradiol, and pituitary content of gonadotropins in beef cows. I. Changes during the estrous cycle *Domestic Animal Endocrinology* **4** 123–132

Padmanabhan V, Convey EM, Roche JF and Ireland JJ (1984) Changes in inhibin-like bioactivity in ovulatory and atretic follicles and utero–ovarian venous blood and prostaglandin-induced luteolysis in heifers *Endocrinology* **115** 1332–1340

Parker KL and Schimmer BP (1997) Steroidogenic factor 1: A key determinant of endocrine development and function *Endocrine Reviews* **18** 361–377

Quirk CC (1997) *Cloning and Characterization of the Murine and Ovine Gonadotropin-releasing Hormone Receptor Genes* PhD Dissertation, Colorado State University, Fort Collins

Richards MW, Wetteman RP and Schoenemann HM (1989) Nutritional anestrus in beef cows: body weight change, body condition, luteinizing hormone in serum and ovarian activity *Journal of Animal Science* **67** 1520–1526

Roberts V, Meunier H, Vaughn J, Rivier J, Rivier C, Vale W and Sawchenko P (1988) Production and regulation of inhibin subunits in pituitary gonadotropes *Endocrinology* **124** 552–554

Rodriguez RE and Wise ME (1991) Advancement of postnatal pulsatile luteinizing hormone secretion in the bull calf by pulsatile administration of gonadotropin-releasing hormone during infantile development *Biology of Reproduction* **44** 432–439

Sakurai H, Adams BM and Adams TE (1997) Concentration of gonadotropin-releasing hormone receptor messenger ribonucleic acid in pituitary tissue of orchidectomized sheep: effect of passive immunization against gonadotropin-releasing hormone *Journal of Animal Science* **75** 189–194

Schoenemann HM, Humphrey WD, Crowder ME, Nett TM and Reeves JJ (1985) Pituitary luteinizing hormone-releasing hormone receptors in ovariectomized cows after challenge with ovarian steroids *Biology of Reproduction* **32** 574–583

Sealfon SC, Laws SC, Wu JC, Gillo B and Miller WL (1990) Hormonal regulation of gonadotropin-releasing hormone receptors and messenger RNA activity in ovine pituitary culture *Molecular Endocrinology* **4** 1980–1987

Sealfon SC, Weinstein H and Millar RP (1997) Molecular mechanisms of ligand interaction with the gonadotropin-releasing hormone receptor *Endocrine Reviews* **18** 180–205

Turzillo AM and Nett TM (1997) Effects of bovine follicular fluid and passive immunization against gonadotropin-releasing hormone (GnRH) on messenger ribonucleic acid for GnRH receptor and gonadotropin subunits in ovariectomized ewes *Biology of Reproduction* **56** 1537–1543

Turzillo AM, Campion CE, Clay CM and Nett TM (1994) Regulation of gonadotropin-releasing hormone (GnRH) receptor messenger ribonucleic acid and GnRH receptors during the early preovulatory period in the ewe *Endocrinology* **135** 1353–1358

Turzillo AM, DiGregorio GB and Nett TM (1995a) Messenger ribonucleic acid for gonadotropin-releasing hormone receptor and numbers of gonadotropin-releasing hormone receptors in ovariectomized ewes after hypothalamic–pituitary disconnection and treatment with estradiol *Journal of Animal Science* **73** 1784–1788

Turzillo AM, Juengel JL and Nett TM (1995b) Pulsatile gonadotropin-releasing hormone (GnRH) increases concentrations of GnRH receptor messenger ribonucleic acid and numbers of GnRH receptors during luteolysis in the ewe *Biology of Reproduction* **53** 418–423

Turzillo AM, Quirk CC, Juengel JL, Nett TM and Clay CM (1997) Effects of ovariectomy and hypothalamic–pituitary disconnection on amounts of steroidogenic factor-1 mRNA in the ovine anterior pituitary gland *Endocrine* **6** 251–256

Turzillo AM, Clapper JA, Moss GE and Nett TM (1998) Regulation of ovine GnRH receptor gene expression by progesterone and oestradiol *Journal of Reproduction and Fertility* **113** 251–256

Turzillo AM, Nolan TE and Nett TM (1998) Regulation of gonadotrophin-releasing hormone (GnRH) receptor gene expression in sheep: interaction of GnRH and estradiol *Endocrinology* **139** 4890–4894

Vizcarra JA, Wetteman RP, Braden TD, Turzillo AM and Nett TM (1997) Effect of gonadotropin-releasing hormone (GnRH) pulse frequency on serum and pituitary concentrations of luteinizing hormone and follicle-stimulating hormone, GnRH receptors, and messenger ribonucleic acid for gonadotropin subunits in cows *Endocrinology* **138** 594–601

Wise ME, Nieman D, Stewart J and Nett TM (1984) Effect of number of receptors for gonadotropin-releasing hormone on the release of luteinizing hormone *Biology of Reproduction* **31** 1007–1013

Wu JC, Sealfon SC and Miller WL (1994) Gonadal hormones and gonadotropin-releasing hormone (GnRH) alter messenger ribonucleic acid levels for GnRH receptors in sheep *Endocrinology* **134** 1846–1850

Young LS, Naik SI and Clayton RN (1984) Adenosine 3′,5′-monophosphate derivatives increase gonadotropin-releasing hormone receptors in cultured pituitary cells *Endocrinology* **114** 2114–2122

Zhou W and Sealfon SC (1994) Structure of the mouse gonadotropin-releasing hormone receptor gene: variant transcripts generated by alternative processing *DNA and Cell Biology* **13** 605–614

Journal of Reproduction and Fertility Supplement **54**, 87–99

Follicle-stimulating isohormones: regulation and biological significance

V. Padmanabhan, J. S. Lee and I. Z. Beitins*

Department of Pediatrics, Reproductive Sciences Program, The University of Michigan, Ann Arbor, MI 48109, USA

Follicle-stimulating hormone (FSH) is a key hormone in the regulation of follicular development. Although the existence of FSH heterogeneity is well established, the physiological significance of this pleomorphism remains unknown. Observed changes in circulating FSH heterogeneity during critical reproductive events such as puberty and reproductive cyclicity suggest that different combinations of FSH isoforms reach the target sites during different physiological states to influence a variety of biological end points such as cellular growth, development, steroidogenesis and protein synthesis. Considering that these FSH isoforms have different physicochemical properties and potential to bind not only their cognate receptors but also structurally related, non-FSH receptors with various affinities, the regulatory implications of FSH heterogeneity in modulating the various FSH-induced functions are enormous. However, assigning functional significance to FSH heterogeneity has been hampered because of (1) difficulties associated with procurement of highly purified, naturally occurring, circulating FSH isoforms; (2) absence of reference standards that contain the entire repertoire of FSH isoforms present in biological fluids; and (3) specificity issues inherent to the detection systems used. If particular FSH isoforms do possess selective biological functions, specific combinations of FSH isoforms could be generated to regulate fertility in farm animals and humans.

Introduction

Although the existence of protein heterogeneity is now uncontested, the intracellular mechanisms governing protein synthesis, post-translational modifications and post-secretion fates are poorly understood. Proteins emerging from the ribosome have a great variety of potential fates, depending on the presence or absence of critical amino acid sequences. These amino acid sequences constitute target signals that determine spatial configuration, folding and post-translational modifications (including the addition and trimming of carbohydrate units, stability, packaging in granules and secretion). Structural features (such as tertiary configuration, carbohydrate composition) determine the metabolic fate of the protein in the peripheral circulation. Although the structure of the oligosaccharide chains of glycoproteins is an important determinant of the circulating half-life, protein structure and exposed epitopes are important for target cell receptor binding and signal transduction. For these and other reasons, it appears that the resulting mix of heterogeneous hormonal isoforms exert effects at the target site that cannot be interpreted properly by conventional radioimmunoassays alone.

Existence, Origin and Functional Attributes of FSH Isoforms

Existence of FSH isoforms

Use of various separation techniques, such as chromatofocusing, isoelectric focusing and gel electrophoresis, has unequivocally established that FSH occurs in multiple forms both in the

* Current address: National Center for Research Resources, National Institute of Health, Bethesda, MD.

pituitary and circulation (reviewed in Chappel *et al.*, 1983; Beitins and Padmanabhan, 1991; Ulloa-Aguirre *et al.*, 1995). The more discriminating the separation technique the more complex the distribution profiles of FSH isoforms. For instance, Burgon *et al.* (1993), using isoelectric focusing and high performance anion exchange chromatography, identified more than 20 human pituitary isoforms. A complete array of FSH isoforms ranging from basic to strongly acidic pI values is found both in the pituitary and circulation (reviewed in Chappel *et al.*, 1983; Beitins and Padmanabhan, 1991; Ulloa-Aguirre *et al.*, 1995). In some physiological and pharmacological conditions, the distribution of circulating and pituitary FSH isoforms appears to be similar, while in others they differ.

Origin of FSH heterogeneity

There are several potential sites at which changes in FSH heterogeneity can originate. At the pituitary, the array of FSH isoforms that exist during different physiological states may represent merely FSH at different stages in the biosynthetic pathway. Alternatively, changes in FSH heterogeneity may be the outcome of alterations in post-translational processing and a function of glycosylation changes. In this context, it is of interest that the activities of glycosyltransferases and sialotransferases, enzymes involved in sulphation and sialylation of luteinizing hormone (LH) and FSH, vary with the physiological state in the pituitary (Dharmesh and Baenziger, 1993; Damián-Matsumura *et al.*, 1998). This may contribute to potential differences in glycosylation of pituitary gonadotrophins.

At the circulatory level changes in FSH heterogeneity may originate from regulated preferential secretion, peripheral modification, or differences in metabolic clearance. *In vitro* studies (reviewed in Chappel *et al.*, 1983; Ulloa Aguirre *et al.*, 1995) and recent *in vivo* studies in sheep (Lee *et al.*, 1998) suggest that the pituitary has the ability to secrete different combinations of FSH isoforms depending on the physiological status of the animal.

Functional attributes of FSH isoforms

The nature of the oligosaccharide side chains varies considerably and may be manifested as changes in carbohydrate content, length and branching of the side chain and the associated charge. These structural differences appear to dictate the receptor binding ability, *in vitro* and *in vivo* biological activities, and circulatory half-lives of the FSH isoforms. Most of the earlier work that used biological to immunological ratios to estimate FSH biopotencies concluded that *in vitro* biological activity of the FSH isoform is positively correlated with the pI value of the isohormone; the FSH isoforms with higher pI values have higher *in vitro* biological activities than those with lower pI values (reviewed in Chappel *et al.*, 1983; Beitins and Padmanabhan, 1991; Ulloa-Aguirre *et al.*, 1995). An exception to this rule is that the basic FSH isoforms that elute in the void volume of the chromatofocusing column (pH gradient 7.4–4.0) appear to act as antagonists of FSH (Dahl *et al.*, 1988; Timossi *et al.*, 1998).

The assumption in the use of B:I ratios for estimating FSH biopotency has been that the immunopotencies are similar between different FSH isoforms. More recent studies that used highly purified FSH isoforms of defined mass (not adjusted for carbohydrate content and oxidative losses) have raised questions about this assumption and shown the contrary, namely that FSH immunopotencies of different FSH isoforms vary (Burgon *et al.*, 1993; Stanton *et al.*, 1996). It then follows that earlier biopotency estimates that used immunoassays to determine gonadotrophin protein mass may be inaccurate. On the basis of the high correlation between immuno-, radioreceptor and *in vitro* bioactivity of 15 purified human FSH isoforms (> 90% purity) in the pI value range of 3.63–5.13, Burgon *et al.* (1993) concluded that: (1) current immunoassays are measuring mass in combination with some measure of bioactivity and (2) an immunoassay directed to the 'invariant region of the molecule' not affected by glycosylation differences is required to assess the true bioactivity of the FSH isoform. Interestingly, in spite of the high correlation between the immuno-, radio-receptor and *in vitro* bioactivities of individual isoforms, these investigators also noted significant differences in ratios of activities between *in vitro* bioassays and other methods (Burgon *et al.*, 1993). For example, the *in vitro* bio- to radio-receptor assay ratio of the human FSH isoform with a pI value of 4.23 was four times greater than those with pI values of 3.63, 3.88, 4.07, 4.85, and 5.13. Isoforms with higher pI values, such as those reported to have antagonistic properties (Dahl *et al.*, 1988; Timossi *et al.*, 1998) or the less acidic FSH isoforms (pI > 5.4) seen during puberty (Padmanabhan *et al.*, 1992), have not been evaluated. Differences in assay ratios from the carefully characterized study of Burgon *et al.* (1993) and the 5–8-fold differences in immuno-, radio-receptor and *in vitro* bioactivities of the various human FSH isoforms indicate that structural heterogeneity of FSH does contribute to functional differences.

In terms of the circulatory half-life, many studies using chemically or enzymatically modified FSH have shown that the less acidic FSH isoforms have shorter half-lives than the more acidic FSH isoforms (reviewed in Chappel *et al.*, 1983; Beitins and Padmanabhan, 1991; Ulloa-Aguirre *et al.*, 1995). Differences in half-life in these instances appear to relate to the degree of sialylation (Morell *et al.*, 1971). Although most studies follow this general rule, others have failed to observe such relationships (Robertson *et al.*, 1991). Assuming that the FSH isohormones with shorter circulatory half-lives have higher biological activity *in vitro*, the issue is to determine whether measures of *in vitro* biological activity are meaningful in interpreting consequences *in vivo*. Studies by Stanton *et al.* (1996) show a 16-fold range in *in vivo* activities between various LH isoforms. These changes parallel estimates of bioactivity *in vitro*. This finding suggests that differences in bioactivity *in vitro* and not the large differences in circulatory half-life are the key determinants of bioactivity *in vivo* as assessed in short-term assays (long-term consequences were not assessed).

The effector mechanisms by which FSH elicits its target cell action appear also to be conducive for rapid functional interactions between FSH and its receptor. For example, minutes after addition, FSH elicits increases in intracellular calcium (Flores *et al.*, 1990). Furthermore, effective activation of adenylate cyclase appears possible under conditions of intermittent receptor activation (Spiegel *et al.*, 1992). In this context it is of interest that FSH is secreted both in a basal and pulsatile mode (Padmanabhan *et al.*, 1997a). Some evidence also indicates that what is secreted in pulses is of different molecular nature than that secreted in the basal mode (reviewed in Ulloa-Aguirre *et al.*, 1995). It still remains to be addressed whether the pulsatile inputs are perceived by the target site as such, and if so, do they respond differently to this intermittent activation as opposed to FSH secreted in the basal mode.

Although many issues still remain to be addressed, it is clear that a combination of circulating gonadotrophin isoforms reaches target tissues and influences a variety of biological end points such as cellular growth, development, steroidogenesis and synthesis of proteins. FSH isoforms possessing prolonged circulation times may have the potential to provide a long-acting stimulus for the progression of maturational events. In contrast, isoforms with shorter half-lives that are secreted intermittently may provide an acute yet potent stimulus (Chappel *et al.*, 1983; Beitins and Padmanabhan, 1991). Relative proportions of the various gonadotrophin isoforms within the circulation, therefore, have the potential to exert qualitatively different effects on target tissues.

Methodological Considerations in Assessing FSH Heterogeneity

Utility of B:I ratios

Initial studies of FSH heterogeneity relied heavily upon comparative bioactivity (B) and immunoreactivity (I) estimates. In these instances changes in B:I ratios were taken to imply that there was a change in the constituent mix of FSH isoforms. More recent studies have questioned this concept (Jaakkola *et al.*, 1990, Simoni *et al.*, 1994). Conclusions on hormonal heterogeneity derived from studies of B:I ratios of FSH must be drawn with caution and may suffer from two major drawbacks. First, it needs to be proven that the differences in measured bioactive FSH and immunoreactive FSH are not simply the result of differences in the effects of interfering substances on the measurement system but result from an ability to discriminate among isoforms. For example, RIAs using different combinations of standards and antisera (with different epitopic recognition) can differentially recognize gonadotrophin mixtures (Simoni *et al.*, 1994). Bioassays suffer in that substances present in the test material can interfere (Simoni *et al.*, 1991) or modulate the effectiveness of FSH (reviewed in Beitins and Padmanabhan, 1991; Chappel, 1995) leading to erroneous estimates.

Second, the FSH standards used in various immuno- and bioassays add another source of error when estimating changes in FSH B:I ratios. The distribution patterns of FSH isoforms of pituitary and urinary FSH standards appear to have sparser representation of less acidic FSH isoforms than those found in circulation and unpurified pituitary extracts (Simoni *et al.*, 1993a, 1993b). This finding suggests that the standards that are currently in use in the various assays do not have the full array of FSH isoforms that are present in biological fluids. Although it is ideal to have a reference preparation that contains the repertoire of FSH isoforms present in the biological fluid being measured, this does not appear practical in view of the high variability associated with its composition during different physiological states. At the very minimum, documentation of changes in FSH heterogeneity using FSH B:I ratios requires careful characterization of the assay systems and use of the same reference preparations. Unfortunately, many of the earlier studies reporting changes in FSH B:I ratios used different FSH standards to estimate the immuno- and biopotencies. In fact, the first study that challenged the utility of B:I ratios itself used two different standards in their immuno- and bioassays (Jakkola *et al.*, 1990). The specificity of the monoclonal antibodies also adds an additional problem in that they fail to recognize all variants (Pettersson and Soderholm, 1991).

In terms of bioassays, the biopotency estimate of the FSH isoform may also vary with the biological end point or assay system chosen. When cAMP is used as the endpoint in the Sertoli cell bioassay, deglycosylated ovine FSH appears to be inactive (Fig. 1) (Padmanabhan, 1995). On the contrary, when oestradiol is chosen as the end point, deglycosylated FSH stimulates oestradiol production. Species specificity of the bioassay is also a consideration. For instance, Ding *et al.* (1991) have shown that the bioactive LH estimates and the B:I ratios vary markedly depending on whether the mouse Leydig cell, rat interstitial cell or human granulosa cell bioassay is used to measure LH bioactivity. Bioassays of FSH that use cells transfected with recombinant FSH receptors (Gudermann *et al.*, 1994; Christine-Maitre *et al.*, 1996) and cAMP as the end point also pose a problem because they do not measure a biological end point and ignore involvement of other second messenger systems involved in FSH signal transduction.

Taking all aforementioned caveats into consideration, it appears that differences in FSH B:I ratios reported by different investigators during similar physiological situations are a direct function of the assay, the standard and end point used. For instance, FSH B:I ratios are constant throughout the cycle (Jia *et al.*, 1986), highest during the preovulatory period (Padmanabhan *et al.*, 1988a; Wide and Bakos, 1993; Zambrano *et al.*, 1995), or maximal in the luteal phase (Christine-Maitre *et al.*, 1996) of the human menstrual cycle. Therefore, when comparing biological potency estimates, it is important to consider the species, the cellular types, as well as the relative endpoints of the bioassay used by the various investigators. Assuming that assay systems are controlled carefully, changes in B:I ratio at best will point to possible changes in heterogeneity but will require confirmation with other chromatographic approaches.

Chromatographic separation

Several fractionation approaches have been used effectively to characterize changes in FSH heterogeneity and have been reviewed extensively (Ulloa-Aguirre *et al.*, 1995). Although a broader classification of FSH isoforms can be achieved by gel permeation chromatography, higher resolution has been achieved with techniques that separate on the basis of charge. Some charge-based techniques that have been used extensively are isoelectric focusing, chromatofocusing and analytical zone electrophoresis. These approaches while providing a general assessment of changes, do not separate isoforms to purity which requires multiple chromatographic steps (Burgon *et al.*, 1993). An inherent caveat in these approaches is that estimates of FSH isoform distribution after fractionation are based on immunoassay systems that have their inherent problems. Recent studies of Simoni *et al.*

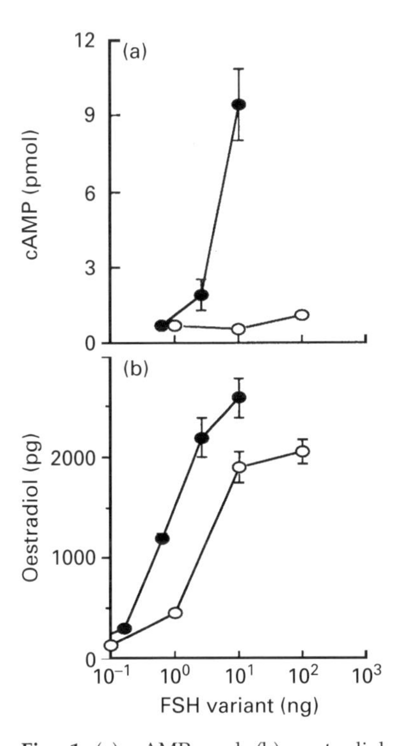

Fig. 1. (a) cAMP, and (b) oestradiol responses of immature rat Sertoli cells to increasing concentrations of native (oFSH) (●) or deglycosylated oFSH (DG-oFSH) (○). Sertoli cells from 7–10-day-old rats were cultured using previously validated methods (Padmanabhan *et al.*, 1987). Cell monolayers were exposed to increasing concentrations of oFSH or DG-oFSH for 24 h. Thirty minutes after addition of FSH, a 200 μl aliquot of medium was removed from each well for cAMP determinations. Medium concentrations of oestradiol (24 h incubates) and cAMP (30 min incubates) were measured using previously validated assays. Note the differences in cAMP and oestradiol responses to the native and deglycosylated FSH (modified from Padmanabhan *et al.*, 1993).

(1994) have shown that the distribution profiles of FSH after chromatofocusing vary considerably between immunoassays because each assay recognizes individual isoforms of FSH differently.

Regulation of FSH Heterogeneity

A central theme emerging from investigations of FSH heterogeneity is that the endocrine status regulates the proportion of various FSH isoforms and that these isoforms show distinct actions.

Although the factors affecting the final distribution of gonadotrophin isoforms within the circulation are multifaceted and complex, it is clear that endocrine changes regulate the proportions of FSH isoforms both within the pituitary and in the peripheral circulation.

Pituitary FSH heterogeneity

Endocrine regulation of pituitary FSH microheterogeneity has been studied widely in numerous species. Several reviews have addressed the regulation of pituitary heterogeneity in great detail (reviewed in Chappel *et al.*, 1983; Beitins and Padmanabhan, 1991; Ulloa-Aguirre *et al.*, 1995). Qualitative differences in pituitary FSH content have been correlated with age, sex and stage of the oestrous cycle in several species by using various techniques (Blum and Gupta, 1980; Chappel *et al.*, 1983; Ulloa-Aguirre *et al.*, 1995). In general, pituitary FSH is less acidic in female than male pituitaries and young than old animals. Changes in pituitary FSH heterogeneity (predominance of less acidic FSH isoforms) have also been found during onset of puberty in rats (Chappel *et al.*, 1983), but not in heifers (Stumpf *et al.*, 1992).

Circulating FSH heterogeneity

Regulation of circulating FSH heterogeneity has not received such intense investigation as pituitary heterogeneity, in part due to limitations imposed by relatively low concentrations of circulating FSH. Development of sensitive, *in vitro* FSH bioassays which use oestradiol as the end point (granulosa cell bioassay: Jia *et al.*, 1986; Sertoli cell bioassay: Padmanabhan *et al.*, 1987) have allowed characterization of changes in circulating bioactive FSH during different physiological states (reviewed in Beitins and Padmanabhan, 1991). Because of the caveats discussed above in using B:I ratios as an index of changes in FSH heterogeneity, in this report, only those studies that used chromatographic approaches either alone or in conjunction with bioactivity measures to assess changes in FSH heterogeneity will be considered. In general, increases in oestradiol and GnRH in gonad-intact models lead to increases in less acidic FSH isoforms and corresponding increases in bioactive FSH in the circulation (Fig. 2) (reviewed in Beitins and Padmanabhan, 1991; Ulloa-Aguirre *et al.*, 1995). Administration of GnRH to prepubertal boys (Phillips and Wide, 1994), men (Simoni *et al.*, 1996), and women (Zambrano *et al.*, 1995) increases the release of less acidic FSH isoforms. Treatment of women with GnRH antagonist, on the other hand, leads to an increase in the circulation of a highly basic FSH isoform (Dahl *et al.*, 1988). This basic FSH isoform was shown to antagonize the action of FSH.

Few studies have also addressed the role of oestradiol in modulating circulating FSH heterogeneity. Studies in humans documenting changes in FSH heterogeneity (shift to a less acidic side) following treatment of postmenopausal women (Wide and Naessén, 1994) or a Turner's girl (Padmanabhan *et al.*, 1988a) with oestrogens provide corroborative evidence in support of a role for oestradiol in modulating circulating FSH heterogeneity. In these instances, since oestrogens can induce their effects via mediation of hypothalamic GnRH secretion, it is difficult to separate the direct effects of oestradiol from mediation via alterations in GnRH input. Studies using ovariectomized, nutritionally growth-retarded (hypogonadotrophic) sheep have shown that pulsatile GnRH administration alone, in the absence of oestradiol, increases LH but does not alter the distribution profile of FSH or bioactive FSH secretion (Fig. 3) (Hassing *et al.*, 1993). In contrast, oestradiol administration to nutritionally growth-retarded lambs leads to an increase in less acidic FSH isoforms (Padmanabhan *et al.*, 1997b) (Fig. 4). Therefore, these findings support a role for oestradiol in mediating FSH heterogeneity.

Progesterone and androgens appear to have an opposite effect to that of oestradiol since they appear to increase the presence of more acidic FSH isoforms (Fig. 2) (reviewed in Ulloa-Aguirre *et al.*, 1995). In the presence of high progesterone, oestradiol fails to increase the presence of less acidic FSH isoforms in the circulation (Wide *et al.*, 1996a). Furthermore, the predominant circulating form of FSH is acidic during the luteal phase of the human menstrual cycle (Padmanabhan *et al.*, 1988a) and

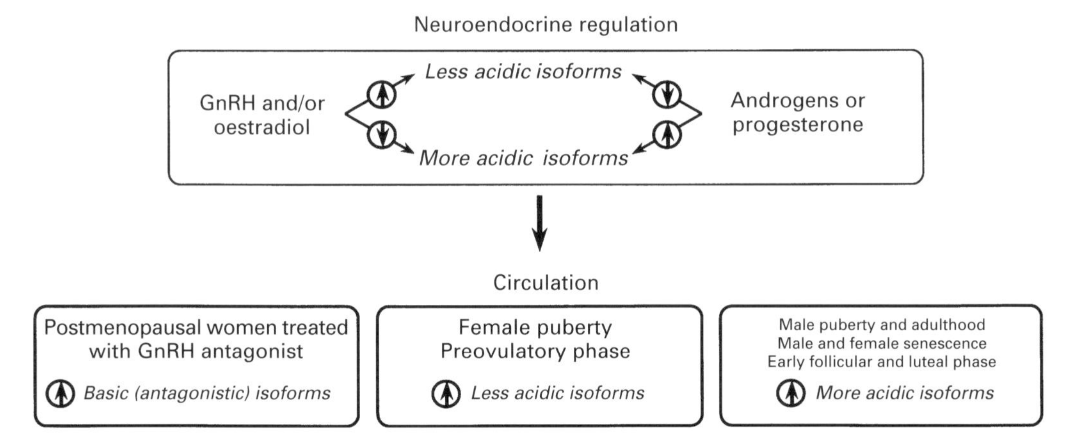

Fig. 2. Schematic diagram summarizing the neuroendocrine regulation of FSH heterogeneity and the nature of circulating FSH isoforms during different physiological states (reviewed in Beitins and Padmanabhan, 1991; Ulloa-Aguirre *et al.*, 1995). GnRH (particularly in ovary-intact models) and oestradiol appear to increase the proportion of less acidic FSH isoforms and decrease the presence of more acidic FSH isoforms. In contrast, androgens and progesterone appear to increase the concentration of more acidic FSH isoforms. The distribution of FSH isoforms also varies with the physiological state. For example, during the onset of puberty in females and the preovulatory phase greater proportions of less acidic FSH isoforms circulate. In contrast, circulating FSH in males, during female senescence, and the early follicular and luteal phases of the reproductive cycle appear predominantly acidic in nature. Interestingly, treatment of post-menopausal women with a GnRH antagonist increases the release of basic FSH isoforms with antagonistic properties (Dahl *et al.*, 1988).

the prepartum period in cattle (Crowe *et al.*, 1998), when concentrations of progesterone and oestradiol are both high.

In all these studies it is difficult to ascertain whether an increase in less acidic FSH isoforms such as those induced by GnRH or oestradiol is due to selective secretion or metabolic alteration. Studies of Harsch *et al.* (1993) suggest that metabolic deglycosylation can occur in circulation. More recent studies characterizing FSH distribution near the site of secretion show that oestradiol selectively increases the secretion of less acidic FSH isoforms (Lee *et al.*, 1998). Evidence is also accumulating to show that oestradiol alters the activity of pituitary glycosyl- and sialotransferases (Dharmesh and Baenziger, 1993; Damian-Matsumura *et al.*, 1998), thereby contributing to glycosylation differences. Overall changes in circulating FSH heterogeneity appear to be the sum effect of secretory changes, metabolic alterations and metabolic clearance.

Biological Significance

When assessing biological significance, it is not sufficient only to show FSH heterogeneity is present and is regulated. It is essential to assess whether such changes are biologically meaningful and to determine whether the naturally occurring FSH isoforms are different functionally. In general, increased release of less acidic FSH isoforms occurs at the onset of puberty and the preovulatory period (Fig. 2). In contrast, more acidic FSH isoforms predominate in males, during senescence in both males and females, and during the early follicular and luteal phases of the oestrous or menstrual cycles (Fig. 2) (reviewed in Beitins and Padmanabhan, 1991; Ulloa-Aguirre *et al.*, 1995).

Changes in FSH heterogeneity may be important in the pubertal process

Experimental induction of puberty in female lambs increases the release of circulating bioactive FSH in pubertal lambs as compared with prepubertal lambs; this increase is not evident when

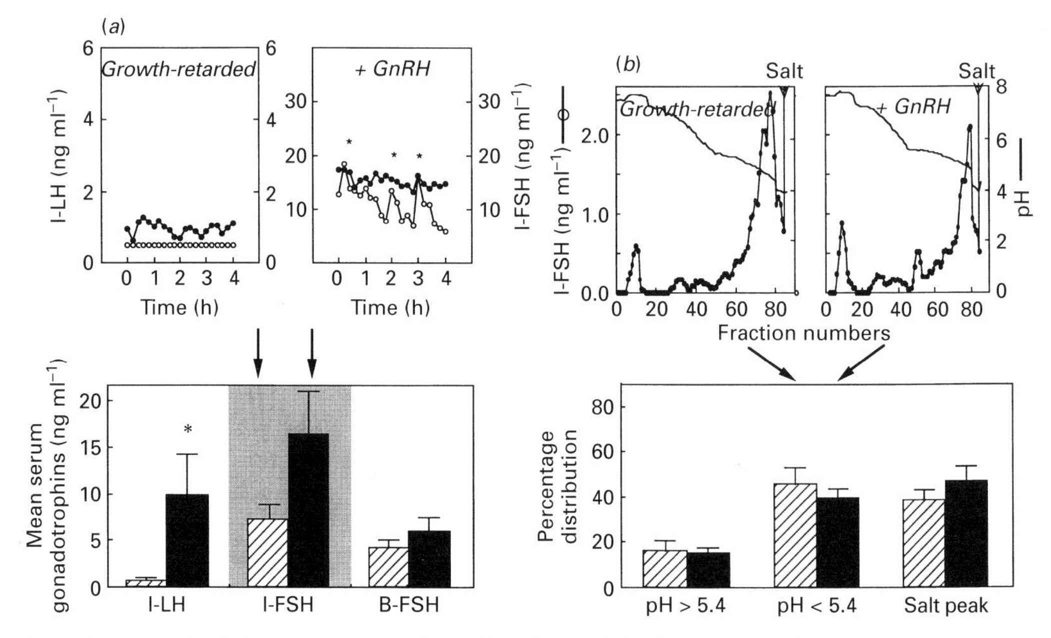

Fig. 3. Changes in circulating concentrations of gonadotrophins and FSH heterogeneity after pulsatile administration of GnRH to ovariectomized nutritionally growth-retarded (hypogonadotrophic lamb) (Hassing *et al.*, 1993). (a) Circulating patterns of immunoreactive (I) FSH in a growth-retarded and GnRH-treated lamb (top) and mean concentrations of I-FSH and bioactive FSH (B-FSH) measured by the Sertoli cell aromatase bioassay (bottom). Circulating patterns and mean concentrations of I-LH are provided for comparison. (b) Distribution pattern of immunoreactive FSH (closed circles) after separation of serum from a representative growth-retarded (left) and GnRH-treated (right) lamb by chromatofocusing (top). Percentage distribution of circulating FSH isoforms that eluted at pH above 5.4, below pH 5.4 and the components bound at the lower limiting pH are shown in the bottom. Values represent mean ± SEM ($n = 4$ for each group).

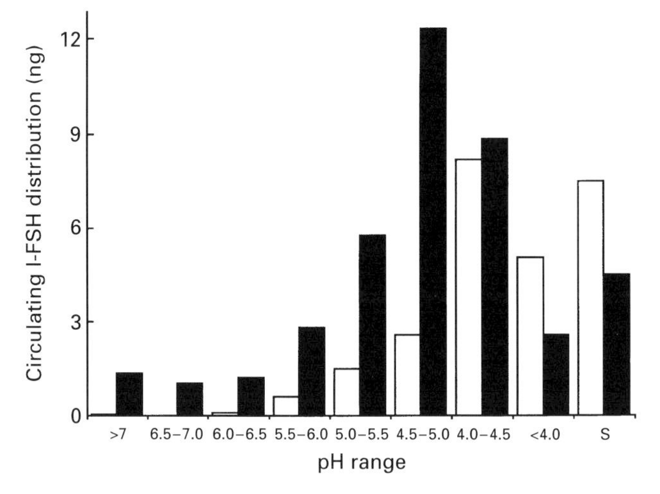

Fig. 4. Distribution pattern of immunoreactive FSH after separation of serum from a representative ovariectomized hypogonadotrophic (□) or oestradiol-treated (■) lamb by chromatofocusing. Ovariectomized hypogonadotrophic lambs were treated with follicular phase concentrations of oestradiol (via implants for 18 h). Note the shift to the less acidic side of FSH distribution in the oestradiol-treated lamb (Padmanabhan *et al.*, 1997b).

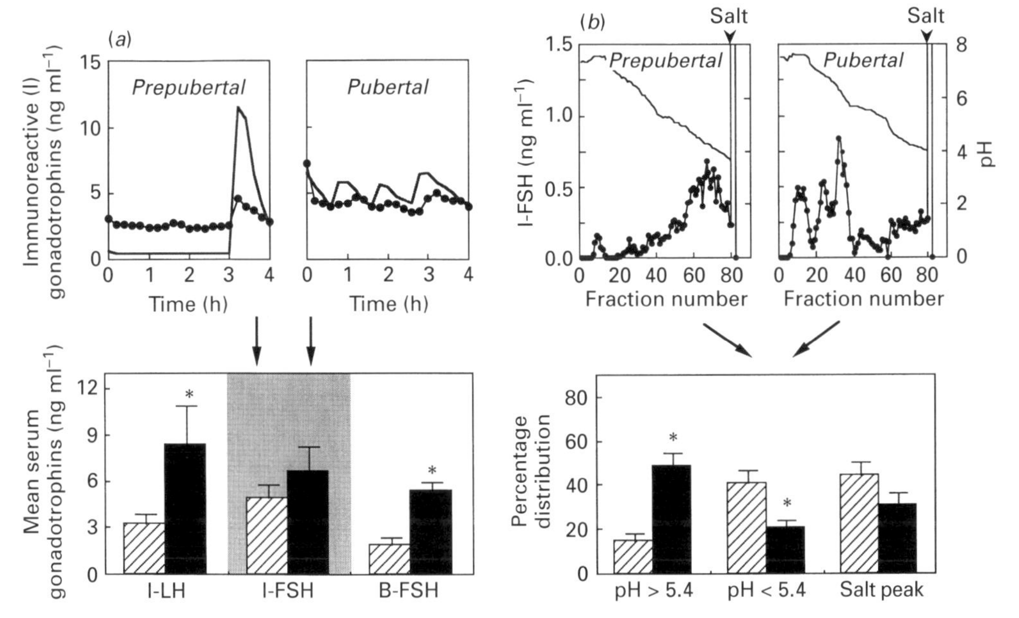

Fig. 5. Changes in circulating concentrations of gonadotrophins and FSH heterogeneity during experimental induction of puberty in female lambs (Padmanabhan *et al.*, 1992). (a) Shown are circulating patterns of immunoreactive (I) FSH in a prepubertal and pubertal lamb (top) and mean concentrations of I-FSH and bioactive FSH (B-FSH) measured by the Sertoli cell aromatase bioassay (SAB) and granulosal cell aromatase bioassay (GAB) in the prepubertal and pubertal lambs (bottom). Circulating patterns and mean concentrations of I-LH are also provided for comparison. Note the selective increase in B-FSH as measured by the SAB and GAB assays at the onset of puberty. (b) Distribution pattern of I-FSH (closed circles) after separation of serum from a representative prepubertal (left) and pubertal (right) lamb by chromatofocusing (top). Note the elution of greater amounts of immunoreactive FSH isoforms in the less acidic region in the pubertal lamb. Percentage distribution of circulating FSH isoforms that eluted at pH above 5.4, below pH 5.4 and the components bound at the lower limiting pH are shown in the bottom. Values represent mean ± SEM (n = 4 for each group).

measured by a radioimmunoassay (Fig. 5) (Padmanabhan *et al.*, 1992). The increased release of bioactive FSH in pubertal lambs is evident whether measured by the Sertoli or the granulosa cell bioassays. However, quantitative differences in bioactive FSH estimates are evident depending on the bioassay used (reviewed in Ulloa-Aguirre *et al.*, 1995). The increased release of B-FSH in pubertal lambs also is accompanied by a change in the distribution pattern of circulating FSH isoforms (Fig. 5) with increased release of less acidic (pI > 5.4) serum FSH isoforms. Considering that pubertal onset is associated with increased gonadal activity, the less acidic FSH isoforms in conjunction with increased secretion of bioactive FSH indicate that the less acidic isoforms may be biologically meaningful and have the potential to provide a potent and acute signal to the ovary. Such changes may be the consequence of increased oestradiol secretion from the ovary.

Changes in FSH heterogeneity may be important in the ovulatory process

Detailed chromatographic studies characterizing the changes in circulating FSH heterogeneity during the ovulatory cycles have been addressed in humans (Padmanabhan *et al.*, 1988a; Wide and Bakos, 1993; Zambrano *et al.*, 1995) and cattle (Cooke *et al.*, 1997). These studies reveal that increases in less acidic circulating FSH isoforms occur during the preovulatory period. Studies using concanavalin chromatography show that the complexity of the oligosaccharide chains are also altered during different phases of the human menstrual cycle and that less complex FSH isoforms are found during mid-cycle than in early follicular or late luteal periods (Anobile *et al.*, 1998). The marked shift in FSH distribution profile favouring less acidic isoforms of FSH coincides with the timing of preovulatory follicular development.

FSH isoforms may differ in their functional attributes

Before attributing physiological significance to heterogeneity, it is essential to determine whether the various isoforms differ in their functionalities. To be biologically meaningful, changes in FSH heterogeneity need to be of sufficient magnitude to alter the net potency of the hormone or have various functions. Estimations of the biopotencies of FSH isoforms on the basis of FSH mass predict a 5–8-fold difference in radio-receptor and biopotency of purified human pituitary-derived FSH isoforms as well as a fourfold discrepancy between radio-receptor and *in vitro* biopotencies (Burgon *et al.*, 1993). Considering that subtle increases in immunoreactive FSH can induce ovarian responses (Ben-Rafael *et al.*, 1995), potency differences such as those reported by Burgon *et al.* (1993) and the magnitude of changes such as those seen during mid-cycle (Padmanabhan *et al.*, 1988a) and puberty (Padmanabhan *et al.*, 1992) have the potential to have meaningful biological consequences. In this context it is of interest that pulsatile administration of GnRH leads to fast clearing FSH signals in patients with idiopathic hypogonadotropic–hypogonadism (Padmanabhan *et al.*, 1988b) and a less negatively charged (less acidic) form of FSH in children with pubertal disorders (Wide *et al.*, 1996b). Similarly treatment with a GnRH antagonist leads to production of a basic form of FSH that is capable of antagonizing FSH action (Dahl *et al.*, 1988).

Less acidic (pI 5.0–5.6) isoforms of human recombinant FSH have been shown to be more potent than those in the mid- (pI 3.6–4.6) and acidic (pI 4.5–5.0) ranges in inducing mouse follicular development *in vitro* (Vitt *et al.*, 1997). In addition to inducing follicles of large final size, the less acidic FSH isoforms, even at the lowest concentration tested (25 mIU ml^{-1}), induced antral formation in 70% of follicles and oestrogen production by day 2 of exposure to FSH. A similar degree of follicle development was achieved only with 100 mIU ml^{-1} of the mid- pI value and 500 mIU ml^{-1} of acidic FSH isoforms. Furthermore, oestradiol secretion was evident only after 4 days of culture even with the 500 mIU acidic FSH ml^{-1}. These studies indicate that the physicochemical nature of FSH isoforms may lead to quantitative and qualitative differences in ovarian function.

An exciting possibility is that the different FSH isoforms encode different functions. On the basis of the multitude of functions FSH mediates at the gonads, involvement of different second messenger systems and potential for receptor cross-talk (Fig. 6), the signal transduction pathway of FSH isoforms may differ and culminate in altered responses. There is precedence with other glycoprotein hormones for such altered signal transduction cascades. FSH and thyroid-stimulating hormone (TSH) have been shown to bind each others receptors (Dobozy *et al.*, 1985). Interaction of TSH with human FSH receptor is a possible mechanism by which some children with juvenile hypothyroidism exhibit unexplained precocious puberty (Anasti *et al.*, 1995). Asialo human chorionic gonadotrophins (hCGs) have been shown to have higher TSH-like activity in human thyroid follicles (Yamazaki *et al.*, 1995). In this context, it is of interest that there is a positive correlation between serum free thyroxine and asialo hCG in patients with gestational thyrotoxicosis (Tsuruta *et al.*, 1995). Similarly, different isoforms of TSH have been shown to have different biological activities: the basic isoforms promoting iodide and thymidine uptake and acidic isoforms increasing intracellular cAMP (Pickles *et al.*, 1992). Certain LH isoform(s) have also been shown to possess renotrophic activity but weak steroidogenic potential (Nomura *et al.*, 1988). Studies with deglycosylated and native mixes of ovine FSH also show isoform specific cAMP and oestradiol responses (Fig. 1) (Padmanabhan, 1995). These observations provide support for the possibility that endocrine-induced changes in FSH heterogeneity can have functional consequences.

Conclusion

Our understanding of mechanisms controlling the pubertal process and ovulatory cyclicity may be flawed, in as much as the basis for such concepts is derived mainly from immunological

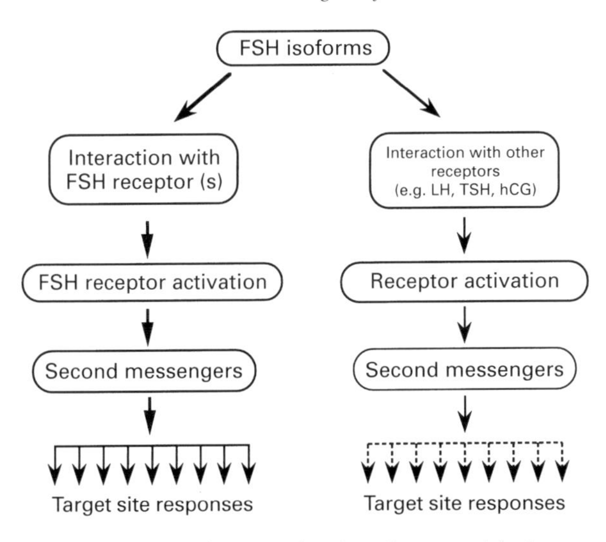

Fig. 6. Schematic diagram showing the potential sites at which the effector mechanisms may vary for the different FSH isoforms leading ultimately to differential target site responses. Different FSH isoforms may show differential affinity to a given FSH receptor, or selectively bind variant FSH receptor population or even structurally unrelated receptors (receptor cross-talk). Changes in receptor binding in turn can lead to changes in signalling cascades involving different second messenger systems and culminate in differential target cell responses.

characterizations of circulating patterns of gonadotrophic hormones that may not recognize the entire repertoire of gonadotrophin variants present in circulation. Mounting evidence indicates that, through an interaction with the hypothalamus and the gonad, the anterior pituitary gland can secrete different types of FSH that vary in biological potency and circulatory half-life. This opens up very interesting questions such as whether target cells differentially respond to the pattern of the imposed FSH signals, much as the T cell is known to differentially respond to small changes in ligand (Marx, 1995). In addition, the molecular modifications responsible for these observed changes in the distribution of FSH isoforms have the potential to lead to changes in affinity for classical FSH receptors, cross-reactivity with non-FSH receptors as well as with any other FSH receptor forms that might have been induced by the same endocrine conditions that led to the changes in FSH. From these perspectives, changes in FSH heterogeneity have the potential to provide an exquisitely fine-tuned mechanism to control gonadal function.

Therefore, it is a matter of great importance: (1) to determine details of the linkage between changes in isoform distribution and modifications in the biological attributes of the FSH signal delivered to the gonad during different physiological and pathological states; such efforts need to use well characterized assay systems and reference preparations that are reflective of the repertoire of FSH isoforms present in biological fluids, and (2) to reveal whether the various isoforms differentially act to initiate distinct functions in target cells. Only when the structure and importance of the various naturally occurring FSH isoforms are determined will it be possible for us to obtain a complete understanding of the mechanisms regulating reproductive processes. If heterogeneity proves to be biologically important, it should be possible to design FSH isoforms for desired functions and to use them too regulate fertility.

This work was supported by USPHS grants HD 23812.

References

Anasti JN, Flack MR, Froehlich J, Nelson LM and Nisula BC (1995) A potential novel mechanism for precocious puberty in juvenile hypothyroidism *Journal of Clinical Endocrinology and Metabolism* **80** 276–279

Anobile CJ, Talbot JA, McCann JA, Padmanabhan V and Robertson WR (1998) Glycoform composition of serum gonadotrophins through the normal menstrual cycle and in post-menopausal state *Human Reproduction* **4** 641–647

Beitins IZ and Padmanabhan V (1991) Bioactivity of gonadotropins *Endocrinology and Metabolism Clinics of North America* **20** 85–120

Ben-Rafael Z, Levy T and Schoemaker J (1995) Pharmacokinetics of follicle-stimulating hormone: clinical significance *Fertility and Sterility* **63** 689–700

Blum WFP and Gupta D (1980) Age and sex-dependent nature of the polymorphic forms of rat pituitary FSH: the role of glycosylation *Neuroendocrinology Letters* **2** 357–365

Burgon PG, Robertson DM, Stanton PG and Hearn MTW (1993) Immunological activities of highly purified isoforms of human FSH correlate with *in vitro* bioactivities *Journal of Endocrinology* **139** 511–518

Chappel SC (1995) Heterogeneity of follicle stimulating hormone: control and physiological function *Human Reproduction Update* **1** 479–487

Chappel SC, Ulloa-Aguirre A and Coutifaris C (1983) Biosynthesis and secretion of follicle-stimulating hormone *Endocrine Reviews* **4** 179–211

Christine-Maitre S, Taylor AE, Khoury RH, Hall JE, Martin KA, Smith PC, Albanese C, Jameson JL, Crowley WF, Jr and Sluss PM (1996) Homologous *in vitro* bioassay for follicle-stimulating hormone (FSH) reveals increased FSH biological signal during the mid-phase to late-luteal phase of the human menstrual cycle *Journal of Clinical Endocrinology and Metabolism* **81** 2080–2088

Cooke DJ, Crowe MA and Roche JF (1997) Circulating FSH isoform patterns during recurrent increases in FSH throughout the oestrous cycle of heifers *Journal of Reproduction and Fertility* **110** 339–345

Crowe MA, Padmanabhan V, Mihm M, Beitins IZ and Roche JF (1998) Resumption of follicular waves in beef cows is not associated with peri-parturient changes in FSH heterogeneity despite major changes in steroid and gonadotropin concentrations *Biology of Reproduction* **58** 1445–1450

Dahl KD, Bicsak TA and Hsueh AJW (1988) Naturally occurring antihormones: secretion of FSH antagonists by women treated with a GnRH analog *Science* **239** 72–74

Damián-Matsumura P, Zaga V, Sánchez-Hernández C, Maldonado A, Timossi C and Ulloa-Aguire A (1998) The changes in α-2,3, sialyltransferase mRNA levels during the rat estrous cycle and after castration correlate with variations in charge distribution of intrapituitary follicle-stimulating hormone (FSH) *Program and Abstracts of the 80th Annual Meeting of the Endocrine Society, New Orleans* p90 (Abstract OR 28-6)

Dharmesh SM and Baenziger JU (1993) Estrogen modulates expression of the glycosyltransferases that synthesize sulfated oligosaccharides on lutropin *Proceedings of the National Academy of Sciences USA* **90** 11127–11131

Ding Y-Q, Ranta T, Nikkanen I and Huhtaniemi I (1991) Discordant levels of serum bioactive LH in man as measured in different *in-vitro* bioassay systems using rat and mouse interstitial cells and human granulosa–luteal cells *Journal of Endocrinology* **128** 131–137

Dobozy O, Csaba G, Hetenyi G and Shahin M (1985) Investigation of gonadotropin–thyrotropin overlapping and hormonal imprinting in the rat testis *Acta Physiologica Hungarica* **66** 169–175

Flores JA, Veldhuis JD and Leong DA (1990) Follicle-stimulating hormone evokes an increase in intracellular free calcium ion concentrations in single ovarian (granulosa) cells *Endocrinology* **127** 3172–3179

Gudermann T, Brockman H, Simoni M, Gromoll J and Nieschlag E (1994) *In vitro* bioassay for serum follicle-stimulating hormone (FSH) based on L cells transfected with recombinant rat FSH receptor: validation of a model system *Endocrinology* **135** 2204–2213

Harsch IA, Simoni M and Nieschlag E (1993) Molecular heterogeneity of serum follicle-stimulating hormone in hypogonadal patients before and during androgen replacement therapy and in normal men *Clinical Endocrinology* **39** 173–180

Hassing JM, Kletter GB, I'Anson H, Woods RI, Beitins IZ, Foster DL and Padmanabhan V (1993) Pulsatile administration of gonadotropin-releasing hormone does not alter the follicle-stimulating hormone (FSH) isoform distribution pattern of pituitary or circulating FSH in nutritionally growth-restricted ovariectomized lambs *Endocrinology* **132** 1527–1536

Jaakkola T, Ding YQ, Kellokumpu-Lehtinen P, Valavaara R, Martikainen H, Tapanainen J, Ronnberg L and Huhtaniemi I (1990) The ratios of serum bioactive/immunoreactive luteinizing hormone and follicle-stimulating hormone in various clinical conditions with increased and decreased gonadotropin secretion: reevaluation by a highly sensitive immunometric assay *Journal of Clinical Endocrinology and Metabolism* **70** 1496–1505

Jia XC, Kessel B, Yen SSC, Tucker EM and Hsueh AJW (1986) Serum bioactive follicle-stimulating hormone during the human menstrual cycle and in hyper- and hypogonadotrophic states: application of a sensitive granulosa cell aromatase bioassay *Journal of Clinical Endocrinology and Metabolism* **62** 1243–1249

Lee JS, Manning JM, Foster DL and Padmanabhan V (1998) Estrogen increases the *secretion* of less-acidic FSH isoforms in ovariectomized prepubertal and peripubertal lambs *Program and Abstracts of the 80th Annual Meeting of the Endocrine Society,* New Orleans p. 348 (Abstract P2–474)

Marx J (1995) The T cell receptor begins to reveal its many facets *Science* **267** 459–460

Morell AG, Gregoriadis G and Scheinberg IH (1971) The role of sialic acid in determining the survival of glycoproteins in circulation *Journal of Biological Chemistry* **246** 1461–1467

Nomura K, Tsunasawa S, Ohmura K, Sakiyama F and Shizume K (1988) Renotropic activity in ovine luteinizing hormone isoform(s) *Endocrinology* **123** 700–712

Padmanabhan V (1995) Neuroendocrine control and physiologic relevance of FSH heterogeneity *Journal of Reproduction and Fertility Abstract Series* **15** Abstract S3

Padmanabhan V, Chappel SC and Beitins IZ (1987) An improved *in vitro* bioassay for follicle-stimulating hormone (FSH): suitable for measurement of FSH in unextracted human serum *Endocrinology* **121** 1089–1098

Padmanabhan V, Lang LL, Sonstein J, Kelch RP and Beitins IZ

(1988a) Modulation of serum follicle-stimulating hormone bioactivity and isoform distribution by estrogenic steroids in normal women and in gonadal dysgenesis *Journal of Clinical Endocrinology and Metabolism* **67** 465–473

Padmanabhan V, Kelch RP, Sonstein J, Foster CM and Beitins IZ (1988b) Bioactive follicle-stimulating hormone responses to intravenous gonadotropin-releasing hormone in boys with idiopathic hypogonadotropic hypogonadism *Journal of Clinical Endocrinology and Metabolism* **67** 793–800

Padmanabhan V, Mieher CD, Borondy M, I'Anson H, Wood RI, Landefeld TD, Foster DL and Beitins IZ (1992) Circulating bioactive follicle-stimulating hormone and less acidic follicle-stimulating hormone isoforms increase during experimental induction of puberty in the female lamb *Endocrinology* **131** 213–220

Padmanabhan V, McFadden K, Mauger DT, Karsch FJ and Midgley AR, Jr (1997a) Neuroendocrine control of follicle-stimulating hormone (FSH) secretion: I. Direct evidence for separate episodic and basal components of FSH secretion *Endocrinology* **138** 424–432

Padmanabhan V, I'anson H, Foster DL and Beitins IZ (1997b) Estradiol increases the release of less acidic FSH isoforms in nutritionally growth-retarded lambs *Program and Abstracts of the 79th Annual Meeting of the Endocrine Society* Abstract P3–336

Pettersson KSI and Soderhölm JR-M (1991) Individual differences in lutropin immunoreactivity revealed by monoclonal antibodies *Clinical Chemistry* **37** 333–340

Phillips DJ and Wide L (1994) Serum gonadotropin isoforms become more basic after an exogenous challenge of gonadotropin-releasing hormone in children undergoing pubertal development *Journal of Clinical Endocrinology and Metabolism* **79** 814–819

Pickles AJ, Peers N, Robertson WR and Lambert A (1992) Different isoforms of human pituitary thyroid-stimulating hormone have different relative biological activities *Journal of Molecular Endocrinology* **9** 251–256

Robertson DM, Foulds LM, Fry RC, Cummins JT and Clarke I (1991) Circulating half-lives of follicle-stimulating hormone and luteinizing hormone in pituitary extracts and isoform fractions of ovariectomized and intact ewes *Endocrinology* **129** 1805–1813

Simoni M, Khan SA and Nieschlag E (1991) Serum bioactive follicle-stimulating hormone-like activity in human pregnancy is a methodological artifact *Journal of Clinical Endocrinology and Metabolism* **73** 1118–1122

Simoni M, Jockenhövel F and Nieschlag E (1993a) Biological and immunological properties of the international standard for FSH 83/575: isoelectrofocusing profile and comparison with other FSH preparations *Acta Endocrinologica* **128** 281–288

Simoni M, Weinbauer GF and Nieschlag E (1993b) Molecular composition of two different batches of urofollitropin: analysis by immunofluorimetric assay, radioligand receptor assay and *in vitro* bioassay *Journal of Endocrinological Investigations* **16** 21–27

Simoni M, Jockenhövel F and Nieschlag E (1994) Polymorphism of human pituitary FSH: analysis of immunoreactivity and *in vitro* bioactivity of different molecular species *Journal of Endocrinology* **141** 359–367

Simoni M, Peters J, Behre HM, Kliesch S, Leifke E and Nieschlag E (1996) Effects of gonadotropin-releasing hormone on bioactivity of follicle-stimulating hormone (FSH) and microstructure of FSH, luteinizing hormone and sex-hormone binding globulin in a testosterone-based contraceptive trial: evaluation of responders and non-responders *European Journal of Endocrinology* **135** 433–439

Spiegel AM, Shenker AM and Weinstein LS (1992) Receptor–effector coupling by G proteins: implications for normal and abnormal signal transduction *Endocrine Reviews* **13**: 536–565

Stanton PG, Burgon PG, Hearn MTW and Robertson DM (1996) Structural and functional characterization of hFSH and hLH isoforms *Molecular and Cellular Endocrinology* **125** 133–141

Stumpf TT, Roberson MS, Wolfe MW, Zalesky DD, Cupp AS, Werth LA, Kojima N, Hejl K, Kittok RJ, Grotjan HE and Kinder JE (1992) A similar distribution of gonadotropin isohormones is maintained in the pituitary throughout sexual maturation in the heifer *Biology of Reproduction* **46** 442–450

Timossi CM, de Tomasi JB, Zambrano E, González R and Ulloa-Aguirre A (1998) A naturally occurring basically charged human follicle-stimulating hormone (FSH) variant inhibits FSH-induced androgen aromatization and tissue-type plasminogen activator enzyme activity *in vitro. Neuroendocrinology* **67** 153–163

Tsuruta E, Tada H, Tamaki H, Kashiwai T, Asahi K, Takeoka K, Mitsuda N and Amino N (1995) Pathogenic role of asialo human chorionic gonadotropin in gestational thyrotoxicosis *Journal of Clinical Endocrinology and Metabolism* **80** 350–355

Ulloa-Aguirre A, Midgley AR, Jr, Beitins IZ and Padmanabhan V (1995) Follicle-stimulating isohormones: characterization and physiological relevance *Endocrine Reviews* **16** 765–787

Vitt UA, Kloosterbocr HJ, Rose UM, Kiesel PS, Bete A and Nayudu PL (1997) Differential effects of 3 hFSH isoforms on ovarian follicle development *in vitro. Journal of Reproduction and Fertility Abstract Series* **20** Abstract 22

Wide L and Bakos O (1993) More basic forms of both human follicle-stimulating hormone and luteinizing hormone in serum at midcycle compared with the follicular and luteal phase *Journal of Clinical Endocrinology and Metabolism* **76** 885–889

Wide L and Naessén T (1994) 17β-Oestradiol counteracts the formation of the more acidic isoforms of follicle-stimulating hormone and luteinizing hormone after menopause *Clinical Endocrinology (Oxford)* **40** 783–789

Wide L, Naessén T, Eriksson K and Rune C (1996a) Time-related effects of progestogen on the isoforms of serum gonadotrophins in 17β-oestradiol treated post-menopausal women *Clinical Endocrinology* **44** 651–658

Wide L, Albertsson-wikland K and Phillips DJ (1996b) More basic isoforms of serum gonadotropins during gonadotropin-releasing hormone therapy in pubertal children *Journal of Clinical Endocrinology and Metabolism* **81** 216–221

Yamazaki K, Sato K, Shizume K, Kanaji Y, Ito Y, Obara T, Nakagava T, Koizumi T and Nishimura R (1995) Potent thyrotropic activity of human chorionic gonadotropin variants in terms of ^{125}I incorporation and de novo synthesized thyroid hormone release in human thyroid follicles *Journal of Clinical Endocrinology and Metabolism* **80** 473–479

Zambrano E, Olivares A, Mendez JP, Guerrero L, Diaz-Cueto L, Veldhuis JD and Ulloa-Aguirre A (1995) Dynamics of basal and gonadotropin-releasing hormone-releasable serum follicle stimulating hormone charge isoform distribution throughout the human menstrual cycle *Journal of Clinical Endocrinology and Metabolism* **80** 1647–1656

Journal of Reproduction and Fertility Supplement **54**, 101–114

Endocrine and reproductive responses of male and female cattle to agonists of gonadotrophin-releasing hormone

M. J. D'Occhio and W. J. Aspden

Animal Sciences and Production Group, Centre for Primary Industries Research, Central Queensland University, Rockhampton, Queensland 4702, Australia

The pituitary response in cattle to treatment with GnRH agonist has two phases. In the acute phase secretion of LH is increased, while the chronic phase is characterized by a downregulation of GnRH receptors and insensitivity of gonadotrophs to natural sequence GnRH. After long-term treatment with GnRH agonist, cattle do not have pulsatile secretion of LH but maintain basal LH. This is associated with reduced pituitary contents of LH, LH mRNA, FSH and FSH mRNA. Long-term treatment of bulls with GnRH agonist results in an increase in testicular LH receptors and high plasma testosterone. Heifers treated with a GnRH agonist from early in the oestrous cycle develop a larger corpus luteum and secrete more progesterone. Increased steroidogenesis is reflected in increased steroid acute regulatory (StAR) protein and steroidogenic enzymes in the testes and corpus luteum. GnRH agonists have potential as novel strategies for reproductive management in cattle. A GnRH agonist bioimplant was recently used to block the LH surge after FSH stimulation of follicle growth in heifers. Ovulation was induced by injection of LH, and heifers were inseminated relative to the LH injection. This GnRH agonist–LH protocol provides a model for studying the gonadotrophin requirements for follicular growth and oocyte maturation in cattle, and will enable controlled *in vivo* maturation of oocytes before recovery for *in vitro* procedures.

Introduction

Gonadotrophin releasing hormone (GnRH) is a ten amino acid neuropeptide that initiates the cascade of reproductive hormones in mammals. Major reproductive events including the onset of puberty and ovulation rely on increases in GnRH secretion. Increased release of GnRH into hypothalamo–hypophyseal portal vessels results in greater secretion of LH and FSH and enhanced gonadal function. The importance of GnRH for reproductive function has focused attention on this neuropeptide for reproductive therapies in cattle (Thatcher *et al.*, 1993).

Agonists of GnRH were developed initially to treat hypogonadism resulting from insufficient endogenous secretion of GnRH. Typical structural features of GnRH agonists that distinguish them from natural sequence GnRH are substitution of glycine at position 6 of the peptide with a D-amino acid (e.g. D-tryptophan) and removal of glycine from the amino terminus. Substitution with a D-amino acid at position 6 increases the half-life of GnRH agonists in circulation, and removal of the amino terminal glycine increases affinity for the GnRH receptor (Karten and Rivier, 1986).

It was recognized early in the development of GnRH agonists that the reproductive response was dependent on dose of agonist and duration of treatment. The acute response, irrespective of dose, was characterized by increased gonadotrophin secretion (Chenault *et al.*, 1990). Chronic treatment with relatively high doses of GnRH agonist resulted in suppressed gonadotrophin secretion (Lahlou *et al.*, 1987). The latter was due to downregulation of GnRH receptors on

Table 1. Anterior pituitary contents of LH and FSH (ng mg^{-1} anterior pituitary) in intact bulls and castrated bulls (steers) treated with GnRH agonist

	LH		FSH	
	Control	GnRH agonist	Control	GnRH agonist
Bulls[a] ($n = 4$)	≈450	≈25 ($P < 0.01$)	≈950	≈50 ($P < 0.05$)
Bulls[b] ($n = 4$–5)	553	33 ($P < 0.001$)	–	–
Steers[c] ($n = 8$–11)	228	26 ($P < 0.001$)	1515	1390 ($P > 0.05$)

Values are means.
[a]Melson *et al.*, 1986; [b]Aspden *et al.*, 1997a; [c]Aspden *et al.*, 1996.

gonadotrophs (Hazum and Conn, 1988) and an uncoupling (desensitisation) of second messenger systems within these cells (Huckle and Conn, 1988; Hawes *et al.*, 1992).

This review provides an overview of the endocrine and molecular features of the responses of the anterior pituitary and gonads in cattle to treatment with GnRH agonists. Applications for GnRH agonists in the reproductive management of cattle also are considered.

Neuroendocrine Response to GnRH Agonist Treatment

Release of endogenous GnRH into hypothalamo–hypophyseal portal vessels in bull calves occurred in a pulsatile manner (Rodriguez and Wise, 1989). From this observation in male cattle and general patterns of LH release in female cattle, it can be assumed that GnRH release in cattle is pulsatile. In male sheep, treatment with GnRH agonist did not influence GnRH secretory patterns (Caraty *et al.*, 1990). It would appear, therefore, that GnRH agonist treatment does not influence the activity of hypothalamic GnRH secreting neurones. This conclusion is consistent with the major known effects of GnRH agonists at the anterior pituitary gland, which are discussed below.

Anterior Pituitary Response to GnRH Agonist Treatment

The major direct effects of GnRH agonists within the reproductive axis appear to be actions at pituitary gonadotrophs (Hazum and Conn, 1988; Huckle and Conn, 1988). Bulls treated with the GnRH agonist nafarelin had reduced pituitary GnRH receptors after treatment for 15 days (Melson *et al.*, 1986). This was consistent with the classical downregulation of GnRH receptors induced by GnRH agonists (Hazum and Conn, 1988). Studies on second messenger uncoupling (desensitization) (Huckle and Conn, 1988; Hawes *et al.*, 1992) have not been conducted in cattle. However, these studies should be considered, as bulls and heifers treated with GnRH agonists maintain basal secretion of LH (discussed below). This result differs from the significant reduction in circulating LH seen in most species during agonist treatment (D'Occhio and Aspden, 1996).

Pituitary contents of LH (Table 1) and LH mRNA (Table 2) were reduced in intact bulls treated with GnRH agonist. This was associated with a lack of pulsatile secretion of LH (Melson *et al.*, 1986) but maintenance of basal LH secretion (D'Occhio and Aspden, 1996; Aspden *et al.*, 1997a,b; Aspden *et al.*, 1998). In an intensive study of LH secretory patterns in bulls treated with the GnRH agonist nafarelin, it was confirmed that basal LH secretion in bulls is slightly but significantly increased during agonist treatment (Jimenez-Severiano *et* al., 1998). Similar increases in basal plasma LH were observed in heifers treated with leuprolide (Evans and Rawlings, 1994) and buserelin (Gong *et al.*, 1995).

An effect of GnRH agonist on pituitary LH in cattle was demonstrated clearly using castrated bulls (Aspden *et al.*, 1996), which have an increased secretion of LH and provide a useful experimental model to demonstrate any decreases in circulating LH (Fig. 1), and pituitary contents of LH and LH mRNA, which might occur during treatment with deslorelin. Although plasma LH

Table 2. Anterior pituitary contents of LH mRNA and FSH mRNA (mean arbitrary relative units) in intact bulls and castrated bulls (steers) treated with the GnRH agonist deslorelin

	LH β-subunit mRNA		FSH β-subunit mRNA	
	Control	GnRH agonist	Control	GnRH agonist
Bulls[a] (n = 4–5)	0.65	0.22 (P = 0.003)	–	–
Steers[b] (n = 8–11)	1.56	0.08 (P < 0.001)	1.01	0.34 (P < 0.001)

Values are means.
[a]Aspden *et al.*, 1997a; [b]Aspden *et al.*, 1996.

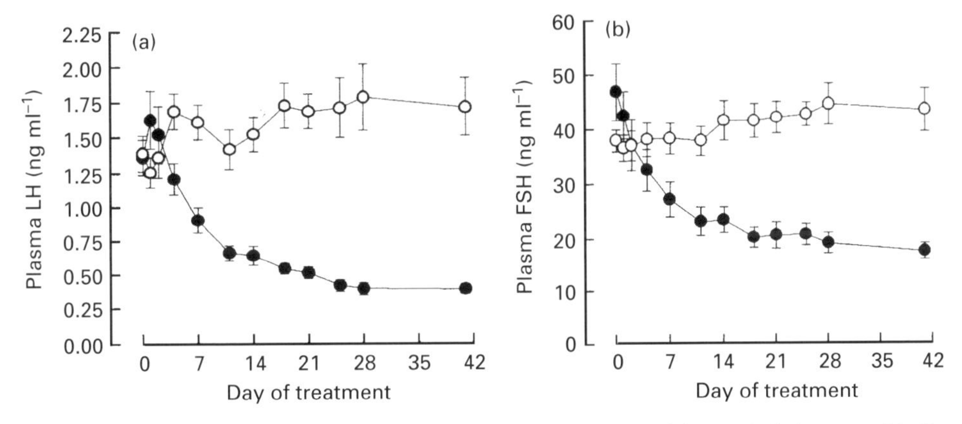

Fig. 1. Longitudinal profiles of plasma (a) LH and (b) FSH in control (open circles) castrated bulls (steers) and steers treated with the GnRH agonist deslorelin (solid circles) commencing on day 0. Results are means ± SEM (n = 8) (Aspden *et al.*, 1996).

was reduced in castrated bulls treated with agonist, basal secretion was maintained, similar to the finding in intact bulls (Fig. 1). The understanding that has emerged from GnRH agonist studies in cattle, therefore, is that pulsatile secretion of LH is blocked, which is consistent with GnRH receptor downregulation, but basal LH secretion is tonically increased (Evans and Rawlings, 1994; Jimenez-Severiano *et al.*, 1998).

The mechanism(s) that promotes increased basal secretion of LH in cattle treated with GnRH agonist has not been defined. The basal secretion could be constitutive and does not require typical second messenger pathways (Huckle and Conn, 1988). Alternatively, GnRH agonists may stimulate second messenger pathways in cattle to maintain increased basal LH secretion. The latter suggestion might be considered inconsistent with the downregulation of GnRH receptors in cattle during agonist treatment. Endogenous GnRH is not required for continued secretion of LH in bulls treated with GnRH agonist, as bulls treated with agonist, and simultaneously actively immunized against GnRH, maintained basal secretion of LH (Aspden *et al.*, 1997b).

Treatment of intact bulls with the GnRH agonist nafarelin caused a decrease in pituitary FSH content (Table 1). In castrated bulls, treatment with deslorelin was associated with a decrease in FSH mRNA (Table 2) but there was no apparent change in pituitary FSH content (Table 1). The latter observation was inconsistent with the reduction in plasma FSH in castrated bulls during treatment with deslorelin (Fig. 1).

The response of the pituitary to natural sequence GnRH is reinstated over several weeks after treatment with GnRH agonist in bulls (Bergfeld *et al.*, 1996a) and heifers (Bergfeld *et al.*, 1996b). It is not known whether the gradual recovery of pituitary responsiveness to GnRH is related to a gradual replenishment of GnRH receptors on gonadotroph cells, or to a gradual re-establishment of second

Table 3. Volume fractions and absolute volume (ml) per testis of seminiferous epithelium, lumen and interstitium for control bulls and bulls treated with deslorelin.

	Volume fraction			Absolute volume per testis		
	Seminiferous epithelium	Lumen	Interstitium	Seminiferous epithelium	Lumen	Interstitium
Control	0.693 ± 0.024[a]	0.116 ± 0.018[a]	0.192 ± 0.015[a]	119.5 ± 5.9[a]	19.8 ± 7.0[a]	33.3 ± 3.6[a]
Deslorelin	0.707 ± 0.014[a]	0.110 ± 0.013[a]	0.183 ± 0.007[a]	168.0 ± 12.5[b]	27.1 ± 4.1[a]	43.4 ± 3.4[a]

Results are means ± SEM (n = 6).
[a, b]Means within columns without a common superscript are significantly different ($P < 0.01$).
[a]Means within columns with a common superscript are not significantly different ($P > 0.05$).

messenger pathways within gonadotroph cells (Gorospe and Conn, 1988). Consistent with a gradual return to normal pituitary function after GnRH agonist treatment, post-pubertal heifers treated with a deslorelin bioimplant for 10, 28 or 56 days ovulated approximately 20 days after treatment (D'Occhio and Kinder, 1995; D'Occhio *et al.*, 1996). Heifers infused with buserelin for 48 days displayed oestrus 8–11 days after treatment, but a preovulatory surge release of LH had not occurred 22 days after treatment (Gong *et al.*, 1996). In young bulls, treatment with leuprolide from 6 to 20 weeks of age delayed the occurrence of a prepubertal rise in plasma LH and testosterone by 4 weeks, from 20 weeks to 24 weeks (Chandolia *et al.*, 1997).

Gonadal Responses to GnRH Agonist Treatment

Male cattle

Testosterone secretion, LH receptors and testis growth. The outstanding feature of the response in bulls to treatment with GnRH agonist is an increase in the secretion of testosterone which is maintained for the duration of treatment, irrespective of the dose of agonist (D'Occhio and Aspden, 1996). Increased testosterone secretion in bulls during agonist treatment was consistently demonstrated for deslorelin (Bergfeld *et al.*, 1996a; D'Occhio and Aspden, 1996; Aspden *et al.*, 1997a,b; Aspden *et al.*, 1998), nafarelin (Melson *et al.*, 1986), buserelin (Rechenberg *et al.*, 1986) and leuprolide (Ronayne *et al.*, 1993). However, relatively young bulls treated with leuprolide had reduced plasma concentrations of testosterone (Chandolia *et al.*, 1997).

In a long-term study, 20-month-old Brahman and Brahman × Hereford Shorthorn bulls treated with deslorelin maintained high plasma testosterone for more than 120 days (D'Occhio and Aspden, 1996). Over this period, bulls receiving GnRH agonist showed a faster rate of testis growth and at the end of treatment had larger testes (Aspden *et al.*, 1998). Increased testis size was associated with greater absolute volume per testis of seminiferous epithelium (Table 3). The numerical density of round spermatids in the testis was not increased in bulls treated with GnRH agonist, but the absolute number of round spermatids per testis was greater in bulls treated with agonist ($23.5 \pm 2.5 \times 10^9$) than in control bulls ($16.2 \pm 1.4 \times 10^9$). Increases in rate of testis growth in bulls during GnRH agonist treatment were not observed consistently in shorter-term studies (Melson *et al.*, 1986; Ronayne *et al.*, 1993). Notwithstanding the latter studies, chronic treatment with GnRH agonist may provide practical applications to influence testis growth and function in bulls, as discussed below.

Bulls treated with GnRH agonist had increased numbers of testicular LH receptors which might explain, in part, increased testosterone secretion (Melson *et al.*, 1986). However, in *in vitro* studies using rat Leydig cells, only approximately 1% LH receptor occupancy was required for a maximal testosterone response to gonadotrophin stimulation (Mendelson *et al.*, 1975). It is possible, therefore, that increased testosterone secretion in bulls treated with GnRH agonist is due to a combination of

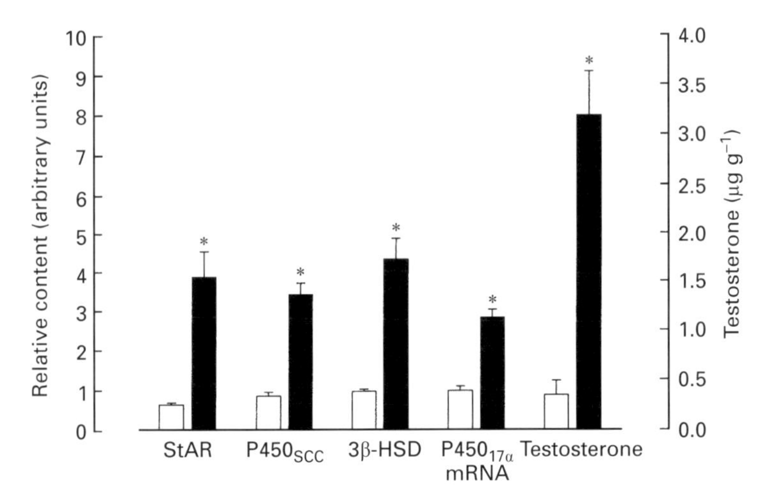

Fig. 2. Relative contents of testicular steroid acute regulatory (StAR) protein, P450scc, 3β-HSD, $P450_{17\alpha}$ mRNA (arbitrary units) and testosterone (μg g^{-1} tissue) in control bulls (open squares) and bulls treated with the GnRH agonist deslorelin (solid squares) for 10 days. Results are means ± SEM ($n = 4$ to 7); *$P < 0.001$ for all data (Aspden *et al.*, 1998).

increased LH receptors and tonically increased basal concentrations of LH. The increase in the number of LH receptors in bulls treated with agonist could be interpreted to mean that pulses of LH typically observed in bulls downregulate LH receptors in Leydig cells. The requirement of LH for increased testosterone secretion during GnRH agonist treatment was confirmed by the observation that plasma testosterone declined in bulls treated with agonist and simultaneously actively immunized against LH (Aspden *et al.*, 1997b).

StAR protein and steroidogenic enzymes in bulls. Steroid acute regulatory (StAR) protein facilitates the transport of cholesterol from the cytoplasm to the inner mitochondrial membrane, which is the rate-limiting step in steroid biosynthesis (Stocco, 1997). StAR protein mRNA was demonstrated in bovine testis tissue (Pilon *et al.*, 1997). Testicular content of StAR protein was increased in bulls treated with GnRH agonist, and which had increased testicular testosterone content and enhanced testosterone secretion (Aspden *et al.*, 1998). Major steroidogenic enzymes were similarly increased in bulls treated with GnRH agonist (Fig. 2). It would appear, therefore, that increased testosterone synthesis and secretion in bulls treated with GnRH agonist results from the stimulation of normal steroidogenic mechanisms in Leydig cells.

Bulls that had an increased plasma concentration of testosterone in response to chronic treatment with deslorelin showed a further acute increase in testosterone secretion subsequent to injection of human chorionic gonadotrophin (Table 4). This finding was interpreted as indicating that StAR protein and presumably steroidogenic enzymes are not maximally stimulated during GnRH agonist treatment in bulls.

Female cattle

Follicles. The acute increase in plasma LH that occurs at initiation of GnRH agonist treatment (Chenault *et al.*, 1990) can induce ovulation of growing preovulatory follicles (Macmillan and Thatcher, 1991). Luteinization, without ovulation, also can be induced by treatment with GnRH agonist (Macmillan and Thatcher, 1991; Rettmer *et al.*, 1992a). In Brahman heifers, ovulation was

Table 4. Plasma concentrations of testosterone at commencement of treatment with deslorelin (day 0) and on day 56 of treatment, and changes in plasma concentrations of testosterone after intramuscular injection of hCG (2 000 iu; 0 min) on day 56 for control bulls and bulls treated with deslorelin

		Testosterone (ng ml^{-1})			
			Day 56		
	n	Day 0	0 min	120 min	Δ Testosterone*
Control	6	1.6 ± 0.5^{a,x}	1.6 ± 0.4^{a,x}	3.7 ± 0.6^{a,y}	2.1 ± 0.7^{a}
Deslorelin	8	1.9 ± 1.0^{a,x}	5.3 ± 1.4^{b,y}	11.1 ± 1.0^{b,z}	5.8 ± 1.4^{b}

Results are means ± SEM (M. J. D'Occhio, unpublished).
*Increase in plasma concentration of testosterone after injection of hCG.
^{a,b}Means within columns without a common superscript are significantly different ($P < 0.05$).
^{x,y,z}Means within rows without a common superscript are significantly different ($P < 0.05$).

induced by GnRH agonist treatment on day 4 and day 6 of the oestrous cycle, but not day 2 or day 8 (Table 5). These findings were consistent with previous observations in which ovulation was consistently induced by treatment with GnRH agonist at about day 5 to day 6 of the oestrous cycle (Schmitt *et al.*, 1996b). Practical applications of the ovulation-inducing response to acute GnRH agonist treatment are discussed below.

In post-pubertal heifers, the absence of pulsatile secretion of LH associated with continued treatment with GnRH agonist prevented the development of follicles (Gong *et al.*, 1996) beyond the stage at which follicles acquire significant LH receptors and become LH-dependent (7–9 mm) (Bao *et al.*, 1997). Subsequently, plasma FSH was suppressed and follicle growth was restricted to early stages of development (≤ 4 mm) (Gong *et al.*, 1996). The GnRH agonist-treated heifer therefore provides an experimental model to study gonadotrophin requirements for growth, development and maturation of ovarian follicles.

In prepubertal heifers, treatment with deslorelin for 28 days was associated with increased plasma concentrations of oestradiol (Bergfeld *et al.*, 1996b). This finding was analogous to increased testosterone secretion in bulls treated with GnRH agonist. There may be a common response mechanism in cattle that leads to increased gonadal steroidogenesis during GnRH agonist treatment. This suggestion is supported by increased progesterone secretion by the corpus luteum in heifers treated with GnRH agonist commencing early in the oestrous cycle (Thatcher *et al.*, 1993; D'Occhio *et al.*, 1996; Pitcher *et al.*, 1997). Increased gonadal steroidogenesis in cattle receiving GnRH agonist could be related to the absence of pulsatile secretion of LH, but maintenance of tonically increased basal LH secretion (Evans and Rawlings, 1994; Jimenez-Severiano *et al.*, 1998).

Since bulls had increased testicular LH receptors during GnRH agonist treatment (Melson *et al.*, 1986), increased secretion of oestradiol in prepubertal heifers receiving agonist (Bergfeld *et al.*, 1996b) may be due to increased stimulation of thecal cells by LH and a greater supply of androgen for follicular aromatization to oestradiol (Berndtson *et al.*, 1995). However, this contention is not supported by the finding that post-pubertal heifers implanted with deslorelin, and undergoing superstimulation of ovarian follicle growth with FSH tended to have lower plasma concentrations of oestradiol (Fig. 3; heifers nos. 23, 50, 146 and 904). In the latter study, reduced secretion of oestradiol may have been due to reduced androgen precursor synthesis by thecal cells, due to a lack of pulsatile secretion of LH during stimulation of follicular growth with FSH.

It was suggested that the acute increase in LH that occurs when GnRH agonist treatment is initiated at the mid-luteal phase of the oestrous cycle (day 12–13) causes luteinization of follicles, which accounts for reduced secretion of oestradiol (Thatcher *et al.*, 1993; Rettmer *et al.*, 1992a). The steroidogenic response of follicles to GnRH agonist would appear to be influenced by age, ovarian status and other endocrine factors.

Heifers treated chronically with GnRH agonist cannot initiate an endogenous preovulatory surge release of LH and ovulation does not occur (D'Occhio *et al.*, 1997). Chronic GnRH agonist

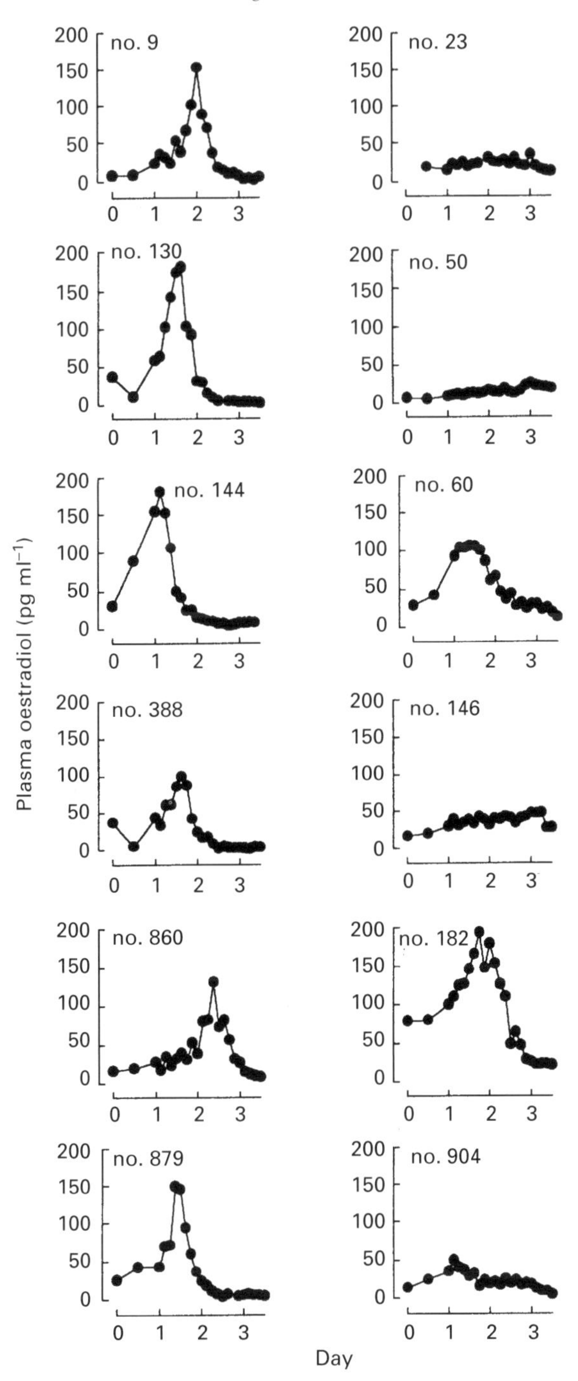

Fig. 3. Plasma concentrations of oestradiol for individual control heifers (left panel) and heifers treated with the GnRH agonist deslorelin (right panel). The latter heifers were implanted with a deslorelin bioimplant one week before treatment with FSH. FSH was used to stimulate ovarian follicle growth in both groups of heifers over 4 days, with FSH treatment ending on day 2 of the profiles (M. J. D'Occhio and J. E. Kinder, unpublished).

Table 5. Proportion of Brahman heifers that ovulated in response to treatment with a GnRH agonist bioimplant commencing on different days of the oestrous cycle

Day of oestrous cycle	Diameter of the largest follicle (mm)	Proportion of heifers that ovulated
2	4.7 ± 0.3	0/4
4	6.5 ± 0.7	2/4
6	10.0 ± 0.0	4/4
8	8.6 ± 0.9	0/4

Values are mean ± SEM (M. J. D'Occhio and F. Cremonesi, unpublished).

Table 6. Size of the corpus luteum and plasma concentrations of progesterone on day 13 of the oestrous cycle in control heifers and heifers treated with deslorelin from day 3 of the cycle[†]

	n	Corpus luteum (g)	Plasma progesterone (ng ml^{-1})
Control	16	3.1 ± 0.2[a]	9.1 ± 1.3[a]
Deslorelin	16	4.2 ± 0.4[b]	18.9 ± 3.5[b]

Results are means ± SEM.
[†]Pitcher *et al.* (1997).
[a, b]Means within columns without a common superscript are significantly different ($P < 0.05$).

treatment therefore has potential as a contraceptive approach in heifers and this is discussed below. The absence of an endogenous LH surge in heifers treated with agonist also provides the opportunity to control the time of ovulation with exogenous gonadotrophin. The value of this approach is discussed below in relation to multiple ovulation and embryo transfer.

Structure and function of the corpus luteum. Heifers treated chronically with GnRH agonist commencing early in the oestrous cycle (day 3) had a larger corpus luteum and secreted more progesterone than did untreated heifers (Table 6). Treatment with deslorelin was associated with higher basal plasma concentrations of LH (Pitcher *et al.*, 1997) which could have contributed, in part, to increased size and function of the corpus luteum. The corpus luteum in heifers treated with deslorelin had a greater content of StAR protein and the steroidogenic enzyme, cytochrome P450 side-chain cleavage enzyme (P450scc; Pitcher *et al.*, 1997). Treatment of heifers with buserelin early in the oestrous cycle caused an increase in the relative numbers of large luteal cells in the corpus luteum (Twagiramungu *et al.*, 1995).

As noted above, heifers treated with GnRH agonist from about day 4 to day 6 of the oestrous cycle can ovulate and develop an accessory corpus luteum. In this situation, increased progesterone secretion during the oestrous cycle is contributed by the existing spontaneous corpus luteum and the accessory corpus luteum. The potential application of GnRH agonist treatment early in the oestrous cycle to increase plasma progesterone and enhance the likelihood of pregnancy recognition and conception is discussed below.

Application of GnRH Agonists in Cattle

Male cattle

Puberty and sperm production. Pubertal (20-month-old) Brahman (*Bos indicus*) bulls treated chronically with deslorelin for 120 days had a faster rate of testis growth and at the end of treatment had a greater sperm production capacity than untreated bulls (Aspden *et al.*, 1998). These findings indicate that reproductive development during the prepubertal period in bulls might be accelerated by chronic treatment with GnRH agonist. However, 5-month-old Freisian (*Bos taurus*) bulls treated with leuprolide for 56 days did not show enhanced testis growth (Ronayne *et al.*, 1993). In another study, Holstein bulls (*Bos taurus*) treated with natural sequence GnRH from 1.5 to 3.0 months of age showed a delay in puberty (Miller and Amann, 1986). In addition, Hereford (*Bos taurus*) bulls treated with a slow-release formulation of leuprolide at approximately 1.5, 2.5 and 3.5 months of age had smaller testes and fewer spermatids at about 12 months of age (Chandolia *et al.*, 1997). The present information therefore indicates that treatment with natural sequence GnRH or GnRH agonist occurring relatively early (1–3 months of age) may be detrimental to pubertal development, but treatment during the peripubertal period or after puberty may enhance testicular function.

Semen quality. Bulls treated with GnRH agonist had increased intratesticular testosterone and high plasma concentrations of testosterone (Aspden *et al.*, 1998). It is possible that GnRH agonist treatment may be beneficial in situations in which testosterone production is reduced and is limiting to sperm production, semen quality or both factors. Increased testosterone synthesis in response to GnRH agonist treatment may enhance semen quality by effects within the testes, epididymides or accessory sex glands.

Female cattle

Treatment of cystic follicles. The acute increase in plasma LH that occurs at initiation of GnRH agonist treatment in female cattle (Chenault *et al.*, 1990) can induce ovulation or luteinization of a cystic follicle (Thatcher *et al.*, 1993). GnRH agonists are used for the treatment of cystic follicles, particularly in post-partum dairy cows.

Synchronization of oestrus. The capacity of acute treatment with GnRH and GnRH agonists to induce ovulation and formation of a corpus luteum has been used to develop oestrous synchronization protocols. A particularly successful protocol involves injections of GnRH or GnRH agonist, in combination with prostaglandin, to induce the emergence of a new dominant preovulatory follicle, and to control the time of ovulation for fixed-time insemination (Twagiramungu *et al.*, 1995; Burke *et al.*, 1996; Schmitt *et al.*, 1996a; Pursley *et al.*, 1997).

Induction of accessory corpus luteum. The relationship of conception rates in heifers to absolute concentrations of progesterone in circulation remains equivocal (Rettmer *et al.*, 1992b; Macmillan and Peterson, 1993). If it is demonstrated that increased plasma concentrations of progesterone are associated with increased pregnancy rates, treatment with a GnRH agonist bioimplant commencing at about day 6 of the oestrous cycle may prove practical for inducing an accessory corpus luteum and increasing plasma progesterone, in combination with artificial insemination, embryo transfer, or both procedures. Treatment with GnRH agonist at about day 12 of the oestrous cycle after mating may extend the luteal phase and increase the opportunity for maternal recognition of pregnancy (Thatcher *et al.*, 1993).

Post-partum anoestrus. The use of GnRH agonist treatment to induce ovulation during the post-partum period was examined extensively in beef and dairy cattle (D'Occhio *et al.*, 1989; Roberge *et al.*, 1992; Twagiramungu *et al.*, 1995). Chronic treatment with agonist induced ovulation in post-partum anoestrous cows, but the life-span of the resulting corpus luteum was reduced. A corpus luteum of normal life-span could be induced if GnRH agonist treatment was preceded by progesterone priming (D'Occhio *et al.*, 1989). Most previously anoestrous animals induced to ovulate with GnRH or GnRH agonist treatment did not initiate regular oestrous cycles. The application of

GnRH agonist treatment to induce ovulation in post-partum anoestrous cows would therefore appear to be restricted to situations in which it can be combined with progestagen priming and artificial insemination.

Superovulation

Sexually mature heifers and cows. A GnRH agonist bioimplant was recently used to develop a new protocol for multiple ovulation and embryo transfer (MOET) in cattle (D'Occhio *et al.*, 1997). In the new protocol, donor heifers or cows are implanted with GnRH agonist approximately one week before commencement of FSH treatment. The preovulatory surge release of LH that typically occurs subsequent to FSH treatment is blocked due to the desensitizing actions of GnRH agonist described above and, accordingly, ovulation does not occur. However, ovulation can be induced by injection of exogenous LH (D'Occhio *et al.*, 1997).

An outstanding feature of the GnRH agonist–LH protocol for MOET is that the time of ovulation is determined by programming the injection of exogenous LH (D'Occhio *et al.*, 1998b). In a time-course study, it was found that optimal fertilization and recovery of embryos occurred when injection of exogenous LH was delayed by about 12 h, relative to normal occurrence of a preovulatory LH surge after stimulation with FSH (D'Occhio *et al.*, 1997). It was considered that a delay in the occurrence of an LH surge allowed additional follicles to develop and acquire sufficient LH receptors for an ovulatory response.

The observation of oestrus subsequent to FSH stimulation is not required in the GnRH agonist–LH protocol, and donors are inseminated relative to the injection of exogenous LH (D'Occhio *et al.*, 1997). In a recent study, typical rates of fertilization and embryo recovery were obtained after donor heifers received one insemination, 12 h after injection of exogenous LH (D'Occhio *et al.*, 1998a).

Heifer calves

The ability to recover viable oocytes from heifer calves provides opportunities for genetic improvement in cattle (Davis *et al.*, 1997). However, current production of viable embryos *in vitro* from oocytes of heifer calves remains relatively low and variable. Ovarian follicular waves are initiated early in the life span of heifer calves (Evans *et al.*, 1994) and there is evidence to indicate that follicular dominance is also established early. These factors may contribute to variability in the follicular response of young heifers to stimulation with gonadotrophins. In addition, heifer calves can have a surge release of LH during stimulation of follicle growth, which may further contribute to the recovery of a heterogeneous population of oocytes (Maclellan *et al.*, 1997).

A GnRH agonist bioimplant was recently used to block pulsatile secretion of LH and prevent the occurrence of a surge release of LH during stimulation of follicle growth with FSH in heifer calves (Maclellan *et al.*, 1997, 1998). Calves treated with deslorelin had more follicles after stimulation with FSH, and this translated into a greater number of oocytes and embryos (Maclellan *et al.*, 1997). This result was not observed in a subsequent study (Maclellan *et al.*, 1998) and additional studies are required in larger numbers of heifer calves to determine whether treatment with a GnRH agonist confers an advantage in follicle growth, oocyte recovery and oocyte developmental competency.

In vivo *oocyte maturation*

The GnRH agonist–LH protocol for MOET described above (D'Occhio *et al.*, 1997) provided an experimental model for examining whether it would be possible to achieve *in vivo* maturation of oocytes before recovery for *in vitro* procedures. Post-pubertal heifers were implanted with GnRH agonist, stimulated with FSH, and injected with LH 24 h after the final injection of FSH (Lindsey *et al.*, 1998). Oocytes were recovered 12 h after injection of LH (i.e. before ovulation) and immediately exposed to spermatozoa for *in vitro* fertilization. The oocytes showed a fertilization potential and

Table 7. Fertilization rate (cleavage) and embryo development (blastocyst) for oocytes recovered after superstimulation of follicle growth with FSH in heifers treated with deslorelin, and heifers treated with deslorelin and treated with exogenous LH

	n	Total number of oocytes	Cleavage	Blastocyst
Deslorelin	9	163	61 (42.3%)[a]	9 (5.5%)[a]
Deslorelin + LH	9	112	61 (54.5%)[a]	25 (22.3%)[b]

Oocytes were recovered 12 h after injection of exogenous LH (Lindsey *et al.*, 1998).
[a, b]Percentages within columns without a common superscript are significantly different ($P < 0.05$).

embryo developmental competency that were typical of oocytes exposed to a conventional 24 h *in vitro* maturation before fertilization (Table 7). Oocytes obtained from heifers treated with GnRH agonist and FSH, but not injected with LH, had normal rates of fertilization but embryo developmental competency was compromised (Table 7). On the basis of these preliminary findings, the capacity for fertilization and embryo developmental competency would appear to have a differential requirement for exposure to pulsatile LH, a surge release of LH, or both. It is possible that oocyte cytoplasmic maturation may be more dependent on exposure to an LH surge than is nuclear maturation.

It would be of interest to examine whether *in vivo* maturation of oocytes using the GnRH agonist–LH protocol could be applied to heifer calves to increase embryo development rates. The GnRH agonist–LH protocol could be further applied to examine fundamental questions relating to the gonadotrophin requirements for normal follicle and oocyte growth and maturation in cattle.

Contraception

The absence of surge releases of LH in heifers and cows treated with GnRH agonist led to studies on the potential of a long-acting GnRH agonist bioimplant as a new contraceptive approach in female cattle (D'Occhio *et al.*, 1996). In a recent study, a substantial contraceptive response was achieved over a period of approximately 12 months with a prototype GnRH agonist bioimplant (Table 8). The use of GnRH agonists provides opportunities for achieving a controlled, reversible suppression of fertility in female cattle.

Conclusion

The response of the anterior pituitary in cattle to treatment with GnRH agonist involves an acute phase (0 to 24 h) during which LH secretion is increased and a chronic phase during which pulsatile secretion of LH is suppressed, but basal release of LH is slightly but significantly increased. The mechanism(s) for continued release of LH during agonist treatment has not been elucidated, but may reflect non-stimulated constitutive release, or an action of the GnRH agonist on gonadotrophs to maintain functional second messenger pathways. An outstanding feature of the gonadal response in cattle treated with GnRH agonists is the increase in steroidogenic activity. This was demonstrated for increased testosterone secretion in bulls, increased oestradiol secretion in pre-pubertal heifers, and enhanced progesterone secretion in post-pubertal heifers treated with agonist commencing early in the oestrous cycle. The basis for the increase in steroidogenesis during agonist treatment is not understood but is likely to be related to tonically elevated basal secretion of LH which appears to be associated with an increase in gonadal LH receptors.

Both the acute and chronic phases of the LH response in cattle to treatment with GnRH agonist provide opportunities for practical applications of agonists. The acute increase in plasma LH that

Table 8. Ovarian activity in female cattle treated with a long-acting GnRH agonist bioimplant

	n	Duration of trial (days)	Number showing ovarian activity	Approximate duration of anoestrus in heifers and cows that showed a return of ovarian function (days)
Trial 1	76	387	20 (26%)	231 ± 19
Trial 2	84	376	8 (10%)	244 ± 13
Trial 3	99	394	9 (9%)	336 ± 3

Heifers and cows were grazed in the presence of bulls, and a return to cyclic ovarian activity was extrapolated from the time of conception (M. J. D'Occhio, G. Fordyce, T. Jubb and T. Whyte, unpublished). Contemporary untreated heifers and cows were introduced at regular intervals and conceived progressively during the trials (data not shown).

occurs at the beginning of agonist treatment has been used in the treatment of cystic follicles, in oestrous synchronisation protocols, and in new superovulation programmes. The lack of preovulatory surge releases of LH in heifers treated chronically with GnRH agonist will lead to the development of a long-acting contraceptive GnRH agonist bioimplant for female cattle. This will have application in the management of fertility in extensively managed beef herds and in the prevention of conception in pre-feedlot heifers. Other possible applications of GnRH agonists could be to increase progesterone secretion in combination with artificial insemination and embryo transfer to enhance conception rates, and to increase testosterone secretion chronically in bulls that have poor semen quality due to reduced testosterone synthesis.

The GnRH agonist-treated heifer will continue to provide an experimental model for studying gonadotrophin requirements for normal follicle and oocyte growth and development in cattle. These studies should lead to GnRH agonist-based protocols for *in vivo* maturation of oocytes before collection for *in vitro* procedures.

The studies undertaken by the authors were made possible by the generous provision of deslorelin bioimplants by Peptech Animal Health Pty Ltd, Sydney, Australia and the collaboration of T. E. Trigg, P. Shober and J. Walsh; D. Miller (Vetrepharm (A/Asia) Pty Ltd, Melbourne, Australia) generously provided FSH (Folltropin-V ®) and LH (Lutropin ®) for the superovulation studies. Other support was provided by Central Queensland University, CSIRO Tropical Agriculture, and Meat and Livestock Australia Ltd.

References

Aspden WJ, Rao A, Scott PT, Clarke IJ, Trigg TE, Walsh J and D'Occhio MJ (1996) Direct actions of the luteinizing hormone-releasing hormone agonist, deslorelin, on anterior pituitary contents of luteinizing hormone (LH) and follicle-stimulating hormone (FSH), LH and FSH subunit messenger ribonucleic acid, and plasma concentrations of LH and FSH in castrated male cattle *Biology of Reproduction* **55** 386–392

Aspden WJ, Rao A, Rose K, Scott PT, Clarke IJ, Trigg TE, Walsh J and D'Occhio MJ (1997a) Differential responses in anterior pituitary luteinizing hormone (LH) content and LHβ- and α-subunit mRNA, and plasma concentrations of LH and testosterone, in bulls treated with the LH-releasing hormone agonist deslorelin *Domestic Animal Endocrinology* **14** 429–437

Aspden WJ, van Reenen N, Whyte TR, Maclellan LJ, Scott PT, Trigg TE, Walsh J and D'Occhio MJ (1997b) Increased testosterone secretion in bulls treated with a luteinizing hormone releasing hormone (LHRH) agonist requires endogenous LH but not LHRH *Domestic Animal Endocrinology* **14** 421–428

Aspden WJ, Rodgers RJ, Stocco DM, Scott PT, Wreford NG, Trigg TE, Walsh J and D'Occhio MJ (1998) Changes in testicular steroidogenic acute regulatory (StAR) protein, steroidogenic enzymes and testicular morphology associated with increased testosterone secretion in bulls receiving the luteinizing hormone releasing hormone agonist deslorelin *Domestic Animal Endocrinology* **15** 227–238

Bao B, Garverick HA, Smith GW, Smith MF, Salfen BE and Youngquist RS (1997) Changes in messenger ribonucleic acid encoding luteinizing hormone receptor, cytochrome

P450-side chain cleavage, and aromatase are associated with recruitment and selection of bovine ovarian follicles *Biology of Reproduction* **56** 1158–1168

Bergfeld EGM, D'Occhio MJ and Kinder JE (1996a) Continued desensitisation of the pituitary gland in young bulls after treatment with the luteinizing hormone-releasing hormone agonist deslorelin *Biology of Reproduction* **54** 769–775

Bergfeld EGM, D'Occhio MJ and Kinder JE (1996b) Pituitary function, ovarian follicular growth, and plasma concentrations of 17β-oestradiol and progesterone in prepubertal heifers during and after treatment with the luteinising hormone-releasing hormone agonist deslorelin *Biology of Reproduction* **54** 776–782

Berndtson AK, Vincent SE and Fortune JE (1995) Effects of gonadotrophin concentration on hormone production by theca interna and granulosa cells from bovine preovulatory follicles *Journal of Reproduction and Fertility Supplement* **49** 527–531

Burke JM, de la Sota RL, Risco CA, Staples CR, Schmitt EJ-P and Thatcher WW (1996) Evaluation of timed insemination using a gonadotropin-releasing hormone agonist in lactating dairy cows *Journal of Dairy Science* **79** 1385–1393

Caraty A, Locatelli A, Delaleu B, Spitz IM, Schatz B and Bouchard P (1990) Gonadotropin-releasing hormone (GnRH) agonists and GnRH antagonists do not alter endogenous GnRH secretion in short-term castrated rams *Endocrinology* **127** 2523–2529

Chandolia RK, Evans ACO and Rawlings NC (1997) Effect of inhibition of increased gonadotrophin secretion before 20 weeks of age in bull calves on testicular development *Journal of Reproduction and Fertility* **109** 65–71

Chenault JR, Kratzer RA, Rzepkowski RA and Goodwin MC (1990) LH and FSH response of holstein heifers to fertirelin acetate, gonadorelin and buserelin *Theriogenology* **34** 81 (Abstract)

Davis GP, D'Occhio MJ and Hetzel DJS (1997) *SMART* breeding: Selection with markers and advanced reproductive technologies *Proceedings of the 12th Conference of the Association for the Advancement of Animal Breeding and Genetics* 6–10th April, Dubbo, Australia **12** 429–432

D'Occhio MJ and Aspden WJ (1996) Characteristics of luteinizing hormone (LH) and testosterone secretion, pituitary responses to LH-releasing hormone (LHRH) and reproductive function in young bulls receiving the LHRH agonist deslorelin: effect of castration on LH responses to LHRH *Biology of Reproduction* **54** 45–52

D'Occhio MJ and Kinder JE (1995) Failure of the LH-releasing hormone agonist, deslorelin, to prevent development of a persistent follicle in heifers synchronized with norgestomet *Theriogenology* **44** 849–857

D'Occhio MJ, Gifford DR, Earl CR, Weatherly T and von Rechenberg W (1989) Pituitary and ovarian responses of post-partum acyclic beef cows to continuous long-term GnRH and GnRH agonist treatment *Journal of Reproduction and Fertility* **85** 495–502

D'Occhio MJ, Aspden WJ and Whyte TR (1996) Controlled, reversible suppression of oestrous cycles in beef heifers and cows using agonists of luteinizing hormone-releasing hormone *Journal of Animal Science* **74** 218–225

D'Occhio MJ, Sudha G, Jillella D, Whyte T, Maclellan LJ, Walsh J, Trigg TE and Miller D (1997) Use of a GnRH agonist to prevent the endogenous LH surge and injection of exogenous LH to induce ovulation in heifers superstimulated with FSH: a new model for superovulation *Theriogenology* **47** 601–613

D'Occhio MJ, Jillella D, Whyte T, Trigg TE and Miller D (1998a) Tight synchrony of ovulation in superstimulated heifers induced to ovulate with LH: embryo recovery after a single insemination *Theriogenology* **49** 376 (Abstract)

D'Occhio MJ, Sudha G, Jillella D, Whyte T, Maclellan LJ, Walsh J, Trigg TE and Miller D (1998b) Close synchrony of ovulation in superstimulated heifers that have a downregulated anterior pituitary gland and are induced to ovulate with exogenous LH *Theriogenology* **49** 637–644

Evans ACO and Rawlings NC (1994) Effects of a long-acting gonadotrophin-releasing hormone agonist (leuprolide) on ovarian follicular development in prepubertal heifer calves *Canadian Journal of Animal Science* **74** 649–656

Evans ACO, Adams GP and Rawlings NC (1994) Follicular and hormonal development in prepubertal heifers from 2 to 36 weeks of age *Journal of Reproduction and Fertility* **102** 463–470

Gong JG, Bramley TA, Gutierrez CG, Peters AR and Webb R (1995) Effects of chronic treatment with a gonadotrophin-releasing hormone agonist on peripheral concentrations of FSH and LH, and ovarian function in heifers *Journal of Reproduction and Fertility* **105** 263–270

Gong JG, Campbell BK, Bramley TA, Gutierrez CG, Peters AR and Webb R (1996) Suppression in the secretion of follicle-stimulating hormone and luteinizing hormone, and ovarian follicle development in heifers continuously infused with a gonadotropin-releasing hormone agonist *Biology of Reproduction* **55** 68–74

Gorospe WC and Conn PM (1988) Restoration of the LH secretory response in desensitized gonadotropes *Molecular and Cellular Endocrinology* **59** 101–110

Hawes BE, Waters SB, Janovick JA, Bleasdale JE and Conn PM (1992) Gonadotropin-releasing hormone-stimulated intracellular Ca^{2+} fluctuations and luteinizing hormone release can be uncoupled from inositol phosphate production *Endocrinology* **130** 3475–3483

Hazum E and Conn PM (1988) Molecular mechanism of gonadotropin releasing hormone (GnRH) action I. The GnRH receptor *Endocrine Reviews* **9** 379–386

Huckle WR and Conn PM (1988) Molecular mechanisms of gonadotropin releasing hormone action II. The effector system *Endocrine Reviews* **9** 387–395

Jimenez-Severiano H, Mussard M, Ehnis L, Koch J, Zanella E, Lindsey B, Enright W, D'Occhio M and Kinder J (1998) Secretion of LH in bull calves treated with analogs of GnRH *Journal of Reproduction and Fertility Supplement* **54**

Karten MJ and Rivier JE (1986) Gonadotropin-releasing hormone analog design. Structure–function studies towards the development of agonists and antagonists: rationale and perspectives *Endocrine Reviews* **7** 44–66

Lahlou N, Roger M, Chaussain J-L, Feinstein M-C, Sultan C, Toublanc JE, Schally AV and Scholler R (1987) Gonadotropin and α-subunit secretion during long term pituitary suppression by D-Trp^6-luteinising hormone-releasing hormone microcapsules as treatment of precocious puberty *Journal of Clinical Endocrinology and Metabolism* **65** 946–953

Lindsey BR, Maclellan LJ, Cremonesi S, Cremonesi F, Whyte TR, Aspden WJ, Kinder JE, Trigg TE, Miller DR and D'Occhio MJ (1998) Differential requirement for exposure of bovine oocytes to the pre-ovulatory LH surge for fertilisation and embryo development. In *Gametes: Development and Function* Eds A Lauria, F Gandolfi, G Enne and L Gianaroli, Serono Symposium, Tabloid S.r.l., Rome, 559 (Abstract)

Maclellan LJ, Bergfeld EGM, Earl CR, Fitzpatrick LA, Aspden WJ, Kinder JE, Walsh J, Trigg TE and D'Occhio MJ (1997) Influence of the luteinizing hormone-releasing hormone agonist, deslorelin, on patterns of estradiol-17β and luteinizing hormone secretion, ovarian follicular responses to superstimulation with follicle-stimulating hormone, and recovery and *in vitro* development of oocytes in heifer calves *Biology of Reproduction* **56** 878–884

Maclellan LJ, Whyte TR, Murray A, Fitzpatrick LA, Earl CR, Aspden WJ, Kinder JE, Grotjan HE, Walsh J, Trigg TE and D'Occhio MJ (1998) Superstimulation of ovarian follicular growth with FSH, oocyte recovery, and embryo production from Zebu (*Bos indicus*) calves: effects of treatment with a GnRH agonist or antagonist *Theriogenology* **49** 1317–1329

Macmillan KL and Peterson AJ (1993) A new intravaginal progesterone releasing device for cattle (CIDR-B) for oestrous synchronisation, increasing pregnancy rates and the treatment of post-partum anoestrus *Animal Reproduction Science* **33** 1–25

Macmillan KL and Thatcher WW (1991) Effects of an agonist of gonadotropin-releasing hormone on ovarian follicles in cattle *Biology of Reproduction* **45** 883–889

Macmillan KL, Day AM, Taufa VK, Gibb M and Pearce MG (1985) Effects of an agonist of gonadotrophin releasing hormone in cattle I. Hormone concentrations and oestrous cycle length *Animal Reproduction Science* **8** 203–212

Melson BE, Brown JL, Schoenemann HM, Tarnavsky GK and Reeves JJ (1986) Elevation of serum testosterone during chronic LHRH agonist treatment in the bull *Journal of Animal Science* **62** 199–207

Mendelson C, Dufau ML and Catt KJ (1975) Gonadotropin binding and stimulation of cyclic adenosine 3'-5'-monophosphate and testosterone production in isolated Leydig cells *Journal of Biological Chemistry* **250** 8818–8823

Miller CJ and Amann RP (1986) Effects of pulsatile injection of GnRH into 6- to 14-wk-old Holstein bulls *Journal of Animal Science* **62** 1332–1339

Pilon N, Daneau I, Brisson C, Ethier J-F, Lussier JG and Silversides DW (1997) Porcine and bovine steroidogenic acute regulatory protein (StAR) gene expression during gestation *Endocrinology* **138** 1085–1091

Pitcher DJ, Aspden WJ, Scott PT, Rodgers RJ and D'Occhio MJ (1997) Pituitary desensitisation, increased progesterone secretion and changes in corpus luteum weight and steroidogenic enzyme content in heifers treated with the GnRH agonist deslorelin *Proceedings of the Australian Society for Reproductive Biology* **28** 28 (Abstract)

Pursley JR, Kosorok MR and Wiltbank MC (1997) Reproductive management of lactating dairy cows using synchronization of ovulation *Journal of Dairy Science* **80** 301–306

Rechenberg WV, Sandow J and Klatt P (1986) Effect of long-term infusion of an LH-releasing hormone agonist on testicular function in bulls *Journal of Endocrinology* **109** R9–R11

Rettmer I, Stevenson JS and Corah LR (1992a) Endocrine responses and ovarian changes in inseminated dairy heifers after an injection of a GnRH agonist 11 to 13 days after estrus *Journal of Animal Science* **70** 508–517

Rettmer I, Stevenson JS and Corah LR (1992b) Pregnancy rates in beef cattle after administering a GnRH agonist 11 to 14 days after insemination *Journal of Animal Science* **70** 7–12

Roberge S, Schramm RD, Schally AV and Reeves JJ (1992) Reduced postpartum anestrus of suckled beef cows treated with microencapsulated luteinizing hormone-releasing hormone analog *Journal of Animal Science* **70** 3825–3830

Rodriguez RE and Wise ME (1989) Ontogeny of pulsatile secretion of gonadotropin-releasing hormone in the bull calf during infantile and pubertal development *Endocrinology* **124** 248–256

Ronayne E, Enright WJ and Roche JF (1993) Effects of continuous administration of gonadotropin-releasing hormone (GnRH) or a potent GnRH analogue on blood luteinizing hormone and testosterone concentrations in prepubertal bulls *Domestic Animal Endocrinology* **10** 179–189

Schmitt EJ, Diaz T, Drost M and Thatcher WW (1996a) Use of a gonadotropin-releasing hormone agonist or human chorionic gonadotropin for timed insemination in cattle *Journal of Animal Science* **74** 1084–1091

Schmitt EJ-P, Diaz T, Barros CM, de la Sota RL, Drost M, Fredriksson EW, Staples CR, Thorner R and Thatcher WW (1996b) Differential response of the luteal phase and fertility in cattle following ovulation of the first-wave follicle with human chorionic gonadotropin or an agonist of gonadotropin-releasing hormone *Journal of Animal Science* **74** 1074–1083

Stocco DM (1997) A StAR search: implications in controlling steroidogenesis *Biology of Reproduction* **56** 328–336

Thatcher WW, Drost M, Savio JD, Macmillan KL, Entwistle KW, Schmitt RL, de La Sota RL and Morris GR (1993) New clinical uses of GnRH and its analogues in cattle *Animal Reproduction Science* **33** 27–49

Twagiramungu H, Guilbault LA and Dufour JJ (1995) Synchronization of ovarian follicular waves with a gonadotropin-releasing hormone agonist to increase the precision of estrus in cattle: a review *Journal of Animal Science* **73** 3141–3151

Journal of Reproduction and Fertility Supplement **54**, 115–126

Control of parturition in ruminants

C. E. Wood

Department of Physiology, University of Florida College of Medicine, Gainesville, FL 32610–0274, USA

Parturition is a process which, when set into motion, occurs to completion. This review concerns the control of parturition in ruminants. Parturition is an endocrine event, dependent upon the activation of the fetal hypothalamus–pituitary–adrenal (HPA) axis. In sheep and other ruminants, increases in plasma concentrations of cortisol induce the activity of 17-hydroxylase and 17,20 lyase in the placenta, increasing the biosynthesis of oestrogen relative to progesterone. The increase in the so-called E:P ratio increases myometrial activity and culminates in labour and delivery. Much work has been done to identify the mechanism of the endogenous activation of the fetal HPA axis. Recent work suggests that production of prostanoids within the fetal brain influences fetal ACTH secretion, and that induction of prostanoid biosynthesis at the end of gestation might be important in the process of parturition. Oestrogen and androgens, secreted by the placenta at the end of gestation, augment activity of the fetal HPA axis by increasing fetal ACTH secretion and by decreasing negative feedback sensitivity to cortisol. Although significant progress has been made concerning the neuroendocrinology of parturition, many significant questions remain. Is parturition regulated or simply programmed? Is parturition the ultimate result of neuronal maturation within the fetal hypothalamus, or is there a complex interplay between the placenta and fetal hypothalamus? Answers to these and other important questions await further research, but may provide key information which will prove useful in understanding general principles of parturition in many mammalian species.

Introduction

Parturition is an all or none event. It must be timed to match the degree of fetal maturation with the ability of the fetus to survive outside the uterus. Once initiated, the process of parturition is difficult to interrupt or delay. The importance of understanding parturition is understood easily in terms of the consequences of prematurity. The fetus, *in utero*, is on a kind of life-support and that life-support is maintained until the fetus can survive and thrive outside the uterine environment. Premature birth is complicated usually by immature pulmonary function and sometimes by incomplete transition of the cardiovascular system from fetal to neonatal anatomy.

Parturition is an endocrine event: this was first demonstrated by G. Liggins, studying the process of parturition in sheep. He found that electrocoagulation of the fetal pituitary indefinitely delayed parturition in sheep (Liggins *et al.*, 1967) and went on to demonstrate that the endocrine axis that is critical for this process in the sheep is the hypothalamus–pituitary–adrenal axis, because infusion of glucocorticoid (Liggins, 1969) or adrenocorticotropin (ACTH) (Liggins, 1968) initiated premature parturition. This observation, which demonstrated a fundamental endocrine process, would have proven to be of immediate importance in human medicine if not for the observation that the endocrinology of parturition in sheep is not identical to that in humans. These differences, detailed in this review, have confounded the basic premise that the ruminant is a good model of parturition for humans, and have focused many research efforts away from the basic neuroendocrinology of the hypothalamus–pituitary–adrenal axis and towards local effectors in the myometrium and other intrauterine tissues.

The debate concerning the control of parturition can be characterized as focusing on two alternative views. While it is well known that the fetus initiates parturition, it is not clear whether the critical event that initiates the process is a fetal neuroendocrine event or whether the fetal neuroendocrine system is responding to, or otherwise augmented by, external stimuli. Perhaps the development of critical synapses within the fetal paraventricular nuclei of the hypothalamus allows an increase in fetal neuroendocrine function which, in turn, alters fetal oestrogen and progesterone biosynthesis and initiates labour. The second possibility is that, while the presence of the fetus appears to be critical, the augmentation of fetal neuroendocrine activity which ultimately initiates parturition might be the result of other stimuli. These stimuli might be placental hormones or increasing fetal stress (changes in fetal blood gases, glucose, or blood pressure). This review will focus on the endocrinology of parturition, particularly the fetal hypothalamus–pituitary–adrenal (HPA) axis which plays a central role. However, the review will also explore the endocrine and other factors that affect the function of the fetal HPA axis, including the secretion of placental hormones and the responsiveness of the fetus to stress.

The Fetal Hypothalamus–Pituitary–Adrenal (HPA) Axis

The fetal HPA axis is functionally similar in most respects to the adult HPA axis (Fig. 1). However, there are some important differences: (1) the fetal brain is still developing late in gestation; and (2) the fetal HPA axis communicates with the maternal HPA axis via the placenta. However, the fetal axis functions in a manner similar to that of the adult axis in that the fetal HPA axis responds to various stimuli (which have been termed 'stresses'). However, the activity of the fetal HPA axis can be altered via 'non-stressful' inputs (for example ontogenetic changes in the activity of the HPA axis).

Several 'stresses' have been investigated in terms of their effects on fetal ACTH secretion. These stimuli include haemorrhage (Wood *et al.*, 1989a), arterial hypotension (Wood *et al.*, 1982; Tong *et al.*, in press), hypoxia, hypercapnia, or asphyxia, and acidaemia (Cudd and Wood, 1996). The ACTH responses to these stimuli are probably considered as the efferent limbs of reflex loops which are subserved by identifiable receptors. For example, ACTH response to hypotension is, in part, mediated by baroreceptors in the fetal carotid sinus or aortic arch (Wood, 1989b) (Fig. 2). The reflexes mediating fetal ACTH responses to the various stimuli, or 'stresses', are only partially identified, making one of the challenges of the future the provision of a more complete understanding of the neuroendocrinology of fetal stress. It has been proposed that repeated fetal hypoxia might alter the timing of parturition secondary to increases in fetal ACTH and cortisol concentrations in plasma. It may be assumed that the ACTH response to hypoxia is mediated by the arterial chemoreceptors, in a similar way to the control of fetal heart rate by these receptors. However, denervation of the carotid sinus chemoreceptors has no measurable effect on fetal ACTH secretion (Giussani *et al.*, 1996). It is possible that the ACTH response to hypoxia could be mediated by the generation of a paracrine or autocrine mediator in the brain of the fetus: it is known that the vasopressin response to hypoxia is blocked by the administration of an adenosine receptor blocker (Koos *et al.*, 1994), and also by pretreatment with indomethacin to block the biosynthesis of prostanoids (Tong *et al.*, 1998).

In adult animals of many species, plasma ACTH concentrations change spontaneously in a 24 h pattern, secondary to a so-called 'circadian' rhythm. This rhythm in rodents and in humans is 'entrained' or reset each day by the timing of exposure to light (the 'light-cycle'). Other factors, such as feeding, can also set a diurnal rhythm in animals maintained in constant light. These changes in ACTH secretion rate are not considered to reflect stress or response to any noxious stimulus in the environment. In fact, responses to 'stressors' have been defined as the increase in activity of the HPA axis above the activity that would otherwise be expected at that time of day. However, fetal ruminants are different, in that there is no endogenous circadian rhythm in ACTH secretion (Bell and Wood, 1991; Simonetta *et al.*, 1991). Although it is sometimes assumed that the axis exhibits circadian activity, it is clear that sheep lack circadian variation in plasma ACTH concentrations, even in adult life (Bell and Wood, 1991). Apparent 24 h rhythms in plasma concentrations of ACTH in the

(a) The hypothalamus–pituitary–adrenal (HPA) axis

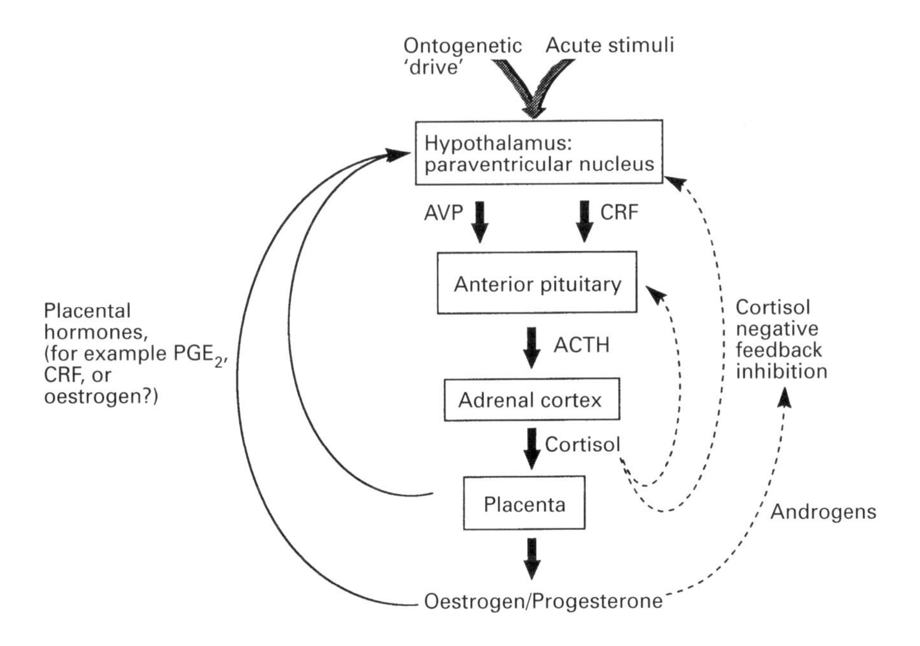

(b) The ontogeny of the fetal HPA axis

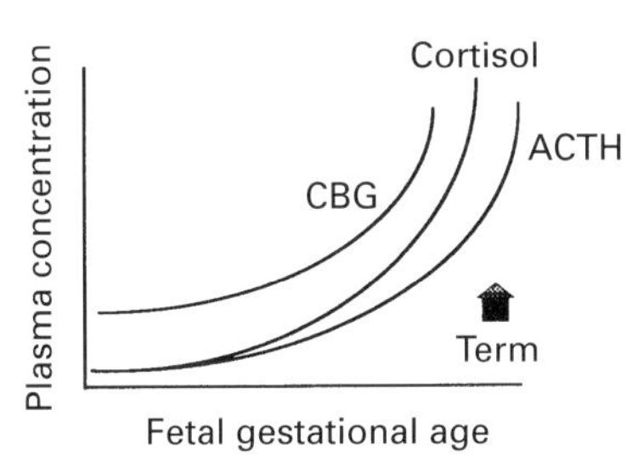

Fig. 1. (a) Schematic diagram of the ovine fetal hypothalamus–pituitary–adrenal (HPA) axis. Solid arrows (left) represent stimulatory interactions and dashed arrows (right) represent inhibitory interactions. Note that cortisol from the fetal adrenal inhibits fetal ACTH secretion by an inhibitory action both at the fetal hypothalamus and at the fetal pituitary, and that androgen from the placenta interrupts the cortisol negative feedback. (b) Schematic representation of the ontogeny of the fetal hypothalamus–pituitary–adrenal axis.

fetus have been secondary to feeding regimens other than *ad libitum* (Simonetta *et al.*, 1991) and, presumably, fluctuations in blood concentrations of metabolic substrates.

Although the fetal HPA axis does not exhibit true circadian rhythmicity, it does follow an ontogenetic pattern of activity. At the end of gestation, fetal ACTH and cortisol secretion rates increase in a semilogarithmic pattern (Wood., 1989a) (Fig. 1). This spontaneous increase in activity might be considered as analogous to the circadian changes in axis activity in other species, since it might reflect an endogenous 'programme' within the fetal hypothalamus and not a response to external or even internal 'stresses'. By analogy to the adult rat or human, therefore, it is possible to define fetal 'stress' as an increase in fetal HPA axis activity above the level that would otherwise be expected at that time in gestation (Wood, 1989a). An alternate view of the ontogenetic increase in

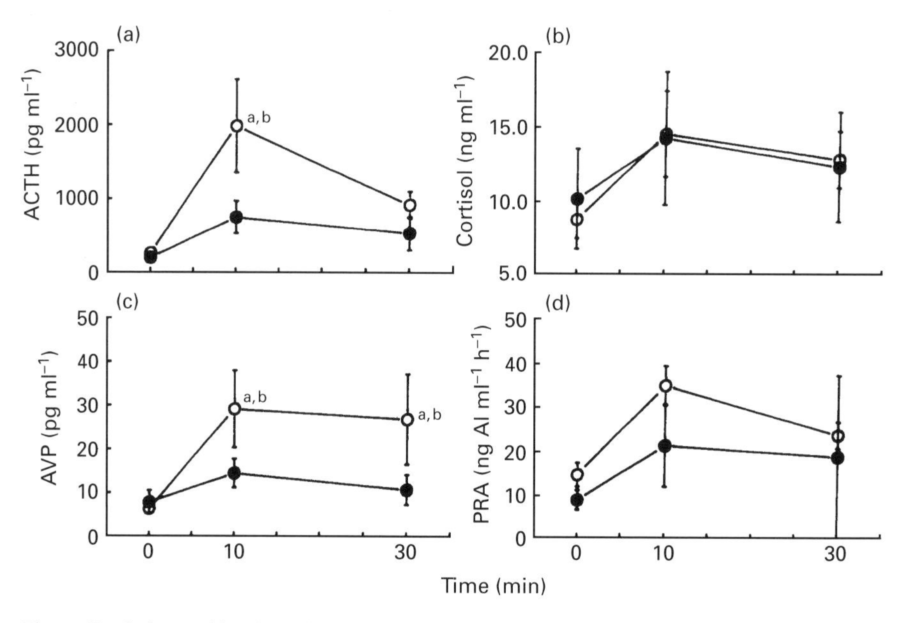

Fig. 2. Fetal plasma (a) ACTH, (b) cortisol, (c) arginine vasopressin (AVP), and (d) plasma renin activity (PRA) responses to hypotension (reduction of fetal blood pressure approximately 50% by vena caval obstruction from 0 to 10 min) in intact fetuses (open symbols) and in sinoaortic denervated fetuses (filled symbols). Sinoaortic denervation significantly reduced the ACTH, AVP, and PRA responses to hypotension, demonstrating the importance of arterial baroreceptors or chemoreceptors in mediating the hormonal responses to hypotension in the fetus. (Reprinted from Wood, 1989b, with permission.)

activity of the fetal HPA axis is that it is the result of increasing sensitivity to fetal stressors (McMillen *et al.*, 1995). According to this hypothesis, the fetus at term recognizes that factors such as the 'fetal' concentrations of glucose and oxygen, and blood pressure are low compared with the values that would be observed in adult animals. It is as though the fetus begins to respond to these variables as it would after birth, driving the activity of the HPA axis higher until parturition is initiated.

Biosynthesis of Progesterone, Oestrogen and the Initiation of Parturition

Parturition could not proceed without uterine contraction. Indeed, it is the contraction of the uterus which ultimately defines labour and results in delivery of the fetus to the extrauterine environment. The ability of the uterine smooth muscle to contract depends on the membrane potential of the smooth muscle cells and the ability of the cells to communicate. A unifying hypothesis that addresses the mechanism of the 'final common pathway' involved in parturition involves the spontaneous changes in the secretion of oestrogen and progesterone at the end of gestation (reviewed in Wood, 1989a). According to this hypothesis, the activity of the uterine myometrium is influenced by the placental production of the steroid hormones progesterone and oestrogen. Throughout much of pregnancy, the placenta synthesizes and secretes large amounts of progesterone. The resulting high concentrations of progesterone maintain uterine quiescence, mainly by hyperpolarizing the myometrial cells. At the end of gestation in many species, there is an increase in the rate of production of oestrogen relative to the rate of production of progesterone. The oestrogen produces a relative depolarization of the uterine myometrial cells, tending to augment their activity. The changes in plasma and tissue concentrations of oestrogen and progesterone also stimulate the formation of gap junctions (Ou *et al.*, 1997). Smooth muscle activation is associated with the synthesis and release prostaglandin $F_{2\alpha}$, which acts in an autocrine and paracrine manner to augment the force of contraction. Indeed, knockout studies in mice have demonstrated that the knockout of the receptor for prostaglandin $F_{2\alpha}$ completely abolishes the process of labour and delivery (Sugimoto *et al.*, 1997).

When comparing most mammalian species, the changes in plasma oestrogen and progesterone are not very uniform (Holtan *et al.*, 1991), and this has led some to question the fundamental importance of this construct. Nevertheless, it is clear that in all species studied to date, the maintenance of pregnancy depends on the continued presence of progesterone, and it is therefore possible that there are significant differences among species in the biochemical and molecular event(s) within the myometrium which alter the responsiveness of the tissue to oestrogen and progesterone.

If the notion that spontaneous changes in oestrogen and progesterone are integral to the final common pathway of parturition is accepted, it is important to understand the endocrine mechanisms which alter placental steroidogenesis. In sheep and other ruminants, before the final stages of pregnancy, the placenta contains the enzymatic machinery for the biosynthesis of progesterone, but lacks the critical enzyme needed for 17α-hydroxylase activity (cytochrome $P450_{c17}$) (reviewed in Wood, 1989a). Therefore, in the preterm sheep fetus, the placenta secretes copious amounts of progesterone, but very little oestrogen. At the end of gestation, increases in plasma cortisol concentration induce cytochrome $P450_{c17}$ in the placenta, allowing an increase in the rate of synthesis of oestrogen at the expense of the rate of synthesis of progesterone. For this reason, the induction of parturition in sheep and in other ruminants depends on the activation of the fetal HPA axis which occurs ontogenetically at the end of gestation.

The increase in oestrogen biosynthesis at the end of gestation is a common theme among mammalian species. However, the strategy for producing this increase varies. In humans and other primates, cytochrome $P450_{c17}$ is not inducible (by cortisol or by any other circulating hormones) (Challis *et al.*, 1974). In the fetuses of these species, the adrenal cortices contain an identifiable pattern of zonation. The adrenal cortex contains a so-called 'fetal' zone and a so-called 'adult' or 'definitive' zone (reviewed in Wood, 1989a). The fetal zone contains the enzymes necessary to synthesize dehydroepiandrosterone (DHEA) and dehydroepiandrosterone sulfate (DHAS), but cannot synthesize cortisol, corticosterone, aldosterone, testosterone, or oestradiol because it lacks 3β-hydroxysteroid dehydrogenase (3β-HSD). However, the adult zone of the fetal adrenal contains all of the enzymes necessary to produce cortisol, corticosterone and aldosterone, the major products of the adrenal gland of the postnatal animal. Both the fetal and the adult zones are under the trophic control of ACTH from the fetal pituitary. When fetal ACTH secretion is increased, the fetal zone increases its production of DHEA and DHAS, and the adult or definitive zone increases its production of cortisol and corticosterone. The mass of the fetal zone greatly exceeds the mass of the adult zone; the mass increases throughout the last half of gestation, and then decreases after birth. The DHEA and DHAS produced by the fetal zone during fetal life is important as a supply of substrate to the placenta for biosynthesis of oestrogens. The placenta, because it lacks cytochrome $P450_{c17}$, cannot synthesize oestrogen from cholesterol. However, the placenta overcomes this deficiency through the use of fetal (17α-hydroxylated) steroidogenic precursors as substrate for the production of oestrogen. For this reason, an increase in the rate of secretion of fetal ACTH increases the placental production of oestrogen, and decreases in the rate of ACTH secretion decrease the rate of secretion of oestrogen in these species.

The horse fetus provides an interesting comparison to those of sheep and primates. If the contention that the final common pathway for increasing uterine contractility is an increase in the oestrogen-to-progesterone ratio is accepted, it becomes apparent that the variants in schema for oestrogen biosynthesis all result in the same final endpoint. The horse is a species without an inducible placental cytochrome $P450_{c17}$ (Mason *et al.*, 1993). Therefore, oestrogen biosynthesis occurs using a fetoplacental unit, a strategy which is in general similar to that of primates. In the horse, the gonads synthesize and secrete large amounts of DHEA, and the initiation of parturition is interrupted by the process of fetal gonadectomy (Pashen and Allen, 1979). This finding suggests that an increase in the activity of the fetal gonad at the end of gestation initiates parturition in this species and that the gonads of the horse are under trophic control by the fetal pituitary.

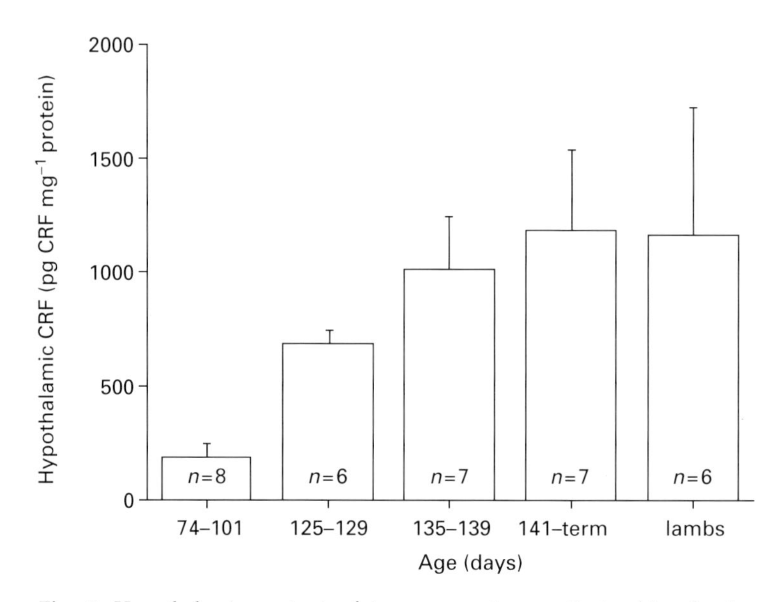

Fig. 3. Hypothalamic content of immunoreactive corticotrophin-releasing hormone (CRF). The tissue concentrations were significantly increased as a function of age, as tested by Dunn's test. The numbers of animals per group are shown within each bar. Values are means ± SEM. (Modified from Saoud and Wood (1996a) with permission from Elsevier Science).

The Ontogeny of the Fetal HPA Axis and the Initiation of Parturition

In sheep, parturition is initiated by a spontaneous increase in activity of the fetal HPA axis. From both a practical and theoretical viewpoint, it is important to understand the mechanism of the increase in fetal ACTH secretion. For example, is the increase in fetal ACTH secretion the result of neuronal maturation within the fetal hypothalamus, or is the increase in fetal ACTH secretion the result of stimulation by another hormone (for example, from the placenta)? Although the mechanism of the increase in fetal ACTH secretion is not understood completely, there are several important facts that have provided a key understanding of several control mechanisms that stimulate or allow increases in cortisol concentration of fetal plasma. These mechanisms will be addressed individually.

The initiation of parturition in sheep requires an intact HPA axis. As mentioned previously, hypophysectomy or bilateral adrenalectomy interrupts the process of parturition. However, bilateral ablation of the paraventricular nuclei or implantation of dexamethasone crystals in the paraventricular nuclei bilaterally also interrupts the process of parturition (Myers *et al.*, 1992). In the final stages of fetal development, the activity of the fetal HPA axis increases spontaneously. Changes at every level of the axis can be demonstrated during the preparturient increase in activity of the HPA axis.

In the hypothalamus, there is an increase in the content of CRH (Saoud and Wood, 1996b) as well as an abundance of mRNA for CRH (Myers *et al.*, 1993) (Fig. 3). Some studies have identified an increase in hypothalamic content of arginine vasopressin (AVP) mRNA (Matthews and Challis, 1995). However, the major increase in hypothalamic AVP content appears to occur after birth (Saoud and Wood, 1996b). In the anterior pituitary, there is an increase in the concentration of ACTH, as well as an augmentation of pituitary post-translational processing of pro-opiomelanocortin (POMC) (Saoud and Wood, 1996c). An increase in the abundance of mRNA for POMC has been demonstrated (McMillen *et al.*, 1990). Histologically, the corticotropes of the anterior pituitary undergo developmental changes. Before day 120 of gestation, the corticotropes in the ovine fetal anterior pituitary appear to be larger than the corticotropes in late gestation (Antolovich *et al.*, 1989). The apparent development of corticotropes in late gestation has been referred to as a change from 'fetal'- to 'adult'-type corticotropes. It is not known whether the function of these morphologically different corticotropes are different from each other. However, it is possible that at least some of the maturation of the anterior pituitary content of POMC and its response to releasing hormones might be explained by the maturation of the corticotrope itself.

Throughout the last 2–3 weeks of development *in utero*, the fetal adrenal gland increases in size relative to fetal body weight, and the cellular sensitivity to ACTH increases (reviewed in Wood, 1989). The increase in sensitivity to ACTH is partially the result of increased adrenal cortical mass and partially the result of increased cellular responsiveness to ACTH. In the adrenal cortical cells, there is an increase in the density of ACTH receptors with a resultant increase in the adrenal cAMP response to ACTH (Durand *et al.*, 1981). This increase in the size and sensitivity of the fetal adrenal, combined with the increased circulating concentrations of ACTH, account for the increase in cortisol secretion rate which ultimately triggers parturition.

Increases in adrenal sensitivity to ACTH at the end of gestation are also, in part, the result of accelerated processing of ACTH from POMC in the anterior pituitary corticotropes. The increase in POMC processing is reflected in an increase in the plasma concentration of fully processed ACTH relative to larger molecular weight forms of immunoreactive ACTH (iACTH) (Thorburn *et al.*, 1991). Measurements of plasma concentrations of iACTH are made using radioimmunoassay, a technique which is dependent upon the binding characteristics of the antiserum used in the assay. An antiserum that binds to an epitope within $ACTH_{1-39}$ would be expected to crossreact with precursor forms of ACTH (such as POMC) or partially processed forms of the precursor (such as the 22 kDa fragment of POMC which is known as 'pro-ACTH') (Eipper and Mains, 1980). In all studies of ACTH that can be immunoassayed in unchromatographed plasma, the value obtained from the radioimmunoassay represents a mixture of unprocessed and partially processed POMC and fully processed ACTH. Thorburn and colleagues (1991) demonstrated that the concentration of fully processed ACTH in plasma increases in late gestation, and that the relative proportion of fully processed to larger molecular weight forms increases. In agreement with this observation was the report by Castro *et al.* (1993) that the ratio of biologically active to immunologically reactive ACTH was increased in fetal plasma in late gestation. The study of adrenal responsiveness to plasma ACTH or ACTH-like peptides using dispersed adrenal cells has provided an understanding of how changes in POMC processing might increase adrenal sensitivity at the end of gestation (Schwartz *et al.*, 1995). In adrenal cells from adult sheep, only fully processed ACTH is biologically active. POMC and pro-ACTH are not active either as agonists or antagonists in the adult cells. In fetal cells, however, these larger peptides act as competitive antagonists at the adrenal gland. In fact, pro-ACTH inhibits ACTH action in nearly equimolar concentrations (Schwartz *et al.*, 1995). Because of the action of ACTH as stimulator of adrenal secretion and the action of POMC and pro-ACTH as inhibitors of adrenal secretion, an increase in the ratio of ACTH:POMC or ACTH:pro-ACTH, as occurs spontaneously in late gestation, will augment adrenal sensitivity to ACTH.

The fetal adrenal requires ACTH for maintenance of its mass, and increases in ACTH concentrations stimulate adrenal growth. It is possible that the sensitivity of the fetal adrenal is dynamically regulated. In sheep fetuses, changes in plasma ACTH and cortisol concentrations sometimes appear to be dissociated from each other. A good example of this dissociation in relation to chronic changes is that the ontogenetic increase in fetal plasma cortisol at the end of gestation has been observed to increase before the increases in fetal plasma ACTH concentration (reviewed in Wood, 1989). A good example in relation to acute changes is that the increase in cortisol concentration of fetal plasma in response to hypotension (Wood *et al.*, 1982) and hypoxia (Giussani *et al.*, 1994) does not always mirror changes in fetal plasma ACTH concentration. Particularly interesting is the observation that carotid sinus denervation blunts the cortisol, but not the iACTH, response to acute hypoxia in sheep fetuses (Giussani *et al.*, 1994). This finding suggests that a component of the HPA axis response to acute stimulation in the sheep fetus involves a dynamic change in adrenal sensitivity to ACTH. In support of this notion is the observation that bilateral splanchnic nerve resection decreased the magnitude of the fetal plasma cortisol response to acute hypoxia (Myers *et al.*, 1990). The proposal that splanchnic nerves alter adrenal responsiveness or

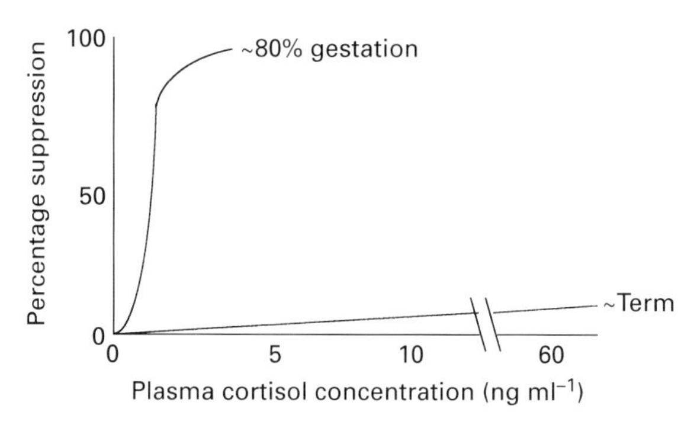

Fig. 4. Schematic representation of the reduction in sensitivity of cortisol negative inhibition of fetal ACTH secretion. In fetuses of between 117 and 131 days of gestation, cortisol effectively inhibits the ACTH response to acute hypotension. In term fetuses, supraphysiological increases in plasma cortisol concentration do not inhibit ACTH responses to hypotension. Data are redrawn and adapted from Wood (1986, 1988).

sensitivity to ACTH is not new and is not restricted to the sheep fetus. Clearly, a better understanding of dynamic changes in adrenal sensitivity in sheep fetuses is needed, because it might be an important part of the puzzle of the mechanism of parturition in this species.

Accompanying the increase in cortisol secretion rate is an increase in the circulating concentrations of corticosteroid-binding globulin (CBG: transcortin). This increase in plasma concentrations of CBG masks the biological activity of the circulating cortisol. However, the binding activity of CBG in sheep fetuses *in vivo* is relatively weak: approximately 25% of the circulating cortisol is unbound, and therefore biologically active (Wood, 1988; Wood, 1986). This is because the concentration of CBG in fetal sheep is low compared with comparable concentrations in the blood of adult rats or humans, and probably because of the large number of steroids circulating in fetal plasma which also bind to (and displace cortisol from) CBG. CBG has been shown to decrease cortisol action. For example, CBG decreases the negative feedback action of corticosterone in adult rats (Kawai and Yates, 1966). It has been proposed that increasing concentrations of CBG significantly interfere with cortisol action (especially in terms of cortisol negative feedback control of ACTH secretion) (Challis and Brooks, 1989). Although this is theoretically possible, the binding efficiency of ovine fetal CBG for cortisol *in vivo* is likely to be too low for dynamic changes in CBG concentration to play a significant role in adjusting cortisol action at target tissues (Wood, 1986).

What Drives Activity of the Preparturient Fetal HPA Axis?

Perhaps the most important question that is critical to our understanding of ovine parturition is the mechanism of the increased activity of the fetal HPA axis. Functionally, the increase in activity of the fetal HPA axis at the end of gestation appears to be different from the increase in activity of the fetal HPA axis in response to a specific stimulus. It is unlikely that the preparturient increase in activity represents the response to a chronic stress, since chronic stimuli often result in an attenuation of ACTH secretion after the initial response (Harvey *et al.*, 1993). For example, acute hypoxia stimulates vigorous ACTH responses in sheep fetuses(Giussani *et al.*, 1994), but long-term hypoxia produced by high altitude residence produces an adaptation of the HPA axis, so that long-term changes in fetal plasma cortisol and ACTH are not measurable (Harvey *et al.*, 1993). Preparturient increases in fetal plasma ACTH and cortisol concentrations, which become progressively pronounced in the last 2–3 weeks of ovine fetal life, are more likely to be caused by a 'programmed' maturation in the fetal hypothalamus or elsewhere in the fetal central nervous system, or by an external influence on the fetal hypothalamus or pituitary, such as the progressive increase in the secretion of a placental hormone.

One important link in the chain of events that initiates parturition in sheep is the interruption of cortisol negative feedback inhibition of fetal ACTH secretion (Fig. 4). This reduction in negative feedback sensitivity occurs in the last few days of fetal life, and functions to allow simultaneous increases in fetal ACTH and cortisol secretion (Wood, 1988). As important as this process might be, it cannot fully explain the process of parturition by itself. It is logical to assume that the preparturient increases in fetal ACTH and cortisol secretion require some stimulation of the axis, as well as interruption of negative feedback.

Recent evidence suggests that at least some of the increase in the activity of three fetal HPA axis at the end of gestation is the result of positive feedback cycle involving oestrogen secreted by the placenta. Physiological, chronic increases in oestradiol concentrations in fetal plasma greatly augment ACTH concentrations in fetal plasma, in unstimulated and stimulated conditions (Saoud and Wood, 1997). The increases in oestradiol concentrations in fetal plasma in that study were within the range of plasma concentrations that are measured endogenously at the end of gestation. We have also reported that chronic physiological increases in androstenedione concentrations in fetal plasma interrupt cortisol negative feedback inhibition of fetal ACTH secretion (Saoud and Wood, 1997). Chronic increases in both oestradiol and androstenedione significantly advance the day of spontaneous parturition (that is, promote premature parturition), indicating that the changes in placental secretion of these two hormones might be an integral part of the process of parturition in sheep. This apparent positive feedback loop might be a feature of the endocrinology of parturition in several species, because simultaneous treatment of pregnant baboons with androstenedione promotes premature parturition (Farber *et al.*, 1997).

Another hypothesis that has been advanced to explain the increase in activity of the fetal HPA axis is that prostaglandin E_2 (PGE_2), secreted by the placenta, circulates in fetal plasma and acts as a hormone to stimulate fetal ACTH secretion (Thorburn *et al.*, 1991). What makes this idea particularly attractive is the fact that circulating concentrations of PGE_2 increase in fetal plasma at the end of gestation, apparently mirroring the changes in ACTH concentration in fetal plasma (Thorburn *et al.*, 1991). The activity of prostaglandin endoperoxide synthase (PGHS) in the cotyledons increases at the end of gestation (Rice *et al.*, 1990), and intravenous infusions of PGE_2 into fetal sheep increase circulating concentrations of iACTH (Thorburn *et al.*, 1991). However, the observation that intracarotid arterial infusions of PGE_2, large enough to increase predicted carotid arterial plasma concentrations of PGE_2 well above the physiological range, are not sufficient to increase fetal ACTH secretion casts some doubt on this hypothesis (Cudd and Wood, 1992). Although it is unlikely that PGE_2 from a placental source stimulates preparturient increases in fetal ACTH secretion, it is still possible that PGE_2 has an important influence on the neuroendocrine mechanisms governing ACTH release in the fetus. The fetal central nervous system contains significant amounts of PGHS, and immunoreactive PGHS has been localized within hypothalamic regions known to be important for controlling activity of the HPA axis (Breder *et al.*, 1992). Recent evidence suggests that the abundance of immunoreactive enzyme increases within these areas before the normal time of parturition (Deauseault and Wood, 1998). It is therefore conceivable that endogenous production of PGE_2 within the brain significantly augments the activity of the HPA axis before birth. It is well known that inhibition of prostaglandin biosynthesis can delay parturition. This is thought to be the result of a reduction in prostaglandin biosynthesis within the myometrium, particularly the result of a reduction in $PGF_{2\alpha}$ production. However, it is possible that general treatment with prostaglandin synthesis inhibitors might also functionally impair the augmentation of the fetal HPA axis which is critical to the initiation of parturition.

It has been proposed that the release of ACTH, CRH or both hormones into the fetal blood is a physiologically important mechanism by which the placenta affects the timing of parturition in primates (Keller-Wood and Wood, 1991a). Ovine placenta contains measurable amounts of immunoreactive ACTH, but little CRH (Keller-Wood and Wood, 1991a,b, 1991b). There are no

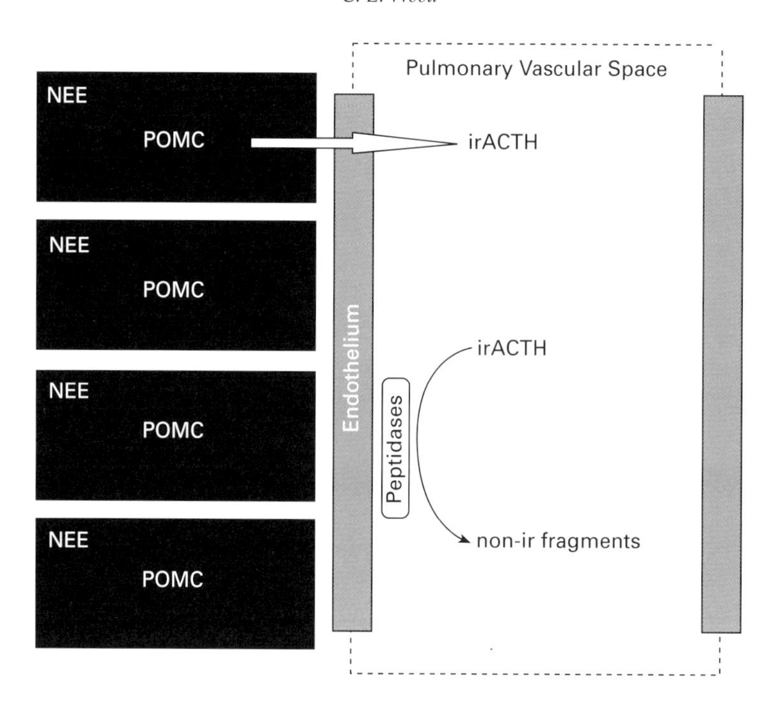

Fig. 5. Schematic representation of a conceptual model of the process of uptake and secretion of immunoreactive ACTH by the fetal lung. Pro-opiomelanocortin (POMC) is the precursor molecule for ACTH, and is synthesized in the fetal pulmonary neuroendocrine cells (NEE). This model assumes simultaneous secretion of immunoreactive ACTH from lung and clearance of approximately 48% of immunoreactive ACTH in pulmonary plasma.

arteriovenous differences across the ovine placenta (either in the fetal circulation or in the maternal circulation) for either peptide (Keller-Wood and Wood, 1991a, 1991b), suggesting that there is no net release of either peptide.

Although the placenta is not a source of either ACTH or CRH in the sheep fetus, the fetal lung contains significant amounts of immunoreactive ACTH (Cudd *et al.*, 1993). The concentration of the peptide in pulmonary tissue, when expressed as ng iACTH per mg protein, is highest in fetuses at mid-gestation and decreases as the fetus matures (Cudd *et al.*, 1993). The peptide is released from the lung into the fetal bloodstream under basal conditions and during acute surgical stress (Cudd *et al.*, 1993; Cudd and Wood, 1995). Indeed, under basal conditions, the lung appears to both secrete iACTH into the bloodstream and metabolize iACTH which reaches the lung via the arterial blood (Cudd and Wood, 1995) (Fig. 5). We have recently demonstrated that the cells within the lung that contain iACTH are pulmonary neuroendocrine cells, both in the form of neuroendocrine epithelial cells (NEEs) and in the form of neuroendocrine bodies (NEBs) (Wood *et al.*, 1998). The decrease in concentration of iACTH in pulmonary tissue coincides with the reported decrease in density of NEEs and NEBs within the lung throughout development (Scheuermann, 1991). The release of iACTH from these cells during acute surgical stress is consistent with the presumed function of these cells as chemosensitive tisssue (Adriaensen and Scheuermann, 1993). Immunoblot analysis of the pulmonary iACTH demonstrates the presence of a form with a molecular weight which is consistent with POMC, and processing products that are different from those identified in fetal pituitary (Saoud and Wood, 1996c). The products of post-translational processing within the lung appear to be consistent with the processing of POMC in adrenal chromaffin tissue (Wood *et al.*, 1998).

Conclusions

Much progress has been made since the original observations by G. C. Liggins concerning the effects of interruption of the HPA axis on the process of parturition in sheep. Yet, it is also undeniable that the answers to the larger questions still elude us. We still do not know, for example, why the activity of the HPA axis increases at the end of gestation. Is this a function of neuronal maturation within the paraventricular nuclei, or is this a function of changing responsiveness to stimuli or changing circulating concentrations of placental hormones, or is this some combination of these factors? Is parturition regulated, and if so, how? Is the process simply programmed, or is there sensation of one or more variables which signal a readiness for birth? Providing the answers to these questions will require answers to more specific questions. For example, how important are dynamic changes in adrenal sensitivity to ACTH? If these changes are important, what causes them: changes in neural efferent activity to the adrenal, or changes in ACTH-like peptides from the pituitary and extrapituitary sites? How important is the positive feedback cycle between oestrogens, androgens, and ACTH? If this is important, we must determine whether conjugated oestrogens (the most abundant form of oestrogen in fetal plasma) are biologically active at the pituitary and hypothalamus.

References

Adriaensen D and Scheuermann DW (1993) Neuroendocrine cells and nerves of the lung *Anatomical Record* **236** 70–85

Antolovich GC, Perry RA, Trahair JF, Silver M and Robinson PJ (1989) The development of corticotrophs in the fetal sheep pars distalis: the effect of adrenalectomy or cortisol infusion *Endocrinology* **124** 1333–1339

Bell ME and Wood CE (1991) Do plasma ACTH and cortisol concentrations in the late gestation fetal sheep vary diurnally? *Journal of Developmental Physiology* **15** 277–282

Breder CD, Smith WL, Raz A, Masferrer J, Seibert K, Needleman P and Saper CB (1992) Distribution and characterization of cyclooxygenase immunoreactivity in the ovine brain *Journal of Comparative Neurology* **322** 409–438

Castro MI, Valego NK, Zehnder T and Rose JC (1993) The ratio of plasma bioactive to immunoreactive ACTH-like activity changes with severity of stress in the late gestation ovine fetus *American Journal of Physiology* **265** E68–E73

Challis JRG and Brooks AN (1989) Maturation and activation of hypothalamic–pituitary–adrenal function in fetal sheep *Endocrine Reviews* **10** 182–204

Challis JR, Davies IJ, Benirschke K, Hendrickx AG and Ryan K J (1974) The effects of dexamethasone on plasma steroid levels and fetal adrenal histology in the pregnant rhesus monkey *Endocrinology* **95** 1300–1305

Cudd TA and Wood CE (1992) Prostaglandin E2 releases ovine fetal ACTH from a site not perfused by the carotid vasculature *American Journal of Physiology* **263** R136–R140

Cudd TA and Wood CE (1995) Secretion and clearance of immunoreactive ACTH by fetal lung *American Journal of Physiology* **268** E845–E848

Cudd TA and Wood CE (1996) Does thromboxane mediate the fetal ACTH response to acidemia? *American Journal of Physiology* **270** R594–R598

Cudd TA Castro MI and Wood CE (1993) Content *in vivo* release and bioactivity of fetal pulmonary immunoreactive adrenocorticotropin *American Journal of Physiology* **265** E667–E672

Deauseault D and Wood CE (1998) Ontogeny of immunoreactive prostaglandin endoperoxide synthase-1 and -2 in ovine fetal and postnatal brainstem, hypothalamus, and pituitary *Journal of the Society for Gynecologic Investigation* **5** 154A(Abstract)

Eipper B A and Mains R E (1980) Structure and biosynthesis of pro-adrenocorticotrophin/endophin and related peptides *Endocrine Reviews* **1** 1–27

Farber DM, Giussani DA, Jenkins SI, Mecenas CA, Winter JA, Wentworth RA and Nathanielsz PW (1997) Timing of the switch from myometrial contractures to contractions in late-gestation pregnant rhesus monkeys as recorded by myometrial electromyogram during spontaneous term and androstenedione-induced by labor *Biology of Reproduction* **56** 557–562

Giussani DA, McGarrigle HHG, Moore PJ, Bennet L, Spencer JAD and Hanson MA (1994) Carotid sinus nerve section and the increase in plasma cortisol during acute hypoxia in fetal sheep *Journal of Physiology* **477** 81–87

Giussani DA, Riquelme RA, Moraga FA, McGarrigle HH, Gaete CR, Sanhueza EM, Hanson MA and Llanos AJ (1996) Chemoreflex and endocrine components of cardiovascular responses to acute hypoxemia in the llama fetus *American Journal of Physiology* **271** R73–R83

Harvey LM, Gilbert RD, Longo LD and Ducsay CA (1993) Changes in ovine fetal adrenocortical responsiveness after long-term hypoxemia *American Journal of Physiology* **264** E741–E747

Holtan DW, Houghton E, Silver M, Fowden AL, Ousey J and Rossdale P D (1991) Plasma progestagens in the mare fetus and newborn foal *Journal of Reproduction and Fertility* **44** 517–528

Kawai A and Yates FE (1966) Interference with feedback inhibition of adrenocorticotropin release by protein binding of corticosterone *Endocrinology* **79** 1040–1046

Keller-Wood M and Wood CE (1991a) Corticotropin-releasing factor in the ovine fetus and pregnant ewe: role of the placenta *American Journal of Physiology* **261** R995–R1002

Keller-Wood M and Wood CE (1991b) Does the ovine placenta secrete ACTH under normoxic or hypoxic conditions? *American Journal of Physiology* **260** R389–R395

Koos BJ, Mason BA and Ervin M G (1994) Adenosine mediates hypoxic release of arginine vasopressin in fetal sheep *American Journal of Physiology* **266** R215–R220

Liggins GC (1968) Premature parturition after infusion of corticotrophin or cortisol into foetal lambs *Journal of Endocrinology* **42** 323–329

Liggins GC (1969) Premature delivery of foetal lambs infused with glucocorticoids *Journal of Endocrinology* **45** 515–523

Liggins GC, Kennedy PC and Holm LW (1967) Failure of initiation of parturition after electrocoagulation of the pituitary of the fetal lamb *American Journal of Obstetrics and Gynecology* **98** 1080–1086

McMillen IC, Antolovich GC , Mercer JE, Perry RA and Silver M (1990) Proopiomelanocortin messenger RNA levels are increased in the anterior pituitary of the sheep fetus after adrenalectomy in late gestation *Neuroendocrinology* **52** 297–302

McMillen IC, Phillips ID, Ross JT, Robinson JS and Owens JA (1995) Chronic stress – the key to parturition? *Reproduction Fertility Development* **7** 499–507

Mason JI, Hinshelwood MM, Murry BA and Swart P (1993) Tissue-specific expression of steroid 17α-hydroxylase/C-17,20 lyase, 3β-hydroxysteroid dehydrogenase, and aromatase in the fetal horse *Proceedings of the Society for Gynecologic Investigation* 357(Abstract)

Matthews SG and Challis JR (1995) Regulation of CRH and AVP mRNA in the developing ovine hypothalamus: effects of stress and glucocorticoids *American Journal of Physiology* **268** E1096–E1107

Myers DA, Robertshaw D and Nathanielsz PW (1990) Effect of bilateral splanchnic nerve section on adrenal function in the ovine fetus *Endocrinology* **127** 2328–2335

Myers DA, McDonald TJ and Nathanielsz PW (1992) Effect of bilateral lesions of the ovine fetal hypothalamic paraventricular nuclei at 118–122 days of gestation on subsequent adrenocortical steroidogenic enzyme gene expression *Endocrinology* **131** 305–310

Myers DA Myers TR, Grober MS and Nathanielsz PW (1993) Levels of corticotropin-releasing hormone messenger ribonucleic acid (mRNA) in the hypothalamic paraventricular nucleus and propiomelanocortin mRNA in the anterior pituitary during late gestation in fetal sheep *Endocrinology* **132** 2109–2116

Ou CW Orsino A and Lye SJ (1997) Expression of connexin-43 and connexin-26 in the rat myometrium during pregnancy and labor is differentially regulated by mechanical and hormonal signals *Endocrinology* **138** 5398–5407

Pashen RL and Allen WR (1979) The role of the fetal gonads and placenta in steroid production, maintenance of pregnancy and parturition in the mare *Journal of Reproduction and Fertilty Supplement* **27** 499–509

Rice GE, Wong MH, Hollingworth SA and Thorburn GD (1990) Prostaglandin G/H synthase activity in ovine cotyledons: a gestational profile *Eicosanoids* **3** 231–236

Saoud CJ and Wood CE (1996a) Ontogeny and molecular weight of immunoreactive arginine vasopressin and corticotropin-releasing factor in the ovine fetal hypothalamus *Peptides* **17** 55–61

Saoud CJ and Wood CE (1996b) Ontogeny of proopiomelanocortin posttranslational processing in the ovine fetal pituitary *Peptides* **17** 649–653

Saoud CJ and Wood CE (1997) Modulation of ovine fetal adrenocorticotropin secretion by androstenedione and 17β-estradiol *American Journal of Physiology* **272** R1128–R1134

Scheuermann DW (1991) Neuroendocrine cells. In *The Lung: Scientific Foundations* pp 289–299 Eds RG Crystal and JB West Raven Press, New York

Schwartz J, Kleftogiannis F, Jacobs R, Thorburn GD, Crosby SR and White A (1995) Biological activity of adrenocorticotropic hormone precursors on ovine adrenal cells *American Journal of Physiology* **268** E623–E629

Simonetta G, Walker DW and McMillen IC (1991) Effect of feeding on the diurnal rhythm of plasma cortisol and adrenocorticotrophic hormone concentrations in the pregnant ewe and sheep fetus *Experimental Physiology* **76** 219–229

Sugimoto Y, Yamasaki A, Segi E, Tsuboi K, Aze Y, Nishimura T, Oida H, Yoshida N, Tanaka T, Katsuyama M, Hasumoto K, Murata T, Hirata M, Ushikubi F, Negishi M, Ichikawa A and Narumiya S (1997) Failure of parturition in mice lacking the prostaglandin F receptor *Science* **277** 681–683

Thorburn GD, Hollingworth SA and Hooper SB (1991) The trigger for parturition in sheep: fetal hypothalamus or placenta? *Journal of Developmental Physiology* **15** 71–79

Tong H, Lakhdir F and Wood CE (1998) Endogenous prostanoids modulate the adrenocorticotropin and vasopressin responses to hypotension in late-gestation fetal sheep *American Journal of Physiology (Reg Int Comp Physiol)* **275** R735–R741

Wood CE (1986) Sensitivity of cortisol-induced inhibition of ACTH and renin in fetal sheep *American Journal of Physiology* **250** R795–R802

Wood CE (1988) Insensitivity of near-term fetal sheep to cortisol: possible relation to the control of parturition *Endocrinology* **122** 1565–1572

Wood CE (1989a) Development of adrenal cortical function. In *Handbook of Human Growth and Developmental Biology* Vol. II: Part A pp 81–94 Eds P Timiras and E Meisami. CRC Press, Boca Raton FL

Wood CE (1989b) Sinoaortic denervation attenuates the reflex responses to hypertension in fetal sheep *American Journal of Physiology* **256** 1103–1110

Wood CE, Keil LC and Rudolph AM (1982) Hormonal and hemodynamic responses to vena caval obstruction in fetal sheep *American Journal of Physiology* **243** E278–E286

Wood CE, Chen H-G and Bell ME (1989) Role of vagosympathetic fibers in the control of adrenocorticotropic hormone, vasopressin, and renin responses to hemorrhage in fetal sheep *Circulation Research* **64** 515–523

Wood CE, Barkoe D, The A, Newman H, Cudd TA, Purinton S and Castro, MI (1998) Fetal pulmonary immunoreactive adrenocorticotropin: molecular weight and cellular localization *Regulatory Peptides* **73** 191–196

COMPARATIVE REPRODUCTIVE FUNCTION: IMPLICATIONS FOR MANAGEMENT

Chair
L. Zarco

Journal of Reproduction and Fertility Supplement **54**, 129–142

Implications of recent advances in reproductive physiology for reproductive management of goats

P. Chemineau[1*], G. Baril[1], B. Leboeuf[2], M. C. Maurel[1], F. Roy[1], M. Pellicer-Rubio[1], B. Malpaux[1] and Y. Cognie[1]

[1]*Institut National de la Recherche Agronomique, Station de Physiologie de la Reproduction des Mammifères Domestiques, URA CNRS 1291, 37380 Nouzilly, France;* [2]*Institut National de la Recherche Agronomique, Station Experimentale d'Insémination Artificielle 86480 Rouillé, France*

The control of reproduction in goats is interesting for technical reasons (synchronization of kiddings, adjustment to forage availability or to economy), and for genetic reasons (identification and dissemination of improved genotypes). The use of short-light rhythms leads to markedly increased production of semen per buck and prevents occurrence of a 'resting' season. Recent identification of a bulbourethral lipase in goat spermatozoa opens new perspectives in sperm preservation. Light plus 'short day' treatments also allow induction of out-of-season oestrous cycles and ovulations leading to enhanced fertility. Repeated use of eCG provokes the production of antibodies, delays the timing of ovulation and causes a reduction in fertility after fixed-time artificial insemination. All steps of embryo production, freezing and transfer are now controlled and allow the attainment of satisfactory numbers of kids born per donor female, which are compatible with the development of the technique for exchanging genotypes between countries. *In vitro* production of embryos allows high development rates to be achieved after *in vitro* maturation and fertilization of oocytes, and will ensure the production of synchronous populations of one-cell zygotes at the stage required by new biotechnologies.

Introduction

As in other domestic species, the control of reproduction in goats offers advantages at the farm and at the level of the population where genetic improvements can be made. The first advantage is the choice of a kidding period at a given time of the year (adjustment to favourable external conditions imposed by the season of forage growth or by marketing of the products). The second advantage is synchronization of kiddings over a reduced period leading to a reduction in kid mortality, constitution of homogeneous groups of mothers and allowing kids to be fed more adequately to their requirements, and optimization of labour for care of the animals. The third advantage of controlling reproduction in goats is that it allows manipulation and storage of the genetic material. Artificial insemination (AI), even used on a small scale, allows links between herds and this increases the efficiency of indexation of sires. Early and accurate estimation of the genetic value of young bucks is feasible. Once identified, the improved males can be used in a large number of herds. Embryo transfer increases the number of progeny from a genetically superior female and is a method for exchanging genotypes without transmitting diseases. Finally, *in vitro* production of embryos, in the near future will give access to the genome of the one-cell embryo.

In this review only a limited number of techniques that have undergone marked progress in recent years are discussed. These techniques are recommended for use in intensive systems in which the income per goat per year is very high, generally because of the price of goat milk.

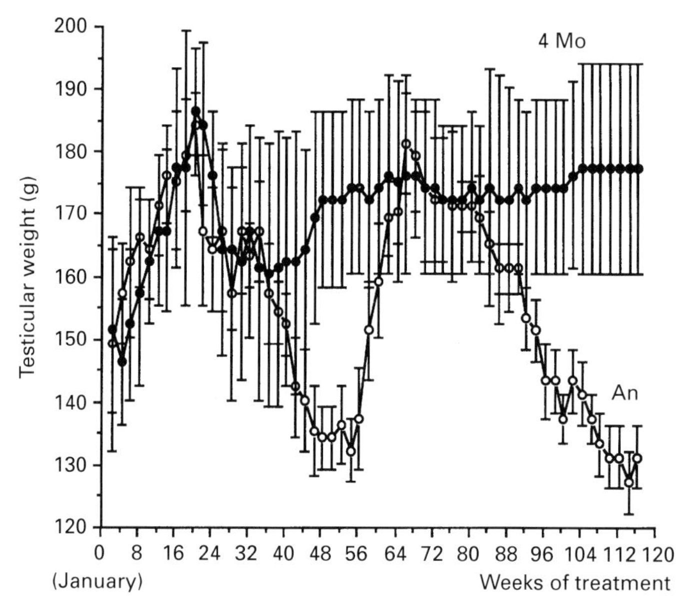

Fig. 1. Testicular weight of Alpine and Saanen bucks treated or not with an accelerated light rhythm of 4 months. Values are monthly means ± SEM. An: natural photoperiodic variations at 46° N latitude (○, *n* = 6). 4 Mo: alternation between two months of long days (16 h light: 8h dark) and two months of short days (8 h light:16 h dark) (●, *n* = 6) (B. Leboeuf and P. Chemineau, unpublished)

Sperm Production and Processing

The application of photoperiodic treatments to bucks of seasonal breeds alleviates the problem of seasonality of sperm production. Initially developed in rams, short light rhythms (that is alternations between 1 or 2 months long days (16 h light: 8 h dark; 16L:8D; LD) and 1 or 2 months short days (8L:16D; SD)) overcome seasonal variations in testis size and sperm production. Alpine and Saanen bucks, subjected for 3 consecutive years to photoperiodic treatments showed a marked increase in all parameters of sperm production, compared with control bucks under natural photoperiod (Delgadillo *et al.*, 1993). When collected twice a week, the total number of spermatozoa produced was improved by 61% (Delgadillo *et al.*, 1991). Semen quality after deep-freezing no longer exhibited the marked seasonal changes observed in untreated bucks. The total number of AI doses produced during the first 2 years of the treatment was much higher (62%) than that produced by control males. Fertility of the semen was not significantly altered by such treatments, in spite of a slight decrease in fertility rate in one group of bucks (Delgadillo *et al.*, 1992).

From these results, it was also apparent that the collection rate (twice a week) could be increased in treated males. It was therefore decided to compare overall sperm production of treated bucks collected four times a week all year round, with sperm production of control bucks collected four times a week from September to February only, as is normal practice. During the 24 months of treatment, as expected, testicular weight of bucks remained constant, at the maximal weight of the full sexual season, while testicular weight of control males underwent the normal seasonal variations (Fig. 1).

As a consequence, sperm production either in terms of total number of spermatozoa produced, or in terms of AI doses, was significantly improved by the treatment (2212 versus 3111 doses per buck). Fertility of AI doses was slightly, although not significantly, lower for light-treated bucks (Table 1; B. Leboeuf and P. Chemineau, unpublished).

Such a high production probably originates from unexpected changes in spermatogenic

Table 1. Number of bucks, collection rhythm, number of AI doses produced per year and fertility after artificial insemination of control bucks (subjected to natural lighting), or treated with short light rhythms of alternation of long days (16 h of light per day) and short days (8 h of light per day); artificial inseminations are done in each flock after distribution of the females to be inseminated in each group of control or treated bucks (Data from Delgadillo *et al.* 1992; and B. Leboeuf and P. Chemineau unpublished results).

<table>
<tr><th></th><th colspan="3">Experimental groups</th></tr>
<tr><th></th><th>Control groups (natural lighting)</th><th>1 month LD/ 1 month SD</th><th>2 months LD/ 2 months SD</th></tr>
<tr><td>Experiment 1 (Delgadillo et al., 1992) (1599 goats in 58 herds)</td><td></td><td></td><td></td></tr>
<tr><td>Number of bucks</td><td>6</td><td>6</td><td>6</td></tr>
<tr><td>Collection rhythm</td><td>2 ejaculates/week</td><td>2 ejaculates/week</td><td>2 ejaculates/week</td></tr>
<tr><td>Number of AI doses (at 200 × 10⁶ sperm/doses) produced per year and per buck</td><td>253</td><td>427</td><td>391</td></tr>
<tr><td>Fertility (% producing kids)</td><td>62.5</td><td>57.9</td><td>57.8</td></tr>
<tr><td>Experiment 2 (B. Leboeuf and O. Chemineau, unpublished) (785 goats in 25 herds)</td><td></td><td colspan="2"></td></tr>
<tr><td>Number of bucks</td><td>6</td><td colspan="2">6</td></tr>
<tr><td>Collection rhythm</td><td>4 ejaculates per week from Sept to Feb</td><td colspan="2">4 ejaculates per week, all the year round</td></tr>
<tr><td>Number of AI doses (at 100 × 10⁶ sperm/doses) produced per year and per buck</td><td>1106</td><td colspan="2">1556</td></tr>
<tr><td>Fertility (% producing kids)</td><td>69.5</td><td colspan="2">61.2</td></tr>
</table>

processes. Light-treated bucks had significantly increased numbers of spermatogonia (the stem cell of the spermatogenic line) while maintaining spermatogenic divisions at the high rate of the full sexual season (Delgadillo *et al.*, 1995). By allowing sperm collection all the year round rather than for 6 out of 12 months, these photoperiodic treatments may accelerate the production of AI doses in young bucks during the 2.5 years of progeny testing. This photoperiodic treatment is now used to improve sperm production of one-year-old bucks in the French national selection programme.

The most recent data obtained in the field of semen technology have been the identification of a seminal plasma enzyme that decreases sperm survival *in vitro*. Egg yolk or skim milk is widely used in extenders for mammalian semen because of their protective role against cold shock of spermatozoa. However, the cryopreservation of goat semen in these media requires that most of the seminal plasma be removed before sperm dilution (washing method) to improve the survival of spermatozoa after freezing and thawing. The bulbourethral gland secretion (BUS) is the fraction of goat seminal plasma responsible for deterioration of sperm viability in egg yolk (Roy, 1957) and milk-based diluents (Nunes *et al.*, 1982). The egg-yolk coagulating enzyme (EYCE) from goat BUS displays phospholipase A activity and may hydrolyse egg yolk lecithin into fatty acids and lysolecithin (Roy, 1957; Iritani and Nishikawa, 1972) which are toxic to goat spermatozoa (Aamdal *et al.*, 1965). The BUS component has been recently purified and identified as a 55–60 kDa glycoprotein (BUSgp60) with triglyceride lipase activity (Pellicer-Rubio *et al.*, 1997). Indeed, BUSgp60 provokes a decrease in the percentage of motile spermatozoa, a deterioration in the quality of movement, breakage of acrosomes and cellular death of goat epididymal spermatozoa diluted in skim milk (Fig. 2).

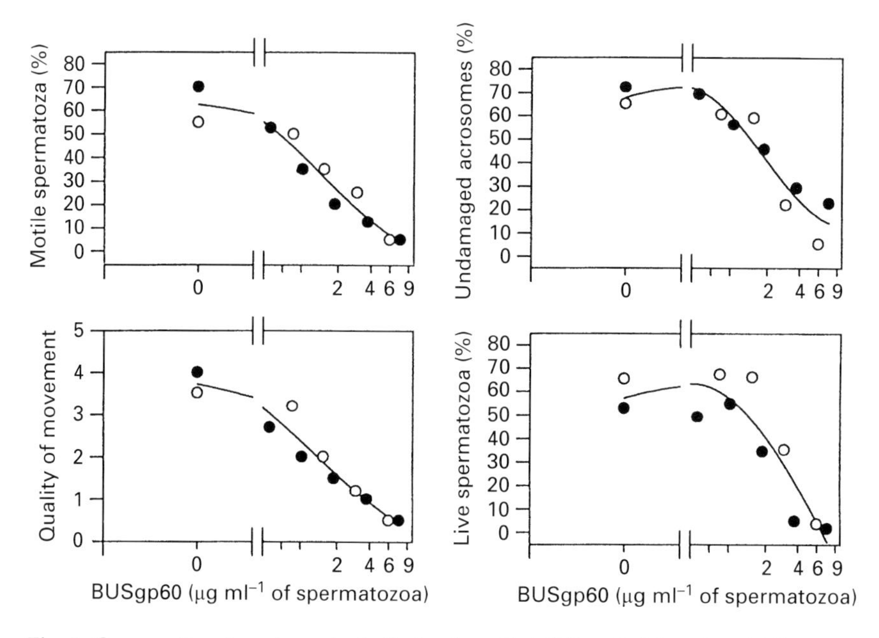

Fig. 2. Comparative dose-dependent effects of purified BUSgp60 (○) and BUS (●) on the quality parameters of goat epididymal spermatozoa diluted in skim milk after incubation for 60 min at 37°C. Values are the mean of two observations. Doses of BUS are expressed as equivalent doses of BUSgp60 (Adapted from Pellicer-Rubio *et al.*, 1997).

The catalysis of oleic acid formation from residual milk triglycerides by BUSgp60 appears responsible for these effects (Pellicer-Rubio and Combarnous, 1998). Interestingly, BUSgp60 has been classified as a novel lipase most probably belonging to the pancreatic lipase-related protein 2 (PLRP2) family (Pellicer-Rubio *et al.*, 1997). Since PLRP2 enzymes are known to display both phospholipase A and lipase activities (Carriere *et al.*, 1994), it has been suggested that BUSgp60 and EYCE are related or even identical enzymes (Pellicer-Rubio and Combarnous, 1998). These results allow the possibility of specifically inhibiting BUSgp60 lipase in milk-based extenders, and avoiding the harmful step of washing goat semen before deep-freezing. Moreover, the use of BUSgp60 inhibitors for better cryopreservation of unwashed goat semen in egg-yolk diluents should be considered.

Induction of Out-of-Season Cyclicity in the Female Goat by Using Photoperiodic Treatments

Appropriate treatment of animals with melatonin could be used to mimic short days while their visual system perceived long days (Chemineau *et al.*, 1992; Deveson *et al.*,1992a; Malpaux *et al.*, 1993), to induce an advance of ovulatory and oestrous activities. However, when used alone in highly seasonal breeds, melatonin treatment provides a maximum advance of only 1.5 months. This is not satisfactory for many farmers, especially in the dairy goat industry in France, who wish to induce a complete out-of-season breeding (that is from April to July). Under such conditions, melatonin treatment should be preceded by at least 2 months of a light treatment composed daily either of long days (Deveson *et al.*, 1992a), or of two periods of supplementary light (Fig. 3; Chemineau *et al.*, 1992). Such long day (LD) treatment probably provides the photoperiodic signal for the onset of the annual breeding season and also restores sensitivity to melatonin (Chemineau *et al.*, 1992; Malpaux *et al.*, 1993). In French dairy goats maintained in open barns, the use of this succession LD + melatonin followed by a 'buck effect' with 'light'-treated bucks, induces ovulatory and oestrous activities that

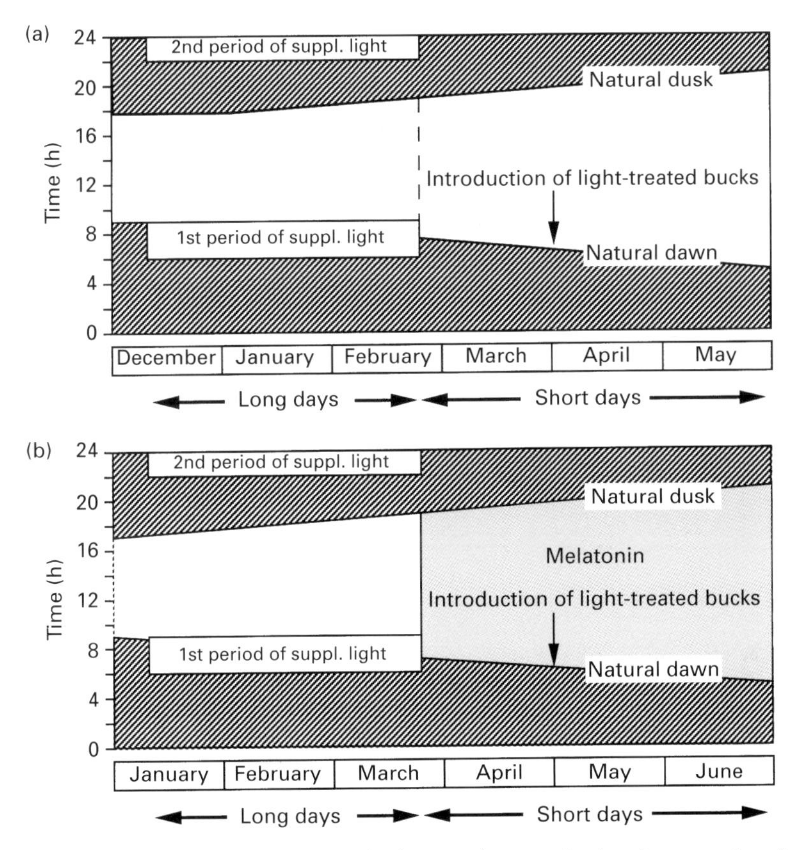

Fig. 3. Photoperiodic treatments applied in open barns and using the succession of long day and short day treatment followed by a 'buck-effect'. The upper treatment (a) is applied early in the year when natural day lenght is short, and the lower treatment (b), using melatonin after the end of the long day period is applied later in the year.

are sufficient to achieve a fertility and prolificacy close to those of the normal annual breeding season (Chemineau *et al.*, 1996).

The LD treatment must be for longer than 2 months; the melatonin concentration provided by the implants should be optimized; and bucks treated with LD + melatonin should be introduced for natural mating 35–70 days after the onset of melatonin treatment (Chemineau *et al.*, 1996). If these conditions are met, peak rates of conception generally occur about 10 days after introduction of bucks and some females conceive at the return to oestrus one cycle later. Such photoperiodic treatments may change the speed of hair growth (Gebbie, 1993) and light-treatment during pregnancy was shown to delay the onset of puberty by about 4 weeks in young female goats born from light-treated mothers (Deveson *et al.*, 1992b).

More recently, it was demonstrated that when applied early in the season (ending before the end of March), the melatonin treatment was not necessary and the return to natural lighting after LD allowed satisfactory fertility rates (Table 2).

Hormonal Synchronization of Oestrus

Hormonal treatment of female goats to induce a synchronous onset of oestrous behaviour and ovulation within a limited time after the end of the treatment is a prerequisite to the use of AI. The

Table 2. Fertility of goats after photoperiodic treatment alone in various flocks and comparison within flock of the treatment with and without melatonin.

	No females	Fertility (No. producing kids)	Litter size (No. kids born/ lambing
Photoperiodic treatment alone Natural mating, 20 herds (GRC, 1996)	3236	76.8%	1.83
Photoperiodic treatment with and without melatonin Natural mating 1996+1997, 1 single herd (B. Lebeouf and P. Chemineau, unpublished)			
With melatonin	126	75.3%	2.07
Without melatonin	115	73.0%	2.00

association between a progestagen (delivered by a vaginal sponge or by a subcutaneous implant), a prostaglandin analogue and PMSG (now called equine chorionic gonadotrophin, eCG) remains the most efficient tool to achieve this objective. These treatments are now widely and successfully used all over the world to control reproduction in female goats. Their use in association with AI on a fixed-time basis in thousands of goats has led to high levels of fertility (Leboeuf *et al.*, 1998). This treatment could also be applied to young goats if specific conditions are respected.

Recent experiments were performed to test modifications to reduce the variability in the interval between the end of the sponge–eCG treatment and onset of oestrus. Neither increase in the quantity of fluorogestone acetate (FGA) delivered by the sponge, nor the use of subcutaneous ear implants reduced this variability (Freitas *et al.*, 1996a, 1997a). Neither the number of corpora lutea, nor the number and size of the follicles observed on the ovary before and during the FGA treatment strongly influenced the response (Freitas *et al.*, 1996b). Finally, it was observed that during natural cycles, the variability in the interval between luteolysis and the onset of oestrus or onset of the LH surge was higher than after FGA–prostaglandin treatment (Freitas *et al.*, 1997b). Thus, it was concluded that further improvements of the 'classic' hormonal treatment would be difficult to obtain.

Paradoxically, when eCG is used repeatedly on the same females, its efficiency decreased. In a single Saanen herd of 169 females in which breeding takes place each year out of season after FGA and eCG treatment, the percentage of goats showing oestrus and producing kids was significantly lower for multiparous than for nulli- and primiparous goats (64 versus 99, and 34 versus 67%, respectively). When goats were treated for the second time during the same year, the percentage showing oestrus was lower than after the first treatment (45 versus 71%; Baril *et al.*, 1992). This situation is due to the appearance of antibodies against eCG (Roy *et al.*, 1995; see later). When eCG binding of the serum was calculated by radioimmunoassay, and expressed as percentage of bound radioactive eCG with plasma (Baril *et al.*, 1992), this percentage was associated with fertility results. Before the treatment, it was higher in multiparous than in nulli- and primiparous goats (18 versus < 1%), and higher in non-pregnant than in pregnant goats (26 versus 7%) (Baril *et al.*, 1992). These results obtained in a single herd have prompted large-scale surveys in private flocks, using FGA/eCG treatments, associated with 'classic' AI with deep-frozen semen. In the first survey, oestrous behaviour was induced in almost all treated goats (98.1% of the 368 Alpines and 272 Saanens goats of 19 private herds) between 24 and 72 h after sponge removal. The distribution of the onset of oestrus after sponge removal did not differ between breeds or with age but was affected by the number of treatments previously received by the females and seemed to increase markedly after the second treatment (Fig. 4). Fertility but not prolificacy after AI was negatively correlated with the interval between sponge removal and onset of oestrus ($R = 0.92$). Fertility of goats that came into

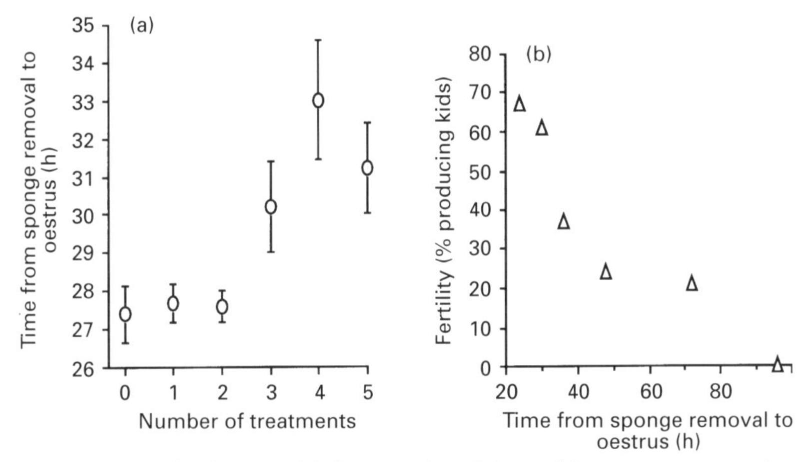

Fig. 4. Relationship between (a) the mean (± SEM) interval from sponge removal to onset of oestrus and the number of treatments previously received by the female goat and (b) between the interval from sponge removal to onset of oestrus and fertility (Adapted from Baril *et al.*, 1993).

oestrus later than 30 h after sponge removal was significantly lower than for those that were first observed in oestrus 24 or 30 h after sponge removal (33 versus 65% respectively, Fig. 4; Baril *et al.*, 1993). This delay in the onset of oestrous behaviour is associated with a delay in the LH preovulatory surge (Maurel *et al.*, 1992) and with a delay in the time of ovulation (Leboeuf *et al.*, 1993, 1996). In the second survey, eCG binding (measured in 524 dairy goats of 17 private herds) before the onset of treatment was significantly lower in herds in which treatments were never used than that measured in samples of the other goats and was not dependent on the age of the female. Binding was increased in the females that had previously received from two to five treatments, compared with that in females that had received no treatment or one treatment (3 versus 10%). On an individual basis, the percentage of goats showing onset of oestrus behaviour more than 30 h after sponge removal was higher (38 versus 7%) and fertility was decreased (51 versus 66% on 166 versus 353 females) when eCG binding was higher (more than 10% of radioactive eCG binding versus less than 5%). When measured 25 days after eCG injection, eCG binding was increased (7% before injection versus 28% after injection), and correlated with binding detected before treatment (Baril *et al.*, 1996b).

Complementary studies were conducted to evaluate the induction of an anti-eCG humoral immune response after a 500 iu eCG injection. An ELISA was developed to quantify the plasma concentration of anti-eCG antibodies and compare kinetics of antibody secretion between individuals (F. Roy *et al.*, 1999). For this experiment 15 goats were treated for the first time with eCG and exhibited an increasing concentration of anti-eCG antibody 10 days (day 10) after eCG injection. Maximum values were reached between day 10 and day 17; thereafter antibody concentration showed a progressive decline over 2 months. Goats previously treated (one or more times) with eCG ($n = 29$) displayed similar kinetics of humoral immune response, except that they exhibited an earlier increase in antibody concentration at day 7 and a longer decreasing phase of the antibody secretion (Fig. 5). Within both treatment groups, all goats had an identical immune response but differed markedly in their anti-eCG concentrations, regardless of the number of previous treatments. Indeed, maximal anti-eCG antibody concentrations varied from 0.7 to 102 μg ml^{-1} in goats treated for the first time and from 3.0 to 219 μg ml^{-1} in goats treated several times. Nevertheless, in spite of the heterogeneity of antibody secretion, results showed that mean antibody concentration measured before treatment increased significantly ($P < 0.05$) as a function of the number of previous eCG treatments. The antibody concentration measured before treatment was defined as residual anti-eCG antibodies. These antibodies resulted from the previous immune response induced by the last eCG injection (about one year before). High residual antibody concentration resulted in decreased

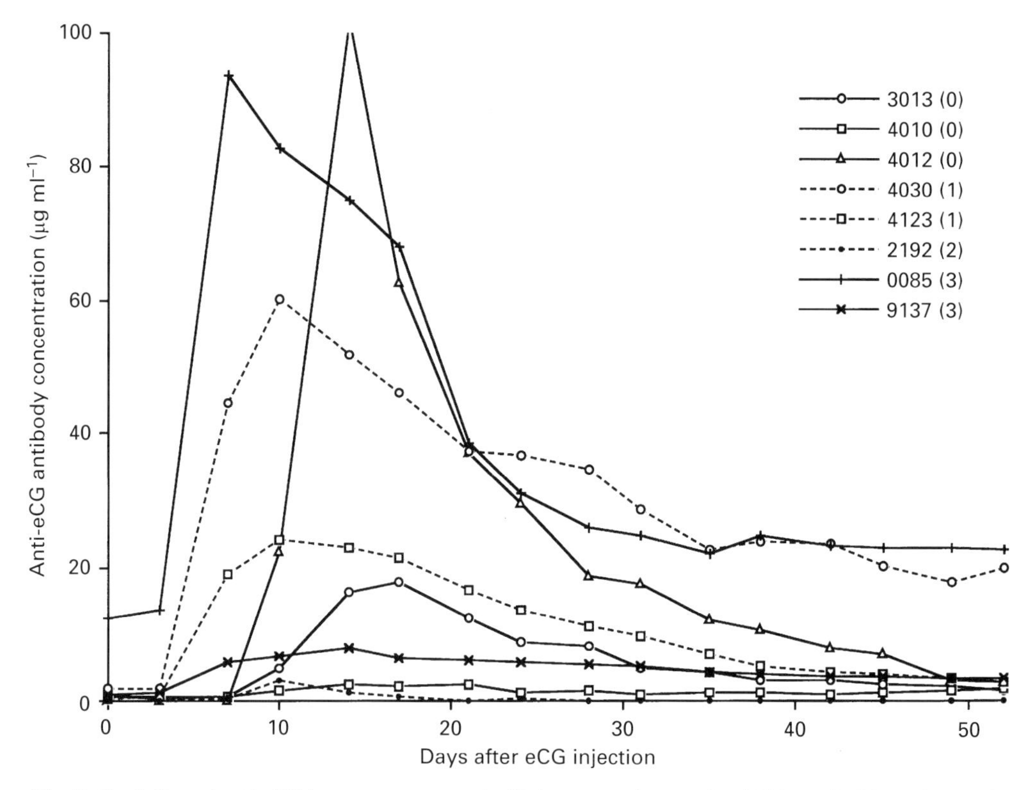

Fig. 5. Evolution of anti-eCG immune response in Alpine goats that received 500 iu of eCG at day 0. The eight goats were considered as representative of the entire group. Anti-eCG antibody concentration was determined by ELISA in plasma samples. Number of previous eCG treatments are indicated in parenthesis. Each point is the mean of duplicate determinations (Adapted from from Roy *et al.*, 1995).

fertility on subsequent treatment for inseminated females, in contrast to antibody concentration measured during the following immune response (at day 10 and day 25).

Fertility of female goats that exhibited oestrous behaviour more than 30 h after sponge removal (representing only 18% of the sample in the previous experiments) is low probably because of their delayed ovulation and because of the use of deep-frozen semen which has a limited lifespan. When these females are artificially inseminated later, adequately with oestrous detection, their fertility was not altered (B. Leboeuf and G. Baril, unpublished). Thus, we recommend artificial insemination only of the females that are detected to be in oestrus 30 h after sponge removal.

Pseudopregnancies

The fertility of goats after artificial insemination can be reduced by pseudopregnancy at the time of induction of oestrus by progestagen or eCG or by other means. Several field trials using ultrasonography have shown that pseudopregnancy appeared in 3–4% of does, sometimes in 20% in some herds (Mialot *et al.*, 1991; Hesselink, 1993; Leboeuf *et al.*, 1994). Pseudopregnancy was related to breed in some trials (Leboeuf *et al.*, 1994) but not in others (Mialot *et al.*, 1991), with reproduction method (3.8% in 1 493 FGA/eCG treated goats versus 2.5% in 3 774 naturally mated goats; Mialot *et al.*, 1991), with sire (20% of 125 daughters from five sires versus 0% of 326 daughters from 12 sires in the same herd; Soulière 1991), with parity (1% of nulliparous versus 18% of primiparous or

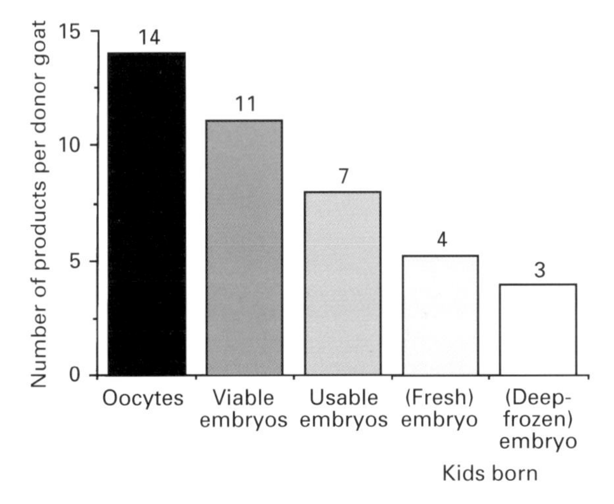

Fig. 6. Number of products per donor female goat at each step of *in vivo* embryo production, collection and transfer, using fresh or deep-frozen embryos (Adapted from Baril, 1995).

multiparous; Hesselink 1993), and with age (10% of 280 does < 5 years old versus 32% of 34 does > 6 years old; Hesselink, 1993).

A recent study perfomed in artifically inseminated dairy goats has demonstrated that more than 50% of the pseudopregnancies identified by ultrasonography at about 45 days after AI followed a late embryonic mortality. Among 67 dairy goats that were diagnosed pseudopregnant 40–60 days after AI, 54% had detectable concentrations of PSPB, a placental hormone, demonstrating the presence of fetal tissue, 30–36 days after AI (Humblot *et al.*, 1995). Thus, at least 50% of so-called pseudopregnancies were the consequence of late embryonic mortality. The reason for this is unknown.

Treatment with a prostaglandin analogue stopped pseudopregnancy (Hesselink 1993), and restored fertility. However, after one or two prostaglandin analogue injections (100 μg), followed 10 days later by progestagen/eCG treatment, the fertility after AI was only 45% (n = 286, Leboeuf *et al.*, 1998).

Embryo Production, Collection, Freezing and Transfer

Embryo transfer is used less frequently in goats than it is in bovine species and is used in goats mainly for the international exchange of genetic material between countries with a concomittant marked reduction in the risk of disease transmission, when international protocols for embryo manipulations are respected. Embryo transfer is also used in transgenic goat programmes to maximize the number of day 3 embryos to be micro-injected (Gootwine *et al.*, 1997).

Donor female goats receive a progestagen treatment ending with gonadotrophic preparation injections to stimulate follicular growth and induce superovulation. The use of FSH is now widely accepted (in fact more or less purified pituitary extracts) rather than eCG to achieve high rates of superovulation. If collection is to be repeated on the same donor females, ovine or caprine FSH (o or cFSH) should be used instead of porcine FSH (pFSH) because of the rapid appearance of antibodies against pFSH which limits the superovulatory response of the females (Remy *et al.*,1991). oFSH can be injected 6–8 times, at intervals of 12 h during the last 3–4 days of the progestagen treatment, with a total dose (for Alpine and Saanen goats) from 16 to 21 mg (standard Armour units). This dose should be adapted to genotype. The use of constant versus decreasing doses may be dependent on the origin of the preparation (Baril, 1995). An FSH:LH ratio increased with 40% LH seems adequate

(Nowshari *et al.*, 1995). On average, the number of ovulations induced by such treatments ranges from 12 to 16 ovulations per goat. However, it should be noted that there is a large variability between females (from 0 to 40, Baril *et al.*, 1995) and that a seasonal effect was described in seasonal breeds (Gootwine *et al.*, 1997).

One of the limitations of these superovulatory treatments comes from the early regression of corpora lutea in 10–35% of the treated females, about 6–8 days after oestrus. The associated decrease in plasma progesterone led to a marked decrease in collection rate (Borque *et al.*, 1993). Even accounting for low body condition score, which could be one of the reasons for such luteal regression, the main causes remain unknown. Use of an antiluteolytic compound or progesterone injections has been described with varying success rates (review Baril,1995).

Successful fertilization of donor females depends on synchronization of ovulation and on the method used to inseminate the females. A reduction in the range in ovulation timing (that is, the time elapsed between the first and the last ovulation) and an increase in ovulation rate was obtained using GnRH injections at a fixed time after the end of the progestagen treatment (Akinlosotu and Wilder, 1993; Krisher *et al.*, 1994). Another alternative is the use of a GnRH antagonist, 12 h after sponge removal, followed by an intravenous injection of 3 mg pLH 24 h later, which mimics the preovulatory LH surge and allows the artificial insemination of the females only once, 16 h after LH injection (Baril *et al.*, 1996a). Natural insemination (mating) can be used satisfactorily (fertility about 80%), but fertility can be reduced during the anoestrous season. If AI with deep-frozen or liquid semen from improved males is used, classic deposition of the semen via the cervix leads to reduced fertilization rates, especially for high ovulation rates. Intra-uterine deposition of the semen after laparoscopy allows the achievement of fertilization rates equivalent to those obtained after natural mating (Vallet *et al.*, 1991).

Embryos can be collected at days 6, 7 and 8 by laparotomy which allows for high collection rates but only once or twice per animal. Collection under laparoscopic control should be used for repetitive collections on the same females (up to 7; Baril 1995). Collection via the cervix should be discounted because penetration into the uterine horns is difficult and collection rates remain low (Soonen *et al.*, 1991; Flores-Foxworth *et al.*, 1992).

Deep freezing of goat embryos is feasible using classic techniques derived from those used in the bovine species. *In vitro* development of frozen–thawed blastocysts was higher than that of frozen morulae whatever the cryoprotectant, glycerol or ethylene glycol, (blastocyst 40.8% $n = 129$, and morula 14.3% $n = 161$; $P < 0.01$). However, *in vivo*, frozen–thawed morulae developed as well as blastocysts did. But for both stages, more embryos developed to term when embryos were frozen with ethylene glycol (51%, $n = 100$), than with glycerol (30%, $n = 83$; $P < 0.08$, Le Gal *et al.*, 1993). Successful vitrification of goat embryos has also been described (Yuswiati and Holtz, 1990). A preliminary result indicated that a similar pregnancy rate was obtained after embryo transfer (ultrasound diagnosis on day 43) for the two methods of cryopreservation, vitrified versus deep-frozen embryos (vitrified 5 pregnancies/7 recipient goats versus deep-frozen 7/8; Traldi *et al.*, 1997).

Transfer should be carried out via laparoscopy which gives equivalent or higher fertilization rates than laparotomy (Baril, 1995) and higher fertility than via the cervix (Flores-Foxworth *et al.*, 1992). Nutrition of recipient goats before and after transfer should be adequate to reach a high fertility (25 versus 67% of kiddings in restricted versus normal-fed Angora goats; Mani *et al.*, 1994).

The number of kids born per donor goat (collected once) varied from three to four, depending on whether embryos were deep frozen or not (Fig. 5; Baril, 1995).

In Vitro Production of Embryos

It is now possible to achieve development to term after transfer of blastocysts produced completely *in vitro* to recipient females(Crozet *et al.*, 1993; Keskintepe *et al.*, 1994). For generation of blastocysts, different steps must be achieved *in vitro*: the maturation of ovarian oocytes (IVM), the capacitation of spermatozoa and fertilization events (IVF) and early cleavage and development to the blastocyst stage in culture (IVC). However, for producing oocytes *in vitro* with full developmental capacity, it is necessary to select oocytes at the end of their growth phase when they became competent for supporting meiotic maturation and embryonic development. Oocytes from small (2–3 mm diameter) and medium follicles (3.1–5.0 mm diameter) yielded a significantly lower proportion of blastocysts than those from large follicles (> 5 mm diameter) (24 versus 39 versus 53%, respectively; Cognié *et al.*, 1996). Ovulated oocytes, fertilized and cultured *in vitro* under the same conditions, yielded 70% blastocysts (G. Baril, N. Poulin and Y. Cognie, unpublished) indicating that the conditions of maturation (*in vivo* or *in vitro*) may also influence the developmental potential of the oocyte. Important progress has been made regarding the development of the optimal medium for maturation of oocytes which consists of caprine follicular fluid (10%) and FSH (100 ng ml^{-1}) in medium M199 under 5% CO_2 allowing a simplification (omitting co-culture with granulosa cells) and better efficiency of the IVM method (Poulin *et al.*, 1996; Cognié *et al.*, 1996).

The age of the donor female may also influence the quality of the oocyte. Oocytes collected from prepubertal goats demonstrated a lower percentage of normal fertilization after IVM than oocytes from adult goats (Martino *et al.*, 1995).

Collection at the abattoir of oocytes from ovaries by aspiration or dissection of follicles provides 1.5–2.1 oocytes per ovary (Martino *et al.*, 1994; Pawshe *et al.*, 1994). Slicing the goat ovary was found to be more efficient for recovering a large number of cumulus–oocyte complexes (six complexes per ovary; Martino *et al.*, 1994), but the extra oocytes, obtained essentially from small follicles, are less competent to develop after IVF (Keskintepe *et al.*, 1994). Ultimately, these three collection techniques seem to be equivalent in terms of embryo yield (Pawshe *et al.*, 1994).

An average of nine cumulus–oocyte complexes per ovary (including four complexes from follicles larger than 5 mm) can be obtained with FSH-primed goats (Crozet *et al.* 1995). When recovery is to be done on improved females, oocytes can also be collected repeatedly(once a week) by laparoscopic aspiration which allows the recovery after FSH priming of 3– 4 cumulus–oocyte complexes per ovary (Todini *et al.*, 1994; Graff *et al.*,1995, respectively). High fertilization rates (about 85%) are achieved using culture media supplemented with serum from oestrous sheep to induce capacitation in spermatozoa (De Smedt *et al.*, 1992). These conditions are also efficient for frozen semen (Cognié *et al.*, 1992). After discarding polyspermic eggs (10–20%), an average of 70% *in vitro* fertilized eggs can be routinely obtained. Procedures for sperm capacitation and IVF conditions will ensure synchronized sperm–egg penetration and, consequently, the production of synchronous populations of one-cell embryos at the stage required for gene injection in pronuclei, performed 14–18 h after insemination. Heparin was shown to increase sperm–egg penetration when added to IVF medium containing sheep serum (Cox *et al.*, 1994) but the quality of the embryos produced with heparin treatment is questionable (Poulin *et al.*, 1996).

For IV development, culture of early embryos (2- to 4-cell embryos) in the presence of oviduct cells leads to significantly more blastocysts and hatched blastocysts than culture with uterine cells or culture in medium alone (Prichard *et al.*, 1992). With the continued refinement of culture techniques, an alternative system, a simple balanced salt solution (SOF: synthetic oviduct fluid) supplemented with amino acids and serum and incubated under an atmosphere of 5% 0_2, 5% $C0_2$, 90% N_2, is being used. Under these conditions, the developmental ability of blastocysts to term after transfer is close to the developmental rate of their *in vivo* counterparts (61% of *in vitro* produced blastocysts gave birth to live young kids; Poulin *et al.*, 1996). Promising results have now been obtained in survival rate of vitrified–thawed and transferred embryos produced *in vitro* (A. Traldi *et al.*, 1998).

Conclusion

Rapid and significant progress has been made in the control of reproduction in goats, in the study of various treatments applied to animals at farm and AI centres, as well as in the field of *in vitro* treatment of caprine gametes. However, in each of these areas, new research results are needed.

In the study of semen production and processing and of AI, additional progress is needed for improving the efficiency of deep-freezing techniques . The use of BUSgp60 lipase inhibitors in seminal plasma for improving sperm viability in milk-based or egg-yolk diluents should be tested.

One of the major problems to be addressed regarding hormonal control of oestrus is reducing the effects of repeated use of eCG, which reduces the fertility of artificially inseminated females. The reasons for the large inter-individual variability in animal response and the development of new products to be administered to female goats in replacement of eCG are the two main directions that should be followed.

Major advances have been made in the area of *in vivo* embryo production, collection, freezing and transfer, and this is now a technique that can be used for exchanging improved breeds with a reduced risk of disease transmission.

In vitro production of embryos has undergone major and rapid progress in recent years, but significant progress in the yields of the different steps still need to be made, to increase the commercial value of the technique. It is reasonable to expect that we will soon obtain the same yield as for *in vivo* production, but at a lower price than current *in vitro* production costs.

A part of the experiments presented here were supported by grants from Région Centre and Région Poitou-Charentes and by grants of the French Ministry of Agriculture and of SANOFI Santé Nutrition Animale.

References

Aamdal J, Lyngset O and Fossum K (1965) Toxic effect of lysolecithin on sperm. A preliminary report *Nordic Veterinary Medicine* **17** 633–634

Akinlosotu, BA and Wilder CD (1993) Fertility and blood progesterone levels following LHRH-induced superovulation in FSH-treated anestrous goats. *Theriogenology* **40** 895–904

Baril G (1995) Possibilidades atuais da transferência de embriões em caprinos. *(Present possibilities of goat embryo transfer)* Proceedings of XI Congreso Brasileiro de Reproduçao Animal, Belo Horizonte, pp. 110–120

Baril G, Remy B, Vallet JC and Beckers JF (1992) Effect of repeated use of progestagen–PMSG treatment for estrus control in dairy goats out of breeding season *Reproduction in Domestic Animals* **27** 161–168

Baril G, Leboeuf B and Saumande J (1993) Synchronization of oestrus in goats: the relationship between time of occurrence of estrus and fertility following artificial insemination *Theriogenology* **40** 621–628

Baril G, Pougnard JL, Freitas VJF, Leboeuf B and Saumande J (1996a) A new method for controlling the precise time of occurrence of the preovulatory gonadotropin surge in superovulated goats *Theriogenology* **45** 697–706

Baril G, Remy B, Leboeuf B, Beckers JF and Saumande J (1996b) Synchronization of estrus in goats : the relationship between eCG binding in plasma, time of occurrence of estrus and fertility following artificial insemination *Theriogenology* **45** 1553–1559

Borque C, Pintado B, Perez B, Gutierrez A, Monoz I and Mateos E (1993) Progesterone levels in superovulated Murciana goats with or without successful embryo collection *Theriogenology* **39** 192 (Abstract)

Carriere F, Thirstrup K, Boel E, Verger R, and Thim L (1994) Structure–function relationships in naturally occurring mutants of pancreatic lipase *Protein Engineering* **7** 563–569

Chemineau P, Malpaux B, Delgadillo JA, Guérin Y, Ravault JP, Thimonier J and Pelletier J (1992) Control of sheep and goat reproduction: use of light and melatonin *Animal Reproduction Science* **30** 157–184

Chemineau P, Malpaux B, Pelletier J, Leboeuf B, Delgadillo JA, Deletang F, Pobel T and Brice G (1996) Emploi des implants de mélatonine et des traitements photopériodiques pour maîtriser la reproduction saisonnière chez les ovins et les caprins *(Use of melatonin implants and photoperiodic treatments to control seasonal reproduction in sheep and goats) INRA Productions Animales* **9** 45–60

Cognié Y, Guérin Y, Poulin N and Crozet N (1992) Successful use of frozen buck semen for *in vitro* fertilization of ovulated goat oocytes *Proceedings of the 5th International Conference on Goats, New Delhi, India* **3** 1248–1252 Ed. RR Lokeshwar, International Goat Association. Little Rock, Ark, USA

Cognié Y, Poulin N and Lamara A (1996) The *in vitro* production of goat embryos using individual oocytes from different sources *Proceedings 12th Association Européenne de Transfert Embryonnaire Lyon* p 122 (Abstract)

Cox JF, Avila J, Saravia F and Santa Maria A (1994) Assessment of fertilizing ability of goat spermatozoa by *in vitro* fertilization of cattle and sheep intact oocytes *Theriogenology* **41** 1621–1629

Crozet N, De Smedt V, Ahmed-Ali M and Sevellec C (1993) Normal development following *in vitro* oocyte maturation and fertilization in the goat *Theriogenology* **39** 206 (Abstract)

Crozet N, Ahmed-Ali M and Dubos MP (1995) Developmental competence of goat oocytes from follicles of different size categories following maturation, fertilization and culture *in vitro Journal of Reproduction and Fertility* **103** 293–298

De Smedt V, Crozet N., Ahmed-Ali M, Martino A and Cognié Y (1992) *In vitro* maturation and fertilization of goat oocytes *Theriogenology* **37** 1049–1060

Delgadillo JA, Leboeuf B and Chemineau P (1991) Decrease of seasonality of sexual behaviour and sperm production in bucks by short photoperiodic cycles *Theriogenology* **36** 755–770

Delgadillo JA, Leboeuf B and Chemineau P (1992) Abolition of seasonal variations in semen quality and maintenance of sperm fertilizing ability by short photoperiodic cycles in he-goats *Small Ruminant Research* **9** 47–59

Delgadillo JA, Leboeuf B and Chemineau P (1993) Maintenance of sperm production in bucks using a third year of short photoperiodic cycles *Reproduction Nutrition Développement* **33** 609–617

Delgadillo JA, Hochereau-de Reviers MT, Daveau A and Chemineau P (1995) Effect of short photoperiodic cycles on male genital tract and testicular parameters in male goats

(*Capra hircus*) *Reproduction Nutrition Développement* **35** 549–558

Deveson S, Forsyth IA and Arendt J (1992a) Induced out-of-season breeding in British Saanen dairy goats : use of artificial photoperiods and/or melatonin administration *Animal Reproduction Science* **29** 1–5

Deveson S, Forsyth IA and Arendt J (1992b) Retardation of pubertal development by prenatal long days in goat kids born in autumn *Journal of Reproduction and Fertility* **95** 629–637

Flores-Foxworth G, McBride BM, Kraemer DC and Nuti LC (1992) A comparison between laparoscopic and transcervical embryo collection and transfer in goats *Theriogenology* **37** 213 (Abstract)

Freitas VJF, Baril G and Saumande J (1996a) Induction and synchronization of oestrus in goats : the relative efficiency of one versus two fluorogestone acetate-impregnated vaginal sponges *Theriogenology* **46** 1251–1256

Freitas VJF, Baril G, Bosc M and Saumande J (1996b) The influence of ovarian status on response to estrus synchronization treatment in dairy goats during the breeding season *Theriogenology* **45** 1561–1567

Freitas VJF, Baril G and Saumande J (1997a) Estrus synchronization in dairy goats : use of fluorogestone acetate vaginal sponges or norgestomet ear implants *Animal Reproduction Science* **46** 237–244

Freitas VJF, Baril G, Martin GB and Saumande J (1997b) Physiological limits to further improvement in the efficiency of oestrous synchronization in goats *Reproduction, Fertility and Development* **9** 551–556

Gebbie F (1993) Control of seasonal breeding and coat development in the goat *PhD Thesis, University of Surrey, UK* pp 205

Gootwine E, Barash I., Bor, A, Dekel I, Friedler A, Heller M, Zaharoni U, Zenue A and Shani M (1997) Factors affecting success of embryo collection and transfer in a transgenic goat program *Theriogenology* **48** 485–499

Graff KJ, Meintjes M, Paul JB, Dyer VW, Denniston RS, Ziomek C and Godke RA (1995) Ultrasound-guided transvaginal oocyte recovery from FSH-treated goats for IVF *Theriogenology* **43** 223 (Abstract)

Hesselink JW (1993) Hydrometra in dairy goats : reproductive performance after treatment with prostaglandins *Veterinary Record* **133** 186–187

Humblot P, Brice G, Chemineau P and Broqua B (1995) Mortalité embryonnaire chez la chèvre laitière après synchronisation des chaleurs et insémination artificielle à contre saison *(Embryonic mortality in dairy goats after out-of-season hormonal synchronisation of oestrus and artifical insemination) Rencontres Recherches Ruminants INRA-IE Paris* **2** 387–390

Iritani A and Nishikawa Y (1972) Studies on the egg yolk coagulation enzyme (phospholipase) in goat semen. IX. Enzyme concentration in the semen collected from the Cowper's gland removed goat *Memories College Agriculture Kyoto University* **101** 57–63

Keskintepe L, Darwish GM, Kenimer A.T and Brackett BG (1994) Term development of caprine embryos derived from immature oocytes *in vitro Theriogenology* **42** 527–535

Krisher RL, Gwazdauskas FC, Page RL, Russel CG, Caneseco RS, Sparks AET, Velander WH, Johnson JL and Pearson RE (1994). Ovulation rate, zygote recovery and follicular populations in FSH- superovulated goats treated with $PGF_{2\alpha}$ and/or GnRH *Theriogenology* **41** 491–498

Le Gal F, Baril G, Vallet JC and Leboeuf B (1993) *In vivo* and *in vitro* survival of goat embryos after freezing with ethylene glycol or glycerol *Theriogenology* **40** 771–777

Leboeuf B, Bernelas D, Pougnard JL, Baril G, Maurel MC, Boué P and Terqui M (1993) Time of ovulation after LH peak in dairy goats induced to ovulate with hormonal treatment *Proceedings 9th Association Européenne de Transfert Embryonnaire Lyon* p 226 (Abstract)

Leboeuf B, Renaud G, De Fontaubert Y, Broqua B and Chemineau P (1994) Echographie et pseudogestation chez la chèvre *(Ultrasonography and pseudopregnancies in the goat) Proceedings 7th International Meeting on Animal Reproduction Murcia, Spain* 251–255

Leboeuf B, Baril G, Maurel MC, Bernelas D, Marcheteau J, Berson Y, Broqua B and Terqui M (1996) Effect of progestagen/PMSG repeated treatments in goats on fertility following artificial insemination (A.I.) *Proceedings 6th International Conference on Goats, Bejing, China* **2** 827 International Academy Publishers, Beijing

Leboeuf B, Manfredi E, Boué P, Piacère A, Brice G, Baril G, Broqua C, Humblot P and Terqui M (1998) Artificial insemination of dairy goats in France *Livestock Production Science* **55** 193–203

Malpaux B, Chemineau P and Pelletier J (1993) Melatonin and reproduction in sheep and goats. In *Melatonin, Biosynthesis, Physiological Effects and Clinical Applications* pp 253–287 Eds HS Yu and RJ Reiter CRC Press, Boca Raton

Mani AU, Watson ED and McKelvey WAC (1994) The effects of subnutrition before or after embryo transfer on pregnancy rate and embryo survival in does *Theriogenology* **41** 1673–1678

Martino A, Palomo MJ, Mogas T and Paramio MT (1994) Influence of the collection technique of prepubertal goat oocytes on *in vitro* maturation and fertilization *Theriogenology* **42** 859–873

Martino A, Mogas T, Palomo MJ and Paramio MT (1995) *In vitro* maturation and fertilization of prepubertal goat oocytes *Theriogenology* **43** 473–485

Maurel MC, Leboeuf B, Baril G and Bernelas D (1992) Determination of the preovulatory LH peak in dairy goats using an ELISA kit on farm *Proceedings 8th Association Européenne de Transfert Embryonnaire, Lyon* p 186 (Abstract)

Mialot JP, Saboureau L, Gueraud JM, Prengere E, Parizot D, Pirot G, Duquesnel R, Petat M and Chemineau P (1991) La pseudogestation chez la chèvre: observations *préliminaires (Pseudopregnancy in the goat: preliminary observations) Recueil de Médecine Vétérinaire, Spécial Reproduction Ruminants* **1** 383–390

Nowshari MA, Beckers JF and Holtz W (1995) Superovulation of goats with purified pFSH supplemented with defined amounts of pLH *Theriogenology* **43** 797–802

Nunes JF, Corteel JM, Combarnous Y and Baril G (1982) Rôle du plasma séminal dans la survie *in vitro* des spermatozoïdes de bouc *(Contribution of the seminal plasma in in vitro survival of goat spermatozoa) Reproduction Nutrition Développement* **22** 611–620

Pawshe CH, Totey SM and Jain SK (1994) A comparison of three methods of recovery of goat oocytes for *in vitro* maturation and fertilization *Theriogenology* **42** 117–125

Pellicer-Rubio MT, Magallon T and Combarnous Y (1997) Deterioration of goat sperm viability in milk extenders is due to a bulbourethral 60-kilodalton glycoprotein with triglyceride lipase activity *Biology of Reproduction* **57** 1023–1031

Pellicer-Rubio MT and Combarnous Y (1998). Deterioration of goat spermatozoa in skimmed milk-based extenders as a result of oleic acid released by the bulbourethral lipase BUSgp60 *Journal of Reproduction and Fertility* **112** 95–105

Poulin N, Guler A, Pignon P and Cognié Y (1996) *In vitro* production of goat embryos : heparin in IVF medium affects developmental ability *Proceedings 6th International Conference on Goats, Bejing, China* **2** 838–840 International Academy Publishers, Beijing

Prichard JF, Thibodeaux JK, Pool SH, Blackewood EG, Menezo Y and Godke RA (1992) *In vitro* co-culture of early stage caprine embryos with oviduct and uterine epithelial cells *Human Reproduction* **7** 553–557

Remy B, Baril G, Vallet JC, Chouvet C, Saumande J, Chupin D and Beckers JF (1991) Are antibodies responsible for a decreased superovulatory response in goats which have been treated repeatedly with porcine follicle-stimulating hormone? *Theriogenology* **36** 389–399

Roy A (1957) Egg yolk-coagulating enzyme in the semen and Cowper's glands of the goat *Nature* **179** 318–319

Roy F, Maurel MC, Combes B, Vaiman D, Cribiu EP, Lantier I, Pobel T, Deletang F, Cambarnous Y and Guillou F (1999) The negative effect of repeated equine chorionic gonadotrophin treatment on subsequent fertility in Alpine goats is due to a humoral immune response involving the major histocompatibility complex *Biology of Reproduction* **60** 805–813

Soonen AH, Lewalski S, Meinecke-Tilman S and Meinecke B (1991) Transcervical collection of ovine and caprine embryos *Proceedings 7th Scientific Meeting European Embryo Transfer Association Cambridge* **1** 208 (Abstract)

Soulière I (1991) La pseudogestation chez la chèvre. Aspects physiologiques et zootechniques *(Pseudopregnancy in goats. Physiological and zootechnical aspects) Mémoire de fin d'Etudes Ecole Nationale Supérieure Féminine d'Agronomie, Rennes* INRA-PRMD Nouzilly ed. 28pp

Todini L, Cognié Y, Poulin N and Guérin Y (1994) Recupero in vivo di oociti di capra mediante aspirazione follicolare per via laparoscopica *(In vivo goat oocyte collection via follicle aspiration under laparoscopy) Proceedings Congresso Società Caprini, Perugia* 331–334

Traldi AS, Leboeuf B, Pougnard JL, Baril G and Mermillod P (1997) Vitrification: a cryopreservation method for embryos produced *in vivo* in the goat. *Arquiros da Facildade de Veterinária Universidade Federal do Rio Grande do Sul UFRGS, Porto Alegre,* Vol.25 no. 1 (Abstract)

Traldi AS, Leboeuf B, Baril G, Cognié Y, Poulin N, Evans G and Mermillod P (1998) Comparative results after transfer of vitrified *in vitro* produced goat and sheep embryos *Proceedings of the 14th Scientific Meeting of the European Embryo Transfer Association Venezia* 258 (Abstract)

Vallet JC, Casamitjana P, Brebion P and Perrin J (1991) Techniques de production, de conservation et de transfert d'embryons chez les petits ruminants *(Production, conservation and embryo transfer techniques in small ruminants) Recueil de Médecine Vétérinaire, Spécial Reproduction Ruminants* **167** 293–301

Yuswiati E and Holtz W (1990) Successful transfer of vitrified goat embryos *Theriogenology* **34** 629–632

Journal of Reproduction and Fertility Supplement **54**, 143–156

Comparative reproductive function in cervids: implications for management of farm and zoo populations

G. W. Asher[1], S. L. Monfort[2] and C. Wemmer[2]

[1]*AgResearch, Invermay Agricultural Centre, Private Bag 50034, Mosgiel, New Zealand;*
[2]*National Zoological Park, Conservation and Research Center, Smithsonian Institution, Front Royal, VA 22630, USA*

The cervids represent a complex assemblage of taxa characterized by extreme diversity in morphology, physiology, ecology and geographical distribution. Farmed species (for example red deer and fallow deer) are usually the common larger-bodied, gregarious and monotocous species that express marked reproductive seasonality in their temperate environment. Their commercial importance has facilitated considerable research into reproductive physiology and the development of assisted reproductive technologies (ART). In contrast, the remaining species, including many of tropical origin, show wide diversity in reproductive patterns, have generally received little scientific scrutiny, and include a number of endangered taxa that are reliant on *ex situ* conservation efforts (such as captive breeding) to ensure their survival. Domestication and *ex situ* management programmes have been associated with widespread translocation of various cervid species around the world, often placing the animals in environments that are not compatible with their evolved reproductive patterns. For example, the summer calving/lactation pattern of red deer, attuned to northern continental climatic patterns, is frequently misaligned with seasonal changes in feed availability in the Australasian pastoral environment. Similarly, seasonal or aseasonal calving patterns of tropical species translocated to temperate regions are usually associated with increased perinatal mortality of calves born in cool seasons. Conversely, temperate species in tropical zones may exhibit aberrant reproductive patterns in the absence of biologically significant photoperiod fluctuations. ARTs, which presently include artificial insemination, embryo transfer and *in vitro* embryo production, have potential application to the genetic management and population growth of various cervid species. Although application to some farmed cervid species is widespread, these technologies are rarely directly transferable from farmed to endangered species. Even within species, ART protocols developed successfully for one genotype (i.e. subspecies) may be ineffective in another (for example superovulation of red deer and wapiti). Therefore, application to genetic management of endangered species necessitates prior research into their reproductive patterns. This is often difficult because of the rarity of the animals, a lack of suitable handling facilities for the particular species, and the timid nature of the deer. More recently, however, non-invasive reproductive profiling, based on remote collection and monitoring of excreted steroid metabolites, has facilitated such research.

Introduction

The ‘cervids’ include 43 species and 206 subspecies (Whitehead 1993) characterized by extreme diversity in morphology, physiology, ecology and geographical distribution. It is important to recognize that in any consideration of reproductive function no one species represents a ‘typical’ deer. For example, some cervids exhibit highly seasonal patterns of births in cool temperate climes, while others are completely aseasonal in equatorial regions. Furthermore, many species are strictly monovular and bear single offspring annually, whereas others are normally polyovular and bear multiple offspring. Even embryonic development and placentation vary considerably among species (Lincoln, 1985).

An increased understanding of reproductive function in cervids over the last few decades stems largely from the growing importance of various deer species in agriculture (for example deer farming) and *ex situ* conservation efforts (for example captive breeding of endangered species). In general, the research has been directed towards successful application of assisted reproductive technologies to manipulate the genetic constitution of populations (for example selection for improved agricultural productivity; avoidance of inbreeding depression in captive populations) or to increase rates of propagation of specific genetic lineages. However, we emphasize the fact that studies to date have focused on a few key species that do not necessarily reflect the full range of cervid physiology.

The international deer farming industry is based broadly on one of two groups of cervids: those of cool temperate northern origin such as European red deer (*Cervus elaphus* spp *scoticus, hippelaphus*), North American and Asiatic wapiti (*Cervus elaphus* spp *nelsoni, roosevelti, manotobensis, xanthopygus*), European fallow deer (*Dama dama*) and Asiatic sika deer (*Cervus nippon*); and those of tropical equatorial origin such as chital deer (*Axis axis*), rusa deer (*Cervus timorensis*) and sambar deer (*Cervus unicolor*). All tend to be genotypes that exhibit notable gregariousness (cf numerous highly territorial species that live solitary lives) and are adapted to mixed browsing/grazing (cf species that are highly selective browsers). Farmed species, typically, can adjust to exclusive grazing and open-range environments even though they probably evolved to live on the forest-pasture interface. In these respects, they represent a specific subset of the entire cervid genome biased towards the larger-bodied, monovulatory species with a high degree of behavioural plasticity. They also represent species that have remained relatively common in the wild.

In contrast, a significant proportion of cervid species have not survived particularly well over the last few centuries. The 1996 IUCN Red List of Threatened Animals identifies approximately 20 species of cervid, representing nine genera, that are either threatened or endangered with extinction world-wide. Species decimated by over-hunting and habitat destruction are often reliant on *ex situ* conservation (ie captive breeding) for any chance of survival. A number of species, such as Pere David's deer (*Elaphurus davidanus*), Mesopotamian fallow deer (*Dama dama mesopotamica*) and Eld's deer (*Cervus eldi eldi*), are represented almost entirely by captive individuals or intensively managed populations. The relevance of captive populations to conservation increases as the risk of extinction increases in *in situ* populations. Several threatened species of deer have internationally registered studbooks, which facilitate demographic and genetic management. These programmes continue without linkage to *in situ* conservation, but nonetheless augment the ability of conservationists to respond to emergencies, should wild populations diminish to critical numbers or become victims of stochastic events. Reproductive management strategies for such species are focused primarily in two areas: first, optimizing reproductive performance within artificial habitats and, second, judicious genetic management to reduce the impact of inbreeding. Traditional captive breeding/management programmes for wild ungulates often have failed to integrate even regional zoological collections into 'single' populations for genetic and demographic management. However, traditional husbandry-management techniques now can be combined with reproductive biotechnology, including germplasm cryopreservation and assisted reproductive technologies (ART) to ensure sufficient genetic diversity without exceeding limited captive animal holding space in zoos (Wildt *et al.*, 1997). With the use of modern management principles, preferred pairings can be co-ordinated by genetic specialists to facilitate natural breedings when feasible, and ART when most practical to avoid long-distance and expensive transport of individual animals. Although this approach does not come without substantial costs in terms of time, energy and expense, the long-term rewards hold the promise for (1) stabilizing the number of captive animals, (2) improving genetic diversity, and (3) establishing viable genome resource banks (GRB) for frozen germplasm. It is interesting to note that while farmed deer species represent a biased 'type' of cervid, captive breeding programmes are in place for a wide range of cervid 'types'. Reproductive management principles and specific technologies developed for the farmed species are, therefore, not always applicable to captive breeding programmes for endangered species. A notable example would be *ex situ* conservation of pudu (*Pudu* sp) and muntjac (*Muntiacus* sp), both being small-bodied, solitary, forest-dwelling taxa.

Seasonality of Reproduction

The necessity for most cervid species to give birth at an appropriate time of year for optimal survival and growth of offspring has exerted considerable influence on their reproductive physiology (Lincoln and Short, 1980). Species of northern temperate origin typically conceive in autumn and calve in summer, whereas species of tropical origin often exhibit limited seasonality or are completely aseasonal (Lincoln, 1985). The endogenous mechanisms governing seasonal reproductive patterns in temperate species are robust, being manifest rigorously when animals are transferred between localities despite subtle regional variations in seasonal feed supply. Furthermore, transference across the equator results in an exact 6 month phase change (Marshall, 1937), even though the relationship between season and feed production differs considerably between continental northern hemisphere and insular southern hemisphere environments (Asher *et al.*, 1993a). Tropical species transferred from equatorial (0–15°) zones may exhibit either pronounced 'reverse seasonality' (i.e. spring conceptions and autumn/winter calving, for example Eld's deer), wide breeding seasons (i.e. calving mostly spread over 3–6 months of the year; for example chital deer in Australia) or complete aseasonality (i.e. calving year round; for example Reeve's muntjac) (Chapman *et al.*, 1984; Loudon and Curlewis, 1988; Monfort *et al.*, 1991; Mylrea, 1992). It should be noted, however, that there is a general paucity of information on the birth season of tropical species in their native tropical environment (Lincoln, 1985).

It is generally accepted that entrainment of seasonal reproductive cycles is effected by endogenous recognition of photoperiodic changes, with the majority of temperate species initiating mating activity during decreasing daylength of late summer and autumn (Lincoln and Short, 1980). Variations in the actual mating season between species of up to 8 weeks usually offsets genetically determined species differences in duration of gestation, such that parturition generally occurs in mid-summer (Lincoln, 1985). A notable exception amongst temperate species is the Pere David's deer, which initiates mating activity in early summer and calves in late spring despite an unusually long gestation of about 280 days (Wemmer *et al.*, 1989; Monfort *et al.*, 1991). There is growing evidence that some tropical species may also perceive, and respond to, changes in photoperiod, although in a markedly different manner than temperate species. For example, rusa deer living in the tropics do not exhibit well-synchronized antler cycles and reproductive seasons, indicating that photoperiod does not effectively modulate reproductive seasonality (van Mourik and Stelmasiak, 1990). Because rusa deer probably migrated from outside the tropical zone, it has been suggested that, once translocated to temperate latitudes, they re-established an evolutionarily-based 'long-day' reproductive rhythm (van Mourik and Stelmasiak, 1990). Although Eld's deer exhibit strong reproductive seasonality in their subtropical native range, this pattern persists after translocation to temperate latitudes (48°N latitude), even though this strategy results in the birth of offspring during the harsh winter months (Monfort *et al.*, 1991). The persistence of such an apparently maladaptive strategy indicates that there is an endogenous rhythm linked to its adaptive significance in their native sub-tropical habitat. Alternatively, Eld's deer may respond to the relatively subtle low amplitude photoperiodic oscillations experienced in their native sub-tropical latitudes; translocation to a temperate latitude may simply serve to reinforce or strengthen seasonal rhythms because the direction of photoperiodic change is similar and the amplitude of seasonal daylength changes is more pronounced. Similarly, treatment of chital deer stags with exogenous melatonin has been shown to hasten antler casting (Mylrea, 1992), indicating possible photoperiodic responsiveness in a species considered to exhibit limited seasonality in both tropical and temperate environments (Loudon and Curlewis, 1988). In contrast, however, tropical sambar deer acclimatized to temperate zones in New Zealand exhibit a loose pattern of 'reverse seasonality' in calving but fail to show any appreciable seasonal patterns in prolactin secretion (Asher *et al.*, 1997a). Thus, some tropical species may not perceive photoperiodic cues, if they are perceived at all, in the same manner as their temperate relatives. Because tropical cervid species are not exposed to strong circannual photoperiodic rhythms, reproductive patterns may have evolved in response to a variety of factors including (1) seasonal fluctuations in the availability of specific food items, (2) local resource competition among sympatric species, or (3) predation pressures; and many of these factors may be directly related to rainfall patterns. For example, predation might increase during seasons when adequate cover is unavailable. Alternatively, selective pressures in some species may have acted more strongly on the ability to conceive rather than on the timing of the birth season; in such cases, optimal forage availability might be predicted during the mating season rather than the birthing season.

Several studies have demonstrated a relationship between annual rainfall patterns and timing of the fawning season of sub-tropical and tropical deer species. It has been asserted that conceptions are timed in white-tailed deer (*Odocoileus virginianus*) living in Columbia (5°N latitude) and the Everglades National Park (25°N latitude) so that births occur during the dry season, which may reduce mortalities associated with the inability of the neonate to thermoregulate properly under wet conditions (Blouch, 1987; Smith *et al.*, 1996). The fawning season in hog deer (*Axis porcinus*, 27°N latitude, Nepal) coincides with new vegetative growth that occurs after seasonal grass fires caused by lightning strikes or agricultural burning, believed to have been occurring for thousands of years (Mishra and Wemmer, 1987). Although other tropical species including red brocket deer (*Mazama americana*, 4–6°N latitude, Suriname) also have fawning seasons that coincide with the end of the rainy season (Branan and Marchinton, 1987), others such as pampas deer (*Ozotocerso bezoarticus bezoarticus*; 18°S latitude, Argentina) and sambar deer (27°N latitude, Nepal) exhibit peak fawning during the wet season (Jackson and Langguth, 1987; Mishra and Wemmer, 1987, respectively). Even species like the sambar deer and chital deer sharing the same habitat (27°N, Nepal) fawn 6 months out-of-phase to one another (Mishra and Wemmer, 1987). Thus, no clear pattern emerges to explain the variance in reproductive patterns exhibited by tropical cervids.

It is precisely this lack of reproductive uniformity among sympatric species from which we can infer differing adaptations to the environment based on a variety of regulatory mechanisms. In Nepal's Royal Chitawan National Park (27°N latitude) four deer species co-exist, including two browsers (sambar deer and muntjac), one grazer (hog deer) and one mixed feeder (chital deer) (Mishra and Wemmer, 1987). Within these guilds each species differs in annual reproductive cycle, habitat selection and feeding ecology. Although the muntjac deer breeds all through the year, the birth season of the sambar deer coincides with the onset of the monsoon; hog deer give birth when new grasses flush after burning, and chital give birth during the dry season. Although photoresponsiveness in tropical cervid species may persist as an evolutionary relic from a time when this trait was advantageous to survival, the tight linkage between photoperiod and seasonal reproduction observed in temperate species may not completely explain the existence of reproductive seasonality in many tropical species. It is important to recognize that we know little about the factors that govern reproductive rhythms in tropical cervid species, and even less about how reproduction is adapted to the ecological setting in which it evolved. Further studies are needed to determine whether tropical cervid species are truly photoresponsive. In addition detailed studies of tropical cervids in native environments are important for determining the impact of uniform photoperiod, seasonal rainfall patterns and food availability on reproductive seasonality.

Implications of Seasonality for Farmed Deer Productivity

Red deer/wapiti (*Cervus elaphus* spp) and fallow deer represent the main species farmed in temperate environments in Australasia, North America and Europe. Both species have reproductive patterns attuned to the northern continental environment. Their autumn mating–summer calving pattern has been strictly conserved when the species have been transferred to the southern hemisphere. Such reproductive seasonality has obvious beneficial consequences for the species within their traditional environment. However, translocation to more temperate, insular Austral environments has been associated with a degree of misalignment between reproductive seasonality and seasonal changes in feed availability (reviewed by Asher *et al.*, 1993a). For example, the spring flush of growth of high quality pasture can occur as early as August–October in the northern regions of New Zealand, yet farmed red deer calve naturally in November–December. Their peak lactational demands and the demands of the young calf for high quality feed are in January–March, when the pasture growth rate and quality are limiting because of seasonal moisture deficits and natural pasture senescence. The challenge for deer farmers is to meet the high energy demands of the lactating female and the calf during such periods. The consequences of failing to do so include increased calf mortality, decreased calf growth rates and depressed hind liveweights. This may further impact on ovulatory activity of the females in the subsequent breeding season. Srategies for reducing the possible impacts of such misalignment include (i) shifting the seasonal peak of pasture growth by using different pasture cultivars or by preventing reproductive senescence of grasses; (ii) selecting deer genotypes with the genetic propensity to calve earlier; and (iii) artificially manipulating the reproductive seasonality of the breeding herd (reviewed by Asher *et al.*, 1993a).

Uncontrolled breeding management of tropical species in temperate zones creates a number of production problems. First, and perhaps most obvious, is the high mortality of neonates born in winter, for example with rusa deer farmed in the southern latitudes of Australia (van Mourik, 1986). Second, predation of neonates has greater overall impact on unsynchronized herds owing to the higher predator:prey ratio than in herds with concise calving patterns (i.e. prey sativation), as exemplified by fox predation of chital deer fawns in Australia (Mylrea, 1992). Generally, however, judicious control of male:female joining dates allows tight control of the season and synchrony of births of tropical species in seasonal environments. Such easy options are not available for temperate species.

There have been recent attempts to establish temperate species, such as red deer and fallow deer, in equatorial zones (despite the presence of locally adapted cervid species). Anecdotal evidence indicates that removal of photoperiodic signals has had a marked effect on the overall physiology of these deer. For example, fallow deer translocated from New Zealand to southeast Asia exhibited apparent 'de-synchronization' to the extent that 70% of does become anovulatory at the time of joining during their 'normal' breeding season. This situation was subsequently corrected by strategic administration of melatonin implants for periods of 2–3 months. The bucks maintained a high within-herd synchrony in their antler cycles with or without melatonin treatment (G Christie, personal communication, 1993).

Seasonal (or aseasonal) constraints encountered in extensively managed farmed populations are seldom considered important in zoo populations owing to high levels of attention given to individuals and the provisions made for intensive feeding and housing. However, many zoo populations consist of non-native species that are maladapted to local climates. Inappropriate birth seasons (for example winter fawning of Eld's deer in North American zoos) can lead to high neonatal mortality due to the inability of the fawn to thermoregulate appropriately (Monfort *et al.*, 1993a).

Ovulatory Cycle

With few exceptions, female cervids are polyoestrous, and non-pregnant animals are capable of exhibiting either continuous oestrous cycles (for example some tropical species) or, more commonly, alternating periods of oestrous cyclicity and anoestrus (Fig. 1). Oestrous cycles have been characterized for a number of species from studies on oestrous behaviour and luteal secretion of progesterone. In non-pregnant females of temperate species, such as red deer and fallow deer, the onset and termination of oestrous cyclicity occur in autumn and spring respectively, with 5–8 cycles expressed. Anoestrus is characterized by low concentrations of progesterone indicative of complete ovulatory arrest, and may persist for 4–6 months from spring to early autumn (Fig. 1). In marked contrast, the tropical brow-antlered Eld's deer initiates oestrous cycles in spring and enters

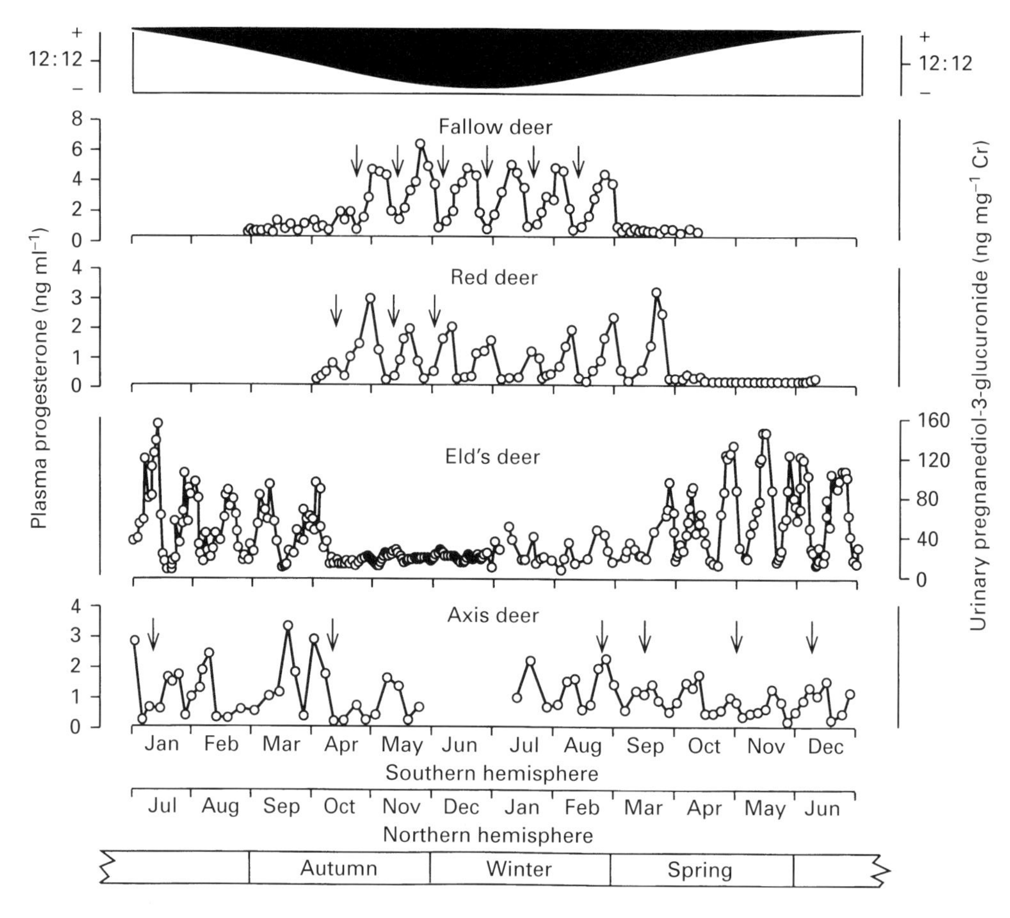

Fig. 1. Oestrous/luteal cycles in temperate (fallow, red) and tropical (Eld's, axis) species of deer in temperate environments, as defined by peripheral plasma progesterone concentrations or urinary pregnanediol metabolite concentrations during the annual cycle of non-pregnant females. The data have been normalized about common hemispheres and placed in relation to relative annual changes in photoperiod. Arrows indicate overt oestrous behaviour (data from Asher, 1985; Asher and Fisher, 1991; Monfort *et al.*, 1991; Mylrea, 1992).

anoestrus in the following autumn in temperate regions, this being effectively a 6 month phase shift from temperate species (i.e. 'reverse seasonality'). Studies on tropical axis deer in temperate regions of Australia indicate oestrous cyclicity throughout the year (Fig. 1), although success of artificial oestrous synchronization techniques appears to vary with season (Mylrea, 1992).

The transition into the breeding season is characterized by 'silent ovulations' (i.e. ovulations not preceded by overt oestrus) and short-lived (8–10 days) corpora lutea in most cervid species studied (Thomas and Cowan, 1975; Asher, 1985; Curlewis *et al.*, 1988; Monfort *et al.*, 1991). In fallow deer, multiple successive silent ovulations leading up to the start of the breeding season have been observed. The transient nature of the preliminary corpora lutea may actually serve to promote within-herd synchrony of first overt oestrus of the season (Asher, 1985). Subsequent oestrous cycles are generally of 'normal' duration, as determined genetically for each species, although occasional 'long cycles' (2–3 times normal duration) have been observed in some Pere David's deer and Eld's deer (Curlewis *et al.*, 1988; Monfort *et al.*, 1991). Average duration of the normal oestrous cycle ranges from 17 days in sambar deer and axis deer, 18–20 days in red deer and Eld's deer, 21–23 days in fallow deer, and 24–27 days for moose (*Alces alces*) and black-tailed deer (*Odocoileus hemionus*). The mean duration of the oestrous cycle tends to increase progressively during the breeding season in

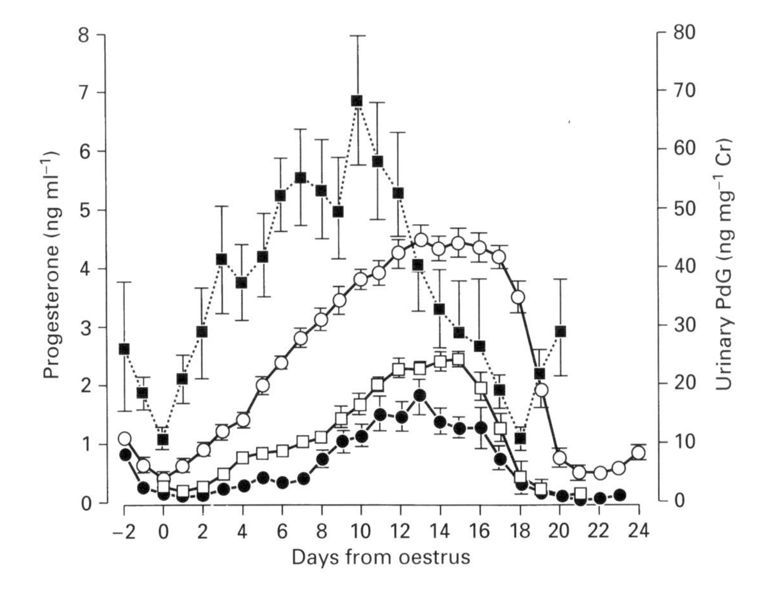

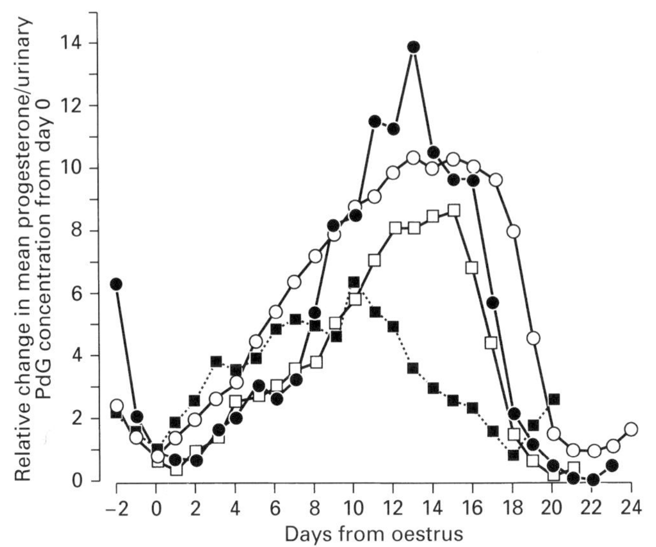

Fig. 2. Profiles of mean peripheral plasma concentrations of progesterone (solid line) or urinary concentrations of PdG (dashed line) during the cervid oestrous cycle for fallow deer (○, *n* = 20 cycles), red deer (□, *n* = 8 cycles), axis deer (●, *n* = 12 cycles) and Eld's deer (■, *n* = 12 cycles). (a) Mean (± SEM) values; (b) the same data presented as relative change in mean concentrations from day 0 (data from Asher, 1985; Asher and Fisher, 1991; Monfort *et al.*, 1991; Mylrea, 1992).

red, fallow and black-tailed deer (Guinness *et al.*, 1971; Asher, 1985; Curlewis *et al.*, 1988; Monfort *et al.*, 1991; Chapple *et al.*, 1993; Schwartz and Hundertmark, 1993; Wong and Parker, 1993).

Luteal events during the cervid oestrous cycle are similar between species (Fig. 2). Luteinization of post-ovulatory follicles is associated with increased secretion of progesterone, with maximal concentrations of peripheral blood or urinary metabolites occurring between days 10 and 16 of the

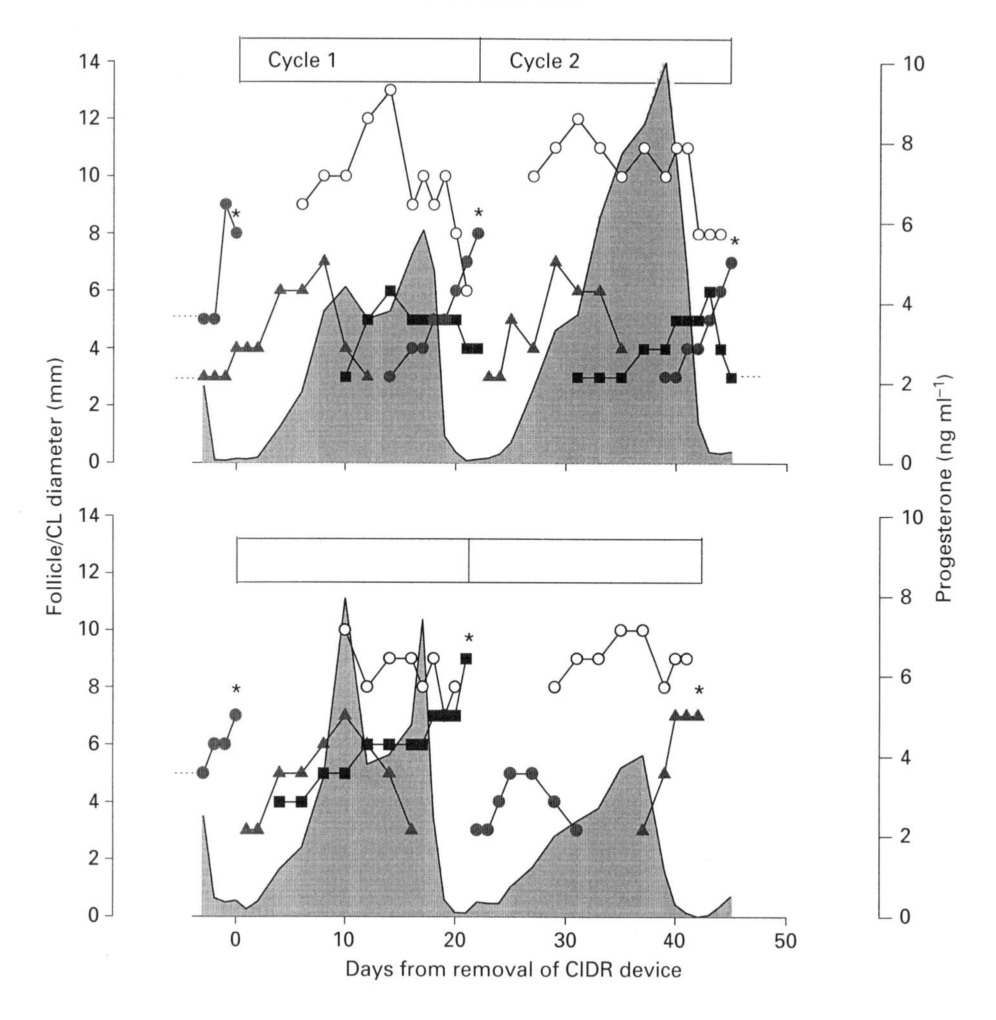

Fig. 3. Representative profile for a fallow deer doe of plasma concentrations of progesterone (shaded profiles), diameters of corpora lutea (○) and diameters of dominant follicles (●, ▲, ■) during two consecutive luteal cycles following synchronization with a CIDR device. Asterisks denote abrupt follicular disappearance indicative of ovulation (G.W. Asher *et al.*, unpublished).

oestrous cycle (day 0 = oestrus). Although absolute plasma concentrations vary between species (Fig. 2a), the relative changes from day 0 are quite similar between species (Fig. 2b). Follicular dynamics during the oestrous cycle have been described from ultrasonographic studies on red deer (Asher *et al.*, 1997b) and, more recently, fallow deer (G.W. Asher *et al.*, unpublished). These studies have demonstrated discrete, non-random patterns of antral follicular growth and regression during the oestrous cycle. The cycle is characterized by a variable number (1–3) of dominant follicular waves (Fig. 3), similar to that observed in cattle. However, these studies relate to species that are highly seasonal breeders and strictly monovular, and may not reflect ovarian dynamics across the range of cervids.

In many respects, the cervid oestrous cycle is similar to that of other domestic ruminant species. Thus methods of artificial oestrous synchronization have tended to follow similar procedures used for sheep and cattle. These include, principally, the use of progestagens (particularly progesterone) to simulate the luteal cycle, and prostaglandins to control luteal longevity (reviewed by Asher *et al.*,

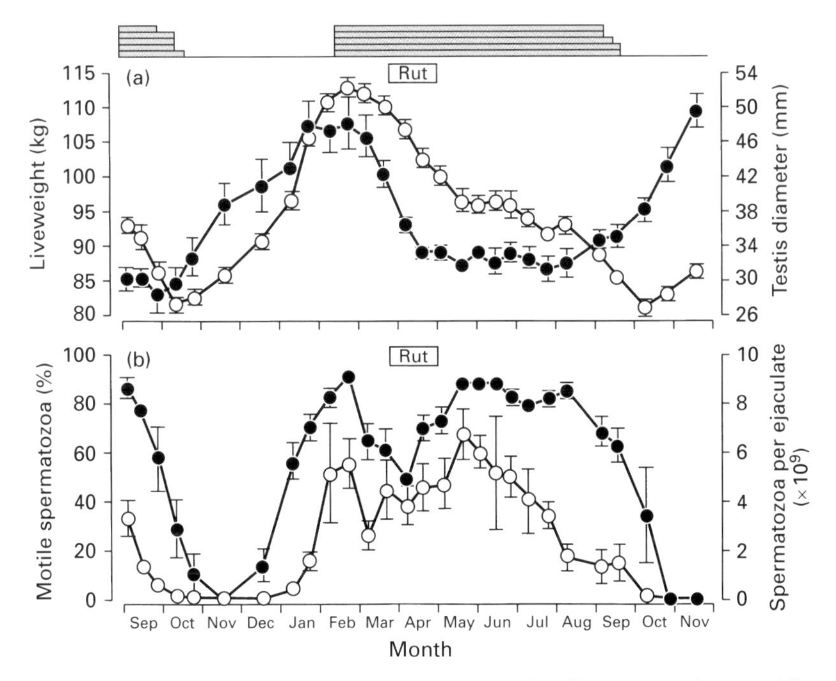

Fig. 4. Seasonal profiles of (a) mean (± SEM) live weight (●), testicular diameter (○), (b) motile spermatozoa (●) and spermatozoa per ejaculate (○) for five F_1 hybrid Mesopotamian × European fallow deer bucks (data from Asher *et al.*, 1996).

1993b). As a generalization, these techniques have been efficacious, leading to the development of a number of successful technologies such as artificial insemination and embryo transfer (see later).

Testicular Cycles

Circannual variation in testicular function is evident in most deer species, reflecting changes in fertility, gross morphometry and behaviour. Of the temperate species, these changes have been best described for red deer (Lincoln, 1971, 1985) and fallow deer (Asher *et al.*, 1989; Schnare and Fischer, 1987). In the adult fallow deer buck, marked annual changes in testicular function (Fig. 4) are controlled primarily by dynamic changes in pituitary secretion of LH. LH is secreted in pulses that vary in amplitude and frequency during the year, being of low amplitude and frequency during the non-breeding season (early summer) and of high amplitude and frequency leading up to the onset of the breeding season in autumn. A similar pattern of secretion has been observed for Eld's deer, a tropical but seasonally breeding species (Monfort *et al.*, 1993a).

Testosterone mediated changes in secondary sexual characteristics are very pronounced in both temperate and tropical cervid species. The antler cycle is linked closely to the testis cycle, with antlers cast annually when testes regress to their minimal dimensions. Casting is in response to a marked decline in testosterone secretion and can be induced by castration (Goss, 1983). New antler growth occurs during the quiescent phase of the testes, in the relative absence of testosterone secretion. Mineralization of antlers coincides with increasing testicular activity and increasing testosterone secretion, and hard antlers are generally retained while testicular tissue is actively secreting modest to high amounts of testosterone. This relationship holds for most antlered species studied (Goss, 1983), although the seasonal patterns vary between species and antler–testis cycles may not be synchronized amongst individuals of some tropical species (Loudon and Curlewis, 1988).

Circannual patterns of testicular regression and recrudescence occurring in temperate cervid species relate to annual cycles of sperm production. Unlike most seasonally breeding domestic ruminants, which show moderate fluctuations in testicular activity (for example sheep), temperate cervids exhibit alternating periods of sperm production and complete spermatogenic arrest that reflect about fivefold changes in seasonal testicular volume (Lincoln, 1971; Haigh *et al.*, 1984; Gosch and Fischer, 1989). In fallow deer, for example, as testicular size increases towards the rut, there is a concomitant increase in spermatogenic activity, such that by the onset of the rut large numbers of viable spermatozoa are present in ejaculates (Fig. 4). The testes remain active throughout winter, secreting modest amounts of testosterone and producing large numbers of spermatozoa. However, towards the onset of spring, LH and testosterone secretion diminish and the testes regress in size. Spermatogenesis is arrested completely by early summer, and the bucks become infertile. They remain in this condition for about two months, gradually regaining fertility towards the end of summer (Fig. 4).

Although tropical cervid species often exhibit circannual cycles of reproductive development, either in a synchronous (for example Eld's deer, chital deer in Australia, Reeve's muntjac in UK) or non-synchronous manner (for example chital deer in UK), males appear to retain a degree of fertility throughout the year (Loudon and Curlewis, 1988; Chapman and Harris, 1991; Mylrea, 1992; Monfort *et al.*, 1993a). Seasonal fluctuations in spermatogenesis are related to the stage of the antler cycle, with reduced production of spermatozoa and minimal testis size during velvet antler growth. However, at no stage is there complete arrest in spermatogenesis. For example, while male Eld's deer exhibit 80% abnormal spermatozoa as well as significantly reduced motility and total numbers of spermatozoa during the nadir of the testicular cycle, they remain effectively fertile.

The implications of the seasonal nature of spermatogenesis or fertility in male deer of most species relate mainly to the collection of semen for artificial insemination programmes and other reproductive technologies. The season of collection is clearly limited to the period of fertility, being 5–7 months in red deer and fallow deer (Asher *et al.*, 1993b) but potentially longer in some tropical species.

Assisted Reproductive Technologies (ART)

Rapid expansion of the deer farming industry around the world within the last 20 years has been accompanied by equally rapid development and adoption of a number of assisted reproductive technologies. These technologies have not only facilitated increased rates of genetic improvement on individual farms, but also allowed widespread movement of genetic material around the world, have been implicated in genetic 'rescue' of rare genotypes or individuals, and have allowed farmers and researchers to cross species boundaries in the production of potentially useful hybrids. The principal tools that have been used include artificial insemination (AI), multiple ovulation–embryo transfer (MOET) and *in vitro* embryo production (IVEP). These technologies have essentially been adapted from those developed for more traditional agricultural species such as sheep and cattle. However, subtle differences in reproductive physiology between livestock species often result in major obstacles in successful application of artificial breeding techniques across species, and considerable research has been required to fine tune these technologies for deer. Even within cervids, there are considerable species and subspecies differences in the effectiveness of standardized techniques, necessitating refinements tailor-made for individual genotypes.

Artificial insemination

The development of successful AI protocols for red deer and fallow deer, including the components of semen collection and cryopreservation, oestrous synchronization and insemination technique, have been reviewed in detail by Asher *et al.* (1993b). In summary, most inseminations performed for these two species have involved fixed-time, laparoscopic intra-uterine deposition of semen that has been collected by electroejaculation. The AI procedures (with the exception of method of semen collection) are essentially similar to those used for sheep in Australasia (Killeen and Caffery, 1982). However, there are significant procedural differences for the two cervid species even though their overall reproductive physiology appears similar. For example, oestrous synchronization in red deer is improved by delivery of equine chorionic gonadotropin (eCG) towards the final phase of progesterone treatment. However, eCG is contra-indicated in fallow deer as it depresses ovulatory response in a high proportion of treated individuals (Asher *et al.*, 1993b). North American wapiti, a subspecies of red deer, have also received considerable attention in the development of AI protocols. Wapiti are considerably larger in body size than red deer, and it is routine practice to perform transcervical intra-uterine AI aided by rectal manipulation of the reproductive tract.

Major biological limitations to the application of AI in these species relate mainly to seasonal constraints. Efficacy of oestrous synchronization techniques improves markedly over the short (1–2 week) period representing the transition from the non-breeding season to the breeding season and remains optimal throughout the rut period (Morrow *et al.*, 1995). Effective synchronization is achieved for the first time during the natural rut, although the additional use of eCG in sychronization programmes for red deer appears to extend this period earlier by 1–2 weeks (Asher *et al.*, 1993b). As there are biological penalties of late calving (for example high winter mortality of calves), the consequences of a failed AI programme can be quite severe, as it inevitably delays the calving period by about 3 weeks (i.e. oestrous cycle duration). Attempts to perform AI programmes earlier in the season to compensate for this often result in poor conception rates, due to high levels of ovulatory failure.

As mentioned earlier, semen harvest from valuable sires is subject to major seasonal constraints which often conflict with the need to use their genetic material in a given season. Usable semen is seldom collected from stags or bucks more than 1–2 weeks before the natural rut. However, effective cryopreservation techniques have overcome this shortfall (Asher *et al.*, 1993b).

Multiple ovulation – embryo transfer (MOET)

MOET technology has been developed mainly for a few cervid species that are farmed, namely red deer and fallow deer (Fennessy *et al.*, 1994). In red deer, the efficiency of MOET approximates that of cattle in terms of surrogate pregnancies per donor. However, efficiency of MOET for fallow deer is very low by comparison (Morrow *et al.*, 1994; Jabbour *et al.*, 1994). This finding represents major species differences in superovulatory response to exogenous gonadotrophins and in fertilization rates of ova. Such diversity in outcomes between two species of apparently similar physiology highlights the difficulties in transferring technology across a range of cervid species. Even subspecies differences, such as that between red deer and wapiti, are associated with major differences in responses to specific exogenous gonadotrophins (Fennessy *et al.*, 1994).

Although MOET would appear to have potential application in *ex situ* conservation of endangered cervid species, allowing increased numbers of progeny to be produced from a small female donor base, there are few examples amongst cervids in which a suitable surrogate (i.e. recipient species) is available. One potential case, however, is the feasibility of using European fallow deer (or hybrids) as surrogates for Mesopotamian fallow deer embryos. These two subspecies represent the extremes in conservation status, and yet are reproductively and genetically compatible (Asher *et al.*, 1996).

In vitro *embryo production (IVEP)*

To date, studies have been largely confined to red deer (Fukui *et al.*, 1991; Berg *et al.*, 1995) and wapiti (Pollard *et al.* 1995). Observed difficulties in oocyte maturation, *in vitro* fertilization and subsequent embryo development using standard bovine and ovine systems highlight the difficulties of direct transference of technology across species. However, it is recognized that the *in vitro* technologies will eventually open up a wider range of potential applications in population and genetic management of farmed and zoo populations of cervids.

Application of ART to zoo populations

Although AI, MOET and IVEP have been heralded as tools for enhancing the captive breeding of endangered species, these assisted techniques have not yet proved to be consistently useful for producing offspring or have not been integrated effectively into captive management schemes (Wildt, 1992). One reason that these biotechniques have not yet become routine is because success is sporadic at best, due, in part, to the failure of scientists to first develop strong biological databases before artificial breeding is attempted. For AI to be successful, detailed prerequisite information must include understanding the reproductive cycle of the female to allow for identification or manipulation of oestrus or ovulation, safe and reliable methods for collecting and storing spermatozoa, and methods for proper deposition of spermatozoa at the optimal time and site in the female. Even after the capacity to produce offspring in an endangered species routinely using AI is developed, a challenge remains: the practical demonstration that such a strategy can be implemented on a 'real-life' basis, for the betterment of regional collections and species conservation. Most zoos are ill-equipped to perform routine animal manipulations and procedures as simple as blood sampling without anaesthetizing animals. Successful production of Eld's deer offspring using AI with frozen–thawed spermatozoa provides an excellent example of how ART can be applied to an endangered species (Monfort *et al.*, 1993b). However, it is important to emphasize that the AI study was preceded by a pre-emptive, 3 year effort to establish an integrative database for both female and male Eld's deer (Monfort *et al.*, 1991; 1993a–d). For example, detailed information on seasonality ensured that hinds were not inseminated too early or late in the breeding season, and seasonal evaluations of ejaculates indicated the time of year most likely to result in peak sperm quality and freezability. Semen collection and cryopreservation studies were conducted to ensure that spermatozoa could be collected routinely by electroejaculation, and various cryodiluents, cryoprotectants and freezing protocols were tested to ensure that excellent motility post-thaw could be achieved. Our research verified that CIDR devices were effective for synchronizing oestrus and demonstrated that the basic AI protocol commonly used in farmed fallow deer (Asher *et al.*, 1990) was effective in the Eld's deer. But, because most zoos do not have adequate animal handling systems, or management schemes that permit routine animal restraint or repeated anaesthesia, blood collection and ultrasonography are generally impractical. Development of biological databases for female and male Eld's deer would not have been feasible without the use of noninvasive endocrine monitoring techniques. In fact, for most species maintained in zoos, urinary or faecal steroid monitoring are the only alternatives for assessing longitudinal endocrine rhythms (Lasley and Kirkpatrick, 1991; Schwarzenberger *et al.*, 1996). These approaches provide considerable promise for improving the success rates of ART in species maintained in zoos.

References

Asher GW (1985) Oestrous cycle and breeding season of farmed fallow deer, *Dama dama. Journal of Reproduction and Fertility* **75** 521–529

Asher GW, Peterson AJ and Bass JJ (1989) Seasonal pattern of LH and testosterone secretion in adult male fallow deer, *Dama dama. Journal of Reproduction and Fertility* **85** 657–665

Asher GW, Kraemer DC, Magyar SJ, Brunner M, Moerbe R and Giaquinto M (1990) Intrauterine insemination of farmed fallow deer (*Dama dama*) with frozen–thawed semen via laparoscopy *Theriogenology* **34** 569–577

Asher GW, Fisher MW, Fennessy PF, Suttie JM and Webster JR (1993a) Manipulation of reproductive seasonality of farmed red deer (*Cervus elaphus*) and fallow deer (*Dama dama*) by strategic administration of exogenous melatonin *Animal Reproduction Science* **33** 267–287

Asher GW, Fisher MW, Fennessy PF, Mackintosh CG, Jabbour HN and Morrow CJ (1993b) Oestrous synchronization, semen collection and artificial insemination of farmed red deer (*Cervus elaphus*) and fallow deer (*Dama dama*) *Animal Reproduction Science* **33** 241–265

Asher GW, Berg DK, Beaumont S, Morrow CJ, O'Neill KT and Fisher MW (1996) Comparison of seasonal changes in reproductive parameters of adult male European fallow deer (*Dama dama dama*) and hybrid Mesopotamian × European fallow deer (*D d mesopotamica* × *D d dama*) *Animal Reproduction Science* **45** 201–216

Asher GW, Muir PD, Semiadi G, O'Neill KT, Scott IC and Barry TN (1997a) Seasonal patterns of luteal cyclicity in young red deer (*Cervus elaphus*) and sambar deer (*Cervus unicolor*) *Reproduction, Fertility and Development* **9** 587–596

Asher GW, Scott IC, O'Neill KT, Smith JF, Inskeep EK and Townsend EC (1997b) Ultrasonographic monitoring of antral follicle development in red deer (*Cervus elaphus*) *Journal of Reproduction and Fertility* **111** 91–99

Asher BW, Scott IC, Mockett BG, O'Neill KT, Diverio S, Inskeep EK and Townsend EC (1998) Ultrasonographic monitoring of ovarian function during the oestrous cycle of fallow deer (*Dama dama*). *Proceedings of the Fourth International Congress on the Biology of Deer*, Kaposvár, Hungary (in press)

Berg DK, Asher GW, Pugh PA, Tervit HR and Thompson JG (1995) Pregnancies following the transfer of *in vitro* matured and fertilised red deer (*Cervus elaphus*) oocytes *Theriogenology* **43** 166 (Abstract)

Blouch RA (1987) Reproductive seasonality of the white-tailed deer on the Columbian Llanos. In *Biology and Management of the Cervidae* pp 339–343 Ed. CM Wemmer. Smithsonian Institution Press, Washington

Branan WV and Marchinton RL (1987) Reproductive ecology of white-tailed and brocket deer in Suriname. In *Biology and Management of the Cervidae* pp 344–351 Ed. CM Wemmer. Smithsonian Institution Press, Washington

Chapman NG and Harris S (1991) Evidence that the seasonal antler cycle of adult Reeve's muntjac (*Muntiacus reevesi*) is not associated with reproductive quiescence *Journal of Reproduction and Fertility* **92** 361–369

Chapman DI, Chapman NG and Dansie O (1984) The periods of conception and parturition in feral Reeves' muntjac (*Muntiacus reevesi*) in southern England based upon age of juvenile animals *Journal of Zoology* **204** 575–578

Chapple RS, English AW and Mulley RC (1993) Characteristics of the oestrous cycle and duration of gestation in chital hinds (*Axis axis*) *Journal of Reproduction and Fertility* **98** 23–26

Curlewis JD, Loudon ASI and Coleman AMP (1988) Oestrous cycles and the breeding season of the Père David's deer hind (*Elaphurus davidianus*) *Journal of Reproduction and Fertility* **82** 119–126

Fennessy PF, Asher GW, Beatson NS, Dixon TE, Hunter JW and Bringans MJ (1994) Embryo transfer in deer *Theriogenology* **41** 133–138

Fukui Y, McGowan LT, James RW, Asher GW and Tervit HR (1991) Effects of culture duration and time of gondadotrophin addition on *in vitro* maturation and fertilization of red deer (*Cervus elaphus*) oocytes *Theriogenology* **35** 499–512

Gosch B and Fischer K (1989) Seasonal changes of testis volume and sperm quality in adult fallow deer (*Dama dama*) and their relationship to the antler cycle *Journal of Reproduction and Fertility* **85** 7–17

Goss RJ (1983) *Deer Antlers: Regeneration, Function and Evolution*, Academic Press, New York

Guinness FE, Lincoln GA and Short RV (1971) The reproductive cycle of the female red deer, *Cervus elaphus*. *Journal of Reproduction and Fertility* **27** 427–438

Haigh JC, Cates WF, Glover GJ and Rawlings NC (1984) Relationships between seasonal changes in serum testosterone concentrations, scrotal circumference and sperm morphology of male wapiti (*Cervus elaphus*) *Journal of Reproduction and Fertility* **70** 413–418

Jabbour HN, Marshall VS, Argo C McG, Hooton J and Loudon ASI (1994) Successful embryo transfer following artificial insemination of superovulated fallow deer (*Dama dama*) *Reproduction, Fertility and Development* **6** 181–185

Jackson JE and Langguth A (1987) Ecology and status of the pampas deer in the Argentinean pampas and Uruguay. In *Biology and Management of the Cervidae* pp 403–409, Ed. CM Wemmer. Smithsonian Institution Press, Washington

Killeen ID and Caffery MGJ (1982) Uterine insemination of ewes with the aid of a laparoscope *Australian Veterinary Journal* **59** 95–96

Lasley BL and Kirkpatrick JF (1991) Monitoring ovarian function in captive and free-ranging wildlife by means of urinary and fecal steroids *Journal of Zoological Wildlife Medicine* **22** 23–31

Lincoln GA (1971) The seasonal reproductive changes in the red deer stag (*Cervus elaphus*) *Journal of Zoology* **163** 105–123

Lincoln GA (1985) Seasonal breeding in deer. In *The Biology of Deer Production* Eds PF Fennessy and KR Drew. Bulletin No 22, Royal Society of New Zealand, Wellington 165–179

Lincoln GA and Short RV (1980) Seasonal breeding: Nature's contraceptive *Recent Progress in Hormone Research* **36** 1–52

Loudon ASI and Curlewis JD (1988) Cycles of antler and testicular growth in an aseasonal tropical deer (*Axis axis*) *Journal of Reproduction and Fertility* **83** 729–738

Marshall FHA (1937) On the change over in the oestrous cycle in animals after transference across the equator, with further observations on the incidence of the breeding seasons and the factors controlling sexual periodicity *Proceedings of the Royal Society of London Series B* **122** 413–428

Mishra HR and Wemmer C (1987) The comparative breeding ecology of four cervids in Royal Chitwan National Park. In *Biology and Management of the Cervidae* pp 259–271 Ed. CM Wemmer. Smithsonian Institution Press, Washington

Monfort SL, Wemmer C, Kepler TH, Bush M, Brown JL and Wildt DE (1991) Monitoring ovarian function and pregnancy in Eld's deer (*Cervus eldi thamin*) by evaluating urinary steroid metabolite excretion *Journal of Reproduction and Fertility* **88** 271–281

Monfort SL, Brown JL, Bush M, Wood TC, Wemmer C, Vargas A, Williamson LR, Montali R and Wildt DE (1993a) Circannual inter-relationships among reproductive hormones, gross morphometry, behaviour, ejaculate characteristics and testicular histology in Eld's deer stags (*Cervus eldi thamin*) *Journal of Reproduction and Fertility* **98** 471– 480

Monfort SL, Asher GW, Wildt DE, Wood TC, Schiewe MC, Williamson LR, Bush M and Rall WF (1993b) Successful intrauterine insemination of Eld's deer (*Cervus eldi thamin*) with frozen–thawed spermatozoa *Journal of Reproduction and Fertility* **99** 459–465

Monfort SL, Brown JL and Wildt DE (1993c) Episodic and seasonal rhythms of cortisol secretion in male Eld's deer (*Cervus eldi thamin*) *Journal of Endocrinology* **138** 41–49

Monfort SL, Williamson LR, Wemmer CM and Wildt DE (1993d) Intensive management of the Burmese brow-antlered deer *Cervus eldi thamin* for effective captive breeding and conservation *International Zoology Yearbook* **32** 44–56

Morrow CJ, Asher GW, Berg DK, Tervit HR, Pugh PA, McMillan WH, Beaumont S, Hall DRH and Bell ACS (1994) Embryo transfer in fallow deer (*Dama dama*): superovulation, embryo recovery and laparoscopic transfer of fresh and cryopreserved embryos *Theriogenology* **42** 579–590

Morrow CJ, Asher GW and MacMillan KL (1995) Oestrous synchronisation in farmed fallow deer (*Dama dama*): effects of season, treatment duration and the male on the efficacy of the intravaginal CIDR device *Animal Reproduction Science* **37** 159–174

Mylrea GE (1992) *Natural and artificial breeding of farmed chital deer* (Axis axis) *in Australia* PhD thesis, University of Sydney, NSW, Australia

Pollard JW, Bringans MJ and Buckrell B (1995) *In vitro* production of wapiti and red deer (*Cervus elaphus*) embryos *Theriogenology* **43** 301–304

Schnare H and Fischer K (1987) Secondary sex characteristics and connected physiological values in male fallow deer (*Dama dama*) and their relationship to changes of the annual photoperiod: doubling the frequency *Journal of Experimental Zoology* **224** 463–471

Schwartz CC and Hundertmark KJ (1993) Reproductive characteristics of Alaskan moose *Journal of Wildlife Management* **57** 454–468

Schwarzenberger F, Mostl E, Palme R and Bamberg E (1996) Faecal steroid analysis for non-invasive monitoring of reproductive status in farm, wild and zoo animals *Animal Reproduction Science* **42** 515–526

Smith TR, Hunter CG, Eisenberg JF and Sunquist ME (1996) Ecology of white-tailed deer in eastern Everglades National Park *Florida Museum of Natural History* **39** 141–172

Thomas DC and Cowan IMcT (1975) The pattern of reproduction in female Columbian black-tailed deer, *Odocoileus hemionus columbianus. Journal of Reproduction and Fertility* **44** 261–272

van Mourik S (1986) Reproductive performance and maternal behaviour in farmed rusa deer, (*Cervus rusa timorensis*) *Applied Animal Behaviour Science* **15** 147–159

van Mourik SA and Stelmasiak T (1990) Endocrine mechanisms and antler cycles in rusa deer, *Cervus (Rusa) timorensis*. In *Horns, Pronghorns and Antlers* pp 265–297 Eds GA Bubenik and AB Bubenik. Springer-Verlag, New York

Wemmer C, Halverson T, Rodden M and Portillo, T (1989) The reproductive biology of female Pere David's deer (*Elaphurus davidianus*) *Zoo Biology* **8** 49–55

Whitehead GK (1993) *The Whitehead Encyclopaedia of Deer* Swan Hill Press, Shrewsbury, UK

Wildt DE (1992) Genetic resource banks for conserving wildlife species: justification, examples and becoming organized on a global basis *Animal Reproduction Science* **28** 247–257

Wildt DE, Rall WF, Critser JK, Monfort SL and Seal US (1997) Genome resource banks: 'Living collection' for biodiversity conservation *Bioscience* **47** 689–698

Wong B and Parker KL (1993) Estrus in black-tailed deer *Journal of Mammology* **69** 168–171

Journal of Reproduction and Fertility Supplement **54**, 157–168

Reproduction in water buffalo: comparative aspects and implications for management

B. M. A. Oswin Perera

Animal Production and Health Section, Joint FAO/IAEA Division of Nuclear Techniques in Agriculture, International Atomic Energy Agency, PO Box 100, A-1400 Vienna, Austria

The domestic buffalo occupies an important niche in many ecologically disadvantaged agricultural systems, providing milk, meat and draught power. Although buffalo can adapt to harsh environments and live on low quality forage, their reproductive efficiency is often compromised by such conditions. Climatic stress depresses ovarian cyclicity, oestrous expression and conception rates. Poor nutrition, usually related to seasonal fluctuations in availability and quality of feed, delays puberty and increases the duration of postpartum anoestrus. Management factors such as the system of grazing (free, tethered or none) and sucking by calves (restricted or *ad libitum*) also modulate reproductive functions. Finally, the skills and capabilities of farmers as well as the quality of support services such as artificial insemination and disease control also influence fertility. The relative importance of these factors vary greatly depending on ecological conditions and production systems. Improvement of reproductive efficiency therefore requires the identification of specific limiting factors under a given situation and the development and field testing of strategies for improvements and interventions that are sustainable with available local resources. The application of modern reproductive technologies in buffaloes requires an appreciation of their biology and reproductive physiology as well as the potentials and limitations under each specific production system.

Introduction

The domestic water buffalo (*Bubalus bubalis*) is an important livestock resource in many countries of Asia, the Mediterranean region and Latin America. The world population of buffaloes is estimated to be 140–150 × 10^6, of which over 90% are in developing countries. They have been classified into three main types: river, swamp and Mediterranean (Cockrill, 1974).

The buffalo has been traditionally regarded as a poor breeder because in most conditions under which it is usually raised, which are small-holder farming systems in harsh environments with minimal managerial inputs, they have low fertility (Bhattacharya, 1974). This is manifested mainly as late maturity, long postpartum anoestrus, poor expression of oestrous signs, low conception rates (CR) and long calving intervals (CI). However, studies on feral buffaloes in Australia (Tulloch, 1979), as well as on domestic buffaloes in Pakistan (Usmani *et al.*, 1985), Sri Lanka (Perera *et al.*, 1987) and Brazil (Vale *et al.*, 1990) prove that they can have excellent fertility provided that genotypes are matched to the environment and they are managed and fed properly. A unique advantage of the buffalo is its ability to utilize poor forages more efficiently than do cattle (Sebastian *et al.*, 1970). However, the buffalo cannot dissipate heat as efficiently as cattle through physiological methods such as sweating (Ranawana *et al.*, 1984) and has therefore developed the strategy of wallowing in order to maintain homeostasis when environmental temperature is high.

In early attempts to improve fertility of buffaloes through reproductive technologies such as artificial insemination (AI), oestrous synchronization and embryo transfer, researchers and livestock personnel often presumed that the reproductive biology of the buffalo was identical to that of cattle. This direct extrapolation of methods resulted in poor fertility, particularly when applied under practical farming conditions. More recently, however, the application of modern research technologies has provided valuable insights into the physiology of buffaloes, and this has allowed a more informed approach to the improvement of their reproductive performance.

Comparative aspects of reproduction in cattle and buffaloes have been the subject of comprehensive reviews a decade ago (Dobson and Kamonpatana, 1986; Jainudeen and Hafez, 1987) and subsequent advances in knowledge on buffalo reproduction have been reviewed by Madan *et al.* (1996); Vale (1997); and Zicarelli (1997). This review will therefore not attempt to cover all aspects of buffalo reproduction. Its main objectives are to: (i) selectively review certain aspects that are important from the viewpoint of practical field applications under prevailing farming systems; (ii) highlight some important species differences compared with cattle; and (iii) discuss opportunities and constraints to the application of modern reproductive technologies for improving buffalo production in situations in which they make an important economic contribution.

Buffalo Farming Systems

Systems of buffalo production vary widely through the different regions of the world (Mahadevan, 1992) and are determined by factors such as climate (tropical or temperate, humid or arid), type of operation (small or large farm, subsistence or commercial), integration with cropping systems (rain-fed or irrigated, annual or perennial) and the primary purpose (milk, meat, draught, capital or mixed). Because the potential for application of improved reproductive technologies is dictated by the characteristics and available resources in specific production systems (Perera, 1994), a brief overview of types of buffalo and husbandry conditions in different regions of the world is provided.

In South Asia, most buffaloes are of the river type and are raised mainly by small-holders in subsistence or semi-commercial farms. Dairy breeds such as Murrah, Surti and Nili-Ravi are kept in irrigated and rain-fed intensive cropping areas under confinement, with stall-feeding on cut grass, crop residues and concentrates. In the less intensively farmed areas they are managed extensively, with grazing on communal lands supplemented by crop residues and byproducts. Urban areas around major cities have large herds raised purely for milk production under a unique intensive system, in which high producing cows are brought in, kept for the duration of their lactation and disposed of.

In East and South-East Asia, the swamp type predominates and is reared mainly on small farms with integrated crop–livestock husbandry. The major usage is for draught associated with rice cultivation. The buffalo is considered a sign of wealth and status in many communities, provides companionship to the family and is featured in religious and cultural events. The production systems are mostly extensive, with free or tethered grazing on communal lands and little or no supplementary feeding except perhaps rice straw and leguminous tree leaves during periods of drought.

Most of the buffalo in North Africa and the Middle East are of the river and Mediterranean types and are concentrated around the Nile delta. They are kept mainly by small-holders for producing milk and eventually beef. There are a few commercial operations in urban areas, for specialized production of milk and beef under intensive systems. In Europe buffaloes of the Mediterranean type predominate and are kept on large commercial farms under semi-intensive or intensive systems for milk and meat production.

Buffalo production has expanded markedly over the past decade in the Latin American region. Buffalo are raised mainly in areas with poor soil or periodic flooding and are able to thrive in conditions that even Zebu cattle cannot cope with. Production systems vary depending on the purpose, but herd sizes are generally large compared with those in other regions. Beef production is usually under extensive conditions while milk production is more intensively managed, but dual-purpose systems are also found.

Reproduction in Female Water Buffalo

Anatomical aspects

The mature ovaries are smaller in water buffalo than in cattle, and weigh about 2.5 g when inactive and 4 g when active; there are fewer primordial and tertiary follicles (Danell, 1987; Zicarelli, 1997). During the oestrous cycle follicular growth occurs in two waves as in cattle. Mature Graafian follicles rarely exceed 8 mm in diameter (Jainudeen *et al.*, 1983), whereas in cattle they often exceed 12 mm. Mature follicles tend to protrude from the surface of the ovary and can be mistaken for an early corpus luteum when palpated *per rectum*. The corpus luteum is soft at first and becomes larger and firmer by mid-cycle. It is smaller than in cattle, and often does not protrude markedly from the surface of the ovary and it sometimes lacks a clear crown. These attributes make accurate identification of ovarian structures by rectal palpation in buffaloes more difficult than they are in cattle (Sharifuddin and Jainudeen, 1983; Perera *et al.*, 1987). It is therefore necessary that veterinarians wishing to gain proficiency in ovarian palpation of buffaloes subsequently visualize the palpated structures using ultrasonography or laparoscopy or, in animals destined for slaughter, *post mortem*.

The uterus and cervix of water buffaloes are smaller than they are in cattle and lie wholly within the pelvic cavity and the uterus is usually tightly coiled (Perera *et al.*, 1977). The cervical canal is narrower than in cattle and presents a greater challenge for AI technicians to negotiate with the insemination gun or pipette, particularly in heifers.

Puberty

Buffalo heifers usually attain puberty when they reach about 55–60% of their adult body weight, but the age at which they attain puberty can be highly variable, ranging from 18 to 46 months (Bhattacharya, 1974). The factors that influence this are genotype, nutrition, management, social environment, climate, year or season of birth, and diseases. The body weight at which puberty is attained is strongly influenced by genotype and is about 200–300 kg for swamp types and 250–400 kg for river types. Although buffaloes attain puberty later than cattle, they have a longer reproductive life, which tends to compensate for this early economic disadvantage.

The best methods for advancing the age at puberty are provision of good management and feeding from birth, alleviation of heat stress where necessary and control of parasites and diseases.

Oestrous cycle

The mean duration of the ovarian cycle in water buffaloes is similar to that in cattle. However, there is greater variability and a higher incidence of both abnormally short and long cycles. This finding can be attributed to various factors, including adverse environmental conditions and nutrition (Kaur and Arora, 1982; Kanai and Shimizu, 1984). There is considerable variability in the duration of oestrus (Luktuke and Ahuja, 1961; Danell, 1987), timing of ovulation (Kanai *et al.*, 1990) and the interval from peak LH concentration to ovulation (Zicarelli *et al.*, 1997) (Table 1). Moioli *et al.* (1998) showed that the interval from peak LH concentration to ovulation was 25±13 h in animals that conceived after AI, while it was 46±18 h in those that did not. Clearly, this high variability imposes major limitations to the wider use of AI, but is probably not of major consequence in natural breeding systems.

Another common abnormality in the ovarian cycle and ovarian activity in buffaloes is anoestrus or acyclicity as discussed later in the section on the postpartum period. Cystic ovaries, either follicular or luteal, are rare in buffaloes.

Oestrous behaviour and external manifestations

Buffalo farmers and researchers agree that detection of oestrus in buffalo is more difficult than it is in cattle and this has been a major impediment to the use of AI and other improved breeding

Table 1. Duration and timing of events associated with the oestrous cycle in buffaloes

Characteristic	Mean	Range
Duration of oestrous cycle (days)	21	17–26
Duration of oestrus (h)	20	5–27
Onset of oestrus to ovulation (h)	34	24–48
Duration of LH surge (h)	10	7–12
LH peak to ovulation (h)	27	12–60
End of oestrus to ovulation (h)	14	6–21

(Sources: Luktuke and Ahuja, 1961; Danell, 1987; Jainudeen and Hafez, 1987; Kanai *et al.*, 1990; Zicarelli *et al.*, 1997).

techniques. Homosexual activity is not as common as it is in cattle: only about 20% of females exhibit it (Danell, 1987; Vale *et al.*, 1990; Barkawi *et al.*, 1993). In a herd, the females that are higher in the social dominance order tend to exhibit such behaviour. Other behavioural signs include restlessness, bellowing and frequent voiding of small quantities of urine (Danell, 1987; Vale *et al.*, 1990), but these behaviours are not consistently exhibited in most animals. During periods of high ambient temperature, the duration of oestrus may be shorter and the signs exhibited only during the night or early morning.

Externally detectable physical changes around the time of oestrus include swelling of the vulva and reddening of the vestibular mucosa (Kanai and Shimizu, 1984; Danell, 1987). Swelling of the vulva results in effacement of the horizontal wrinkles that are present on its external surface and this, together with vestibular reddening, can be detected by regular examination of individual animals under confined conditions. Mucus secreted from the cervix during oestrus is less copious than it is in cattle, and does not usually hang as strands from the vulva, but tends to accumulate on the floor of the vagina and be discharged either when the animal is lying down (Perera *et al.*, 1977) or with the urine (Kanai *et al.*, 1990). Thus in buffaloes that are housed or tethered, a good practice is to examine the floor near the rear of the animal each morning and evening for signs of mucus. Studies using intravaginal probes to detect either electrical conductivity or resistance (Narasimha Rao and Venkataramiah, 1989) appeared promising as aids to detection of oestrus, but their applicability under practical farming situations has not been reported subsequently.

Improvement in the detection of oestrus needs education and motivation of farmers, proper identification and records for individual animals and regular close observation for the occurrence of the behavioural and physical signs described above. In many conditions the main method of detection of oestrus is the use of teaser animals such as vasectomized bulls.

Endocrinology

The changes occurring in the main reproductive hormones during the oestrous cycle, pregnancy, parturition and the postpartum period are now well documented for both river and swamp buffaloes (see Dobson and Kamonpatana, 1986; Jaiundeen and Hafez, 1987).

Progesterone. The temporal changes of progesterone in blood and milk during the ovarian cycle are similar to those in cattle, but the concentration of progesterone is lower, particularly in swamp buffaloes (Kamonpatana *et al.*, 1979; Kanai and Shimizu, 1984; Perera *et al.*, 1987). In river buffaloes, the concentrations are higher than those in swamp buffaloes but still below those in cattle (Danell, 1987; Chohan *et al.*, 1992). In both types of buffalo, the concentration of progesterone in milk parallels that in blood; the concentrations of progesterone in whole milk are four to six times higher than in blood. Concentrations of progesterone in skim (defatted) milk are similar to or slightly lower than those in blood. The main practical application of the progesterone assay is in discriminating between the presence and absence of a corpus luteum, which can be used for monitoring the onset of puberty and postpartum ovarian activity, for diagnosis of absence of pregnancy and for investigation of infertility (Perera and Abeyratne, 1979; Garcia *et al.*, 1995). However, given the variability in concentrations of progesterone in buffaloes that are attributable to genotype and sampling procedures, it is necessary that each laboratory establishes standardized methods as well as the levels of discrimination used for diagnosing the presence or absence of luteal function.

Oestradiol. Results from oestradiol measurements in buffaloes are difficult to interpret because of variations in assay methods and limitations of assay sensitivity. In general, concentrations of oestradiol during the follicular phase of the ovarian cycle appear to be relatively lower than they are in cattle (Batra and Pandey, 1982; Kanai and Shimizu, 1984; Avenell *et al.*, 1985) and this may be a reason for the lower intensity of oestrous signs exhibited by buffaloes (Zicarelli, 1997).

Gonadotrophins. Owing to the lack of availability of purified LH and FSH from buffaloes, most studies have relied on heterologous assays with results calculated against a bovine standard. The available data indicate that the temporal pattern of both LH and FSH in buffaloes is basically similar to that in cattle (Avenell *et al.*, 1985; Kanai *et al.*, 1990); a preovulatory LH surge occurs on the day of oestrus and lasts for 7–12 h.

Pregnancy

The fertilized ovum reaches the uterus by day 4–5 after oestrus and the blastocyst hatches by day 6–8, the latter event occurring 2–3 days earlier than in cattle (Jainudeen, 1990). However, subsequent development of the conceptus and its features that are palpable *per rectum* occur at comparatively later gestational ages than in cattle. This appears to be related to the longer duration of gestation in the buffalo, in which it ranges from 300 to 330 days (Perera and de Silva, 1985).

Diagnosis of pregnancy

Clinical methods for pregnancy diagnosis include rectal palpation and ultrasonography, while the main laboratory methods are based on measurement of progesterone or pregnancy-specific protein B (PSPB). The techniques are basically similar to those used routinely in cattle.

Rectal palpation. Owing to the longer gestation in buffalo, each palpable feature is first discernible about 2–4 weeks later during pregnancy than it is in cattle (Perera, unpublished). Enlargement of the gravid uterine horn and thinning of the uterine wall can be detected in heifers at about 35 days and in cows at 40 days. Presence of fluid and the amniotic vesicle are usually palpable by 40–50 days, when slipping of fetal membranes is also possible. By 60 days the gravid horn is nearly double the diameter of the non-gravid horn but usually the uterus is still within the pelvic cavity. The stage of pregnancy at which it descends to the abdomen, as well as the disposition of the fetus within the pelvis and abdomen, are more variable than they are in cattle. This appears to be influenced by the degree of rumen fill by altering the space available within the abdomen. The fetus can usually be ballotted from the fourth to the sixth months of gestation and can be directly palpated thereafter until term. Development of fremitus in the middle uterine arteries on the gravid and non-gravid sides occurs relatively later than it does in cattle. There is a greater tendency for buffaloes to resist palpation and to strain when the arm is in the rectum, so adequate restraint of the animal and patience are required.

Ultrasonography. As in cattle, the most practical type of instrument is one based on real-time B-mode ultrasonography, equipped with an intra-rectal transducer operating at 5.0 or 7.5 MHz. The embryonic vesicle (non-echogenic) and the embryo proper (echogenic) can be imaged by about 19–21 days after mating. The heartbeat of the embryo is discernible at 30–35 days and structures such as the allantois, amnion, spinal cord and limbs can be imaged at 35–40 days (Perera, unpublished). The stage of gestation can be deduced from the heart rate and the length of the embryo.

Hormone assay. Absence of pregnancy can be diagnosed by measuring progesterone in blood or milk samples collected at 20–23 days after mating with an accuracy approaching 100%, whereas the accuracy for diagnosing pregnancy is only 65–80% (Perera *et al.*, 1980). Early diagnosis of non-pregnancy has particular advantages in buffaloes in which returns to oestrus are more easily missed, in that animals requiring closer attention can be identified and appropriate action taken. For positive confirmation of pregnancy the measurement of PSPB in blood can be performed from 30–40 days onwards (Debenedettia *et al.*, 1997).

Parturition

The process, stages and duration of parturition in buffaloes are similar to those in cattle. Complications such as dystocia are less common in buffaloes but, if they do occur, are more severe and difficult to correct than they are in cattle (L. N. A. de Silva, personal communication). The changes occurring in hormones such as progesterone, oestrone and the main metabolite of prostaglandin $F_{2\alpha}$ (PGFM) around the period of parturition are similar to those in cattle (Perera *et al.*, 1981).

Postpartum period

Uterine involution. Uterine involution is usually completed by 25–35 days after calving (Jainuden and Hafez, 1987; Perera *et al.*, 1987), which is similar to the period required in cattle. The stimulus of suckling has been shown to shorten the involution time (Usmani *et al.*, 1990). Postpartum infections and complications are generally uncommon in buffaloes, except in certain management conditions in which poor hygiene in the calving environment, wallowing in contaminated water or prevalence of specific infections of the reproductive tract lead to endometritis and pyometra.

Ovarian activity. The period of postpartum anoestrus or acyclicity is usually longer in buffaloes than in cattle under comparative management conditions (see Dobson and Kamonpatana, 1986; Jainudeen and Hafez, 1987). Studies in Sri Lanka (de Silva *et al.*, 1985) have shown that indigenous buffaloes raised by village farmers have a highly variable period of postpartum anoestrus, depending on the management system. In one traditional system where they were raised under free-grazing conditions with abundant natural feed, restricted suckling and presence of intact bulls, ovarian activity commenced by 30–60 days after calving and most cows conceived at the first ovulation, resulting in calving intervals of 12–13 months (Perera *et al.*, 1987). However, in a similar system but with poor feed availability and free suckling by the calves, ovarian activity was delayed until 150–200 days after calving, resulting in calving intervals of 18–20 months (Perera *et al.*, 1988). In both of these situations the incidence of short ovarian cycles at the commencement of ovarian activity was about 12%, which is much lower than that observed in dairy cattle. However, dairy buffaloes under confined management systems have a much higher incidence of short luteal phases before the first oestrus (Agarwal and Purbey, 1983; Usmani *et al.*, 1990). Recent studies (Zicarelli, 1997; Moioli *et al.*, 1998) indicate that the presence of bulls has a biostimulatory effect on postpartum ovarian activity of buffalo cows, which can reduce cyclic irregularities and also advance the time of ovulation.

Buffalo cows that are freely suckled by their calves have a longer period of postpartum anoestrus than do those that are subjected to limited or no sucking (Mohan *et al.*, 1990; Tiwari and Pathak, 1995). In milked dairy buffaloes, sucking for 2 min twice a day delays follicular development, first rise in progesterone, first palpable corpus luteum and first oestrus (Usmani *et al.*, 1990). A study on indigenous buffaloes in Sri Lanka (Abeygunawardena *et al.*, 1996a) showed that under extensive management systems, in which weaned calves are not provided with supplementary feed, early weaning has adverse effects on their growth and survivability. In contrast, restricted sucking once or twice a day improved pregnancy rates without increasing mortality of calves (Table 2).

Other factors that have important influences on postpartum ovarian activity are climate and nutrition (Ahmed *et al.*, 1981), which are often interrelated in extensive grazing systems (Lundstrom *et al.*, 1982; Kaur and Arora, 1982), and the presence of parasitism and diseases.

Reducing the duration of the postpartum anoestrous period is perhaps the most important consideration in attempts to improve reproductive efficiency of buffaloes. This can be achieved by

Table 2. Effects of different sucking regimens on pregnancy rates in indigenous buffalo cows at 90 and 150 days postpartum and on mortality of their calves up to 6 months of age

Sucking regimen	*n*	Pregnancy rate (%) at: 90 days	Pregnancy rate (%) at: 150 days	Calf deaths (%)
Free sucking (Control)	74	19	23	5.4
Weaned at 45 days	11	54	73	55.0
Weaned at 60 days	16	6	37	31.0
Weaned at 90 days	18	5	27	33.0
Once per day	29	38	59	6.8
Twice per day	22	31	59	4.5

(Data from Abeygunawardena *et al.*, 1996a)

ensuring that the inhibitory effects of climate, management, nutrition and diseases discussed above are eliminated or ameliorated. In small-holder systems this demands a holistic approach from researchers and livestock support services to develop and transfer a package of appropriate technologies to farmers through close interaction with them (Abeygunawardena *et al.*, 1996b). The use of hormones and other therapeutic interventions to overcome anoestrus are unlikely to be successful unless these basic constraints are resolved.

Endocrinology. LH secretion in buffaloes remains low during the early postpartum period and episodic pulses become detectable a few weeks before ovarian activity commences (Mohan *et al.*, 1990). The timing of this change depends on the factors mentioned above and, for example, occurs earlier in animals that are subjected to better nutrition or restricted suckling than in those under poor nutrition or free sucking. Progesterone profiles during the postpartum period reflect luteal function (Perera, 1980; Jainudeen *et al.*, 1983) and are useful for monitoring ovarian activity, detecting causes of poor reproductive efficiency, improving reproductive management and comparing the responses to interventions aimed at improving fertility (Perera *et al.*, 1984; Garcia *et al.*, 1995).

Measurement of reproductive efficiency

For assessing reproductive efficiency, of an individual, herd or population, reproductive indices commonly used in cattle, for example interval from calving to first oestrus and conception, conception rate to first or all services, overall pregnancy rate, services per conception, calving rate and calving interval, are applied to buffaloes. Values for these characteristics available from different countries for various buffalo genotypes and their crosses, and under varying agro-climatic and management conditions, confirm the highly variable nature of such measures depending on the specific situation (Jainudeen and Hafez, 1987).

Therefore, in setting acceptability criteria and standards for reproductive indices it is important to consider these factors as well as the main purpose for which the animals are raised. In modern dairy cattle systems, the optimum calving interval is considered to be one year and, since the average duration of gestation is 280 days, cows must become pregnant by 85 days after calving. With the longer duration of gestation of 300–330 days in buffaloes a one year calving interval is often impossible to achieve. Furthermore, although a short calving interval is desirable in purely dairy systems, it may not be advantageous in situations in which draught power and other outputs are also important. Indeed, in conditions with limited and highly seasonal feed resources buffaloes may calve only every other year. Attempts to reduce this interval could result in losses due to deaths of calves born during unfavourable seasons or to low survivability of the dam. Thus a more realistic and achievable set of criteria needs to be developed for such production systems, based on the optimum economic benefits that can be derived by farmers using the available resources in a sustainable manner.

Reproduction in Male Water Buffalo

Puberty and sexual maturity

The age at puberty is influenced by breed and nutrition, and varies from 18 to 46 months (Jainudeen and Hafez, 1987). The growth of the testis can be assessed by measuring scrotal circumference and is related to body weight and age (Bongso *et al.*, 1984). Cell division in the spermatogenic series commences by about 12 months of age and active spermatogenesis can be detected by 15 months. Fully formed spermatozoa have been detected in the seminiferous tubules by 17–19 months, but the ejaculate contains motile spermatozoa only after 24–30 months of age (Jainudeen and Hafez, 1987). Thus there appears to be a longer period of delay in buffalo than in cattle between the establishment of spermatogenesis in the testis and the appearance of motile spermatozoa in the ejaculate. Sexual maturity or full potential is usually reached about one year after puberty.

The adult male

Although buffalo bulls can reproduce throughout the year, under certain conditions (as is also the case with buffalo cows) seasonal changes are observed in libido and fertility. In feral or wild buffaloes a period of annual sexual activity similar to the rutting behaviour seen in other wild ungulates is common (Tulloch, 1979). Under domestication these seasonal changes in reproductive functions are less marked and appear to be related to factors such as rainfall through fluctuations in the quality and availability of feed and to temperature and humidity through effects on spermatogenic functions.

The secretion of LH and testosterone is basically similar to that in cattle, with a typical episodic or pulsatile pattern, but with the exception that the concentration of testosterone in blood is significantly lower than it is in cattle (Perera *et al.*, 1979; Ohashi *et al.*, 1996). This has been implicated as a reason for lower libido in buffalo bulls compared with cattle, but requires further study for confirmation.

In many South-East Asian countries, swamp type buffalo cows are mated to improved river type bulls in order to increase milk production while retaining draught ability and climatic adaptability. Since the number of diploid chromosomes in swamp types is 48 whereas it is 50 in river types, the F1 generation will have a proportion of animals with an unbalanced karyotype of 49 chromosomes (Bongso *et al.*, 1983). Such animals can have a higher rate of degeneration of spermatogenic cells and it is therefore important that cross-breeding programmes include the detection and culling of such infertile animals.

Reproductive Technologies

A major obstacle to improved breeding technologies in buffaloes is the difficulty in detection of oestrus. This has resulted in a much lower usage of AI in buffaloes than in cattle in most countries where both species are raised. Improvement of buffalo production in these countries will undoubtedly benefit from wider use of AI and, where appropriate, other improved breeding technologies. Their rational use, however, requires an understanding of the reproductive functions of this species as well as an appreciation of the potential and limitations of the farming system under which they are to be applied.

Artificial insemination (AI)

The procedures for collection, evaluation, dilution, preservation and storage of buffalo semen are now well established (see Vale, 1997) and are based mainly on techniques developed for cattle with some modifications, the major differences being in semen diluents, Tris buffers, skim milk and coconut water are more commonly used.

There is evidence that buffalo spermatozoa subjected to freezing and thawing may have a shorter fertile lifespan inside the female tract than fresh semen (Moioli *et al.*, 1998). Thus accurate detection of oestrus and timing of AI become critical when frozen semen is used and may be one reason for the difference seen in conception rates between AI and natural mating. Another factor that can contribute to poor conception rates is the narrow cervix of the buffalo, which makes AI more difficult.

The timing of AI commonly used in buffalo is based on the AM/PM rule derived from cattle (that is cows first seen at oestrus in the morning are served in the afternoon and those seen at oestrus in the afternoon are served the next morning). However, as discussed earlier, the temporal relationships between luteolysis, oestrus, LH peak and ovulation are highly variable in buffalo. The optimum time for AI appears to be the end of standing oestrus, when the cow refuses further mounting by teaser males (Moioli *et al.*, 1998), which is later than that in cattle. In most buffalo production systems, however, the use of teaser bulls is not practical. The option may be to inseminate cows about 24 h after the first detection of signs of oestrus, but this needs further research for confirmation.

Apart from these technical problems, the main obstacles, particularly in less developed countries, are lack of awareness among farmers of the advantages of AI and inadequacies in infrastructure for efficient delivery of the service.

Synchronization of oestrus

The methods that have been used to synchronize oestrus in buffaloes are based on those developed in cattle using progestational compounds with or without oestradiol, prostaglandins (PG), or combinations of these (see Perera, 1987). Most studies have shown that the percentage of treated buffalo cows that come into oestrus is acceptable (70–90%), but the conception rates to AI are about 20–35% (Pathiraja *et al.*, 1979; Rajamahendran *et al.*, 1980). However, use of natural service can give better results approaching those in cattle.

After withdrawal of exogenous progestagens or treatment with PG the interval to ovulation is more variable than in cattle, resulting in poor fertility when AI is done at predetermined times (Zicarelli *et al.*, 1997). This can be overcome to some extent by doing AI after detection of oestrus, but this defeats one of the main purposes of synchronization, that of being able to dispense with detection of oestrus. There is also the possibility that some ovulating animals will either show very mild or no signs of oestrus and therefore be missed.

The selection of animals to be used in synchronization of oestrus is very important. For regimens using PG the animals must be non-pregnant and cyclic. Ensuring that this occurs is difficult under situations with free range grazing and presence of intact bulls. Use of PG in pregnant animals will cause embryonic death. Although progestagens will not have this effect, the subsequent act of AI can. Animals must also be in good nutritional condition and free from disease. It is also important to ensure that they are not stressed during procedures such as injection of hormones, implantation of devices or AI, particularly under tropical conditions and if they need to be herded together or moved for this purpose. In such situations adequate water, feed and shade must be provided to prevent stress and increase in body temperature which can cause conception failure or early embryonic death. Finally, when there are seasonal differences in breeding activity, synchronization programmes should be scheduled for the more favourable periods when most animals can be expected to be cyclic.

Embryo transfer

The early experiences in buffaloes with methods of embryo transfer developed for cattle were disappointing but, with recent advances in knowledge on buffalo reproductive physiology, the success rate is clearly improving (Drost *et al.*, 1983; Jainudeen, 1990; Madan *et al.*, 1996; Zicarelli, 1997). In addition, techniques have now been developed in buffaloes for ultrasound-guided ovum pick-up (Boni *et al.*, 1996), *in vitro* fertilization and embryo culture (Madan *et al.*, 1996).

In general, the superovulatory responses obtained, recovery of transferable embryos and pregnancy rates in buffaloes have been lower than those in cattle. This has been attributed to differences in ovarian follicular populations, reproductive hormones and general fertility discussed earlier. Given the current success rates, this technology is likely to be applicable only under specific situations for improving buffalo production. Two such opportunities might be in providing the alternative of importing frozen embryos instead of live animals when nucleus breeding stocks need to be established or in the conservation of rare genetic resources. At present it is difficult to see practical applications for embryo transfer and related technologies in tropical small-holder systems.

Conclusion

The domestic buffalo has unique biological characteristics that distinguish it from cattle. Although some of these attributes have made it an indispensable livestock resource, particularly to small-holder farmers in developing countries, others have been a hindrance to greater exploitation of its potential. Under most small-holder production systems the reproductive efficiency of buffaloes is compromised by factors related to climate, management, nutrition and diseases. Thus the main thrust in improving fertility under such situations is likely to be based on simple but innovative improvements and interventions that alleviate these stresses and, where relevant, on the rational application of modern technologies.

The basic technical problems associated with AI in buffaloes were largely overcome a decade ago but it has failed to have the expected impact, due largely to infrastructural and logistic problems. More advanced approaches such as use of hormones for treating anoestrus and synchronizing oestrus are likely to succeed only after the more basic problems have been rectified. They also demand a good knowledge of the production system in order to be used in a cost-effective manner. Embryo transfer and related technologies will, for the foreseeable future, be applicable only under very specific conditions as in the establishment and multiplication of elite institutional herds, as an alternative to importation of live animals, or for the conservation of germplasm.

References

Abeygunawardena H, Kuruwita VY and Perera BMAO (1996a) Effects of different suckling regimes on postpartum fertility of buffalo cows and growth rates and mortality of calves. In *Role of the Buffalo in Rural Development in Asia* pp 321–336 Eds BMAO Perera, JA deS Siriwardene, NU Horadagoda and MNM Ibrahim. Natural Resources, Energy and Science Authority for Sri Lanka, Colombo

Abeygunawardena H, Subasinghe DHA, Perera ANF, Ranawana SSE, Jayatilake MWAP and Perera BMAO (1996b) Transfer of technology in smallholder intensive buffalo farming: results from a pilot study in Mahaweli System "H". In *Role of the Buffalo in Rural Development in Asia* pp 67–94 Eds BMAO Perera, JA deS Siriwardene, NU Horadagoda and MNM Ibrahim. Natural Resources, Energy and Science Authority for Sri Lanka, Colombo

Agarwal SK and Purbey LN (1983) Aberrations of the oestrous cycle in rural buffaloes *Indian Veterinary Journal* **60** 989–991

Ahmed N, Chaudry RA and Khan BB (1981) Effect of month and season of calving on the length of subsequent calving interval in Nili-Ravi buffaloes *Animal Reproduction Science* **3** 301–306

Avenell JA, Seepudin Y and Fletcher IC (1985) Concentrations of LH, oestradiol 17β and progesterone in the peripheral plasma of swamp buffalo cows (*Bubalus bubalis*) around the time of oestrus *Journal of Reproduction and Fertility* **74** 419–424

Barkawi AK, Bedeir LH and El Wardani MA (1993) Sexual behaviour of Egyptian buffaloes in postpartum period *Buffalo Journal* **9** 225–236

Batra SK and Pandey RS (1982) Luteinizing hormone and oestradiol 17β in blood plasma and milk during the oestrous cycle and early pregnancy in Murrah buffaloes *Animal Reproduction Science* **5** 147–157

Bhattacharya P (1974) Reproduction. In *The Husbandry and Health of the Domestic Buffalo* pp 105–158 Ed. WR Cockrill. FAO, Rome

Bongso TA, Hilmi M and Basrur PK (1983) Testicular cells in hybrid water buffaloes *Research in Veterinary Science* **35** 253–260

Bongso TA, Hassan MD and Nordin W (1984) Relationship of scrotal circumference and testis volume to age and body weight in swamp buffalo *Theriogenology* **22** 127–134

Boni R, Roviello S and Zicarelli L (1996) Repeated ovum pick-up in Italian Mediterranean buffalo cows *Theriogenology* **46** 899–909

Chohan KR, Chaudhry RA, Khan NU and Chaudhry MA (1992) Serum progesterone profile during oestrus cycle and early pregnancy in normal and synchronized Nili-Ravi buffaloes *Buffalo Journal* **1** 77–82

Cockrill WR (1974) *The Husbandry and Health of the Domestic Buffalo* pp 48–56 Ed. WR Cockrill FAO, Rome

Danell B (1987) *Oestrous Behaviour, Ovarian Morphology and Cyclical Variation in Follicular System and Endocrine Pattern*

In Water Buffalo Heifers PhD Thesis Swedish University of Agricultural Sciences, Uppsala

Debenedettia A, Malfatti A, Borghese A, Barile VL and Humblot P (1997) Pregnancy specific protein B (PSPB) detection by RIA in buffalo cows. In *Proceedings of the Fifth World Buffalo Congress* pp 771–775 Caserta, Italy

de Silva LNA, Perera BMAO, Tilakaratne N and Edqvist LE (1985) *Production Systems and Reproductive Performance of Indigenous Buffaloes in Sri Lanka* Swedish University of Agricultural Sciences, Uppsala

Dobson H and Kamonpatana M (1986) A review of female cattle reproduction with special reference to a comparison between buffaloes, cows and zebu *Journal of Reproduction and Fertility* **77** 1–36

Drost M, Wright JM, Cripe WS and Richter AR (1983) Embryo transfer in water buffalo (*Bubalus bubalis*) *Theriogenology* **20** 579–584

Garcia M, Jayasuriya MCN and Perera BMAO (1995) Improving animal productivity by nuclear techniques. In *IAEA Yearbook 1995* pp B17–B32 International Atomic Energy Agency, Vienna

Jainudeen MR (1990) A review of embryo transfer technology in the buffalo. In *Domestic Buffalo Production in Asia* pp 103–112 International Atomic Energy Agency, Vienna

Jainudeen MR and Hafez ESE (1987) Cattle and water buffalo. In *Reproduction in Farm Animals* (5th Edn) pp 297–314 Ed. ESE Hafez. Lea and Febiger, Philadelphia

Jainudeen MR, Sharifuddin W and Bashir Ahmad F (1983) Relationship of ovarian contents to plasma progesterone concentration in the swamp buffalo (*Bubalus bubalis*) *Veterinary Record* **113** 369–372

Kamonpatana M, van de Wiel DFM, Koops W, Leenanuraksha LD, Ngramsuriyaroj C and Usanakornkul S (1979) Oestrus control and early pregnancy diagnosis in swamp buffalo: comparison of EIA and RIA for plasma progesterone *Theriogenology* **11** 399–409

Kanai Y and Shimizu H (1984) Plasma concentrations of LH, progesterone and oestradiol during the oestrous cycle in swamp buffalo (*Bubalus bubalis*) *Journal of Reproduction and Fertility* **70** 507–510

Kanai Y, Abdul Latief T, Ishikawa N and Shimizu H (1990) Behavioural and hormonal aspects of the oestrous cycle in swamp buffaloes reared under temperate conditions. In *Domestic Buffalo Production in Asia* pp 113–120 International Atomic Energy Agency, Vienna

Kaur H and Arora SP (1982) Influence of level of nutrition and season on the oestrous cycle rhythm and on fertility in buffaloes *Tropical Agriculture (Trinidad)* **59** 274–278

Luktuke SN and Ahuja LD (1961) Studies on ovulation in buffaloes *Journal of Reproduction and Fertility* **2** 200–201

Lundstrom K, Abeygunawardena H, de Silva LNA and Perera BMAO (1982) Environmental influence on calving interval and estimates of its repeatability in the Murrah buffalo in Sri Lanka *Animal Reproduction Science* **5** 99–109

Madan ML, Das SK and Palta P (1996) Application of reproductive technology to buffaloes *Animal Reproduction Science* **42** 299–306

Mahadevan P (1992) Distribution, ecology and adaptation. In *Buffalo Production* World Animal Science Series C6 pp 1–12 Eds NM Tulloh and JHG Holmes. Elsevier, Amsterdam

Mohan V, Kuruwita VY, Perera BMAO and Abeygunawardena H (1990) Effects of suckling on the resumption of post-partum ovarian activity in buffaloes *Tropical Agricultural Research* **2** 306–315

Moioli BM, Napolitano F, Puppo S, Barile VL, Terzano GM, Borghese A, Malfatti A, Catalano A and Pilla AM (1998) Patterns of oestrus, time of LH release and ovulation and effects of time of artificial insemination in Mediterranean buffalo cows *Animal Science* **66** 87–91

Narasimha Rao AV and Venkataramiah P (1989) Luteolytic effect of a low dose of cloprostenol monitored by changes in vaginal resistance in suboestrous buffaloes *Animal Reproduction Science* **21** 149–152

Ohashi OM, Oba E and Nogueira JC (1996) Levels of testosterone and androstenedione in male buffaloes of different ages *Buffalo Journal* **3** 313–320

Pathiraja N, Abeyratne AS, Perera BMAO and Buvanendran V (1979) Fertility in buffaloes after oestrus synchronisation with cloprostenol and fixed time insemination *Veterinary Record* **104** 279–281

Perera BMAO (1980) Hormonal profiles and synchronisation of oestrus in river buffaloes in Sri Lanka. In *Animal Production and Health in the Tropics* pp 431–434 Eds MR Jainudeen and AR Omar. Penerbit Universiti Pertanian Malaysia, Serdang

Perera BMAO (1987) A review of experiences with oestrous synchronization in buffaloes in Sri Lanka *Buffalo Journal, Supplement* **1** 105–114

Perera BMAO (1994) Current buffalo production systems and future strategies for improvement. In *Proceedings of Fourth World Buffalo Congress* pp 27–38 Sao Paulo, Brazil

Perera BMAO and Abeyratne AS (1979) The use of nuclear techniques in improving reproductive performance of farm animals *World Animal Review (FAO)* **32** 2–8

Perera BMAO and de Silva LNA (1985) Gestation length in indigenous (Lanka) and exotic (Murrah) buffaloes in Sri Lanka *Buffalo Journal* **1** 83–87

Perera BMAO, Pathiraja N, Kumaratilake WLJS, Abeyratne AS and Buvanendran V (1977) Synchronisation of oestrus and fertility in buffaloes using a prostaglandin analogue *Veterinary Record* **101** 520–521

Perera BMAO, Pathiraja N, Motha MXJ and Weerasekera DA (1979) Seasonal differences in plasma testosterone profiles in buffalo bulls *Theriogenology* **12** 33–38

Perera BMAO, Pathiraja N, Abeywardena SA, Motha MXJ and Abeygunawardena H (1980) Early pregnancy diagnosis in buffaloes from plasma progesterone concentration *Veterinary Record* **106** 104–106

Perera BMAO, Abeygunawardena H, Thamotharam A, Kindahl H and Edqvist LE (1981) Peripartal changes of oestrone, progesterone and prostaglandin in the water buffalo *Theriogenology* **15** 463–467

Perera BMAO, de Silva LNA and Karunaratne AM (1984) Studies on reproductive endocrinology and factors influencing fertility in dairy and draught buffaloes of Sri Lanka. In *The Use of Nuclear Techniques to Improve Domestic Buffalo Production in Asia* pp 13–28 International Atomic Energy Agency, Vienna

Perera BMAO, de Silva LNA, Kuruwita VY and Karunaratne AM (1987) Postpartum ovarian activity, uterine involution and fertility in indigenous buffaloes at a selected village location in Sri Lanka *Animal Reproduction Science* **14** 115–127

Perera BMAO, Kuruwita VY, Mohan V, Chandratillake D and Karunaratne AM (1988) Effects of some managerial factors on postpartum reproduction in buffaloes and goats *Acta Veterinaria Scandinavica Supplementa* **83** 91–100

Rajamahendran R, Jayatilake KN, Dharmawardena J and Thamotharam M (1980) Oestrus synchronization in buffaloes (*Bubalus bubalis*) *Animal Reproduction Science* **3** 107–112

Ranawana SSE, Tilakratne N and Srikandakumar A (1984) Utilization of water by buffaloes in adapting to a wet-tropical environment. In *The Use of Nuclear Techniques to Improve Domestic Buffalo Production in Asia* pp 171–187 International Atomic Energy Agency, Vienna

Sebastian L, Mudgal VD and Nair PG (1970) Comparative efficiency of milk production by Sahiwal cattle and Murrah buffalo *Journal of Animal Science* **30** 253–256

Sharifuddin W and Jainudeen MR (1983) The accuracy of rectal diagnosis of corpora lutea in water buffalo (*Bubalus bubalis*) *Animal Reproduction Science* **6** 185–189

Tiwari SR and Pathak MM (1995) Influence of suckling on postpartum reproduction performance of Surti buffaloes *Buffalo Journal* **2** 213–217

Tulloch DG (1979) The water buffalo (*Bubalus bubalis*) in Australia: reproductive and parent–offspring behaviour *Australian Wildlife Research* **6** 265–287

Usmani RH, Ahmad M, Inskeep EK, Dailey RA, Lewis PE and Lewis GS (1985) Uterine involution and postpartum ovarian activity in Nili-Ravi buffaloes *Theriogenology* **24** 435–448

Usmani RH, Dailey RA and Inskeep EK (1990) Effects of limited suckling and varying prepartum nutrition on postpartum reproductive traits of milked buffaloes *Journal of Dairy Science* **73** 1564–1570

Vale WG (1997) News on reproduction biotechnology in males. In *Proceedings Fifth World Buffalo Congress* pp 103–123 Caserta, Italy

Vale WG, Ohashi OM, Sousay JS and Ribeiro HFL (1990) Studies on the reproduction of water buffalo in the Amazon basin. In *Livestock Reproduction in Latin America* pp 201–210 International Atomic Energy Agency, Vienna

Zicarelli L (1997) News on buffalo cow reproduction. In *Proceedings Fifth World Buffalo Congress* pp 124–141 Caserta, Italy

Zicarelli L, de Filippo C, Francillo M, Pacelli C and Villa E (1997) Influence of insemination technique and ovulation time on fertility percentage in synchronized buffaloes. In *Proceedings Fifth World Buffalo Congress* pp 732–737 Caserta, Italy

Journal of Reproduction and Fertility Supplement **54**, 169–178

Reproduction in female South American domestic camelids

J. B. Sumar

IVITA Research Institute, San Marcos University, Av. De Los Incas 1412, Cusco, Perú.

Alpacas and llamas are induced ovulators. They show marked reproductive seasonality in the Andean region, but under Northern Hemisphere conditions of feeding and management, they are non-seasonal breeders. Puberty is attained when they reach 50% of adult body weight. When they are not exposed to a male, females show successive waves of follicular maturation and atresia. Growth, maintenance and regression of a follicle each require an average of 4 and 6 days in alpacas and llamas, respectively. After sterile mating, progesterone concentrations in blood were increased from day 5, reached maximum concentrations on day 7–8, and declined rapidly at 9–10 days after mating. A fertile mating results in formation of a corpus luteum that remains functional throughout gestation. The duration of gestation is 340–346 days. Almost all fetuses were found to occupy the left uterine horn, even though ovulation occurs from both ovaries with equal frequency. Several methods of pregnancy diagnosis have been described. Mating is recommended within 15–20 days after parturition to obtain good fertility rates and one offspring per year. The factors that contribute to high rates of embryonic mortality are unknown. Reproductive technologies, such as AI, superovulation, embryo transfer and IVF, have not been used very extensively in these species but can be successfully applied.

Introduction

The family Camelidae comprises six species, and they are believed to have originated in western North America. Two of the species migrated through the Bering Strait into Asia, and four into South America where the llama (*Lama glama*) and the alpaca (*Lama pacos*) were domesticated 4000 to 5000 years ago and the guanaco (*Lama guanicoe*) and vicuña (*Lama vicugna*) are still found in the wild.

Puberty

Young female alpacas of 12–13 months of age show behavioural oestrus similar to that of adult alpacas (Novoa *et al.*, 1972); ovarian activity begins at 10 months of age, with the presence of follicles of 5 mm or more in diameter. In a study carried out in southern Perú using 280 yearling female alpacas, a relationship was found ($P < 0.001$) between body weight at mating and subsequent birth rates (Leyva and Sumar, 1981). For each kilogram increase in body weight, there was a 5% increase in birth rate, but when body weight exceeded 33 kg, the percentage of non-pregnant females was relatively independent of body weight. In traditional Peruvian production systems, 50% or fewer of yearling alpacas reach 33 kg of body weight at mating time (1 year); therefore, breeding age is postponed until 2 years of age in alpacas and until after 3 years of age in llamas. It has also been shown that with better nutrition after weaning (7–8 months of age), almost 100% of yearling alpacas can reach 33 kg of body weight (Sumar, 1985).

Seasonality

Studies with alpacas and llamas in their natural habitat in the highlands of southern Perú (peasant community farms), where males and females are together all year, showed that the mating activities are seasonal, and last from December to March (summer months). These are the warmest months of the year, with sufficient rain and abundant green forage (San Martín *et al.*, 1968). In addition, the wild species of camelids, the vicuña and guanaco, show this marked seasonality of reproduction (Sumar and García1986).

Alternatively, when females are kept separately from males and copulation is allowed only once a month, both sexes are sexually active throughout the year. Ovulation and fertilization rates, together with embryo survival, were not affected significantly by the season of the year (Fernández-Baca *et al.*, 1972).

Continuous association of females and males inhibits the sexual activity of the male. Factors responsible for the onset and cessation of sexual activity under natural conditions are unknown. Environmental factors, in addition to visual and olfactory stimulation, could be of influence via the central nervous system (Fernández-Baca *et al.*, 1972).

Observations in different zoological parks indicate that camelids, both domestic and wild, are year-round breeders (Schmidt, 1973). In North America, where llamas are kept under continuous good feeding conditions, llamas are considered non-seasonal breeders. An analysis of the birthing season for llamas in the Rocky Mountain area of the USA (LR Johnson, personal communication) indicated that births occurred all year, but most of them (73%) occurred between June and November, i.e. during the warmest months.

Pattern of Ovarian Events

As copulation is a necessary prelude to ovulation in alpaca and llama, they have been classified as reflex or induced ovulators, rather than spontaneous ovulators (San Martín *et al.*, 1968; England *et al.*, 1969). Oestrus and ovulation do not occur in a repetitive, cyclic, and predictable fashion.

Follicular development

When they are not exposed to a male, female alpacas show long periods of sexual receptivity, and short periods of non-receptivity to the male that can last for 48 h (San Martín *et al.*, 1968). These changes may be correlated with rhythmic increases and decreases in serum oestrogen concentrations that reflect successive waves of follicular maturation and atresia.

On the basis of ovarian laparoscopic examination of alpacas, it was found that growth, maintenance and regression of a follicle each required an average of 4 days (total 12 days; range 9–17 days) (Bravo and Sumar, 1989). Transrectal ultrasonography of llamas, in their natural habitat of Peru (Adams *et al.*, 1990) showed that successive dominant follicles emerge at intervals of 19.8 ± 0.7 days in unmated and vasectomized–mated llamas and at intervals of 14.8 ± 0.6 days in pregnant llamas. Lactation was associated with an interwave interval that was shortened by 2.5 ± 0.05 days. Differences between species (alpaca versus llama), management conditions (poor natural grassland in Perú at 4300 m altitude versus intensive management and feeding conditions in USA at sea level), seasonal variation, method of examination, and number of animals evaluated may contribute to variation among studies. However, considerable variability between individuals has been detected regardless of reproductive status (parous versus nonparous). There is variability in both duration of sexual receptivity and regularity of its occurrence presumably because the follicular phase is not terminated by ovulation at a predetermined time in unmated females, and there is no luteal phase to delineate the timing of events after the end of oestrus.

Ovulation

The minimum time to ovulation in alpaca was estimated to be 26 h after natural mating and 24 h after treatment with hCG (San Martín *et al.*, 1968). In receptive female alpacas allowed a single mating, 50% ovulated between 26 and 30 h, 24% ovulated between 30 and 72 h, and 26% failed to ovulate after mating (Sumar, 1991). Forty per cent of the animals that failed to ovulate were yearlings and 15% were adults. Single service by an intact or vasectomized male resulted in ovulation in 77–82% of alpacas, and an increase in the number of services by intact males to three within a period of 24 h did not significantly affect ovulation rate (Fernández-Baca *et al.*, 1970c). Ovulation was also induced successfully after treatment with 1 mg LH, and a dose of 4–8 μg of GnRH was also necessary to provide an adequate stimulus for ovulation (Sumar, 1985). Use of an ultrasonographic technique in llamas allowed the detection of ovulation, on average, 2 days (range 1 to 3 days) after a single mating (Adams *et al.*, 1989, 1990).

A significant increase in serum LH concentration was observed 15 min after the onset of copulation, and the preovulatory peak surge of LH occurred at 2 h; concentrations reached basal values by 7 h after copulation (Bravo *et al.*, 1990). A second release of LH was not detected after a second copulatory period within 24 h of the first release. The LH surge observed subsequent to copulation is consistent with the contention that alpaca and llama are induced ovulators.

There are some indications that females can ovulate in the absence of coital stimulation or exogenous hormones. This is particularly evident when the female is initially isolated from, and then reintroduced to, a male. The rate of spontaneous ovulation was reported to be approximately 5–10% in alpacas (Fernández-Baca *et al.*, 1970a; Sumar and García, 1986) and 9–15% in llamas (Adams *et al.*, 1989, 1990; England *et al.*, 1969).

In addition, it has been reported that deposition of semen in the vagina of the bactrian camel will induce ovulation (Chen *et al.*, 1985), which has led to the proposition that there may be an 'ovulation inducing factor' (OIF) in camel and bull semen. Because of the close phylogenetic relationship between Old and New World Camelidae, it was considered likely that similar effects would be observed in alpacas and llamas. Similarly, ovulation in alpacas and llamas can be induced by deposition of alpaca, llama or bull semen in receptive alpacas or llamas (Sumar, 1994). In the case of camels, Chinese investigators proposed the presence of a GnRH-like substance in camel and bull semen. Conversely, Paolicchi *et al.* (1996) found that alpaca semen contains a factor or factors, other than GnRH, that contributes to the LH secretion mechanism in this species.

Function of the corpus luteum

The function of the corpus luteum after sterile and fertile mating has been studied in domestic camelids. The concentration of progesterone in blood of female alpacas and llamas that were mated with infertile (vasectomized) males was increased from day 5, reached maximum concentrations of 10–20 nmol l^{-1} on day 7–8, and declined rapidly at 9–10 days after mating in association with repeated surge releases of prostaglandin $F_{2\alpha}$ (Sumar *et al.*, 1988). Oestradiol concentrations were > 100–200 pmol l^{-1} during oestrus when the animals were mated. A temporary increase in oestradiol was detected that was related to the rise in progesterone concentrations in the early luteal phase. Otherwise, oestradiol concentrations remained low, 20–40 pmol l^{-1}, during the luteal phase, and increased in most animals to 40–60 pmol l^{-1} after luteolysis. During the 3–4 days after coitus, when the corpus luteum is forming and progesterone concentrations are low, most females remain receptive to males. Plasma concentrations of progesterone in females that displayed sexual receptivity at this time were between 0.06 and 0.28 ng ml^{-1}(Sumar *et al.*, 1993). In ultrasonography studies in llamas, the corpus luteum was first detected on day 3.1 ± 0.2 (day 0 = ovulation) in females mated to vasectomized males and reached maximum diameter (12.8 ± 0.3 mm) on mean day 5.9 ± 0.3 (Adams *et al.*, 1991). Luteal diameter and plasma concentration of progesterone were highly correlated ($r = 0.83$, $P < 0.0001$). A prolonged luteal phase was not observed in any sterile-mated (non-pregnant) alpaca or llama.

A fertile mating results in formation of a corpus luteum that remains functional throughout gestation. Luteal diameter was monitored by transrectal ultrasonography and plasma progesterone concentrations were determined in 68 pregnant llamas (Adams *et al.*, 1991). The corpus luteum was detected on mean day 3.1 ± 0.2, and reached maximum diameter (16 mm) on mean day 21.4 ± 1.2 (day 0 = day of ovulation). There was a decrease in mean plasma progesterone concentration between days 8 and 10, as well as a transient decrease in luteal diameter during this period. A similar decrease in progesterone between days 8 and 11 has been reported in alpacas (Sumar *et al.*, 1993). The transient fall in progesterone is coincident with the initiation of uterine-induced luteal regression in non-pregnant animals. It has been suggested that the rescue and resurgence of the corpus luteum between 8 and 10 days after mating represents the luteal response to pregnancy (maternal recognition). Plasma concentrations of progesterone remained high until about 2 weeks before parturition (Leon *et al.*, 1990). Thereafter progesterone concentrations began to decline and decreased markedly during the final 24 h before parturition. In addition, progesterone concentrations during sterile and fertile mating have been measured in the milk of alpacas and llamas (Sumar, 1991).

Sexual Receptivity and Mating Behaviour

The receptive female will assume the prone position (ventral recumbence) after a short period of pursuit by a male, or she may approach a male that is copulating with another female and adopt the prone position (San Martin *et al.*, 1968; England., 1971). Some receptive females may occasionally display mounting behaviour with other females of the herd, although such behaviour is much less common than in cattle. A non-receptive female runs away from and spits at the male to show rejection. During the very short courting phase and during mating, the males make blowing, grunting, laryngeal–nasal sounds. Copulation takes place in a recumbent position, with the male mounted above and just behind the female (England *et al.*, 1971). The female assumes a very passive attitude during copulation, and in some instances when copulation is prolonged, will appear to tire and may change positions so that she is lying on her side. Compared with other domestic species, coitus is remarkably prolonged in camelids (10–50 min) (Sumar, 1985).

Pregnancy

Duration of gestation

The duration of gestation in alpacas of the Huacaya and Suri breeds was reported to be 341 and 345 days, respectively (San Martin *et al.*, 1968). In llamas, the duration of gestation was 346 ± 8 days (327–357), and neither parity nor sex of the cria was found to influence the duration of gestation (Sumar, 1985).

Site of pregnancy and role of the corpus luteum

Almost all alpaca and llama fetuses were found to occupy the left uterine horn (based on position of the conceptus and site of umbilical attachment), even though ovulation occurs from both ovaries with equal frequency (Table 1) (Fernández-Baca *et al.*, 1973; Sumar, 1985). Thus embryos originating in the right side, migrate to the left horn for attachment. The reason for the right-to-left migration, which is apparently unique to Camelidae is not well known. There may be a differential luteolytic effect between the left versus right uterine horns. The right horn effects luteolysis via a local pathway, whereas the left horn effects luteolysis via both systemic and local pathways (Fernández-Baca *et al.*, 1979), which may be implicated in the reduction of twin pregnancies to single pregnancies (Sumar, 1991, 1996). The role of the corpus luteum during pregnancy in llama and alpaca has been studied by Sumar (1983), and results indicate that the corpus luteum is necessary for maintenance of pregnancy during the entire gestation period in both species.

Pregnancy diagnosis

Several methods of pregnancy diagnosis have been described for llamas and alpacas.

Sexual behaviour. Females that showed behavioural receptivity to teaser alpaca males (vasectomized) 20 or more days after a previous service were found not to be pregnant (Fernandez-

Table 1. Location of corpora lutea (CL) and embryos in alpacas and llamas

Species	Number of animals	Number with CL in right ovary (%)	Number with CL in left ovary (%)	Number with CL in both ovaries (%)	Right horn pregnancy (%)	Left horn pregnancy (%)
Alpaca	928	472 (50.9)	440 (47.4)	16 (1.7)	15 (1.6)	913 (98.4)
Llama	110	60 (54.5)	49 (44.5)	1 (0.9)	1 (0.9)	109 (99.1)

Data from Fernández-Baca *et al.* (1973); Sumar (1985); Sumar (1986).

Baca *et al.*, 1970a, b). However, not all females that rejected the male were found to be pregnant. In another study, the accuracy of pregnancy diagnosis was 84% and 95% in the alpaca and llama, respectively, within 70–125 days of gestation, and this is comparable to the accuracy obtained in sheep (Alarcón *et al.*, 1990).

External palpation. This method is still used in the traditional breeding system in southern Perú, with an accuracy of about 80%. Pregnancy diagnosis is done by external palpation or ballotement at approximately 8 months of gestation (Sumar, 1985).

Rectal palpation. In alpacas rectal palpation is possible as early as 30 days of gestation, but this method is limited because of the small pelvis and fat deposition in the pelvic inlet, particularly in yearling animals (Alarcón *et al.*, 1990). Seventy per cent of yearlings and 90% of adult alpacas can be palpated rectally. In llamas, almost 100% can be palpated, given adequate restraint, lubrication and a skilled veterinarian, with a glove size no greater than 7. The accuracy of pregnancy diagnosis by rectal palpation at 2 months after mating was 100% in alpacas and llamas (Alarcón *et al.*, 1990).

Circulating hormones. Progesterone concentrations can be determined during gestation by standard analytical methods such as radioimmunoassay and enzyme immunoassay. Alpacas or llamas that failed to ovulate, as well as those that failed to conceive, showed basal concentrations of progesterone on day 12 and 30 after mating, whereas animals were considered pregnant when progesterone concentrations were > 1.8 ng ml^{-1}. The occurrence of false positive predictions, related to high blood progesterone concentrations, is probably due to early embryonic loss (Sumar *et al.*, 1993).

Oestrone sulfate concentrations varied between 0.02 and 1.2 nmol l^{-1} (0.26 ± 1.9) from mating until about days 246 and 262 of gestation in alpacas and llamas (Sumar *et al.*, 1990). Thereafter, oestrone sulfate concentrations increased very rapidly, and the highest concentrations were observed at 3 days before parturition at 19.82 ± 5.74 and 15.6 ± 2.7 nmol l^{-1} in alpacas and llamas, respectively. A sharp decrease of oestrone sulphate occurred on day 1 postpartum in both species, reaching basal concentrations at 6–7 days postpartum. Therefore oestrone sulfate can be used for diagnosis of advanced stages of pregnancy and for the well being of the fetus.

Milk progesterone. Differences in concentrations of progesterone in milk can be observed between lactating non-pregnant and pregnant alpacas at 12 days after mating and may provide an early pregnancy test (Sumar, 1991; Sumar *et al.*, 1993).

Fecal progestagens. Schwarzenberger *et al.* (1995) reported that concentrations of progestagens in faeces can be a non-invasive method for investigating reproductive events in wild camelids such as the vicunas.

Ultrasound techniques. Ultrasound techniques that were developed specifically for sheep have been used in alpacas and llamas (Alarcón *et al.*, 1990). In alpacas, the highest accuracy (92%) was recorded at a mean fetal age of 80 days. In llamas, 100% accuracy was obtained at 75 days of gestation. External ultrasound diagnosis of pregnancy can be performed with a 3 MHz probe from 50 days until term, by applying the probe to the bare-skinned area just medial to the stifle (Johnson, 1989). Recent reports showed that transrectal ultrasonography is well suited to the study of the fetal development from 15 days after mating (Adams *et al.*, 1989, 1990).

Parturition

Unassisted labour in alpacas at 4 250 m above sea level in Perú lasts a mean of 203 ± 129 min for primiparous females and 193 ± 122 min for multiparous females (Sumar, 1985). In llamas, a mean of 176 min was reported for the three different stages of labour (Sumar, 1991). Camelids do not lick their offspring at birth, nor do they abandon the cria, even if they are of extremely poor nutritional status. More than 90% of births in alpacas and llamas occur between 07.00 and 13.00 h. This adaptation gives the cria the best chance to get warm and dry before the coldness of the night, where even in the summer, freezing temperatures are common at altitudes higher than 4 000 m (Sumar, 1985).

Puerperium

Up to day 4 after parturition the female alpaca is submissive and will allow herself to be mounted by a male (Sumar *et al.*, 1972). However, luteal regression, follicular growth and uterine involution are not complete, and the female will not ovulate or become pregnant from such early matings. Occasionally fertilization occurs subsequent to mating at 5 days post-partum. By day 10 post-partum the follicles are 8–10 mm in diameter, the corpus luteum has regressed considerably, and the uterus has involuted substantially (weighing only a fifth of its weight 24 h after birth). Mating in alpaca is recommended within 15–20 days after the female gives birth to obtain good fertility rates and one offspring per year (Sumar *et al.*, 1972).

Embryonic and fetal loss

In an early study, more that 80% of the ova recovered 3 days after mating were in the process of dividing, and only 50% of the fertilized ova survived for more that 30 days of gestation in alpacas (Fernández-Baca *et al.*, 1970a). The factors that are responsible for this high rate of embryonic mortality are unknown, but nutritional constraints, hormonal imbalances, and chromosomal aberrations may be principal aetiologies. These studies were conducted in alpacas living in their natural habitat, affected by a harsh natural environment, deteriorating feed supply and the presence of infectious and parasitic diseases. Measurement of mean concentrations of plasma progesterone (Table 2) in female alpacas and llamas that were mated with fertile males indicate that progesterone concentration on day 8 after service are higher in pregnant than in non-pregnant animals ($P < 0.05$) (Sumar *et al.*, 1993). Whether these levels are higher in pregnant animals, owing to the presence of a live embryo, or lower in non-pregnant animals, because of the incapability of the corpus luteum to secrete progesterone, is still unknown.

Twinning

Multiple ovulations occur in 3–10% of alpacas after natural mating and in 9–20% after treatment with gonadotrophins. However, the birth of twins is very rare. Cases of twin pregnancies in alpacas, seen in the early stages of gestation (< 40 days of gestation) are not uncommon (Sumar, 1986). It is believed that there is a reduction in the number of embryos and that probably a single embryo continues to develop. Twin births in llamas may be somewhat more common than in alpacas, and a few cases of twin births in llamas have been reported in the USA (Fowler, 1989). Nearly all alpaca and llama twins are of the fraternal or dizygotic type. Monozygotic twin pregnancies are rare. Similar to the mare, in 95% of alpacas or llamas with twin ovulations, one or both ova or embryos are

Table 2. Mean plasma progesterone concentrations (nmol l–1) in female alpacas and llamas mated with fertile males

	Alpacas (n = 12)		Llamas (n = 10)	
Day after mating	Pregnant	Non-pregnant	Pregnant	Non-pregnant
1	0.32	0.38	0.53	0.45
5	2.46	1.46	1.93	1.38
8	18.50	12.03	16.41	10.90
9	16.34	3.20	17.81	14.10
10	13.70	0.76	20.70	6.90
11	12.84	–	25.13	2.90
12	16.00	–	23.28	0. 28

Values are means. Reproduced with permission from Sumar *et al* (1993).

lost early in the gestation period. In twin pregnancies proceeding to term, one is well at birth and the other is frequently small or very weak.

Artificial insemination

A number of studies have examined the feasibility of artificial insemination in alpacas and llamas, involving deposition of fresh semen into the corpus uteri or into the right uterine horn with the use of the recto-palpation method (Fernández-Baca and Novoa, 1968). Inter-species crosses have also been tested between alpaca and vicuña (the F1 is known as paco-vicuña), and between llama and vicuña (llama-vicuña). In one study conducted to determine the most appropriate time for insemination, the highest proportion of fertilized ova occurred 35–45 h after induction of ovulation (Calderón *et al.*, 1968). The fertility rates were higher when vasectomized males were used to induce ovulation than when hCG was used.

Further studies were conducted using vicuña semen (V) and paco-vicuña semen with female alpacas (A) and llamas (Ll) (Leyva *et al.*, 1977). The birth rate obtained when vicuña was crossed with llama was 16.7% (½V–½A), and when vicuña was crossed with alpaca was 22% (½V–½A). Crossing the paco-vicuña with llama produced a 60% birth rate (½A–½V–½Ll), and 31.1% birth rate for paco-vicuña semen with alpaca (3/4A–½V). Domestic and wild camelids offer advantages compared with other animal species in the potential use of artificial insemination because females are in continuous oestrus during the breeding season, ovulation can be induced with vasectomized males, and semen can be inseminated into the uterus (Calderón *et al.*, 1968).

Embryo Transfer

Pugh and Montes (1994) and Del Campo (1997) used advanced reproductive technologies, such as superovulation, cryopreservation of embryos, embryo transfer, *in vitro* maturation, and *in vitro* fertilization of oocytes, in domestic South American camelids.

Superovulation of donors

The first attempt to superovulate three alpacas was in 1968 (Sumar, 1985) with the use of 1 200 iu PMSG given subcutaneously (Gestyl, Organon) as three consecutive daily doses of 400 iu. At 24 h after the last PMSG injection, the females were given an intravenous injection of hCG (750 iu; Pregnyl, Organon), and the female was mated immediately with a fertile male. At 72 h after mating, the females were sedated and three embryos from each female were collected surgically. Follicular growth and number of corpora lutea were not recorded. More recent studies in alpacas and llamas reported successful superstimulation with eCG or pituitary FSH in the presence of an induced corpus luteum or a progestagen treatment with ovulation being induced by GnRH, hCG or mating (Bourke *et al.*, 1991, 1992, 1995a; Del Campo *et al.*, 1995). Co-ordination of ovarian superstimulation with follicle development was achieved with eCG administered when follicles were 3 mm in diameter and after a complete follicle wave had occurred as determined by ultrasonography (Bravo *et al.*, 1995). ECG doses of 500 and 1000 iu are appropriate for inducing follicular growth in llamas. In contrast a dose of 2000 iu eCG caused hyperstimulation with an increasing incidence of cystic follicles. As evident in other species, there was wide variation in individual animal responses to eCG and embryo recovery rate was very low (Del Campo *et al.*, 1995).

Recovery procedures

Embryos from donor alpacas have been collected either surgically or non-surgically. Collection of zygotes from the oviduct of alpacas by abdominal laparotomy has been reported (Sumar, 1986). Zygotes were flushed from the oviduct to the uterus and collected through an incision made in the uterine horn. The presence of a valve in the utero–tubal junction precludes flushing from the uterus into the oviduct. The recovery rate in single ovulating alpacas was about 80% and no data were reported for superovulated animals.

Non-surgical techniques used in other domestic animals for the recovery of embryos have been adapted successfully for use in llamas and alpacas (Wiepz and Chapman, 1985; Correa *et al.*, 1992; Bourke *et al.*, 1992, 1995a; Del Campo, 1997). Embryo recovery is usually attempted 7 days after mating or GnRH injection. It is important to note that alpaca or llama embryos enter the uterus at about day 4 or 5 (day 0 = day of mating). To date, embryo recovery rate has not been higher than 50% (Del Campo *et al.*, 1995).

Ultrasound-guided transvaginal follicle aspiration has been used to collect oocytes in llamas with a 64% collection rate (Brogliatti *et al.*, 1966).

Recipient management and embryo transfer

Recipient females were synchronized with a single injection of hCG in receptive alpaca females or GnRH at the time of mating the female llama donor (Bourke *et al.*, 1995b; Sumar; 1986, 1996). Embryos were loaded into an inseminating pipette and transferred surgically to the left uterine horn.

A nonsurgical transfer approach was made transcervically via a Cassou-gun, into the tip of the uterine horn ipsilateral to the corpus luteum. The embryo was loaded into a 0.25 ml straw (Bourke *et al.*, 1995b). It is recommended that embryos always be placed into the left uterine horn since embryos in the left uterine horn migrate to the right uterine horn and this may contribute to embryo mortality (Fernández-Baca *et al.*, 1979).

Over the past 26 years, 11 crias have been born throughout the world as a result of embryo transfer techniques (Del Campo, 1997). Researchers from Perú reported a successful surgical embryo transfer and live birth of one alpaca and three late abortions (Sumar, 1985, 1986). The first llama born by a non-surgical collection and transfer technique was in the USA (Weipz and Chapman, 1985). Later, six live crias were born in the United Kingdom between 1992 and 1995 (Bourke *et al.*, 1995b). In Chile, the birth of one llama cria after two non-surgical embryo transfers was reported in 1994 (Del Campo, 1997).

IVF technology

Llama oocytes were collected either by mincing the ovary with a razor blade or by aspiration from ovarian follicles 2–11 mm in diameter (Del Campo *et al.*, 1992). Del Campo and coworkers (1994) conducted the first *in vitro* maturation of oocytes, *in vitro* fertilization with epididymal spermatozoa and *in vitro* co-culture of embryos with oviductal cells. From a large number of oocytes examined for signs of fertilization, 29.2% were penetrated by spermatozoa, 57.1% of the penetrated oocytes had male and female pronuclei. There is also evidence to suggest that a longer period is necessary for oocyte maturation in alpacas and llamas than in other species such as cows (36 h versus 24 h).

According to Del Campo *et al.* (1995), llama embryo–trophoblast expansion ranged from a mean of 1.2 mm in diameter on day 6.5–7.5 to 83 mm on day 13–14. This accelerated rate of embryo development may be related to the early maternal recognition of pregnancy in these species (Adams *et al.*, 1991).

Conclusions

The results obtained to date indicate that domestic South American camelids have particular and unique reproductive characteristics, but there is a need for more detailed investigation and careful observation. Maximizing reproductive performance is highly desirable in monotocous species, such as alpacas and llamas, that have a long pregnancy.

The ovarian follicular dynamics, hormone secretion and ovulation have not been very well established. Determining the follicular wave stage and follicular maturation will be a great advance in reproductive physiology of these species, and will help the reproductive management of mating or insemination. These studies require the development of advanced ultrasonographic techniques or rapid laboratory methods to measure the hormonal status of the animals. As alpacas and llamas are induced or reflex ovulators, studies of the different oestrogenic hormones and rapid quantification would be particularly helpful.

In addition, for embryo transfer, studies on the control of the follicular wave and on obtaining multiple ovulations, without damaging the ovaries and other organs, are urgently needed. Superovulation and embryo transfer in South American camelids are still in their early stages, and better results will be obtained, when appropriate techniques for induced ovulators that do not have an oestrous cycle, but do have follicular waves, are available.

References

Adams GP, Griffin PG and Ginther OJ (1989) *In situ* morphologic dynamic of ovaries, uterus and cervix in llamas *Biology of Reproduction* **41** 551–558

Adams GP, Sumar J and Ginther OJ (1990) Effects of lactational and reproductive status on ovarian follicular waves in llamas (*Lama glama*) *Journal of Reproduction and Fertility* **90** 535–545

Adams GP, Sumar J and Ginther OJ (1991) Form and function of the corpus luteum in llamas *Animal Reproduction Science* **24** 127–138

Alarcón V, Sumar J, Riera GS and Foote WC (1990) Comparison of three methods of pregnancy diagnosis in alpacas and llamas *Theriogenology* **34** 1119–1127

Bourke DA, Adam CL and Kyle CE (1991) Successful pregnancy following non-surgical embryo transfer in llamas *Veterinary Record* **128** 68

Bourke DA, Adam CL, Kyle CE, McEvoy TG and Young P (1992) Ovulation, superovulation and embryo recovery in llamas. In *Proceedings of the 12th International Congress on Animal Reproduction* **57** 193–195

Bourke DA, Kyle CE, McEvoy TG, Young P and Adam CL (1995a) Superovulatory responses to eCG in llamas (*Lama glama*) *Theriogenology* **44** 255–268

Bourke DA, Kyle CE, McEvoy TG, Young P and Adam CL (1995b) Recipient synchronization and embryo transfer in South American camelids *Theriogenology* **43** 171–177

Bravo PW and Sumar J (1989) Laparoscopic examination of the ovarian activity in alpaca *Animal Reproduction Science* **21** 271–281

Bravo PW, Fowler ME, Stabenfeldt GH and Lasley B (1990) Endocrine responses in the llama to copulation *Theriogenology* **33** 891–899

Bravo PW, Tsutsui T and Lasley BL (1995) Dose response to equine chorionic gonadotropins and subsequent ovulation in llamas *Small Ruminant Research* **18** 157–163

Brogliatti GM, Palasz AT and Adams GP (1996) Ultrasound-guided transvaginal follicle aspiration and oocyte collection in llamas (*Lama glama*) *Theriogenology* **45** 249

Calderón W, Sumar J and Franco E (1968) Avances en la inseminación artificial de las alpacas (*Lama pacos*) *Revista de la Facultad de Medicina Veterinaria Universidad Nacional Mayor de San Marcos* **22** 19–35

Chen BX, Yuen ZX and Pan GW (1985) Semen-induced ovulation in the Bactrian camel (*Camelus bactrianus*) *Journal of Reproduction and Fertility* **74** 335–339

Correa JE, Gatica R, Ratto M, Ladrix R and Schuler C (1992) Studies on non-surgical recovery of embryos from South American camelids. In *Proceedings of the 12th International Congress on Animal Reproduction*, The Hague, 788–790

Del Campo MR (1997) Reproductive technologies in South American camelids. In *Current Therapy in Large Animal Theriogenology* Ed. R. S. Youngquist. W. B. Saunders Company, Philadelphia

Del Campo MR, Donoso MX, Del Campo CH, Rojo R, Barros C, Parrish JJ and Mapletoft RJ (1992) *In vitro* maturation of llama (*Lama glama*) oocytes. In *Proceedings of the 12th International Congress on Animal Reproduction* The Hague, Vol. 1 324–326

Del Campo MR, Donoso MX, Del Campo CH, Berland M and Mapletoft RJ (1994) *In vitro* fertilization and development of llama (*Lama glama*) oocytes using epididymal spermatozoa and oviductal cell co-culture *Theriogenology* **41** 1219–1229

Del Campo MR, Del Campo CH, Adams GP and Mapletoft RJ (1995) The application of new reproductive technologies to South American camelids *Theriogenology* **43** 21–30

England BG, Foote WC, Matthews AG, Cardozo SG and Riera S (1969) Ovulation and corpus luteum function in the llama (*Lama glama*) *Journal of Endocrinology* **45** 505–513

England BG, Foote WC, Cardozo AG, Matthews DH and Riera S (1971) Oestrous and mating behaviour in the llama (*Lama glama*) *Animal Behavior* **19** 722–726

Fernández-Baca S and Novoa C (1968) Primer ensayo de inseminación artificial de alpacas (*Lama pacos*) con semen de vicuña (*Vicugna vicugna*). *Revista de la Facultad de Medicina Veterinaria, Universidad Nacional Mayor de San Marcos* **22** 9–18

Fernández-Baca S, Hansel W and Novoa C (1970a) Embryonic mortality in the alpaca *Biology of Reproduction* **3** 243–251

Fernández-Baca S, Hansel W and Novoa C (1970b) Corpus luteum function in the alpaca *Biology of Reproduction* **3** 252–261

Fernández-Baca S, Madden DHL and Novoa C (1970c) Effects of different mating stimuli on induction of ovulation in the alpaca *Journal of Reproduction and Fertility* **22** 261–267

Fernández-Baca S, Sumar J and Novoa C (1972) Comportamiento de la alpaca macho frente a la renovación de las hembras *Revista de Investigaciones Pecuarias (IVITA) Universidad Nacional Mayor de San Marcos* **1** 115–128

Fernández-Baca S, Sumar J, Novoa C and Leyva V (1973) Relación entre la ubicación del cuerpo luteo y la localización del embrión en la alpaca *Revista de Investigaciones Pecuarias (IVITA) Universidad Nacional Mayor de San Marcos* **2** 131–135

Fernández-Baca S, Hansel W, Saatman R, Sumar J and Novoa C (1979) Differential luteolytic effect on right and left uterine horns in the alpaca *Biology of Reproduction* **20** 586–595

Fowler ME (1989) *Medicine and Surgery of South American Camelids. Llama, Alpaca, Vicuña, Guanaco.* Iowa University Press, Ames

Johnson LW (1989) Llama reproduction. In *The Veterinary Clinics of North America. Food Animal Practice. Llama Medicine*, Vol. 5 (1) W. B. Saunders Company, Philadelphia

Leon JB, Smith BB, Timm KI and Le Cren G (1990) Endocrine changes during pregnancy, parturition and the early post-partum period in the llama (*Lama glama*) *Journal of Reproduction and Fertility* **88** 503–511

Leyva V and Sumar J (1981) Evaluación del peso corporal al empadre sobre la capacidad reproductiva de hembras alpaca de un año de edad. *Memorias de la IV Convención Internacional sobre Camélidos Sudamericanos, Corporación Nacional e Instituto de la Patagonia, Chile, Punta Arenas* (Abstract 1)

Leyva V, Franco J and Sumar J (1977) Inseminación Artificial en Camélidos Sudamericanos. In *Memorias de la I Reunión Asociacion Peruana de Producción Animal (APPA). Lima, Perú.*

Novoa C, Fernández-Baca S, Sumar J and Leyva V (1972) Pubertad en la alpaca. *Revista de Investigaciones Pecuarias (IVITA) Universidad Nacional Mayor de San Marcos* **1** 29–35

Paolicchi F, Urquieta B, Del Valle L and Bustos-Obregón E (1996) Actividad Biológica del Plasma Seminal de Alpaca: Estímulo para la Producción de LH por Células gonadotropas *Revista Argentina de Produccion Animal* **16** 351–356

Pugh DG and Montes AJ (1994) Advanced reproductive technologies in South American camelids. In *Veterinary Clinics of North America: Food Animal Practice. Update on Llama Medicine* Vol. 10 (2). W. B. Saunders Company, Philadelphia

San Martín M, Copaira M, Zúñiga J, Rodríguez R, Bustinza G and Acosta L (1968) Aspects of reproduction in the alpaca *Journal of Reproduction and Fertility* **16** 395–399

Schmidt CR (1973) Breeding season and notes on some other aspects of reproduction in captive camelids *International Zoo Yearbook* **13** 387–390

Schwarzenberger F, Speckbacher G and Bamberg E (1995) Plasma and fecal progestagen evaluations during and after the breeding season of the female vicuña (*Vicugna vicugna*) *Theriogenology* **43** 625–634

Sumar J (1983) Removal of the ovaries or ablation of the corpus luteum and its effect on the maintenance of gestation in the alpaca and llama *Acta Veterinaria Scandinavica Supplementum* **83** 133–141

Sumar J (1985) Reproductive physiology in South American camelids. In *Genetics of Reproduction in Sheep* Eds RB Land and DW Robinson. Butterworths, London.

Sumar J (1991) Contribution of the radioimmunoassay technique to knowledge of the reproductive physiology of South American Camelids. In *Isotope and Related Techniques in Animal Production and Health* pp 353–379 FAO/IAEA, Vienna

Sumar J (1994) Effect of various ovulation induction stimuli in alpacas and llamas *Journal of Arid Environments* **26** 39–45

Sumar J (1996) Reproduction in llamas and alpacas *Animal Reproduction Science* **42** 405–415

Sumar J and García M (1986) Fisiología de la reproducción de la alpaca. In *Proceedings of the Symposium on Nuclear and Related Techniques in Animal Production and Health* pp 169–177 IAEA, Vienna

Sumar J, Novoa C and Fernández-Baca S (1972) Fisiología reproductiva post-parto en la alpaca *Revista de Investigaciones Pecuarias (IVITA) Universidad Nacional Mayor de San Marcos* **1** 21–27

Sumar J, Fredriksson G, Alarcón V, Kindahl H and Edqvist L-E (1988) Levels of 15-keto-13,14-dihydro-PGF2α(progesterone and oestradiol-17β, after induced ovulations in llamas and alpacas *Acta Veterinaria Scandinavica* **29** 339–346

Sumar J, Edqvist L-E, Kindahl H, Fredriksson G and Alarcón V (1990) Niveles de Sulfato de Estrona periférica durante la gestación y puerperio en la alpaca y llama. In *X Congreso Nacional de Ciencias Veterinarias. Cusco*

Sumar J, Alarcón V and Echevarría L (1993) Niveles de progesterona periférica en alpacas y llamas y su aplicación en el diagnóstico precoz de gestación y otros usos clínicos *Acta Andina* **2** 161–167

Wiepz DW and Chapman RJ (1985) Non-surgical embryo transfer and live birth in a llama *Theriogenology* **24** 251–257

THE CORPUS LUTEUM

Chair
J. Kotwica

Journal of Reproduction and Fertility Supplement **54**, 181–191

Growth and development of the corpus luteum

L. P. Reynolds and D. A. Redmer

Department of Animal & Range Sciences,
North Dakota State University, Fargo, ND 58105–5727, USA

The mammalian corpus luteum, which plays a central role in the reproductive process because of its production of hormones such as progesterone, is an exceptionally dynamic organ. Growth and development of the corpus luteum are extremely rapid, and even when the corpus luteum is functionally mature cellular turnover remains high. Associated with this high rate of cell turnover, the mature corpus luteum receives the greatest blood supply per unit tissue of any organ, and also exhibits a relatively high metabolic rate. Central to the growth and development of the corpus luteum, therefore, is luteal vascular growth, which appears to be regulated primarily by the angiogenic growth factors, basic fibroblast growth factor and vascular endothelial growth factor. In addition, the corpus luteum is a complex tissue composed of parenchymal (small and large steroidogenic) and nonparenchymal (for example fibroblasts, vascular smooth muscle, pericytes and endothelial) cells. Recent studies evaluating the expression, location and regulation of gap junctions in the corpus luteum indicate an important role of gap junctional intercellular communication in the coordination of function among these diverse cell types during luteal growth and development. These studies will lead to an improved understanding not only of luteal function but also of tissue growth and development in general.

Introduction

The corpus luteum is a transient endocrine gland that is formed from the cells of the ovarian follicle after ovulation (Niswender and Nett, 1988; Reynolds *et al.*, 1994). In mammals, the principal function of the corpus luteum is to secrete progesterone, during the nonpregnant cycle as well as during pregnancy (Niswender and Nett, 1988; Reynolds *et al.*, 1994). In fact, the concentration of progesterone in systemic blood is used as an index of luteal function (Niswender and Nett, 1988; Reynolds *et al.*, 1994). In the nonpregnant animal, progesterone inhibits pituitary gonadotrophin secretion and thereby regulates the duration of the oestrous or menstrual cycle (Niswender and Nett, 1988; Reynolds *et al.*, 1994). During pregnancy, progesterone relaxes the uterine smooth muscle, stimulates uterine growth and secretory activity, influences maternal metabolism and mammary development, and is a precursor for other gestational steroids, all of which serve to support the developing embryo or fetus as well as the resulting offspring (Niswender and Nett, 1988; Reynolds *et al.*, 1994). Inadequate luteal function, therefore, leads to death of the embryo or fetus in most mammalian species (Niswender *et al.*, 1985; Niswender and Nett, 1988; Reynolds *et al.*, 1994).

The corpus luteum is one of the few adult tissues that exhibits regular periods of growth and development (Hudlicka, 1984; Jablonka-Shariff *et al.*, 1993; Reynolds *et al.*, 1994). In addition, growth and development of the corpus luteum are extremely rapid. For example, immediately after ovulation the corpus luteum of the ewe weighs about 30–40 mg. By day 12 of the oestrous cycle (that is, 12 days after ovulation), the corpus luteum reaches a maximum weight of about 750 mg, which represents a 20-fold increase in tissue mass over 12 days. On the basis of several studies of luteal growth and development, the doubling time for luteal tissue mass during this rapid growth phase is about 60–70 h (Reynolds *et al.*, 1994). Such rapid growth is equalled only by the fastest growing tumours (Baserga, 1985; Reynolds *et al.*, 1994). However, in contrast to tumour growth, growth of the corpus luteum is a self-limiting and highly ordered process. Thus, whereas growth of the corpus luteum has stopped by mid-cycle, luteal blood flow and progesterone secretion are maximal (Reynolds *et al.*, 1994).

An improved understanding of luteal function obviously has important implications for the regulation of fertility in mammals. In addition, because the corpus luteum is so dynamic, it provides an ideal model for studying the regulation of tissue growth and development, thereby leading to an improved understanding of these processes not only in the normal state but also in abnormal conditions such as in tumour growth or other pathologies (Reynolds *et al.*, 1992, 1994).

Luteal Growth and Development

General aspects

The corpus luteum is a complex tissue composed of parenchymal (small and large steroidogenic) and nonparenchymal (for example fibroblasts, vascular smooth muscle, pericytes and endothelial) cells (Niswender and Nett, 1988). In a complex tissue, the various cell types must interact to ensure normal growth and development (Hudlicka, 1984; Baserga, 1985). For example, tissue growth depends upon growth of new blood vessels and establishment of a functional blood supply (Hudlicka, 1984; Reynolds *et al.*, 1992). Growth of the vascular bed, in turn, must be highly coordinated with the metabolic demands of the tissue, since either rampant or insufficient vascular growth invariably leads to a pathological condition (Hudlicka, 1984; Reynolds *et al.*, 1992). Coordination among the various cell types is accomplished through several means, including humoral mechanisms involving endocrine or paracrine factors, as well as direct communication via gap junctions (Grazul-Bilska *et al.*, 1997, 1998a). However, before the role of these various factors can be determined, the normal patterns of tissue growth must be understood.

The corpus luteum is formed from the remnant of the ovarian follicle after ovulation and exhibits rapid growth until reaching its mature size around mid-cycle (Fig. 1; Reynolds *et al.*, 1994). At least in domestic ruminants, if pregnancy is established successfully the size of the corpus luteum remains relatively constant throughout most of gestation (Reynolds *et al.*, 1994), and thus as mentioned previously the growth of the corpus luteum appears to be self-limiting (Reynolds *et al.*, 1994). In fact, in mammals the total luteal weight is species specific and related to body size (Fig. 2). Although size of the mature corpus luteum is relatively constant within a species under normal circumstances (Fig. 2; Reynolds *et al.*, 1994), it does not seem to be absolutely fixed, but rather is reduced when increased numbers of corpora lutea are present such as in superovulated cows and ewes, and is increased when total numbers of corpora lutea are decreased such as after unilateral ovariectomy of pregnant pigs (Reynolds *et al.*, 1994).

Cellular aspects

During the luteal growth phase, luteal weight and DNA content increase exponentially (Fig. 1). The labelling index (rate of DNA synthesis) of the early, rapidly growing corpus luteum is extremely high compared with that of other tissues and is comparable with that of growing tumours and regenerating tissues (for example regenerating liver; Baserga, 1985; Jablonka-Shariff *et al.*, 1993; Zheng *et al.*, 1994; Reynolds *et al.*, 1994).

Even though the labelling index of the corpus luteum decreases markedly by mid-cycle, the number of proliferating luteal cells remains relatively constant because of the increase in the number of cells (Table 1). In spite of the continued proliferation of large numbers of luteal cells, however, weight, DNA content or total numbers of cells do not continue to increase in the mature corpus luteum (Fig. 1; Niswender *et al.*, 1985; Farin *et al.*, 1986; Jablonka-Shariff *et al.*, 1993; Zheng *et al.*, 1994; Reynolds *et al.*, 1994; Ricke, 1995). After mid-cycle, therefore, the mature corpus luteum appears to remain a very dynamic tissue, which exhibits a high rate of cell turnover (Reynolds *et al.*, 1994). Whether this high rate of cell turnover continues during pregnancy is not known; however,

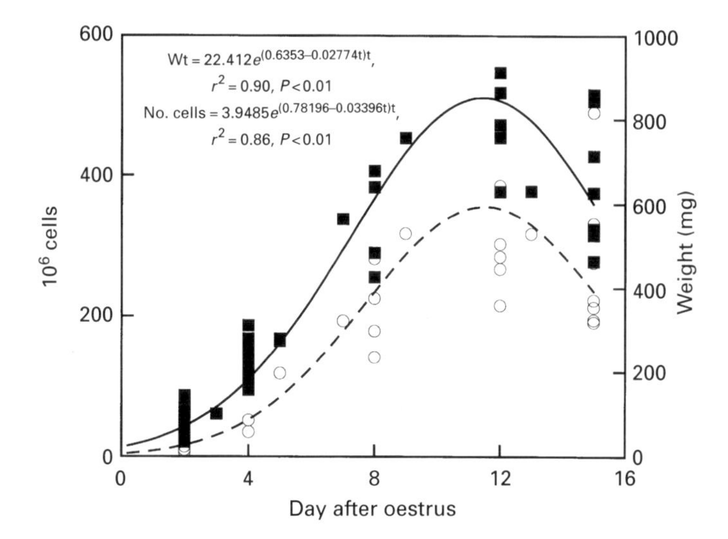

Fig. 1. Regression analysis of weight (■) and number of cells (○) for sheep corpora lutea throughout the oestrous cycle. The exponential growth model was of the form $y = ae^{(k1-k2t)t}$, where y = weight or number of cells at time = t, a = weight or number of cells at time = 0, and k_1 and k_2 are exponential growth constants. Data are adapted from Jablonka-Shariff *et al.* (1993).

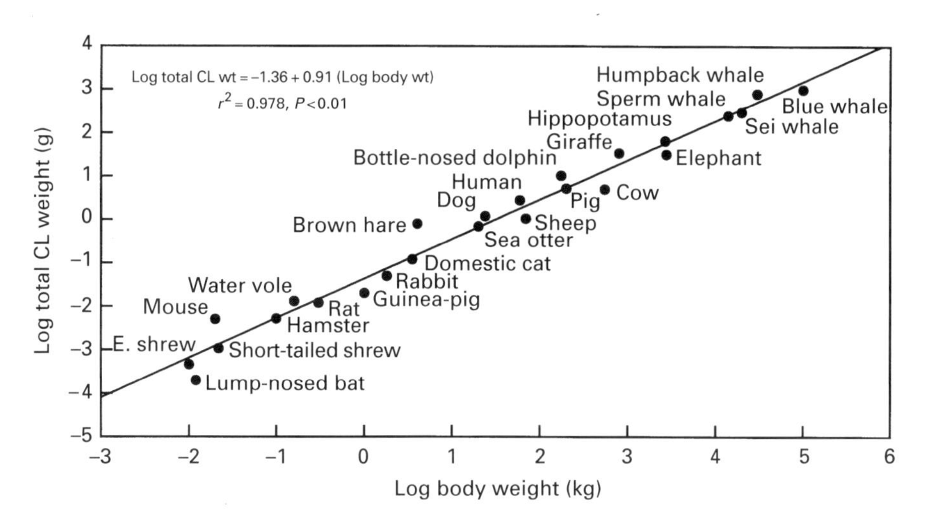

Fig. 2. Regression analysis of log of total weight of corpus luteum versus log of body weight in various mammals. Data for body weights are taken from *Walker's Mammals of the World* (1991) 5th Edn (Ed. RM Nowak, Johns Hopkins University Press, Baltimore). Data for total weight of corpus luteum are taken from various sources and represent maximal weights attained during the nonpregnant cycle or early pregnancy; when weight was not given it was estimated by assuming that 1 mm^3 = 1 mg and 1 cm^3 = 1 g.

morphological descriptions as well as quantitative morphometry have indicated that this may be the case (Reynolds *et al.*, 1994; Jablonka-Shariff *et al.*, 1997).

Luteal differentiation also occurs rapidly, and luteinization of the follicular cells that will form the corpus luteum and the associated progesterone production by these cells begin before ovulation (Gore-Langton and Armstrong, 1988). In addition, expression of 3β-hydroxysteroid dehydrogenase, a key enzyme in progesterone synthesis, is high in a large proportion of luteal cells soon after

Table 1. Kinetics of luteal cell proliferation in domestic livestock[a]

Species	Day after oestrus	No. of animals	No. of cells (× 10^6)	Dividing cells (× 10^6/day)
Sheep	2	12	16	12
	4	12	91	42
	8	6	223	38
	12	6	296	31
	15	7	274	4
Cow	2	5	177	78
	8	5	1394	106
	14	5	1631	117
	20	5	328	24
Pig	2	6	67	35
	4	7	97	38
	8	7	150	33
	12	6	155	27
	15	7	203	32
	18	7	163	30

[a]Values are means and are adapted from Jablonka-Shariff *et al.* (1993) for sheep, Zheng *et al.* (1994) for cow, and Ricke (1995) for pig. Number of cells was estimated based on DNA content, and number of dividing cells was estimated by multiplying the number of cells by the *in vivo* labelling index (nuclear incorporation of bromodeoxyuridine during 1 h of pulse labelling for sheep and pig; nuclear localization of proliferating cell nuclear antigen for cow) and assuming the duration of the S-phase *in vivo* is approximately 11 h (Baserga, 1985).

ovulation (Conley *et al.*, 1995). Although concentrations of progesterone in systemic blood do not reach maximal values until mid-cycle, the early corpus luteum is capable of producing progesterone, on a per unit tissue or per cell basis, at a rate similar to that of the mature corpus luteum (Reynolds *et al.*, 1994). This finding suggests that low systemic concentrations of progesterone early in the oestrous cycle may be due primarily to the relatively small tissue mass or low vascularity of the developing corpus luteum rather than to its inability to produce progesterone (Reynolds *et al.*, 1994).

Consistent with their early differentiation, a large proportion of the steroidogenic parenchymal cells of the corpus luteum, and especially the large luteal cells, exhibit little proliferation (Farin *et al.*, 1986; Jablonka-Shariff *et al.*, 1993). For example, during growth of the sheep corpus luteum, large steroidogenic luteal cells do not increase in number but rather increase about three-fold in size, whereas the small steroidogenic luteal cells and luteal capillary (endothelial and pericytes) cells do not increase in size but increase about five- to six-fold in number (Fig. 3). In agreement with these observations, the luteal cells that continue to proliferate after mid-cycle are principally non-parenchymal cells, such as endothelial cells, fibroblasts, or possibly undifferentiated parenchymal 'stem' cells (Reynolds *et al.*, 1994).

The rapid and maintained proliferation of luteal endothelial cells is associated with a large increase in the rate of blood flow to the corpus luteum throughout its growth (Reynolds, 1986; Reynolds *et al.*, 1994). The mature corpus luteum is highly vascular and receives the greatest rate of blood flow, per unit tissue, of any organ (Reynolds, 1986; Zheng *et al.*, 1993; Redmer and Reynolds, 1996; Reynolds and Redmer, 1998). In addition, in the mature corpus luteum nearly every parenchymal cell is in contact with a capillary (Reynolds *et al.*, 1992; Redmer and Reynolds, 1996). Not surprisingly, in conjunction with its extremely high vascularity and blood flow, the mature corpus luteum also exhibits a high metabolic rate (Swann and Bruce, 1987), perhaps because of its high rates of steroid production and cell turnover. These observations again emphasize the importance of intercellular communication in coordinating cellular proliferation and function in a complex tissue such as the corpus luteum, and also emphasize the importance of evaluating function of the individual cell types rather than the tissue as a whole.

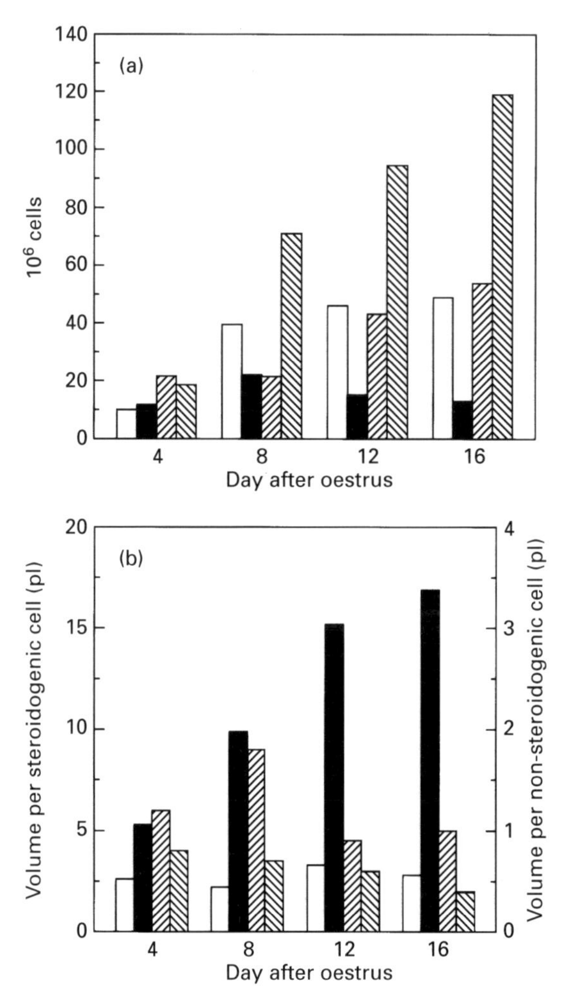

Fig. 3. Number of cells (a) and volume per cell (b) for individual cell types (□ small; ■ large; ▨ fibroblast; and ▧ endothelial and pericyte) of sheep corpus luteum throughout the oestrous cycle. Volume was estimated by assuming that 1 μm^3 = 1 femtolitre. Data are adapted from Farin *et al.* (1986).

Angiogenesis in Luteal Growth and Development

General aspects

Because the corpus luteum grows so rapidly, and because tissue growth depends on concomitant vascular growth, or angiogenesis (Hudlicka, 1984), it has long been appreciated that angiogenesis is a critical aspect of growth and function of the corpus luteum (for review, see Reynolds *et al.*, 1992, 1994, and Redmer and Reynolds, 1996). Conversely, inadequate luteal function has been associated with decreased luteal vascularity (Reynolds *et al.*, 1994, and Redmer and Reynolds, 1996), and several investigators have suggested that reduced ovarian blood flow plays a critical role in luteal regression (Reynolds, 1986; Niswender and Nett, 1988). Although a portion of the capillary bed degenerates during luteal regression, some of the capillaries and the larger microvessels are maintained, and these probably play an important role in resorption of the luteal tissue (Zheng *et al.*, 1993; Ricke, 1995; Reynolds and Redmer, 1998).

Luteal angiogenic factors

Because of the importance of vascular development in the growth, development and function of ovarian tissues, much of our research has focused on evaluating the activity and expression of luteal angiogenic factors (Redmer *et al.*, 1988; Reynolds *et al.*, 1992, 1994; Redmer and Reynolds, 1996). An important finding from our initial studies was that the angiogenic activity produced by bovine follicles and bovine and ovine corpora lutea binds relatively strongly to heparin-affinity columns (Reynolds *et al.*, 1992, 1994; Redmer and Reynolds, 1996).

Thus, we hypothesized that these ovarian angiogenic factors belong to one of the families of heparin-binding angiogenic factors, namely the fibroblast growth factors (FGF) or the vascular endothelial growth factors (VEGF), rather than to one of the other families of growth factors found in the ovary (Reynolds *et al.*, 1992, 1994; Redmer and Reynolds, 1996). This hypothesis is consistent with the suggestion that FGF and VEGF are probably key mediators of the angiogenic process in a variety of tissues (Klagsbrun and D'Amore, 1991; Reynolds *et al.*, 1992, 1994; Redmer and Reynolds, 1996; Ferrara and Davis-Smyth, 1997; Reynolds and Redmer, 1998).

Basic fibroblast growth factor

Basic FGF (bFGF) protein is present in bovine and ovine corpora lutea (Grazul-Bilska *et al.*, 1992; Zheng *et al.*, 1993; Schams *et al.*, 1994; Jablonka-Shariff *et al.*, 1997). Basic FGF also stimulates proliferation of bovine and ovine luteal endothelial cells, especially those from early in the cycle (Gospodarowicz *et al*, 1986; Grazul-Bilska *et al.*, 1995). In addition, in the bovine corpus luteum the pattern of expression of bFGF mRNA closely follows that of production of angiogenic activity, and both increase throughout the oestrous cycle and in response to LH (Redmer *et al.*, 1987, 1988; Stirling *et al.*, 1991). Moreover, the majority (approximately 82%) of the endothelial mitogenic activity produced by bovine, ovine and porcine corpora lutea was neutralized by antibodies against bFGF (Ricke, 1995; Redmer and Reynolds, 1996; Reynolds and Redmer, 1998). This last observation is important because endothelial proliferation is one of the critical components of an angiogenic response (Hudlicka, 1984; Klagsbrun and D'Amore, 1991).

However, if bFGF is one of the major angiogenic factors in the corpus luteum, it was not clear why the amounts of its mRNA and protein remain constant or even increase throughout the oestrous cycle, even though most luteal vascular development occurs in the early corpus luteum (Zheng *et al.*, 1993; Reynolds *et al.*, 1994; Ricke, 1995; Doraiswamy, 1998). To attempt to solve this conundrum, we recently evaluated luteal expression of the two major FGF receptors (FGFR) in ovine corpus luteum (Doraiswamy, 1998).

FGFR-1 was present in the luteal vasculature throughout the oestrous cycle; in parenchymal cells, it was present during early and mid-cycle, but was barely detectable late in the oestrous cycle. Conversely, FGFR-2 was present in the parenchymal cells at all stages, but localized to the larger microvessels only late in the oestrous cycle. We therefore hypothesized that bFGF may affect the function of not only luteal endothelial cells but also other luteal cell types, which is consistent with the ability of bFGF to stimulate progesterone production by bovine and ovine luteal cells (Grazul-Bilska *et al.*, 1995; Reynolds and Redmer, 1998). In addition, as mentioned above, some of the capillaries and larger microvessels of the corpus luteum are maintained late in the oestrous cycle to enable resorption of the regressing luteal tissue. The FGFs have been shown to inhibit cell death in granulosal cells (Tilly *et al.*, 1992) as well as in other types of cell (Gospodarowicz *et al.*, 1976; Yasuda *et al.*, 1995). Thus, the maintenance of FGFR in the luteal microvessels could explain why they are maintained while the other luteal tissues regress (Doraiswamy *et al.*, 1998; Reynolds and Redmer, 1998).

Vascular endothelial growth factor

It has become clear during the last decade that the VEGF family of proteins play a pivotal role in the angiogenic process (Ferrara and Davis-Smyth, 1997). The VEGF are specific stimulators of vascular permeability, and endothelial cell protease production, migration and proliferation, all of which are critical components of angiogenesis, and VEGF receptors localize exclusively to endothelial cells. Exogenous VEGF will stimulate vascular growth in a variety of *in vivo* models, and VEGF mRNA is markedly increased in numerous tissues during normal and pathological angiogenesis. In addition, a number of hormones, growth factors and cytokines that may be involved in the angiogenic process have been shown to stimulate VEGF expression (Reynolds and Redmer, 1998). Moreover, hypoxia, which is thought to be a primary stimulus for tissue angiogenesis, strongly induces VEGF gene expression (Ferrara and Davis-Smyth, 1997).

More recently, treatment with VEGF-neutralizing monoclonal antibodies *in vivo* was shown to block the growth and reduce the vascularity of a variety of tumours in mice, even though these same antibodies did not affect the growth of tumour cells *in vitro* (Kim *et al.*, 1993). In addition, disruption of the genes for VEGF receptors resulted in embryonic death by about day 8 of pregnancy in mice (Fong *et al.*, 1995; Shalaby *et al.*, 1995). Similarly, homozygous or heterozygous gene knockouts for VEGF were lethal by about day 11 of pregnancy in mice (Carmeliet *et al.*, 1996; Ferrara *et al.*, 1996). In all of these embryos, marked cardiovascular defects were observed, such as delayed or abnormal development of the heart, aorta, major vessels and extraembryonic vasculature, including the yolk sac and placenta. These observations indicate a central role for VEGF not only in angiogenesis but also in organization and maintenance of the microvasculature.

Although VEGF mRNA is present in the ovine corpus luteum throughout the oestrous cycle, its abundance is two- to threefold higher early in the cycle, when luteal vascularization is occurring, compared with mid- or late cycle (Redmer and Reynolds, 1996; Doraiswamy, 1998). Although only about 30% of the endothelial mitogenic activity produced by the early ovine corpus luteum was neutralized with an anti-VEGF antibody, 65% of the endothelial migration-stimulating activity was neutralized (Doraiswamy, 1998). On the basis of these observations, we have suggested that VEGF plays a major role in luteal vascular development (Redmer and Reynolds, 1996; Reynolds and Redmer, 1998).

Consistent with the mRNA and immunoneutralization studies, localization of VEGF protein appeared to be greatest during early luteal development and least late in the oestrous cycle, during luteal regression (Doraiswamy, 1998). In addition, VEGF was expressed by connective tissue cells in the luteal capsule and connective tissue tracts, in cells within the luteal parenchymal lobules, and in luteal arterioles. The VEGF-expressing cells within the parenchymal lobules were associated with the luteal capillaries and, based on their anatomical location, appeared to be capillary pericytes. Recently, we confirmed that these cells are capillary pericytes by co-localization of VEGF and smooth muscle α-actin, which is a specific marker of cells of the smooth muscle lineage, including pericytes (Hirschi and D'Amore, 1996; Doraiswamy, 1998). Thus, VEGF protein is localized to thecal-derived connective tissue tracts, arteriolar smooth muscle, and capillary pericytes of the developing corpus luteum. In agreement with these observations, we have shown that VEGF is expressed exclusively in the theca and not the granulosa of preovulatory bovine and ovine follicles. In addition, the thecal-derived VEGF-expressing cells invade the granulosa within hours after ovulation, before invasion of the granulosa by the thecal-derived endothelial cells (Doraiswamy, 1998).

These observations are consistent with several seemingly disparate observations made previously. First, the thecal-derived cells are responsible for vascularization of the developing corpus luteum (Zheng *et al.*, 1993; Reynolds *et al.*, 1994). Second, the thecal cells are mesodermal in origin, and the pattern of vascularization as well as VEGF expression of the developing corpus luteum closely resembles that of embryonic organs and placenta, which are vascularized by mesodermally-derived cells (Reynolds *et al.*, 1994; Hirschi and D'Amore, 1996; Redmer and Reynolds, 1996; Doraiswamy, 1998). Third, within the last decade it has been proposed that capillary pericytes are the primary vascular cells that initiate the process of angiogenesis and subsequently interact closely with the endothelial cells to establish the mature vascular bed (Folkman and D'Amore, 1996; Hirschi and D'Amore, 1996).

These observations, taken together with studies showing that VEGF is critical for normal vascular development, led us to propose a novel model for vascularization of the corpus luteum (Fig. 4; Doraiswamy, 1998; Reynolds and Redmer, 1998). In this model, the pericytes of the thecal capillaries produce VEGF (Fig. 4a). Just before ovulation, the theca prepares to invade the membrana

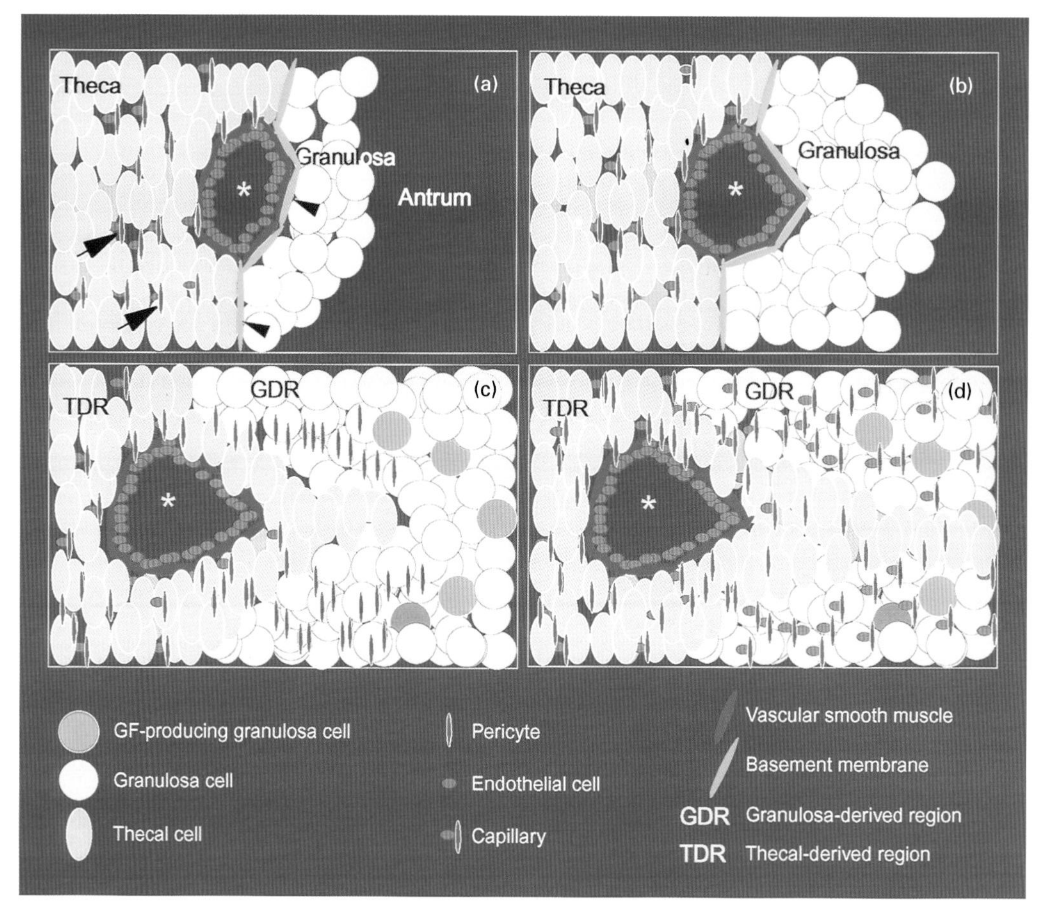

Fig. 4. Model for neovascularization of the corpus luteum. In the large preovulatory follicle (a), note the extensive network of thecal capillaries (arrows), composed of endothelial cells and pericytes, and also the larger thecal microvessels (*) separated from the avascular membrana granulosa by a basement membrane (arrowheads). A few hours before ovulation (b), the granulosal layer expands, and the larger thecal microvessels increase in size and invaginate the basement membrane. At or shortly after ovulation (c), thecal-derived pericytes migrate into the granulosal-derived region under the influence of growth factor (GF), perhaps fibroblast growth factor 2 (FGF-2) or platelet-derived growth factor (PDGF), produced by the granulosal cells. As the corpus luteum develops (d), pericytes within the granulosal-derived region produce vascular endothelial growth factor (VEGF), which stimulates migration of thecal-derived endothelial cells; subsequently, the pericytes and endothelial cells form the mature capillary bed of the luteal parenchymal lobule. Figure modified from Doraiswamy (1998).

granulosa, which is undergoing expansion (Fig. 4b). After ovulation and breakdown of the basement membrane, the thecal-derived pericytes are the initial vascular cells that invade the granulosal-derived region (Fig. 4c). Furthermore, we propose that proliferation and migration of thecal pericytes is stimulated by growth factors (GF) produced by the granulosa (Fig. 4c and d), most likely FGF-2 or platelet-derived growth factor (PDGF), both of which have been shown to stimulate pericyte migration or proliferation and may be present in the granulosa (Neufeld *et al.*, 1987; VanWezel *et al.*, 1995; Folkman and D'Amore, 1996). The pericytes then produce VEGF, which stimulates migration of the thecal-derived endothelial cells into the granulosal-derived region (Fig. 4d). Subsequently, the pericytes and endothelial cells form the mature capillary bed of the luteal parenchymal lobule (Fig. 4d).

Gap Junctions in Luteal Growth and Development

General aspects

Another important means of coordinating function among the various populations of luteal cells is by contact-dependent mechanisms, which involve direct coupling of the cells via gap junctions (Grazul-Bilska *et al.*, 1997, 1998a). Coordination of cellular functions is critical since, for example, growth of the vascular beds must be coordinated closely with tissue growth to ensure normal function of the mature corpus luteum (Klagsbrun and D'Amore, 1991; Reynolds *et al.*, 1992).

The gap junction, nexus, or junctio communicans, is an intercellular junction that allows communication and electrical coupling between adjacent cells, and which can be open or closed (that is, a gated channel; Grazul-Bilska *et al.*, 1997, 1998a). Gap junctions are composed of protein subunits termed connexins and are ubiquitous in multicellular organisms (Grazul-Bilska *et al.*, 1997, 1998a). In fact, gap junctions are present in almost all mammalian tissues and have been found in ovarian follicles and corpora lutea of numerous species (Grazul-Bilska *et al.*, 1997, 1998a).

Gap junctions in luteal function

The presence of functional gap junctions as well as gap junctional connexins in bovine and ovine luteal tissues and cultured luteal cells has recently been shown (Grazul-Bilska *et al.*, 1997, 1998a). In addition, the concentrations and function of luteal gap junctions vary with the stage of luteal development. For example, luteal expression of connexin 43 is greater during the early and mid-luteal phases compared with the late luteal phase (Grazul-Bilska *et al.*, 1997, 1998a; Khan-Dawood *et al.*, 1996). Luteal gap junction expression and function are regulated by luteotrophic and luteolytic hormones and second messengers (Grazul-Bilska *et al.*, 1997, 1998a). Moreover, the rate of gap-junctional intercellular communication varies with luteal cell type; it is greatest between small luteal cells, least between large luteal cells, and intermediate between large and small luteal cells (Grazul-Bilska *et al.*, 1997, 1998a).

On the basis of these observations and the known roles of gap junctions in other tissues, we proposed that gap junctions play a critical role in luteal development and function (Grazul-Bilska *et al.*, 1997, 1998a). In support of this proposal, we have now shown that transfection of bovine luteal cells with an antisense oligonucleotide corresponding to the sequence of bovine connexin 43 not only reduced the rate of gap junctional communication but also reduced LH-induced progesterone secretion, while not affecting basal progesterone secretion (Grazul-Bilska *et al.*, 1998b).

By using dual immunostaining techniques, we have localized connexin 43 to the cellular borders between luteal steroidogenic and endothelial cells (Grazul-Bilska *et al.*, 1997). Thus, gap junctions may be important in coordinating growth and function among the various types of luteal cell, including parenchymal cells, endothelial cells and pericytes (Grazul-Bilska *et al.*, 1997, 1998a). This type of coordination may be especially important in light of our observations cited above concerning the potential importance of pericytes in luteal growth and vascularization.

Conclusion

In the last decade, we have learned much about cellular growth and cellular interactions in the corpus luteum. For example, we have determined the patterns of cell proliferation and identified the major angiogenic factors that regulate luteal vascular development. In addition, we have shown that gap junctional intercellular communication probably plays an important role in coordinating cellular functions during luteal growth and development.

However, there are many questions remaining concerning the mechanisms regulating luteal growth and development. For example, what are the roles of angiogenic factors in differentiated luteal function or luteal regression? Do additional factors play a role in maturation of the luteal microvasculature, as proposed for other systems (Folkman and D'Amore, 1996)? What are the other factors involved in intercellular communication among the various types of luteal cell, and do gap junctions play a role in this? What roles do gap junctions play during luteal regression: for example, do they transduce the luteolytic signal from the large luteal cells, which contain the majority of PGF receptors, to the small luteal cells? Thus, even though we have made substantial progress in our understanding of cellular interactions during luteal growth and development, much remains to be learned.

Work from our laboratories has been supported by grants from the US Department of Agriculture (National Research Initiative), National Institutes of Health, and National Science Foundation. The authors would like to thank colleagues (Anna Grazul-Bilska, Derek Killilea, and Robert Moor), technicians (James Kirsch and Kim Kraft), and former students (Vinayak Doraiswamy, Paul Fricke, Albina Jablonka-Shariff, Mary Lynn Johnson, William Ricke, and Jing Zheng) without whom it would not have been possible to accomplish this work.

References

Baserga R (1985) *The Biology of Cell Reproduction* Harvard University Press, Cambridge, MA

Carmeliet P, Ferreira V, Breier G, Pollefeyt S, Kieckens L *et al.* (1996) Abnormal blood vessel development and lethality in embryos lacking a single VEGF allele *Nature* **380** 435–439

Conley AJ, Kaminski MA, Dubowsky SA, Jablonka-Shariff A, Redmer DA and Reynolds LP (1995) Immunohistochemical localization of 3β-hydroxysteroid dehydrogenase and P450 17α-hydroxylase during follicular and luteal development in pigs, sheep and cows *Biology of Reproduction* **52** 1081–1094

Doraiswamy V (1998) Angiogenesis in the ovine ovary: expression of vascular endothelial growth factor (VEGF) and fibroblast growth factor (FGF) *PhD Dissertation* North Dakota State University, Fargo

Farin CE, Moeller CL, Sawyer HR, Gamboni F and Niswender GD (1986) Morphometric analysis of cell types in the ovine corpus luteum throughout the estrous cycle *Biology of Reproduction* **35** 1299–1308

Ferrara N and Davis-Smyth T (1997) The biology of vascular endothelial growth factor *Endocrine Reviews* **18** 4–25

Ferrara N, Moore KC, Chen H, Dowd M, Lu L, O'Shea KS, Braxton LP, Hillian KJ and Moore MW (1996) Heterozygous embryonic lethality induced by targeted inactivation of the VEGF gene *Nature* **380** 439–442

Folkman, J and D'Amore PD (1996) Minireview. Blood vessel formation: What is its molecular basis? *Cell* **87** 1153–1155

Fong GH, Rossant J, Gertsenstein M and Breitman ML (1995) Role of the flt-1 receptor tyrosine kinase in regulating the assembly of vascular endothelium *Nature* **376** 66–70

Gore-Langton RE and Armstrong DT (1988) Follicular steroidogenesis and its control. In *The Physiology of Reproduction* pp 331–385 Ed. E Knobil and J Neill, Raven Press, New York

Gospodarowicz D, Moran J, Braun DL and Birdwell CR (1976) Clonal growth of bovine vascular endothelial cells: fibroblast growth factor as a survival agent *Proceedings of the National Academy of Sciences USA* **73** 4120–4124

Gospodarowicz D, Massoglia S, Cheng J and Fujii DK (1986) Effect of fibroblast growth factor and lipoproteins on the proliferation of endothelial cell derived from bovine adrenal cortex, brain cortex and corpus luteum capillaries *Journal of Cell Physiology* **127** 121–136

Grazul-Bilska AT, Redmer DA, Killilea SD, Kraft KC and Reynolds LP (1992) Production of mitogenic factor(s) by ovine corpora lutea throughout the estrous cycle *Endocrinology* **130** 3625–3632

Grazul-Bilska AT, Redmer DA, Jablonka-Shariff A, Biondini ME and Reynolds LP (1995) Proliferation and progesterone production of ovine luteal cells from several stages of the estrous cycle: effects of fibroblast growth factors and luteinizing hormone *Canadian Journal of Physiology and Pharmacology* **73** 491–500

Grazul-Bilska AT, Reynolds LP and Redmer DA (1997) Minireview: gap junctions in the ovaries *Biology of Reproduction* **57** 947–957

Grazul-Bilska AT, Reynolds LP and Redmer DA (1998a) Cellular interactions in the corpus luteum *Seminars in Reproductive Endocrinology* **15** 383–393

Grazul-Bilska AT, Reynolds LP and Redmer DA (1998b) Transfection of bovine luteal cells with gap junctional connexin 43 (Cx43) antisense oligonucleotide affects progesterone secretion *Biology of Reproduction* **58** (Supplement 1) 78

Hirschi KK and D'Amore PA (1996) Pericytes in the microvasculature *Cardiovascular Research* **32** 687–698

Hudlicka O (1984) Development of microcirculation: capillary growth and adaptation. In *Handbook of Physiology, Section 2: The Cardiovascular System, Vol IV Microcirculation, Part 1* pp 165–216 Ed. EM Renkin and CC Michel. American Physiological Society, Waverly Press, Baltimore

Jablonka-Shariff A, Grazul-Bilska AT, Redmer DA and Reynolds LP (1993) Growth and cellular proliferation of ovine corpora lutea throughout the estrous cycle *Endocrinology* **133** 1871–1879

Jablonka-Shariff A, Grazul-Bilska AT, Redmer DA and Reynolds LP (1997) Cellular proliferation and fibroblast growth factors in the corpus luteum during early pregnancy in ewes *Growth Factors* **14** 15–23

Khan-Dawood FS, Yang J and Dawood Y (1996) Expression of gap junction protein connexin-43 in the human and baboon (*Papio anubis*) corpus luteum *Journal of Clinical Endocrinology and Metabolism* **81** 835–842

Kim KJ, Li B, Winer J, Armanini M, Gillett N, Phillips HS and Ferrarra N (1993) Inhibition of vascular endothelial growth factor-induced angiogenesis suppresses tumour growth *in vivo. Nature* **362** 841–844

Klagsbrun MA and D'Amore PA (1991) Regulators of angiogenesis *Annual Review of Physiology* **53** 217–239

Neufeld G, Ferrara N, Schweigerer L, Mitchell R and

Gospodarowicz D (1987) Bovine granulosa cells produce basic fibroblast growth factor *Endocrinology* **121** 597–603

Niswender GD and Nett TM (1988) The corpus luteum and its control. In *The Physiology of Reproduction* pp 489–525 Ed. E Knobil and J Neill. Raven Press, New York

Niswender GD, Schwall RH, Fitz TA, Farin CE and Sawyer HR (1985) Regulation of luteal function in domestic ruminants: new concepts *Recent Progress in Hormone Research* **41** 101–150

Nowak RM (1991) *Walker's Mammals of the World*. The Johns Hopkins University Press, Baltimore and London

Redmer DA and Reynolds LP (1996) Angiogenesis in the ovary *Reviews of Reproduction* **1** 182–192

Redmer DA, Kirsch JD and Grazul AT (1987) *In vitro* production of angiotropic factor by bovine corpus luteum: partial characterization of activities that are chemotatic and mitogenic for endothelial cells. In *Regulation of Ovarian and Testicular Function, Advances in Experimental Medicine and Biology Vol. 219* pp 683–688 Ed. VB Mahesh *et al.* Plenum Press, New York

Redmer DA, Grazul AT, Kirsch JD and Reynolds LP (1988) Angiogenic activity of bovine corpora lutea at several stages of luteal development *Journal of Reproduction and Fertility* **82** 627–634

Reynolds LP (1986) Utero–ovarian interactions during early pregnancy: role of conceptus-induced vasodilation *Journal of Animal Science* **62** (Supplement 2) 47–61

Reynolds LP and Redmer DA (1998.) Expression of the angiogenic factors, basic fibroblast growth factor (bFGF) and vascular endothelial growth factor (VEGF), in the ovary *Journal of Animal Science* **76** 1671–1681

Reynolds LP, Killilea SD and Redmer DA (1992) Angiogenesis in the female reproductive system *FASEB Journal* **6** 886–892

Reynolds LP, Killilea SD, Grazul-Bilska AT and Redmer DA (1994) Mitogenic factors of corpora lutea *Progress in Growth Factor Research* **5** 159–175

Ricke WA (1995) Cellular growth of porcine corpora lutea throughout the estrous cycle *MS Thesis* North Dakota State University, Fargo

Schams D, Amselgruber W, Einspanier R, Sinowatz F and Gospodarowicz D (1994) Localization and tissue concentration of basic fibroblast growth factor in the bovine corpus luteum *Endocrine* **2** 907–912

Shalaby F, Rossant J, Yamaguchi TP, Gertsenstein M and Schuh AC (1995) Failure of blood-island formation and vasculogenesis in flk-1-deficient mice *Nature* **376** 62–66

Stirling D, Waterman MR and Simpson ER (1991) Expression of mRNA encoding basic fibroblast growth factor, bFGF, in bovine corpora lutea and cultured luteal cells *Journal of Reproduction and Fertility* **91** 1–8

Swann RT and Bruce NW (1987) Oxygen consumption, carbon dioxide production and progestagen secretion in the intact ovary of the day-16 pregnant rat *Journal of Reproduction and Fertility* **80** 599–605

Tilly JL, Billig H, Kowalski KI and Hsueh AJW (1992) Epidermal growth factor and basic fibroblast growth factor suppress the spontaneous onset of apoptosis in cultured rat ovarian granulosa cells and follicles by a tyrosine kinase-dependent mechanism *Molecular Endocrinology* **6** 1942–1950

Van Wezel IL, Umapathysivam K, Tilley WD and Rodgers RJ (1995) Immunohistochemical localization of basic fibroblast growth factor in bovine ovarian follicles *Molecular and Cellular Endocrinology* **115** 133–140

Yasuda T, Grinspan J, Stern J, Franceschini B, Bannerman P and Pleasure D (1995) Apoptosis occurs in the oligodendroglial lineage, and is prevented by basic fibroblast growth factor *Journal of Neuroscience Research* **40** 306–317

Zheng J, Redmer DA and Reynolds LP (1993) Vascular development and heparin-binding growth factors in the bovine corpus luteum at several stages of the estrous cycle *Biology of Reproduction* **49** 1177–1189

Zheng J, Fricke PM, Reynolds LP and Redmer DA (1994) Evaluation of growth, cell proliferation, and cell death in bovine corpora lutea throughout the estrous cycle *Biology of Reproduction* **51** 623–632

Journal of Reproduction and Fertility Supplement **54**, 193–205

Molecular regulation of luteal progesterone synthesis in domestic ruminants

J. L. Juengel and G. D. Niswender

Animal Reproduction and Biotechnology Laboratory, Colorado State University, Fort Collins, CO 80523–1683, USA

Regulation of progesterone secretion from the corpus luteum during the oestrous cycle requires the integration of multiple signals to achieve the appropriate amount of progesterone to maximize reproductive efficiency. Development of a mature corpus luteum capable of secreting sufficient amounts of progesterone is dependent upon the pituitary hormones LH and growth hormone (GH). Continued secretion of progesterone from the mature corpus luteum is also dependent upon pituitary hormones. If pregnancy does not occur, prostaglandin $F_{2\alpha}$ ($PGF_{2\alpha}$) of uterine origin causes a precipitous decrease in progesterone secretion and demise of the corpus luteum. A major point of regulation of progesterone secretion by both luteotrophic and luteolytic hormones appears to be regulation of transport of cholesterol through the mitochondrial membranes to cytochrome P450scc. It is likely that both luteotrophic and luteolytic hormones regulate steroidogenic acute regulatory protein (StAR), which facilitates transport. Regulation may be occurring through increases or decreases in gene transcription, translation efficiency or post-translational modifications such as phosphorylation. Thus, although synthesis of progesterone is a complex process, both positive and negative regulation of the process appears to occur primarily at a single step (transport of cholesterol to the inner mitochondrial membrane) in the pathway.

Introduction

The corpus luteum, which secretes progesterone, is a transient endocrine gland formed from follicular cells following ovulation. Progesterone is necessary for maintenance of pregnancy in all domestic ruminants. Inadequate luteal secretion of progesterone is a major cause of early embryonic mortality (Nancarrow, 1994; Zavy, 1994). However, if fertilization does not occur, the corpus luteum must stop producing progesterone to allow the complex series of events that results in another ovulation.

In domestic ruminants, the primary luteotrophic hormones, which support the development and function of the corpus luteum, are LH and GH. In addition, locally produced prostaglandins (PG) of the E and I series, and insulin-like growth factors (IGF) probably support luteal function. The luteolytic hormone that causes decreased secretion of progesterone and demise of luteal cells is $PGF_{2\alpha}$. This review will focus on how luteotrophic and luteolytic hormones regulate progesterone synthesis at the cellular and molecular levels.

Luteal Steroidogenic Pathway

For a better understanding of the way in which these hormones regulate secretion of progesterone from the corpus luteum it is first important to understand how steroidogenic cells produce progesterone (Fig. 1). Progesterone is made from the precursor cholesterol. Under normal conditions *in vivo*, the majority of the cholesterol used for synthesis of all steroid hormones is obtained from high density lipoprotein (HDL) or low density lipoprotein (LDL) (Gwynne and Strauss, 1982). Once

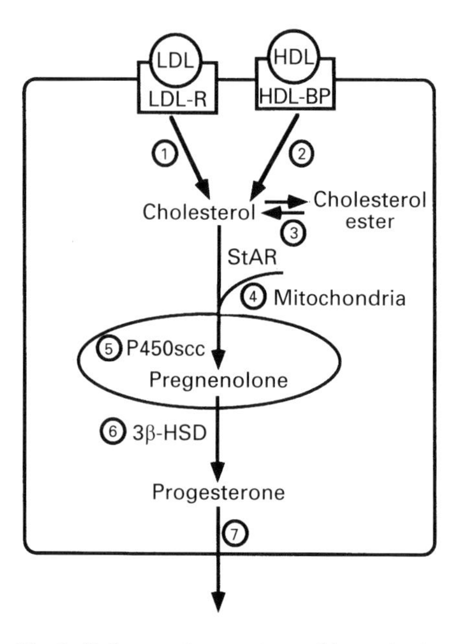

Fig. 1. Pathway of progesterone biosynthesis in a generic luteal cell. Three sources of cholesterol can be utilized for substrate: (1) low density lipoprotein (LDL); (2) high density lipoprotein (HDL); or (3) hydrolysis of stored cholesterol esters by cholesterol esterase. The free cholesterol is transported to the mitochondria apparently with cytoskeletal involvement. Cholesterol is then transported from the outer to inner mitochondrial membrane (4), which appears to involve steroidogenic acute regulatory protein (StAR). Cholesterol is converted to pregnenolone by cytochrome P450 side-chain cleavage enzyme (P450scc); (5), transported out of the mitochondria and converted to progesterone by 3β-hydroxysteroid dehydrogenase/Δ^5,Δ^4 isomerase (3β-HSD; (6), which is present in the smooth endoplasmic reticulum. Progesterone appears to diffuse from the luteal cell (7).

cholesterol is taken into the cell, it can be shuttled into the steroid biosynthetic pathway or stored as cholesterol esters in the form of lipid droplets (Gwynne and Strauss, 1982). Cholesterol can be released from these stores by cholesterol esterase, when demand for cholesterol exceeds the supply. During steroidogenesis, cholesterol moves through the cytoplasm to the mitochondria, where the side chain is cleaved. This cytoplasmic translocation is dependent upon the cytoskeleton (Jefcoate *et al.*, 1992) and probably involves sterol-binding proteins. Cholesterol must then pass through the outer to the inner mitochondrial membrane where the enzyme complex involved in cholesterol side-chain cleavage is located. Transport of cholesterol across the mitochondrial membranes appears to be the rate-limiting step of steroidogenesis (Stocco and Clark, 1996). It has been proposed that the recently identified steroidogenic acute regulatory (StAR) protein facilitates transport of cholesterol to cytochrome P450 side-chain cleavage enzyme (P450scc). The hypothesis that StAR is crucial to transport of cholesterol is supported by several lines of evidence. First, there is preliminary evidence that StAR is capable of binding cholesterol (Liu *et al.*, 1996). Second, StAR has been detected in the inner mitochondrial membrane where cholesterol side-chain cleavage occurs (King *et al.*, 1995). Third, naturally occurring mutations in the StAR gene severely limit steroidogenesis in adrenal and gonadal tissues (Lin *et al.*, 1995). Finally, gonadal tissue collected from patients with mutated StAR protein is capable of producing normal amounts of steroids if provided with membrane permeable cholesterol analogues which by-pass the usual route of cholesterol transport (Miller, 1996).

Another protein that appears to facilitate transport of cholesterol to the side chain cleavage enzyme complex is the peripheral benzodiazepine receptor (PBR; Papadopoulos *et al.*, 1997a). Targeted deletion of PBR resulted in loss of steroidogenic capacity of Leydig cells, which was restored when the cells were supplied with membrane permeable cholesterol analogues (Papadopoulos *et al.*, 1997b). In addition, PBR appears to associate with a voltage-dependent anion channel (VDAC) and molecular modelling indicates that the PBR–VDAC complex will form a pore permeable to cholesterol which spans the mitochondrial membranes (Papadopoulos *et al.*, 1997a). However, little information is available regarding regulation of this system in luteal tissue and thus, the role of this complex in control of luteal progesterone synthesis in domestic ruminants is not known. Once in the mitochondria, cholesterol is converted to pregnenolone by the side chain cleavage enzyme complex which comprises cytochrome P450scc, adrenodoxin and adrenodoxin reductase proteins. Pregnenolone is then converted to progesterone by 3β-hydroxysteroid dehydrogenase/Δ^5, Δ^4 isomerase (3β-HSD) in the smooth endoplasmic reticulum.

Steroidogenic Cells of the Corpus Luteum

The corpus luteum contains at least two types of steroidogenic cell that differ in morphological as well as biochemical characteristics. In mature ovine luteal tissue, small luteal cells are 8–22 μm in diameter, have lipid droplets, abundant smooth endoplasmic reticulum and are spindle shaped (Niswender *et al.*, 1985). These cells have receptors for LH and respond to this hormone with a 6–12-fold increase in steroid production (Fitz *et al.*, 1982; Alila *et al.*, 1988a). Treatment of small luteal cells with PGI_2 also increases progesterone synthesis (Fitz *et al.*, 1984; Alila *et al.*, 1988b). In the ewe, small luteal cells do not have high-affinity receptors for $PGF_{2\alpha}$ or PGE_2 (Fitz *et al.*, 1982); however, in the cow, mRNA encoding $PGF_{2\alpha}$- receptor (Mamluk *et al.*, 1998) as well as $PGF_{2\alpha}$ and PGE_2 (Chegini *et al.*, 1991) binding sites have been detected in small luteal cells.

Large luteal cells are more spherical, greater than 22 μm in diameter, contain abundant quantities of smooth and rough endoplasmic reticulum, have protein containing secretory granules and in cows, but not sheep, contain lipid droplets (Niswender *et al.*, 1985). Although large luteal cells contain receptors for LH (Harrison *et al.*, 1987; Chegini *et al.*, 1991), binding of LH to its receptor does not increase secretion of progesterone from these cells (Rodgers *et al.*, 1983; Hoyer *et al.*, 1984; Alila *et al.*, 1988a). Large luteal cells also contain receptors for prostaglandins $F_{2\alpha}$, E_2 and I_2 (Fitz *et al.*, 1982; Chegini *et al.*, 1991) as well as GH (Lucy *et al.*, 1993) and IGF-I (Perks *et al.*, 1995). Basal secretion of progesterone is 10–40 fold higher in large than in small cells but treatment with stimulatory hormones, such as PGE_2 or PGI_2 only increases secretion of progesterone from ovine large luteal cells 2–4 fold (Fitz *et al.*, 1984). Prostaglandin $F_{2\alpha}$ decreases secretion of progesterone from these cells (Wiltbank *et al.*, 1991). It has been calculated that the majority of the progesterone (< 80%) secreted from the mature ovine corpus luteum is derived from large luteal cells (Niswender *et al.*, 1985) although this may be controversial (Rodgers *et al.*, 1983).

Regulation of Progesterone Secretion During the Oestrous Cycle

Changes in the concentration of progesterone in sera normally observed during the oestrous cycle are due to changes in luteal blood flow, size and number of steroidogenic cells and changes in the

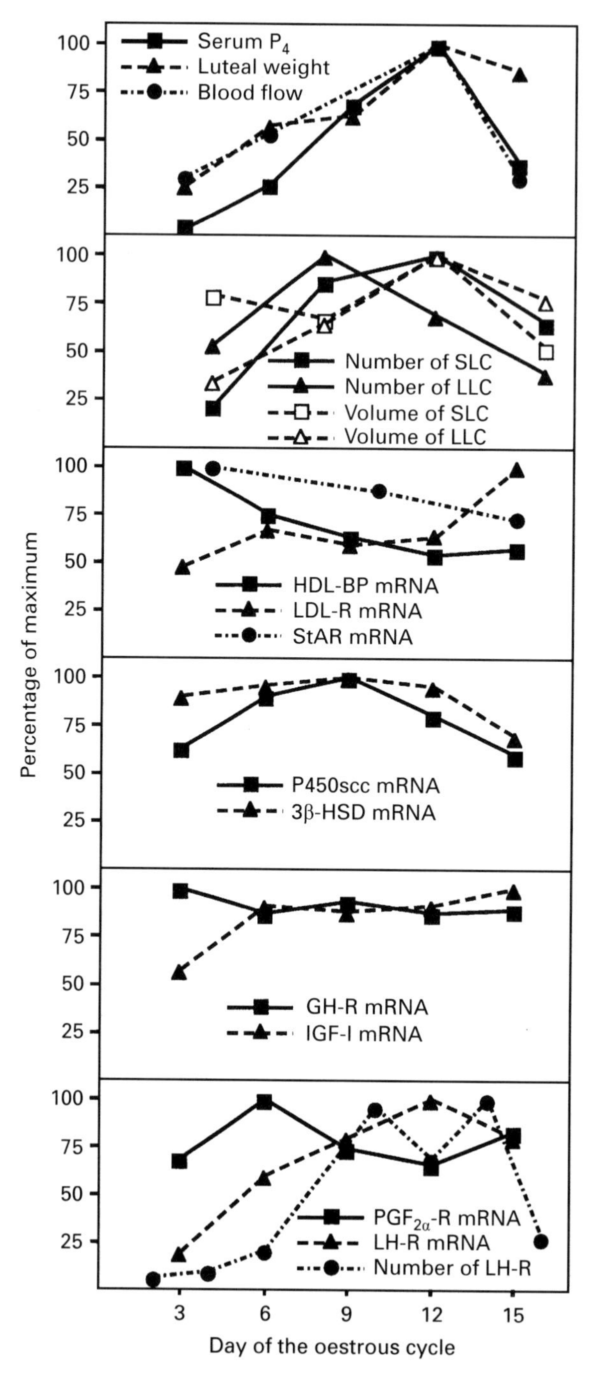

Fig. 2. Changes in concentrations of progesterone in sera (serum P_4), luteal weight, blood flow to the ovary bearing a corpus luteum, and numbers and sizes of small (S) and large (L) luteal cells (LC) over the ovine oestrous cycle are presented in the top two panels. Changes in mRNA encoding proteins important for cholesterol uptake and transport to the mitochondria (high density lipoprotein-

steroidogenic capacity of luteal cells (Fig. 2). Increases in concentrations of progesterone in sera during the early luteal phase are paralleled by increases in blood flow (Niswender *et al.*, 1976), luteal weight and numbers and sizes of steroidogenic luteal cells (Fig. 2; Farin *et al.*, 1986). In contrast, mRNAs encoding many proteins important for steroidogenesis are already expressed at high concentrations by day 3 of the ovine oestrous cycle (Fig. 2; Juengel *et al.*, 1994, 1995a; Tandeski *et al.*, 1996). For this communication, data regarding mRNA encoding important proteins that regulate the functions of luteal cells are expressed, in all cases, as steady-state concentrations. Although the total amount of mRNA encoding most of these proteins in the corpus luteum would increase as the numbers and size of steroidogenic cells increase during luteal development, it is the concentrations of each species of mRNA versus other mRNAs which should control the activity of individual cells. Therefore, steady-state concentrations of mRNA should reflect the ability of an individual mRNA to influence cellular function. It does not appear that expression of any of these mRNAs severely limits secretion of progesterone during the early luteal phase.

Both mRNA and the numbers of receptors for LH (LH-R) increase as luteal development progresses (Fig. 2, Diekman *et al.*, 1978a; Spicer *et al.*, 1981; Garverick *et al.*, 1985; Guy *et al.*, 1995). Thus, it is possible that cellular growth and division and increased progesterone secretion are driven by increasing the responsiveness of individual cells to LH. Progesterone increases expression of LH-R in luteal cells from bovine corpora lutea collected during the early luteal phase (Jones *et al.*, 1992). In addition, luteal concentrations of mRNA encoding IGF-I, IGF-II and the receptor for IGF-I increase during luteal development (Einspanier *et al.*, 1990; Perks *et al.*, 1995; Juengel *et al.*, 1997b) providing additional potential mechanisms for hormonal support of normal luteal growth. During luteal regression, the decrease in concentrations of progesterone in serum is closely associated with reduced blood flow to the corpus luteum (Niswender *et al.*, 1976) and decreased steroidogenic capacity of luteal cells (McGuire *et al.*, 1994; Juengel *et al.*, 1995a; Tandeski *et al.*, 1996). The number of steroidogenic cells decreases later during luteal regression (Braden *et al.*, 1988) and is not associated with the acute downregulation of progesterone synthesis seen during luteal regression. The focus of the remainder of this review will be regulation of steroidogenic capacity of luteal cells as well as their ability to respond to luteotrophic and luteolytic hormones.

Acquisition of maximum steroidogenic capability

The preovulatory LH surge causes release of the oocyte from the follicle and differentiation of follicular cells into luteal cells (luteinization). Luteinization is characterized by increased steroid production and a switch from producing oestradiol to progesterone. Not surprisingly concentrations of LDL-R, StAR, P450scc and 3β-HSD increase during this time (mRNA, protein, or both; Rodgers *et al.*, 1986, 1987; Couet *et al.*, 1990; Voss and Fortune, 1993a; Juengel *et al.*, 1994; Pescador *et al.*, 1996). In addition, enzymes important for oestrogen synthesis (aromatase and 17α-hydroxylase cytochrome P450) were greatly decreased in the newly formed bovine corpus luteum (Rodgers *et al.*, 1986, 1987; Voss and Fortune, 1993b). Cells also become more responsive to LH (Diekman *et al.*, 1978a) and GH (Lucy *et al.*, 1993). Receptors for $PGF_{2\alpha}$ also increase during luteinization to reach maximum numbers per large luteal cell shortly after ovulation (Wiltbank *et al.*, 1995). Thus, during luteinization, the newly formed corpus luteum gains increased ability to synthesize progesterone and respond to regulatory hormones.

---

binding protein (HDL-BP); low density lipoprotein-receptor (LDL-R) and steroidogenic acute regulatory protein (StAR)) are shown in the third panel, whereas patterns of expression of mRNA encoding cytochrome P450 side-chain cleavage (P450scc) and 3β-hydroxysteroid dehydrogenase/Δ^5,Δ^4 isomerase (3β-HSD) over the oestrous cycle are displayed in the fourth panel. The last two panels contain information about expression of luteotrophic and luteolytic hormones or their receptors (receptor (R) for growth hormone (GH), LH and prostaglandin $F_{2\alpha}$ ($PGF_{2\alpha}$) and insulin-like growth factor I (IGF-I)) during the ovine oestrous cycle. Notice the close association between luteal blood flow, luteal weight and amount of progesterone in serum. Data obtained from Niswender *et al.*, 1976; Diekman *et al.*, 1978a; Farin *et al.*, 1986; Juengel *et al.*, 1994, 1995a, 1996, 1997b; Guy *et al.*, 1995 and Tandeski *et al.*, 1996.

The corpus luteum continues to secrete increasing amounts of progesterone for several days following luteinization, and continued development of the corpus luteum is dependent upon pituitary support (Farin *et al.*, 1990; Juengel *et al.*, 1995b). Supply of cholesterol probably does not limit secretion of progesterone from the corpus luteum during development, as steady-state concentrations of mRNA encoding HDL-binding protein (HDL-BP; Tandeski *et al.*, 1996) and LDL-R (Rodgers *et al.*, 1987; Tandeski *et al.*, 1996) were maximal early in the oestrous cycle (Fig. 2), and hypophysectomy of ewes did not decrease expression of these mRNAs (Tandeski *et al.*, 1996). Steady-state concentrations of mRNA encoding StAR or StAR protein also peak early in the oestrous cycle (Juengel *et al.*, 1995a; Pescador *et al.*, 1996). However, removal of pituitary support during luteal development severely reduces mRNA encoding StAR and replacement of LH or GH prevented this decrease (Juengel *et al.*, 1995a). Thus, expression of normal amounts of mRNA encoding StAR, and presumably protein since these two variables are highly correlated (Pescador *et al.*, 1996), appears essential for maximal secretion of progesterone. Conversion of cholesterol to pregnenolone by the P450scc enzyme complex may limit secretion of progesterone during development of the corpus luteum as mRNA, or protein, or both for P450scc and adrenodoxin increase during luteal development (Rodgers *et al.*, 1986; 1987; Juengel *et al.*, 1994). Removal of the pituitary gland prevents the normal increases in mRNA encoding P450scc observed during the ovine oestrous cycle, and replacement of LH or GH supports normal expression of this mRNA (Juengel *et al.*, 1995b). Conversion of pregnenolone to progesterone by 3β-HSD does not appear to limit secretion of progesterone from the corpus luteum. Concentrations of mRNA encoding 3β-HSD, as well as 3β-HSD protein and enzyme activity reach maximum values early in the oestrous cycle and appear to be in great excess (Couet *et al.*, 1990; Wiltbank *et al.*, 1993; Juengel *et al.*, 1994). In addition, maximal expression of 3β-HSD mRNA, which is dependent upon LH, was not required for maximal secretion of progesterone (Juengel *et al.*, 1995b).

The increase in the number of receptors for LH observed during luteal development (Diekman *et al.*, 1978a; Spicer *et al.*, 1981, Garverick *et al.*, 1985) was preceded by an increase in mRNA encoding this receptor (Guy *et al.*, 1995). However, removal of the pituitary gland during luteal development did not prevent expression of normal concentrations of LH-R in luteal tissue (Farin *et al.*, 1990; Juengel *et al.*, 1995b). Thus, reduced expression of the LH-R was not limiting secretion of progesterone in hypophysectomized ewes. In contrast, while mRNA encoding the receptor for GH (GH-R) is expressed at maximal concentrations in corpora lutea by day 3 of the ovine oestrous cycle, removal of the pituitary gland caused a decrease in expression of this mRNA (Juengel *et al.*, 1997b). Somewhat surprisingly, LH, but not GH, supported normal expression of GH-R mRNA (Juengel *et al.*, 1997b). However, concentrations of progesterone in sera of GH-treated, hypophysectomized ewes were not different from those in pituitary-intact control ewes. Thus, secretion of progesterone in hypophysectomized ewes was not probably limited by a lack of GH binding. Similarly, although concentrations of mRNA encoding IGF-I increase during luteal development, this increase in expression of IGF-I mRNA was not necessary for normal progesterone biosynthesis (Juengel *et al.*, 1997b). In contrast, increases in luteal weight were associated with increased expression of IGF-I mRNA (Juengel *et al.*, 1997b). Thus, concentrations of progesterone in sera are not tightly linked to luteal expression of IGF-I mRNA, but increases in luteal weight were associated with increases in this mRNA.

Maintenance of maximal steroidogenic capacity

Once the corpus luteum is fully formed, it does not appear to require pulsatile LH release to maintain secretion of progesterone at normal, mid-luteal phase values in either sheep or cattle (McNeilly *et al.*, 1992; Peters *et al.*, 1995). However, basal amounts of LH are necessary to maintain normal serum concentrations of progesterone and luteal weights in sheep (Haworth *et al.*, 1998). An additional pituitary hormone may also be necessary for normal luteal function during the mid-luteal phase, as hypophysectomy decreased concentrations of progesterone in sera and luteal weights more severely than specific removal of LH with an antiserum (Haworth *et al.*, 1998). Both specific removal of LH and hypophysectomy decreased luteal weight and luteal concentrations of mRNAs encoding StAR, P450scc and 3β-HSD (Haworth *et al.*, 1998). Thus, removal of the pituitary decreases both the amount of luteal tissue and the capability of that tissue to synthesize progesterone. However, since hypophysectomy during the mid-luteal phase only reduced concentrations of progesterone in sera 4 days later by approximately 60% (Haworth *et al.*, 1998), it seems likely that the corpus luteum of the ewe is somewhat independent of pituitary support at this time.

Cellular regulation of enhanced synthesis of progesterone

In small luteal cells, LH-induced increases of progesterone synthesis were associated with a slight increase in release of cholesterol from cholesterol esters but not with increased uptake of cholesterol or activity of P450scc or 3β-HSD (Wiltbank *et al.*, 1993). However, the modest increase in cholesterol esterase activity was not sufficient to account for the marked increase in progesterone secretion in these cells. Therefore, it was postulated that LH was increasing transport of cholesterol through the cell or across the mitochondrial membranes. Since StAR facilitates the transport of cholesterol to P450scc (Stocco and Clark, 1996), regulation of this molecule appears to be crucial in control of secretion of progesterone. Chronic removal of LH decreases concentrations of mRNA encoding StAR (Juengel *et al.*, 1995a); however, whether LH acutely increases luteal concentrations of StAR mRNA is not clear. Intra-ovarian infusion of LH did not increase steady-state concentrations of mRNA encoding StAR 4, 12 or 24 h later (Juengel *et al.*, 1997a); thus, if LH acutely increases luteal concentrations of StAR mRNA it must do so in a very rapid and transient manner. However, a direct effect of activation of PKA on transcription of the StAR gene has been demonstrated as the human (Sugawara *et al.*, 1997), mouse (Caron *et al.*, 1997) and sheep (J. L. Juengel, C. M. Clay and G. D. Niswender; unpublished observations) StAR promoters respond directly to PKA activation with modest increases in activity. In addition, the ability of StAR to stimulate steroidogenesis was increased by PKA-dependent phosphorylation (Arakane *et al.*, 1997). Thus, the acute luteotrophic effects of LH are probably due to increased transcription, translation or phosphorylation of StAR.

Other hormones such as PGE_2 (Fitz *et al.*, 1984; Alila *et al.*, 1988b), PGI_2 (Fitz *et al.*, 1984; Alila *et al.*, 1988b), GH (Liebermann and Schams, 1994) and IGF-I (McArdle and Holtorf, 1989) have also been shown to increase secretion of progesterone from luteal cells; however, the mechanisms by which these hormones stimulate progesterone synthesis are not known. Prostaglandin E_2 and I_2 have been reported to increase cAMP (Marsh, 1975; Bennegard *et al.*, 1990) and thus, in small luteal cells, would be expected to increase progesterone secretion in a manner similar to LH. However, both of these hormones also increase progesterone synthesis in large luteal cells (Fitz *et al.*, 1984; Alila *et al.*, 1988b), which do not respond to increased cAMP with an increase in progesterone secretion (Fitz *et al.*, 1984; Hoyer *et al.*, 1984). Therefore, binding of these hormones to receptors in large luteal cells must regulate progesterone secretion in another manner. A better understanding of how both basal and luteotropin stimulated progesterone synthesis is controlled in small, and particularly, large luteal cells is crucial if these processes are to be manipulated effectively.

Cellular regulation of reduced secretion of progesterone

Progesterone secretion is maintained for several days, and then rapidly declines if pregnancy does not occur. This rapid decline in the concentration of progesterone in sera is followed by a slower decrease in luteal weight. Prostaglandin $F_{2\alpha}$ induces both the rapid decrease in progesterone secretion and loss of luteal tissue (Niswender and Nett, 1994). Binding of $PGF_{2\alpha}$ to its receptor causes activation of PKC and an influx of calcium (Wiltbank *et al.*, 1991). Pharmacological activation of PKC *in vivo* decreases secretion of progesterone without causing luteal cell death (McGuire *et al.*, 1994). Similarly, a low dose of $PGF_{2\alpha}$, which transiently decreased secretion of progesterone and induced oligonucleosome formation, does not cause luteolysis (J. L. Juengel, J. D. Haworth, M. K. Rollyson, P. J. Silvia, E. W. McIntush, H. R. Sawyer and G. D. Niswender, unpublished observations). Thus, destruction of the corpus luteum is not required for reduced progesterone secretion from this tissue.

In fact, it seems clear that the anti-steroidogenic effects of $PGF_{2\alpha}$ are mediated via the PKC second messenger pathway, while the cytotoxic effects are mediated via calcium influx (Wiltbank *et al.*, 1991).

Some mechanisms whereby activation of PKC by $PGF_{2\alpha}$ could decrease synthesis of progesterone at the cellular level include (1) influencing uptake or transport of cholesterol, (2) regulating conversion of cholesterol to pregnenolone or pregnenolone to progesterone, or (3) reducing the ability of the corpus luteum to respond to luteotrophic hormones by reducing receptors for LH, GH or both hormones, or interfering with the ability of the hormone receptor complex to activate their second messenger systems. The cellular mechanisms by which $PGF_{2\alpha}$ decreases synthesis of progesterone are rapidly being elucidated.

Treatment of ewes with $PGF_{2\alpha}$ decreased concentrations of mRNA encoding LDL-R; however, concentrations of HDL-BP mRNA actually increased during the first 12 h after treatment with $PGF_{2\alpha}$ (Fig. 3; Tandeski *et al.*, 1996). Suppression of LDL-R mRNA to values similar to those seen after treatment with $PGF_{2\alpha}$ did not affect secretion of progesterone (T.R. Tandeski, J. L. Juengel, and G. D. Niswender, unpublished observations). Therefore, it seems unlikely that suppression of LDL-R mRNA is important for reduced secretion of progesterone during luteolysis. Furthermore, lipoprotein uptake was not limiting progesterone secretion in cultures of either ovine or bovine luteal cells after treatment with $PGF_{2\alpha}$ (Grusenmeyer and Pate 1992; Wiltbank *et al.*, 1993). Finally, treatment of ovine luteal cells with $PGF_{2\alpha}$ did not affect release of cholesterol from cholesterol esters (Wiltbank *et al.*, 1993). Thus, $PGF_{2\alpha}$ does not appear to reduce synthesis of progesterone by decreasing the availability of cholesterol for steroidogenesis.

In cultures of ovine and bovine luteal cells, $PGF_{2\alpha}$ appeared to decrease transport of cholesterol through the cytoplasm, or from the outer to inner mitochondrial membrane, or both (Grusenmeyer and Pate 1992; Wiltbank *et al.*, 1993). Transport of cholesterol through the cell to the mitochondria is facilitated by the cytoskeleton (Jefcoate *et al.*, 1992) and disruption of the cytoskeleton will decrease secretion of progesterone from luteal cells (Niswender and Nett, 1994). Treatment of ewes with $PGF_{2\alpha}$ rapidly disrupts the microtubular network of ovine corpora lutea (Murdoch, 1996). This disruption occurred before concentrations of progesterone in sera or luteal tissues decreased (Murdoch, 1996). In addition, in rat corpora lutea, administration of $PGF_{2\alpha}$ rapidly decreases concentrations of sterol carrier protein-2 (SCP-2), which facilitates transport of cholesterol through the cytoplasm (McLean *et al.*, 1995). Whether $PGF_{2\alpha}$ has a similar effect on SCP-2 in luteal tissue of domestic ruminants is not known. Thus, one mechanism that $PGF_{2\alpha}$ may use to reduce progesterone synthesis is disruption of cholesterol transport to the mitochondria.

Prostaglandin $F_{2\alpha}$ also appears to disrupt transport of cholesterol across the mitochondrial membranes, potentially through regulation of StAR. Concentrations of StAR mRNA or protein in ovine and bovine corpora lutea were reduced within 12 h of treatment with $PGF_{2\alpha}$ (Fig. 3; Juengel *et al.*, 1995a; Pescador *et al.*, 1996). In addition, since the activity of StAR is modified by phosphorylation (Arakane *et al.*, 1997), and StAR contains several potential PKC phosphorylation sites (Hartung *et al.*, 1995; J. L. Juengel and G. D. Niswender, unpublished observations), its activity may be modified by $PGF_{2\alpha}$ to reduce progesterone synthesis more rapidly. Of particular interest are three potential PKC phosphorylation sites in the mitochondrial targeting sequence, the modification of which could be proposed to prevent targeting of StAR to the mitochondria. However, removal of the mitochondrial targeting sequence did not appear to affect the ability of StAR to stimulate steroidogenesis (Arakane *et al.*, 1996). It should be pointed out that marked over-expression of the mutated form of StAR without its mitochondrial targeting sequence may have saturated the normal cholesterol transport mechanisms. Thus, the potential for modifications of the mitochondrial targeting sequence of StAR to interfere with cholesterol transport is unclear. The PBR system has been shown to be essential for transport of cholesterol to P450scc in other steroidogenic cells (Papadopoulus *et al.*, 1997b); however, little is known about its potential regulation in luteal tissue of domestic ruminants. Thus, whether $PGF_{2\alpha}$-induced downregulation of progesterone synthesis is mediated partially through regulation of the PBR system remains to be determined. It is clear that one of the mechanisms by which $PGF_{2\alpha}$ decreases synthesis of progesterone in luteal cells is disruption of transport of cholesterol across the mitochondrial membranes.

For some time, conversion of cholesterol to pregnenolone was thought to be the rate-limiting

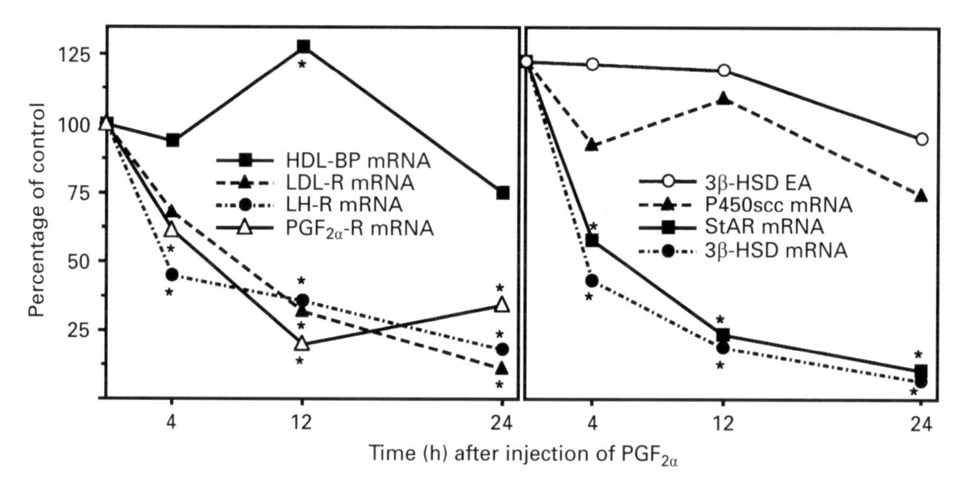

Fig. 3. Pattern of expression of mRNA encoding high density lipoprotein-binding protein (HDL-BP), low density lipoprotein-receptor (LDL-R), luteinizing hormone-receptor (LH-R) and prostaglandin $F_{2\alpha}$-receptor ($PGF_{2\alpha}$-R) 4, 12 and 24 h after injection of $PGF_{2\alpha}$ are shown in the left panel. Values are expressed as a percentage of the control value. Pattern of expression of mRNA encoding steroidogenic acute regulatory protein (StAR), cytochrome P450 side-chain cleavage enzyme (P450scc) and 3β-hydroxysteroid dehydrogenase/Δ^5,Δ^4 isomerase (3β-HSD) as well as 3β-HSD enzyme activity (EA) after $PGF_{2\alpha}$ administration are given in the right panel. Within a variable, values that differ ($P < 0.05$) from controls (time of injection) are indicated with an asterisk. Data obtained from Guy *et al.* 1995; Juengel *et al.*, 1995a, 1996, 1998; and Tandeski *et al.*, 1996.

step in steroid synthesis. However, the activity, protein concentration and concentrations of mRNA encoding P450scc are not acutely reduced following treatment with $PGF_{2\alpha}$ (Fig. 3; Grusenmeyer and Pate, 1992; Wiltbank *et al.*, 1993; McGuire *et al.*, 1994; Tian *et al*, 1994; Rodgers *et al.*, 1995). Thus, reduced progesterone synthesis is not due to a decreased capacity to convert cholesterol to pregnenolone. Induction of luteolysis with $PGF_{2\alpha}$ rapidly decreased mRNA encoding 3β-HSD in both ewes and cows (McGuire *et al.*, 1994; Tian *et al.*, 1994). However, in the same tissue that had markedly less mRNA encoding 3β-HSD, amounts of 3β-HSD protein or enzyme activity were not reduced (Fig. 3; Rodgers *et al.*, 1995; Juengel *et al.*, 1998). Thus, $PGF_{2\alpha}$ -induced downregulation of luteal progesterone secretion does not appear to be mediated by decreased ability to convert pregnenolone to progesterone.

Administration of $PGF_{2\alpha}$ did not decrease concentrations of mRNA encoding GH-R or IGF-I (Juengel *et al.*, 1997b). In fact, IGF-I mRNA and protein increased during the later stages of luteolysis (Perks *et al.*, 1995). Binding of IGF-I to luteal tissue has also been shown to increase during luteolysis; however, this appeared to be due to increased expression of IGF-I binding proteins and not to an increased number of IGF-I receptors (Perks *et al.*, 1995). Since IGF-I binding proteins can decrease IGF-I availability, decreased free IGF-I may be important in luteolysis. Injection of $PGF_{2\alpha}$ decreased mRNA encoding LH-R (Fig. 3; Guy *et al.*, 1995); however, numbers of LH receptors were not decreased until after concentrations of progesterone in sera had declined (Diekman *et al.*, 1978b; Spicer *et al.*, 1981). Thus, the decrease in progesterone secretion following $PGF_{2\alpha}$ is not likely due to an inability of the corpus luteum to respond to tropic hormonal stimuli; however, $PGF_{2\alpha}$ could affect the ability of the hormone–receptor complex to activate its second messenger. Indeed, $PGF_{2\alpha}$ decreases adenylate cyclase activity and increases phosphodiesterase, thus decreasing the amount of cAMP available to activate PKA (Agudo *et al.*, 1984; Garverick *et al.*, 1985) which would decrease secretion of progesterone particularly from small luteal cells.

One of the most interesting aspects of the changes that occur after $PGF_{2\alpha}$ treatment of ewes is the marked decline in concentrations of mRNA encoding 3β-HSD, StAR, LDL-R, LH-R and $PGF_{2\alpha}$-R

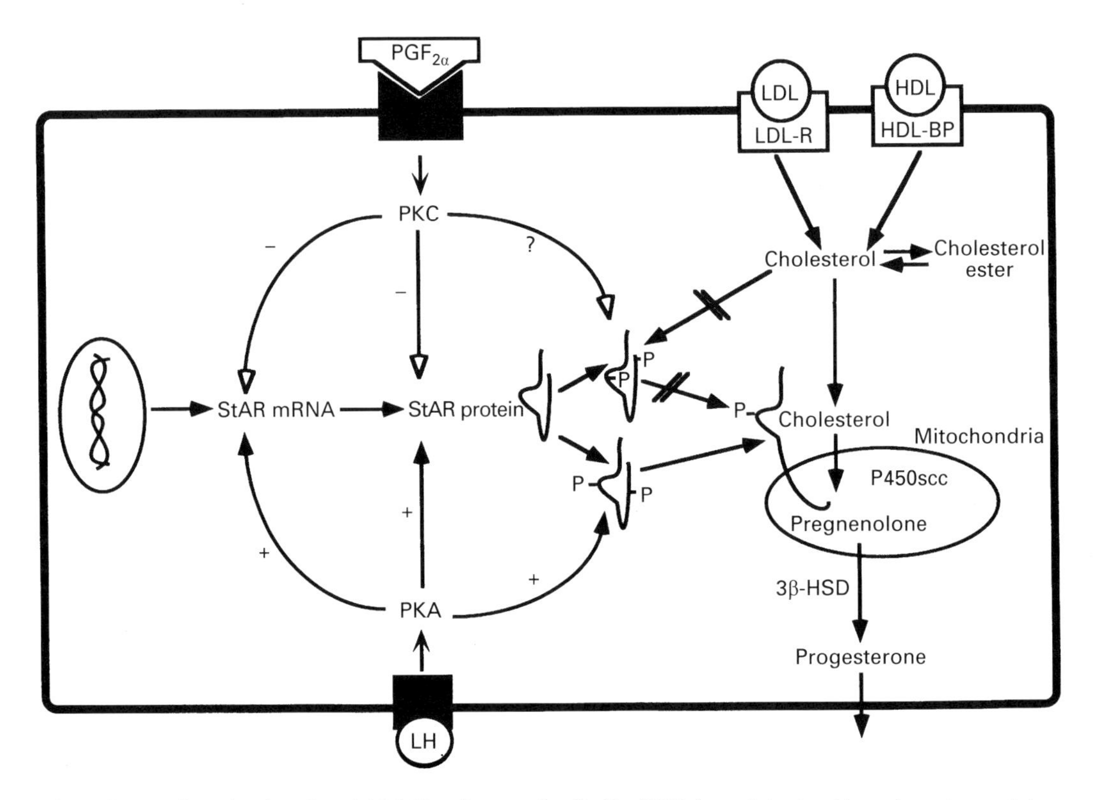

Fig. 4. Potential mechanisms by which LH and prostaglandin $F_{2\alpha}$ ($PGF_{2\alpha}$) regulate steroidogenic acute regulatory protein- (StAR) facilitated transport of cholesterol to cytochrome P450 side-chain cleavage enzyme (P450scc). Binding of LH to its receptor activates protein kinase A (PKA), which increases mRNA encoding StAR through transcriptional activation of the StAR gene but not through stabilization of the StAR mRNA (Kiriakidou *et al.*, 1996). Increases in mRNA encoding StAR, and potentially increased translational efficiency of the StAR mRNA, would increase the amount of StAR protein. In addition, PKA phosphorylates StAR thereby increasing its ability to transport cholesterol to cytochrome P450scc. The anti-steroidogenic effects of $PGF_{2\alpha}$ have been shown to be mediated through activation of protein kinase C (PKC). Prostaglandin $F_{2\alpha}$ decreases steady-state concentrations of mRNA encoding StAR, potentially by decreasing transcription of the StAR gene, destabilizing the StAR mRNA, or by both mechanisms. Concentrations of StAR protein decrease and it is possible that activity of the StAR protein is reduced through post-translational modifications, such as phosphorylation, resulting in decreased ability to bind cholesterol or decreased StAR-facilitated transport of cholesterol to P450scc. In addition, activation of PKC decreases cAMP concentrations, thus potentially reducing the PKA-activated increases in StAR activity. LDL-R: low density lipoprotein receptor; HDL-BP: high density lipoprotein-binding protein; 3β-HSD: 3β-hydroxysteroid dehydrogenase/ Δ^5,Δ^4 isomerase.

while concentrations of mRNA encoding HDL-BP and P450scc were unaffected (Fig. 3). This may indicate that a very specific mechanism is activated by the PKC second messenger pathway which results in enhanced degradation or inhibited synthesis of some mRNAs with no effect on others. This mechanism may be specific to luteal cells as activation of PKC in other steroidogenic cell types has very different effects on several of the mRNAs encoding steroidogenic proteins.

Conclusion

Development and maintenance of the steroidogenic machinery in luteal cells is dependent upon pituitary hormones, LH and GH. These hormones not only support increases in luteal weight but also support normal chronic expression of StAR, P450scc and 3β-HSD mRNA, which encode three proteins essential for progesterone synthesis. Luteotrophic and luteolytic hormones also acutely regulate progesterone synthesis. The primary effect of LH appears to be regulation of pregnenolone production by increasing transport of cholesterol to cytochrome P450scc. One protein likely to facilitate this process is StAR. The precise mechanisms by which StAR facilitates transport of cholesterol to cytochrome P450scc are not known. However, phosphorylation of StAR by PKA is likely one mechanism by which LH acutely enhances the activity of StAR and thus progesterone synthesis (Fig. 4). In addition, activation of PKA may also increase transcription of the StAR gene, or increase translation efficiency of StAR mRNA (or both), thereby increasing the amount of StAR protein available to facilitate transfer of cholesterol (Fig. 4). Prostaglandin $F_{2\alpha}$ reduces transport of cholesterol across the mitochondrial membranes, potentially by decreasing amounts of StAR protein or its ability to transport cholesterol. Prostaglandin $F_{2\alpha}$ may reduce StAR protein by decreasing transcription of the StAR gene, interfering with stability of the StAR transcript or reducing translation efficiency of the StAR mRNA (Fig. 4). Once StAR is inserted into the mitochondria, it is likely that it is no longer able to transport cholesterol. Therefore, only newly synthesized protein transports cholesterol giving StAR an effective half life of 3–5 min. Thus, one mechanism that both luteotrophic and luteolytic hormones appear to use to regulate progesterone secretion from the corpus luteum acutely is regulation of StAR-facilitated transport of cholesterol. This may be accomplished through regulation of transcription of the StAR gene, stability of the StAR mRNA, synthesis of StAR protein, or post-translational modifications of StAR such as phosphorylation (Fig. 4).

This research was supported by NIH grant HD 11590 and the Colorado Agricultural Experiment Station. The authors appreciate the invaluable technical expertise of M. Gallegos, B. Meberg, C. Moeller and K. Sutherland. They would also like to thank Kathy Thomas for her assistance with preparation of manuscripts.

References

Agudo SPL, Zahler WL and Smith MF (1984) Effects of prostaglandins $F_{2\alpha}$ on the adenylate cyclase and phosphodiesterase activity of ovine corpora lutea *Journal of Animal Science* **58** 955–962

Alila HW, Dowd JP, Corradino RA, Harris WV and Hansel W (1988a) Control of progesterone production in small and large bovine luteal cells separated by flow cytometry *Journal of Reproduction and Fertility* **82** 645–655

Alila HW, Corradino RA and Hansel W (1988b) A comparison of the effects of cyclooxygenase prostanoids on progesterone production by small and large bovine luteal cells *Prostaglandins* **36** 259–270

Arakane F, Sugawara T, Nishino H, Liu Z, Holt JA, Pain D, Stocco DM, Miller WL and Strauss JF III (1996) Steroidogenic acute regulatory protein (StAR) retains activity in the absence of its mitochondrial import sequence: implications for the mechanism of StAR action *Proceedings of the National Academy of Sciences USA* **93** 13 731–13 736

Arakane F, King SR, Du Y, Kallen CB, Walsh LP, Watari H, Stocco DM and Strauss JF, III (1997) Phosphorylation of steroidogenic acute regulatory protein (StAR) modulates its steroidogenic activity *Journal of Biological Chemistry* **272** 32 656–32 662

Bennegard B, Hahlin M and Hamberger L (1990) Luteotrophic effects of prostaglandins I_2 and D_2 on isolated human corpora lutea *Fertility and Sterility* **54** 459–464

Braden TD, Gamboni F and Niswender GD (1988) Effects of prostaglandin $F_{2\alpha}$-induced luteolysis on the populations of cells in the ovine corpus lutea *Biology of Reproduction* **39** 245–253

Caron KM, Ikeda Y, Soo S-C, Stocco DM, Parker KL and Clark BJ (1997) Characterization of the promoter region of the mouse gene encoding the steroidogenic acute regulatory protein *Molecular Endocrinology* **11** 138–147

Chegini N, Lei ZM, Rao Ch V and Hansel W (1991) Cellular distribution and cycle phase dependency of gonadotropin and eicosanoid binding sites in bovine corpora lutea *Biology of Reproduction* **45** 506–513

Couet J, Martel C, Dupont E, Luu-The V, Sirard M-A, Zhao H-F, Pelletier G and Labrie F (1990) Changes in 3β-hydroxysteroid dehydrogenase/Δ^5–Δ^4 isomerase messenger ribonucleic acid, activity and protein levels during the estrous cycle in the bovine ovary *Endocrinology* **127** 2141–2148

Diekman MA, O'Callaghan P, Nett TM and Niswender GD (1978a) Validation of methods and quantification of luteal receptors for LH throughout the estrous cycle and early pregnancy in ewes *Biology of Reproduction* **19** 999–1009

Diekman MA, O'Callaghan P, Nett TM and Niswender GD (1978b) Effect of prostaglandin $F_{2\alpha}$ on the number of LH receptors in ovine corpora lutea *Biology of Reproduction* **19** 1010–1013

Einspanier R, Miyamoto A, Schams D, Muller M and Brem G (1990) Tissue concentration, mRNA expression and stimulation of IGF-I in luteal tissue during the oestrous cycle and pregnancy of cows *Journal of Reproduction and Fertility* **90** 439–445

Farin CE, Moeller CL, Sawyer HR, Gamboni F and Niswender GD (1986) Morphometric analysis of cell types in the ovine corpus luteum throughout the estrous cycle *Biology of Reproduction* **35** 1299–1308

Farin CE, Nett TM and Niswender GD (1990) Effects of luteinizing hormone on luteal cell populations in hypophysectomized ewes *Journal of Reproduction and Fertility* **88** 61–70

Fitz TA, Mayan MH, Sawyer HR and Niswender GD (1982) Characterization of two steroidogenic cell types in the ovine corpus luteum *Biology of Reproduction* **27** 703–711

Fitz TA, Hoyer PB and Niswender GD (1984) Interactions of prostaglandins with subpopulations of ovine luteal cells I. Stimulatory effects of prostaglandins E^1, E_2 and I_2 *Prostaglandins* **28** 119–126

Garverick HA, Smith MF, Elmore RG, Morehouse GL, Agudo SPL and Zahler WL (1985) Changes and interrelationships among luteal LH receptors, adenylate cyclase activity and phosphodiesterase activity during the bovine estrous cycle *Journal of Animal Science* **61** 216–223

Grusenmeyer DP and Pate JL (1992) Localization of prostaglandins $F_{2\alpha}$ inhibition of lipoprotein use by bovine luteal cells *Journal of Reproduction and Fertility* **94** 311–318

Guy MK, Juengel JL, Tandeski TR and Niswender GD (1995) Steady-state concentrations of mRNA encoding the receptor for luteinizing hormone during the estrous cycle and following prostaglandin $F_{2\alpha}$ treatment of ewes *Endocrine* **3** 585–589

Gwynne JT and Strauss JF, III (1982) The role of lipoproteins in steroidogenesis and cholesterol metabolism in steroidogenic glands *Endocrine Reviews* **3** 299–329

Harrison LM, Kenny N and Niswender GD (1987) Progesterone production, LH receptors, and oxytocin secretion by ovine luteal cell types on days 6, 10 and 15 of the oestrous cycle and day 25 of pregnancy *Journal of Reproduction and Fertility* **79** 539–548

Hartung S, Rust W, Balvers M and Ivell R (1995) Molecular cloning and *in vivo* expression of the bovine steroidogenic acute regulatory protein *Biochemical and Biophysical Research Communications* **215** 646–653

Hoyer PB, Fitz TA and Niswender GD (1984) Hormone-independent activation of adenylate cyclase in large steroidogenic ovine luteal cells does not result in increased progesterone secretion *Endocrinology* **114** 604–608

Jefcoate CR, McNamara BC, Artemenko I and Yamazaki T (1992) Regulation of cholesterol movement to mitochondrial cytochrome P450scc in steroid hormone synthesis *Journal of Steroid Biochemistry and Molecular Biology* **43** 751–767

Jones LS, Ottobre JS and Pate JL (1992) Progesterone regulation of luteinizing hormone receptors on cultured bovine luteal cells *Molecular and Cellular Endocrinology* **85** 33–39

Juengel JL, Guy MK, Tandeski TR, McGuire WJ and Niswender GD (1994) Steady-state concentrations of messenger ribonucleic acid encoding cytochrome P450 side-chain cleavage and 3β-hydroxysteroid dehydrogenase/Δ^5, Δ^4 isomerase in ovine corpora lutea during the estrous cycle *Biology of Reproduction* **51** 380–384

Juengel JL, Meberg BM, Turzillo AM, Nett TM and Niswender GD (1995a) Hormonal regulation of messenger ribonucleic acid encoding steroidogenic acute regulatory protein in ovine corpora lutea *Endocrinology* **136** 5423–5429

Juengel JL, Nett TM, Tandeski TR, Eckery DC, Sawyer HR and Niswender GD (1995b) Effects of luteinizing hormone and growth hormone on luteal development in hypophysectomized ewes *Endocrine* **3** 323–326

Juengel JL, Wiltbank MC, Meberg BM and Niswender GD (1996) Regulation of steady-state concentrations of messenger ribonucleic acid encoding prostaglandin $F_{2\alpha}$-receptor in ovine corpus luteum *Biology of Reproduction* **54** 1096–1102

Juengel JL, Larrick TL, Meberg BM and Niswender GD (1997a) Luteal expression of steroidogenic factor-1 mRNA during the estrous cycle and in response to luteinizing hormone (LH) stimulation *Biology of Reproduction* **56 (Supplement 1)** Abstract 145

Juengel JL, Nett TM, Anthony RV and Niswender GD (1997b) Effects of luteotrophic and luteolytic hormones on expression of mRNA encoding insulin-like growth factor-1 and growth hormone receptor in the ovine corpus luteum *Journal of Reproduction and Fertility* **110** 291–298

Juengel JL, Meberg BM, McIntush EW, Smith MF and Niswender GD (1998) Concentration of mRNA encoding 3β-hydroxysteroid dehydrogenase/Δ^5,Δ^4 isomerase (3β-HSD) and 3β-HSD enzyme activity following treatment of ewes with prostaglandin $F_{2\alpha}$ *Endocrine* **8** 45–50

King SR, Ronen-Fuhrmann T, Timberg R, Clark BJ, Orly J and Stocco DM (1995) Steroid production after *in vitro* transcription, translation, and mitochondrial processing of protein products of complementary deoxyribonucleic acid for steroidogenic acute regulatory protein *Endocrinology* **136** 5165–5176

Kiriakidou M, McAllister JM, Sugawara T and Strauss JF, III (1996) Expression of steroidogenic acute regulatory protein (StAR) in the human ovary *Journal of Clinical Endocrinology and Metabolism* **81** 4122–4128

Liebermann J and Schams D (1994) Actions of somatotrophin on oxytocin and progesterone release from the microdialysed bovine corpus luteum *in vitro*. *Journal of Endocrinology* **143** 243–250

Lin D, Sugawara T, Strauss JF, III, Clark BJ, Stocco DM, Saenger P, Rogol A and Miller WL (1995) Role of steroidogenic acute regulatory protein in adrenal and gonadal steroidogenesis *Science* **267** 1828–1831

Liu Z, Frolov AA, Schroeder F and Stocco DM (1996) Does cholesterol bind to the steroidogenic acute regulatory (StAR) protein? *Biology of Reproduction* **54 (Supplement 1)** Abstract 194

Lucy MC, Collier RJ, Kitchell ML, Dibner JJ, Hauser SP and Kriui GG (1993) Immunohistochemical and nucleic acid analysis of somatotropin receptor populations in the bovine ovary *Biology of Reproduction* **48** 1219–1227

McGuire WJ, Juengel JL and Niswender GD (1994) Protein kinase C second messenger system mediates the antisteroidogenic effects of prostaglandin $F_{2\alpha}$ in the ovine corpus luteum *in vivo*. *Biology of Reproduction* **51** 800–806

McLean MP, Billheimer JT, Warden KJ and Irby RB (1995) Prostaglandin $F_{2\alpha}$ mediates ovarian sterol carrier protein-2 expression during luteolysis *Endocrinology* **136** 4963–4972

McNeilly AS, Crow WJ and Fraser HM (1992) Suppression of pulsatile luteinizing hormone secretion by gonadotrophin-releasing hormone antagonist does not affect episodic progesterone secretion or corpus luteum function in ewes *Journal of Reproduction and Fertility* **96** 865–874

Mamluk R, Chen D-b, Greber Y, Davis JS and Meidan R (1998) Characterization of messenger ribonucleic acid expression for prostaglandin $F_{2\alpha}$ and luteinizing hormone receptors in various bovine luteal cell types *Biology of Reproduction* **58** 849–856

Marsh JM (1975) The role of cyclic AMP in gonadal function. In *Advances in Cyclic Nucleotide Research* Vol. 6 pp 137–200 Eds P Greengard and GA Robison. Raven Press, New York

McArdle CA and Holtorf A-P (1989) Oxytocin and progesterone release from bovine corpus luteal cells in culture: effects of insulin-like growth factor I, insulin, and prostaglandins *Endocrinology* **124** 1278–1286

Miller WL (1996) Mitochondrial specificity of the early steps in steroidogenesis *Journal of Steroid Biochemistry and Molecular Biology* **55** 607–616

Murdoch WJ (1996) Microtubular dynamics in granulosa cells of periovulatory follicles and granulosa-derived (large) lutein cells of sheep: relationships to the steroidogenic folliculo–luteal shift and functional luteolysis *Biology of Reproduction* **54** 1135–1140

Nancarrow CD (1994) Embryonic mortality in the ewe and doe. In *Embryonic Mortality in Domestic Species,* pp 79–97 Eds MT Zavy and RD Geisert. CRC Press, Boca Raton, FL

Niswender GD and Nett TM (1994) Corpus luteum and its control in infraprimate species. In *The Physiology of Reproduction* Vol. 1 pp 781–816 Eds E Knobil and JD Neill. Raven Press, Ltd, New York

Niswender GD, Reimers TJ, Diekman MA and Nett TM (1976) Blood flow: a mediator of ovarian function *Biology of Reproduction* **14** 64–81

Niswender GD, Schwall RH, Fitz TA, Farin CE and Sawyer HR (1985) Regulation of luteal function in domestic ruminants: new concepts *Recent Progress in Hormone Research* **41** 101–151

Papadopoulos V, Amri H, Boujrad N, Cascio C, Culty M, Garnier M, Hardwick M, Li H, Vidic B, Brown AS, Reversa JL, Bernassau JM and Drieu K (1997a) Peripheral benzodiazepine receptor in cholesterol transport and steroidogenesis *Steroids* **62** 21–28

Papadopoulos V, Amri H, Li H, Boujrad N, Vidic B and Garnier M (1997b) Targeted disruption of the peripheral-type benzodiazepine receptor gene inhibits steroidogenesis in the R2C Leydig tumor cell line *Journal of Biological Chemistry* **272** 32 129–32 135

Perks CM, Denning-Kendall PA, Gilmour RS and Wathes DC (1995) Localization of messenger ribonucleic acid for insulin-like growth factor I (IGF-I), IGF-II and the type 1 IGF receptor in the ovine ovary throughout the oestrous cycle *Endocrinology* **136** 5266–5273

Pescador N, Soumano K, Stocco DM, Price CA and Murphy BD (1996) Steroidogenic acute regulatory protein in bovine corpora lutea *Biology of Reproduction* **55** 485–491

Peters KE, Bergfeld EG, Cupp AS, Kojima FN, Mariscal V, Sanchez T, Wehrman ME, Grotjan HE, Hamernick DL, Kittok RJ and Kinder JE (1995) Luteinizing hormone has a role in development of fully functional corpora lutea (CL) but is not required to maintain CL function in heifers *Biology of Reproduction* **51** 1248–1254

Rodgers RJ, O'Shea JD and Findlay JK (1983) Progesterone production *in vitro* by small and large ovine luteal cells *Journal of Reproduction and Fertility* **69** 113–124

Rodgers RJ, Waterman MR and Simpson ER (1986) Cytochromes P-450^{scc}, $P450^{scc17\alpha}$, adrenodoxin, and reduced nicotinamide adenine dinucleotide phosphate-cytochrome P-450 reductase in bovine follicles and corpora lutea *Endocrinology* **118** 1366–1374

Rodgers RJ, Waterman MR and Simpson ER (1987) Levels of messenger ribonucleic acid encoding cholesterol side-chain cleavage cytochrome P-450, 17α-hydroxylase cytochrome P-450, adrenodoxin, and low density lipoprotein receptor in bovine follicles and corpora lutea throughout the ovarian cycle *Molecular Endocrinology* **1** 274–279

Rodgers RJ, Vella CA, Young FM, Tian XC and Fortune JE (1995) Concentrations of cytochrome P450 cholesterol side-chain cleavage enzyme and 3β-hydroxysteroid dehydrogenase during prostaglandin $F_{2\alpha}$-induced luteal regression in cattle *Reproduction Fertility Development* **7** 1213–1216

Spicer LJ, Ireland JJ and Roche JF (1981) Changes in serum LH, progesterone, and specific binding of ^{125}I-hCG to luteal cells during regression and development of bovine corpora lutea *Biology of Reproduction* **25** 832–841

Sugawara T, Kiriakidou M, McAllister JM, Kallen CB and Strauss JF III (1997) Multiple steroidogenic factor 1 binding elements in the human steroidogenic acute regulatory protein gene 5′-flanking region are required for maximal promoter activity and cyclic AMP responsiveness *Biochemistry* **36** 7249–7255

Stocco DM and Clark BJ (1996) Regulation of the acute production of steroids in steroidogenic cells *Endocrine Reviews* **17** 221–244

Tandeski TR, Juengel JL, Nett TM and Niswender GD (1996) Regulation of mRNA encoding low density lipoprotein receptor and high density lipoprotein-binding protein in ovine corpora lutea *Reproduction Fertility and Development* **8** 1107–1114

Tian XC, Berndtson AK and Fortune JE (1994) Changes in levels of messenger ribonucleic acid for cytochrome P450 side-chain cleavage and 3β-hydroxysteroid dehydrogenase during prostaglandin $F_{2\alpha}$-induced luteolysis in cattle *Biology of Reproduction* **50** 349–356

Voss AK and Fortune JE (1993a) Levels of messenger ribonucleic acid for cholesterol side chain cleavage cytochrome P-450 and 3β hydroxysteroid dehydrogenase in bovine preovulatory follicles decreases after the luteinizing hormone surge *Endocrinology* **132** 888–894

Voss AK and Fortune JE (1993b) Levels of messenger ribonucleic acid for cytochrome P450 17 alpha-hydroxylase and P450 aromatase in preovulatory bovine follicles decrease after the luteinizing hormone surge *Endocrinology* **132** 2239–2245

Wiltbank MC, Diskin MG and Niswender GD (1991) Differential actions of second messenger systems in the corpus luteum *Journal of Reproduction and Fertility Supplement* **43** 65–75

Wiltbank MC, Belfiore CJ and Niswender GD (1993) Steroidogenic enzyme activity after acute activation of protein kinase (PK) A and PKC in ovine small and large luteal cells *Molecular and Cellular Endocrinology* **97** 1–7

Wiltbank MC, Shiao TF, Bergfelt DR and Ginther OJ (1995) Prostaglandin $F_{2\alpha}$ receptors in the early bovine corpus luteum *Biology of Reproduction* **52** 74–78

Zavy MT (1994) Embryonic mortality in cattle. In *Embryonic Mortality in Domestic Species* pp 99–140 Eds MT Zavy and RD Geisert. CRC Press, Boca Raton, FL

Journal of Reproduction and Fertility Supplement **54**, 207–216

Luteal peptides and their genes as important markers of ovarian differentiation

R. Ivell, R. Bathgate and N. Walther

Institute for Hormone and Fertility Research, University of Hamburg, Grandweg 64, 22529 Hamburg, Germany

Secreted peptide hormones and components of the steroidogenic machinery are molecules that are expressed usually in high amounts and in a time- and cell-specific fashion within the cells that give rise to the bovine corpus luteum. They thus serve as useful markers for the events occurring within the nuclei of these cells that result in differentiation and the expression of the specific luteal phenotype. We have studied the bovine genes of three such luteal products: oxytocin, the new relaxin-like factor (RLF), and the steroidogenic acute regulatory protein (StAR). The oxytocin gene is expressed in the granulosal cells of the preovulatory follicle and in the large luteal cells of the immediately resulting early corpus luteum. The RLF gene is a major thecal cell product in antral and atretic follicles. It is also transcribed in luteal cells, but only in the mid- to late ovarian cycle and in pregnancy, following a temporal pattern of expression very similar to that of relaxin in pigs. The StAR gene appears to be upregulated only in the mid- to late ovarian cycle, several days after the increase in steroidogenic enzymes associated with luteinization and progesterone production. All three genes make use of the transcription factor SF-1 (Ad4BP) and, although they all respond to LH activation of adenylate cyclase, none utilize CRE-linked systems. Specific transcriptional activation must involve other factors to encode the information for the widely diverse temporal and cellular patterns of gene expression for these three genes.

Introduction

The bovine derivatives of the mesonephric mesenchymal cells of the embryonic female gonad offer an ideal model system in which to investigate differentiation processes in a steroidogenic tissue. In a synchronous fashion through the course of embryonic and postnatal development, and then later in each ovarian cycle and in pregnancy, these cells exhibit a marked switching on and off of genes, which are important for the changing phenotype of the ovary. We have been especially interested in the major switching systems involved in luteinization, that is the differentiation of cells of the membrana granulosa and theca interna of the preovulatory follicle into a functional corpus luteum, and in luteal maintenance in the event of pregnancy. In order to access the molecular systems involved in such gene-switching events one makes use of genes that encode molecules central for the specifically changing phenotype of the ovary. In the case of the bovine model, suitable genes are those for components of the steroidogenic pathway, and those encoding secreted peptide hormones. The reason for their usefulness is that they show a high degree of temporal and cell-type specificity, are expressed synchronously in the granulosal, thecal or luteal cells, are expressed in high amounts (often up to 1% of all transcripts), and their expression can be studied in a regulated fashion at different levels (mRNA, protein, function) in primary cell cultures. Although a large number of genes fall into this category, our studies have focussed on three genes representing different cell or temporal specificity. In this way, we have been able to dissect some of the molecular mechanisms occurring in the cell nuclei involved in the formation of a healthy corpus luteum from a preovulatory follicle. These are the genes for the secreted peptide hormones oxytocin and the new relaxin-like factor (RLF), and for the intracellular signalling peptide StAR (steroidogenic acute regulatory peptide).

Oxytocin

The production of oxytocin by the ruminant corpus luteum has been described by a number of authors (for example Flint and Sheldrick, 1982; Wathes and Swann, 1982; Fields *et al.*, 1983; Ivell and Richter, 1984; Ivell *et al.*, 1985, 1990; Meidan *et al.*, 1992; Fortune and Voss, 1993). The important features of oxytocin biosynthesis are first that the nonapeptide hormone is made as part of a larger precursor polypeptide, together with neurophysin I. Thus post-translational processing, particularly within the Golgi and dense-core secretory vesicles, must form an important part of any regulatory pathway, as also the regulated release of those granules by specific secretagogues, such as prostaglandin (PG)$F_{2\alpha}$. Second, ruminants express much more oxytocin in the ovary than do non-ruminant species. Whereas in the latter (for example primates), the function of oxytocin is thought to be paracrine within the ovary, in ruminants oxytocin forms part of an endocrine positive feedback loop involving oxytocin receptors on the epithelium of the uterine endometrium; upon stimulation of these receptors at oestrus large amounts of $PGF_{2\alpha}$ are released which in the absence of pregnancy lead to luteolysis and further secretion of luteal oxytocin. From the point of view of gene regulation, in the cow oxytocin mRNA is exclusively produced within the luteinizing granulosal cells of the preovulatory follicle and in the large luteal cells of the early corpus luteum (Fehr *et al.*, 1987). The oxytocin gene is upregulated only in highly oestrogenic follicles following the LH surge (Holtorf *et al.*, 1989), and appears to be downregulated 3–5 days after ovulation (Fig. 1; Ivell *et al.*, 1985; Fehr *et al.*, 1987). Only very small amounts of oxytocin mRNA can be detected in the corpus luteum of pregnancy, although an upregulation of the oxytocin gene appears to occur again in the luteal cells of the corpus luteum after the onset of labour at term of pregnancy (Ivell *et al.*, 1995). Thus, the bovine oxytocin gene can act as an indicator gene for regulatory events occurring at these times and in the granulosa–large luteal cell lineage.

There have been numerous studies on oxytocin secretion and gene expression in primary cell cultures. Earlier studies using luteal cells showed that, in general, any oxytocin detected was being secreted from premade stores, and that under most conditions the oxytocin gene was permanently downregulated in such cell cultures (McArdle and Holtorf, 1989; Furuya *et al.*, 1990). Granulosal cells derived from large preovulatory follicles with high oestrogen concentration in the follicular fluid have yielded most information on regulation of the oxytocin gene. These cells appear to luteinize spontaneously in culture, producing high concentrations of progesterone. However, whereas the presence of serum has little effect on steroidogenesis, it appears to be quite inhibitory for the production of oxytocin. For optimal induction of the oxytocin gene, less than 1% fetal calf serum is necessary, together with either insulin or insulin-like growth factor I (IGF-I), acting via independent receptors (Holtorf *et al.*, 1989). At submaximal concentrations of insulin or IGF-I (for example 10–100 ng ml^{-1}), further addition of either LH or FSH to stimulate adenylate cyclase leads to a further increase of gene expression. However, at higher insulin or IGF-I concentrations, activation of adenylate cyclase by these hormones has little further influence. In the absence of insulin or IGF-I, no amount of LH or FSH can induce oxytocin gene expression. Recently, we have been able to show that an essential part of the upregulation of the oxytocin gene at luteinization involves a local positive feedback loop comprising the luteal steroid progesterone (Lioutas *et al.*, 1997). Addition of the competing antigestagens RU486 or onapristone to insulin/forskolin-stimulated bovine granulosal cells completely inhibited the transcription of the oxytocin gene and consequently the production of the peptide hormone (Lioutas *et al.*, 1997). However, oestrogens have no effect on oxytocin production by cultured granulosa–luteal cells (Furuya *et al.*, 1990; McArdle *et al.*, 1991).

The switching on and off of genes is a complex resultant of a number of different processes, beginning with local changes in chromatin structure, the activation of the DNA by single or combined transcription factors and cofactors, and finally by inhibition or reversal of these processes. We have studied the bovine oxytocin gene extensively in the context of its activation in luteinizing

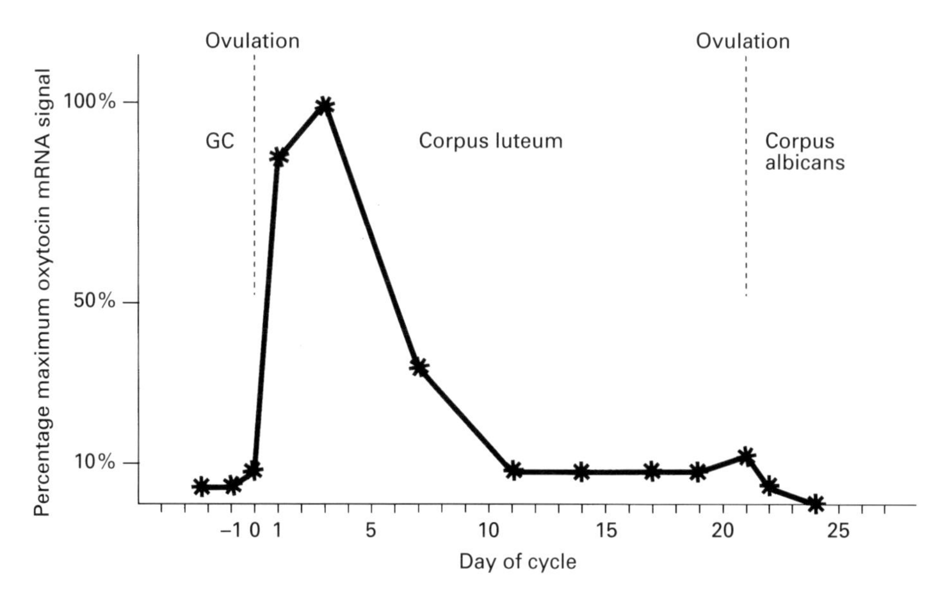

Fig. 1. Specific oxytocin-mRNA levels in the bovine corpus luteum of the oestrous cycle (GC, freshly prepared granulosal cells). 10 mg of total RNA from bovine corpora lutea collected at the times indicated were subjected to dot blot hybridization to a bovine oxytocin-gene-specific probe radiolabelled using ^{32}P. This was followed by autoradiography and densitometric estimation of the resulting specific signals, which were expressed as percentage of the maximum signal measured. Redrawn from Ivell *et al.* (1985).

bovine granulosal cells. An investigation of the *in vivo* pattern of methylation in the upstream promoter region of the gene indicated regions that are specifically hypomethylated when the gene is being actively transcribed (Kascheike *et al.*, 1997). As might be expected, there is hypomethylation in the proximal region of the transcription start site up to about –500 bp; however, this does not appear to be absolutely tissue specific. The second hypomethylated region is further upstream in the distal promoter at –1700 to –1900 bp (Fig. 2). This hypomethylation appears to be specific to luteinizing granulosal cells actively transcribing the oxytocin gene. Although several studies using human and rat oxytocin gene promoter–reporter constructs in heterologous transfections have suggested that different members of the steroid receptor superfamily (for example ERα, TR, RARα) can interact with the oxytocin gene promoter under these artificial conditions, we have been able to show that for active transcription of the bovine oxytocin gene in luteinizing granulosal cells, only the orphan receptor SF-1 binds within this proximal promoter region (Wehrenberg *et al.*, 1994a, b). SF-1 appears to be replaced by another related factor, COUP-TF, when the gene is being downregulated in the mid-cycle corpus luteum (Wehrenberg *et al.*, 1992, 1994a). Using protein factors extracted from bovine cells and tissues wherein the oxytocin gene is endogenously up- or downregulated, we have further been able to show that in the distal region of the promoter, there is first binding of luteal-specific nuclear factors to two artiodactyl-specific repeat regions (Fig. 2; Walther *et al.*, 1991). Second, our results indicate that another novel transcription factor, which is similar though not identical to SF-1, also binds in the distal promoter (about –1800 bp; Kascheike *et al.*, 1997), a region which, as noted above, appears to be specifically hypomethylated in tissues in which the gene is actively transcribed. Although the oxytocin gene is expressed only in follicles containing high oestrogen, there is no evidence to date for an interaction of a nuclear oestrogen receptor with the oxytocin gene promoter, nor is there any evidence for an interaction with the progesterone receptor. This is surprising in view of the marked inhibition of oxytocin gene transcription in luteinizing granulosal cells by antigestagens (Lioutas *et al.*, 1997).

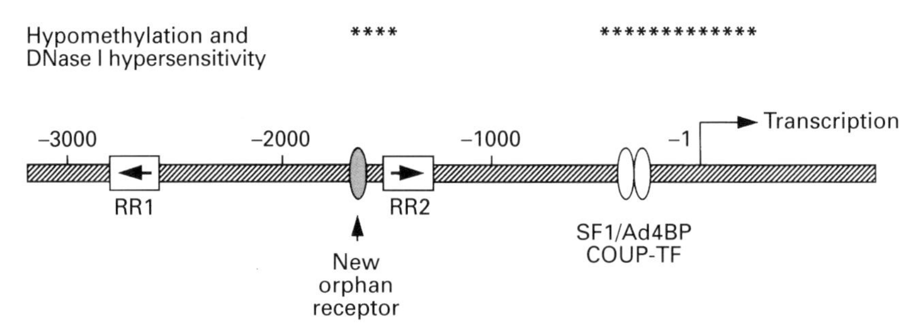

Fig. 2. Schematic representation of the bovine oxytocin gene promoter indicating the locations of luteal-specific nuclear factor binding and sites of specific genomic hypomethylation and *in vivo* DNaseI hypersensitivity (***), both implying regions of open chromatin. RR1 and RR2 (artiodactyl-specific repeat sequences; Walther *et al.*, 1991). For further details, see text and Kascheike *et al.* (1997).

Together, these data imply that genetic information within the first 2000 upstream nucleotides of the transcription start site is sufficient to regulate oxytocin gene expression. However, attempts to express the bovine oxytocin gene transgenically in mice using comparable promoter–reporter constructs have led to curious results. When a short promoter including the region –1 to –500 is used, there is no ovarian expression, nor expression in the hypothalamus, where high levels of expression would be expected. However, there is substantial expression of the transgene specifically in the Sertoli cells of the mouse testis (Ang *et al.*, 1991, 1994). Longer promoter regions failed to yield significant expression. However, when a part of the genetically linked bovine vasopressin gene is included in a mini-locus construct, specific hypothalamic expression is detected, although there is still no expression in the ovary of these transgenic mice (Ang *et al.*, 1993). When the bovine vasopressin gene promoter is used to drive the transgene, there is expression in the ovary (Ho *et al.*, 1995), although this gene is not expressed under normal conditions in the ovary of the cow. These unexpected results would imply that probably there are interactions between nuclear factors and the promoters of both the vasopressin and oxytocin genes, which lie very close together (about 8 kb apart) within the bovine genome, before one of these genes is expressed in specific cell types.

Relaxin-like Factor

It is well known that there is a relaxin-type of physiology associated with parturition in cows (Anderson *et al.*, 1995). Furthermore, application of porcine relaxin can induce cervical dilatation in cows, just as in species such as the pig and the rat (Anderson *et al.*, 1995). However, all attempts to characterize a bovine relaxin at the molecular level have failed (Hartung *et al.*, 1995a), and in the sheep it has been shown that there is a large deletion in the relaxin genomic locus (Roche *et al.*, 1993). A novel member of the relaxin–insulin–IGF family of genes was identified as being expressed exclusively in the testicular Leydig cells of pigs, humans, rats, mice, sheep and bulls (reviewed in Ivell, 1997). The cloned cDNAs from these species encoded a molecule with a very similar B–C–A peptide domain structure to that of relaxin, including a motif in the B-domain which shows high similarity to the proposed receptor-binding motif in relaxin. Indeed, a chemically synthesized peptide was shown to bind to mouse relaxin receptors (Büllesbach and Schwabe, 1995). For this reason the new gene product, originally referred to as Leydig-insulin-like peptide (Adham *et al.*, 1993), is now more generally referred to as relaxin-like factor (RLF). Having cloned the RLF cDNA from the bovine testis (Bathgate *et al.*, 1996), we were able to show that in cows, unlike in all non-ruminant species, the RLF gene is expressed at a very high level also in the ovary. *In situ* hybridization analysis showed that it was expressed exclusively in the theca interna of antral follicles, but also of atretic follicles, as well as in the corpus luteum (Bathgate *et al.*, 1996). A more

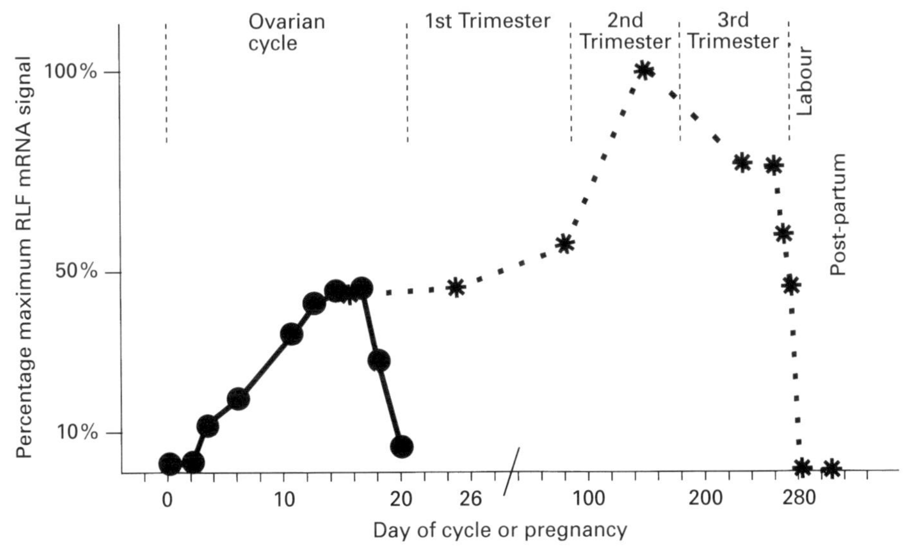

Fig. 3. RLF-mRNA profile in the bovine corpus luteum through the oestrous cycle (solid line) and pregnancy (dashed line). Briefly northern blot hybridizations were performed with 10 mg total RNA extracted from bovine corpora lutea at the times indicated; these were hybridized to a bovine RLF-specific ^{32}P-labelled cDNA probe, and the resulting signals measured using a phosphorimager. Data derived from R. Bathgate, M. Fields and R. Ivell (unpublished).

detailed analysis of expression in the corpus luteum was obtained using northern blots of luteal RNA from various times of the cycle and pregnancy (summarized in Fig. 3; Bathgate, Fields and Ivell, unpublished). Although it is expressed highly in the theca interna of the preovulatory follicle, it is evidently downregulated upon ovulation. Within the corpus luteum there is only low expression in the early cycle; however, the RLF gene is evidently upregulated in the mid–late luteal phase in what appear to be the large luteal cells (Fig. 4). In the event of pregnancy the maximum values of the cycle are maintained and even increased reaching a maximum at the end of the second trimester. Amounts of RLF mRNA fall markedly at term, and transcripts are undetectable during labour. This pattern of gene expression is very similar to that reported for the related hormone relaxin in the pig ovary, and supports the notion that in the cow, RLF might be substituting for the missing relaxin. To date it has not been possible to purify the endogenous RLF peptide so that physiological experiments to confirm this idea cannot yet be performed. We have succeeded in raising antibodies against peptide fragments as well as bacterially produced human and mouse RLF. These antibodies appear to work well in an immunohistochemical context (Ivell *et al.*, 1997; Balvers *et al.*, 1998); however, except for one of these (Fig. 4), they fail to work in the cow. Neither do they appear to function in the context of western blots or immunoassays. It is clearly of great importance to produce better antibodies so that a study of this very interesting molecule can be pursued at protein and functional levels.

Given the high endogenous RLF gene expression in the thecal cells of bovine antral follicles, we have begun to study the regulation of this expression using primary thecal cell cultures. Preliminary experiments show that in all culture conditions the RLF gene is initially downregulated with a kinetic which varies somewhat under different experimental conditions. Of great interest, however, is the observation that under continuous IGF-I or insulin stimulation, RLF mRNA increases again after about 6 days in culture, maintaining a high level for several days (Bathgate, Moniac and Ivell, unpublished). Addition of fetal calf serum appears to irrevocably downregulate the RLF gene. Thus, the reappearance of RLF mRNA upon insulin or IGF-I stimulation appears to follow a time course in cell culture, which is parallel to the upregulation of RLF in the corpus luteum *in vivo*.

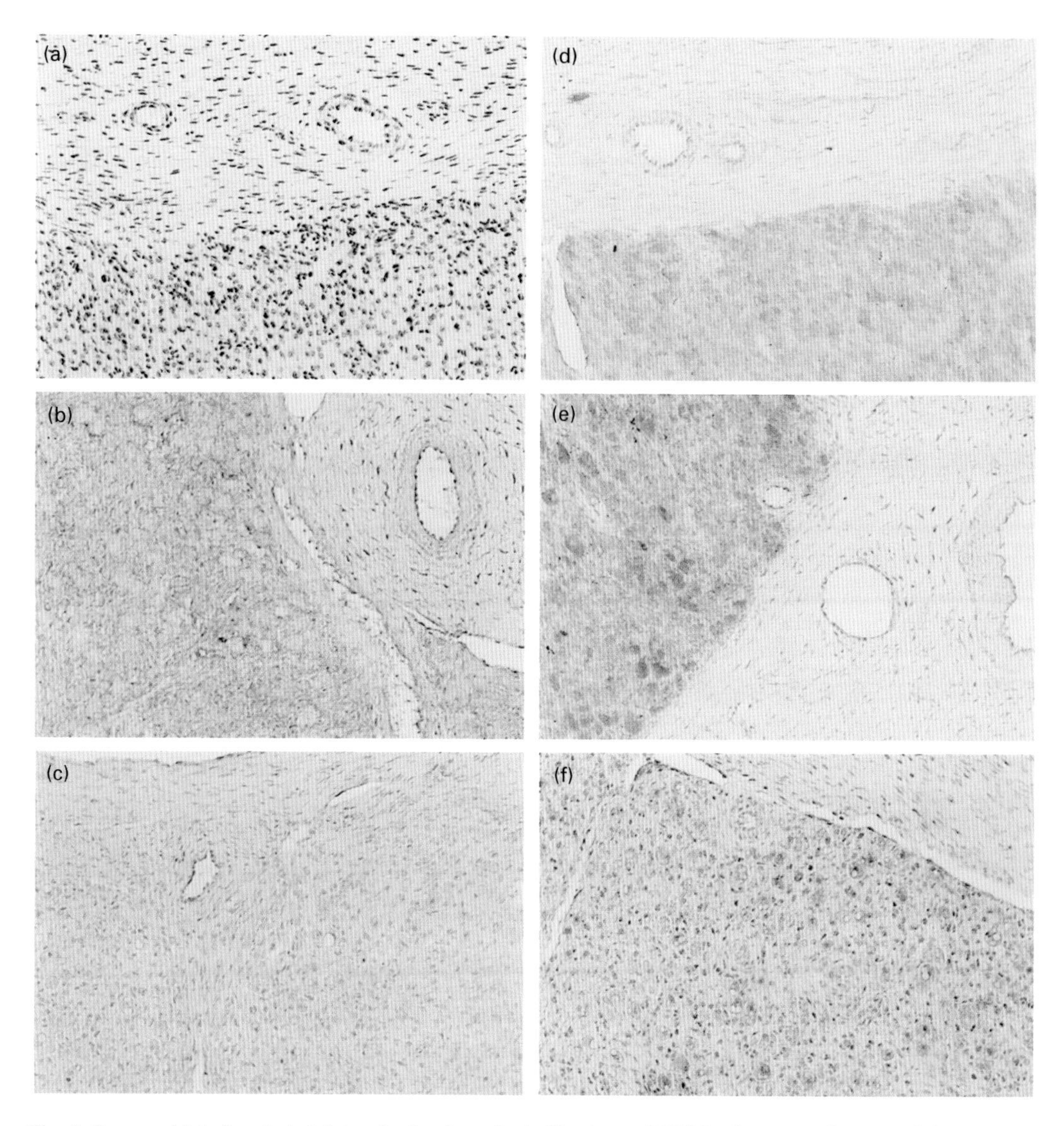

Fig. 4. Immunohistochemical staining for bovine relaxin-like factor (RLF) in the corpus luteum of the oestrous cycle. (a)–(c) Controls replacing the primary antibodies by preimmune serum from the same animals. (d)–(f), as (a)–(c), anti-RLF antiserum M1 (Ivell *et al.*, 1997). The antibody M1 was raised against a peptide epitope of the RLF B-domain which is identical in human and bovine RLF sequences. (d) Early–mid-cycle; (e) mid–late cycle; (f) late cycle (day 21). Conditions for immunohistochemistry were as in Ivell *et al.* (1997)

The bovine RLF gene has not yet been studied. However, there is some information already on the gene for mouse RLF (Zimmerman *et al.*, 1997; Koskimies *et al.*, 1997). In this species, in which RLF is highly expressed in the testis, and expressed at a very low level only in the ovary (Balvers et al., 1998), it appears that only a few hundred nucleotides of the promoter region upstream of the transcription start site are sufficient to confer a degree of cell specificity, as well as high activity in promoter–reporter transfection experiments (Koskimies *et al.*, 1997). This region includes two recognition motifs for the transcription factor SF-1. An interesting feature of the mouse RLF gene,

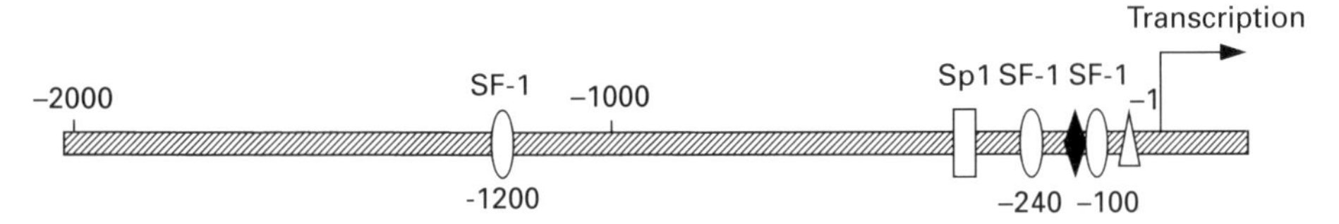

Fig. 5. Schematic representation of the bovine StAR gene promoter indicating the locations of the putative binding motifs for SF-1 (Ad4BP; ellipses), DAX-1 (triangle), C/EBPβ (filled rhombus), Sp1 (rectangle). See Rust *et al.* (1998) for further details.

which appears to be true also for the human RLF gene (Safford et al., 1997), is that it appears to be wholly included within an intron of a quite different gene, namely that for the signal transduction molecule JAK3 (Koskimies *et al.*, 1997).

Steroidogenic Acute Regulatory Protein

The steroidogenic acute regulatory (StAR) peptide is a small intracellular protein, which is considered to be responsible for the acute regulation of steroidogenesis in the testis, ovary and adrenal gland (reviewed in Stocco, 1998). It functions by interacting with the mitochondrion and, presumably, via the transient establishment of contact points between the outer and the inner mitochondrial membrane, allows the transfer of cholesterol to the steroidogenic enzyme complex on the inner mitochondrial surface. As part of a differential cloning project to identify genes expressed in the bovine corpus luteum of the late ovarian cycle and pregnancy, we were able to clone the full-length bovine StAR transcript for the longer of the two mRNA species evident in northern hybridizations (Hartung *et al.*, 1995b; Pescador *et al.*, 1996). The difference in size of the two commonly seen transcripts lies in a differential polyadenylation (Hartung *et al.*, 1995b). The steady state levels of StAR mRNA in the bovine corpus luteum indicate that these basal levels begin to increase only in the mid- to late cycle, as has recently also been shown in pigs (LaVoie *et al.*, 1997), and remain high throughout pregnancy. This result is somewhat confusing, since the initial upregulation occurs about 5 days after the upregulation of P450scc mRNA and the increase in progesterone production. Thus, if the StAR peptide is involved in acute regulation of steroidogenesis also in the bovine ovary, this is evidently not reflected by the basal levels of the specific transcript. However, from preliminary studies assessing StAR gene expression in bovine adrenal cells (Nicol *et al.*, 1997), as well as in gonadotrophin-stimulated rat ovaries (Ronen-Fuhrmann *et al.*, 1998), it would appear that there is a rapid transient increase in StAR mRNA upon activation of adenylate cyclase by ACTH or LH, respectively.

We have recently succeeded in cloning the complete genomic sequence of the bovine StAR gene, including approximately 3 kb of the 5′ promoter region upstream of the transcription start site (Rust *et al.*, 1998). Computer analysis identified three motifs in this region corresponding to the SF-1 (Ad4BP) transcription factor binding site (Fig. 5). Use of heterologous transfection systems, whereby a bovine SF-1 expression vector was co-transfected into the cells together with appropriate bovine StAR promoter-deletion reporter constructs, indicated that the two proximal SF-1 motifs at –100 and –240 interacted with SF-1 to induce an upregulation of the gene. Furthermore, additional cotransfection of a constitutive protein kinase A catalytic subunit indicated that the effect of cAMP was probably mediated by SF-1. However, upon performing electrophoretic mobility shift assays to detect specific DNA–nuclear protein binding, it was found that bovine SF-1 is incapable of binding to the most proximal (–100) motif, and only very weakly to the second (–240) site. Instead SF-1 showed excellent binding to the most distal (–1000) motif. This discrepancy between the two types of experiment suggests that *in vivo* the SF-1 mediated regulation of the bovine StAR gene is probably more complex and may involve a number of additional cofactors or interactions with other as yet unidentified transcription factors (Rust *et al.*, 1998). Computer analysis has identified putative motifs also for the single-strand DNA-binding DAX-1, for SpI, and for C/EBPβ (Fig. 5), although whether these can be utilized either *in vitro* or *in vivo* has yet to be demonstrated.

The role of SF-1 (Ad4BP) in Follicular/Luteal Cell Differentiation

At present there are three different genes that are expressed in different cell types and at different times within the differentiation pathway of the bovine corpus luteum. The oxytocin gene is expressed in granulosa–lutein cells of the early cycle; the RLF gene is expressed in theca–lutein cells of the follicle and in the corpus luteum of the late cycle and pregnancy; the StAR gene is probably expressed at a low basal level in unstimulated granulosal and thecal cells of the follicle, but at increased basal levels only in the corpus luteum of the mid- to late cycle and in pregnancy. However, whereas the upregulation *in vitro* of the oxytocin gene requires stimulation for about 16 h of adenylate cyclase activation before an increase is evident, the StAR gene – at least in bovine adrenal cells – increases already after only 1 h. The RLF gene in contrast appears to be upregulated in the corpus luteum first after 4–6 days of stimulation.

All of these genes include motifs for the transcription factor SF-1 (Ad4BP) in their 5′ promoter regions. Equally, although they exhibit a pattern of expression *in vivo* which would correlate with adenylate cyclase activation (either by LH or by $PGF_{2\alpha}$), none includes the classic response element for such activation which could interact with the CREB or related transcription factors. These genes thus differ from the bovine inhibin-α gene which is upregulated in granulosa–lutein cells via the mediation of a typical CRE motif (Ungefroren *et al.*, 1994). The bovine oxytocin, RLF and StAR genes are thus also like the genes for some of the steroidogenic enzymes (for example CYP11A and CYP21B; Lauber *et al.*, 1993) which also include an important SF-1 motif, are regulated by adenylate cyclase activation, but in a CRE-independent fashion (Lauber et al. 1993). The conclusion that must be drawn from all these studies is that while SF-1 is an essential permissive factor targetting gene expression to the cells of the mesonephric mesenchymal lineage, and possibly also directly mediating some of the activating information from the cAMP-linked second messenger system, it is only one component in a complicated transcriptional regulatory complex. There must be further components that discriminate between granulosal and thecal cell-specific expression, and which determine the timing of luteal gene expression. Such dynamic features probably involve not simply activating molecules but probably are the resultants of complex interactions between activating and inactivating cofactors. It has recently been shown that SF-1 mediated gene transcription can be activated by the interaction of SF-1 with the transcription factor WT-1 (Nachtigal *et al.*, 1998). In contrast, SF-1 may interact with the factor DAX-1 to recruit the inactivating cofactor complex which includes the N-CoR protein (Crawford *et al.*, 1998). We have shown for the bovine oxytocin gene that binding of SF-1 can be inhibited by competitive binding of another related factor, COUP-TF, to the same *cis* element (Wehrenberg *et al.*, 1994a,b). It has also been suggested for the StAR gene in the rat testis (Zazopoulos *et al.*, 1997) that SF-1 interaction can be sterically blocked by a DNA-dependent binding of DAX-1 to a single-strand region of a hairpin loop in the immediate neighbourhood of one of the proximal SF-1 *cis* elements. It has also been shown that there may be a synergistic activating interaction between SF-1 and other DNA-dependent transcription factors, such as C/EBPβ (Nalbant *et al.*, 1998). Thus there is probably a wide variety of different molecular mechanisms which acting through the mediation of SF-1 can lead to the type of coded response needed to explain the highly specific expression patterns of luteal genes.

Conclusion

We are probably only just beginning to unravel the complexity of the molecular codes responsible for specific gene expression in the cells of the granulosa–theca–luteal lineages. In this context, the study of hormone genes, and in particular those involved in the production of luteal peptides as well as steroids, together with the power of the bovine ovarian system, should help us to reach a better understanding of ovarian function and its pathology.

Appendix: List of Abbreviations

Transcription factors are indicated in the text by their common acronyms: C/EBPβ, CCAAT/ enhancer-binding protein β; COUP-TF, chicken ovalbumin upstream promoter–transcription factor; CREB, cAMP-responsive element (CRE) binding protein; CREM, CRE-modulator protein; DAX-1, dosage-sensitive sex-reversal, X-linked gene; ERα and ERβ, oestrogen receptor α and β; PR, progesterone receptor; RAR, retinoic acid receptor; SpI, Simian virus 40 specific transcription factor 1; TR, thyroid hormone receptor.

The authors are very grateful to all the members of the Hamburg team, who over the years have helped us to unravel some of the complexity of luteal gene expression. They would particularly like to thank Ms Nicole Moniac for allowing us to refer to some of her unpublished data. They also thank the Deutsche Forschungsgemeinschaft for continued financial support of this project (Iv7/1 and Iv7/5).

References

Adham IM, Burkhardt E, Benahmed M and Engel W (1993) Cloning of a cDNA coding for a novel insulin-like peptide of the testicular Leydig cells *Journal of Biological Chemistry* **268** 26668–26672

Anderson LL, Gazal OS, Dlamini B and Li Y (1995) The role of relaxin in ruminants. In *Progress in Relaxin Research* pp 428–438 Eds AH McLennan, G Tregear and G Bryant-Greenwood. World Scientific Publishing, Singapore

Ang HL, Ungefroren H, De Bree F, Foo NC, Carter D, Burbach JPH, Ivell R and Murphy D (1991) Testicular oxytocin gene expression in seminiferous tubules of cattle and transgenic mice *Endocrinology* **128** 2110–2117

Ang HL, Carter DA and Murphy D (1993) Neuron-specific expression and physiological regulation of bovine vasopressin transgenes in mice *EMBO Journal* **12** 2397–2409

Ang HL, Ivell R, Walther N, Nicholson H, Ungefroren H, Millar M, Carter D and Murphy D (1994) Over-expression of oxytocin in the testes of a transgenic mouse model *Journal of Endocrinology* **140** 53–62

Balvers M, Spiess AN, Domagalski R, Hunt N, Kilic E, Mukhopadhyay A, Hanks E, Charlton HM and Ivell R (1998) Relaxin-like factor expression as a marker of differentiation in the mouse testis and ovary *Endocrinology* **139** 2960–2970

Bathgate R, Balvers M, Hunt N and Ivell R (1996) Relaxin-like factor gene is highly expressed in the bovine ovary of the cycle and pregnancy: sequence and messenger ribonucleic acid analysis *Biology of Reproduction* **55** 1452–1457

Büllesbach E and Schwabe C (1995) A novel Leydig cell-derived protein is a relaxin-like factor *Journal of Biological Chemistry* **270** 16011–16015

Crawford PA, Dorn C, Sadovsky Y and Milbrandt J (1998) Nuclear receptor Dax-1 recruits nuclear corepressor N-CoR to Steroidogenic Factor 1 *Molecular and Cellular Biology* **18** 2949–2956

Fehr S, Ivell R, Koll R, Schams D, Fields M and Richter D (1987) Expression of the oxytocin gene in the large cells of the bovine corpus luteum *FEBS Letters* **210** 45–50

Fields PA, Eldridge RK, Fuchs AR, Roberts RF and Fields MJ (1983) Human placental and bovine corpora luteal oxytocin *Endocrinology* **112** 1544–1546

Flint APF and Sheldrick EL (1982) Ovarian secretion of oxytocin is stimulated by prostaglandin *Nature* **297** 587–588

Fortune JE and Voss AK (1993) Oxytocin gene expression and action in bovine preovulatory follicles *Regulatory Peptides* **45** 257–261

Furuya K, McArdle CA and Ivell R (1990) The regulation of oxytocin gene expression in early bovine luteal cells *Molecular and Cellular Endocrinology* **70** 81–88

Hartung S, Kondo S, Abend N, Hunt N, Rust W, Balvers M, Bryant-Greenwood G and Ivell R (1995a) The search for ruminant relaxin. In *Progress in Relaxin Research* pp 439–456 Eds AH McLennan, G Tregear and G Bryant-Greenwood. World Scientific Publishing, Singapore

Hartung S, Rust W, Balvers M and Ivell R (1995b) Molecular cloning and *in vivo* expression of the bovine steroidogenic acute regulatory protein *Biochemical and Biophysical Research Communications* **215** 646–653

Ho MY, Carter DA, Ang HL and Murphy D (1995) Bovine oxytocin transgenes in mice: hypothalamic expression, physiological regulation and interactions with the vasopressin gene *Journal of Biological Chemistry* **270** 27199–27205

Holtorf AP, Furuya K, Ivell R and McArdle CA (1989) Oxytocin production and oxytocin messenger ribonucleic acid levels in bovine granulosa cells are regulated by insulin and insulin-like growth factor-I: dependence on developmental status of the ovarian follicle *Endocrinology* **125** 2612–2620

Ivell R (1997) Biology of the relaxin-like factor (RLF) *Reviews of Reproduction* **2** 133–138

Ivell R and Richter D (1984) The gene for the hypothalamic peptide hormone oxytocin is highly expressed in the bovine corpus luteum: biosynthesis, structure and sequence analysis *EMBO Journal* **3** 2351–2354

Ivell R, Brackett KH, Fields MJ and Richter D (1985) Ovulation triggers oxytocin gene expression in the bovine ovary *FEBS Letters* **190** 263–267

Ivell R, Hunt N, Abend N, Brackmann B, Nollmeyer D, Lamsa JC and McCracken JA (1990) Structure and ovarian expression of the oxytocin gene in sheep *Reproduction Fertility and Development* **2** 703–711

Ivell R, Rust W, Einspanier A, Hartung S, Fields M and Fuchs AR (1995) Oxytocin and oxytocin receptor gene expression in the reproductive tract of the pregnant cow: rescue of luteal oxytocin production at term *Biology of Reproduction* **53** 553–560

Ivell R, Balvers M, Domagalski R, Ungefroren H, Hunt N and Schulze W (1997) Relaxin-like factor: a highly specific and constitutive new marker for Leydig cells in the human testis *Molecular Human Reproduction* **3** 459–466

Kascheike B, Ivell R and Walther N (1997) Alterations in the chromatin structure of the distal promoter region of the bovine oxytocin gene correlate with ovarian expression *DNA and Cell Biology* **16** 1237–1248

Koskimies P, Spiess AN, Lahti P, Huhtaniemi I and Ivell R (1997) The mouse relaxin-like factor gene and its promoter are located within the 3′ region of the JAK3 genomic sequence *FEBS Letters* **419** 186–190

Lauber ME, Kagawa N, Waterman MR and Simpson ER (1993) cAMP-dependent and tissue-specific expression of genes encoding steroidogenic enzymes in bovine luteal and granulosa cells in primary culture *Molecular and Cellular Endocrinology* **93** 227–233

LaVoie HA, Benoit AM, Garmey JC, Dailey RA, Wright DJ and Veldhuis JD (1997) Coordinate developmental expression of genes regulating sterol economy and cholesterol side-chain cleavage in the porcine ovary *Biology of Reproduction* **57** 402–407

Lioutas C, Einspanier A, Kascheike B, Walther N and Ivell R (1997) An autocrine progesterone positive feedback loop mediates oxytocin upregulation in bovine granulosa cells during luteinization *Endocrinology* **138** 5059–5062

McArdle CA and Holtorf AP (1989) Oxytocin and progesterone release from bovine corpus luteum cells in culture: effects of insulin-like growth factor I, insulin and prostaglandins *Endocrinology* **124** 1278–1286

McArdle CA, Kohl C, Rieger K, Gröner I and Wehrenberg U (1991) Effects of gonadotrophins, insulin and insulin-like growth factor I on ovarian oxytocin and progesterone production *Molecular and Cellular Endocrinology* **78** 211–220

Meidan R, Altstein M and Girsh E (1992) Biosynthesis and release of oxytocin by granulosa cells derived from preovulatory bovine follicles: effects of forskolin and insulin-like growth factor-I *Biology of Reproduction* **46** 715–720

Nachtigal MW, Hirokawa Y, Enyeart-VanHouten DL, Flanagan JN, Hammer GD and Ingraham HA (1998) Wilms' tumor 1 and Dax-1 modulate the orphan nuclear receptor SF-1 in sex-specific gene expression *Cell* **93** 445–454

Nalbant D, Williams SC, Stocco DM and Khan SA (1998) Luteinizing hormone-dependent gene regulation in Leydig cells may be mediated by CCAAT/enhancer-binding protein-β *Endocrinology* **139** 272–279

Nicol MR, Morley SD, Stirling D, Ivell R, Walker SW and Mason JI (1997) The expression of steroidogenic acute regulatory protein (StAR) mRNA in bovine adrenocortical cells *Journal of Endocrinology* **152** (Supplement) P7

Pescador N, Soumano K, Stocco DM, Price CA and Murphy BD (1996) Steroidogenic acute regulatory protein in bovine corpora lutea *Biology of Reproduction* **55** 485–491

Roche PJ, Crawford RJ and Tregear GW (1993) A single copy relaxin-like gene sequence is present in sheep *Molecular and Cellular Endocrinology* **91** 21–28

Ronan-Fuhrmann T, Timberg R, King SR, Hales KH, Hales DB, Stocco DM and Orly J (1998) Spatio–temporal expression patterns of steroidogenic acute regulatory protein (StAR) during follicular development in the rat ovary *Endocrinology* **139** 303–315

Rust W, Stedronsky K, Tillmann G, Morley S, Walther N and Ivell R (1998) The role of SF-1/Ad4BP in the control of the bovine gene for the Steroidogenic Acute Regulatory (StAR) protein *Journal of Molecular Endocrinology* **21** 189–200

Safford MG, Levenstein M, Tsifrina E, Amin S, Hawkins AL, Griffin CA, Civin CI and Small D (1997) JAK3: expression and mapping to chromosome 19p12–13.1 *Experimental Hematology* **25** 374–386

Stocco D (1998) A review of the characteristics of the protein required for acute regulation of steroid hormone biosynthesis: the case for the steroidogenic acute regulatory (StAR) protein *Proceedings of the Society for Experimental Biology and Medicine* **217** 123–129

Ungefroren H, Wathes DC, Walther N and Ivell R (1994) Structure of the alpha-inhibin gene and its regulation in the ruminant gonad: inverse relationship to oxytocin gene expression *Biology of Reproduction* **50** 401–412

Walther N, Wehrenberg U, Brackmann B and Ivell R (1991) Mapping of the oxytocin gene control region: identification of binding sites for luteal nuclear proteins in the 5′ non-coding region of the gene *Journal of Neuroendocrinology* **3** 539–549

Wathes DC and Swann RW (1982) Is oxytocin an ovarian hormone? *Nature* **297** 225–227

Wehrenberg U, Ivell R and Walther N (1992) The COUP transcription factor (COUP-TF) is directly involved in the regulation of oxytocin gene expression in luteinizing bovine granulosa cells *Biophysical and Biochemical Research Communications* **189** 496–503

Wehrenberg U, Ivell R, Jansen M, Von Goedecke S and Walther N (1994a) Two orphan receptors binding to a common site are involved in the regulation of the oxytocin gene in the bovine ovary *Proceedings of the National Academy of Sciences USA* **91** 1440–1444

Wehrenberg U, Von Goeddecke S, Ivell R and Walther N (1994b) The orphan receptor SF-1 binds to the COUP-like element in the promoter of the actively transcribed oxytocin gene *Journal of Neuroendocrinology* **6** 1–4

Zazopoulos E, Lalli E, Stocco DM and Sassone-Corsi P (1997) DNA-binding and transcriptional repression by DAX-1 blocks steroidogenesis *Nature* **390** 311–315

Zimmermann S, Schöttler P and Adham I (1997) Mouse Leydig insulin-like (Ley-I-L) gene structure and expression during testis and ovary development *Molecular Reproduction and Development* **47** 30–38

Journal of Reproduction and Fertility Supplement **54**, 217–228

Intraovarian regulation of luteolysis

R. Meidan[1], R. A. Milvae[2], S. Weiss[3], N. Levy[1] and A. Friedman[3]

Department of Animal Sciences, Sections of [1]Reproduction and [3]Immunology, Faculty of Agriculture, Food and Environmental Quality Sciences, The Hebrew University of Jerusalem, Rehovot 76100, Israel; [2]Department of Animal Science, University of Connecticut, Storrs, CT 06269–4040, USA

The corpus luteum is a transient gland, which is only functional for 17–18 days in the cyclic cow or for up to 200 days in the pregnant cow. Regression of the corpus luteum is essential for normal cyclicity as it allows the development of a new ovulatory follicle, whereas prevention of luteolysis is necessary for the maintenance of pregnancy. Evidence acquired over the past three decades indicated that $PGF_{2\alpha}$ is the luteolytic hormone in ruminants. Nevertheless, the detailed mechanisms of $PGF_{2\alpha}$ action are just beginning to be clarified. A pivotal role for an endothelial cell product endothelin 1 (ET-1) has been documented in $PGF_{2\alpha}$-induced luteal regression. ET-1 inhibited progesterone production by luteal cells in a dose-dependent manner via selective ET-1 binding sites (ET_A). The inhibitory action of $PGF_{2\alpha}$ on progesterone secretion (*in vivo* and *in vitro)* was blocked by a selective ET_A receptor antagonist. This implied that ET-1 (through ET_A receptors present on steroidogenic cells) may have mediated the inhibitory effect of $PGF_{2\alpha}$. The involvement of ET-1 in luteal regression was also suggested by the observation that the highest concentrations of ET-1 coincide with uterine $PGF_{2\alpha}$ surges. Furthermore, $PGF_{2\alpha}$ administration upregulated ET-1 expression within the corpus luteum. Later stages of luteal regression, which involve programmed cell death (PCD), are presumably mediated by immune cells. ET-1 may also be involved in this process by promoting leukocyte migration and stimulating macrophages to release tumour necrosis factor α (TNFα). The TNFα receptor type 1 (p55) is present on luteal cells (endothelial and steroidogenic cells) and could initiate PCD and the structural demise of the corpus luteum.

Introduction

In cattle and other species, the corpus luteum plays a central role in the regulation of cyclicity and in the maintenance of pregnancy (Hansel and Blair, 1996). This gland undergoes dynamic changes throughout its life span; its formation, induced by the luteotrophic hormone LH, involves a complex process of cell differentiation and neovascularization (Jablonka-Shariff *et al.*, 1993; Fields and Fields, 1996). This highly vasculated endocrine organ functions to prepare the female reproductive system to receive and maintain a conceptus. In the absence of an embryonic signal, the corpus luteum will regress. Since the early 1970s substantial evidence has accumulated to indicate that the primary luteolysin in domestic ruminants is $PGF_{2\alpha}$ of uterine origin (McCracken *et al.*, 1972; Ellinwood *et al.*, 1979; Pate, 1994).

Administration of $PGF_{2\alpha}$ between day 5 and day 15 of the bovine oestrous cycle triggers a sequence of irreversible changes in the corpus luteum, similar to the spontaneously occurring events (McCracken *et al.*, 1972; Pate, 1994). Initially, there is reduction in progesterone release, and later an infiltration of macrophages and morphological changes consistent with apoptotic cell death (Juengel *et al.*, 1993; Hahnke *et al.*, 1994). Corpus luteum regression is necessary for normal cyclicity, as a functional corpus luteum suppresses the final stages of follicular development which leads to ovulation. Not only is the suppression of corpus luteum function (that is progesterone production) required, but the gland must also be physically eliminated, to keep the ovary at its proper size. Luteal regression may be regarded as consisting of two processes – functional and structural luteolysis – that differ in their temporal and mechanistic features. Functional luteolysis refers to the rapid decline in luteal progesterone, while structural luteolysis describes the events leading to the structural demise of the corpus luteum and is accomplished within a few days.

In contrast to the well documented consistent luteolytic effects of $PGF_{2\alpha}$ *in vivo*, studies examining its direct effects on luteal cells produced controversial data, especially when using isolated luteal cell populations in which it had paradoxically increased progesterone production (Davis *et al.*, 1989; Meidan *et al.*, 1991; Miyamoto *et al.*, 1993). This discrepancy had led us to postulate that a non-steroidogenic cell, such as the endothelial cell, may mediate the actions of $PGF_{2\alpha}$. The possible involvement of endothelial cells in physiology of the corpus luteum is supported by their abundance (> 50% of total cells of the corpus luteum; O'Shea *et al.*, 1989, and their associated contact with steroidogenic cells (Grazul-Bilska *et al.*, 1992). Endothelial cells have the unique ability to sense changes in blood flow, blood pressure and oxygen tension to which they respond by the appropriate upregulation of endothelin 1 (ET-1) expression. As a result of its strategic location, the endothelium layer can integrate a myriad of physical and biochemical signals within an organ; therefore, it is not surprising that recent evidence indicates that these cells are key players in many biological functions (Inagami *et al.*, 1995). At the time of luteal regression, macrophages invade the corpus luteum; these cells might participate in apoptotic events by secreting cytokines such as tumour necrosis factor α (TNFα). Athough the role of TNFα in the regression of the corpus luteum is presently unknown, it should be noted that peak amounts of TNFα within the corpus luteum correspond to the initiation of the apoptotic process (Shaw and Britt, 1995).

This review summarizes studies documenting the involvement of endothelial cells and macrophages in $PGF_{2\alpha}$-induced regression of the corpus luteum.

Role of Endothelial Cells in $PGF_{2\alpha}$-induced Anti-steroidogenic Action

The corpus luteum is a heterogeneous tissue and besides endothelial cells and steroidogenic large luteal (LLC) and small luteal (SLC) cells it also consists of fibroblasts, smooth muscle cells and immune cells (O'Shea *et al.*, 1989). Therefore, two experimental models were used to study the involvement of endothelial cells in the anti-steroidogenic actions of $PGF_{2\alpha}$: (1) luteal slices in which the integrity and communication between the various cells is preserved, and (2) isolated luteal steroidogenic cells co-cultured in the presence of endothelial cells (Girsh *et al.*, 1995).

The accumulation of progesterone secretion from 2- to 4-day-old corpus luteum slices is shown in Fig. 1a. $PGF_{2\alpha}$ (1 μg ml^{-1}) and LH (100 ng ml^{-1}) increased progesterone secretion to values that were 60% higher than control values; however, this increase was not statistically significant (Fig. 1a). Incubation with LH + $PGF_{2\alpha}$ did not alter progesterone secretion in these young corpora luteal slices, in relation to LH (Fig. 1a). In contrast, in mature corpora luteal slices (6- to 12-day-old), LH stimulated progesterone secretion by 1.6 times after incubation for 30 min (11.2 versus 6.7 ng mg^{-1} corpus luteum). The LH-stimulated progesterone curve rose steadily until 5 h of incubation, when it reached concentrations that were 2.8-fold higher than control values (Fig. 1b). In contrast to results with slices obtained from young corpus luteum, $PGF_{2\alpha}$ significantly reduced progesterone stimulation by LH in the mature corpus luteum ($P < 0.02$; Fig. 1b). The finding that the effect of $PGF_{2\alpha}$ on luteal tissue was age dependent is in agreement with other reports demonstrating that $PGF_{2\alpha}$ can induce luteolysis only after day 5 of the cycle (Pate, 1994). When a model with isolated luteal steroidogenic and endothelial cells was used, $PGF_{2\alpha}$ did not inhibit progesterone secretion when incubated with steroidogenic cells; the anti-steroidogenic effect of $PGF_{2\alpha}$ was apparent only when luteal cells were co-cultured together with endothelial cells (Girsh *et al.*, 1995). This was the first indication that the anti-steroidogenic effect of $PGF_{2\alpha}$ may be mediated by factors released from the resident endothelial cells in corpus luteum tissue. On the basis of these findings, we investigated the effects of an endothelial cell product ET-1 on $PGF_{2\alpha}$-induced luteal regression.

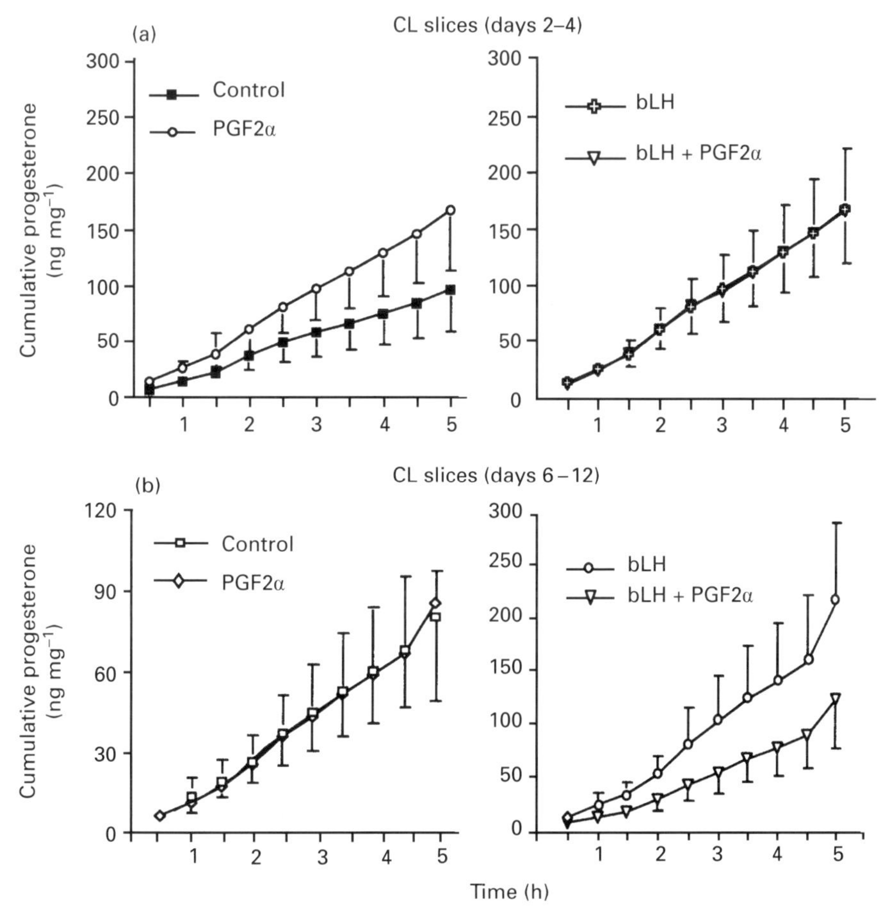

Fig. 1. Cumulative progesterone production (ng mg^{-1} tissue) by 2–4-day-old (a), and 6–12-day-old (b) corpora lutea slices. Slices were preincubated in DMEM/HEPES containing 5% FCS for 2 h at 37.5°C. Slices were then incubated for an additional 5 h in the same media (control) or with the addition of $PGF_{2\alpha}$ (1 µg ml^{-1}), LH (100 ng ml^{-1}) or LH + $PGF_{2\alpha}$. During the preincubation and incubation periods media were replaced every 30 min. Results are the means ± SEM of three corpora lutea.

Inhibition of Luteal Steroidogenesis by ET-1

Endothelin 1 is synthesized and secreted by endothelial cells. This peptide was originally isolated from porcine aortic endothelial cells (Yanagisawa *et al.*, 1988), and subsequently has been found in a wide range of tissues including ovarian granulosa, endometrial and placental (Yanagisawa and Masaki, 1989). ET-1 – a 21 amino acid peptide– is one of the most potent vasoconstrictor peptides known; it belongs to a structurally homologous peptide family which includes ET-2, ET-3 and sarafotoxins (Yanagisawa and Masaki, 1989; Luscher *et al.*, 1992). It is synthesized as a 203 amino acid prepropeptide, which is proteolytically cleaved to produce big ET-1, which is finally processed to the mature form of ET-1 by an endothelin-converting enzyme (Luscher *et al.*, 1992; Opgenorth *et al.*, 1992). Induction of ET-1 secretion results from an enhanced preproET-1 gene expression since ET-1 is not present in storage pools (Yanagisawa and Masaki, 1989; Opgenorth *et al.*, 1992). Although the three ETs are products of three distinct genes, they share extensive sequence homology and a

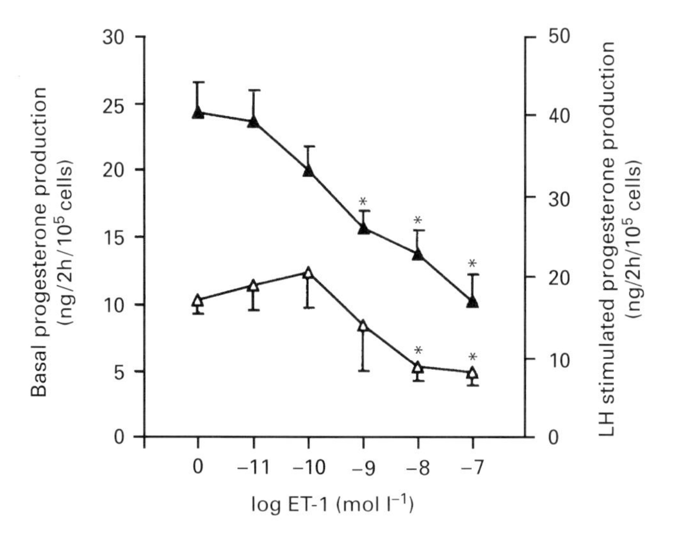

Fig. 2. Dose–response effect of endothelin 1 (ET-1) on progesterone secretion (ng 10^{-5} cells) from dispersed bovine luteal cells. After collagenase dispersion, luteal cells were incubated with various doses of ET-1 in the presence (▲) or absence (△) of bLH (5 ng ml^{-1}; USDA b5). Progesterone concentrations represent the concentration in media after incubation for 2 h. Data are the means ± SEM, $n = 5$, $*P < 0.05$.

common structural design (Yanagisawa and Masaki, 1989). Vascular endothelial cells produce mainly ET-1, and much less of the other two peptides (Yanagisawa and Masaki, 1989)

The effects of ET-1 on cells derived from the corpus luteum are shown in Fig. 2. Both basal- and bLH (5 ng ml^{-1})-stimulated progesterone secretion from luteal cells were inhibited by ET-1 in a dose dependent manner (Fig. 2; Girsh *et al.*, 1996a). Using an *in vitro* microdialysis system (MDS), Miyamoto *et al.* (1997) demonstrated that $PGF_{2\alpha}$ acutely stimulated the release of progesterone from bovine corpus luteum tissue. Preincubation with $PGF_{2\alpha}$ potentiated the inhibitory action of ET-1 on progesterone production (Miyamoto *et al.*, 1997). Of the two steroidogenic luteal cells, only the granulosa-derived cells responded to this peptide with an inhibition of basal and cAMP-stimulated progesterone production (Girsh *et al.*, 1996a). Similarly, when endothelial–luteal co-cultures were used, there was a ($PGF_{2\alpha}$-induced) reduction in progesterone, which was observed only in the large (granulosa-derived) cells (Girsh *et al.*, 1995). However, in luteal–endothelial co-cultures, the inhibitory effect of $PGF_{2\alpha}$ was attained after a longer incubation. The difference in time scale between these two experimental models may have been due to the time required for endogenous ET-1 to be released before a reduction in progesterone could be observed. The mechanism whereby ET-1 affects steroidogenesis warrants further research.

Expression of ET-1 Receptors in Luteal Cells

Two distinct receptor subtypes of the ET family have been cloned (Arai *et al.*, 1990; Sakurai *et al.*, 1990). These receptors belong to the seven transmembrane G-coupled superfamily and have been termed ET_A (for aorta) and ET_B (for bronchus). The order of potency of the different endothelins binding to the ET_A receptor is ET-1,ET-2 > ET-3 (Arai *et al.*, 1990). The other receptor subtype, ET_B, exhibits equipotent affinity for these three peptides (Sakurai *et al.*, 1990). Endothelin, via binding to these receptors, activates numerous transmembrane signalling systems in various tissues and cell types. The multiple ET-stimulated signal transduction pathways probably contribute to the diversity

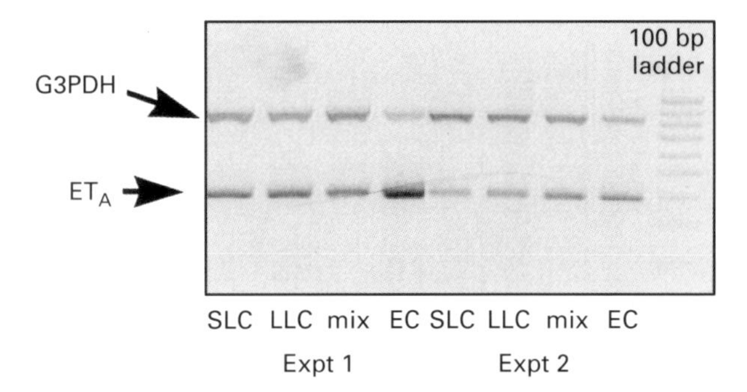

Fig. 3. Expression of endothelin 1 binding sites (ET_A) in bovine luteal cells. Endothelial cells (EC) and small and large luteal cells (SLC and LLC, respectively) were enriched from bovine corpora lutea by elutriation. Mixed cells (mix) – non-separated dispersed cells. Input of 100 ng total RNA of each sample was reverse transcribed and amplified for 21 and 26 cycles (with G3PDH and ET_A primers, respectively). PCR products were subjected to electrophoresis on 2% agarose gel and stained with ethidium bromide. An inverse image is presented.

of the biological responses induced by this peptide. Besides its well described effects on vascular smooth muscle cells and myocytes, ET-1 modulates steroidogenesis in several cell types including ovarian rat granulosa cells. Nevertheless, the identification of ET-1 receptors type(s) expressed by these cells was inconclusive (Iwai *et al.*, 1991; Tedeschi *et al.*, 1992; Iwai *et al.*, 1993; Flores *et al.*, 1995). Saturable, high-affinity ET-1 binding sites were identified in mid-cycle bovine corpus luteum and in the two steroidogenic luteal cell types (Girsh *et al.*, 1996a). These binding sites are of the ET_A subtype as demonstrated by competition with BQ123 (a selective ET_A antagonist) and mRNA expression. ET_A mRNA was detected in SLC, LLC and endothelial cells enriched from bovine corpus luteum (Fig. 3).

In agreement with binding data, ET-1 induced inhibition in progesterone release was prevented when cells were preincubated with a BQ compound. In addition, the addition of ET-3 (up to 10^{-6} mol l^{-1}) was ineffective in terms of progesterone secretion from cells derived from the corpus luteum, either under basal or bLH-stimulated conditions (Girsh *et al.*, 1996a).

The anti-steroidogenic effect of $PGF_{2\alpha}$ was retained in slices of corpus luteum incubated *in vitro* (Girsh *et al.*, 1995). Therefore, the involvement of ET-1 in $PGF_{2\alpha}$-induced progesterone reduction was studied using luteal slices (Girsh *et al.*, 1996a). Incubation with $PGF_{2\alpha}$ significantly inhibited LH-stimulated progesterone secretion (Girsh *et al.*, 1996). When BQ610 was added to the incubation medium it prevented the decline in progesterone secretion. Similarly, a single intraluteal injection of BQ123 to ewes at the mid-luteal phase was sufficient to attenuate the luteolytic action of exogenous $PGF_{2\alpha}$ compared with saline pretreatment (Fig. 4), implying that ET-1 (acting via ET_A receptors) mediated the inhibitory effect of $PGF_{2\alpha}$. The involvement of ET-1 in luteal regression is further suggested by the findings demonstrating that $PGF_{2\alpha}$ could upregulate ET_A expression in small and large luteal-like cells, obtained after *in vitro* luteinization (R. Meidan and N. Levy, unpublished data). However, whether the content of ET-1 receptors in the corpus luteum fluctuates throughout the oestrous cycle is unknown.

Induction of ET-1 Gene Expression by PGF_{2a}

The increase in ET-1 production suggested by luteal–endothelial co-cultures (Girsh *et al.*, 1995) can be explained, at least in part, by a direct effect of $PGF_{2\alpha}$ on endothelial cells. Indeed, we found that

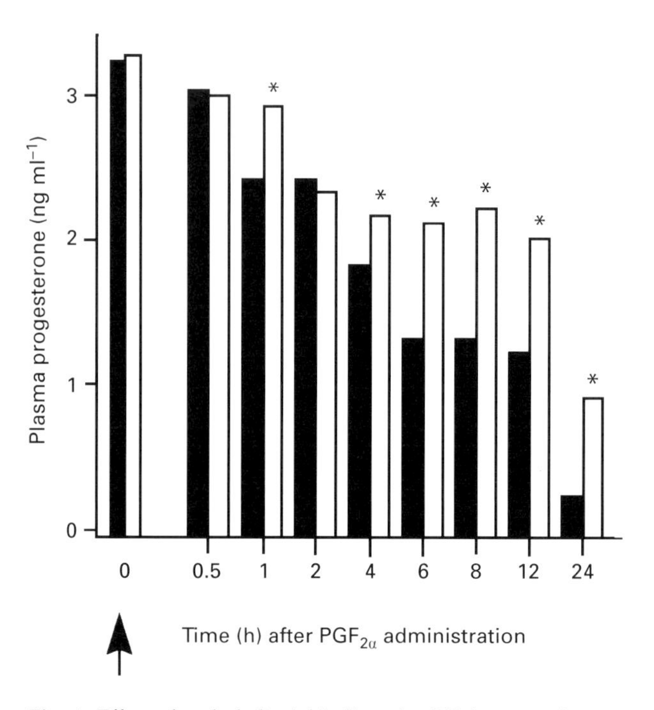

Fig. 4. Effect of endothelin 1 binding site (ET_A) antagonist on plasma progesterone concentrations. Normally cyclic ewes were injected with either saline (control ■) or 100 µg BQ123 (□), 30 min before the administration of 10 mg PGF$_{2\alpha}$ (Lutalyse, i. m.). Ewes underwent flank laparotomy on day 8 or 9 of the oestrous cycle and each corpus luteum per animal was injected with the appropriate treatment. Data are the means ± SEM, $n = 5$. *Significantly different from control ($P < 0.05$).

PGF$_{2\alpha}$ stimulated ET-1 secretion and its mRNA expression in luteal endothelial cells (Girsh *et al.*, 1996b). Under basal conditions the amount of ET-1 produced after incubation for 3 h was 30.2 ± 6.69 pg 10^{-5} cells and the addition of PGF$_{2\alpha}$ (1 µg ml^{-1}) increased ET-1 production approximately threefold. PGF$_{2\alpha}$ induced ET-1 expression in a time-dependent manner. An initial rise in ET-1 mRNA was detected within 15 min after addition of PGF$_{2\alpha}$ and ET-1 mRNA content continued to rise and reached a value 2.5 times higher than those of non-treated cells after 3 h (Girsh *et al.*, 1996). These effects are exerted via specific PGF$_{2\alpha}$ receptors (of the FP type), expressed by luteal endothelial cells (Mamluk *et al.*, 1998). This study was the first to report the presence of PGF$_{2\alpha}$ receptors in luteal endothelial cells.

These data imply that, at least under *in vitro* conditions, PGF$_{2\alpha}$ can rapidly enhance ET-1 expression. However, for ET-1 to have a physiological role in luteal regression, ET-1 concentrations should be high during this phase of the cycle and should vary in a PGF$_{2\alpha}$-dependent fashion.

ET-1 and ET-1 mRNA content were determined in corpus luteum obtained at different stages of the bovine oestrous cycle (Girsh *et al.*, 1996b). ET-1 content was highest in corpus luteum collected on days 17–21 of the cycle (30 times higher than on days 5–6). Similarly, on day 18 there was an increase in ET-1 mRNA expression compared with days 5 and 10 of the oestrous cycle (Fig. 5; Girsh *et al.*, 1996b). Ohtani *et al.* (1998) also observed that highest ET-1 concentrations were found in peripheral plasma during luteolysis. These findings suggest that enhancement of ET-1 expression in the aged corpus luteum could result from uterine PGF$_{2\alpha}$ secretion which occurs at this stage of the bovine cycle. This contention was examined by injecting a luteolytic dose of PGF$_{2\alpha}$ into heifers during mid-

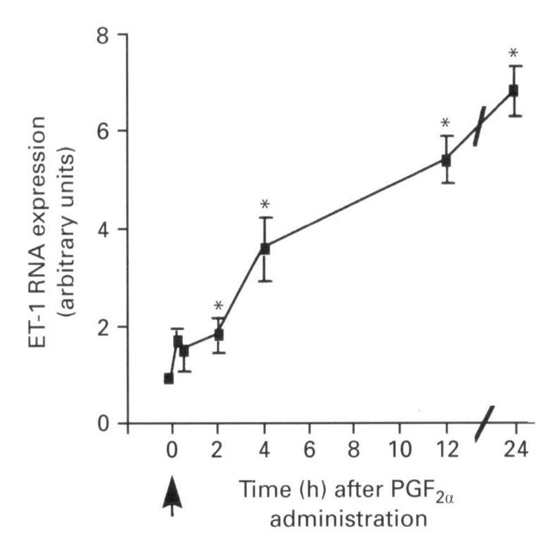

Fig. 5. Endothelin 1 (ET-1) mRNA expression after administration of PGF$_{2\alpha}$ to heifers on day 10 of the oestrous cycle. Corpora lutea were collected at various times after administration of PGF$_{2\alpha}$ (25 mg Lutalyse) as indicated. Data (densitometric units) were normalized to the value of the untreated control (day 10 corpora lutea). Data are means ± SEM from three corpora lutea at each time point. *Significantly different from control ($P < 0.05$).

cycle, resulting in a marked increase in the expression of ET-1 (Fig. 5). Likewise, incubation of luteal slices with PGF$_{2\alpha}$ resulted in a four-fold increase in their ET-1 concentrations (Girsh *et al.*, 1996b). These findings have been corroborated and extended by Miyamoto *et al.* (1997), who reported that PGF$_{2\alpha}$ significantly stimulated release of ET-1 from corpus luteum pieces *in vitro* after infusion for 2 h. Ohtani *et al.* (1998) have documented real time, intraluteal changes in ET-1, oxytocin and progesterone with an MDS device implanted *in vivo*. After administration of PGF$_{2\alpha}$ there was a rapid increase in ET-1 and oxytocin which was accompanied by simultaneous inhibition in progesterone produced by luteal tissue *in vivo*.

Clearly, these data show that PGF$_{2\alpha}$, both under *in vitro* and *in vivo* conditions, quickly augments luteal expression of ET-1 mRNA and protein content.

Role of Endothelial Cells and ET-1 in Structural Luteolysis

ET-1 may also be involved in later stages of luteal regression – namely in structural luteolysis. At present this is a poorly understood process, which is characterized by atrophy of corpus luteum tissue followed by the formation of scar tissue (Fields and Fields, 1996). Atrophy of corpus luteum tissue, which includes both vascular and endocrine epithelial cells, is a selective process involving programmed cell death (PCD) (Juengel *et al.*, 1993; Shikone *et al.*, 1996). Structural luteolysis is also accompanied by the influx of leukocytes, mainly macrophages (Hahnke *et al.*, 1994) and the local secretion or expression of several inflammatory cytokines, such as TNFα (Shaw and Britt, 1995; Wuttke *et al.*, 1997) and monocyte chemoattractant protein 1 (MCP-1) (Tsai *et al.*, 1997; Haworth *et al.*, 1998). Migration of leukocytes is a complex process that involves direct interaction of the migrating cells with endothelium (Mantovani *et al.*, 1997). Thus, any process leading to recruitment of

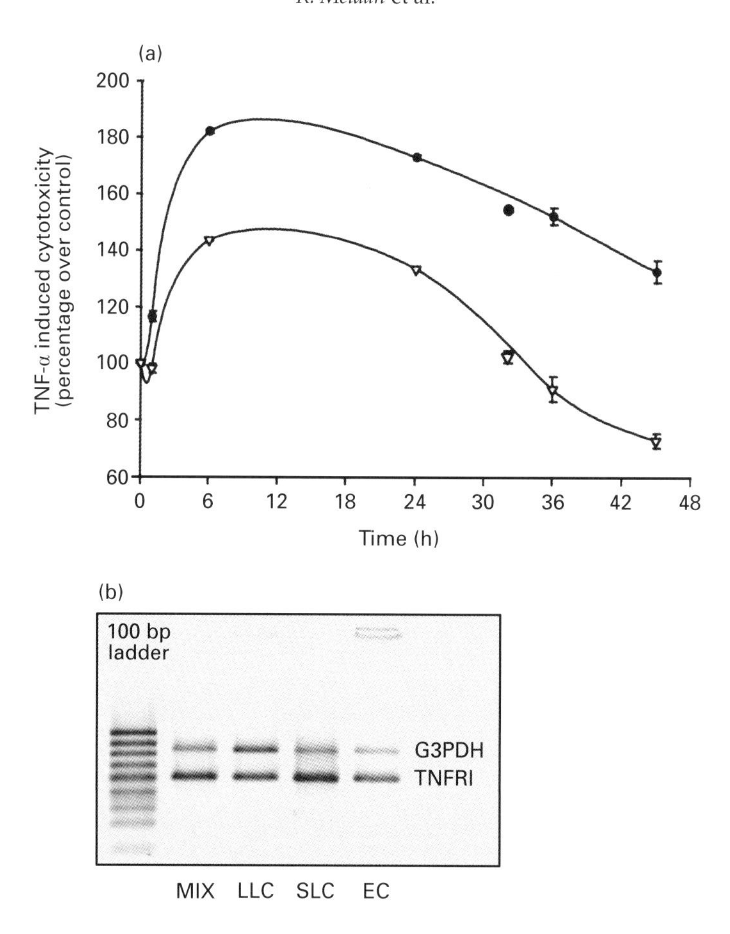

Fig. 6. (a) Peripheral blood derived bovine macrophages were cultured in the presence of increasing doses of endothelin 1 (ET-1) (● 10^{-7} mol ET-1 l^{-1}; ▽ 10^{-9} mol ET-1 l^{-1}). Supernatants were collected at different time intervals and assayed for tumour necrosis factor α (TNFα) activity in a bioassay using a TNFα sensitive WEHI cell line. (b) Expression of p55 TNFR1 in bovine luteal cells. Endothelial cells (EC), small and large luteal cells (SLC and LLC, respectively) were enriched from bovine corpora lutea by elutriation. Mixed cells (MIX) are non-separated dispersed cells. Input of 100 ng total RNA of each sample was reverse transcribed and amplified for 23 and 28 cycles (with G3PDH and TNFR1 primers, respectively). PCR products were subjected to electrophoresis on a 2% agarose gel and stained with ethidium bromide. An inverse image is presented.

leukocytes to the corpus luteum would have to involve endothelial cells. Such a possibility is indicated by our observations that, in response to $PGF_{2\alpha}$ stimulation, corpus luteum-derived endothelial cells secrete ET-1, a cytokine previously shown to be important for leukocyte migration (Boros *et al.*, 1998). To establish further a functional link between macrophages, endothelium and PCD in the corpus luteum, we determined responses of macrophages to endothelial cytokines (ET-1) and investigated expression of receptors for TNFα in different cell populations of the bovine corpus luteum.

ET-1 induced secretion of TNFα by bovine macrophages: 10^{-7} mol l^{-1} was more effective than 10^{-9} mol l^{-1}, and peak secretion was measured between 6 and 24 h of incubation (Fig. 6a). Dispersed

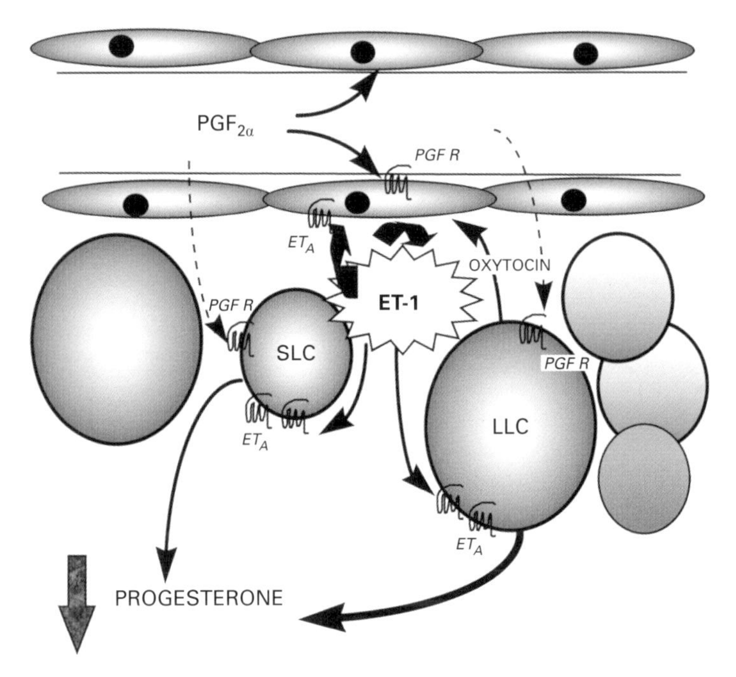

Fig. 7. Model of steroidogenic–endothelial cell interactions during functional luteolysis in ruminants. $PGF_{2\alpha}$ directly enhances endothelin 1 (ET-1) production by the resident endothelial cells, acting through the FP type receptor present in these cells. Large luteal cells may also respond to $PGF_{2\alpha}$ by secreting oxytocin, and increasing ET-1 expression. ET-1 is preferentially released towards the basal cell surface rather than towards the apical surface, and may thus reach nearby luteal steroidogenic cells to reduce their progesterone output. $PGF_{2\alpha}$ and ET-1 both induce vasoconstriction in the corpus luteum, and subsequently hypoxic conditions may develop and further augment ET-1 secretion via a positive feedback loop. ET_A: type A ET-1 receptor; PGFR: $PGF_{2\alpha}$ receptor; SLC: small luteal cells; LLC: large luteal cells.

total corpus luteum cells, SLC, LLC and endothelial cells express the p55 type receptor for TNFα (receptor for secreted form TNFα (Mantovani *et al.*, 1997); Fig. 6b).

Collectively, these results indicate that endothelial cells might have a pivotal role in structural luteolysis via $PGF_{2\alpha}$-induced secretion of ET-1. This statement is supported by data showing high MCP-1 in the corpus luteum in response to $PGF_{2\alpha}$ (Tsai *et al.*, 1997; Haworth *et al.*, 1998), and although the cells secreting MCP-1 were not identified in these studies, previous investigations showed that endothelial cells are a source of MCP-1 (Mukaida *et al.*, 1992).

Conclusions

We demonstrated that $PGF_{2\alpha}$ stimulates luteal ET-1 production by several mechanisms that are not mutually exclusive (summarized in Fig. 7). Large luteal cells respond to $PGF_{2\alpha}$ by secreting oxytocin, which could increase ET-1 production by resident endothelial cells. $PGF_{2\alpha}$ can also directly enhance ET-1 production in endothelial cells, acting through the FP type receptor present in these cells. ETs are preferentially released towards the basal cell surface rather than towards the apical surface, and may thus reach nearby luteal steroidogenic cells and reduce their progesterone output. In addition,

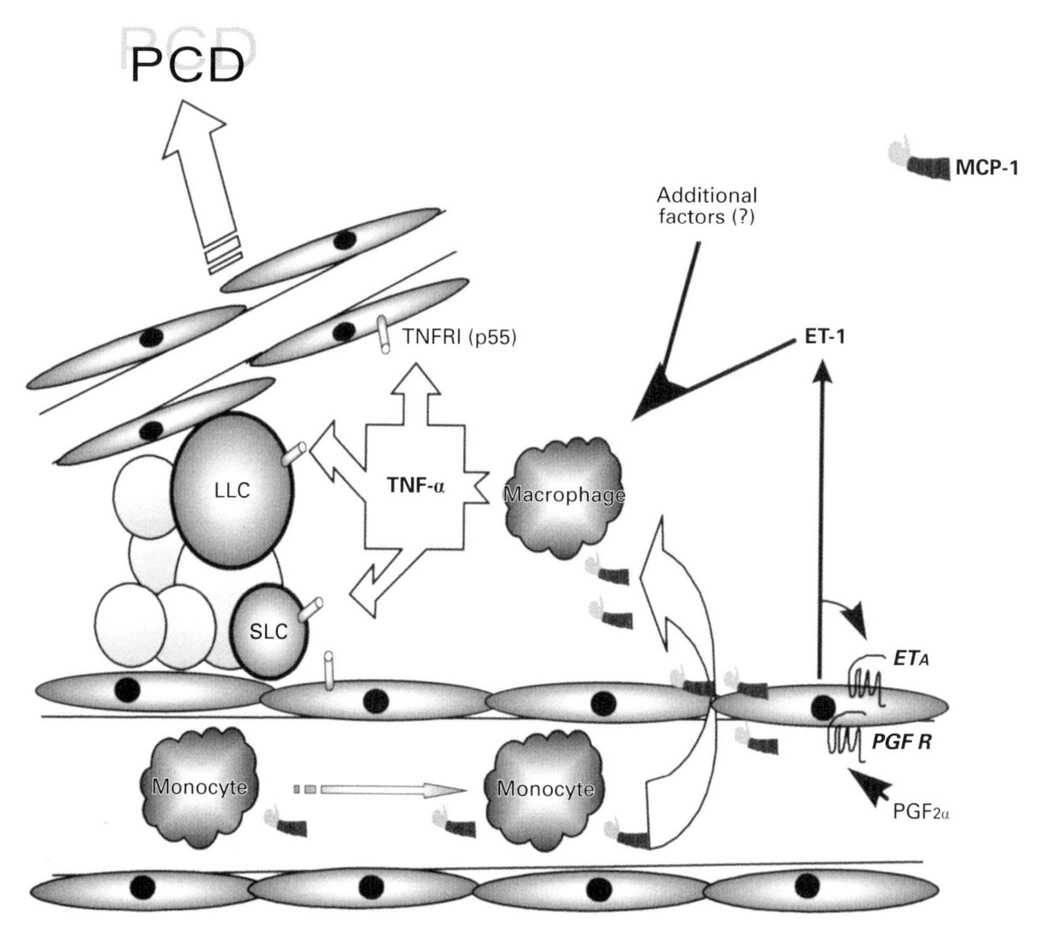

Fig. 8. Model for structural regression of the corpus luteum of ruminants. Conditions established during functional regression, that is endothelial cell activation and cytokine secretion, form the initial phase of structural regression, which assigns a key role to endothelium and macrophages in events leading to programmed cell death (PCD) in the corpus luteum. $PGF_{2\alpha}$ sensitized endothelial cells might initiate a cascade of events leading to the recruitment and transmigration of monocytes to the corpus luteum. (The model proposes that monocyte chemoattractant protein 1 (MCP)-1 is the inducer of migration). Migration and concomitant maturation to macrophages is followed by their activation, which could lead to PCD via secretion of tumour necrosis factor α (TNFα). ET-1: endothelin 1; TNFR1: receptor type 1 (p55) for TNFα

$PGF_{2\alpha}$ may potentiate the inhibitory effect of ET-1 on progesterone release. $PGF_{2\alpha}$, like ET-1 but somewhat weaker, can induce constriction of smooth muscle cells in arterioles. Subsequent to vasoconstriction, hypoxic conditions may develop and further augment ET-1 secretion via a positive feedback loop. The existence of multiple pathways may be instrumental in ensuring the marked increase in luteal endothelin secretion. The quick upregulation of ET-1 mRNA transcription in luteal endothelial cells coupled with lack of peptide storage pools, a unique feature of ET-1, enable acute changes in ET-1 concentrations. This phenomenon undoubtedly facilitates the mediatory role of ET-1 in functional regression.

Conditions established during functional regression, that is endothelial cell activation and cytokine secretion, form the initial phase of structural regression, which assigns a key role to endothelium and macrophages in events leading to PCD in the corpus luteum (Fig. 8). $PGF_{2\alpha}$ sensitized endothelial cells might initiate a cascade of events leading to the recruitment and transmigration of monocytes to the corpus luteum (MCP-1 may be the inducer of migration; Fig. 8). Migration and concomitant maturation to macrophages are followed by their activation, which could lead to PCD via secretion of TNFα. Whether endothelial cells are a primary target for PCD remains to be elucidated. We favour this notion, for regression of corpus luteum vasculature might cause local anoxia, which is an inducer of epithelial cell death.

References

Arai H, Hori S, Aramori I, Ohkubo H and Nakanishi S (1990) Cloning and expression of a cDNA encoding an endothelin receptor *Nature* **348** 730–732

Boros M, Massberg S, Baranyi L, Okada H and Messmer K (1998) Endothelin 1 induces leukocyte adhesion in submucosal venules of the rat small intestine *Gastroenterology* **114** 103–114

Davis J, Alila H, West L, Corradino R, Weakland L and Hansel W (1989) Second messenger systems and progesterone secretion in the small cells of the bovine corpus luteum: effects of gonadotropins and prostaglandin F-2α *Journal of Steroid Biochemistry* **32** 643–649

Ellinwood WE, Nett TM and Niswender GD (1979) Maintenance of the corpus luteum of early pregnancy in the ewe. II. Prostaglandin secretion by the endometrium *in vitro* and *in vivo. Biology of Reproduction* **21** 845–856

Fields M and Fields P (1996) Morphological characteristics of the bovine corpus luteum during the estrous cycle and pregnancy *Theriogenology* **45** 1295–1325

Flores J, Winters T, Knight J and Veldhuis J (1995) Nature of endothelin binding in the porcine ovary *Endocrinology* **136** 5014–5019

Girsh E, Greber Y and Meidan R (1995) Luteotrophic and luteolytic interactions between bovine small and large luteal-like cells and endothelial cells *Biology of Reproduction* **52** 954–962

Girsh E, Milvae R, Wang W and Meidan R (1996a) Effect of endothelin-1 on bovine luteal cell function: role in prostaglandin F2α-induced antisteroidogenic action *Endocrinology* **137** 1306–1312

Girsh E, Wang W, Mamluk R, Arditi F, Friedman A, Milvae R and Meidan R (1996b) Regulation of endothelin-1 expression in the bovine corpus luteum: elevation by prostaglandin F2α *Endocrinology* **137** 5191–5196

Grazul-Bilska A, Reynolds L and Redmer D (1992) Contact-dependent intercellular communication between bovine endothelial cells and early or mid-cycle bovine luteal cells *IX Ovarian Workshop*, Chapel Hill, North Carolina Abstract 31 Serono Symposia

Hahnke KH, Christenson LK, Ford SP and Taylor M (1994) Macrophage infiltration into the porcine corpus luteum during prostaglandin $F_{2\alpha}$-induced luteolysis *Biology of Reproduction* **50** 10–15

Hansel W and Blair R (1996) Bovine corpus luteum: a historic overview and implications for future research. *Theriogenology* **45** 1267–1294

Haworth J, Rollyson M, Silva P, McIntush E and Niswender G (1998) Messenger ribonucleic acid encoding monocyte chemoattractant protein-1 is expressed by the ovine corpus luteum in response to prostaglandin F2α *Biology of Reproduction* **58** 169–174

Inagami T, Naruse M and Hoover R (1995) Endothelium as an endocrine organ *Annual Review of Physiology* **57** 171–189

Iwai M, Hasaaki M, Taii S, Sagawa N, Nakao K, Imura H, Nakanishi S and Mori T (1991) Endothelins inhibit luteinization of cultured porcine granulosa cells *Endocrinology* **129** 1909–1914

Iwai M, Hori S, Shigemoto R, Kanzaki H, Mori T and Nakanishi S (1993) Localization of endothelin receptor messenger ribonucleic acid in the rat ovary and fallopian tube by *in situ* hybridization *Biology of Reproduction* **49** 675–680

Jablonka-Shariff A, Grazul-Bilska A, Redmer D and Reynolds L (1993) Growth and cellular proliferation of ovine corpora lutea throughout the estrous cycle *Endocrinology* **133** 1871–1879

Juengel JL, Garverick HA, Johnson AL, Youngquist RS and Smith MF (1993) Apoptosis during luteal regression in cattle *Endocrinology* **132** 249–254

Luscher T, Boulanger C, Dohi Y and Yang Z (1992) Endothelium-derived contracting factors *Hypertension* **19** 117–130

McCracken J, Carlson J, Glew M, Goding J and Baird D (1972) Prostaglandin $F_{2\alpha}$ identified as a luteolytic hormone in sheep *Nature New Biology* **238** 129–134

Mamluk R, Chen D, Greber Y, Davis J and Meidan R (1998) Characterization of prostaglandin $F_{2\alpha}$ and LH receptor mRNA expression in different bovine luteal cell types *Biology of Reproduction* **58** 849–856

Mantovani A, Bussolino F and Introna M (1997) Cytokine regulation of endothelial cell function: from molecular level to the bedside *Immunology Today* **18** 231–240

Meidan R, Aberdam E and Aflalo L (1991) Steroidogenic enzyme content and progesterone induction by cAMP-generating agents and prostaglandin $F_{2\alpha}$ in bovine theca and granulosa cells luteinized *in vitro. Biology of Reproduction* **46** 786–792

Miyamoto A, Lutzov H and Schams D (1993) Acute actions of prostaglandin $F_{2\alpha}$, E_2 and I_2 in microdialyzed bovine corpus luteum *in vitro. Biology of Reproduction* **49** 423–430

Miyamoto A, Kobayashi S, Arata S, Ohtani M, Fukui Y and Schams D (1997) Prostaglandin F2 alpha promotes the inhibitory action of endothelin-1 on the bovine luteal function *in vitro. Journal of Endocrinology* **152** R7–11

Mukaida NA, Arada K, Yasumoto K and Matsushim K (1992) Properties of pro-inflammatory cell type-specific leukocyte chemotactic cytokines, interleukin 8 (IL-8) and monocyte chemotactic and activating factor (MCAF) *Microbiology and Immunology* **36** 773–789

Ohtani M, Kobayashi S, Miyamoto A, Hayashi K and Fukui Y (1998) Real time relationships between intraluteal and plasma concentrations of endothelin, oxytocin, and progesterone during prostaglandin $F_{2\alpha}$-induced luteolysis in the cow *Biology of Reproduction* **58** 103–108

Opgenorth T, Wu-Wong J and Shiosaki K (1992) Endothelin-converting enzymes *Federation Association FASEB Journal* **6** 2653–2659

O'Shea J, Rodgers R and D'Occhio M (1989) Cellular

composition of cyclic corpus luteum of the cow *Journal of Reproduction and Fertility* **85** 483–487

Pate JL (1994) Cellular components involved in luteolysis *Journal of Animal Science* **72** 1884–1890

Sakurai T, Yanagisawa M, Takuwa Y, Miyazaki H, Kimura S, Goto K and Masaki T (1990) Cloning of a cDNA encoding a nonisopeptide-selective subtype of the endothelin receptor *Nature* **348** 732–735

Shaw D and Britt J (1995) Concentrations of tumour necrosis factor α and progesterone within the bovine corpus luteum sampled by continuous-flow microdialysis during luteolysis *in vivo. Biology of Reproduction* **53** 847–854

Shikone T, Yamoto M, Kokawa K, Yamashita K, Nishimori K and Nakano R (1996) Apoptosis of human corpora lutea during cyclic luteal regression and early pregnancy *Journal of Clinical Endocrinology and Metabolism* **81** 2376–2380

Tedeschi C, Hazum E, Kokia E, Ricciarelli E, Adashi E and Payne D (1992) Endothelin-1 as a luteinization inhibitor: inhibition of rat granulosa cell progesterone accumulation via selective modulation of key steroidogenic steps affecting both progesterone formation and degradation *Endocrinology* **131** 2476–2478

Tsai S, Juengel J and Wiltbank M (1997) Hormonal regulation of monocyte chemoattractant protein-1 messenger ribonucleic acid expression in corpora lutea *Endocrinology* **138** 4517–4520

Wuttke W, Pitzel L, Knoke K, Theiling K and Jarry H (1997) Immune–endocrine interactions affecting luteal functions in pigs *Journal of Reproduction and Fertility Supplement* **52** 19–29

Yanagisawa M and Masaki T (1989) Biochemistry and molecular biology of the endothelins *Trends in Pharmacological Sciences* **10** 374–378

Yanagisawa M, Kurihara H, Kimura S, Mitsui Y, Kobayashi M, Watanabe T and Masaki T (1988) A novel potent vasoconstrictor peptide produced by vascular endothelial cells *Nature* **332** 411–415

MALE FUNCTION AND FERTILITY

Chair
G. B. Martin

Journal of Reproduction and Fertility Supplement **54**, 231–242

Regulation of gonadotrophin-releasing hormone secretion by testosterone in male sheep

S. M. Hileman[1] and G. L. Jackson[2]

[1]*Department of Medicine, Division of Endocrinology, Beth Israel Deaconess Medical Center, Harvard Medical School, Boston, MA 02215, USA;* [2]*Department of Veterinary Biosciences, University of Illinois, Urbana, IL 61802, USA*

In males, including the ram, testosterone, acting via its primary metabolites oestradiol and dihydrotestosterone (DHT), suppresses circulating LH concentrations. This effect is due primarily, although not totally, to decreased frequency of gonadotrophin-releasing hormone (GnRH) pulses. The arcuate–ventromedial region (ARC–VMR) of the mediobasal hypothalamus and possibly the medial preoptic area (mPOA) are sites at which oestradiol acts to suppress GnRH, but the site of DHT action is not known. Given that native GnRH neurones appear to contain few or no oestrogen or androgen receptors, the effects of testosterone metabolites probably are exerted by modulating activity of inhibitory interneurone systems such as β-endorphin, dopamine, and γ-aminobutyric acid (GABA). Although β-endorphin clearly inhibits GnRH secretion, the observation that testosterone treatment during a long-day photoperiod reduced proopiomelanocortin (POMC) mRNA in the arcuate nucleus while coincidentally suppressing GnRH release indicates that β-endorphin does not mediate the inhibitory effect of testosterone on GnRH. Activation of $GABA_A$ receptors in either the mPOA or ARC–VMR suppressed LH, whereas activation of $GABA_B$ receptors in the ARC–VMR increased LH pulse amplitude. Therefore, it is suggested that GABA acts in both regions to regulate LH. Whereas testosterone affects GABA metabolism in the rat hypothalamus, its effect in the ram hypothalamus is yet to be determined. Testosterone treatment activated dopaminergic cells in the retrochiasmatic A15 area in the same animals in which it suppressed POMC mRNA in the arcuate nucleus. This dopaminergic system may partially mediate the negative feedback effect of testosterone in the ram analogous to its role in partially mediating the negative effect of oestrogen in the ewe. Future studies must concentrate on determining how these and other putative inhibitory neuronal systems interact and how they in turn are regulated by environmental factors such as photoperiod.

Introduction

Upon casual observation it may be concluded that there is a relatively simple relationship among the secretory patterns of reproductive hormones in the male. A pulse of GnRH released from the hypothalamus releases a pulse of LH from the pituitary, which in turns elicits a burst of testosterone secretion from the testis. The increased testosterone then suppresses GnRH and LH, completing a typical negative feedback loop. Although it is correct, this simple depiction belies a far more complex relationship. Closer observation reveals that the feedback loop has many components and that the relationships between those components are highly dynamic, being influenced by factors such as age, photoperiod, nutritional status, and social cues. The objective of this review is to describe some of those components, how their function is modulated, and how they affect the efficacy by which testosterone regulates GnRH and LH secretion in one representative animal, the ram.

Is Metabolism of Testosterone Important?

Testosterone treatment reduces LH secretion in males of all species studied so far (Kalra and Kalra, 1989). Several observations indicate that this inhibitory effect of testosterone is mediated primarily by the testosterone metabolites oestrogen and DHT rather than testosterone *per se*. First, very much smaller amounts of oestrogen or DHT than testosterone are necessary for suppression of LH release (Parrott and Davies, 1979). Second, both oestrogen and DHT are produced by peripheral (Hileman *et al.*, 1994) and neuronal aromatization and reduction (Naftolin and Ryan, 1975; Selmanoff *et al.*, 1977) of testosterone. Third, immunization of intact rams against oestrogen greatly increases circulating concentrations of both LH and testosterone (Monet-Kuntz *et al.*, 1988). This result could be due to neutralization of oestrogen produced by the testis as well as oestrogen produced by aromatization from testosterone. However, the observation that treatment of testosterone-treated gonadectomized rams with the aromatase inhibitor aminoglutethimide also significantly increases circulating LH (Scanbacher, 1984) provides strong evidence that oestrogen produced by aromatization from testosterone contributes significantly to normal negative feedback.

Although it was clear that administration of DHT inhibited LH release and blocked a post-castration rise in LH secretion, it was not clear whether conversion of testosterone to DHT is an essential component by which testosterone suppresses LH release. To investigate this question, we infused wethers for 72 h with either testosterone alone, reductase inhibitor alone (L-651-723 supplied by Merck Research Laboratory, Rahway, NJ), or testosterone together with reductase inhibitor (Hileman *et al.*, 1994b). Infusion of inhibitor alone had no effect on either LH secretion or circulating concentrations of testosterone, oestrogen or DHT. In the testosterone-infused males, reduced LH release was associated with increased concentrations of all three steroid hormones. Infusion of the inhibitor with testosterone blocked only DHT formation (by over 80%) and significantly reduced, but did not completely abolish, the ability of testosterone to inhibit LH secretion (Fig. 1). These data led to the suggestion that formation of DHT is an important step in testosterone-induced reduction of pulsatile LH release. Notably, the effect of blocking both aromatase and reductase activity in testosterone-treated castrated rams has not been reported. Given the results of the cited studies, a severe or perhaps total attenuation of testosterone action on LH secretion could be expected.

Site of Testosterone Action on LH

Although the inhibitory action of testosterone on LH release is clearly established, the specific sites of action of testosterone are only partially known. Whether testosterone reduces GnRH secretion, responsiveness of the pituitary to GnRH, or both, remains unclear for some species (Kalra and Kalra, 1989). In sheep, testosterone acts primarily, although not exclusively, on the brain to suppress GnRH pulse frequency. Specifically, castration leads to increased GnRH pulse frequency (Caraty and Locatelli, 1988), whereas testosterone replacement reduces GnRH pulse frequency (Jackson *et al.*, 1991; Tilbrook and Clarke, 1995). Circulating testosterone, at concentrations that severely reduced GnRH pulse frequency, had a marginal effect on pituitary response to exogenous GnRH (Jackson *et al.*, 1991). However, evaluation of pituitary responsiveness across the annual breeding season leads to the conclusion that testosterone also acts directly on the pituitary to modulate the response to GnRH (Rhim *et al.*, 1993). The neural sensitivity to the negative feedback action of testosterone as well as circulating concentrations of testosterone vary greatly with stage of the annual reproductive cycle (for example, Rhim *et al.*, 1993). Thus it is likely that the relative effect of testosterone on the pituitary versus brain also varies with stage of the annual reproductive cycle.

The specific neural sites at which testosterone or testosterone metabolites act to regulate LH are poorly established. Both oestrogen receptor α and androgen receptor distribution in the sheep brain have been described (Lehman *et al.*, 1993; Herbison, 1995). Within the hypothalamus there are high concentrations of oestrogen receptor α and androgen receptor in the preoptic area, arcuate and ventromedial nucleus, and median eminence. Recently, another form of the oestrogen receptor,

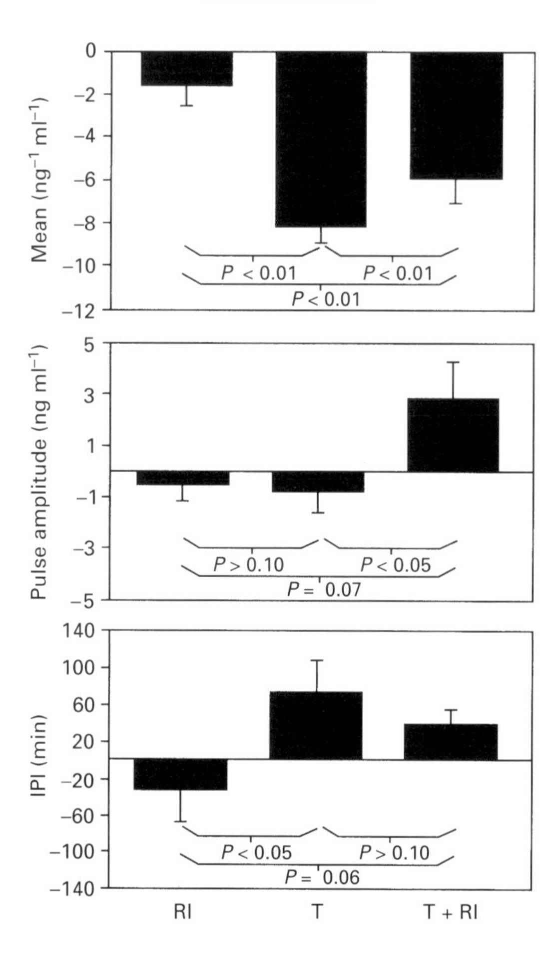

Fig. 1. Changes in LH pulse parameters of castrated male sheep treated with either 0.6 mg kg^{-1} 5α-reductase inhibitor (RI) L-651,723 or 768 μg kg^{-1} day^{-1} testosterone (T), or T + RI for 3 days. IPI: interpulse interval. Bars represent means ± SEM of day 3 values minus day 0 values (n = 5 per treatment). P values are indicated within each graph for comparisons made between groups using tests of least significant differences. (Reproduced from Hileman *et al.*, 1994 with permission.)

oestrogen receptor β, has been found. Although the function of oestrogen receptor β is not clear, its distribution in the male sheep hypothalamus is similar to that described for the rodent (Shugrue *et al.*, 1997) with localization in the medial preoptic area, retrochiasmatic area, bed nucleus stria terminalis, paraventricular nucleus, supraoptic nucleus and dorsomedial hypothalamus. Only sparse labelling for oestrogen receptor β is found in the arcuate nucleus and ventromedial hypothalamus (S. M. Hileman and R. J. Handa, unpublished). Thus, each of these sites, as well as others in the brainstem and amygdala, may be involved in mediating the action of testosterone on GnRH secretion.

In an attempt to delineate specific sites at which testosterone, oestrogen, and DHT act to suppress LH, we (Scott *et al.*, 1997) placed implants of these steroids into the mPOA and ARC–VMR of the ventromedial hypothalamus of long-term castrated rams. Implants of testosterone and DHT at either site were ineffective at suppressing LH. In contrast, implants of oestrogen in the mPOA were marginally effective whereas implants of oestrogen into the ARC–VMR clearly suppressed LH (Fig. 2). These results implicate the ARC–VMR as an important site at which oestrogen acts to suppress LH secretion. The reason for the failure of testosterone and DHT implants to suppress LH is not clear, but may reflect less efficient diffusion, downregulation of androgen receptors following castration (Handa *et al.*, 1996), or the fact that androgens may act at other or additional untested sites.

By Which Neural Pathways Does Testosterone Act?

The specific mechanisms by which testosterone or its metabolites act to suppress GnRH release are not clear. The most parsimonious mechanism would be action of the steroids directly on GnRH neurones. Recent observations indicating that immortalized GT1-7 GnRH-secreting cells contain both androgen and oestrogen receptors support this contention (Belsham *et al.*, 1998; Shen *et al.*, 1998). However, it is debatable whether these cells are fully representative of endogenous GnRH secreting cells *in vivo*. In addition, it is notable that several investigations have found few or no steroid receptors on native GnRH cells *in vivo* (Shivers *et al.*, 1983; Huang and Harlan, 1993; Lehman and Karsch, 1993). Consequently, the prevailing hypothesis is that steroids affect GnRH neurones through actions on interneurones. This idea gains support by observations that numerous neurotransmitter agonists and antagonists affect GnRH secretion and that several neuronal systems concentrate gonadal steroids. Under this concept, steroids could reduce GnRH secretion either by increasing secretion of inhibitory neurotransmitters or by reducing secretion of stimulatory neurotransmitters or by a combination of effects. Owing to the inherent difficulty of such studies, relatively little effort has been given to investigating the second or third possibilities. The first possibility is perhaps the easiest to address and results of several studies lead to the suggestion that gonadal steroids modulate the secretion of at least three neurotransmitters known to inhibit LH secretion: opiates (i.e. β-endorphin), dopamine, and γ-aminobutyric acid (GABA).

β-Endorphin

β-Endorphin neurones are found in high concentrations in the arcuate nucleus, contain oestrogen receptors, and contact GnRH neurones (Leranth *et al.*, 1988; Thind and Goldsmith, 1988; Barb *et al.*, 1991; Lehman and Karsch, 1993). Numerous data show that administration of the opiate agonist morphine suppresses LH, whereas injection of antagonists such as naloxone increases LH (see Barb *et al.*, 1991). Observations that the effects of opiate antagonists are much more robust in intact or testosterone-treated than in castrated animals led to the concept that opiates may mediate the inhibitory action of testosterone (Ebling and Lincoln, 1985; Barb *et al.*, 1991). The observation that peripheral concentrations of β-endorphin were highest during the breeding season (Ssewannyana and Lincoln, 1990) indicates that the influence of endorphin is greatest during that period. Intuitively, this seems inconsistent with the fact that testosterone is relatively ineffective at inhibiting GnRH release at this time. To investigate this issue further, we performed a series of experiments (Hileman *et al.*, 1996; Hileman *et al.*, 1998) to examine the effect of testosterone on POMC mRNA in the arcuate nucleus under the assumption that amounts of the precursor mRNA may reflect the synthesis and secretion of the peptide. In the most recent study, males were either castrated or castrated and implanted with testosterone, and then placed under either an inhibitory long-day or stimulatory short-day photoperiod. Testosterone did not alter either LH release or amounts of POMC mRNA in animals exposed to short days. In contrast, testosterone greatly reduced both mean LH concentrations and the amount of POMC mRNA (Fig. 3), but not of GnRH mRNA, in animals exposed to long days (Fig. 3). This finding was consistent with our previous report (Hileman *et al.*, 1996) that testosterone administered for either 3 days or 3 months reduced POMC mRNA in the

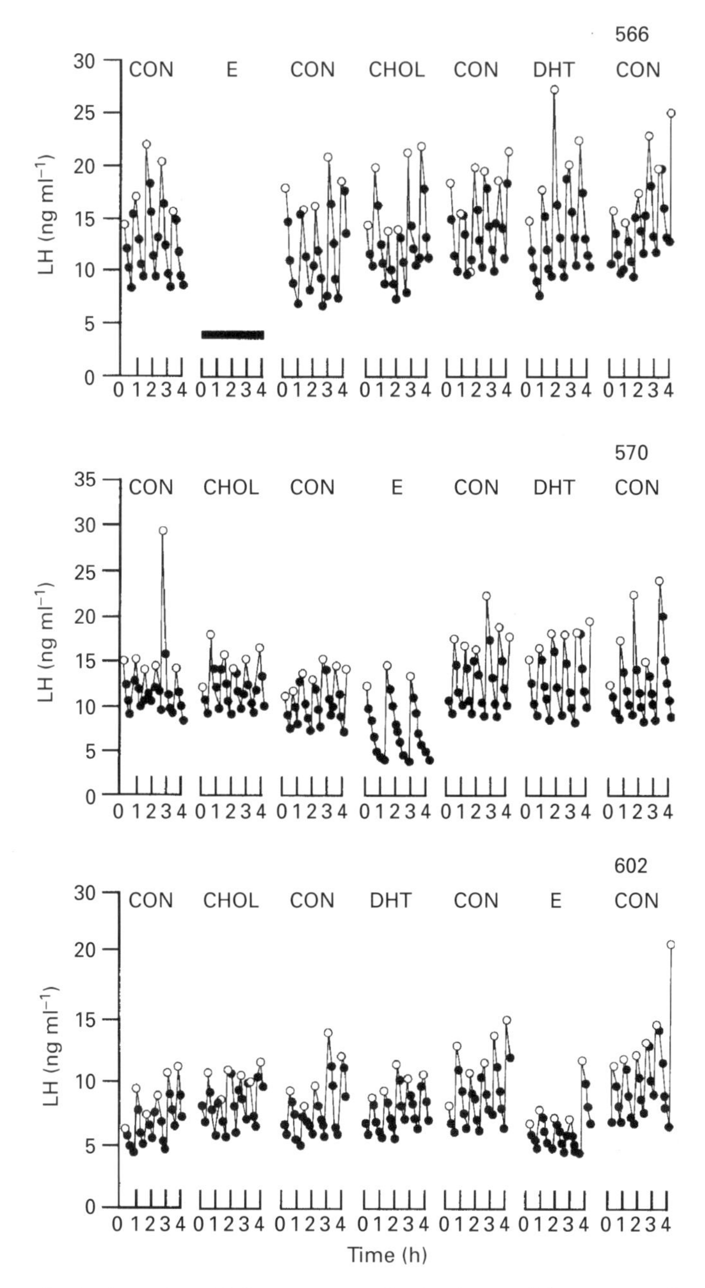

Fig. 2. Examples of LH secretory profiles in plasma of three castrated rams with either stylets or steroid implants placed bilaterally into a site located dorsal–lateral to the arcuate nucleus and medial to the ventromedial nucleus of the hypothalamus (VMH). CON: stylet; CHOL: cholesterol; DHT: dihydrotestosterone; E: oestradiol. Peaks of LH pulses are represented by hollow circles. Sampling periods were 7 days apart. The number in the upper right of each panel refers to an individual animal identification number. (Modified from Scott *et al.*, 1997.)

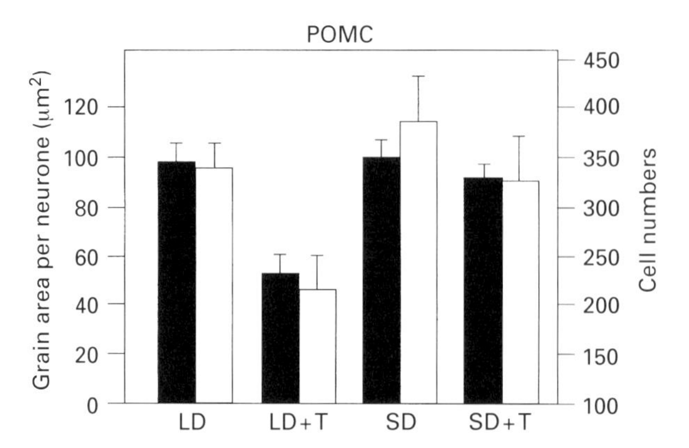

Fig. 3. Effects of testosterone (T) and duration of photoperiod on silver grain area per neurone (solid bars) and numbers of proopiomelanocortin (POMC) mRNA positive cells (■) in the arcuate nucleus of castrated rams. LD: long days (16L:8D); SD: short days (10L:14D). Data are presented as means ± SEM. POMC mRNA grain area in the LD + T group was lower ($P < 0.01$) than in the other groups which did not differ from each other ($P > 0.10$). There was an effect of steroid ($P < 0.05$) and a strong tendency for an effect of photoperiod ($P = 0.06$), but no interaction of treatments ($P > 0.40$), on the number of POMC cells. (Reproduced from Hileman *et al.*, 1998.)

arcuate nucleus of wethers kept under ambient long days. Although it is recognized that mRNA content may not always reflect actual peptide release, studies from two other laboratories independently investigating the effects of feed restriction on POMC mRNA content and β-endorphin release indicate that relative steady-state amounts of POMC mRNA are indicative of relative hypothalamic β-endorphin release rates in the lateral median eminence of sheep (Prasad *et al.*, 1993; McShane *et al.*, 1993). In addition, preliminary data from our recent studies using portal-cannulated males indicate that naloxone effectively increases GnRH and LH release in castrated males exposed to inhibitory long days. This finding suggests that the magnitude of response to naloxone may be dependent more on the basal LH secretory rate than on steroid background. These findings support the concept that the increased synthesis of β-endorphin is not a mechanism whereby testosterone suppresses GnRH release in animals exposed to long days. It is noteworthy that Goodman *et al.* (1995) reached a similar conclusion regarding the relationship between oestrogen, the opiates, and LH secretion in ewes.

Dopamine

Dopaminergic neurones are clumped into nuclei or groups scattered throughout the brain. Small groups designated A12, A14, A15, and located in the anterior part of the hypothalamus appear to have a significant role in regulating LH in sheep. The A14–A15 group is located in the retrochiasmatic area; the A12 group is located in the arcuate nucleus–median eminence region. A large body of evidence from studies using female sheep and a variety of approaches led to the concept that dopaminergic input from these areas inhibits LH secretion, at least when the animals are exposed to long-day photoperiods (Thiery *et al.*, 1995; Lehman *et al.*, 1996; Viguie *et al.*, 1997).

In comparison, there are relatively few studies on the role of the dopaminergic system on LH secretion in rams, but most of these data support an inhibitory role. Although pimozide, a dopamine

D_2 receptor antagonist, did not increase LH in rams (Tilbrook and Clarke, 1992), the more specific D_2 antagonist sulpiride was effective in Soay rams, particularly during the non-breeding season (Tortonese and Lincoln, 1994). In addition, supporting evidence was obtained in our laboratory by monitoring the effect of testosterone administration on expression of the early–intermediate gene c-Fos in the A14–A15 cell groups. We used dual-label immunocytochemistry for c-Fos and tyrosine hydroxylase (the rate-limiting enzyme in dopamine formation) to determine the percentage of dopaminergic cells activated by testosterone. The tissue used was from our previously cited study in which we had shown that infusion of testosterone for 3 days suppressed LH release coincident with reduced POMC mRNA (Hileman *et al.*, 1996). Testosterone treatment significantly increased the percentage of dual-labelled cells in the A15 group and strongly tended ($P < 0.06$) to increase that percentage in the A14 group. No effect of testosterone was noted on either the A13 cell group or the total number of tyrosine hydroxylase positive cells (Lubbers *et al.*, 1995). These results are very similar to those obtained from ewes treated with oestrogen during the non-breeding season (Lehman *et al.*, 1996). In summary, there is evidence to support the concepts (1) that activation of the A14–A15 cell groups is involved in the feedback action of gonadal steroids during the non-breeding season and (2) that during this season gonadal steroids selectively activate this dopaminergic subsystem in both rams and ewes.

GABA

Gamma-aminobutyric acid (GABA) is a widely distributed neurotransmitter, the primary action of which is to inhibit the activation of other neuronal systems. Several observations indicate that it acts in the mPOA of rats to inhibit GnRH release, and results of experiments in male rats support the concept that GABA may mediate the inhibitory effect of testosterone on GnRH and LH release, particularly within the mPOA (Grattan and Selmanoff, 1993; Grattan *et al.*, 1996; Sagrillo and Selmanoff, 1997). Results of studies in ewes also led to the contention that GABA acts in the mPOA to suppress GnRH secretion (Robinson, 1995; Scott and Clarke, 1993a,b).

Work from our laboratory supports the concept that GABA acts in both the mPOA and ARC–VMR to suppress GnRH release and within the ARC–VMR to regulate specifically LH pulse amplitude (Ferreira *et al.*, 1996). Castrated rams had guide tubes stereotaxically placed bilaterally either in the mPOA or ARC–VMR to deliver drugs into these specific sites by microdialysis. Subsequently, these areas were perfused with artificial cerebrospinal fluid for 4 h followed by 4 h of either cerebrospinal fluid, the $GABA_A$ receptor agonist muscimol, or the $GABA_B$ receptor agonist baclofen. In the mPOA, muscimol treatment reduced pulsatile LH release, whereas baclofen was without effect. In the ARC–VMR, muscimol also inhibited pulsatile LH release, but surprisingly baclofen actually increased LH release. This effect was due primarily to an increase in pulse amplitude rather than pulse frequency and probably reflected an increase in GnRH pulse amplitude (Fig. 4). Thus, it appears that GABA may act through the $GABA_A$ receptor to suppress pulsatile GnRH release in both the POA and ARC–VMR. Interpreting the effect of baclofen is not as straightforward, but may be explained by the fact that $GABA_B$ receptors apparently function as autoreceptors. Local stimulation of presynaptic autoreceptors may reduce local secretion of endogenous GABA and thus free GnRH neurones from chronic inhibition.

Although activation of the $GABA_A$ and $GABA_B$ receptor subtypes obviously altered LH release, results from our studies addressing the issue of whether these receptor types mediate testosterone negative feedback are more difficult to interpret. The $GABA_A$ receptor antagonist bicuculline methiodide (BMI) and the $GABA_B$ receptor antagonist CGP 55854A were administered during the breeding season into only the ARC–VMR of castrated males and castrated testosterone-treated males. The expectation was that the $GABA_A$ antagonist would increase LH. However, BMI consistently suppressed LH release in castrated males and failed to increase LH secretion in the testosterone-treated males. CGP 55854A was without effect in either group. The apparently paradoxical effects of BMI might be explained by inhibitory effects of this drug on *N*-methyl-D-aspartic acid (NMDA) receptors or other neurotransmitters (Svenneby and Roberts, 1973; Miller and

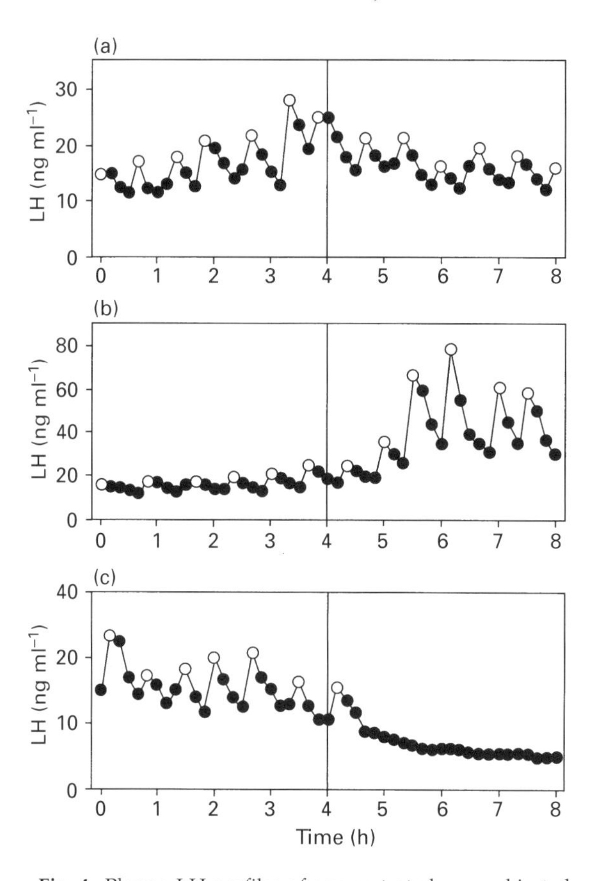

Fig. 4. Plasma LH profiles of one castrated ram subjected to separate sequential bilateral microdialysis infusion of artificial cerebrospinal fluid (aCSF) only (a), or aCSF followed by 1 mmol baclofen l^{-1} (b), or aCSF (c) followed by 1 mmol muscimol l^{-1} into the arcuate–ventromedial region of the hypothalamus. The drug concentrations listed are those of the dialysis solution. It is estimated that the total doses of baclofen and muscimol delivered at each site were 7.9 and 4.5 µg, respectively. Note the enlarged scale of the middle panel. Peaks of LH pulses are represented by hollow circles. Drug delivery started at 4 h, as indicated by the vertical line. (Reproduced from Ferreira *et al.*, 1996.)

McLennan, 1974; Krebs *et al.*, 1994; Musshoff *et al.*, 1994) and in retrospect indictaes that studies using this antagonist are interpreted with caution.

However, it should be noted that although BMI injected into the POA of ovariectomized ewes consistently reduced LH when given during the breeding season, it increased LH in some oestrogen-treated ovariectomized ewes when given during the anoestrous period (Scott and Clarke, 1993b). Thus, the effect of BMI on LH may vary with steroidal background or season of treatment. Possibly, during the breeding season the GABAergic system is relatively inactive, particularly in the absence of gonadal steroids, and the only observable effect of BMI on LH secretion is inhibition due to blockade of other essential stimulatory systems attained by relatively high doses of the drug. In contrast, during the non-breeding season, when GABA activity is postulated to be increased by

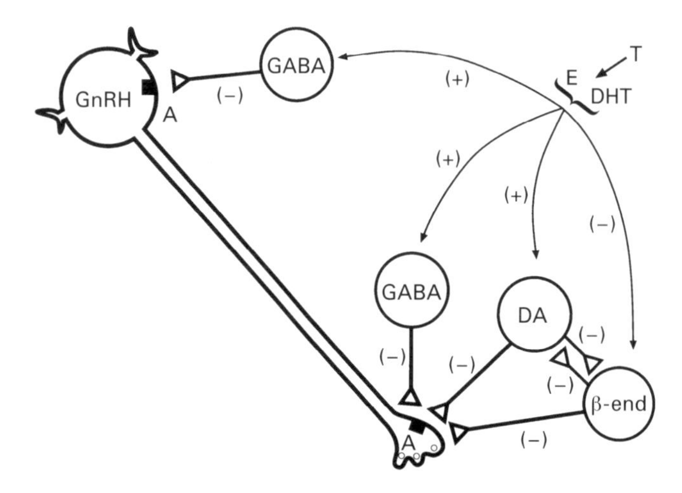

Fig. 5. Schematic illustration of the possible mechanisms by which testosterone (T) controls release of gonadotrophin-releasing hormone (GnRH) in male sheep. (+): stimulatory action; (−): inhibitory action; E: oestradiol, DHT: dihydrotestosterone; β-end: β endorphin secreting neurone; DA: dopamine secreting neurone; GABA: γ-aminobutyric acid secreting neurone; A: $GABA_A$ receptor.

gonadal steroids, the GABAergic system may become more sensitive to blockade. Although it is clear that GABA modulates LH secretion in rams, data are insufficient to conclude whether it partially mediates the action of testosterone on LH. Clearly, additional study will be required to resolve these issues.

Summary Model

Results of studies from several laboratories are incorporated into a summary model (Fig. 5) illustrating some of the possible neurochemical pathways by which testosterone alters GnRH release in rams. The first point is that the effects of testosterone are mediated largely, if not exclusively, by the metabolites oestrogen and DHT. These steroids act in the ARC–VMR and probably other sites to reduce GnRH pulse frequency, but not GnRH synthesis. A second point is that these steroids probably do not act directly on GnRH neurones, but act indirectly by modulating secretion of one or more inhibitory neuromodulators such as β-endorphin, dopamine and GABA.

The specific role of β-endorphin remains unclear, but there is overwhelming evidence that it acts tonically to inhibit GnRH release. On the other hand, it does not appear to mediate the inhibitory action of testosterone. Our results support the postulate that testosterone suppresses, not stimulates, synthesis of β-endorphin, particularly when the animals are exposed to a long-day photoperiod. It appears paradoxical that when testosterone is maximally suppressing GnRH release, it also is suppressing a system that inhibits GnRH release. However, the model deals with this by incorporating the possibility that testosterone also increases local dopamine release, particularly during the non-breeding season. Stimulation of dopamine release probably has two outcomes. First, DA may act directly to suppress GnRH release. Second, DA also may act to inhibit β-endorphin secretion (Tortonese and Lincoln, 1994). In our study in which testosterone suppressed POMC mRNA (Hileman *et al.*, 1996), it also activated A14 and A15 dopamine neurones (Lubbers *et al.*, 1995). As β-endorphin neurones contain oestrogen receptors, there may also be a direct effect of this testosterone metabolite. However, it is not clear that these associated changes reflect cause and effect. As indicated in the model testosterone-induced reduction in β-endorphin secretion could also secondarily lead to increased dopaminergic activity. In either case, if the inhibitory effect of dopamine on GnRH secretion is relatively stronger than that of β-endorphin, activation of this network by testosterone will still suppress GnRH release. In addition, it should be noted that the relationship among testosterone and the neurotransmitters is not static. During the breeding season the effectiveness of testosterone in altering the activity of these systems and in suppressing GnRH is reduced. Although the suppression of β-endorphin may be reduced, resulting in greater β-endorphin release, a parallel reduction in dopamine release would ultimately result in increased GnRH release.

In the model we suggest that GABA inhibits GnRH release by acting via $GABA_A$ receptors in both the mPOA and ARC–VMR. However, the indication that GABA partially mediates the action of testosterone is speculative.

Conclusions

There are several important unanswered questions about how testosterone reduces GnRH and LH secretion. Although oestrogen and DHT appear to mediate the action of testosterone, the relative contribution of these two metabolites is unclear. Suppression of either DHT or oestrogen formation partially blocked the inhibitory effects of testosterone on pulsatile LH release. The effect of simultaneously blocking both reductase and aromatase activities has not been determined. In addition, it is not clear exactly where in the hypothalamus or brainstem oestrogen and DHT act to alter GnRH release. Indeed, it is yet to be clearly demonstrated that DHT alters GnRH release. This issue requires further investigation, perhaps using a more androgen-responsive model than the long-term castrated male.

The specific role of the various neuromodulators remains unresolved. We are currently addressing the question of whether testosterone stimulates GABA release in the hypothalamus. However, the specific physiological roles of the GABA receptor subtypes, and the respective roles of the widely distributed GABA neurones in modulating or mediating the effects of testosterone on GnRH release remain to be elucidated fully in the ram.

The specific role of β-endorphin, and of other opiates, also remains elusive. It is unknown whether testosterone alters hypothalamic β-endorphin release and coincidentally whether hypothalamic release of β-endorphin reflects changes in steady state POMC mRNA. Furthermore, it is not known whether testosterone or environmental factors regulate opiate receptors, nor is it clear exactly where the opiates act. The report by Sanella *et al.* (1997) that GnRH neurones lack opiate receptors leads to the suggestion that still other interneurone systems must be involved – perhaps those secreting nitric oxide (Brann and Mahesh, 1997; Lopez *et al.*, 1997).

Dopamine neurones of the A14 and A15 groups appear to be activated by testosterone during long days. Whether this is specific for an inhibitory photoperiod in males as in females is not known. The relative importance and relationship of this subset, and the A12, dopamine neurones to GnRH release has not been determined nor have efferent and afferent pathways to and from these dopamine groups. Given that the A14–A15 dopaminergic neurones apparently contain little oestrogen receptor-α (Lehman and Karsch, 1993) it is not clear how their function is modulated by either photoperiod or steroids.

In addition, there is a fundamental question as to how photoperiod acts to gate the sensitivity of these and possibly other systems to testosterone. Ultimately this involves melatonin, the pineal hormone which transduces photic information into a chemical signal. Although considerable progress has been made (for example Malpaux *et al.*, 1998; Hileman *et al.*, 1994a) neither the neuroanatomical components nor the identity of neural systems involved in this pathway are completely known. Perhaps even more perplexing is the consistent observation that although testosterone or oestrogen administration suppresses LH pulse frequency, pharmacological manipulations of various neurotransmitters have failed to mimic fully the action of either steroid treatment or removal. Consequently this leads to the suggestion that testosterone, its metabolites or both may act through the coordinated activity of several neurotransmitter systems rather than a single system. Thus, it seems likely that much effort will be required before we fully understand the 'relatively simple' pathways by which testosterone regulates GnRH and LH release in males before those by which environmental factors modulate the action of testosterone can be determined.

The authors gratefully recognize the contributions of our colleagues Suzie Ferreira, David Kuehl, Laura Lubbers, and Chris Scott who contributed to the work from this laboratory. These studies were supported by NIH Grant HD 27453 and USDA Grants AG92-37203-8177 and AG95-37203-2033.

References

Barb CR, Kraeling RR and Rampacek GB (1991) Opioid modulation of gonadotropin and prolactin secretion in domestic farm animals *Domestic Animal Endocrinology* **8** 15–27

Belsham DD, Evangelou A, Roy D, Le DV and Brown TJ (1998) Regulation of gonadotropin-releasing hormone (GnRH) gene expression in GnRH-secreting GT1–7 hypothalamic neurons *Endocrinology* **139** 1108–1114

Brann DW and Mahesh VB (1997) Excitatory amino acids: evidence for a role in the control of reproduction and anterior pituitary hormone secretion *Endocrine Reviews* **18** 678–700

Caraty A and Locatelli A (1988) Effect of time after castration on secretion of LHRH and LH in the ram *Journal of Reproduction and Fertility* **82** 263–269

Ebling FJP and Lincoln GA (1985) Endogenous opioids and the control of seasonal LH secretion in Soay rams *Journal of Endocrinology* **107** 341–353

Ferreira SA, Scott CJ, Kuehl DE and Jackson GL (1996) Differential regulation of luteinizing hormone release by γ-aminobutyric acid receptor subtypes in the arcuate–ventromedial region of the castrated ram *Endocrinology* **137** 3453–3460

Ferreira SA, Hileman SM, Kuehl DE and Jackson GL (1998) Effects of dialyzing γ-aminobutyric acid receptor antagonists into the medial preoptic and arcuate ventromedial region on luteinizing hormone release in male sheep *Biology of Reproduction* **58** 1038–1046

Goodman RL, Parfitt DB, Evans NP, Dahl GE and Karsch FJ (1995) Endogenous opioid peptides control the amplitude and shape of GnRH pulses in the ewe *Endocrinology* **136** 2412–2420

Grattan DR and Selmanoff M (1993) Regional variation in γ-aminobutyric acid turnover: effect of castration on γ-aminobutyric acid turnover in microdissected brain regions of the male rat *Journal of Neurochemistry* **60** 2254–2264

Grattan DR, Rocca MS, Sagrillo CA, McCarthy MM and Selmanoff M (1996) Antiandrogen microimplants into the rostral medial preoptic area decrease γ-aminobutyric acidic neuronal activity and increase luteinizing hormone secretion in the intact male rat *Endocrinology* **137** 4167–4173

Handa RJ, Kerr JE, DonCarlos LL, McGivern RF and Hejna G (1996) Hormonal regulation of androgen receptor messenger mRNA in the medial preoptic area of the male rat *Molecular Brain Research* **39** 57–67

Herbison AE (1995) Neurochemical identity of neurones expressing oestrogen and androgen receptors in sheep hypothalamus *Journal of Reproduction and Fertility Supplement* **49** 271–283

Hileman SM, Kuehl DE and Jackson GL (1994a) Effect of anterior hypothalamic area lesions on photoperiod-induced shifts in reproductive activity of the ewe *Endocrinology* **135** 1816–1823

Hileman SM, Lubbers LS, Kuehl DE, Schaeffer DJ, Rhodes L and Jackson GL (1994b) Effect of inhibiting 5α-reductase activity on the ability of testosterone to inhibit luteinizing hormone release in male sheep *Biology of Reproduction* **50** 1244–1250

Hileman SM, Lubbers LS, Petersen SL, Kuehl DE, Scott CJ and Jackson GL (1996) Influence of testosterone on LHRH release, LHRH mRNA and proopiomelanocortin mRNA in male sheep *Journal of Neuroendocrinology* **8** 113–121

Hileman SM, Kuehl DE and Jackson GL (1998) Photoperiod affects the ability of testosterone to alter proopiomelanocortin mRNA, but not luteinizing hormone-releasing hormone mRNA, levels in male sheep *Journal of Neuroendocrinology* **10** 587–592

Huang X and Harlan RE (1993) Absence of androgen receptors in LHRH immunoreactive neurons *Brain Research* **624** 309–311

Jackson GL, Kuehl D and Rhim TJ (1991) Testosterone inhibits gonadotropin-releasing hormone pulse frequency in the male sheep *Biology of Reproduction* **45** 188–194

Kalra SP and Kalra PS (1989) Do testosterone and estradiol-17β enforce inhibition or stimulation of luteinizing hormone-releasing hormone *Biology of Reproduction* **41** 559–570

Krebs MO, Kemel ML, Gauchy C, Desban M and Glowinski J (1994) Does bicuculline antagonize NMDA receptors? Further evidence in the rat striatum *Brain Research* **634** 345–348

Lehman MN and Karsch FJ (1993) Do gonadotropin-releasing hormone, tyrosine hydroxylase-, and beta-endorphin-immunoreactive neurons contain estrogen receptors? A double-label immunocytochemical study in Suffolk ewes *Endocrinology* **133** 887–895

Lehman MN, Ebling FJP, Moenter SM and Karsch FJ (1993) Distribution of estrogen receptor-immunoreactive cells in sheep brain *Endocrinology* **133** 876–886

Lehman MN, Durham DM, Jansen HT, Adrian B and Goodman RL (1996) Dopaminergic A14/A15 neurons are activated during estradiol negative feedback in anestrous, but not breeding season, ewes *Endocrinology* **137** 4443–4450

Leranth C, MacLusky NJ, Shanabrough M and Naftolin F (1988) Immunohistochemical evidence for synaptic connections between pro-opiomelanocortin-immunoreactive neurons and LH-RH neurons in the preoptic area of the rat *Brain Research* **449** 167–176

Lopez JL, Moretto M, Merchenthaler I and Negro-Vilar A (1997) Nitric oxide is involved in genesis of pulsatile LHRH secretion from immortalized LHRH neurons *Journal of Neuroendocrinology* **9** 647–654

Lubbers LS, Hileman SM, Jansen HT, Lehman MN and Jackson GL (1995) Testosterone-induced activation of tyrosine hydroxylase-containing neurons of the A14 and A15 hypothalamic nuclei in the male sheep *Abstracts 25th Annual Meeting, Society for Neuroscience* Abstract 745 2, p 1897

McShane TM, Petersen SL, McCrone S and Keisler DH (1993) Influence of food restriction on neuropeptide Y, proopiomelanocortin, and luteinizing hormone-releasing hormone gene expression in sheep hypothalami *Biology of Reproduction* **49** 831–839

Malpaux B, Daveau A, Maurice-Mandon F, Duarte G and Chemineau P (1998) Evidence that melatonin acts in the premamillary area to control reproduction in the ewe: presence of binding sites and stimulation of LH secretion by *in situ* microimplant delivery *Endocrinology* **139** 1508–1516

Miller JJ and McLennan H (1974) The action of bicuculline upon acetylcholine-induced excitations of central neurons *Neuropharmacology* **13** 784–785

Monet-Kuntz C, Hochereau-deReviers MT, Pisselet C, Perreau C, Fontaine I and Schanbacher BD (1988) Endocrine parameters, hormone receptors, and functions of the testicular interstitium and seminiferous epithelium in estradiol-immunized Il-de-France rams *Journal of Andrology* **9** 278–283

Musshoff U, Majeda M, Bloms-Funke P and Speckmann E-J (1994) Effects of epileptogenic agent bicuculline methiodide on membrane currents induced by *N*-methyl-D-aspartate and kainate (oocyte:*Xenopus laevis*) *Brain Research* **639** 135–138

Naftolin F and Ryan KJ (1975) The metabolism of androgens in central neuroendocrine tissues *Journal of Steroid Biochemistry* **6** 993–997

Parrott RF and Davies RV (1979) Serum gonadotropin levels in prepubertally castrated male sheep treated for long periods with propionated testosterone, dihydrotestosterone, 19-hydroxytestosterone or oestradiol *Journal of Reproduction and Fertility* **56** 543–548

Prasad BM, Conover CD, Sarkar DK, Rabji J and Advis JP (1993) Feed restriction in prepubertal lambs: effect on puberty onset and on *in vivo* release of luteinizing hormone-releasing hormone, neuropeptide Y and beta-endorphin from the posterior-lateral median eminence *Neuroendocrinology* **57** 1171–1181

Rhim T, Kuehl D and Jackson GL (1993) Seasonal changes in the relationships between secretion of gonadotropin-releasing hormone, luteinizing hormone, and testosterone in the ram *Biology of Reproduction* **48** 197–204

Robinson JE (1995) Gamma amino-butyric acid and the control of GnRH secretion in sheep *Journal of Reproduction and Fertility Supplement* **49** 221–230

Sagrillo CA and Selmanoff M (1997) Castration decreases single cell levels of mRNA encoding glutamic acid decarboxylase in the diagonal band of broca and the sexually dimorphic nucleus of the preoptic area *Journal of Neuroendocrinology* **9** 699–706

Sannella MI and Petersen SL (1997) Dual label *in situ* hybridization studies provide evidence that luteinizing hormone-releasing hormone neurons do not synthesize messenger ribonucleic acid for mu, kappa, or delta opiate receptors *Endocrinology* **138** 1667–1672

Schanbacher BD (1984) Regulation of luteinizing hormone secretion in male sheep by endogenous estrogen *Endocrinology* **115** 944–950

Scott CJ and Clarke IJ (1993a) Inhibition of LH secretion in ovariectomized ewes during the breeding season by gamma-aminobutyric acid (GABA) is effected by $GABA_A$ receptors, but not by $GABA_B$ receptors *Endocrinology* **132** 1789–1796

Scott CJ and Clarke IJ (1993b) Evidence that changes in the function of the subtypes of the receptors for γ-aminobutyric acid may be involved in seasonal changes in the negative-feedback effects of estrogen on gonadotropin-releasing hormone secretion and plasma luteinizing hormone levels in the ewe *Endocrinology* **133** 2904–2912

Scott CJ, Kuehl DE, Ferreira SA and Jackson GL (1997) Hypothalamic sites of action for testosterone, dihydrotestosterone, and estrogen in regulation of luteinizing hormone secretion in male sheep *Endocrinology* **138** 3686–3694

Selmanoff MK, Brodkin LD, Weiner RI and Sitteri PK (1977) Aromatization and 5α-reduction of androgens in discrete hypothalamic and limbic regions of the male and female rat *Endocrinology* **101** 841–848

Shen ES, Meade EH, Perez MC, Deecher D, Negro-Vilar A and Lopez FJ (1998) Expression of functional estrogen receptors and galanin messenger ribonucleic acid in immortalized luteinizing hormone-releasing neurons: estrogenic control of galanin gene expression *Endocrinology* **139** 939–948

Shivers BD, Harlan RE, Morrell JI and Pfaff DW (1983) Absence of oestradiol concentration in cell nuclei of LHRH-immunoreactive neurones *Nature* **304** 345–347

Shugrue PJ, Lane MV and Merchenthaler I (1997) Comparative distribution of estrogen receptor-α and -β mRNA in the rat central nervous system *Journal of Comparative Neurology* **388** 507–525

Ssewannyana E and Lincoln GA (1990) Regulation of photoperiod-induced cycle in the peripheral blood concentrations of β-endorphin and prolactin in the ram: role of dopamine and endogenous opioids *Journal of Endocrinology* **127** 461–469

Svenneby G and Roberts E (1973) Bicuculline and *N*-methylbicuculline-competitive inhibitors of brain acetylcholinesterase *Journal of Neurochemistry* **21** 1025–1026

Thiery J-C, Gayrard V, LeCorre S, Viguie C, Martin GB, Chemineau P and Malpaux B (1995) Dopaminergic control of LH secretion by the A15 nucleus in anoestrous ewes *Journal of Reproduction and Fertility Supplement* **49** 285–296

Thind KK and Goldsmith PC (1988) Infundibular gonadotropin-releasing hormone neurons are inhibited by direct opioid and autoregulatory synapses in juvenile monkeys *Neuroendocrinology* **47** 203–216

Tilbrook AJ and Clarke IJ (1992) Evidence that dopaminergic neurons are not involved in the negative feedback effect of testosterone on luteinizing-hormone in rams in the non-breeding season *Journal of Neuroendocrinology* **4** 365–374

Tilbrook AJ and Clarke IJ (1995) Negative feedback regulation of the secretion and actions of GnRH in male ruminants *Journal of Reproduction and Fertility Supplement* **49** 297–306

Tortonese DJ and Lincoln GA (1994) Photoperiodic modulation of dopaminergic control of pulsatile LH secretion in sheep *Journal of Endocrinology* **143** 25–32

Viguie C, Thibault J, Thiery JC, Tillet Y and Malpaux B (1997) Characterization of the short day-induced decrease in median eminence tyrosine hydroxylase activity in the ewe: temporal relationship to the changes in luteinizing hormone and prolactin secretion and short day-like effect of melatonin *Endocrinology* **138** 499–506

Journal of Reproduction and Fertility Supplement **52**, 243–257

Role of male–female interaction in regulating reproduction in sheep and goats

S. W. Walkden-Brown[1], G. B. Martin[2] and B. J. Restall[3]

[1]*Animal Science, School of Rural Science and Natural Resources, University of New England, Armidale, NSW 2351, Australia;* [2]*Faculty of Agriculture (Animal Science), The University of Western Australia, Nedlands 6907, Australia;* [3]*Capratech Consulting, 822 Teven Rd., Teven, NSW 2478 Australia.*

The induction of synchronous ovulatory activity in anovulatory sheep and goats after the introduction of males, the 'male effect', has probably been used to advantage since these species were domesticated and the underlying physiological and behavioural mechanisms have been progressively elucidated over the past 50 years. Less well understood is the analogous effect of oestrous females on males. This review examines the nature and importance of these male–female interactions in sheep and goats, and describes the most important internal and external factors influencing the reproductive outcomes of such interactions. It is proposed that the male and female effects are both components of a self-reinforcing cycle of stimulation that, under ideal conditions, culminates in the synchronous very rapid onset (within days) of fertile reproductive activity. However, precisely because of the speed of this response, it is suggested that mechanisms have evolved to limit its efficacy, and thus prevent conception at inappropriate times. The complexity of these factors and the interactions between them are highlighted, and a broad conceptual framework for understanding them is proposed based upon an appreciation of variation in both the responsiveness of the target animal and the quality of the signal from the signalling animal.

Introduction

In some production systems for small ruminants, reproduction is strictly controlled by isolating females from males apart from a brief mating period, whereas in other systems, males and females are not physically segregated at all. Irrespective of the mating system used, it is clear that there are times when male–female interaction triggers a cascade of physiological events culminating in successful conception, often highly synchronous, times when no response is evident, and times when an intermediate response is observed. In this review the nature of male–female interactions in two species, sheep and goats, is examined and the most important factors influencing the reproductive outcome of such interactions are described. The complexity of these factors and their interactions is highlighted, and a broad conceptual framework for understanding them is proposed. By virtue of its breadth, the review will not cover all of the detailed physiology underlying the responses.

Nature and Importance of Male–Female Interaction in Sheep and Goats

An evolutionary perspective

Reproduction in wild sheep and goats is characterized by long periods of segregation between mature males and females while the females are anoestrous (before puberty, during gestation and lactation, during photoperiodic or nutritional anoestrus) interspersed with shorter periods during which the male and female herds combine, often accompanied by intense sexual activity (Shackelton and Shank, 1984). Periodic reuniting of the sexes appears to have two important roles in these animals. The first is to initiate breeding activity by inducing females to make the transition from the anoestrous to the oestrous state earlier than in the absence of males, in which case other environmental and physiological factors determine the timing. The second is to ensure synchrony of the female cycles, primarily to enable synchronous parturition as a defence mechanism against predators. This 'fine tuning' of the timing of the end of anoestrus by social cues remains strongly evident in domestic breeds, although the extent to which it can be used to advance the spontaneous onset of fertile oestrus varies greatly among breeds and species.

Female responses to male introduction – the 'male effect'

The induction of synchronous ovulatory activity in anovulatory females after the introduction of males, the so-called 'male effect', has been widely documented for sheep and goats (reviews: Martin *et al.*, 1986; Chemineau, 1987). In responsive ewes and does, the primary response to the male stimulus is an immediate increase in LH pulse frequency (Martin *et al.*, 1980; Chemineau *et al.*, 1986a), reflecting increased pulsatile secretion of GnRH (Hamada *et al.*, 1996). The secretion of LH increases within minutes and, in fully responsive ewes, leads to a preovulatory surge of LH 6–52 h after introduction of the male, followed by ovulation 23–24 h later (Oldham *et al.*, 1978). In ewes, this initial ovulation is invariably 'silent' (that is, not accompanied by oestrus) and may be followed by a short luteal phase and a second silent ovulation 6–7 days later or a normal luteal phase and ovulation with oestrus about 18 days later, or both in sequence. In each case, if the response is maintained, subsequent ovulations follow a luteal phase of normal duration and are associated with oestrus. The situation in responsive female goats differs in that a variable number of does exhibit oestrus at the first induced ovulation, 2–3 days after introduction of the male, and the majority go on to have a short luteal phase followed by a second ovulation 5–7 days later, invariably accompanied by oestrus. As in the sheep, subsequent ovulations follow a normal luteal phase and are accompanied by oestrus. Administration of progesterone or progestagens to ewes and does just before introduction of the male prevents the abnormal luteal phases and can be used to ensure oestrus and fertility at the first or second induced ovulation (Cognié *et al.*, 1982; Chemineau, 1985).

Male responses to exposure to females – the 'female effect'

In both the ram and goat buck, exposure of males to oestrous females may stimulate an immediate increase in LH secretion, analogous to the male effect in females (Sanford *et al.*, 1974; Howland *et al.*, 1985). Consequently, this has been termed the 'female effect' (Fig. 1). Testosterone secretion is stimulated by the increase in LH secretion (Sanford *et al.*, 1974; Schanbacher *et al.*, 1987; Walkden-Brown *et al.*, 1994a) and there are also increases in plasma concentrations of FSH, cortisol and prolactin (Howland *et al.*, 1985; Gonzalez *et al.*, 1988a,b; Borg *et al.*, 1992). Increased secretion of gonadotrophins and androgen is associated with courtship behaviour, whereas increased concentrations of cortisol and prolactin are associated with mounting and intromission (Borg *et al.*, 1992). Increases in cortisol and prolactin are also induced by electroejaculation in the absence of females (Martin *et al.*, 1984) indicating that they represent a non-specific stress response. Homosexual rams fail to exhibit an LH or testosterone response to oestrous ewes (Perkins and Fitzgerald, 1992).

The functional significance of the female effect has yet to be determined. It is possible that the increase in testosterone secretion leads to behavioural changes that improve the success of mating although reproductive behaviour does not appear to be acutely sensitive to testosterone concentrations (Mattner and Braden, 1975). However, exposure to oestrous females clearly enhances the ability of rams and goat bucks to induce ovulation in seasonally anovulatory females (Knight, 1985; Walkden-Brown *et al.*, 1993a) suggesting that the female effect is one component of a self-reinforcing cycle of stimulation that may be initiated by either sex.

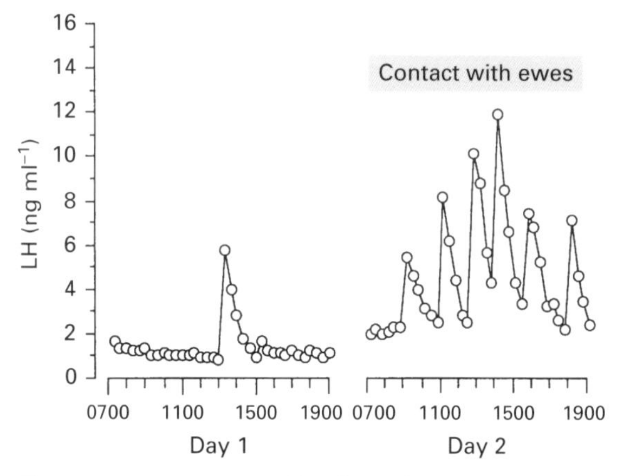

Fig. 1. Representative profile of peripheral LH concentration in a mature Merino ram before and during exposure to oestrous ewes. Samples were collected at 20 min intervals for the same 12 h period on successive days with ewes placed in pens beside rams after the first blood sample on day 2. (From data of Walkden-Brown et al., 1993d)

Signalling mechanisms – the male stimulus

In sheep, the male effect involves a range of sex steroid-dependent stimuli. Long-term castrates (wethers) are ineffective unless they have been treated with androgen or oestrogen and androgen-treated ovariectomized ewes can also induce ovulation (Fulkerson *et al.*, 1981; Signoret *et al.*, 1982). The odour of fleece, but not urine, may induce a full ovulatory response (Knight and Lynch, 1980; Knight *et al.*, 1983a), and anosmic ewes exhibit a depressed ovulatory response to rams (Morgan *et al.*, 1972), but olfactory cues are not the only male stimulus responsible. In a series of experiments, Pearce and Oldham (1988) showed that rams in full contact with ewes invariably induced a greater ovulatory response than rams separated from ewes by clear or opaque fences. They concluded that tactile stimuli were important in the male effect. Anosmic ewes also exhibit a normal LH response to rams, but not rams fleece, suggesting that other cues are involved (Cohen-Tannoudji *et al.*, 1986).

The situation appears to be similar in goats. Bucks in direct contact with does induce a greater ovulatory response than those separated by a fence, a narrow passage or a solid partition (Shelton, 1980; Chemineau, 1987). Exposing does to buck fleece alone may induce a partial ovulatory response (Shelton 1980; Claus *et al.*, 1990; Walkden-Brown *et al.*, 1993b) but urine odours do not appear to be important in mediating the effect (Walkden-Brown *et al.*, 1993b). Anosmia had no effect on the proportion of does exhibiting an increase in LH secretion in response to introduction of a buck, but it reduced by half the number of does ovulating in the study of Chemineau *et al.* (1986b). These observations suggest that the male stimulus is multi-sensory, possibly involving olfactory, visual, tactile and auditory cues and that, as in the sheep, the intensity of the male stimulus is important.

Considerable research effort has been expended on the olfactory component of the stimulus, encouraged by reports that the full ovulatory response of ewes is induced by olfactory cues from rams (Knight and Lynch, 1980; Knight *et al.*, 1983a), and reports that goat bucks and extracts of buck hair are able to induce ovulation in seasonally anovulatory ewes, in some cases as effectively as rams (Knight *et al.*, 1983a; Birch *et al.*, 1989). This interesting interspecies interaction led to suggestions that a pheromone common to both species is a necessary component of the multi-sensory complex needed to induce the male effect (see Martin *et al.*, 1986 for discussion on the use of the term pheromone in this context). Despite some work on ram fleece (review; Signoret, 1991), most work on the isolation of a male pheromone in small ruminants has concentrated on the goat buck because of the pronounced buck odour associated with the seasonal rut in this species, and despite the dangers

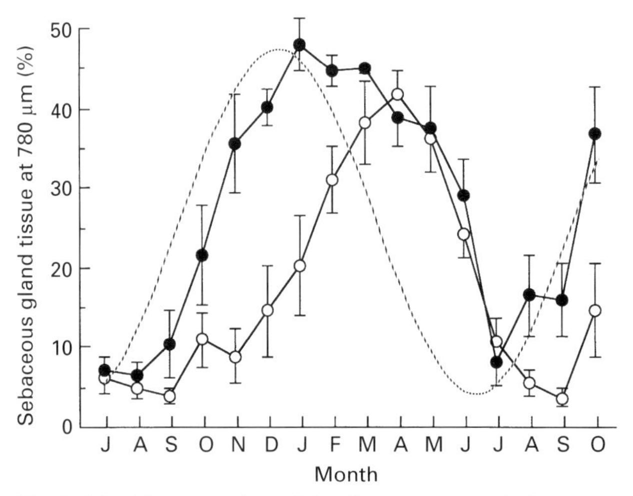

Fig. 2. Monthly means (± SEM) for the percentage of sebaceous gland tissue at a skin depth of 780 µm in samples taken from the occipital region of 3-year-old cashmere bucks fed diets of Low (○, $n = 6$) or High (●, $n = 6$) quality *ad libitum* under natural photoperiod at 29°S for 16 months. The dashed curve represents the annual curve of photoperiod (units not shown, range 10.3–14.0 h). Note the season × nutrition interaction resulting in increased sebaceous gland activity during spring and summer. Adapted from Walkden-Brown *et al.* (1994b).

inherent in such a subjective and anthropomorphic hypothesis. The rutting odour of the goat has been shown to be a testis-dependent component of sebum that varies with the activity of sebaceous glands especially in the head and neck region (Jenkinson *et al.*, 1967; Walkden-Brown *et al.*, 1994b; Hillbrick and Tucker, 1996). This variation is markedly influenced by both season and level of nutrition (Fig. 2). During the breeding season, the mean lipid content of buck fleece may reach 7% by weight (Hillbrick and Tucker, 1996). Both the fatty acid and non-acid components of buck hair extracts exhibit pheromonal activity, although 4-ethyloctanoic acid, which is largely responsible for the characteristic strong odour of bucks, does not (Birch *et al.*, 1989; Claus *et al.*, 1990). Taking these observations together, it is possible that a 'cocktail' of compounds is involved. This idea is supported by the demonstrable chemical complexity of the lipid fraction of buck hair (Jenkinson *et al.*, 1967; Sugiyama *et al.*, 1986) which contains up to 29 specific fatty acids that are not found in the fleece of wethers or does (Hillbrick *et al.*, 1995).

Signalling mechanisms – the female stimulus

There is considerable evidence that olfactory cues from the female are important in sexual communication in sheep, particularly in the detection of oestrus. Rams are able to discriminate between urine from oestrous and non-oestrous ewes, primarily using the main olfactory system, rather than the vomeronasal olfactory system (Blissitt *et al.*, 1990). However, non-oestrous ewes are able to induce LH and testosterone responses in rams, albeit to a lesser degree than oestrous ewes (Gonzalez *et al.*, 1991a), and olfactory cues appear to play a minor role in inducing endocrine responses to either oestrous or non-oestrous ewes in males. Olfactory stimulation of rams with female urine, wool and vaginal secretions fails to induce, and surgical anosmia fails to abolish the LH and testosterone response of experienced rams to oestrous females (Gonzalez *et al.*, 1991b). On

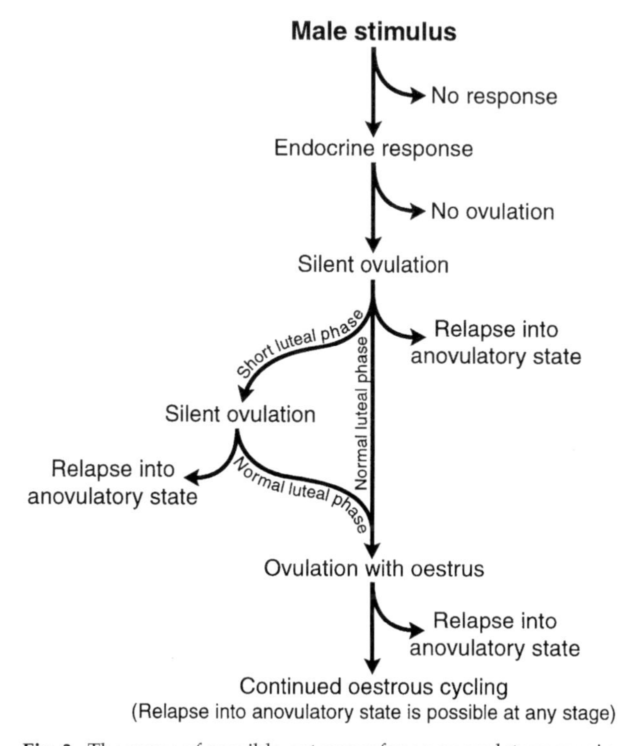

Fig. 3. The range of possible outcomes for an anovulatory ewe in response to a male stimulus. Note that the response can be terminated at any stage providing the ewe with protection against conception at an inappropriate time. Continuation of the response depends upon responsiveness of the ewe and continuation of the male stimulus. The situation in the goat is different in that many does exhibit oestrus at the first ovulation (though usually infertile) and almost all exhibit fertile oestrus at a second ovulation following a short luteal phase.

the basis of these data, olfactory cues are more important in mediating the 'male effect' than the 'female effect' and in both of these, olfactory cues form only part of a multi-sensory stimulus.

Variation in response

A key feature of male–female interaction in sheep and goats is variation in the response after contact with the other sex. This ranges from a failure to elicit any endocrine response to the stimulus, detection of an endocrine response without behavioural or other physiological sequelae, development of a significant physiological or behavioural response that is not sustained, through to a sustained transition from an infertile to a fertile state (Fig. 3). This variation in response may be due to either variation in responsiveness of the target animal or variation in the quality of the stimulus provided by the signalling animal.

Both of these factors appear to be important in determining the type of response obtained after male–female interaction, and both of them seem to be influenced by complex interactions between a number of internal and external environmental factors. This concept is illustrated in Fig. 4 and the major determinants of responsiveness and stimulus quality are discussed below.

Variation in target animal responsiveness. In females, variation in responsiveness to males is generally reported as differences in 'depth of anoestrus'. Although the term was originally used to

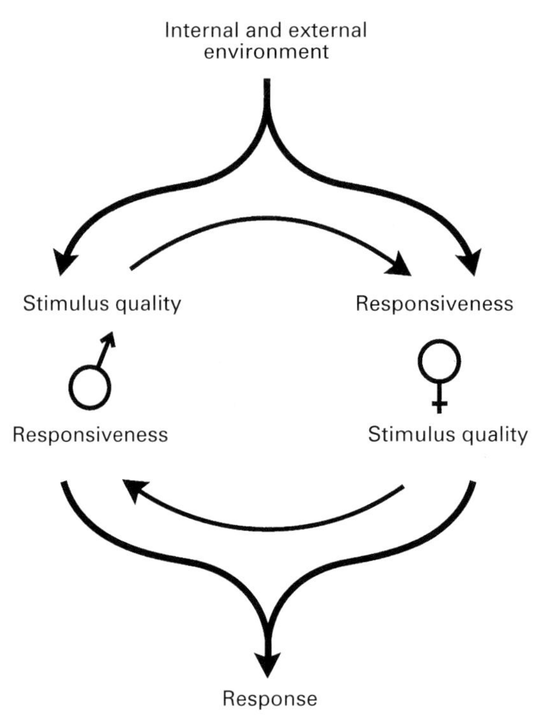

Fig. 4. A conceptual framework for considering male–female interaction. Internal and external factors influence responses by acting on responsiveness and stimulus quality in both sexes. Increases in one stimulate increases in the other in a self-reinforcing cycle that may be damped by environmental and physiological constraints.

describe variation in the extent of photoperiodic inhibition of reproduction in the ewe, its meaning has now been broadened to refer to the extent of inhibition of the reproductive axis in the female irrespective of cause (for example Martin *et al.*, 1986; Chemineau, 1987). If the male effect in goats is used as an example, increasing depth of anoestrus seems to be associated with:

(1) Reductions in the proportion of does ovulating, or total failure to ovulate (Chemineau, 1987; Restall, 1992);
(2) Reductions in the proportion of does exhibiting oestrus at first ovulation (Chemineau, 1983; Walkden-Brown *et al.*, 1993c);
(3) Increases in the proportion of responding does having a short luteal cycle after the first ovulation (Chemineau, 1983);
(4) Increases in the interval (up to a week) between introduction of the buck and the first ovulation (Chemineau 1983, 1987);
(5) Increases in the proportion of does returning to an anovulatory state after one or more ovulations (Chemineau *et al.*, 1986a).

Indeed, the extent of responsiveness to the male effect has been used by Restall (1992) to partition the reproductive cycle of female goats into active, responsive and quiescent periods.

There is no universal measure of 'depth of anoestrus', but because the proportion of anovulatory ewes ovulating in response to introduction of the ram is positively associated with the proportion of ewes ovulating spontaneously in the flock before introduction of the male (Lindsay and Signoret,

1980), either variable is a useful indicator of it. However while the 'depth of anoestrus' concept is a useful means of integrating all of the inhibitory influences on the reproductive axis into a single measure, it has two drawbacks. The first is that the term 'depth of anoestrus' is inappropriate to apply to males although the underlying concept applies in males as well as females. The second drawback is that measures of 'depth of anoestrus' based upon responses to male introduction confound true 'depth of anoestrus' in the female with variation in signal strength from the male and probably result in underestimation of the importance of the latter in explaining variation in responses. For these reasons less ambiguous terminology to describe the extent of responsiveness will be used in the remainder of the review.

Variation in the quality of the stimulus. The quality of the sexual stimulus appears to be a function of the intensity, duration and complexity of the sexual stimulus, with each influencing the type of response obtained. Thus, in both sheep and goats, the degree and type of separation from males influences the ovulatory response obtained (Shelton, 1980; Pearce and Oldham, 1988), and the continued presence of the male is required for maximum ovulatory response or persistence of the induced ovulatory activity (Signoret *et al.*, 1982; Murtagh *et al.*, 1984a). Similarly, the LH response to rams is sensitive to the continued presence of the male stimulus, with short-term stimulation (< 24 h) failing to induce ovulation or maintain the increase in LH pulse frequency (Signoret *et al.*, 1982; Cohen-Tannoudji and Signoret, 1987). The ovulatory response in both sheep and goats has been shown to vary with male libido; high libido males induce the greater response (Fig. 5; Signoret *et al.*, 1982; Perkins and Fitzgerald, 1994). Anosmia presumably reduces the complexity of the stimulus received and this may explain the low ovulatory response of anosmic females to males, as observed in both species (Morgan *et al.*, 1972; Chemineau *et al.*, 1986b).

Less is known about the 'female effect', except that oestrous ewes elicit greater LH responses in rams than do non-oestrous ewes (Gonzalez *et al.*, 1991a) suggesting a difference in stimulus quality. Direct contact between rams and ewes is also required to obtain the acute LH response, and separation by as little as 30 cm is sufficient to prevent it (Gonzalez *et al.*, 1988a).

Internal and External Determinants of Responsiveness and Stimulus Quality

The ways that internal and external factors can modulate male–female interactions and responses are inherently complex because those factors often interact amongst themselves to influence the reproductive axis, they may have similar or different effects on the two sexes, and their actions may influence both the responsiveness to stimuli from the other sex and the quality of the stimulus provided for the other sex.

Novelty of the stimulus

Originally it was considered that about a month of separation between the sexes was required before an efficient 'male effect' could be obtained. However, subsequent work has shown that exposure to rams for 2–3 h has no effect on the responsiveness of ewes to re-exposure to rams as little as 24 h later (Cohen-Tannoudji and Signoret, 1987), and that the introduction of novel rams to ewes that are already in contact with rams will induce a high proportion of anovulatory ewes to ovulate (Pearce and Oldham, 1988; Cushwa *et al.*, 1992). This finding suggests that the novelty of the male stimulus, rather than its presence or absence, is most important and implies that ewes become refractory to an unchanging male stimulus. This contention is supported by findings that ewes maintained continuously with rams exhibit a seasonal pattern of oestrous activity which is similar to that of isolated ewes, whereas ewes that are intermittently exposed to rams show a higher incidence of oestrus during the anoestrous period (Riches and Watson, 1954; Lishman, 1969). On the other hand, Notter (1989) found that anovulatory ewes in continuous association with rams were less responsive to introduction of a novel ram than were isolated ewes. He suggested that a novel ram stimulus cannot totally overcome the effects of refractoriness. In contrast to sheep, female goats do not appear to become refractory to continuous exposure to bucks, because does exhibit the same

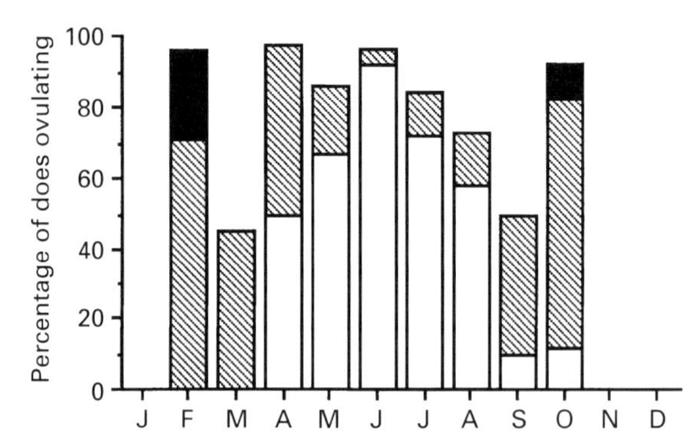

Fig. 5. Effect of varying the male stimulus on responses of Australian cashmere goats to the introduction of males. Data derived from Restall (1992) illustrate the proportion of does ovulating spontaneously (□) and ovulating 14 days after exposure to vasectomized males (▧) throughout the year in a paddock (data adjusted to mid-point of each month). The data from Walkden-Brown *et al.* (1993c) (columns with solid section at the top) are from experiments carried out when does were unresponsive in Restall's experiment (February and October). They illustrate the ovulatory responses of seasonally anovulatory cashmere does 10 days after being exposed to normal (▧) or 'enhanced' (■) males in 100 m^2 enclosures containing a buck and 10 does (threre replicates for each treatment). Bucks were 'enhanced' by improved nutrition or exposure to oestrous females before introduction. In the February experiment the 'enhanced' bucks were exposed to oestrous females for 2 days before introduction, whereas in the October experiment the 'enhanced' bucks had been on a high quality *ad libitum* diet for 16 months before introduction and had been exposed to oestrous females at intervals of 2 months. The high ovulatory response to normal bucks reported by Walkden-Brown *et al.* (1993c) suggests that enclosing males and females in a relatively confined space increases the response.

prolonged breeding season when they are in constant contact with bucks and when they are exposed to bucks intermittently throughout the year (Cameron and Batt, 1989; Restall, 1992). Under both conditions, the breeding season is initiated earlier and terminated later compared with that of does isolated from bucks.

Again, the roles of refractoriness and novelty for the 'female effect' are less well documented but they are likely to be just as important. Sanford *et al.* (1974) found that the LH and testosterone response of rams to oestrous ewes declined to basal values after about 12 h, despite continued sexual activity, suggesting some refractoriness to the female stimulus. Thiéry and Signoret (1978) found that male reproductive behaviour declined rapidly within 5 min of the introduction of an oestrous ewe. Introduction of a new ewe, but not re-introduction of the same ewe, induced reproductive behaviours that were not different from those induced by the initial introduction.

Requirement for prior sexual experience

Prior sexual contact with males does not appear to be an important requirement for the male effect, and ovulatory responses are evident in many experiments using maiden females with limited post-weaning contact with males. However, Murtagh *et al.* (1984b) showed that exposing 11-month-old maiden ewes to rams improved their responsiveness to introduction to a ram 4 months later, albeit by a small amount. In goats, Walkden-Brown *et al.* (1993b) found that the ovulatory response to introduction to a buck in 30-month-old does was the same whether they had been isolated from males since weaning at 4 months of age, or had been exposed to vasectomized bucks for 10 days a year earlier.

The role of sexual experience in the female effect appears to be more important because sexually naive rams exhibit a smaller response in LH and testosterone secretion after exposure to females than do experienced males (Gonzalez *et al.*, 1991a; Borg *et al.*, 1992). Naive rams also show similar responses to both oestrous or anoestrous ewes, whereas experienced rams show a much greater response to oestrous ewes, suggesting a greater ability to detect oestrus (Gonzalez *et al.*, 1991a). These variations in endocrine response may reflect differences in sexual activity (Gonzalez *et al.*, 1991a; Borg *et al.*, 1992; Perkins and Fitzgerald, 1992), although other studies have shown a lack of association between sexual activity and endocrine response (Gonzalez *et al.*, 1988a,b).

Effect of recent sexual stimulation

Exposing rams (Knight, 1985) or goat bucks (Walkden-Brown *et al.*, 1993a) to oestrous females, shortly before or during introduction to anovulatory females, increases the magnitude of the ovulatory response obtained. As discussed above, this is probably a major function of the 'female effect' in the reproductive strategy of the species. In sheep this mechanism probably explains reports that oestrous females induce ovulatory activity in anovulatory females, because direct female–female stimulation does not appear to be responsible (Knight, 1985). In contrast, goat does in oestrus are able to induce ovulation in anovulatory does both directly or via enhancement of the male effect (Walkden-Brown *et al.*, 1993a; Restall *et al.*, 1995).

Stage of reproductive cycle

The reproductive consequences of the male effect are greatest during anoestrus, although there is evidence that introduction of the male can induce a degree of synchronization in cyclic Creole goats (Chemineau, 1983). One mechanism for this may be a shortening of the follicular phase which has been reported after introduction of the male to cyclic ewes (Martin *et al.*, 1986). Whether the introduction of males can influence the duration of the luteal phase of the oestrous cycle has not been clearly established. In acyclic females, introduction of the male advances the onset of puberty and advances the end of post-partum and seasonal anoestrus (see for example Amoah and Bryant, 1984; Geytenbeek *et al.*, 1984; Martin and Scaramuzzi, 1983). Effects on pubertal advancement are constrained by the maturational development of the animal and the photoperiodic environment, but there is little doubt that in many controlled breeding situations involving seasonal breeds, the initial ovulation and oestrus in pubertal ewes and does is triggered by the introduction of rams and bucks. In the case of post-partum anoestrus, Geytenbeek *et al.* (1984) found that the ovulatory responses to males increased with time after parturition in autumn-lambing Merinos, but there was no effect on the timing of the first post-partum oestrus. However, responses are strongly influenced by both the time of year and the nutritional status of the ewes with poor nutritional status able to inhibit ram-induced oestrus (but not ovulation) and facilitate the cessation of male-induced oestrous cycles (Wright *et al.*, 1990).

Photoperiod

Photoperiod is undoubtedly the most important single regulator of reproductive activity in sheep and goats worldwide and is the basis of most of the 'seasonal breeding' observed in these species (for reviews see: Ortavant *et al.*, 1988; Walkden-Brown and Restall, 1996). When considering the effects of photoperiod on male–female interaction the following concepts are important:

(1) Photoperiod acts by selective inhibition of the reproductive axis at various times of the year, to ensure the birth and rearing of the young at the most favourable time of year. Generally speaking, inhibition is maximal during periods of increasing or long daylength (late winter through early summer).
(2) Responsiveness to the effects of photoperiod varies widely and is heritable. This is the basis of breed variation in seasonality. For this reason the effects of photoperiod must be considered both in terms of the current (and preceding) photoperiodic milieu and the photoresponsiveness of the breed in question.
(3) Within a given breed, photoperiodic inhibition of reproduction in females is greater than in males, probably because the adverse consequences of conception at an inappropriate time are much greater for females than for males. Thus, while long periods of seasonal anovulation and anoestrus are common in females, total cessation of sperm production is rare in males. Similarly, in photoresponsive breeds, testicular growth generally commences during increasing daylength in spring–summer and testicular size peaks in late summer–early autumn, well before the peak in spontaneous ovulatory activity in females in late autumn or early winter (Ortavant *et al.*, 1988). However, in practice, male–female interaction ensures that most sheep and goats mate at the peak of the male cycle and early in the female cycle of responsiveness.
(4) The effects of photoperiod interact with other environmental and physiological cues to regulate the timing and magnitude of most reproductive variables and should not be considered in isolation.

Photoperiod and the male effect. Photoperiod is a major cause of variation in the degree of female responsiveness to males, whether it be at the pubertal, post-partum or seasonal transitions from anoestrus. In breeds exhibiting only moderate seasonality such as the Merino sheep and Creole goats, introduction of the male may induce an ovulatory response at any time of the year, although responses will vary with the extent of photoperiodic inhibition (Lindsay and Signoret, 1980; Chemineau, 1983). On the other hand, in breeds that are more strongly seasonal, the male effect may advance the onset of the normal breeding season by as little as a few weeks (Martin and Scaramuzzi, 1983; Chemineau, 1987). Importantly, the limitation on the usefulness of the male effect that is imposed by photoperiodic inhibition of female responsiveness might be overcome under experimental conditions by measures that increase the quality of the male stimulus. For example, in Australian cashmere goats in spring and summer, the ovulatory response to introduction of a buck is low or absent under normal paddock conditions (Cameron and Batt, 1989; Restall, 1992), whereas joining males and females in small groups in small enclosures, under experimental conditions that possibly impose a greater degree of contact between them, induces ovulatory responses in a high proportion of does (Fig. 5). Similarly Chemineau *et al.* (1986a) obtained full ovulatory responses in the markedly seasonal Saanen goat breed during deep anoestrus in late spring–early summer by using androgenized females and bucks treated with a light–melatonin regimen. From these observations, the quality of the male stimulus is an important contributor to the failure of the male effect during seasonal anoestrus, a concept supported by observations that dairy goat bucks used for out-of-season matings with light-treated does also require light treatment to ensure adequate libido (Ashbrook, 1982).

Photoperiod and the female effect. The effects of photoperiod on the female effect are less clear cut than those for the male effect. Reasons for this probably include a lower level of photoperiodic inhibition of responsiveness in males (see above) and the wider use of a standardized stimulus, namely females in induced oestrus. Price *et al.* (1994) showed that the sexual performance of rams is not affected by the mode of induction of oestrus in intact or ovariectomized ewes, although it is not known whether this is true for the endocrine response to ewes. In sheep, the greatest response to females is observed in rams during the non-breeding season or when LH pulse frequencies are low (Schanbacher *et al.*, 1987; Gonzalez *et al.*, 1988a), but significant responses can occur during the breeding season (Gonzalez *et al.*, 1988b). In pygmy goat bucks, Howland *et al.* (1985) found that the responses in LH, FSH and testosterone evoked by oestrous females were maximal in summer, absent

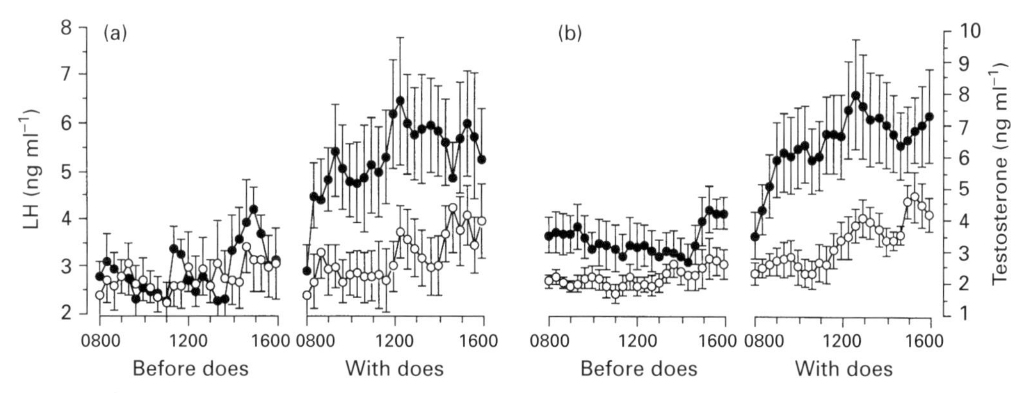

Fig. 6. Effect of oestrous does and diet on mean plasma (a) LH and (b) testosterone concentrations in mature Australian cashmere bucks fed either a low quality diet of pasture hay (○, $n = 6$ or a high quality diet of pelleted lucerne (●, $n = 6$) ad libitum for 16 months. Blood samples were collected at intervals of 20 min between 08.00 and 16.00 h on subsequent days. A doe in oestrus was introduced into each buck pen after the 08.00 h blood sample and was left with the buck until blood sampling ceased. Data for all sampling periods are pooled. Note that an overall effect of diet on LH and testosterone concentration is evident only during exposure to oestrous females (Walkden-Brown *et al.*, 1994a).

during the autumn rut, and small during winter and spring. Australian cashmere bucks exhibit maximal LH and testosterone responses in summer; responses are also evident in autumn and early winter, but not late winter and spring (Walkden-Brown *et al.*, 1994a).

Nutrition

Nutrition is a powerful regulator of reproductive function, but in seasonal breeds of sheep and goat its major effects are on ovulation rate, sperm production and the timing of the pubertal and post-partum transitions. There is little evidence that short-term changes in the nutritional status of females greatly influence their responsiveness to males during seasonal anoestrus (Knight *et al.*, 1983b; Fisher *et al.*, 1993a). However, longer term differences in nutrition may induce differences in the proportion of ewes ovulating during seasonal anoestrus and this in turn may be reflected in altered responsiveness to rams (Fisher *et al.*, 1993a).

In contrast to females, short-term improvements in nutrition in males may induce rapid endocrine and testicular responses (review: Martin and Walkden-Brown, 1995) and these may allow a greater LH response to oestrous ewes (Fisher *et al.*, 1993b). In Australian cashmere goat bucks, longer term improvements in nutrition increased the LH and testosterone response to oestrous females (Fig. 6), but in a seasonally dependent way (Walkden-Brown *et al.*, 1994a). Strong links between the male and female effects were revealed when bucks from this experiment, after 16 months on diets of high and low quality, were placed with does in deep seasonal anoestrus for 10 days. Bucks from the high diet induced more does to ovulate, exhibit oestrus and conceive, than did bucks on the low diet (Walkden-Brown *et al.*, 1993a). The female responses were correlated positively with the serving capacity of the bucks and their testosterone response to oestrous does. In a similar experiment using Merino rams, 19 weeks of differential nutrition did not affect their ability to induce ovulation in late spring, and there was no association between the ovulatory response and male serving capacity (Fisher *et al.*, 1994).

Conclusion

Under optimal conditions, interaction of males with anovulatory but responsive females initiates a self-reinforcing cycle of stimulation that culminates in the synchronous onset of fertile reproductive

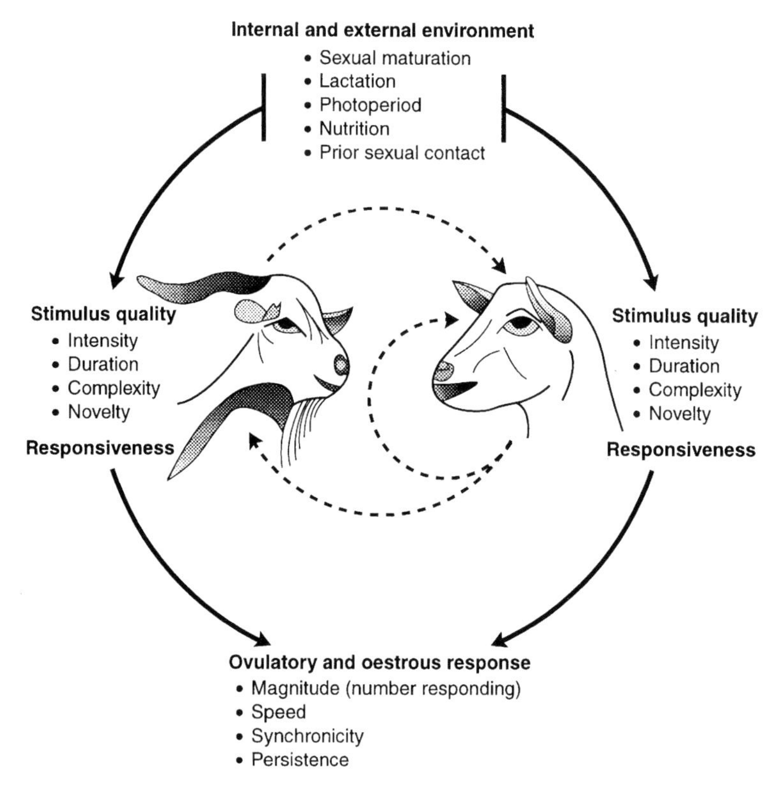

Fig. 7. Schematic representation of male–female interaction in the goat and sheep. Note the self-reinforcing cycle of stimulation that can be initiated by either sex, and which includes female–female stimulation (absent in sheep). A range of factors act on stimulus quality and responsiveness of both males and females to determine the magnitude and persistence of the outcome. Adapted from Walkden-Brown *et al.* (1993c).

activity (Fig. 7). Because this transition can occur so rapidly, it represents a truly opportunistic reproductive response. However, the major metabolic consequences of reproduction in sheep and goats do not occur until 4–6 months after conception, that is late gestation and early lactation, so such opportunism is tempered to limit the extent of inappropriate conception. Thus, the extent to which male–female interaction succeeds in inducing fertile reproductive activity is constrained by a range of interacting factors, particularly by photoperiod and more particularly in females (Fig. 7). This results in a wide range of possible outcomes following the initial stimulation, including termination of the response following ovulation, possibly providing a late escape from an inappropriate opportunistic pregnancy.

In the wild, social facilitation of reproduction probably operates as a means of ensuring synchronous parturition at an appropriate time and, in the longer term, as a means of varying the extent of photoresponsiveness in populations, with less photoresponsive animals continually testing the consequences of advancing the breeding season. This is consistent with the observation that breeding for reduced seasonality advances the onset of the breeding season to a much greater extent than it extends the end of the breeding season. Whether the onset of the breeding season in wild sheep and goats is triggered by a waft of male pheromones or the spontaneous onset of oestrus in a single female genetic outlier is a moot point, although the observation that the breeding season commences earlier in large herds of goats than small herds (Corteel, 1977; Shelton, 1978) lends some credence to the latter.

In domestic sheep and goats, the male effect is a valuable tool for advancing the breeding season of seasonal breeds and for low cost synchronization of reproduction in anovulatory ewes and does. Although some of the variability in response can be removed by appropriate timing of joining, and pretreatment of females with progesterone, our understanding of the importance of the male stimulus has increased in recent years and there is good scope for extending the period during which the male effect can be reliably used by maximizing and standardizing the male stimulus through use of improved male nutrition, exposure of males to oestrous females or use of testosterone-treated castrates.

References

Amoah EA and Byrant MJ (1984) A note on the effect of contact with male goats on the occurrence of puberty in female goat kids *Animal Production* **38** 141–144

Ashbrook PF (1982) Year round breeding for uniform milk production. In *Proceedings of the Third International Conference on Goat Production and Disease Tuscon, Arizona)* pp 153–154 International Goat Association, Rutland, MD

Birch EJ, Knight TW and Shaw GJ (1989) Separation of male goat pheromones responsible for stimulating ovulatory activity in ewes *New Zealand Journal of Agricultural Research* **32** 337–341

Blissitt MJ, Bland KP and Cottrell DF (1990) Olfactory and vomeronasal chemoreception and the discrimination of oestrous and non-oestrous ewe urine odours by the ram *Applied Animal Behaviour Science* **27** 325–335

Borg KE, Esbenshade KL, Johnson BH, Lunstra DD and Ford JJ (1992) Effects of sexual experience, season and mating stimuli on endocrine concentrations in the adult ram *Hormones and Behaviour* **26** 87–109

Cameron AWN and Batt PA (1989) The effect of continuous or sudden introduction of bucks on the onset of the breeding season in female goats *Proceedings of the Australian Society for Reproductive Biology* **21** 109 (Abstract)

Chemineau P (1983) Effect on oestrus and ovulation of exposing Creole goats to the male at three times of the year *Journal of Reproduction and Fertility* **67** 65–72

Chemineau P (1987) Possibilities for using bucks to stimulate ovarian and oestrous cycles in anovulatory goats – a review *Livestock Production Science* **17** 135–147

Chemineau P, Normant E, Ravault JP and Thimonier J (1986a) Induction and persistence of pituitary and ovarian activity in the out of season lactating dairy goat after a treatment combining a skeleton photoperiod melatonin and the male effect *Journal of Reproduction and Fertility* **78** 497–504

Chemineau P, Levy F and Thimonier J (1986b) Effects of anosmia on LH secretion ovulation and oestrous behaviour induced by males in the anovular Creole goat *Animal Reproduction Science* **10** 125–132

Claus R, Over R and Dehnhard M (1990) Effect of male odour on LH secretion and the induction of ovulation in seasonally anoestrous goats *Animal Reproduction Science* **22** 27–38

Cognié Y, Gray SJ, Lindsay DR, Oldham CM, Pearce DT and Signoret JP (1982) A new approach to controlled breeding in the sheep using the 'ram effect' *Animal Production in Australia* **14** 519–522

Cohen-Tannoudji J and Signoret JP (1987) Effect of short exposure to the ram on later reactivity of anoestrous ewes to the male effect *Animal Reproduction Science* **13** 263–268

Cohen-Tannoudji J, Locatelli A and Signoret JP (1986) Non pheromonal stimulation by the male of LH release in the anoestrus ewe *Physiology and Behaviour* **36** 921–924

Corteel JM (1977) Management of artificial insemination of dairy goats through oestrus synchronization and early pregnancy diagnosis. In *Proceedings of the Sheep Industry Development Conference* pp 1–20 Ed. C. Terrill, University of Wisconsin

Cushwa WT, Bradford GE, Stabenfeldt GH, Berger YM and Dally MR (1992) Ram influence on ovarian and sexual activity in anoestrous ewes: effects of isolation of ewes before joining and date of ram introduction *Journal of Animal Science* **70** 1195–1200

Fisher JF, Martin GB, Oldham C and Gray S (1993a) Long term effects of nutrition on spontaneous ovulation in Merino ewes and their responses to the 'ram effect' *Proceedings of the VII World Conference on Animal Production* University of Alberta, Edmonton Alberta Vol. 2 *Short Papers and Abstracts* pp 32–33

Fisher JF, Martin GB, Hughes P, Bouckhliq R and Gray S (1993b) Nutritional effects on luteinizing hormone (LH) secretion in rams after exposure to oestous ewes *Proceedings of the Australian Society for Reproductive Biology* **25** 7 (Abstract)

Fisher JF, Martin GB, Oldham C and Shepherd K (1994) Do differences in nutrition or serving capacity affect the ability of rams to elicit the 'ram effect'? *Animal Production in Australia* **20** 426 (Abstract)

Fulkerson WJ, Adams NR and Gherardi PB (1981) Ability of castrate male sheep treated with oestrogen or testosterone to induce and detect oestrus in ewes *Applied Animal Ethology* **7** 57–66

Geytenbeek PE, Oldham CM and Gray SJ (1984) The induction of ovulation in the postpartum ewe *Animal Production in Australia* **15** 353–356

Gonzalez R, Orgeur P and Signoret JP (1988a) Luteinizing hormone testosterone and cortisol responses in rams upon presentation of estrous females in the non-breeding season *Theriogenology* **30** 1075–1086

Gonzalez R, Poindron P and Signoret JP (1988b) Temporal variation in LH and testosterone responses of rams after the introduction of oestrous females during the breeding season *Journal of Reproduction and Fertility* **83** 201–208

Gonzalez R, Orgeur P, Poindron P and Signoret JP (1991a) Female effect in sheep 1. The effects of sexual receptivity of females and the sexual experience of rams *Reproduction, Nutrition Développement* **31** 97–102

Gonzalez R, Levy F, Orgeur P, Poindron P and Signoret JP (1991b) Female effect in sheep 2. Role of volatile substances from the sexually receptive female; implication of the sense of smell *Reproduction, Nutrition Développement* **31** 103–109

Hamada T, Nakajima M, Takeuchi Y and Mori Y (1996) Pheromone-induced stimulation of hypothalamic gonadotropin-releasing hormone pulse generator in ovariectomized estrogen-primed goats *Neuroendocrinology* **64** 313–319

Hillbrick GC and Tucker DJ (1996) Effect of nutrition on lipid production and composition of cashmere buck fleece *Small Ruminant Research* **22** 225–230

Hillbrick GC, Tucker DJ and Smith GC (1995) The lipid composition of cashmere goat fleece *Australian Journal of Agricultural Research* **46** 1259–1271

Howland BE Sanford LM and Palmer WM (1985) Changes in the serum levels of LH, FSH, prolactin, testosterone and cortisol associated with season and mating in male pygmy goats *Journal of Andrology* **6** 89–96

Jenkinson DM, Blackburn PS and Proudfoot R (1967) Seasonal changes in the skin glands of the goat *British Veterinary Journal* **123** 541–549

Knight TW (1985) Are rams necessary for the stimulation of anoestrus ewes with oestrus ewes? *Proceedings of the New Zealand Society for Animal Production* **45** 49–50

Knight TW and Lynch PR (1980) Source of ram pheromones that stimulate ovulation in ewes *Animal Reproduction Science* **3** 133–136

Knight TW, Tervit HR and Lynch PR (1983a) Effects of boar pheromones, rams wool and presence of bucks on ovarian activity in anovular ewes early in the breeding season *Animal Reproduction Science* **6** 129–134

Knight TW, Hall DRH and Wilson LD (1983b) Effects of teasing and nutrition on the duration of the breeding season in Romney ewes *Proceedings of the New Zealand Society of Animal Production* **43** 17–19

Lindsay DR and Signoret JP (1980) Influence of behaviour on reproduction *Proceedings of the 9th International Congress on Animal Reproduction and Artificial Insemination*, Madrid **1** 83–92

Lishman AW (1969) The seasonal pattern of oestrus amongst ewes as affected by isolation from and joining with rams *Agroanimalia* **1** 95–102

Martin GB and Scaramuzzi RJ (1983) The induction of oestrus and ovulation in seasonally anovular ewes by exposure to rams *Journal of Steroid Biochemistry* **19** 869–875

Martin GB and Walkden-Brown SW (1995) Nutritional influences on reproduction in mature male sheep and goats *Journal of Reproduction and Fertility Supplement* **49** 437–449

Martin GB, Oldham CM, Cognie Y and Pearce DT (1986) The physiological responses of anovulatory ewes to the introduction of rams – a review *Livestock Production Science* **15** 219–247

Martin ICA, Lapwood KR and Elgar HJ (1984) Changes in plasma concentrations of cortisol and prolactin in rams associated with ejaculation of semen. In *Reproduction in Sheep* pp 86–88 Eds DR Lindsay and DT Pearce, Australian Academy of Science, Canberra

Mattner PE and Braden AWH (1975) Studies in the flock mating of sheep 6. Influence of age hormone treatment shearing and diet on the libido of Merino rams *Australian Journal of Experimental Agriculture and Animal Husbandry* **15** 330–336

Morgan PD, Arnold GW and Lindsay DR (1972) A note on the mating behaviour of ewes with various senses impaired *Journal of Reproduction and Fertility* **30** 151–152

Murtagh JJ, Gray SJ, Lindsay DR, Oldham CM and Pearce DT (1984a) The effect of the presence of rams on the continuity of ovarian activity of maiden merino ewes in spring. In *Reproduction in Sheep* pp 37–38 Eds DR Lindsay and DT Pearce. Australian Academy of Science, Canberra

Murtagh JJ, Gray SJ, Lindsay DR and Oldham CM (1984b) The influence of the 'ram effect' in 10–1-month-old Merino ewes on their subsequent performance when introduced to rams again at 15 months of age *Animal Production in Australia* **15** 490–493

Notter DR (1989) Effects of continuous ram exposure and early spring lambing on initiation of the breeding season in yearling crossbred ewes *Animal Reproduction Science* **19** 265–272

Oldham CM, Martin GB and Knight TW (1978) Stimulation of seasonally anovular Merino ewes by rams. I Time from introduction of the rams to the preovulatory LH surge and ovulation *Animal Reproduction Science* **1** 283–290

Ortavant R, Bocquier F, Pelletier J, Ravault JP, Thimonier J and Volland-Nail P (1988) Seasonality of reproduction in sheep and its control by photoperiod *Australian Journal of Biological Science* **41** 69–85

Pearce GP and Oldham CM (1988) Importance of non-olfactory ram stimuli in mediating ram-induced ovulation in the ewe *Journal of Reproduction and Fertility* **84** 333–339

Perkins A and Fitzgerald JA (1992) Luteinizing hormone testosterone and behavioral response of male-oriented rams to estrous ewes and rams *Journal of Animal Science* **70** 1787–1794

Perkins A and Fitzgerald JA (1994) The behavioural component of the ram effect: the influence of ram sexual behavior on the induction of estrus in anovulatory ewes *Journal of Animal Science* **72** 51–55

Price EO, Blackshaw JK, Blackshaw A, Borgwardt R, Dally MR and Bondurant RH (1994) Sexual responses of rams to ovariectomised and intact estrous ewes *Applied Animal Behaviour Science* **42** 67–71

Restall BJ (1992) Seasonal variation in reproductive activity in Australian goats *Animal Reproduction Science* **27** 305–318

Restall BJ, Restall H and Walkden-Brown SW (1995) The induction of ovulation in anovulatory goats by oestrous females *Animal Reproduction Science* **40** 299–303

Riches JH and Watson RH (1954) The influence of the introduction of rams on the incidence of oestrus in Merino ewes *Australian Journal of Agricultural Research* **5** 141–147

Sanford LM Palmer WM and Howland BE (1974) Influence of sexual activity on serum levels of LH and testosterone in the ram *Canadian Journal of Animal Science* **54** 579–585

Schanbacher BD, Orgeur P, Pelletier J and Signoret JP (1987) Behavioural and hormonal responses of sexually-experienced Ile-de-France rams to oestrous females *Animal Reproduction Science* **14** 293–300

Shackelton DM and Shank CC (1984) A review of the social behavior of feral and wild sheep and goats *Journal of Animal Science* **58** 500–509

Shelton M (1960) The influence of the presence of the male on initiation of oestrus cycling and ovulation in Angora does *Journal of Animal Science* **19** 368–375

Shelton M (1978) Reproduction and breeding of goats *Journal of Dairy Science* **61** 994–1010

Shelton M (1980) Goats: influence of various exteroceptive factors on initiation of estrus and ovulation *International Goat and Sheep Research* **1** 156–162

Signoret JP (1991) Sexual pheromones in the domestic sheep: importance and limits in the regulation of reproductive physiology *Journal of Steroid Biochemistry and Molecular Biology* **39** 639–645

Signoret JP, Fulkerson WJ and Lindsay DR (1982) Effectiveness of testosterone treated wethers and ewes as teasers *Applied Animal Ethology* **9** 37–45

Sugiyama T, Matsuura H, Sasada H, Masaki J and Yamashita K

(1986) Characterization of fatty acids in the sebum of goats according to sex and age *Agricultural Biology and Chemistry* **50** 3049–3052

Thiéry JC and Signoret JP (1978) Effect of changing the teaser ewe on the sexual activity of the ram *Applied Animal Ethology* **4** 31–34

Walkden-Brown SW and Restall BJ (1996) Environmental and social factors affecting reproduction. In *Proceedings of the VI International Conference on Goats* (Beijing, 1996) **Vol. 2** pp 762–775 Ed. PJ Host. International Academic Publishers, Beijing

Walkden-Brown SW, Restall BJ and Henniawati (1993a) The male effect in Australian cashmere goats 3. Enhancement with buck nutrition and use of oestrous females *Animal Reproduction Science* **32** 69–84

Walkden-Brown SW, Restall BJ and Henniawati (1993b) The male effect in Australian cashmere goats 2. Role of olfactory cues from the male *Animal Reproduction Science* **32** 55–67

Walkden-Brown SW, Restall BJ and Henniawati (1993c) The male effect in Australian cashmere goats 1. Ovarian and behavioural response of seasonally anovulatory does following the introduction of bucks *Animal Reproduction Science* **32** 41–53

Walkden-Brown SW, Boukhliq R, Fisher JS and Martin GB (1993d) Does the nutrition of the ram influence its behavioural and endocrine response to oestrous ewes? *Proceedings of the Australian Society for Reproductive Biology* **25** 38 (Abstract)

Walkden-Brown SW, Restall BJ, Norton BW and Scaramuzzi RJ (1994a) The 'female effect' in Australian cashmere goats. Effect of season and diet quality on the LH and testosterone response of bucks to oestrous does *Journal of Reproduction and Fertility* **100** 521–531

Walkden-Brown SW, Restall BJ, Norton BW, Scaramuzzi RJ and Martin GB (1994b) Effect of nutrition on seasonal patterns of LH, FSH and testosterone concentration testicular mass, sebaceous gland volume and odour in Australian cashmere goats *Journal of Reproduction and Fertility* **102** 351–360

Wright PJ, Geytenbeek PE and Clarke IJ (1990) The influence of nutrient status of post-partum ewes on ovarian cyclicity and on the oestrous and ovulatory responses to ram introduction *Animal Reproduction Science* **23** 293–303

Journal of Reproduction and Fertility Supplement **54**, 259–269

Sexual behaviour of rams: male orientation and its endocrine correlates

J. A. Resko[1], A. Perkins[2], C. E. Roselli[1], J. N. Stellflug[3] and F. K. Stormshak[4]

[1]*Department of Physiology and Pharmacology, School of Medicine, Oregon Health Sciences University, Portland, OR 97201–3098, USA;* [2]*Department of Psychology, Carroll College, Helena, MAO 59625, USA;* [3]*US Sheep Experimentation Station, Dubois, ID 83425, USA; and* [4]*Department of Animal Sciences, Oregon State University, Corvallis, OR 97331–6702, USA*

The components of heterosexual behaviour in rams are reviewed as a basis for understanding partner preference behaviour. A small percentage of rams will not mate with oestrous females and if given a choice will display courtship behaviour towards another ram in preference to a female. Some of the endocrine profiles of these male-oriented rams differ from those of heterosexual controls. These differences include reduced serum concentrations of testosterone, oestradiol and oestrone, reduced capacity to produce testosterone *in vitro*, and reduced capacity to aromatize androgens in the preoptic–anterior hypothalamus of the brain. Our observation that aromatase activity is significantly lower in the preoptic–anterior hypothalamic area of male-oriented rams than in female-oriented rams may indicate an important neurochemical link to sexual behaviour that should be investigated. The defect in steroid hormone production by the adult testes of the male-oriented ram may represent a defect that can be traced to the fetal testes. If this contention is correct, partner preference behaviour of rams may also be traceable to fetal development and represent a phenomenon of sexual differentiation.

Sexual Behaviour

Sexual differentiation of the brain

Although the effects of androgen on the development of the reproductive tract and the external organs of reproduction were known previously for several species, early experiments using the guinea-pig as an experimental model expanded the 'hormonal theory' of sexual differentiation to include the brain (Phoenix *et al.*, 1959). This landmark publication provided a rationale for behavioural scientists and neuroendocrinologists to study the effects of androgens and their metabolites in the developing fetus to ascertain the relationship between androgen action and the expression of heterosexual behaviour and gonadotrophin release in adulthood.

In general, the principles that govern the 'organizational' effects of androgen on the fetal brain, that is, permanent effects that carry over into the adult period, are similar to those proposed for the reproductive tract and the external organs of reproduction. In males, androgens secreted by the fetal testes act upon undifferentiated target tissues and produce a male phenotype (both anatomical and behavioural). The basic phenotype appears to be female and will remain female without the intervention of endogenous androgen. Females exposed to androgen during a 'critical' period for sexual differentiation will be androgenized permanently. Males deprived of androgen exposure during the 'critical period' regardless of the method, for example castration, use of drugs, or chemicals, will be feminized.

The principles described above for brain organization apply to a wide variety of mammalian species including sheep (Short, 1974). In sheep fetuses, the gonads differentiate into testes or ovaries by day 35 of gestation (Resko, 1985); the period of gestation is approximately 151 days. Testosterone can be quantified in the developing gonad on day 30 of gestation and its concentrations in testes increase with fetal age (Attal, 1969). Plasma obtained from sheep fetuses on day 70 of gestation contained significantly higher quantities of testosterone if it was obtained from male compared with female fetuses (Pomerantz and Nalbandov, 1975). This sex difference in testosterone concentrations probably occurred from day 35 (the day on which the fetal gonads differentiate) to day 60, the period in which sexual differentiation of the brain can be controlled by exogenous androgen administration to the pregnant ewe (Short, 1974). Androgen effects on brains of male fetuses can be determined in at least two ways: by the repertoire of reproductive behaviours that are displayed in adulthood and by the pattern of gonadotrophins released in response to an oestrogen challenge (Short, 1974; Karsch and Foster, 1975). Typically, rams will pursue oestrous females and display courtship behaviour which terminates with ejaculation (Banks, 1964). Reproductive behaviours in rams can be seasonally dependent in temperate climates; therefore, time of year is an important consideration for testing male behaviours in sheep (Pepelko and Clegg, 1965). Neither female- nor male-oriented rams release surge amounts of gonadotrophin in response to an oestrogen challenge, which is an adequate trigger for gonadotrophin release in females (Short, 1974; Karsch and Foster, 1975; Perkins *et al.*, 1995). Adult females, on the other hand, whose brains were not exposed to 'high' levels of androgen during the 'critical period' for sexual differentiation in the fetus, typically display oestrous behaviour and are receptive to the courtship behaviours of the male.

Partner preference behaviour

Partner preference behaviour can be defined as the sex of the partner toward which an individual directs behaviour of a sexual nature. If a male directs his sexual overtures towards a female, the male is classified as female-oriented. If, on the other hand, these behaviours are directed towards another male, the male is classified as male-oriented. In this review, we will not, by definition, consider the behaviour of anatomically feminized males, due to low amounts of androgen exposure during the critical period for sexual differentiation, and which display female (receptive) behaviours towards another male. Rather, we will discuss only males that possess the normal male anatomical phenotype yet choose another male over a female as a sexual partner.

A schematic representation of the phases and interrelationships of heterosexual behaviours is presented in Fig. 1. (Beach, 1976). Heterosexual behaviours of both sexes can be divided into four phases: attractive, appetitive, consummatory and postconsummatory (Beach, 1976). For a more detailed description of these phases, the reader should consult the review by Beach (1976). Generally, however, sexual attractiveness is an abstraction inferred from observations of the behaviour of males toward females. Appetitive behaviour of males such as approaching and investigating females probably depends on olfactory and visual cues. Consummatory responses that is, ejaculations, on the other hand, are associated with contact stimuli and secretions of the vagina. In the postconsummatory phase, the male temporarily loses the capacity to respond to the stimuli that initiated the sexual activity.

Male sexual performance depends upon the various stimuli provided by the female. The behavioural and nonbehavioural characteristics of a female that enable a ram to distinguish an oestrous ewe from one that is not in oestrus have been divided into the following categories: attractivity, proceptivity and receptivity. From observations of mating behaviour of sheep, one might conclude that rams, aggressively, seek out oestrous females and that the female plays only a minor role in this process. However, it has been shown that ewes in oestrus actively solicit males (proceptive behaviour) so that mating can be completed (Lindsay, 1961). Sexual receptivity in females can be defined as the response of the female to stimuli provided by the male. It is part of the consummatory phase. In ovariectomized ewes, receptivity can be induced by multiple injections of progesterone followed by a single dose of oestrogen (Lindsay, 1966) or by treatment with synthetic steroids and prostaglandins (Fitzgerald *et al.*, 1985). It is possible that heterosexual behaviours in mammals follow a bell shaped curve. The sexual activity of most animals falls within two standard deviations of the central point of this curve (the mean). A small number of animals fall within three standard deviations of the mean. Those on the upper end of the curve represent high-libido males; those on the lower end of the curve possess little or no sexual drive.

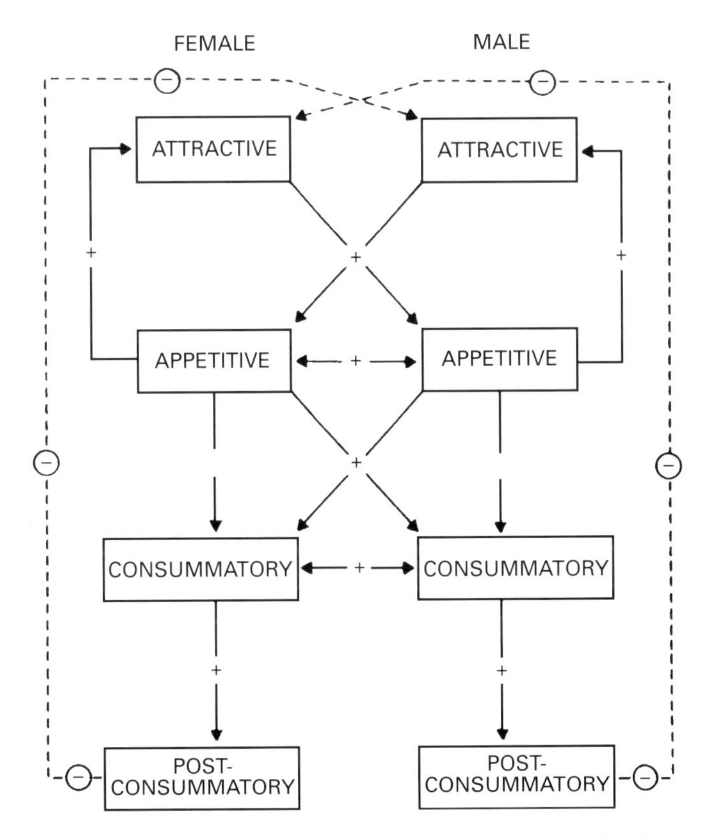

Fig. 1. Phases and interrelationships between a male and a female in heterosexual matings. (–) indicates inhibition; (+) indicates stimulation. (Reprinted from Beach, 1976 *Hormones and Behavior* **7** 105–138, with permission).

Many species of vertebrate including humans with normal to high libido participate in male-oriented behaviour (Dagg, 1984). Male orientation of domestic rams, determined by preferential sexual contact with males over females, has been reported (Perkins and Fitzgerald, 1992; Zenchak *et al.*, 1981). These behaviours are found not only in domesticated sheep but also in nondomesticated Bighorn sheep (Geist, 1971). In wild populations, rams segregate into all male groups after maturity. The dominant male attempts copulation with the subordinate males. Most of the aggression in the group occurs between the subordinates directed towards the dominate ram(s). Subordinate rams that are not aggressive live peacefully within the group. Thus it appears that the submissive behaviour of the subordinate rams has some adaptive significance perhaps to reduce aggression within the all male group (Geist, 1971). The behaviour of the dominant male is not exclusively male-oriented, however, because dominate males, if given the opportunity, will mate with oestrous females. Males that display both male- and female-oriented behaviours have been identified in many animal species. This type of behaviour seems to be confounded with dominance. On the other hand, a small percentage of rams, in any given population, will be exclusively male-oriented. Thus, the domestic ram may be a useful model for investigating the physiological basis of partner preference because these latter males, if given a choice, will exhibit courtship behaviour towards a male rather than an oestrous female (Resko *et al.*, 1996) .

The phenomenon of homosexual behaviour in men is not well understood but, as mentioned above for sheep, only a small percentage of men are truly male oriented, that is, are attracted only to men. Theories of aetiology range from defective parental nurturing (Freud, 1953) to differences in brain structure and chemistry between homosexual and heterosexual individuals (Dorner *et al.*, 1975; Gladue, 1984; Swaab and Hofman, 1990; Le Vay, 1991; Allen and Gorski, 1992) . The influence of individual genetic make-up has also been considered. Homosexual behaviour correlates with inheritance of polymorphic markers on the X chromosome (Xq28 region) in men which has led to speculation regarding maternal inheritance of these behavioural characteristics (Hamer *et al.*, 1993; Hu *et al.*, 1995). The short arm of the Y chromosome contains the sex-determining gene(s) which organizes the fetal testes (Page *et al.*, 1987), the secretions of which play a major role in sexual differentiation. In early prophase I, the sex chromosomes condense to form a heterochromatic structure called the XY body (Solari, 1980) and there is evidence that genetic material can exchange between the X and the Y chromosome at this time (Freije *et al.*, 1992). It is conceivable that in a small number of cases exchange of the sex-determining gene(s) from the Y to the X chromosome occurs. This exchange may affect normal organization of the fetal testes and the quantity and type of hormones that they produce with concomitant effects on brain development. Genetic control of sexual orientation has also been reported in *Drosophila* (Ryner *et al.*, 1996).

Role of the medial preoptic–anterior hypothalamic area in male sexual behaviour

The medial preoptic–anterior hypothalamic area (MPOA) mediates male reproductive behaviours in many species. In rats (Heimer and Larsson, 1966; Giantonio *et al.*, 1970; Ginton and Merari, 1977), cats (Hart *et al.*, 1973), dogs (Hart, 1989), goats (Hart, 1986) and monkeys (Slimp *et al.*, 1978), lesions of the MPOA interfere with normal male copulatory responses. A possible function of this area is that it is the sensory processor of sexual behaviours but not of sexual arousal (Giantonio *et al.*, 1970; Everitt and Stacey, 1987). Animals with brain lesions cannot integrate information of a sexual nature thus resulting in defective copulatory behaviour (Heimer and Larsson, 1966; Everitt and Stacey, 1987). The relationship of the MPOA to partner preference behaviour is not well understood. Male ferrets, however, that have been treated with oestrogen and have bilateral lesions of the MPOA spend more time with stimulus males than do control males (Paredes and Baum, 1995).

Role of in situ oestrogen formation by neural tissue in male sexual behaviour

Aromatase activity (the sum total of biochemical events that convert androgen to oestrogen) is relatively high in the MPOA of rodents compared with other parts of the brain (Roselli *et al.*, 1985). It appears that the concentrations of this enzyme activity within the MPOA are important for the following reasons: (1) testosterone or oestradiol is capable of restoring male copulatory behaviour in castrated rats (Davidson, 1969); (2) testosterone is unable to maintain normal male sexual behaviour in animals treated with inhibitors of aromatization (Beyer *et al.*, 1986). Similarly, sexual differentiation of the male brain during the 'critical period' is mediated by aromatization of androgen to oestrogen in some species (MacLusky and Naftolin, 1981). Treatment of rats during the 'critical period' for sexual differentiation of the brain with compounds that inhibit aromatization or are oestrogen antagonists prevents androgen-induced defeminization of the brain (McEwen *et al.*, 1977; Sodersten, 1978) and differentiation of the sexually dimorphic nucleus in the preoptic area (Dohler *et al.*, 1984). In a similar way, male ferrets require *in situ* oestrogen formation in the MPOA for differentiation of the sexually dimorphic nucleus (Baum *et al.*, 1996). Male rats exposed to an aromatase inhibitor during the neonatal period showed a preference for stimulus males over oestrous females when tested in the early part of the dark phase of the light–dark cycle (Bakker *et al.*, 1993).

Endocrine Correlates of Partner Preference Behaviour in Rams

In previous work (Resko *et al.*, 1996), we studied partner preference behaviour in rams and correlated this behaviour with steroid concentrations in the systemic circulation, capacity of the testes to produce androgens and oestrogens, and brain aromatase activity.

Tests for sexual behaviour

Rams used in this study were of Targhee, Rambouillet, Columbia and Polypay breeds. All rams were born in the spring lambing season (April and May). Rearing conditions have been described by Fitzgerald *et al.* (1993). Briefly, ewes and lambs grazed spring and summer ranges until weaning in August. At weaning, ram and ewe lambs were separated from the dams. Ram lambs were combined into all male groups of approximately 400–500 animals. The ram lambs were kept on autumn range land, grazing for an additional 2 months, and then moved to a feedlot (November until April) where they received rations developed for growing animals. At one year of age, ram lambs were moved as a group onto spring ranges and kept through the summer and autumn. During this time, they were exposed only to natural changes in photoperiod and had no physical contact with females. Sexual behaviour of the rams was tested beginning at approximately 16 to 18 months of age.

Preliminary tests

From August to October of the second year of life, the sexual behaviours of these rams were tested. Rams were placed with three mature ovariectomized ewes in which oestrus was induced by exogenous hormone treatments as described by Fitzgerald *et al.* (1985). Each test was for 30 min and each ram was tested at least six times. Repeated tests revealed three groups of rams: those that mated with oestrous females repeatedly (female-oriented), those that mated with oestrous females occasionally but at least once (low libido males), those that did not mount oestrous females (potentially male-oriented).

After these tests were completed, the size of the male population reared together was reduced to 25–30 rams of mixed sexual preference. These animals were housed together in outdoor pens approximately 12 m^2 and were not permitted physical contact with females, but females were housed in adjacent pens. Rams that would not mount oestrous females in the preliminary tests mentioned above, as well as female-oriented rams, mounted male pen-mates in the group setting.

Preference tests

Only rams that would not mount oestrous females in the preliminary tests satisfied the criterion for entrance into the sexual preference paradigm. A description of the method and apparatus used in the sexual preference tests have been described by Perkins and Fitzgerald (1992) and are shown in Fig. 2.

Briefly, in November and December of the second year of life, rams were exposed, simultaneously, to two restrained oestrous females and two males that were chosen at random for use. Rams that courted and mounted males in preference to females during a 30 min test that was repeated at least three times were classified as male-oriented. Using these procedures, we identified six males that would not mount females in the preliminary sexual tests, mounted males in a group setting, and mounted males in preference to females in the sexual preference tests. Male-oriented rams were given an additional preference test 5 days before they were killed at three and one-half years of age.

Treatment of female-oriented controls

During the breeding season of the second year of life (November and December), the seven, female-oriented rams were used as breeders and were housed with cyclic females in single-sire outdoor pens for this purpose. They remained with the females for 32 days after which they were returned to the all male group which contained male-oriented and low libido males from which they were taken. Female-oriented rams were tested with oestrous females for a 2 week period in August of the third year of life. Afterwards they were returned to their all male groups where they remained until they were killed along with the male-oriented rams in late October of the same year.

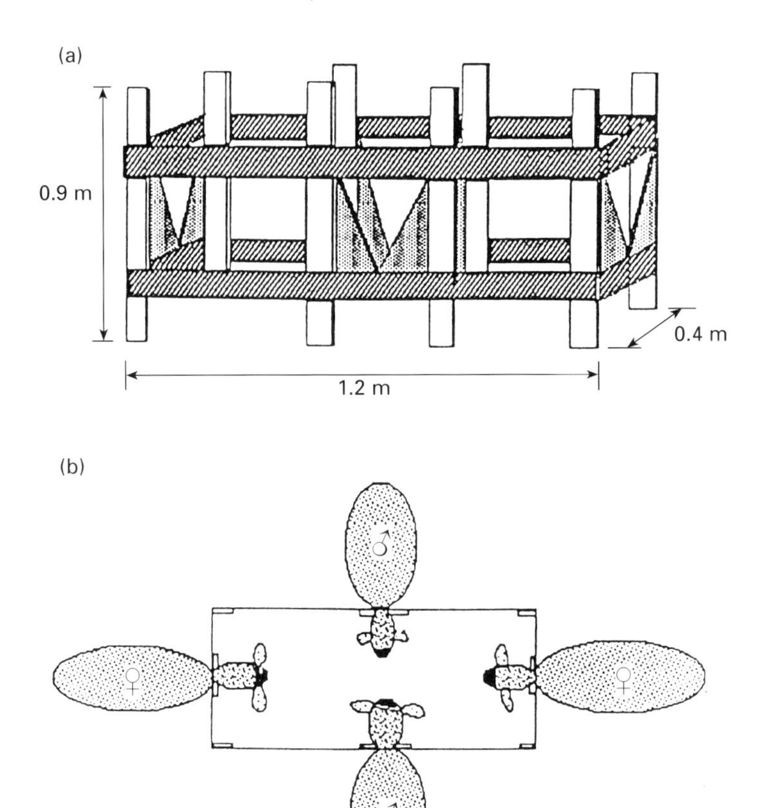

Fig. 2. A diagrammatic representation of the method used for sexual preference testing. The upper panel shows a diagram of a four-way stanchion for restraining test subjects. The lower panel demonstrates that oestrous females (and stimulus males) were restrained in the stanchion facing one another. (Reprinted from Perkins and Fitzgerald, 1992 *Journal of Animal Science* **70** 1787–1794, with permission).

Birth records

Because it has been shown that rams born co-twin with males are sexually more active as adults than those born co-twin with females (Fitzgerald *et al.*, 1993) , we reviewed the birth records of all the rams included in this study. Of the six male-oriented rams studied, two were born as singletons, two as twins, two as triplets. Three of the female-oriented rams were born as singletons, and four as twins. No obvious relationship was found between the number of offspring developing *in utero* or the sex of these offspring to sexual orientation in adulthood.

Behavioural data

The results of the behavioural tests from six rams classified as male-oriented are shown in Table 1. Each ram was given a 30 min test which was repeated at least three times in the preference protocol described previously. Data from two preference tests designated trials 1 and 2 in Table 1 are presented. Behavioural end points of the male-oriented rams that differed significantly ($P < 0.05$) depending on the sex of the stimulus animal were: sniffs, foreleg kicks, vocalizations and mounts.

Table 1. Reproductive behaviours of rams classified as male-oriented in partner preference tests[a]

Partner preference	Number of responses per trial (mean ± SEM) Sniffs	Foreleg kicks	Vocalizations	Flehmen	Mounts	Ejaculations
Male stimulus						
Trial 1	14.8*	16.7*	12.8*	00.6	17.3*	00.2
	±2.5	±7.5	±6.4	±0.6	±7.9	±0.2
Trial 2	7.0*	10.8*	10.3*	0	5.5*	0.4
	±1.7	±1.8	±1.5	0	±1.9	±0.4
Female stimulus						
Trial 1	4.5	0	0.2	0	0	0
	±2.6	0	0.2	0	0	0
Trial 2	0	0	0	0	0	0
	0	0	0	0	0	0

[a]Animals that did not mount females in, at least, six nonpreference behavioural tests (designated Preliminary Tests in the text) were tested further. Behaviours from the last two preference tests (designated Trials) were recorded during exposure for 30 min to both males and females. The last trial was conducted 5 days before the subjects were killed. $n = 6$ rams *$P < 0.05$, response to males compared with response to females. (Data reprinted from Resko *et al.*, 1996 *Biology of Reproduction* **55** 120–126, with permission).

Table 2. Serum steroid concentrations of rams as a function of sexual orientation

Treatment	n	Steroids (pg ml^{-1} serum) T	DHT	Δ^4	E_1	E_2
Female-oriented	7	1559	103	245	46	15
		±228	±45	±67	±2	±3
Male-oriented	6	874*	50	127	40*	8*
		±196	±10	±61	±2	±1

Values are means ± SEM. T, testosterone; DHT, dihydrotestosterone; Δ^4, androstenedione, E_1, oestrone; E_2, oestradiol *Significantly different from female-oriented rams, $P < 0.05$.
(Data reprinted from Resko *et al.*, 1996 *Biology of Reproduction* **55** 120–126, with permission).

The Flehmen response and the ejaculations did not differ significantly. It is important to remember that the six males for which data are provided in this table were given at least six sexual behaviour performance tests with oestrous females over two years before the preference tests. In none of these tests were these rams interested in females.

Tissue collection

Rams were killed in random order 5 days after the last preference test. Brains were quickly removed and the various areas dissected according to previously published procedures (Moss *et al.*, 1980). Brain areas, pituitaries, prostates and testes were rapidly frozen on dry ice. Tissue and serum samples were coded and the code was not broken until after all the samples were assayed.

Steroid radioimmunoassays and gonad incubations

Steroids were quantified by specific radioimmunoassays after chromatography on Sephadex LH-20 columns as described by Resko *et al.* (1980). Steroid biosynthesis by the testes *in vitro* was

Table 3. Biosynthesis of steroids from [^{3}H]progesterone by ram testicular homogenates[a]

Treatment	*n*	µmole mg^{-1} h^{-1}	
		Testosterone	17OH-progesterone
Female-oriented	7	28.8 ± 8.1	416.9 ± 100.8
Male-oriented	6	12.1 ± 2.3*	186.3 ± 30.7*

[a]See Resko *et al.* (1996) for details of incubations and isolation of steroid metabolites. Values are means ± SEM. *Significantly different from female-oriented rams, $P < 0.05$. (Data reprinted from Resko *et al.*, 1996 *Biology of Reproduction* **55** 120–126, with permission).

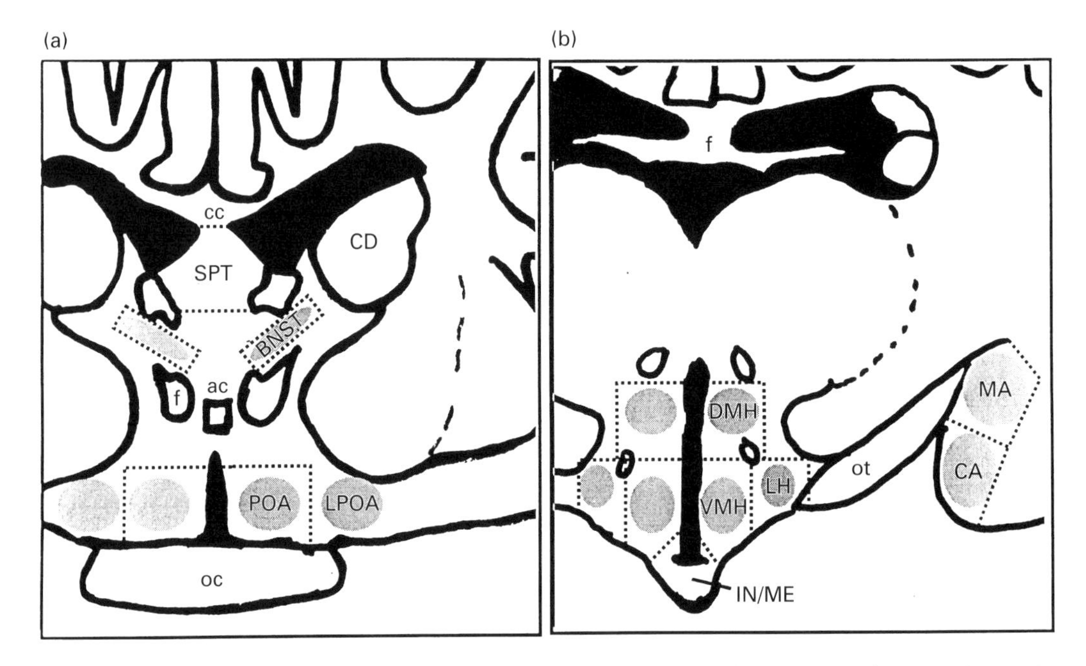

Fig. 3. Tracings of frontal sections of the adult ram brain at two levels. (a) A section at the preoptic–anterior hypothalamus; (b) a section at the medial hypothalamus. cc: corpus callosum; CD: caudate nucleus, SPT: septum; BNST: bed nucleus of the stria terminalis; ac: anterior commissure; f: fornix; POA: preoptic area–anterior hypothalamus; LPOA: lateral preoptic area–anterior hypothalamus; oc: optic chiasma; DMH: dorso–medial hypothalamus; VMH: ventro–medial hypothalamus; LH: lateral hypothalamus; IN/ME: infundibulum/median eminence; ot: optic tract; CA: cortical amygdala; and MA: medial amygdala.

determined by the method described by Resko *et al.* (1996). Serum steroid concentrations of rams as a function of sexual orientation are shown in Table 2. Sera from female-oriented rams contained significantly higher ($P < 0.05$) concentrations of testosterone, oestrone and oestadiol than sera from male-oriented rams. Androstenedione and 5α-dihydrotestosterone did not differ significantly between groups.

The capacity of testicular homogenates from male- and female-oriented rams to biosynthesize testosterone and 17α-hydroxyprogesterone from tritiated progesterone *in vitro* was tested (Table 3). Under the same experimental conditions, the testes of male-oriented rams possessed a reduced capacity for hormone biosynthesis compared with the testes of female-oriented controls. These results taken together indicate that partner preference behaviour is correlated with steroid

Table 4. The effects of castration on aromatase activity in the medial preoptic–anterior hypothalamus of the heterosexual ram

Treatment	n	Aromatase activity ($fmol\ ^3H_2O\ h^{-1}\ mg^{-1}$ protein)
Intact	3	211.4 ± 30.6
Castrated	3	102.9 ± 13.5 *

n, number of animals *Significantly different from intact value, $P < 0.02$.

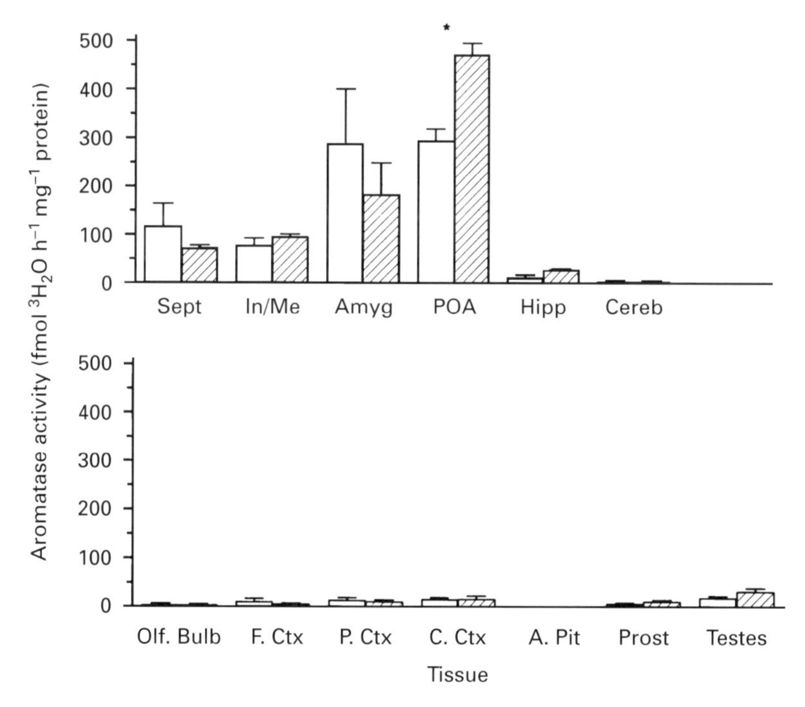

Fig. 4. Distribution of aromatase activity in brain, anterior pituitary gland, prostate and testis of intact (□) male-oriented (n = 6) and (▨) female-oriented (n = 7) rams. Data are presented as means ± SEM. *Significant effect of sexual orientation on aromatase activity ($P < 0.05$). Sept: septum; In/Me: infundibulum/median eminence; Amyg: amygdala; POA: preoptic area; Hipp: hippocampus; Cereb: cerebellum; Olf. Bulb: olfactory bulb; F. Ctx: frontal cortex; P. Ctx: parietal cortex; C. Ctx: cingulate cortex; A. Pit: anterior pituitary gland; Prost: prostate. (Data reprinted from Resko *et al.*, 1996 *Biology of Reproduction* **55** 120–126, with permission).

production by the ram testes and that male-oriented behaviour is displayed by those animals whose testes produce less hormone.

Aromatase assays

Brain aromatase activity was assayed using a ^3H_2O assay as described by Roselli *et al.* (1984) and validated for the ram by Resko *et al.* (1996). A diagrammatic representation of a frontal view of the ram brain at two levels, at the level of the preoptic/anterior hypothalamus and at the level of the medial hypothalamic area (Fig. 3) can be used to locate the areas of the ram brains represented by data depicted in Fig. 4 and Table 4. A comparison of aromatase activity among different brain parts and between male- and female-oriented rams is shown in Fig. 4. The brains of male-oriented rams contained reduced amounts of aromatase activity only in the POA. This is the brain area usually associated with the mediation of sexual behaviour in many mammalian species.

The effects of castration on aromatase activity in the POA of brains from heterosexual rams (Table 4; Roselli *et al.*, 1998) show that castration significantly reduced the amount of aromatase activity in this region ($P < 0.02$) adding to the list of species in which castration or androgen control aromatase activity. A corollary to the above is the fact that castration not only reduces reproductive behaviours in female-oriented rams but also did so in male-oriented rams. (Pinckard *et al.*, 1998). These results seem to imply that male-oriented behaviour in this species is androgen-dependent or alternatively is regulated by a product of the testes.

Conclusions

The rationale for studying male-oriented behaviour in rams is twofold. First, from the point of view of animal husbandry, these rams are 'behaviourally infertile' and, therefore, constitute a liability to breeders who cannot determine this fact before purchase. Second, further research may identify this species as a model for the study of male-oriented behaviour in general. In the former case, an understanding of the role of *in situ* oestrogen formation in the preoptic–anterior hypothalamic area of the brain may lead to endocrine therapies that may be effective for improving the reproductive success of these animals. In the latter case, examination of the defects in testicular androgen production in male-oriented rams may ultimately lead to studies of the fetal testes and a better understanding of the unresolved complexities of the nature versus nurture origins of male-orientation of males in a wide variety of species.

References

Allen LS and Gorski RA (1992) Sexual orientation and the size of the anterior commissure in the human brain *Proceedings of the National Academy of Sciences USA* **89** 7199–7202

Attal J (1969) Levels of testosterone, androstenedione, estrone and estradiol-17β in the testes of fetal sheep *Endocrinology* **85** 280–289

Bakker J, Ophemert JV and Slob AK (1993) Organization of partner preference and sexual behavior and its nocturnal rhythmicity in male rats *Behavioral Neuroscience* **107** 1049–1058

Banks EM (1964) Some aspects of sexual behaviour in domestic sheep, *Ovis aries. Behaviour* **23** 249–279

Baum MJ, Tobet SA, Cherry JA and Paredes RG (1996) Estrogenic control of preoptic area development in a carnivore, the ferret *Cellular and Molecular Neurobiology* **16** 117–128

Beach FA (1976) Sexual attractivity, proceptivity, and receptivity in female mammals *Hormones and Behaviour* **7** 105–138

Beyer C, Morali G, Naftolin F, Larsson K and Perez-Palacios G (1986) Effect of some antiestrogens and aromatase inhibitors on androgen induced sexual behavior in castrated male rats *Hormones and Behaviour* **7** 353–363

Dagg AT (1984) Homosexual behavior and female–male mounting in mammals – a first survey *Mammal Review* **14** 155–185

Davidson JM (1969) Effects of estrogen on the sexual behavior of male rats *Endocrinology* **84** 1365–1372

Dohler KD, Srivastava SS, Shryne JE, Jarzab B, Sipos A and Gorski RA (1984) Differentiation of the sexually dimorphic nucleus in the preoptic area of the rat brain is inhibited by postnatal treatment with an estrogen antagonist *Neuroendocrinology* **38** 297–301

Dorner G, Rohde W, Stahl F, Kreel L and Masius WG (1975) A neuroendocrine predisposition for homosexuality in men *Archives Sexual Behaviour* **4** 1–8

Everitt BJ and Stacey P (1987) Studies of instrumental behavior with sexual reinforcement in male rats (*Rattus norvegicus*): II. Effects of preoptic area lesions, castration, and testosterone *Journal of Comparative Psychology* **101** 407–419

Fitzgerald JA, Ruggles AJ, Stellflug JN and Hansel W (1985) A seven-day synchronization method for ewes using medroxyprogesterone acetate (MPA) and prostaglandin $F_{2\alpha}$ *Journal of Animal Science* **61** 466–469

Fitzgerald JA, Perkins A and Hemenway K (1993) Relationship of sex and number of siblings *in utero* with sexual behavior of mature rams *Applied Animal Behavior Science* **38** 283–290

Freije D, Helms C, Watson MS and Donis-Keller H (1992) Identification of a second pseudoautosomal region near the Xq and Tq telomeres *Science* **258** 1784–1786

Freud S (1953) Three essays on the theory of sexuality. In *Standard Edition of the Complete Psychological Works of Sigmund Freud* pp 125–143 Anonymous Hogarth Press, London

Geist V (1971) *Mountain Sheep: A Study In Behavior And Evolution* p 139 University of Chicago Press, Chicago IL

Giantonio GW, Lund NL and Gerall AA (1970) Effect of diencephalic and rhinencephalic lesions on the male rat's sexual behavior *Journal of Comparative Physiology and Psychology* **73** 38–46

Ginton A and Merari A (1977) Long range effects of MPOA

lesions on mating behavior in the male rat *Brain Research* **120** 158–163

Gladue BA (1984) Neuroendocrine response to estrogen and sexual orientation *Science* **28** 1496–1499

Hamer DH, Hu S, Magnuson VL, Hu N and Pattatucci AML (1993) A linkage between DNA markers on the X chromosome and male sexual orientation *Science* **261** 321–327

Hart BL (1986) Medial preoptic–anterior hypothalamic lesions and sociosexual behavior of male goats *Physiology and Behavior* **36** 301–305

Hart BL (1989) Medial preoptic–anterior hypothalamic area and sociosexual behavior of male dogs: a comparative and neuropsychological analysis *Journal of Comparative Physiology and Psychology* **86** 328–349

Hart BL, Haugen CM and Peterson DM (1973) Effects of preoptic–anterior hypothalamic lesions on mating behavior in male cats *Brain Research* **54** 177–191

Heimer L and Larsson K (1966) Impairment of mating behavior in male rats following lesions in the preoptic–anterior hypothalamus continuum *Brain Research* **3** 248–263

Hu S, Pattatucci AML, Patterson C, Li L, Fulker DW, Cherny SS, Kruglyak L and Hamer DH (1995) Linkage between sexual orientation and chromosome Xq28 in males but not in females *Nature Genetics* **11** 248–256

Karsch FJ and Foster DL (1975) Sexual differentiation of the mechanism controlling the preovulatory discharge of luteinizing hormone in sheep *Endocrinology* **97** 373–379

Le Vay S (1991) A difference in hypothalamic structure between heterosexual and homosexual men *Science* **253** 1034–1037

Lindsay DR (1961) Studies on the efficiency of mating in the sheep II. The effects of freedom of rams, paddock size, and age of ewes *Journal of Agricultural Science* **57** 141–145

Lindsay DR (1966) Modification of behavioral oestrus in the ewe by social and hormonal factors *Animal Behaviour* **14** 73–83

McEwen BS, Lieberburg I, Chaptal C and Krey LC (1977) Aromatization: important for sexual differentiation of the neonatal rat brain *Hormones and Behavior* **9** 249–263

MacLusky NJ and Naftolin F (1981) Sexual differentiation of the central nervous system *Science* **211** 1294–1302

Moss GE, Adams TE, Niswender GD and Nett TM (1980) Effects of parturition and suckling on concentrations of pituitary gonadotropins, hypothalamic GnRH and pituitary responsiveness to GnRH in ewes *Journal of Animal Science* **50** 496–502

Page DC, Mosher R, Simpson EM, Fisher MC, Mardon G, Pollack J, McGillivary B and de la Chapelle A (1987) The sex-determining region of the human Y chromosome encodes a zinc finger protein *Cell* **51** 1091–1104

Paredes RG and Baum MJ (1995) Altered sexual partner preference in male ferrets given exitotoxic lesions of the preoptic area anterior hypothalamus *Journal of Neuroscience* **15** 6619–6630

Pepelko WE and Clegg MT (1965) Influence of season of the year upon patterns of sexual behavior in male sheep *Journal of Animal Science* **24** 633–637

Perkins A and Fitzgerald JA (1992) Luteinizing hormone, testosterone and behavioral response of male-oriented rams to estrous ewes and rams *Journal of Animal Science* **70** 1787–1794

Perkins A, Fitzgerald JA and Moss GE (1995) A comparison of LH secretion and brain estradiol receptors in heterosexual and homosexual rams and female sheep *Hormones and Behavior* **29** 31–41

Phoenix CH, Goy RW, Gerall AA and Young WC (1959) Organizing action of prenatally administered testosterone propionate on the tissue mediating mating behavior in the female guinea pig *Endocrinology* **65** 369–382

Pinckard K, Stellflug J, Williams M and Stormshak F (1998) Influence of castration and estrogen replacement on sexual behavior in asexual, heterosexual and male-oriented rams *Biology of Reproduction (Supplement 1)* 58 (Abstract)

Pomerantz DK and Nalbandov AV (1975) Androgen level in the sheep fetus during gestation *Proceedings of the Society for Experimental Biology and Medicine* **149** 413–416

Resko JA (1985) Gonadal hormones during sexual differentiation in vertebrates. In *Handbook of Behavioral Neurobiology* pp 21–42 Eds N Alder, D Pfaff, and RW Goy. Plenum Publishing Corporation, New York

Resko JA, Ellinwood WE, Pasztor LM and Buhl AE (1980) Sex steroids in the umbilical circulation of fetal rhesus monkeys from the time of gonadal differentiation *Journal of Clinical Endocrinology and Metabolism* **50** 900–905

Resko JA, Perkins A, Roselli CE, Fitzgerald JA, Choate JVA and Stormshak F (1996) Endocrine correlates of partner preference behavior in rams *Biology of Reproduction* **55** 120–126

Roselli CE, Ellinwood WE and Resko JA (1984) Regulation of brain aromatase activity in rats *Endocrinology* **114** 192–200

Roselli CE, Horton LE and Resko JA (1985) Distribution and regulation of aromatase activity in the rat hypothalamus and limbic system *Endocrinology* **117** 2471–2477

Roselli CE, Stormshak F and Resko JA (1998) Distribution and regulation of aromatase activity in the ram hypothalamus and amygdala *Brain Research* **811** 105–110

Ryner LC, Goodwin SF, Castrillon DH, Anand A, Villella A, Baker BS, Hall JC, Taylor BJ and Wasserman SA (1996) Control of male sexual behavior and sexual orientation in *Drosophila* by the *fruitless* gene *Cell* **87** 1079–1089

Short RV (1974) *The Sexual Endocrinology of the Perinatal Period* pp 121 Eds MG Forest and J Bertrand. INSERM Colloque International, Lyon

Slimp JC, Hart BL and Goy RW (1978) Heterosexual, autosexual and social behavior of adult male rhesus monkeys with medial preoptic–anterior hypothalamic lesions *Brain Research* **142** 105–122

Sodersten P (1978) Effects of anti-oestrogen treatment of neonatal male rats on lordosis behaviour and mounting behaviour in the adult *Journal of Endocrinology* **76** 241–249

Solari AJ (1980) Synaptonemal complexes and associated structures in microspread human spermatocytes *Chromosoma* **81** 315–337

Swaab DF and Hofman MA (1990) An enlarged suprachiasmatic nucleus in homosexual men *Brain Research* **537** 141–148

Zenchak JJ, Anderson GC and Schein MW (1981) Sexual partner preference of adult rams (*Ovis aries*) as affected by social experiences during rearing *Applied Animal Ethology* **7** 157–167

Journal of Reproduction and Fertility Supplement **54**, 271–283

The functional integrity and fate of cryopreserved ram spermatozoa in the female tract

L. Gillan and W. M. C. Maxwell

Department of Animal Science, University of Sydney, NSW 2006, Australia

Cryopreservation advances capacitation-like changes in ram spermatozoa. These changes are reflected in an increased fertilizing ability compared with fresh spermatozoa, followed by an accelerated decline in fertilizing ability after incubation *in vitro* or *in vivo*. Furthermore, frozen–thawed spermatozoa are released earlier than fresh spermatozoa after binding to oviduct cells *in vitro*, confirming their physiological readiness to participate in fertilization despite their short lifespan. After insemination large numbers of spermatozoa are lost from the female reproductive tract of the ewe via the vagina. Frozen–thawed spermatozoa are expelled faster than fresh spermatozoa. The advanced membrane status of frozen–thawed spermatozoa may provoke their rapid loss and possibly makes them more vulnerable to attack by uterine leucocytes, or by some other mechanism, as a high proportion of spermatozoa lost from the tract are decapitated. The observed destabilization of the membranes of cryopreserved spermatozoa is accompanied by impaired sperm transport, associated with mitochondrial injury, necessitating intrauterine deposition of frozen–thawed semen to obtain satisfactory fertility after artificial insemination. However, the frozen–thawed spermatozoa that can participate in fertilization may contribute to increased embryonic loss by the advancement of cleavage or through a direct effect of cryopreservation on the male genome.

Introduction

The discovery of cryoprotective properties of glycerol (Bernstein and Petropavlovskij, 1937) led to the routine use of cryopreserved semen for artificial insemination in cattle. However, its application in the artificial insemination of other livestock is limited, due to reduced fertility after non-surgical insemination compared with that of fresh or liquid-stored semen. This reduced fertility results from factors such as poor recovery of viability after cryopreservation (as in pigs and horses) and reduced sperm transport in the female tract, necessitating costly intrauterine insemination (as in sheep and goats). Although considerable progress has been made in some aspects of gamete and embryo cryopreservation, such as pig embryo vitrification (Dobrinsky and Johnson, 1994), there has been little corresponding advance in boar semen cryopreservation (Almlid and Hofmo, 1996). This is due in part to economic considerations, since liquid semen delivery systems are well advanced in the pig industry and cryopreservation is, therefore, required only for the importation of new genes and the banking of valuable germplasm. However, in other livestock, such as sheep, where a low-cost method for utilization of cryopreserved spermatozoa would be of considerable economic advantage, progress has been limited despite considerable research (Salamon and Maxwell, 1995a,b).

The purpose of this review is to consider, with particular reference to the ram, recent findings on the effects of cryopreservation on sperm function and the fate of cryopreserved spermatozoa in the female reproductive tract.

Factors Associated with Poor Fertility of Cryopreserved Ram Spermatozoa

Although motility is preserved in a relatively high proportion (40–60%) of ram spermatozoa after freeze–thawing, only 20–30% remain biologically unchanged (Salamon and Maxwell, 1995a). It is possible that cryopreservation selects the most viable spermatozoa and destroys the rest, leaving a much smaller, but fertile population (Watson, 1995). However, although this theory is supported by the observed improvement in fertility in ewes after cervical insemination with a greater number of frozen–thawed spermatozoa (Salamon, 1977), it does not fully explain the low fertility of frozen semen. Even when very large numbers of motile, frozen–thawed spermatozoa are placed in the cervix, fertility is lower than for fresh semen (Maxwell and Hewitt, 1986). This finding suggests that the spermatozoa regaining motility after thawing must be compromised in some way that prevents them either reaching the site of fertilization or participating in fertilization. Thus, cryogenic changes to spermatozoa must be responsible for a decrease in their functional integrity or effective transport through the female tract to the site of fertilization.

Functional Integrity of Spermatozoa

Ultrastuctural and biochemical damage occurs to many spermatozoa during freeze–thawing but those remaining motile, and presumably intact, are functional, giving high fertilization rates (85–93%) after intrauterine or tubal deposition of even very small numbers of spermatozoa (sheep: Maxwell *et al.*, 1993; cattle: Seidel *et al.*, 1995). Although it has been assumed that the functional condition of this motile subpopulation of cryopreserved spermatozoa would be similar to that of fresh motile spermatozoa, evidence suggests that this is not the case, as a greater proportion of the surviving spermatozoa show evidence of altered membrane responses to physiological stimuli (Watson, 1995; Maxwell and Watson, 1996).

Capacitation status of spermatozoa

Ejaculated spermatozoa cannot fertilize until they undergo capacitation, which is the first stage of membrane destabilization events involving an efflux of cholesterol and redistribution of intrinsic membrane proteins and lipids (Harrison, 1996). This destabilization continues with the acrosome reaction and ultimately results in loss of viability. Cryopreservation suspends the continuum of changes in spermatozoa that begins in the epididymis and ends at fertilization or cell death. However, rather than being revived in the same state as they were before freezing, the spermatozoa may emerge in a state resembling that of capacitation, having bypassed some of this normal maturation (Watson, 1995). Watson proposed that the apparently 'capacitated' cells in frozen–thawed semen would have a reduced life-span and if not exposed to oocytes within a short time would be unable to achieve fertilization. Ageing of functionally capacitated ram spermatozoa in the reproductive tract of the ewe after cervical insemination may further aggravate this situation, and such spermatozoa may die prematurely in the lower part of the reproductive tract.

To test Watson's hypothesis, Gillan *et al.* (1997a) investigated the proposal that cryopreservation may cause ram spermatozoa to undergo membrane changes similar to capacitation. The events of capacitation may involve an influx of calcium ions. The antibiotic chlortetracycline (CTC, a fluorescent compound) accumulates and fluoresces in membrane compartments in which there are high concentrations of calcium ions next to hydrophobic sites (Tsien, 1989). CTC has been used as a fluorescent probe to visualize the course of capacitation and the acrosome reaction in mouse (Saling and Storey, 1979), bull (Fraser *et al.*, 1995) and boar spermatozoa (Wang *et al.*, 1995; Maxwell and Johnson, 1997a). The CTC patterns assessed by Gillan *et al.* (1997a) were F (non-capacitated), B (capacitated) and AR (acrosome reacted). A high proportion of the spermatozoa had either capacitated or acrosome reacted, as determined by these patterns, after incubation at 37°C for 6 h or after freezing and thawing with or without further incubation, whereas most fresh unincubated spermatozoa had not capacitated (Fig. 1). *In vitro* matured oocytes were penetrated earlier by frozen

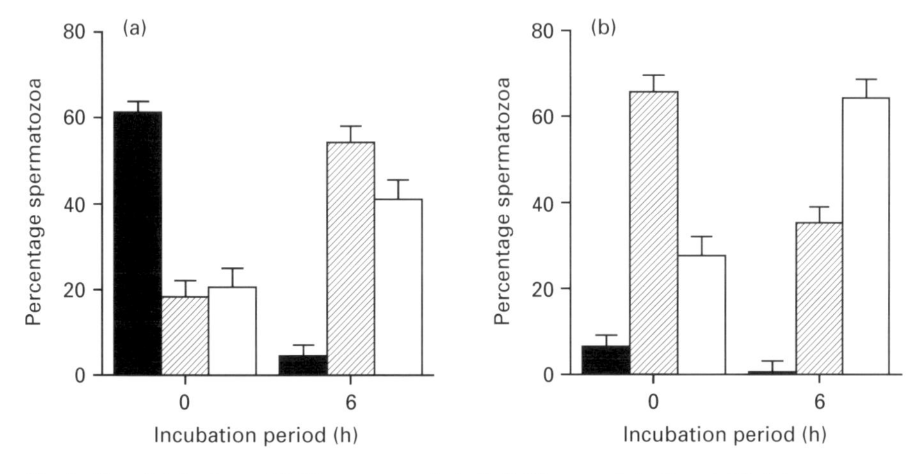

Fig. 1. The distribution of F (■; non-capacitated), B (▨; capacitated) and AR (□; acrosome-reacted) chlortetracycline staining patterns at $t = 0$ h (no incubation) or 6 h (incubation at 37°C) for fresh (a) and frozen–thawed (b) ram spermatozoa. Values are means ± SEM of six replicates. (Adapted from Gillan *et al.*, 1997a.)

Table 1. Proportion of advanced zygotes (cleavage stage or at syngamy with the sperm tail present) 24 h after *in vitro* fertilization of *in vitro* matured sheep oocytes with fresh and frozen–thawed ram spermatozoa

Type of spermatozoa	Incubation time (h at 37°C)	Zygotes assessed	Zygotes advanced	
			Number	%
Fresh	0	88	24	27.3
	6	62	42	67.7
Frozen–thawed	0	76	53	69.7
	6	40	30	75.0

(L. Gillan, G. Evans and W. M. C. Maxwell, unpublished.)

than by fresh spermatozoa, as evidenced by the presence of more advanced zygotes 24 h after insemination (Table 1). Fresh spermatozoa incubated for 6 h at 37°C produced advanced zygotes in similar proportions to frozen spermatozoa.

These findings suggest that both incubation and freeze–thawing advance capacitation-like changes in ram spermatozoa and that further incubation of such cells causes them to acrosome react. Similar findings have been published for bovine spermatozoa (Cormier *et al.*, 1997). The *in vitro* fertilization results also indicate that the cryopreserved spermatozoa are capacitated, confirming earlier findings that frozen ram spermatozoa exhibit the highest penetration of zona-free hamster eggs immediately after thawing in contrast to fresh spermatozoa, which require a period of incubation of several hours to attain the highest penetration rate (Watson, 1995).

Other treatments that might be expected to increase the proportions of capacitated spermatozoa in boar semen (incubation in the presence of bicarbonate, flow cytometric sorting, cooling and freezing) also increase the percentage of B-pattern cells in comparison with fresh semen (Maxwell and Johnson, 1997a). Lipid phase transitions have been implicated as responsible for membrane destabilization during cooling (Holt and North, 1984). It is probably the process of cooling and rewarming that induces capacitation-like changes in membrane function after cryopreservation rather than events associated with ice formation, and thus cells merely cooled to 0–4°C also display this effect (Watson, 1996; Maxwell and Johnson, 1997a,b). However, it is not possible to draw definitive conclusions from these results about the importance of sperm membrane changes associated with frozen storage when the semen is used for artificial insemination. Even after prolonged ageing or freeze–thawing of spermatozoa, the number of acrosome-intact cells in the inseminate may still be above the threshold required to achieve satisfactory fertilization of oocytes.

Integrity of the sperm tail

In addition to changes in capacitation status of the spermatozoa, the integrity of the sperm tail and its mitochondrial function may also be influenced by freezing and thawing. Windsor (1997) demonstrated that mitochondrial respiration was important for cervical penetration by ram spermatozoa and that mitochondrial injury, as assessed by uptake of rhodamine 123 (R123), was a significant contributor to the poor fertility in ewes after cervical compared with intrauterine insemination with frozen–thawed semen. This effect was independent of post-thaw motility of spermatozoa. The same study demonstrated that inhibition of glycolysis by rotenone did not prevent fertilization of ova by ram spermatozoa, but did reduce their ability to traverse the cervix and reach the site of fertilization. Our studies (L. Gillan, G. Evans and W. M. C. Maxwell, unpublished) suggest that mitochondrial function is influenced by freezing and thawing in a similar way to functional membrane status (as illustrated in Fig. 1). Fresh spermatozoa incubated for 6 h at 37°C exhibited the same uptake of R123 as frozen–thawed, unincubated spermatozoa (Fig. 2). There was a more rapid decline in the proportion of frozen–thawed than fresh spermatozoa able to take up R123 during incubation ($P < 0.001$), indicating that fresh spermatozoa maintained their respiratory activity longer than frozen–thawed cells.

The fertility observed by Windsor (1997) in ewes inseminated with rotenone-treated semen (low after cervical but high after intrauterine insemination) is similar to that reported with frozen–thawed ram semen (Lightfoot and Salamon, 1970a; Maxwell and Hewitt, 1986). Windsor (1997) also demonstrated that the cryoprotective procedures currently used for ram semen, including those that enhance post-thaw motility, do not prevent injury to the mitochondria of spermatozoa during the freeze–thawing process. Mitochondrial injury, causing impaired sperm transport, and advanced head membrane destabilization may result in the premature death of large numbers of frozen–thawed spermatozoa in the lower parts of the female tract after cervical or vaginal deposition.

Integrity of the sperm genome and embryonic survival

Early embryonic mortality may contribute to the low fertility rate after artificial insemination of ewes with cryopreserved semen (Salamon and Maxwell, 1995b). An increase in abnormalities of embryonic development associated with ageing of spermatozoa has been observed for several species, and considerable attention has been focused on possible changes in the haploid genome of the sperm cell due to ageing. There is some evidence that embryonic loss may be increased when stored spermatozoa are further aged in the female tract resulting in asynchrony between the age of the spermatozoa and ova (Salamon *et al.*, 1979). Thus, there was a steeper decline in embryonic survival with time of chilled (5°C) storage of ram semen after single than after double insemination, pointing to the importance of time of semen deposition into the cervix to avoid additional ageing of spermatozoa.

A direct effect of freeze–thawing on the sperm nucleus cannot be ruled out, as cryopreservation is known to cause some DNA denaturation, as assessed by the uptake of acridine orange, of human (Royere *et al.*, 1988) and ram spermatozoa (L. Gillan and W. M. C. Maxwell, unpublished; Fig. 3). Spermatozoa weakened or not totally injured or capacitated during freeze–thawing, and those which after thawing and insemination are aged in the female tract may initiate embryos that are not viable and perish at an early stage. The advanced stage of fertilization obtained after *in vitro* fertilization of *in vitro* matured oocytes with frozen spermatozoa (Table 1) may not have resulted in normal embryonic development. The fast cleavage observed also after cervical insemination with liquid stored semen (Lopyrin and Rabocev, 1968) can be considered to presage embryonic mortality. There is some evidence (Gillan *et al.*, 1997) that ewes inseminated into the uterine horns with

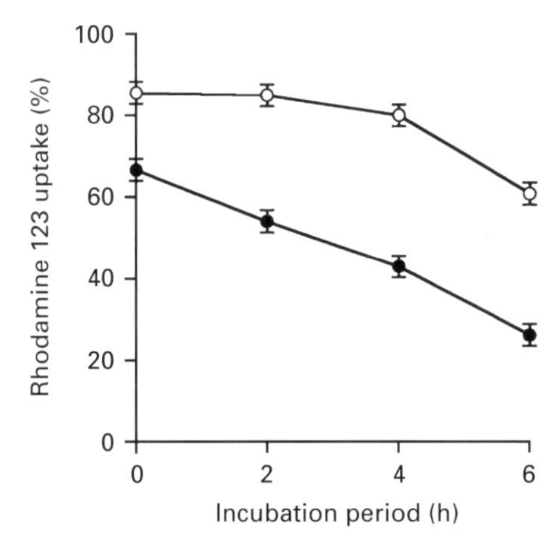

Fig. 2. Uptake of rhodamine 123 by fresh (○) and frozen–thawed ram spermatozoa (●) after incubation at 37°C in air at a concentration of 100×10^{-6} ml^{-1}. Values are means ± SEM of three replicates. (L. Gillan, G. Evans and W. M. C. Maxwell, unpublished.)

frozen–thawed spermatozoa, and at a time closely associated with the expected presence of an oocyte(s), are more likely to produce viable embryos.

Transport of Spermatozoa Through the Female Tract

The initial theory of sperm transport in sheep and cattle was outlined by Hawk (1983). After semen deposition in the anterior vagina, or in the cervix, spermatozoa were thought to travel to the oviducts in two phases – an initial rapid phase in which small numbers of spermatozoa reached the oviducts within a few minutes, followed by a slower ascent of spermatozoa from cervical reservoirs into and through the uterus, with a gradual increase over several hours of numbers of spermatozoa in the oviducts. It was established by Hunter *et al.* (1980), using ligation of segments of the oviduct to prevent post-mortem effects on sperm transport, that viable spermatozoa did not reach the oviducts until several hours after mating. Hunter showed that the main functional sperm reservoir in sheep was the caudal portion of the isthmus, and that spermatozoa did not enter the ampulla of the oviduct until after ovulation (Hunter and Nichol, 1983).

A number of investigators (reviewed by Salamon and Maxwell, 1995b) have reported inadequate transport and reduced viability of frozen–thawed spermatozoa in the reproductive tract of the ewe, which contributed to low fertility after cervical insemination. When the thawed semen was deposited into the uterus or oviducts, high (85–95%) fertilization rates were obtained, indicating that the spermatozoa remaining biologically intact maintained their fertilizing capacity after freeze–thawing. Nevertheless, sperm transport may be impaired even after uterine insemination of cryopreserved semen; thus, fertilization failure in superovulated ewes after late insemination with frozen–thawed spermatozoa was associated with the inability of spermatozoa to pass through the uterotubal junction and isthmus (Jabbour and Evans, 1991). These findings support the original contention of Lightfoot and Salamon (1970a) that an impairment of sperm transport through the female reproductive tract is a major contributor to poor fertility after artificial insemination of sheep with cryopreserved semen. The reduced viability of frozen–thawed spermatozoa in the female reproductive tract is characterized by longevity, which is often half that of fresh spermatozoa (reviewed by Salamon and Maxwell, 1995b) and impaired ability to penetrate the cervix (Lopyrin

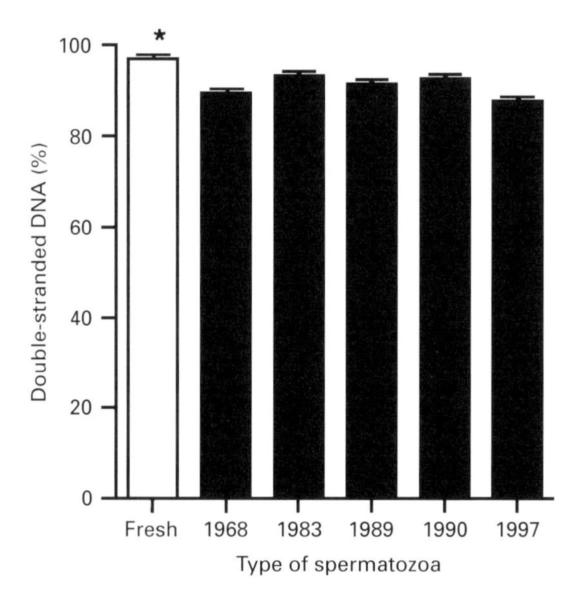

Fig. 3. Percentage of double-stranded DNA, as assessed by uptake of the metachromatic dye acridine orange, in fresh (□) ram spermatozoa compared with spermatozoa frozen (■) in different years. Values are means ± SEM of three replicates. *Significantly different from values for frozen spermatozoa ($P < 0.001$). (L. Gillan and W. M. C. Maxwell, unpublished.)

and Loginova, 1958), utero-tubal junction or both structures (Mattner *et al.*, 1969; Platov, 1983). However, this may be exacerbated by an increased rate of expulsion of such spermatozoa from the tract. It is not known how many spermatozoa, and in what condition, are required to achieve fertilization, what is the fate of spermatozoa deposited in various parts of the tract and how they interact with epithelial cells when they reach the oviducts.

Number of spermatozoa required to achieve fertility

The ultimate purpose of sperm transport is to provide adequate numbers of functional spermatozoa, from a heterologous population, at the site of fertilization over a period that will ensure fertilization of all oocytes. Therefore, sperm transport mechanisms need to be both facilitative and regulatory. Given optimal timing of insemination, most species produce more spermatozoa than are required to achieve this response. Consequently, satisfactory fertility can be achieved by artificial insemination of considerably fewer spermatozoa than are in an ejaculate, and the number that reach the site of fertilization at the appropriate time is extremely small compared with the number deposited. However, there are species differences. For example, small numbers of viable spermatozoa are required to achieve fertilization in cows, while pigs require many millions. This quantitative difference may be an important determinant of the fertility of cryopreserved semen, as those species that require large numbers of spermatozoa for conception are less tolerant of poor survival of spermatozoa after cryopreservation.

For achieving comparable fertility rates with fresh and frozen bull spermatozoa, ten times the number of cryopreserved spermatozoa are required (Shannon, 1978). The situation is similar for artificial insemination of sheep, in which a minimum of 100 million viable fresh spermatozoa are required to be deposited in the cervix for adequate fertility. Although cervical insemination of cryopreserved semen is not recommended because of low and variable fertility (Evans and Maxwell,

1987), published estimates range from 160 (Lightfoot and Salamon, 1970b) to 800 million motile frozen–thawed spermatozoa (Colas, 1979; near the 10:1 ratio for cattle) required to achieve fertility similar to that of fresh semen. However, if in the ewe the cervical barrier to the ascent of spermatozoa is bypassed by surgical or laparoscopic insemination, as few as 100 000 or 10 000 fresh and 500 000 or 100 000 frozen–thawed spermatozoa are sufficient to achieve fertilization after insemination into the uterus or oviducts, respectively (Maxwell *et al.*, 1993). These results further support the hypothesis that a combination of impaired transport and reduced functional integrity of spermatozoa are important limiting factors to improved fertility of sheep after artificial insemination with frozen semen.

Fate of spermatozoa deposited in the female tract

Loss of spermatozoa occurs at all levels of the reproductive tract including the cervix, uterotubal junction and oviducts (reviewed by Drobnis and Overstreet, 1992). The low ratio of spermatozoa to ova (approximately 1:1) in the ampulla at the time of fertilization (sheep: Hunter and Nicol, 1986) indicates that these spermatozoa need to be very fertile. For this fertility to be achieved, the female tract must have highly selective mechanisms to eliminate unsuitable cells, and these may also act preferentially to eliminate many cryopreserved spermatozoa before they can reach the site of fertilization.

The fate of spermatozoa that fail to reach the oviducts after natural mating or artificial insemination has been the subject of speculation for many years. In early reports, only a small proportion of the spermatozoa inseminated into the ewe cervix could be recovered (Quinlivan and Robinson, 1969). Hawk and Conley (1971) reported that most spermatozoa in an inseminate drain from the female reproductive tract within a few minutes or hours after insemination; they suggested that the remaining spermatozoa were removed from the tract by slower drainage or phagocytosis. These authors placed a ligature under the mucosa at the vulvo–vaginal junction at artificial insemination to prevent the loss of spermatozoa to the exterior. Twenty-four hours later, 62% of the spermatozoa in the inseminate were recovered from the ligated reproductive tract. Many of the missing spermatozoa had been removed by phagocytosis, as evidenced by their tails protruding from leucocytes. In contrast, only 0.5% of the inseminate was recovered from unligated control ewes. Dead spermatozoa deposited in the vagina are known to be lost in large numbers between 3 and 9 h after insemination (Tilbrook and Pearce, 1984), but it is not clear whether there is a selective mechanism to remove such non-viable compared with viable cells.

Frozen–thawed spermatozoa need to be deposited in the uterus to achieve adequate fertility after artificial insemination of sheep. There have been few reports on the movement of these spermatozoa from the site of insemination and the number of spermatozoa retained in the tract. It is not certain that insemination into the uterus eliminates the cervix as a critical segment of the tract for sperm transport. In an experiment on intrauterine insemination of oestrous ewes (n = 12), only 8.0 ± 5.75% of fresh and 1.5 ± 0.89% of frozen–thawed ram spermatozoa were recovered from the cervix, uterus and oviducts after deposition in the uterine horns by laparoscopy (L. Gillan, P. F. Watson and W. M. C. Maxwell, unpublished). Overall, more spermatozoa were recovered from the uterus (33.1 ± 5.69%) than from the ampulla (4.0 ± 3.74%) or isthmus (2.9 ± 5.69% of the total cells recovered). The low recovery rate reflects considerable loss of cells by phagocytosis or transport out of the reproductive tract. Large numbers of spermatozoa were located in the cervix of the ewes (67.7 ± 16.16%), and many of these cells were motile, indicating a possible role for the cervix as a sperm reservoir after intrauterine insemination with fresh or frozen semen. However, it was not clear whether such cells could have subsequently ascended to the oviducts and participated in fertilization or whether they remained in the lower part of the reproductive tract until death or expulsion through the vagina. This study, together with that of Hawk and Conley (1971), confirms that a high proportion of the inseminate is normally lost through the vagina to the exterior regardless of the site of semen deposition.

Since a large proportion of frozen–thawed spermatozoa may be capacitated, it is likely that the fresh and frozen–thawed spermatozoa would interact in a different way with the female

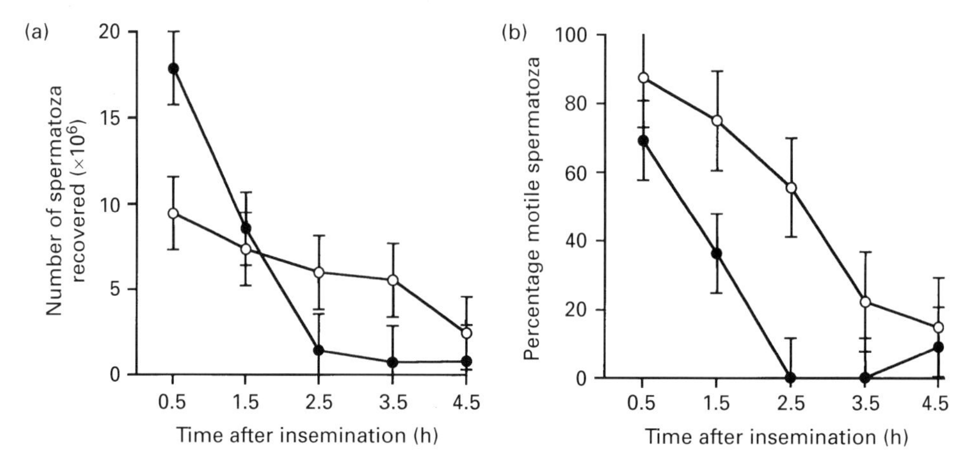

Fig. 4. Number (a) and motility (b) of spermatozoa recovered from the vaginae of ewes after intrauterine insemination with 100×10^6 fresh (○) and frozen–thawed (●) spermatozoa. Values are means ± SEM of 13 animals. (L. Gillan, P. F. Watson, K. Skogvold and W. M. C. Maxwell, unpublished.)

reproductive tract. Moreover, the frozen–thawed spermatozoa may acrosome-react and die earlier, decreasing the fertilizing population and promoting their loss from the reproductive tract. The rate of loss of spermatozoa from ewes in synchronized oestrus was therefore examined after intrauterine insemination by laparoscopy with 100×10^6 motile Percoll-separated fresh or frozen–thawed spermatozoa (L. Gillan, P. F. Watson, K. Skogvold and W. M. C. Maxwell, unpublished). At the time of insemination, 48 h after removal of progestagen sponges and injection of 400 i.u. PMSG, a balloon catheter was inserted into the vagina and inflated to prevent loss of spermatozoa. Spermatozoa were recovered from the vagina by flushing the catheter with phosphate-buffered saline after 30 min and then at hourly intervals after insemination. The overall number of spermatozoa recovered decreased from $13.63 \pm 0.31 \times 10^6$ after 0.5 h to $1.64 \pm 0.36 \times 10^6$ at 4.5 h after insemination ($P < 0.001$). Over the entire recovery period (4.5 h after insemination), equal numbers of fresh and frozen–thawed spermatozoa were recovered from the tract. However, fresh spermatozoa were lost gradually from the tract, whereas large numbers of frozen–thawed spermatozoa were lost within 0.5 h of insemination and these losses decreased as time within the tract increased ($P < 0.05$, Fig. 4).

Spermatozoa recovered from the tract were stained with CTC (see Fig. 5). As time within the tract increased, the proportion of F and B pattern spermatozoa decreased ($P < 0.001$ in each case) and AR pattern increased. However, there were more AR pattern cells recovered when frozen–thawed spermatozoa were inseminated ($P < 0.001$). More motile fresh than frozen–thawed spermatozoa were recovered ($P < 0.05$) and as the collection period increased the overall motility decreased ($P < 0.001$, Fig. 4).

Phagocytosed spermatozoa are often seen in flushings of the reproductive tract (Hawk, 1987), and most of the spermatozoa not lost by drainage or expulsion to the exterior are probably ultimately phagocytosed. However, significant numbers of decapitated or 'tailless' spermatozoa are found in the female reproductive tract after insemination of ewes with both fresh (Mattner, 1964) and frozen spermatozoa (Hawk and Conley, 1971). In the experiment described above, large numbers of decapitated spermatozoa were recovered from the tract (50.3% ± 0.95 decapitated compared with <1 % before insemination), with significantly more frozen–thawed than fresh spermatozoa lacking tails ($P < 0.05$). Moreover, in an earlier experiment (L. Gillan, P. F. Watson and W. M. C. Maxwell, unpublished), more decapitated spermatozoa were found in the cervix (31.1 ± 7.01%), uterus (42.4 ± 1.77%) and isthmus (32.3 ± 1.66%) than in the ampulla (10.55 ± 1.94%). These results indicate that there may be another important, but as yet unknown, mechanism for removal of the tails from

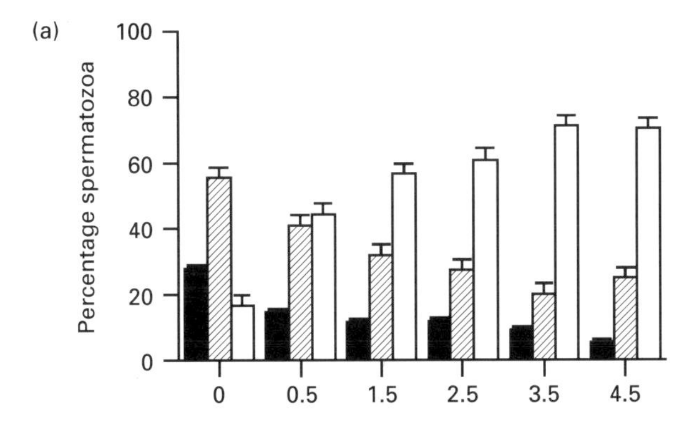

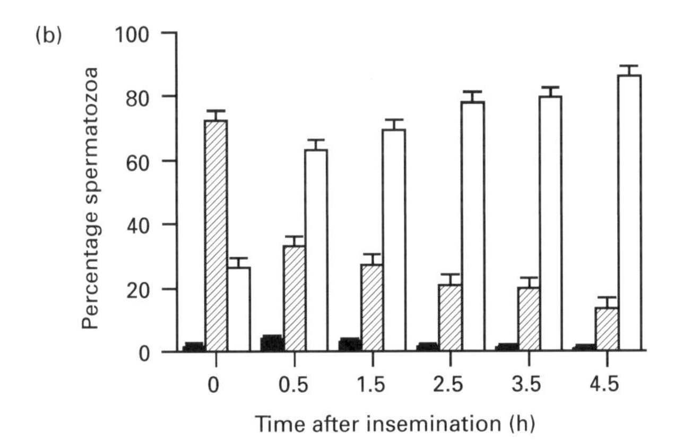

Fig. 5. The distribution of F (■; non-capacitated), B (▨; capacitated) and AR (□; acrosome reacted) chlortetracycline staining patterns of fresh (a) and frozen–thawed (b) ram spermatozoa lost from the tract after intrauterine insemination by laparoscopy. The $t = 0$ h sample was analysed before insemination. Values are means ± SEM of 12 animals.

non-viable spermatozoa in the lower and medium regions of the female reproductive tract, thus ensuring that such cells are unable to participate in the fertilization of oocytes.

Interaction of spermatozoa with the oviducts

In all animal models studied so far, the isthmus of the oviduct acts as a reservoir and conserves the function of spermatozoa if mating or insemination takes place before ovulation. A minimum of 6–8 h is required for a functional population of spermatozoa to be established in the oviducts of sheep and cows mated at oestrus (Hunter and Wilmut, 1984; Hunter and Nicol, 1986). Immediately before ovulation the oviduct activates a proportion of these spermatozoa to progress to the site of fertilization in the ampulla (Hunter and Nicol, 1983).

The oviduct may also regulate the speed of capacitation of spermatozoa. For example, when hamster spermatozoa were recovered from the isthmus in animals mated shortly after the onset of oestrus, they required additional time *in vitro* before they could penetrate eggs, but spermatozoa recovered from females immediately after ovulation could penetrate eggs within 30 min (Smith and Yanagimachi, 1989). Thus, capacitation and activation of cells sequestered in the isthmic reservoir

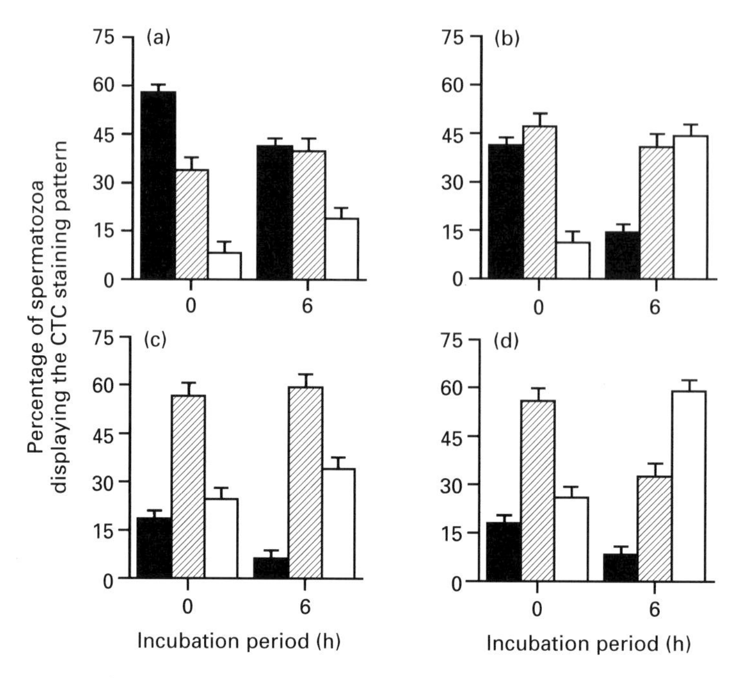

Fig. 6. The percentage of fresh (a and b) and frozen–thawed (c and d) spermatozoa displaying the F (■; uncapacitated), B (▨; capacitated) and AR (□; acrosome reacted) chlortetracycline staining patterns after incubation in the absence (control; a and c) or presence (experimental; b and d) of an oviduct epithelial cell monolayer. Values are means ± SEM of five replicates. (Adapted from Gillan *et al.*, 1997b.)

may not be achieved synchronously but rather be staggered over a period of time (Hunter, 1993). Periods of highly active free swimming appear to be alternated with phases of adhesion to the epithelium. Most uncapacitated hamster spermatozoa deposited in the oviducts attach to the oviductal mucosa, unlike capacitated spermatozoa which do not, indicating that the surface of the spermatozoa regulates these events (Smith and Yanagimachi, 1991). Thus in hamsters, it would appear that spermatozoa remain attached to the isthmic mucosa until they become capacitated, after which they detach and migrate to the ampulla to fertilize the ova.

It is clear that spermatozoa undergo necessary membrane changes during intimate contact with the epithelial cells of the oviduct. As frozen–thawed spermatozoa are likely to be in a more advanced membrane state than freshly ejaculated spermatozoa, these two populations of cells may respond differently when in contact with epithelial cells of the oviduct. Suzuki *et al.* (1997) found that frozen–thawed bull spermatozoa did react in a different way from fresh spermatozoa when coincubated with bovine oviduct epithelial cells, but they did not assess the membrane status of the spermatozoa. To investigate this possibility further, Gillan *et al.* (1997b) incubated fresh and frozen–thawed ram spermatozoa *in vitro* with an oviduct epithelial cell monolayer (OECM) for 6 h, and observed the ability of spermatozoa to attach to the OECM and undergo membrane changes. Control fresh and frozen spermatozoa were incubated in the absence of an OECM. Frozen spermatozoa bound to the OECM immediately and in larger quantities than did fresh spermatozoa. After the removal of unbound spermatozoa, fresh spermatozoa were released from the OECM gradually (59 ± 1.0, 49 ± 1.0 and 27 ± 2.0%), whereas frozen–thawed cells were released rapidly (58 ± 3.7, 16 ± 1.9 and 9 ± 1.0% attached after 2, 4 and 6 h incubation, respectively). A sample of medium containing unattached spermatozoa was also removed to assess membrane status using the CTC fluorescence assay; the majority of the fresh control spermatozoa were initially non-capacitated and over the incubation period gradually became capacitated and acrosome reacted (Fig. 6a). In contrast to the control spermatozoa, fresh spermatozoa incubated with OECM were more likely to be capacitated, but acrosome intact at 0 h (Fig. 6b), perhaps due to the immediate binding of many uncapacitated cells to the OECM. The spermatozoa released from the OECM were mainly capacitated (51.2 ± 3.9%) at 2 h and most were acrosome reacted at the end of the incubation (Fig. 6b). Most of the frozen control and experimental spermatozoa displayed the B pattern immediately upon thawing (Fig. 6c,d). The control spermatozoa gradually became acrosome reacted (Fig. 6c), whereas the experimental cells underwent this transition more rapidly and to a greater extent (Fig. 6d).

The earlier release of cryopreserved spermatozoa from the OECM may indicate that these cells reached the correct physiological state to participate in fertilization earlier than fresh spermatozoa, adding support to the hypothesis that frozen–thawed spermatozoa are already in a state of membrane destabilization similar to capacitation. The observation that cryopreserved spermatozoa underwent a transition to the acrosome-reacted state earlier than fresh spermatozoa, and earlier in the presence of the OECM than in its absence, indicates that the two types of spermatozoa respond differently to oviduct stimuli and that they have a different lifespan in the female reproductive tract. Thus, even when intrauterine insemination is used to ensure fertilization, mistiming of semen deposition with respect to ovulation will lead to lower fertility than insemination just before, or after, ovulation (Salamon and Maxwell, 1995b). This may be the result of premature exhaustion of the limited supplies of 'uncapacitated' cells from the isthmic reservoirs as a result of their early activation.

Conclusion

The studies discussed in this paper demonstrate that fresh and frozen–thawed spermatozoa behave differently when deposited in the female reproductive tract. In comparison with fresh spermatozoa, the frozen–thawed cells exhibit destabilized membranes as assessed by CTC fluorescence patterns, capacitation, impaired sperm transport associated with decreased respiratory activity, increased denaturation of DNA, faster loss from the tract and accelerated binding to and release from oviduct cells. These findings support the hypotheses that cryopreservation bypasses the normal processes of sperm maturation leading to capacitation (Watson, 1995) and results in the transport of fewer spermatozoa to the site of fertilization (Lightfoot and Salamon, 1970a).

Studies to date have not explained the mechanisms of the membrane changes observed after cryopreservation or provided strong clues on how they might be reversed. Detailed information is not yet available on the interaction of cryopreserved spermatozoa with the cells within the functional sperm reservoir in the caudal isthmus. However, it is likely that cryopreservation or thawing methods that incorporate strategies to stabilize sperm membranes and limit the rate of loss of spermatozoa from the tract would also ensure that more fertile cells, in the best functional state, reach and establish themselves in the tubal reservoirs.

The authors are grateful to G. Evans, S. Mortimer, S. Salamon and P.F. Watson for their comments on the manuscript.

References

Almlid T and Hofmo PO (1996) A brief review of frozen semen application under Norwegian AI service conditions. In *Proceedings of the Third International Conference on Boar Semen Preservation*, Mariensee, Germany. Eds D Rath, LA Johnson and KF Weitze *Reproduction in Domestic Animals* **31** 169–174

Bernstein AD and Petropavlovskij VV (1937) Effect of non-electrolytes on viability of spermatozoa *Bjulleten Experimentalnoj Biologii i Mediciny* **3** 41–43 (in Russian)

Colas G (1979) Fertility in the ewe after artificial insemination with fresh and frozen semen at the induced oestrus, and influence of the photoperiod on the semen quality of the ram *Livestock Production Science* **6** 153–166

Cormier N, Sirard M-A and Baley JL (1997) Premature capacitation of bovine spermatozoa is initiated by cryopreservation *Journal of Andrology* **18** 461–468

Dobrinsky JR and Johnson LA (1994) Cryopreservation of porcine embryos by vitrification: a study of *in vitro* development *Theriogenology* **42** 25–35

Drobnis EZ and Overstreet JW (1992) Natural history of mammalian spermatozoa in the female reproductive tract *Oxford Reviews of Reproductive Biology* **14** 1–45

Evans G and Maxwell WMC (1987) *Salamon's Artificial Insemination of Sheep and Goats* Butterworths, Sydney

Fraser LR, Abeydeera LR and Niwa K (1995) Ca^{2+}-regulating mechanisms that modulate bull sperm capacitation and acrosomal exocytosis as determined by chlortetracycline analysis *Molecular Reproduction and Development* **40** 233–241

Gillan L, Evans G and Maxwell WMC (1997a) The capacitation status and fertility of fresh and frozen–thawed ram spermatozoa *Reproduction, Fertility and Development* **9** 481–487

Gillan L, Evans G and Maxwell WMC (1997b) Cryopreservation induced membrane changes in ram spermatozoa and their effect on oviductal interactions *Proceedings of the Australian Society for Reproductive Biology, Canberra* **28** 85 (Abstract)

Gómez MC, Catt JW, Gillan L, Evans G and Maxwell WMC (1997) Effect of culture, incubation and acrosome reaction of fresh and frozen–thawed ram spermatozoa for *in vitro* fertilization and intracytoplasmic sperm injection *Reproduction, Fertility and Development* **9** 665–673

Harrison RAP (1996) Capacitation mechanisms, and the role of capacitation as seen in eutherian mammals *Reproduction, Fertility and Development* **8** 581–594

Hawk HW (1983) Sperm survival and transport in the female reproductive tract *Journal of Dairy Science* **66** 2645–2660

Hawk HW (1987) Transport and fate of spermatozoa after insemination of cattle *Journal of Dairy Science* **70** 1487–1503

Hawk HW and Conley HH (1971) Loss of spermatozoa from the reproductive tract of the ewe and intensification of sperm 'breakage' by progestagen *Journal of Reproduction and Fertility* **27** 339–347

Holt WV and North RD (1984) Partially irreversible cold-induced lipid phase transitions in mammalian sperm plasma membrane domains: freeze–fracture study *Journal of Experimental Zoology* **230** 473–483

Hunter RHF (1993) Sperm:egg ratios and putative molecular signals to modulate gamete interactions in polytocous mammals *Molecular Reproduction and Development* **35** 324–327

Hunter RHF and Nichol R (1983) Transport of spermatozoa in the sheep oviduct: preovulatory sequestering of cells in the caudal isthmus *Journal of Experimental Zoology* **228** 121–128

Hunter RHF and Nichol R (1986) Post-ovulatory progression of viable spermatozoa in the sheep oviduct, and the influence of multiple mating on their pre-ovulatory distribution *British Veterinary Journal* **142** 52–58

Hunter RHF and Wilmut I (1984) Sperm transport in the cow: peri-ovulatory redistribution of cells within the oviduct *Reproduction, Nutrition Development* **24** 597–608

Hunter RHF, Nichol R and Crabtree SM (1980) Transport of spermatozoa in the ewe: timing of the establishment of a functional population in the oviduct *Reproduction, Nutrition Development* **20** 1869–1875

Jabbour HN and Evans G (1991) Fertility of superovulated ewes following intrauterine or oviducal insemination with fresh or frozen–thawed semen *Reproduction, Fertility and Development* **3** 1–7

Lightfoot RJ and Salamon S (1970a) Fertility of ram spermatozoa frozen by the pellet method. I. Transport and viability of spermatozoa within the genital tract of the ewe *Journal of Reproduction and Fertility* **22** 385–398

Lightfoot RJ and Salamon S (1970b) Fertility of ram spermatozoa frozen by the pellet method. II. The effects of method of insemination on fertilization and embryonic mortality *Journal of Reproduction and Fertility* **22** 399–408

Lopyrin AI and Loginova NV (1958) Method of freezing ram semen *Ovtsevodstvo* **No. 8** 31–34 (in Russian)

Lopyrin AI and Rabocev VK (1968) The optimum time to use preserved semen *Ovtsevodstvo* **13 No. 9** 34–36 (in Russian)

Mattner PE (1964) *Studies on Transport and Distribution of Spermatozoa in the Genital Tract of the Ewe* MVSc thesis, University of Sydney

Mattner PE, Entwistle KW and Martin ICA (1969) Passage, survival and fertility of deep-frozen ram semen in the genital tract of the ewe *Australian Journal of Biological Science* **22** 181–187

Maxwell WMC and Hewitt LJ (1986) A comparison of vaginal, cervical and intrauterine insemination of sheep *Journal of Agricultural Science* **106** 191–193

Maxwell WMC and Johnson LA (1997a) Chlortetracycline analysis of boar spermatozoa after incubation, flow cytometric sorting, cooling, or cryopreservation *Molecular Reproduction and Development* **46** 408–418

Maxwell WMC and Johnson LA (1997b) Membrane status of boar spermatozoa after cooling or cryopreservation *Theriogenology* **48** 209–219

Maxwell WMC and Watson PF (1996) Recent progress in the preservation of ram semen *Animal Reproduction Science* **42** 55–65

Maxwell WMC, Evans G, Rhodes SL, Hillard MA and Bindon BM (1993) Fertility of superovulated ewes after intrauterine or oviducal insemination with low numbers of fresh or frozen–thawed spermatozoa *Reproduction, Fertility and Development* **5** 57–63

Platov EM (1983) An important factor affecting ewe fertility *Ovtsevodstvo* **No. 6** 35–37 (in Russian)

Quinlivan TD and Robinson TJ (1969) Number of spermatozoa in the genital tract after artificial insemination of progestagen-treated ewes *Journal of Reproduction and Fertility* **19** 73–86

Royere D, Hamamah S, Nicolle JC, Barthelemy C and Lansac J (1988) Freezing and thawing alter chromatin stability of ejaculated human spermatozoa: fluorescence acridine orange staining and feulgen-DNA cytophotometric studies *Gamete Research* **21** 51–57

Salamon S (1977) Fertility following deposition of equal numbers of frozen–thawed ram spermatozoa by single and double insemination *Australian Journal of Agricultural Research* **28** 477–479

Salamon S and Maxwell WMC (1995a) Frozen storage of ram semen. I. Processing, freezing, thawing and fertility after cervical insemination *Animal Reproduction Science* **37** 185–249

Salamon S and Maxwell WMC (1995b) Frozen storage of ram semen. II. Causes of low fertility after cervical insemination and methods of improvement *Animal Reproduction Science* **38** 1–36

Salamon S, Maxwell WMC and Firth JH (1979) Fertility of ram semen after storage at 5°C *Animal Reproduction Science* **2** 373–385

Saling PM and Storey BT (1979) Mouse gamete interactions during fertilization *in vitro:* chlortetracycline as a fluorescent probe for the mouse sperm acrosome reaction *Journal of Cell Biology* **83** 544–555

Seidel GE, Allen CH, Brink Z, Graham JK and Cattell MB (1995) Insemination of Holstein heifers with very low numbers of unfrozen spermatozoa *Journal of Animal Science Supplement 1* **73** 232 (Abstract 488)

Shannon P (1978) Factors affecting semen preservation and conception rates in cattle *Journal of Reproduction and Fertility* **54** 519–527

Smith TT and Yanagimachi R (1989) Capacitation status of hamster spermatozoa in the oviduct at various times after mating *Journal of Reproduction and Fertility* **86** 255–261

Smith TT and Yanagimachi R (1991) Attachment and release of spermatozoa from the caudal isthmus of the hamster oviduct *Journal of Reproduction and Fertility* **91** 567–573

Suzuki H, Foote RH and Farrell PB (1997) Computerized imaging and scanning electron microscope (SEM) analysis of co-cultured fresh and frozen bovine sperm *Journal of Andrology* **18** 217–226

Tilbrook AJ and Pearce DT (1984) Losses of spermatozoa from the vagina of the ewe *Animal Production in Australia* **15** 759 (Abstract)

Tsien RY (1989) Fluorescent indicators of ion concentrations. In *Fluorescence Microscopy of Living Cells in Culture. Part B. Quantitative Fluorescence Microscopy – Imaging and Spectroscopy.* Methods in Cell Biology Vol. 30 pp 127–156 Eds DL Taylor and Y-L Wang. Academic Press, New York

Wang WH, Abeydeera LR, Fraser LR and Niwa K (1995) Functional analysis using chlortetracycline fluorescence and *in vitro* fertilization of frozen–thawed ejaculated boar spermatozoa incubated in a protein-free chemically defined medium *Journal of Reproduction and Fertility* **104** 305–313

Watson PF (1995) Recent developments and concepts in the cryopreservation of spermatozoa and the assessment of their post-thawing function *Reproduction, Fertility and Development* **7** 871–891

Watson PF (1996) Cooling of spermatozoa and fertilizing capacity. In *Proceedings of the Third International Conference on Boar Semen Preservation, Mariensee, Germany* Eds D Rath, LA Johnson and KF Weitze *Reproduction in Domestic Animals* **31** 135–140

Windsor DP (1997) Mitochondrial function and ram sperm fertility *Reproduction, Fertility and Development* **9** 279–284

EMBRYONIC SURVIVAL

Chair
D. C. Wathes

Journal of Reproduction and Fertility Supplement **54**, 287–302

Uterine differentiation as a foundation for subsequent fertility

F. F. Bartol[1], A. A. Wiley[1], J. G. Floyd[1], T. L. Ott[2], F. W. Bazer[2], C. A. Gray[2] and T. E. Spencer[2]

[1]*Department of Animal and Dairy Sciences, Auburn University, AL 36849–5415; and* [2]*Institute of Biosciences and Technology, Center for Animal Biotechnology, Texas A & M University, College Station, TX 77843–2471, USA*

Uterine differentiation in cattle and sheep begins prenatally, but is completed postnatally. Mechanisms regulating this process are not well defined. However, studies of urogenital tract development in murine systems, particularly those involving tissue recombination and targeted gene mutation, indicate that the ideal uterine organizational programme evolves epigenetically through dynamic cell–cell and cell–matrix interactions that define the microenvironmental context within which gene expression occurs and may ensure adult tissue stability. In the cow and ewe, transient postnatal exposure of the developing uterus to steroids can produce immutable changes in adult uterine tissues that may alter the embryotrophic potential of the uterine environment. Thus, success of steroid-sensitive postnatal events supporting uterine growth and development can dictate the functional potential of the adult uterus. Studies to determine effects of specific steroidal agents on patterns of uterine development during defined neonatal periods, as well as the functional consequences of targeted neonatal steroid exposure in the adult uterus, should enable identification of critical developmental mechanisms and determinants of uterine integrity and function. Extreme adult uterine phenotypes (lesion models) created in cattle and sheep by strategic postnatal steroid exposure hold promise as powerful tools for the study of factors affecting uterine function and the rapid identification of novel uterine genes.

Introduction

The uterus is an essential reproductive organ. Functions of the uterus in domestic ruminants include generation of the luteolytic signal required for ovarian cyclicity, transport and maturation of spermatozoa, recognition and reception of embryos, provision of an embryotrophic environment for conceptus development, and expulsion of the fetus and placenta at parturition (Bartol, 1999). These functions are borne by the uterine mucosa or endometrium and smooth muscle or myometrium. Developmental determinants of uterine function are not well defined. However, studies of laboratory animals (Mori and Nagasawa, 1988), humans (Mori and Nagasawa, 1988; Cooper and Kavlock, 1997), wildlife species (Cooper and Kavlock, 1997), and domestic ungulates, including sheep (Bartol *et al.*, 1988a, 1997), cattle (Hancock *et al.*, 1994; Bartol *et al.*, 1995; King *et al.*, 1995; Bartol *et al.*, 1996; Bartol and Floyd, 1996) and pigs (Bartol *et al.*, 1993), indicate that exposure of developing uterine tissues to agents that disrupt critical organizational events can have lasting effects on reproductive health. Thus, while genetic potential for uterine competence and reproductive success may be defined at conception, success of developmental events regulating uterine growth, morphogenesis and cytodifferentiation ultimately determines phenotypic potential of the uterus to support essential reproductive processes.

The fact that disruptive effects of steroids can have long-term consequences for uterine function, and that the nature of such effects reflects specific conditions of steroid exposure, suggests a strategy

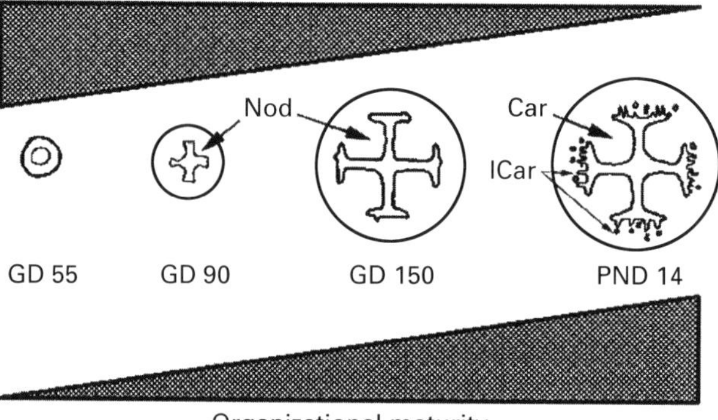

Fig. 1. Schematic representation of ovine uterine morphogenesis between gestational day (GD) 55 and postnatal day 14 (PND 14; birth = PND 0). Significant caruncular morphogenesis occurs prenatally (GD 150 = term), while genesis and proliferation of endometrial glands (adenogenesis) is a postnatal event in both sheep and cattle (Nod: precaruncular nodule; Car: caruncle; ICar: intercaruncular area). In general, the potential for organizational disruption of uterine development is inversely related to tissue maturity.

for identification of critical, steroid-sensitive developmental periods, and a scheme for the creation of extreme adult uterine phenotypes (lesion models) of use in studies of uterine function. The objectives of this review are to: (1) summarize primary developmental events that support uterine organization; (2) present evidence that the functional integrity of adult uterine tissues is determined, in part, by the success of steroid-sensitive postnatal uterine organizational events; (3) describe strategies for creation of extreme adult uterine phenotypes in domestic ruminants, based on the concept of endocrine disruption; and (4) present evidence of the utility of such models for study of uterine development and function. Where possible, emphasis will be placed on uterine development in sheep and cows.

Uterine Organogenesis

The uterus develops as a specialization of the paramesonephric ducts, which give rise to the infundibula, oviducts, uterus, cervix and anterior vagina (Bartol, 1999). Paramesonephric fusion occurs between gestational day (GD) 34 and 55 in sheep (Wiley *et al.*, 1987), and GD 55 and 60 in cattle (Marion and Gier, 1971). Fusion is partial in both species, producing a bicornuate uterus that supports intercornual migration of embryos.

Uterine histogenesis has been described to some extent for both cattle (see Marion and Gier, 1971; Atkinson *et al.*, 1984; and references therein) and sheep (Wiley *et al.*, 1987; Bartol *et al.*, 1988a,b). Differentiation of paramesonephric tissues into histologically discernible zones indicating endometrium and presumptive myometrium is evident in the ovine fetus by GD 55, and in the bovine fetus by GD 70. Definitive uterine tissue layers, including the adluminal zone of densely packed endometrial stroma or stratum compactum, the deeper more loosely arranged stromal cells of the stratum spongiosum, and both inner and outer layers of myometrial smooth muscle, are evident in sheep and cattle by GD 90–100. Caruncles, raised aglandular structures that are macroscopic features of the adult endometrium in sheep and cattle, emerge during fetal life as precaruncular nodules (Fig. 1). Extensive prenatal caruncular morphogenesis defines both the number and distribution of these structures along the uterine wall. Genesis of uterine glands (adenogenesis) begins during the last month of gestation when short epithelial invaginations appear along the uterine mucosa surrounding the base of precaruncular nodules. Endometrial morphogenesis is completed postnatally with continued growth of caruncles, extensive proliferation of endometrial glands, and establishment of definitive aglandular caruncular and intensely glandular intercaruncular endometrial areas.

Organizational Mechanisms

Mechanisms regulating growth and differentiation of the paramesonephric duct axis in ruminants are not well defined. Presently, much is inferred about these mechanisms from studies of laboratory species. Jost (1953) established the paradigm that prenatal urogenital tract development in female mammals is an ovary-independent process. Uterine development is also unaffected for defined periods after ovariectomy at birth in the mouse (Bigsby and Cunha, 1985), rat (Branham and Sheehan, 1995), pig (Tarleton *et al.*, 1998) and sheep (Bartol *et al.*, 1988a,b). Thus, uterine organizational mechanisms are ovary-independent and may be steroid-independent for some period before and after birth.

Roles for ligand-dependent nuclear receptors

The extent to which members of the nuclear receptor superfamily of ligand-regulatable transcription factors are required to support uterine development during pre- and early postnatal life is unclear. However, female mice lacking functional oestrogen receptor-α (ER) or progesterone receptor (PR) genes were born with complete reproductive tracts, indicating that uterine organogenesis, at least in the mouse, does not require ER, PR or their cognate ligands (Korach *et al.*, 1996; Lydon *et al.*, 1996). Uterine hypoplasia in ER-null mice confirmed that an active ER system is required for uterine growth (Korach *et al.*, 1996). Uterine hypoplasia was also observed in weaned mice lacking receptors for 1α, 25-dihydroxy vitamin D3 (Yoshizawa *et al.*, 1997). In addition, the fact that retinyl palmitate administered to neonatal pigs perturbed early postnatal uterine development (Vallet *et al.*, 1995), and that retinoids can affect homeogene expression (see below, and Marshall *et al.*, 1996), indicates that retinoic acid receptors may mediate uterine organizational events.

Ontogeny of steroid receptor expression and function in developing ungulate uterine tissues are incompletely characterized. In the cow, PR mRNA was not detected consistently in dispersed fetal uterine cells from mid- to late gestation, whereas ER mRNA was detected on GD 100–110 and increased on GD 185–200 (Malayer and Woods, 1998). Oestrogen did not bind to or affect DNA synthesis in cultured bovine mesonephric cells obtained on GD 50–59 (Winters *et al.*, 1993). However, oestrogen responsiveness was reported for cultured uterine cells from GD 185–200 (Malayer and Woods, 1998). Thus, neither PR nor functional ER may be present in developing uterine tissues prenatally until after gross uterine morphology and basic histoarchitectural features of the uterine wall have emerged (see above). Since exposure of the immature urogenital tract to steroids can affect the integrity of adult tissues (see below), and steroids are present in the fetal circulation throughout gestation, ontogeny of steroid sensitivity in developing uterine tissues may reflect a natural strategy to ensure organizational success.

Expression of functional steroid receptors during the perinatal period may be necessary for normal uterine development (Bartol *et al.*, 1988b; Malayer and Woods, 1998; Tarleton *et al.*, 1998). Nuclear steroid receptors mediate classical ligand-dependent events and enhance target cell responsiveness to peptide growth factors by coupling with membrane receptor-mediated signal transduction pathways (Smith, 1998). Thus, expression of ER-positive (ER+) character in uterine cells later in development, as described for cows (Malayer and Woods, 1998), may define the point at which oestrogens begin to elicit trophic effects on the fetal uterus and(or) increase the sensitivity of ER+ uterine target cells to paracrine mediators of oestrogen action such as insulin-like growth factor I (IGF-I) or epidermal growth factor (EGF). Like the ER-null mouse (Korach *et al.*, 1996), both IGF-I-null and EGF receptor (EGFR)-null mice have hypoplastic uteri (Baker *et al.*, 1996; Hom *et al.*, 1998).

Uterotrophic effects of oestradiol were limited to epithelium and virtually absent from fibromuscular stroma when EGFR expression, also documented in the fetal bovine uterus (Malayer and Woods, 1998), was eliminated in mice (Hom *et al.*, 1998). Thus, compartment-specific cross-talk between growth factor and steroid receptor signalling pathways may affect uterine growth in the fetus and neonate.

Homeogenes as effectors of uterine organization and tissue stability

Homeogenes are primary effectors of tissue organization. Mammalian homeogenes encode transcription factors that determine regional tissue identities along anteroposterior body axes, and may maintain functional stability of adult tissues (Taylor *et al.*, 1997). Expression of homeogenes Hoxa-9, -10, -11, -13 and Msx1 is uniform along the murine paramesonephric duct prenatally, but becomes restricted spatially during the first two weeks of postnatal life such that oviducts express Hoxa-9, the uterus Hoxa-10, -11 and Msx1, the cervix Hoxa-11 and -13, and the anterior vagina Hoxa-13. This pattern persists in adult murine and human genital tracts and may be subject to steroid regulation (Pavlova *et al.*, 1994; Taylor *et al.*, 1997).

Disruption of homeogene expression is associated with homeotic tissue transformations. In adult ewes, prolonged exposure to oestrogen causes permanent infertility associated with destabilized cervical histoarchitecture and redifferentiation of the cervix toward a more anterior, uterine-like phenotype (Adams and Sanders, 1993; Adams, 1995). This partial homeotic transformation can be accompanied by development of uterine glandular cysts and adenomyotic lesions (Adams, 1995). Similarly, uteri of mice lacking functional Hoxa-10 (Benson *et al.*, 1996) or Hoxa-11 genes (Gendron *et al.*, 1997) displayed partial homeotic transformation toward a more anterior, oviduct-like phenotype. Impaired ability to form uterine glands was observed in Hoxa-11 mutants (Gendron *et al.*, 1997). Mice lacking functional Hoxa-10, Hoxa-11 or Hmx3 genes cannot support embryo development due to aberrant expression of critical uterine proteins (Gendron *et al.*, 1997; Wang *et al.*, 1998). Uterine expression of Wnt-5a was altered in Hmx3-null mice (Wang *et al.*, 1998). Stromal expression of Wnt-5a affects uterine epithelial expression of Msx1, which may be important for maintenance of epithelial receptivity to stromal and conceptus signals (Pavlova *et al.*, 1994). Thus, structural and functional stability of uterine tissues may evolve and be maintained through steroid-sensitive mechanisms operating to ensure spatially unique, tissue-specific patterns of homeotic gene expression yet to be defined in domestic ruminants.

Cell–cell and cell–matrix interactions and selective stabilization

Uterine development requires continuous reciprocal interactions between epithelium and underlying stroma (Bigsby, 1991). The communication network that develops through these interactions involves paracrine-acting factors and their receptors that may include homeogene products and Wnts (see above, and Moon *et al.* 1997,), hepatocyte growth factor (HGF; Sugawara *et al.*, 1997), EGF (Hom *et al.*, 1998), heparin-binding EGF (HB-EGF; Zhang *et al.*, 1998), transforming growth factors (TGF; Takahashi *et al.*, 1994; Godkin and Dore, 1998), IGF (Stevenson *et al.*, 1994), keratinocyte growth factor (KGF; Koji *et al.*, 1994), vascular endothelial growth factor (VEGF; Grant *et al.*, 1995; Torry and Torry, 1997), and others. Specific regulatory interactions that must evolve for successful uterine development reflect communication between epithelial, stromal and endothelial cells within the context of their extracellular matrix (ECM; Ettinger and Doljanski, 1992; Grant *et al.*, 1995). These relationships direct local patterns of gene expression and dictate cellular responsiveness to gene products (Ettinger and Doljanski, 1992). Therefore, the ideal uterine organizational programme may be determined epigenetically through selective stabilization of specific cell–cell and cell–ECM interactions. This organizational model predicts that each configuration of cells and ECM during the course of development increases the likelihood of the next, and decreases the likelihood of others that might be less desirable (Ettinger and Doljanski, 1992). Conversely, aberrations in initial conditions would be amplified through development with severe implications for end-organ integrity. If applicable to domestic ruminants, this model predicts

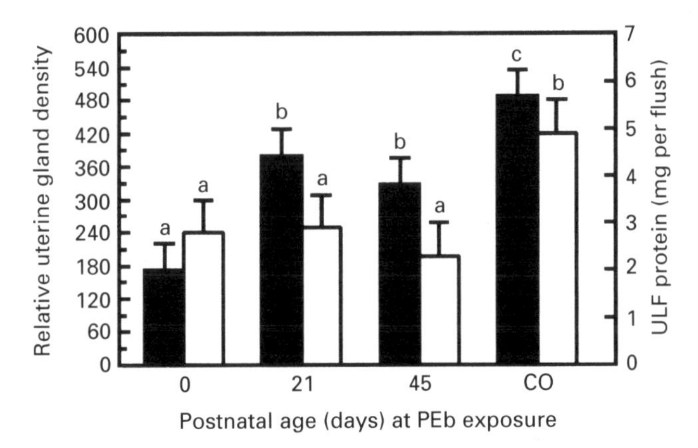

Fig. 2. Effects of chronic (about 200 days) administration of progesterone plus oestradiol benzoate (PEb), beginning on either postnatal day 0 (PND 0 = birth), 21 or 45, on relative endometrial gland density (dark bars) and uterine luminal fluid (ULF) protein content (open bars) in adult beef heifers during dioestrus. Treated heifers ($n = 5$ per group) received a single implant containing progesterone (100 mg) and oestradiol benzoate (10 mg). Control (CO) heifers were not exposed to progesterone and oestradiol benzoate. Uteri were obtained from heifers at approximately 15 months of age on day 12 of a $PGF_{2\alpha}$-induced oestrous cycle. For each response, bars with different letters are significantly different ($P < 0.01$). (Adapted from Bartol *et al.*, 1995.)

that disruption of critical primary conditions that define the programmatic context for uterine development should alter the capacity of uterine tissues to develop and function properly.

Postnatal Disruption of Uterine Development

The idea that disruption of development during specific 'critical' periods could have enduring effects on adult tissues is not new. Perinatal exposure of rodents to steroids, including oestrogens progestins and androgens, can disrupt uterine development, initiate uterine lesions and impair fertility (Sananes *et al.*, 1980; Mori and Nagasawa, 1988; Ohta, 1995). Tissue susceptibility to such organizational effects of steroids tends to be inversely related to age or tissue maturity and directly related to dosage and duration of exposure (Fig. 1).

Postnatal exposure to steroids can have lasting effects in domestic ungulates (Bartol *et al.*, 1993, 1995; Spencer *et al.*, 1993; King *et al.*, 1995). Moreover, the potential for inappropriate exposure of developing tissues in domestic animals to either natural hormones or xenobiotics is real. Exposure can occur: (1) physiologically, as a consequence of aberrant production of hormones during critical periods; (2) by diet, as a consequence of the consumption of bioactive agents such as phytoestrogens or mycotoxins (Adams, 1995); (3) pharmacologically, as a consequence of the intentional use of endocrinologically active agents to enhance performance traits (Hancock *et al.*, 1994; King *et al.*, 1995); and (4) unintentionally, as a consequence of the presence of endocrine-active industrial pollutants in the environment (Cooper and Kavlock, 1997). Compounds that disrupt development by altering critical endocrinological events are categorized as endocrine disruptors (EDs).

Postnatal exposure to steroidal endocrine disruptors in cows

Commercially, beef calves are often exposed to steroidal agents released from implants designed to enhance growth performance. In the United States, implants approved for female calves intended for use as breeding replacements contain either the oestrogenic compound zeranol alone (36 mg) (Schering-Plough Animal Health Corp., Union, NJ), or a combination of progesterone (100 mg) and oestradiol benzoate (10 mg) (Fort Dodge Animal Health, KS; Vetlife, Norcross, GA).

Effects of postnatal exposure to zeranol on bovine reproductive performance were related to both period of exposure and dosage. In numerous trials, pregnancy rates decreased by an average of 35% in yearling heifers given single zeranol implants at birth, but were essentially unaffected by the same treatment initiated between one and 14 months of age, suggesting a critical period for oestrogen sensitivity during the first postnatal month. However, pregnancy rates were depressed by as much as 40% in heifers treated with two or more zeranol implants between one and 11 months of age (Hancock *et al.*, 1994; Bartol and Floyd, 1996). Chronic exposure to zeranol for 300 days from birth delayed puberty and reduced pre- and postpubertal uterine diameter in beef heifers (King *et al.*, 1995). Similarly, exposure of heifers to several compounds for one year from PND 84, including trenbolone acetate, oestradiol, zeranol, or a combination of TBA and oestradiol, had variable but consistently negative effects on adult uterine wet weights (Moran *et al.*, 1990). These anti-uterotrophic effects could reflect lesions of the central nervous system, altered gonadotrophin secretion and lack of uterotrophic support from the ovary. However, abortion frequency increased between GD 25 and 45 in heifers exposed to zeranol from birth to PND 300, indicating that zeranol-induced uterine lesions that affected attachment of the conceptus.

Treatment of beef heifers with an implant designed to release both progesterone and oestradiol benzoate for approximately 200 days beginning on either PND 0, 21 or 45 reduced adult uterocervical wet weight by 35%, myometrial area by 23%, and endometrial area by 27%, regardless of age at first exposure to progesterone and oestradiol benzoate (Bartol *et al.*, 1995). Effects were accompanied by a marked decrease in uterine glandularity that was most severe when exposure began at birth (Fig. 2). Uterine luminal fluid protein content was reduced by approximately 45% in heifers exposed to progesterone and oestradiol benzoate (Fig. 2). Thus, generalized uterine hypoplasia, endometrial aplasia and altered uterine protein content were observed in cyclic adult heifers 13.5 to 15 months after initiation of chronic (about 200 day) exposure to progesterone and oestradiol benzoate on or before PND 45, and a potentially critical period of uterine sensitivity to developmental disruption induced by progesterone and oestradiol benzoate was identified between birth and PND 21 (Bartol *et al.*, 1995).

Subsequently, crossbred beef heifers were assigned to one of five groups at birth (groups I–V; n = 5 or 6 heifers per group). Heifers in groups I–III received a single progesterone and oestradiol benzoate implant at birth, while those in groups IV and V served as unexposed controls (CO). All heifers were laparotomized on PND 21, when each uterus was measured and progesterone and oestradiol benzoate implants were removed from calves in group I. This created a group of uterine-intact adults exposed to progesterone and oestradiol benzoate for 21 days from birth. In addition, on PND 21, heifers in groups II (progesterone and oestradiol benzoate) and V (CO) were hemi-hysterectomized to permit evaluation of short-term treatment effects on uterine histoarchitecture. Jugular blood samples were taken from heifers at 16 months of age during dioestrus, before and after administration of oxytocin (100 iu), and plasma was assayed for 13,14-dihydro-15-keto$PGF_{2\alpha}$ (PGFM) as a reflection of oxytocin-inducible uterine prostaglandin generating ability (Wolfenson *et al.*, 1993). Uteri were obtained at slaughter during dioestrus at 26 months of age. Tissues were processed for histomorphometry, and endometrial samples were assayed for oxytocin receptor concentrations (Spencer *et al.*, 1995; F. F. Bartol and M. A. Mirando, unpublished).

Anti-uterotrophic effects of progesterone and oestradiol benzoate were evident by PND 21, when both uterine horn length (141.6 versus 78.2 ± 5.0 mm) and volume (5019 versus 2020 ± 59 mm^3) were reduced ($P < 0.05$) in treated heifers. Nascent uterine glands were present in both groups on PND 21, but were more frequently branched and appeared less stable structurally in heifers exposed to progesterone and oestradiol benzoate (Fig. 3). Chronic exposure to progesterone and oestradiol benzoate, exceeding 21 days from birth, was required to produce overt effects on adult bovine uterine size and endometrial histoarchitecture. Uterine weights (Fig. 4a) indicated that anti-uterotrophic effects of chronic exposure to progesterone and oestradiol benzoate alone (group III) were approximately equivalent to neonatal hhx (group V). Effects of hhx combined with chronic

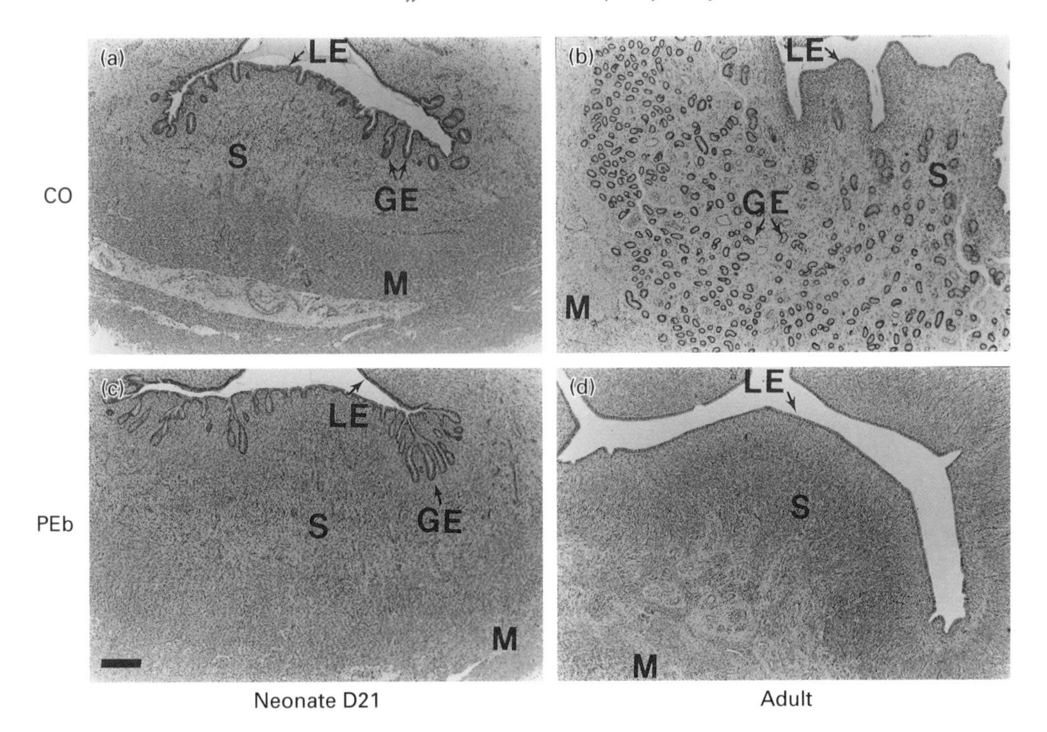

Fig. 3. Postnatal histogenesis of the bovine uterine wall and extreme effects of exposure to progesterone plus oestradiol benzoate (PEb) from birth on adult uterine histoarchitecture. Photomicrographs show histology of the uterine wall in individual animals hemihysterectomized on neonatal day 21 (a and c), and in the contralateral uterine horn of the same animals at 26 months of age during dioestrus (b and d). Micrographs (a) and (b) illustrate normal histogenesis in a representative control heifer (CO) not exposed to progesterone and oestradiol benzoate. Micrographs (c) and (d) illustrate an extreme consequence of chronic exposure to progesterone and oestradiol benzoate from birth in an adult heifer. Nascent endometrial glands (GE) were present in both CO and heifers exposed to progesterone plus oestradiol benzoate on PND 21 (a versus c). Chronic exposure to progesterone plus oestradiol benzoate reduced or eliminated endometrial glands in adults (b versus d). No endometrial glands were found in multiple serial sections of the adult uterus (d) that was exposed neonatally to progesterone plus oestradiol benzoate. However, this neonatally hemihysterectomized heifer displayed oestrous cycles of normal duration. LE: luminal epithelium, GE: glandular epithelium; S: endometrial stroma; M: myometrium. Haematoxylin and eosin staining. Scale bar represents 300 μm.

progesterone and oestradiol benzoate were additive, producing a 52% reduction in uterine mass in group II heifers (Fig. 4a). The reduction in adult endometrial glandularity, expected with chronic exposure to progesterone and oestradiol benzoate, was most severe for group II heifers, in which few or no endometrial glands were found (Fig. 3). Consistently, both peak uterine PGFM response and endometrial OTR concentrations were reduced ($P < 0.07$) in group II heifers (Fig. 4).

Treatment-induced loss of oxytocin-sensitive prostaglandin-generating uterine parenchyma may explain the reduced peak PGFM response observed in group II heifers (Fig. 4). However, all heifers displayed regular oestrous cycles of normal duration, including the group II heifer in which no endometrial glands were found (Fig. 3). Thus, uterine glandular epithelium may not be essential for normal cyclicity in cattle. Results also indicate that evidence of normal cyclicity is not necessarily evidence of normal endometrial integrity.

Results show that transient postnatal exposure to steroids can have specific, extreme and lasting effects on the adult bovine endometrium that could alter the embryotrophic potential of the uterine environment (Martal *et al.*, 1997). Effects reflect particular conditions of exposure and tend to be

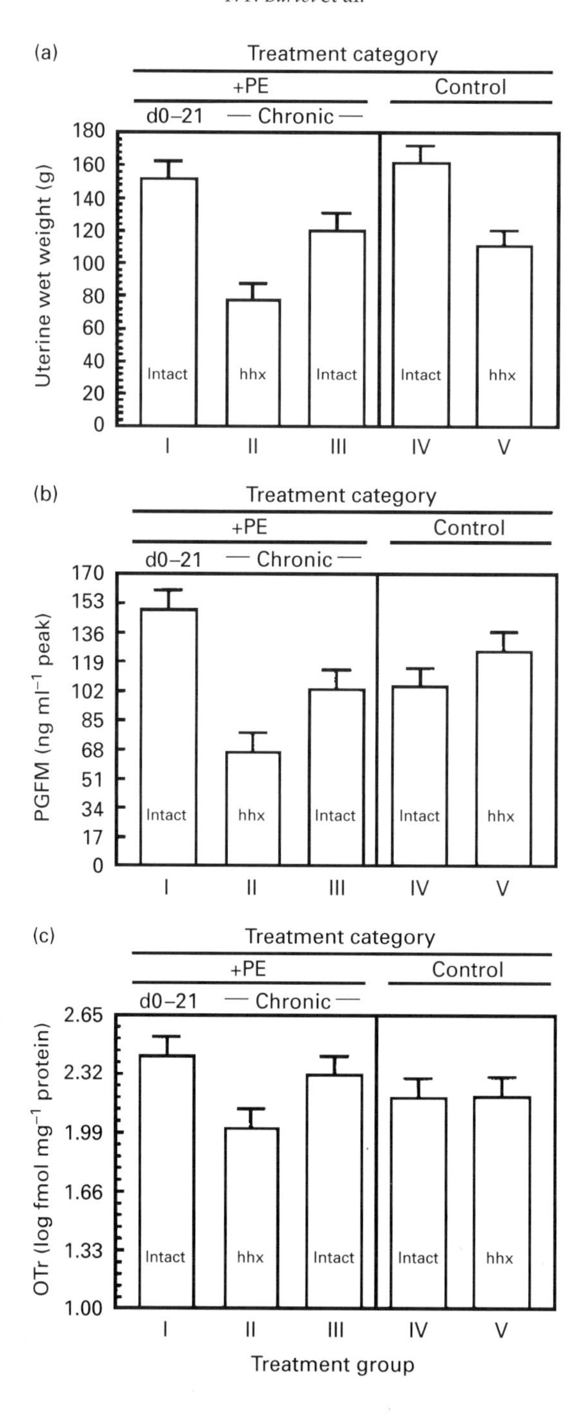

Fig. 4. Effects of neonatal exposure to progesterone plus oestradiol benzoate (PEb) from birth to neonatal day 21 (d0–21), or for approximately 200 days from birth (chronic), and hemihysterectomy (hhx) on neonatal day 21, on specific uterine responses in adult beef heifers (n = 5–6 per treatment group). Responses (least squares means + SEM) illustrated are: (a) uterine wet weight; (b) peak peripheral plasma concentrations of 13,14-dihydro-15-keto $PGF_{2\alpha}$ (PGFM) in response to an oxytocin (100 iu) challenge during dioestrus; and (c) endometrial oxytocin receptor density (OTr). Uterine weight (a) was reduced by neonatal progesterone plus oestradiol benzoate

more pronounced when initiated at birth. Studies also showed that postnatal exposure to steroidal endocrine disruptors of uterine development could be used to create unique adult uterine phenotypes.

Exposure to progestin from birth and uterine development in sheep

Endometrial glands proliferate rapidly between PND 0 and 26 in the ewe (Bartol *et al.*, 1988a,b). Exposure of lambs to a 19-nor-progestin (NOR; 17α-acetoxy-11β-methyl-19-norpreg-4-ene-3,20 dione) from PND 0 to 13 prevented gland development, as reflected by their absence in tissues from PND 13 (Bartol *et al.*, 1988a). Withdrawal of the progestin block permitted genesis of poorly organized uterine glands between PND 13 and 26 (Fig. 5A). Similarly, patterns of *in vitro* uterine protein synthesis characteristic of tissues obtained from NOR-exposed ewes on PND 13 and from NOR-withdrawn ewes on PND 26 were not identical to those observed during the normal morphogenetic transition between PND 0 and PND 13 (Fig. 5b). It was proposed that: (1) withdrawal of uterine tissues from a progestin-dominated prenatal environment at birth provides an endocrine cue for initiation of uterine adenogenesis; (2) this organizational programme could be disrupted by postnatal exposure to NOR; and that (3) prolonged exposure to NOR from birth should disintegrate critical organizational events sufficiently to produce a stable, extreme endometrial phenotype in adult ewes characterized by the absence of uterine glands, an organizationally induced uterine gland 'knock-out' (UGKO).

The 'UGKO' phenotype was created in adult ewes exposed to NOR for 32 weeks from birth (Bartol *et al.*, 1997). Uteri were obtained from NOR-exposed UGKO ewes during follicular ($n = 5$) and luteal ($n = 2$) phases of the ovarian cycle, and from one ewe with inactive ovaries. In striking contrast to intensely glandular endometrium obtained from control ewes during dioestrus (Fig. 6), endometrium from UGKO ewes was aglandular (7/8), or contained a few glandular cysts (1/8, not shown). This extreme phenotype may be induced by transient exposure to NOR for no more than 8 weeks from birth (T. E. Spencer, T. L. Ott, F. W. Bazer and F. F. Bartol, unpublished). Studies to determine whether UGKO ewes can cycle normally and conceive, and whether pregnancy can be established and maintained in an aglandular uterus are underway.

Distinct patterns of endometrial gene expression were identified between control and UGKO endometrium using mRNA differential display PCR. The majority (> 95%) of over 80 cDNAs cloned to date were amplified from control and absent from UGKO samples (Fig. 7). If structural differences between UGKO and control tissues are reflected at the transcriptional level, many differentially expressed mRNAs should be specific to the epithelium. Consistently, an antisense cRNA probe generated from endometrial cDNA DD54, identified as described above, hybridized specifically to uterine luminal and glandular epithelium from normal cyclic and pregnant ewes (Fig. 8a). Northern blot analysis revealed two major endometrial transcripts of approximately 2 kb and 6 kb, and DD54 expression increased during dioestrus, suggesting endocrine regulation of this epithelial gene product (Fig. 8b and unpublished results). The DD54 cDNA lacks sequence homology with known genes as determined using the BLAST algorithm (National Center for Biotechnology Information, NIH, Bethesda, MD). Results illustrate the immediate utility of the UGKO model for discovery of potentially novel genes encoding uterine proteins required for establishment of an embryotrophic uterine microenvironment in domestic ungulates (Martal *et al.*, 1997).

---

exposure (+PE < Control, $P < 0.06$) and hhx ($P < 0.01$). Among groups exposed to progesterone plus oestradiol benzoate (I–III), uterine weight was lower in group II than in groups I and III ($P < 0.01$). Relative to intact controls (group IV), short-term exposure to progesterone plus oestradiol benzoate (d0–21, group I) did not affect uterine weight. Peak PGFM concentrations (b), defined as the maximum value detected for each heifer within 45 min after oxytocin, were not affected by progesterone plus oestradiol benzoate exposure or hhx alone. However, among progesterone plus oestradiol benzoate exposed heifers, peak PGFM values were lower ($P < 0.07$) in group II than in group III. Identical relationships were detected for endometrial OTR density (c).

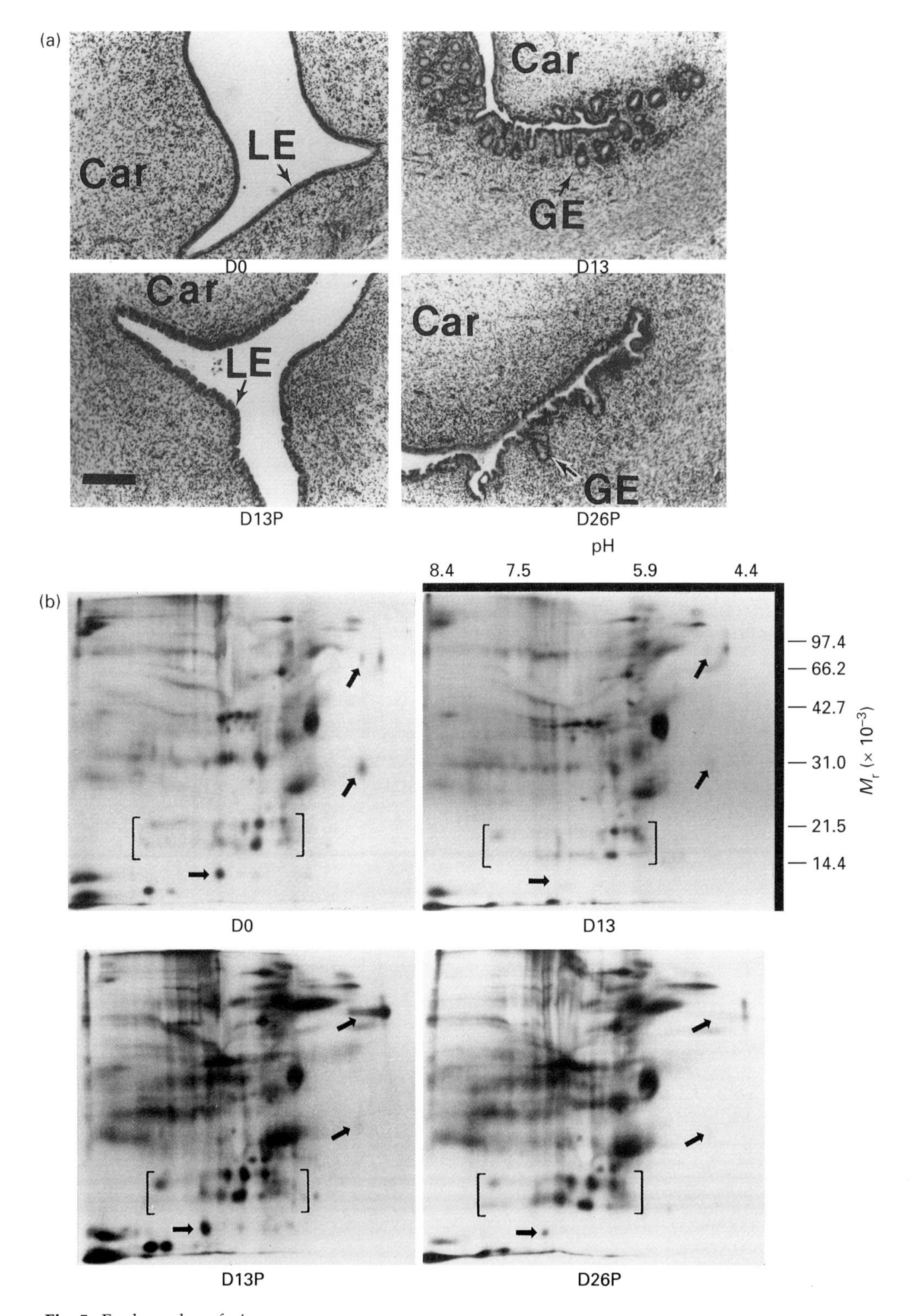

Fig. 5. For legend see facing page.

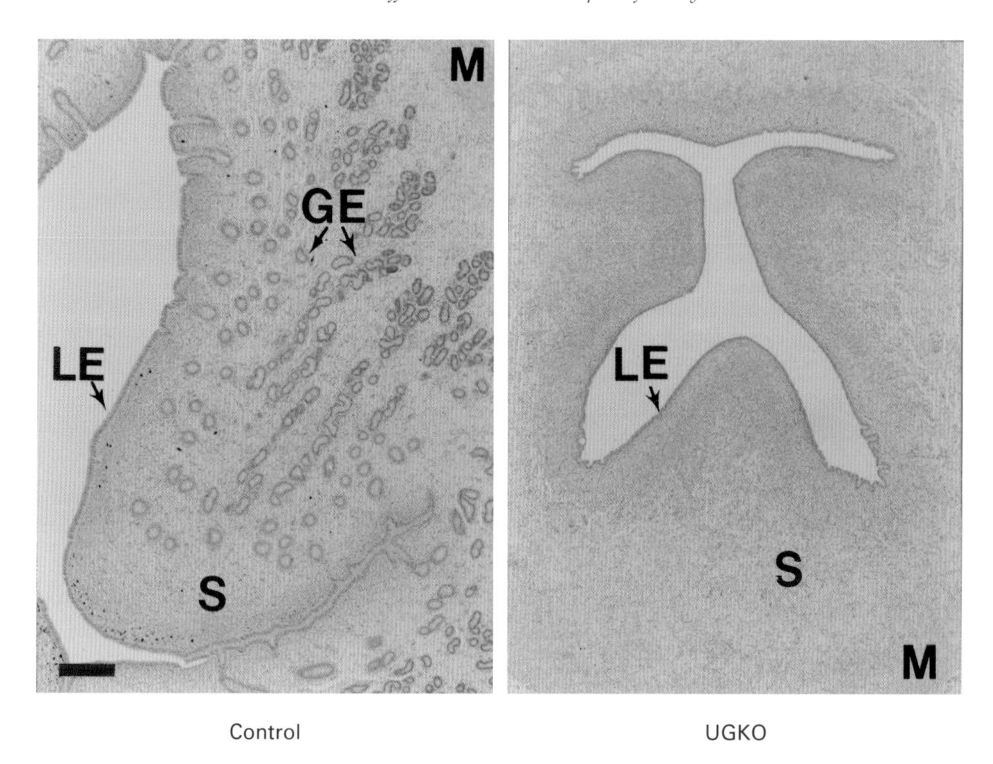

Fig. 6. Histological characterization of the uterine gland 'knock-out' (UGKO) phenotype in adult ewes. Photomicrographs depict normal adult endometrial histology for a typical control ewe, and the UGKO phenotype in an adult ewe that was exposed to a 19-norprogestin (19-norpreg-4-ene-3,20 dione) for 32 weeks from birth. Note the intense endometrial glandularity characteristic of control endometrium (left), compared with the glandless condition found in the UGKO endometrium (right). Tissues were obtained during dioestrus. LE: luminal epithelium; GE: glandular epithelium; S: endometrial stroma; M: myometrium. Scale bar represents 33 µm.

Fig. 5. Effects of chronic neonatal exposure of ewe lambs to 19-norprogestin (P; 19-norpreg-4-ene-3,20 dione) from birth on: (a) postnatal histogenesis of the endometrium (Car: caruncle; LE: luminal epithelium; GE: glandular epithelium); and (b) patterns of uterine protein synthesis *in vitro* in neonatal ewes ovariectomized at birth. (a) Endometrial glands are absent at birth (D0), but present throughout the intercaruncular endometrium by postnatal day (PND) 13 (D13). When ewe lambs are exposed to progestin from birth to PND 13, endometrial adenogenesis is inhibited and uterine glands are absent on PND 13 (D13P). Withdrawal of the progestin block on PND 13 permits some gland development as observed on PND 26 (D26P), although newly formed endometrial glands are structurally abnormal (D26P versus D13). (b) Uterine tissues of the type illustrated in (a) were explanted under defined conditions in the presence of L-4,5-[^{3}H]leucine and labelled proteins in explant medium were identified by fluorography of dried two-dimensional PAGE gels. Labelled uterine products were separated by isoelectric focussing in the first dimension (pH) and SDS-PAGE in the second ($M_r \times 10^{-3}$). Changes in patterns of uterine protein synthesis associated with uterine development between birth (D0) and PND 13 (D13) are illustrated by the top fluorographs. Some proteins produced by normally glandless uterine tissues from postnatal day 0 are no longer produced, or produced in the same relative abundance by tissues from postnatal day 13, in which endometrial glands are normally present (D0 versus D13, arrows). Chronic exposure of ewe lambs to progestin from birth inhibits gland genesis, restores production of some uterine proteins and induces production of others (D0 versus D13P, arrows and brackets). Tissues obtained on PND 26 after withdrawal of the progestin block to adenogenesis on PND 13 (D26P) displayed suppressed production of some proteins normally associated with uterine gland development (arrows), but relatively stable production of other proteins induced by progestin exposure (brackets). Scale bar represents 200 mm.

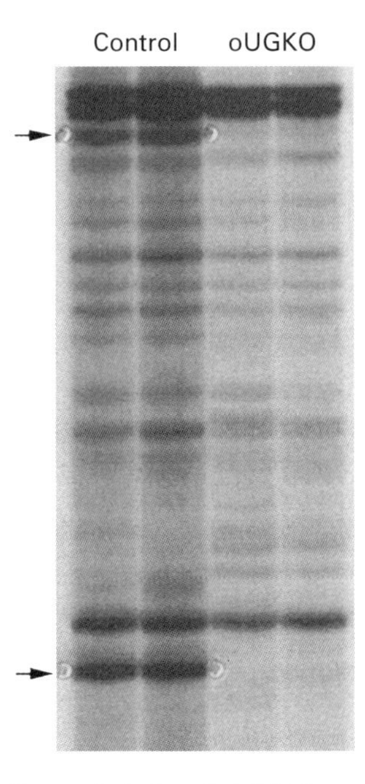

Fig. 7. Progestin-induced inhibition of uterine gland genesis alters transcriptional activity in adult ovine endometrium. An autoradiograph of [^{32}P-α]ATP-labelled cDNAs generated by mRNA differential display PCR (DD-PCR) and separated by electrophoresis on a 4.5% acrylamide–urea sequencing gel is shown. Typical results from duplicate endometrial total RNA samples obtained from normal control (left lanes) and uterine gland 'knock-out' (UGKO; right lanes) endometrium are shown (see Fig. 6). Evidence of differential gene expression is illustrated by the presence of bands in control lanes (arrows) and the absence of these bands in UGKO lanes.

Fig. 8. Expression of endometrial DD54 mRNA. (a) Representative dark-field photomicrographs of DD54 mRNA expression in endometrium obtained from cyclic and pregnant ewes detected by *in situ* hybridization analysis. The cDNA corresponding to DD54 was identified by mRNA DD-PCR as in Fig. 7. Cross-sections of uteri from cyclic and pregnant ewes were hybridized with [^{35}S]UTP-labelled sense and antisense DD54 cRNA probes and hybridization signals visualized by autoradiography. Tissues shown are from a cyclic ewe on day 1 after oestrus (D1C), and a pregnant ewe on day 19 (D19Px). Note intense hybridization signal in luminal epithelium (LE) and glandular epithelium (GE). No signal above background was detected with the labelled sense probe (D1C Sense). (b) Expression of endometrial DD54 mRNA detected by northern hybridization analysis. Two DD54 mRNA transcripts of approximately 2 kb and 6 kb (arrows) were identified in endometrial total RNA (20 mg) from cyclic (days 1, 5, 9 and 15 after oestrus) and pregnant ewes (Px; days 15 and 19) with a radiolabelled antisense cRNA probe generated from DD54 cDNA template. Expression of this epithelial gene product increased during dioestrus. Scale bar represents 40 μm.

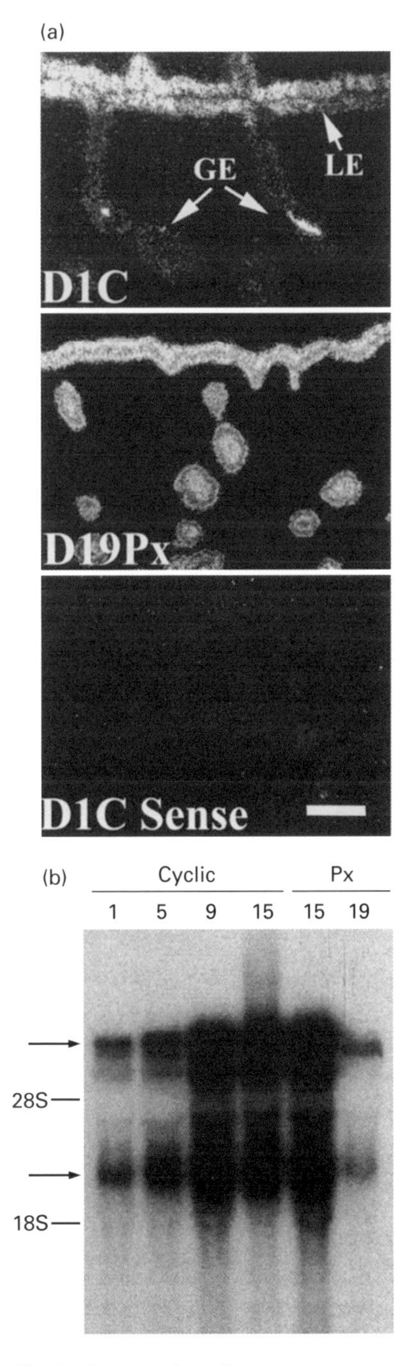

Fig. 8. For legend see facing page.

Conclusions

The extent to which embryo mortality (Martal *et al*, 1997) or uteroplacental dysfunctions associated with fetal growth retardation in domestic ruminants are attributable to uterine lesions induced by disruption of development is unknown. However, the fact that such lesions can be induced indicates that reproductive performance could be affected in this way. Since, in both cattle and sheep, postnatal uterine organizational events are steroid sensitive, mechanisms regulating these events are likely to involve steroid receptors and their co-activators and co-repressors (Hirotaka *et al.*, 1997). How the ideal uterine organizational programme evolves, and the extent to which disruption of organizationally critical cell–cell and cell–matrix interactions may affect the fate of uterine tissues and cells are topics that warrant investigation if factors affecting uterine capacity to support reproduction in domestic ruminants are to be defined.

Studies in cows and ewes indicate that normal and aberrant programmes of uterine organization, and consequences of organizational disruption, can be delineated by comparing patterns of uterine development and function in neonatally steroid exposed and unexposed animals. Physiological, biochemical and molecular comparisons of normal and lesioned adult uterine tissues should enable identification of factors affecting uterine function, developmental determinants of uterine integrity, biological markers of exposure to steroidal disruptors of uterine development, and rapid identification of novel uterine genes. Such studies will facilitate the design of environments and refinement of management guidelines to ensure that genetic potential for reproductive performance is realised, and will aid in efforts to evaluate the potential reproductive impact of environmental exposure to endocrine disruptors during development.

The authors thank Dr William W. Thatcher (University of Florida) and Dr Mark A. Mirando (Washington State University) for performing PGFM and OTR assays; Merial (Athens, GA) for providing Norgestomet implants; Ms Margaret M. Joyce for technical assistance; members of Dr Bazer's laboratory for assistance with surgery and collection of ovine tissues; Dr Shawn W. Ramsey and Mr Todd Taylor of the Texas A & M University Sheep and Goat Center, and Mr Jason Word for care and management of ewes; and Dr Dale A. Coleman and Ms Mabel Robinson (Auburn University) for technical assistance in bovine and early ovine studies. Special thanks are extended to the Center for Animal Biotechnology, Institute of Biosciences and Technology at Texas A&M University, for support leading to establishment and initial molecular characterizations of the ovine UGKO model. Work reported here was supported in part by AAES Project ALA 04–022 and USDA-NRI grants 95–37203–1995 to FFB and 98–35203–6322 to TES. This is AAES Journal No. 4–985927.

References

Adams NR (1995) Organizational and activational effects of phytoestrogens on the reproductive tract of the ewe *Proceedings of the Society for Experimental Biology and Medicine* **208** 87–91

Adams NR and Sanders MR (1993) Development of uterus-like redifferentiation in the cervix of the ewe after exposure to estradiol-17β *Biology of Reproduction* **48** 357–362

Atkinson BA, King GJ and Amoroso EC (1984) Development of the caruncular and intercaruncular regions in the bovine endometrium *Biology of Reproduction* **30** 763–774

Baker J, Hardy MP, Zhou J, Bondy C, Lupu F, Bellve AR and Efstratiadis A (1996) Effects of an Igf1 gene null mutation on mouse reproduction *Molecular Endocrinology* **10** 903–918

Bartol FF (1999) Uterus: Nonhuman. In *Encyclopedia of Reproduction* Vol. 4 Eds E Knobil and JD Neill. Academic Press, San Diego, CA

Bartol FF and Floyd JG (1996) Critical periods, steroid exposure and reproduction *Proceedings for the Annual Meeting of the Society for Theriogenology* 101–111

Bartol FF, Wiley AA, Spencer TE, Vallet JL and Christenson RK (1993) Early uterine development in pigs *Journal of Reproduction and Fertility Supplement* **48** 99–116

Bartol FF, Johnson LL, Floyd JG, Wiley AA, Spencer TE, Buxton DF and Coleman DA (1995) Neonatal exposure to progesterone and estradiol alters uterine morphology and luminal protein content in adult beef heifers *Theriogenology* **43** 835–844

Bartol FF, Floyd JG, Wiley AA, Coleman DA, Wolfe DF and Thatcher WW (1996) Neonatal steroid exposure and hemihysterectomy affect adult bovine uterine weight and response to oxytocin *Biology of Reproduction Supplement* 1 **54** 180

Bartol FF, Wiley AA, Spencer TE, Ing NH, Ott TL and Bazer FW (1997) Progestin exposure from birth: epigenetic induction of a unique adult uterine phenotype in sheep – a glandless endometrium *Biology of Reproduction Supplement* 1 **56** 133

Bartol FF, Wiley AA, Coleman DA, Wolfe DF and Riddell MG (1988a) Ovine uterine morphogenesis: effects of age and progestin administration and withdrawal on neonatal endometrial development and DNA synthesis *Journal of Animal Science* **66** 3000–3009

Bartol FF, Wiley AA and Goodlett DR (1988b) Ovine uterine morphogenesis: histochemical aspects of endometrial development in the fetus and neonate *Journal of Animal Science* **66** 1303–1313

Benson GV, Lim H, Paria BC, Satokata I, Dey SK and Maas RL (1996) Mechanisms of reduced fertility in Hoxa-10 mutant mice: uterine homeosis and loss of maternal Hoxa-10 expression *Development* **122** 2687–2696

Bigsby RM (1991) Reciprocal tissue interactions in morphogenesis and hormonal responsiveness of the female reproductive tract. In *Cellular Signals Controlling Uterine Function* pp 11–20 Ed. LA Lavia. Plenum Press, New York

Bigsby RM and Cunha GR (1985) Effects of progestins and glucocorticoids on deoxyribonucleic acid synthesis in the uterus of the neonatal mouse *Endocrinology* **117** 2520–2526

Branham WS and Sheehan DM (1995) Ovarian and adrenal contributions to postnatal growth and differentiation of the rat uterus *Biology of Reproduction* **53** 863–872

Cooper RL and Kavlock RJ (1997) Endocrine disruptors and reproductive development: a weight-of-evidence overview *Journal of Endocrinology* **152** 159–166

Ettinger L and Doljanski F (1992) On the generation of form by the continuous interactions between cells and their extracellular matrix *Biological Reviews* **67** 459–489

Gendron RL, Paradis H, Hsieh-Li HM, Lee DW, Potter SS and Markoff E (1997) Abnormal uterine stromal and glandular function associated with maternal reproductive defects in Hoxa-11 null mice *Biology of Reproduction* **56** 1097–1105

Godkin JD and Dorè JJE (1998) Transforming growth factor β and the endometrium *Reviews of Reproduction* **3** 1–6

Grant DS, Rose RW, Kinsella JK and Kibbey MC (1995) Angiogenesis as a component of epithelial–mesenchymal interactions. In *Epithelial–Mesenchymal Interactions in Cancer* pp 235–248 Ed ID Goldberg and EM Rosen. Birkhauser Verlag, Basel

Hancock RF, Deutscher GH, Nielsen MK and Colburn DJ (1994) Effects of Synovex C implants on growth rate, pelvic area, reproduction, and calving performance of replacement heifers *Journal of Animal Science* **72** 292–299

Hirotaka S, Spencer TE, Onate SA, Jenster G, Tsai SY, Tsai M-J and O'Malley BW (1997) Role of co-activators and co-repressors in the mechanism of steroid/thyroid receptor action *Recent Progress in Hormone Research* **52** 141–165

Hom YK, Young P, Wiesen JF, Miettinen PJ, Derynck R, Werb Z and Cunha GR (1998) Uterine and vaginal organ growth requires epidermal growth factor receptor signaling from stroma *Endocrinology* **139** 913–921

Jost A (1953) Problems of fetal endocrinology: the gonadal and hypophyseal hormones *Recent Progress in Hormone Research* **8** 379–418

King BD, Bo GA, Lulai C, Kirkwood RN, Cohen RDH and Mapletoft RJ (1995) Effect of zeranol implants on age at onset of puberty, fertility and embryo and fetal mortality in beef heifers *Canadian Journal of Animal Science* **75** 225–230

Koji T, Chedid M, Rubin JS, Slayden OD, Csaky KG, Aaronson SA and Brenner RM (1994) Progesterone-dependent expression of keratinocyte growth factor mRNA in stromal cells of the primate endometrium: keratinocyte growth factor as a progestomedin *Journal of Cell Biology* **125** 393–401

Korach KS, Couse JF, Curtis SW, Washburn TR, Lindzey J, Kimbro KS, Eddy EM, Migliaccia S, Snedeker SM, Lubahn DB, Schomberg DW and Smith EP (1996) Estrogen receptor gene disruption: molecular characterization and experimental and clinical phenotypes *Recent Progress in Hormone Research* **51** 159–188

Lydon JP, DeMayo FJ, Conneely OM and O'Malley BW (1996) Reproductive phenotypes of the progesterone receptor null mutant mouse *Journal of Steroid Biochemistry* **56** 67–77

Malayer JR and Woods VM (1998) Expression of estrogen receptor and maintenance of hormone-responsive phenotype in bovine fetal uterine cells *Domestic Animal Endocrinology* **15** 141–154

Marion GB and Gier HT (1971) Ovarian and uterine embryogenesis and morphology of the non-pregnant female mammal *Journal of Animal Science Supplement* 1 **32** 24–47

Marshall H, Morrison A, Studer M, Popperl H and Krumlauf R (1996) Retinoids and Hox genes *FASEB Journal* **10** 969–978

Martal J, Chene N, Camous S, Huynh L, Lantier F, Hermier P, Haridon RL, Charpigy G, Charlier M and Chaouat G (1997) Recent developments and potentialities for reducing embryo mortality in ruminants: the role of IFN-t and other cytokines in early pregnancy *Reproduction Fertility and Development* **9** 355–380

Moon RT, Brown JD and Torres M (1997) WNTs modulate cell fate and behavior during development *Trends in Genetics* **13** 157–162

Moran C, Prendiville DJ, Quirke JF and Roche JF (1990) Effects of oestradiol, zeranol or trenbolone acetate implants on puberty, reproduction and fertility in heifers *Journal of Reproduction and Fertility* **89** 527–536

Mori T and Nagasawa H (1988) *Toxicity of Hormones in Perinatal Life* CRC Press, Boca Raton, FL

Ohta Y (1995) Sterility in neonatally androgenized female rats and the decidual cell reaction *International Review of Cytology* **160** 1–52

Pavlova A, Boutin E, Cunha G and Sassoon D (1994) Msx1 (Hox-7.1) in the adult mouse uterus: cellular interactions underlying regulation of expression *Development* **120** 335–346

Sananes N, Baulieu EE and Le Goascogne C (1980) Treatment of neonatal rats with progesterone alters the capacity of the uterus to form deciduomata *Journal of Reproduction and Fertility* **58** 271–273

Smith CL (1998) Cross-talk between peptide growth factor and estrogen receptor signaling pathways *Biology of Reproduction* **58** 627–632

Spencer TE, Wiley AA and Bartol FF (1993) Neonatal age and period of estrogen exposure affect porcine uterine growth, morphogenesis, and protein synthesis *Biology of Reproduction* **48** 741–751

Spencer TE, Becker WC, George P, Mirando MA, Ogle TF and Bazer FW (1995) Ovine interferon-t regulates expression of endometrial receptors for estrogen and oxytocin but not progesterone *Biology of Reproduction* **53** 732–745

Stevenson KR, Gilmour RS and Wathes DC (1994) Localization of insulin-like growth factor-I (IGF-I) and -II messenger ribonucleic acid and type 1 IGF receptors in the ovine uterus during the estrous cycle and early pregnancy *Endocrinology* **134** 1655–1664

Sugawara J, Fukaya T, Murakami T, Yoshida H and Yajima A (1997) Hepatocyte growth factor stimulates proliferation, migration, and lumen formation of human endometrial epithelial cells *in vitro*. *Biology of Reproduction* **57** 936–942

Takahashi T, Eitzman B, Bossert NL, Walmer D, Sparrow K, Flanders KC, McLachlan J and Nelson KG (1994) Transforming growth factors beta 1, beta 2, and beta 3

messenger RNA and protein expression in mouse uterus and vagina during estrogen-induced growth: a comparison to other estrogen-regulated genes *Cell Growth and Differentiation* **5** 919–935

Tarleton BJ, Wiley AA, Spencer TE, Moss AG and Bartol FF (1998) Ovary-independent estrogen receptor expression in neonatal porcine endometrium *Biology of Reproduction* **58** 1009–1019

Taylor HS, Vanden Heuvel GB and Igarashi P (1997) A conserved Hox axis in the mouse and human female reproductive system: late establishment and persistent adult expression of the Hoxa cluster genes *Biology of Reproduction* **57** 1338–1345

Torry DS and Torry RJ (1997) Angiogenesis and the expression of vascular endothelial growth factor in endometrium and placenta *American Journal of Reproductive Immunology* **37** 21–29

Vallet JL, Christenson RK, Bartol FF and Wiley AA (1995) Effect of retinyl palmitate, progesterone, oestradiol, and tamoxifen treatment on secretion of a retinol binding protein-like protein during uterine gland development in neonatal swine *Journal of Reproduction and Fertility* **103** 189–197

Wang W, Van De Water T and Lufkin T (1998) Inner ear and maternal reproductive defects in mice lacking the Hmx3 homeobox gene *Development* **125** 621–634

Wiley AA, Bartol FF and Barron DH (1987) Histogenesis of the ovine uterus *Journal of Animal Science* **64** 1262–1269

Winters TA, Febres GF, Fulgham DL, Bertics PJ, Duello TM and Gorski J (1993) Ontogeny of the epidermal growth factor receptor during development of the fetal bovine mesonephros and associated organs of the urogenital tract *Biology of Reproduction* **48** 1395–1403

Wolfenson D, Bartol FF, Badinga L, Barros CM, Marple DN, Cummins KA, Wolfe DF, Lucy MC, Spencer TE and Thatcher WW (1993) Secretion of $PGF_{2\alpha}$ and oxytocin during hyperthermia in cyclic and pregnant heifers *Theriogenology* **39** 1129–1141.

Yoshizawa T, Handa Y, Uematsu Y, Takeda S, Sekine K, Yoshihara Y, Kawakami T, Arioka K, Sato H, Uchiyama Y, Masushige S, Fukamizu A, Masumoto T and Kata S (1997) Mice lacking the vitamin D receptor exhibit impaired bone formation, uterine hypoplasia and growth retardation after weaning *Nature Genetics* **16** 391–396

Zhang Z, Laping J, Glasser S, Day P and Mulholland J (1998) Mediators of estradiol-stimulated mitosis in the rat uterine luminal epithelium *Endocrinology* **139** 961–966

Journal of Reproduction and Fertility Supplement **54**, 303–315

IGF paracrine and autocrine interactions between conceptus and oviduct

A. J. Watson[1], M. E. Westhusin[2] and Q. A. Winger[2]

[1]*Depts of Obstetrics and Gynaecology and Physiology, The University of Western Ontario, London, Ontario, Canada, N6A 5C1;* [2]*Depts of Veterinary Physiology and Pharmacology, Texas A&M University, College Station, Texas, USA*

Development *in vitro* is influenced by embryo density, serum, somatic cell co-culture and the production of 'embryotrophic' paracrine and autocrine factors. Research in our laboratory has focussed principally on the insulin-like growth factor (IGF) family. We have demonstrated that pre-attachment bovine and ovine embryos express mRNAs encoding a number of growth factor ligand and receptor genes including all members of the IGF ligand and receptor family throughout this developmental interval. In addition, early embryos express mRNAs encoding IGF-binding proteins (IGFBPs) 2–5 from the one-cell to the blastocyst stage and IGFBP5 mRNA at the blastocyst stage. Cultured bovine blastocysts release up to 35 pg per embryo in 24 h, whereas release of IGF-I was below detectable values. Analysis extended to bovine oviductal cultures has also demonstrated that mRNAs encoding these IGF family members are present throughout an 8 day culture period. Transcripts encoding IGFBPs 2–6 were also present. Release of both IGFs was recorded over an 8 day culture period. IGF-II release was significantly greater than that observed for IGF-I. Therefore, the IGFs are present throughout the maternal environment during early embryo development. The oocyte, within the follicle, is held in an environment high in IGFs and IGFBPs. The zygote, after fertilization, is maintained in an IGF-rich environment while free-living in the oviduct and the uterus. This review is focused on the IGF family and IGFBPs and their roles in enhancing development up to the blastocyst stage.

Overview

The production of mammalian zygotes *in vitro* is an important approach for studying oocyte maturation, fertilization, and early development. We are still, however, unable to mimic the environment of the female reproductive system for any mammalian species, and zygotes produced *in vitro* generally result in lower pregnancy rates after transfer to the uterus than do zygotes produced *in vivo*. *In vitro* derived embryos often display a developmental lag, contain fewer cells, and are morphologically distinct (Farin and Farin, 1995; Thompson *et al.*, 1995; Walker *et al.*, 1996). Successful progression through the first week (preimplantation phase) of development is essential for implantation and establishment of pregnancy. Studies are required to define and optimize culture environments and technologies for treating various forms of human infertility, to increase developmental frequencies of livestock embryos *in vitro* for commercial applications, and to understand better the genetic programme regulating early development.

Preimplantation development is somewhat autonomous as development to the blastocyst stage can be supported *in vitro* in simple defined conditions for many mammalian species (Chatot *et al.*, 1990; Gardner *et al.*, 1994; Keskintepe *et al.*, 1995; Summers *et al.*, 1995; Walker *et al.*, 1996). Despite this autonomous nature it is clear that development *in vitro* is enhanced by factors such as embryo density (O'Neill, 1997), serum and somatic cell co-culture (Gandolfi and Moor, 1987; Sirard *et al.*, 1988; Xia *et al.*, 1996). The desire to support development of preimplantation mammalian embryos *in vitro* has driven the characterization of 'embryotrophic' factors and unleashed a period of analysing growth factor and cytokine expression patterns during early development (for review see Schultz *et al.*, 1993; Schultz and Heyner, 1993; Kaye and Harvey, 1995; Kane *et al.*, 1997; Stewart and Cullinan, 1997). These types of study have largely outpaced functional studies and to date a convincing foundation illuminating specific roles for these factors in supporting early development has been put forward only for the mouse preimplantation embryo (Kaye and Harvey, 1995; Stewart and Cullinan, 1997). There are, however, compelling data that support similar roles for these factors in influencing early development in other species too.

Research in our laboratories has focussed principally on the insulin-like growth factor (IGF) family. The IGFs are present throughout the maternal environment during early embryo development (for review see Clemmons, 1993; Murphy and Barron, 1993; Jones and Clemmons, 1995). The oocyte, within the follicle, is held in an environment high in IGFs and insulin-like growth factor binding proteins (IGFBPs) (Ling *et al.*, 1993; de la Sota *et al.*, 1996). The zygote, following fertilization, is maintained in an IGF-rich environment while free-living in the oviduct and the uterus (Giudice *et al.*, 1992; Wiseman *et al.*, 1992; Carlsson *et al.*, 1993; Xia *et al.*, 1996; Winger *et al.*, 1997). For these reasons, the current review will focus on the IGF family and IGFBPs and their roles in enhancing development up to the blastocyst stage. Emphasis will be placed on the early mouse embryo (because the majority of information has been generated from this species). We will, however, summarize data supporting roles for these modulators in regulating early development of livestock embryos as well.

Mammalian Embryo Culture

Early embryos from different mammalian species exhibit broad variations in their capacity to complete the first week of development *in vitro*. Simple media capable of supporting mouse development from the two-cell to blastocyst stage *in vitro* have been in use for over two decades. Initially, the persistence of a 'two-cell' culture block impeded development of one-cell zygotes to the blastocyst stage but, recently, media (most notably CZB and KSOMaa) have been designed specifically to circumvent this problem in this species (Chatot *et al.*, 1990; Summers *et al.*, 1995). In this regard, mouse embryos are somewhat atypical as their apparent ability to adapt to culture environments is not equally shared by early embryos of other species.

The development of serum-supplemented media and embryo co-culture methods for supporting bovine development *in vitro* occurred during the mid- to late 1980s (Gandolfi and Moor, 1987; Sirard *et al.*, 1988). These first systems were critical for advancing research investigating early mammalian development and also for applying assisted reproductive technologies to domestic species. Ovine embryo culture systems originated from synthetic oviduct fluid medium (SOFM) which was designed from concentrations of salts and energy metabolites found in sheep oviductal fluid (Tervit *et al.*, 1972; Walker *et al.*, 1996).

The complete removal of serum from culture protocols has been more difficult, but is clearly an important goal as evidenced by increasing reports of the negative effects of serum on early development (Farin and Farin, 1995; Thompson *et al.*, 1995; Walker *et al.*, 1996). Transitional stages from serum supplementation have included replacement with bovine serum albumin (BSA), and most recently the addition of polyvinyl alcohol (PVA) and amino acids (Gardner *et al.*, 1994; Keskintepe *et al.*, 1995; Walker *et al.*, 1996). With these advances it is now possible to record similar developmental frequencies to the blastocyst stage for bovine zygotes placed into culture under complex or simple media systems, as is displayed in Table 1.

The use of simple defined conditions is imperative for studies directed at examining the physiological roles of 'embryotrophic' growth factors on early development. It is very difficult to formulate clear conclusions about effects on development when suboptimal culture conditions are used. We believe this limitation in culture conditions has especially impeded progress in understanding the regulation of development in early livestock embryos.

Table 1. Development of bovine *in vitro* matured and inseminated oocytes in complex versus simple culture media

Treatment	Zygotes[a]	Cleavage[b]	Morulae[c]	Blastocysts[d]
TCM-199+ serum + embryo co-culture[e]	285	73.8 ± 8.3%	43.0 ± 6.8%	31.2 ± 4.6%
cSOFM[f]	108	71.2 ± 7.4%	46.9 ± 8.7%	39.1 ± 5.1%[g]

[a]number of replicates = 3; [b]frequency (mean + SD) of zygotes at two-cell stage or further at 72 h post insemination (p.i.); [c]frequency (mean + SD) of cleaved zygotes reaching the morula stage (> 32 cell zygotes) at day 6 p.i.; [d]frequency (mean + SD) of cleaved zygotes reaching the blastocyst stage at day 8 p.i; [e]culture medium composed of TCM-199 medium + 10% steer serum; zygotes cultured in 50 μl culture drops with primary bovine oviduct vesicle cultures and under a 5% CO_2 in air culture atmosphere; [f]serum-free culture medium composed of modified synthetic oviduct fluid medium (cSOFM; Keskintepe *et al.*, 1995); zygotes cultured in 50 μl culture drops under a 5% CO_2:5%O_2:90%N_2 culture atmosphere; [g]$P < 0.05$, blastocyst formation for cSOFM versus TCM-199 + serum + coculture.

Insulin-like Growth Factors, Receptors and Binding Proteins

IGF ligands

Bovine IGF-I is a 70 amino acid, basic, single chain polypeptide, with a molecular mass of 7649 daltons. The bovine cDNA is 93% identical to the human sequence, and the amino acid sequence is 96% conserved (Fotsis *et al.*, 1990). Three disulfide bridges maintain tertiary structure of the molecule. The IGF-II protein is highly conserved between species (the 180 amino acids that encode the mature bovine and ovine IGF-II clones are identical) and rat, human, bovine and ovine forms differ at only one amino acid (Brown *et al.*, 1990). The precursor molecule contains a 24-residue amino-terminal signal peptide, a 67 amino acid, mature IGF-II polypeptide and an 89 amino acid carboxyl terminal. Bovine IGF-II has over 60% identity with IGF-I (Brown *et al.*, 1990; Fotsis *et al.*, 1990).

IGF-I/type-1 receptor

The actions of IGF-I and IGF-II are mediated largely through the IGF-I receptor (LeRoith *et al.*, 1995). The IGF-I receptor is synthesized as a single chain polypeptide. Post-translational modifications include cleavage of a signal polypeptide and further cleavage into a 707 amino acid, extracellular α-subunit and a 626 amino acid, transmembrane β-subunit. The α- and β- subunits are linked by disulfide bonds. Two αβ complexes are joined by additional disulfide bonds creating the mature $\alpha_2\beta_2$ receptor. Binding of IGF ligands is mediated by the extracellular α-subunit within a cysteine-rich region. Tyrosine kinase activity occurs in the cytoplasmic β-domain. Binding of IGF ligand to the α-subunit stimulates phosphorylation of both tyrosine and serine residues (LeRoith *et al.*, 1995). Autophosphorylation of the IGF-I receptor results in multiple signalling pathway cascades leading to the stimulation of cell growth (LeRoith *et al.*, 1995; Jones and Clemmons, 1995).

IGF-II–mannose-6-phosphate receptor

The IGF-II/M6P receptor is a monomeric 215 kDa glycoprotein with high IGF-II binding affinity, binding IGF-I at 500-fold lower level than IGF-II, with no affinity for insulin (for review see Schultz and Heyner, 1993; Jones and Clemmons, 1995). Sequence comparisons of the IGF-II receptor and the cation-independent mannose-6-phosphate receptor revealed identical molecules. The binding sites for IGF-II and M6P are distinct and both ligands can bind simultaneously to the receptor (Morgan *et al.*, 1987). The IGF-II receptor protein contains a large extracellular domain, comprising 93% of the total receptor, a single transmembrane domain and a small cytoplasmic tail. Fifteen repeat sequences of eight conserved cysteine residues, a single fibronectin type II repeat and 19 *N*-linked glycosylation sites are located on the extracellular domain (Morgan *et al.*, 1987). The binding of IGF-II to the receptor results in internalization and degradation of IGF-II (Morgan *et al.*, 1987). It is still unclear whether the IGF-II receptor has a biological role beyond regulating free concentrations of IGF-II. A soluble form of the IGF-II/M6P receptor generated by proteolytic cleavage of the membrane bound receptor has been identified in rats (for review, see Jones and Clemmons, 1995).

Insulin-like growth factor binding proteins (IGFBPs)

The IGFs are almost entirely bound *in vivo* to high-affinity IGF-binding proteins of which there are at least six members (for review see Clemmons, 1993; Murphy and Barron, 1993; Jones and Clemmons, 1995). All IGFBPs display structural homology, bind IGF-I and IGF-II specifically and have a negligible affinity for insulin. Sequence alignments of IGFBPs reveal regions of homology within the amino- and carboxyl-terminal regions. The positions of 18 cysteines, which participate in the formation of disulfide bridges and contribute to three-dimensional structure, are conserved in IGFBPs 1–5. The rat IGFBP-6 sequence lacks two and the human IGFBP-6 sequence lacks four of the 18 conserved cysteines found in the other IGFBPs. In serum, approximately 75% of the circulating IGF is complexed with IGFBP-3, and an 88 kDa glycoprotein, the acid labile subunit (ALS), forming a 150 kDa protein complex. This 150 kDa complex prolongs the half-life of IGFs in serum to 12–15 h, which is considerably longer than the 10 min half-life of free IGFs. The half-life of free binding proteins is between 30 and 90 min. IGFBPs can inhibit or potentiate IGF action under various conditions (for review see Clemmons, 1993; Murphy and Barron, 1993; Jones and Clemmons, 1995). The characterization of specific proteases for the IGFBPs has further complicated the situation as these proteases cleave binding proteins into forms with altered affinity for the IGFs. IGFBPs are subject to post-translational modifications, and direct cellular effects, in which binding to IGF ligand is not necessary, have been described (Jones and Clemmons, 1995).

Oviductal Growth Factors and Binding Proteins

Oviductal fluid provides an environment in which fertilization and early embryonic growth take place (Leese, 1988). The oviduct may provide an environment rich in 'embryotrophic factors' capable of enhancing development (Wiseman *et al.*, 1992). The precise role these molecules play in supporting early mammalian development is still under investigation but growth factors certainly perform roles expected for 'embryotrophic' factors. We characterized the expression of transcripts encoding basic fibroblast growth factor (bFGF), transforming growth factor α (TGF-α), TGF-β1, TGF-β2; platelet-derived growth factor (PDGF-A), IGF-I, and IGF-II in bovine and ovine oviduct primary cultures by applying reverse transcription–polymerase chain reaction methods (RT–PCR; Watson *et al.*, 1992, 1994).

The production of IGF-I and IGF-II by the oviduct has been established for several species (Giudice *et al.*, 1992; Wiseman *et al.*, 1992; Carlsson *et al.*, 1993; Winger *et al.*, 1997). Insulin, IGF-I and IGF-II of maternal origin have been detected in the murine reproductive tract (Murphy and Barron, 1993). Likewise transcripts and polypeptides encoding IGF-I have been reported in the rat fallopian tube (Carlsson *et al.*, 1993). Our more recent analysis of oviduct growth factors mapped out the distribution of both mRNAs and polypeptides encoding IGF-I and IGF-II in bovine oviduct and in primary cultures (Xia *et al.*, 1996). Bovine primary oviduct cultures released ten times more IGF-II than IGF-I into the culture medium (Winger *et al.*, 1997). Schmidt *et al.* (1994) reported the localization of IGF-I mRNAs in the bovine oviduct during the entire oestrous cycle by RT–PCR and northern blot analysis. IGF-I mRNA expression increased after ovulation, indicating a possible role for IGF-I following ovulation by either influencing oviduct function or early embryo development.

Table 2. Growth factor ligand mRNAs during preimplantation development

Factor	Mouse	Cow	Sheep
Insulin	No	No	No
IGF-I	2C-blast	1C-blast	1C-blast
IGF-II	2C-blast	1C-blast	1C-blast
TGF-α	1C-blast	1C-blast	1C-blast
EGF	No	No	No
NGF	No	No	No
bFGF	No	1C-8/16C	1C-8/16C
TGFβ1	1C-blast	1C-blast	1C-blast
PDGF-α	2C-blast	1C-blast	not investigated

IGF: insulin-like growth factor; TGF-α: transforming growth factor α; EGF: epidermal growth factor; NGF: nerve growth factor; bFGF: basic fibroblast growth factor; PDGF: platelet-derived growth factor; blast: blastocyst; C: cell.
Data from Schultz and Heyner (1993); Kaye and Harvey (1995).

IGF-I protein has also been localized in human oviductal epithelial cells (Giudice *et al.*, 1992) and in porcine oviductal fluid (Wiseman *et al.*, 1992). Furthermore, IGF-I and IGF-II polypeptides may be transported to the oviduct via the circulation, as reported in the mouse (Murphy and Barron, 1993). This source of IGFs could augment the low expression observed within intact oviduct and increase the overall amounts of IGF polypeptides associated with these epithelial cells.

IGFBPs 1–4 have been detected in the human oviduct (Giudice *et al.*, 1992). Experiments conducted in our laboratory detected transcripts encoding IGFBP-2, -3, -4, and -5 in bovine oviduct and primary oviduct cultures by RT–PCR analysis (Winger *et al.*, 1997). The mRNAs encoding IGFBP-1 were not detected and IGFBP-6 mRNAs were not consistently detected in all oviduct sample replicates. Western ligand blot analysis revealed four IGFBPs of approximate molecular masses 24 kDa, 31 kDa, 36 kDa and a broad band extending from 46 to 53 kDa in bovine oviduct conditioned media. We confirmed the identity of the 24, 31 and 36 kDa proteins by western immunoblot as IGFBP -4, -5, and -2, respectively. The 46–53 kDa broad band represented IGFBP-3.

Embryonic Growth Factors and Binding Proteins

Data collected from studies applying RT–PCR and immunolocalization to characterize the expression of a number of growth factor genes during murine preimplantation development are summarized in Table 2 (for review see Schultz *et al.*, 1993; Schultz and Heyner, 1993; Kaye and Harvey, 1995). These include mRNAs encoding TGF-α , TGF-β1, TGF-β2, TGF-β3, PDGF-A, Kaposi's sarcoma-type growth factor (kFGF) , IGF-I and IGF-II. Transcripts encoding cytokines such as interleukin 3 (IL-3), interleukin 6 (IL-6) and leukaemia inhibitory factor (LIF) also are expressed by the blastocyst during early mouse development. The activation of growth factor ligand and receptor genes is selective, as transcripts encoding several factors including EGF, nerve growth factor (NGF) and insulin have not been detected during the first week of development for any mammalian species (Watson *et al.*, 1992, 1994). Until recently it was thought that IGF-I was not produced by the preimplantation mouse embryo until the eight-cell stage. However, the use of RNA extracts from larger embryo pools has led to the detection of IGF-I mRNAs in all murine preimplantation stages (Doherty *et al.*, 1994). Our failure to detect IGF-I mRNAs in rat embryos (Zhang *et al.*, 1994) may reflect this problem of transcript abundance, as we were able to detect IGF-I mRNAs during both bovine and ovine early development (Watson *et al.*, 1992; 1994).

RT–PCR amplicons for TGF-β 2, TGF-α , PDGF-A and IGF-II were detected throughout bovine early development (Watson *et al.*, 1992) and transcripts encoding IGF-II and TGF-α throughout ovine development (Watson *et al.*, 1994). Similar to mouse and rat early embryos, products encoding EGF, NGF, or

Table 3. Growth factor receptor mRNAs during preimplantation development

Factor	Mouse	Bovine	Ovine
Insulin-r	8C-blast	1C-blast	1C-blast
IGF-Ir	8C-blast	1C-blast	1C-blast
IGF-IIr	2C-blast	1C-blast	1C-blast
EGF-r	1C-blast	blast	no reports
PDGF-a	1C-blast	1C-blast	not investigated

IGF: insulin-like growth factor; EGF: epidermal growth factor; PDGF: platelet-derived growth factor; blast: blastocyst; C: cell;
Data from Schultz and Heyner (1993); Kaye and Harvey (1995).

insulin were not detected in bovine or ovine early embryos. However, transcripts encoding bFGF were detected only up to the eight- to16-cell stage in bovine embryos, and then declined markedly following the eight- to16-cell stage during early ovine development (Watson *et al.*, 1992, 1994).

A number of receptor genes are expressed in the early mouse embryo (summarized in Table 3) including the insulin-r, IGF-Ir, IGF-IIr, EGFr, PDGF-α r, and colony-stimulating factor 1r (CSF-I) (for review see Schultz and Heyner, 1993; Kaye and Harvey, 1995). In the mouse, the IGF type 1 receptor was detected by cell surface binding of IGF-I and IGF-II at the morula and blastocyst stages of development (Mattson *et al.*, 1988) and by gold-labelled IGF-I binding as early as the eight-cell stage (Smith *et al.*, 1993). The IGF-II/M6P receptor was first detected in two-cell mouse embryos (for review see Schultz and Heyner, 1993; Kaye and Harvey, 1995). Bovine and ovine early embryos express transcripts encoding PDGF α-r, insulin-r, and IGF-I-r and IGF-II-r throughout the first week of development (Watson *et al.*, 1992,1994).

Transcripts encoding IGFBPs 2–4 were detected throughout early bovine development while IGFBP-5 mRNAs were detected only weakly in bovine blastocysts (Winger *et al.*, 1997). mRNAs encoding IGFBPs 1 and 6 were not detected during this developmental interval in bovine embryos. In the mouse embryo, mRNAs encoding IGFBP-6 were detected only in blastocysts, while transcripts encoding IGFBP-2, -3 and -4 were detected throughout murine preimplantation development (Hahnel and Schultz, 1994). Transcripts encoding IGFBP-5 were not detected in any preimplantation stage in mice (Hahnel and Schultz, 1994).

IGFs in Bovine Parthenogenotes

Parthenogenetic embryos contain only maternal genetic material representing a set of genes derived completely from oogenesis. Imprinting implies that through epigenetic modification a particular genetic allele will become silenced, with expression resulting from the second (non-silenced) allele (DeChiara *et al.*, 1991; Latham *et al.*, 1994). IGF-II is an imprinted gene in which expression stems from only the paternal allele (DeChiara *et al.*, 1991). Imprinting of IGF-II may regulate IGF-II expression, since overexpression of IGF-II can be detrimental (see below). In contrast, expression of IGF-II receptor is the result of the maternal allele only (Baker *et al.*, 1993). Although imprinting of the IGF-II gene is well established by birth, a debate continues regarding the state of the imprint during preimplantation development (Rappolee *et al.*, 1992; Latham *et al.*, 1994). We have measured amounts of IGF-II released into medium by bovine parthenogenotes and contrasted these values with those observed for inseminated embryos. Blastocysts produced following fertilization released significantly greater amounts (mean ± SEM) of IGF-II (36.2 ± 3.9 pg per embryo) compared with parthenogenetic embryos (9.6 ± 2.8 pg per embryo, $P < 0.05$; Fig. 1). Bovine parthenogenetic blastocysts expressed transcripts encoding IGFBP-2, -3, –4 and -5 as was observed for *in vitro* fertilized controls (Fig. 2). The release of IGF-II by bovine parthenogenetic embryos is of interest. If imprinting influences gene expression (Rappolee *et al.*, 1992), it is possible that the IGF-II released is the result of an incomplete imprint, aberrant gene expression due to chromosomal ploidy or perhaps

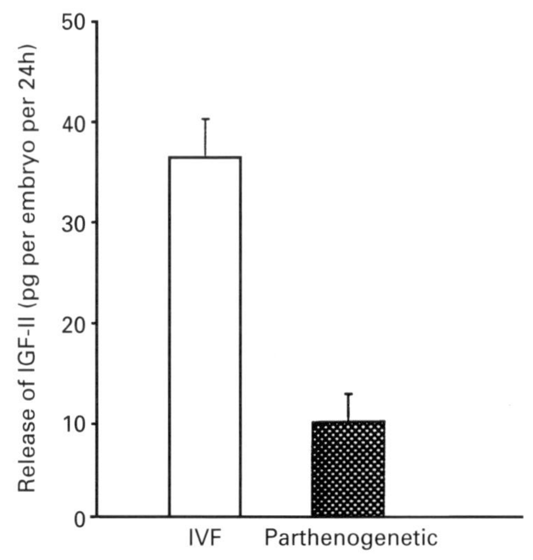

Fig. 1. Release of insulin-like growth factor II (IGF-II) from bovine parthenogenetic and *in vitro* fertilized (IVF) bovine blastocysts. For collection of conditioned media, inseminated and parthenogenetic blastocysts were removed from culture on day 7. They were washed three times in serum-free medium and groups of ten blastocysts (from both groups) were placed in 200 μl of serum-free TCM-199 incubation medium for 24 h (experimental, $n = 6$). Conditioned media were collected, lyophilized and resuspended in 50 μl for radioimmunoassay using recombinant human IGF-I and IGF-II iodinated to specific activities of 150–250 μCi μg^{-1} protein as outlined by Winger *et al.* (1997). Insulin-like growth factor binding proteins (IGFBPs) were extracted to release IGF–IGFBP complexes and precipitate the binding proteins. IVF blastocysts ($n = 6$) released significantly greater ($P < 0.05$) amounts of IGF-II than did parthenogenetic blastocysts. Release of IGF-I was below the limits of detection of the assay for both groups of embryos.

due to release of membrane bound IGF-II from embryos accumulated during culture. Alternatively, the IGF-II gene may not become imprinted until later stages of embryo development (Latham *et al.*, 1994), or the lower IGF-II release may be the result of lower overall transcription in less healthy parthenogenetic embryos. However, it would certainly appear that even bovine parthenogenotes are exposed to autocrine IGF-II during the first week of development.

The presence of gene products encoding the IGFs, their receptors and binding proteins in early embryos from several species indicates that IGFs expressed by the embryo or maternal tissues could exert receptor-mediated actions on the embryo and therefore influence growth and development by supporting the progression of embryos through the first week of development.

Biological Actions of IGFs

The actions of IGF *in vitro* include effects on protein and carbohydrate metabolism, and effects on cell replication and differentiation (Murphy and Barron, 1993; Schultz and Heyner, 1993; Jones and

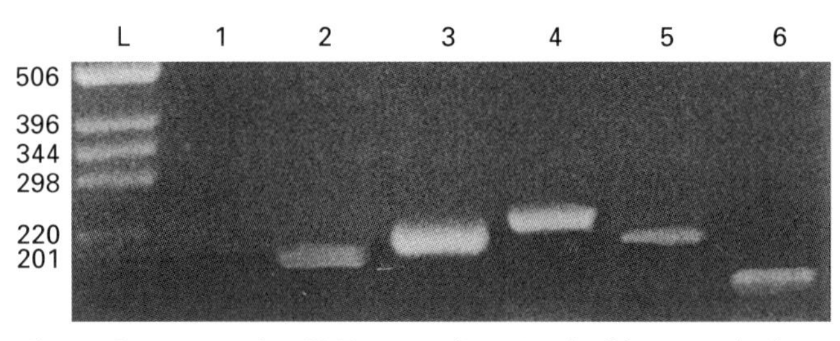

Fig. 2. Detection of mRNAs encoding insulin-like growth factor binding proteins (IGFBPs) in bovine parthenogenetic blastocysts by reverse transcription–polymerase chain reaction (RT–PCR). Bovine oocytes were activated by treatment with 7% ethanol for 5 min after *in vitro* maturation for 24 h. Activated oocytes were further treated with cytochalasin D (5 μg ml^{-1} for 6 h) to produce diploid chromosomal complements. Parthenogenetic zygotes were co-cultured with oviductal vesicles in TCM-199 medium supplemented with 10% steer serum to support development to the blastocyst stage. For RT–PCR analysis, total RNA was isolated from pools of 10–20 parthenogenetic blastocysts, and the RNA was reverse transcribed into cDNA as described by Watson *et al.* (1992, 1994). IGFBP primer sequences were derived from published human and bovine cDNA sequences. Expected sizes of products are: 239 bp (IGFBP-1); 186 bp (IGFBP-2); 210 bp (IGFBP-3); 222 bp (IGFBP-4); 215 bp (IGFBP-5) and 345 bp (IGFBP-6). The identity was confirmed by cloning each amplified product into a pCRII vector by the use of the TA cloning kit (Invitrogen) followed by base-specific termination of enzyme catalysed primer extension reactions using a T7 sequencing kit. Lanes are (L) ladder (bands from top to bottom: 516/506 bp, 396 bp, 344 bp, 298 bp, 220/200 bp, 154/142 bp), while lanes 1–6 represent IGFBPs 1–6. Transcripts encoding IGFBP-2, -3, -4, and -5 were detected in bovine parthenogenetic blastocysts. Transcripts encoding IGFBP-1 and -6 were not detected; however, a smaller than expected product was detected with the IGFBP-6 primers. After sequence analysis, this product was found not to represent IGFBP-6.

Clemmons, 1995). IGF-I acts as a progression factor in the cell cycle. Quiescent cells in G0 when treated with a competence factor (PDGF, bFGF) progress to G1 and arrest. Treatment with IGF-I induces progression through the cell cycle leading to DNA synthesis and cell proliferation.

The functions of IGF-I and IGF-II on fetal development have been studied using gene targeting and transgenic approaches and the outcomes of these experiments are summarized in Table 4. Mice carrying copies of the human IGF-I gene fused to the metallothionein-I promoter have high IGF-I (for review see Baker *et al.*, 1993). These mice have increased body weight largely due to increased muscle, brain, spleen, kidney and pancreas mass. Most mice with a disrupted IGF-I gene died at birth and those that survived displayed growth retardation, reaching only 60% of normal birth weight (Liu *et al.*, 1993). IGF-II homozygous null mutants produce live pups with birth weights 60% of normal size with prenatal growth defects starting at about day 13.5 (DeChiara *et al.*, 1991). IGF-I and IGF-II double 'knock-outs' displayed complete neonatal lethality with birth weights 30% of normal. Type-1 receptor deficient mice displayed fetal growth deficits of 45% of normal birth weight and complete neonatal lethality (Liu *et al.*, 1993). The IGF-I–IGF-Ir mutants displayed the same phenotype as the IGF-Ir (–/–) mice, thus indicating that the essential functions of IGF-I are mediated through the IGF-Ir. IGF-II–IGF-Ir knockouts resulted in lower fetal birth weight than IGF-Ir knockouts alone, indicating that IGF-II must be acting by a route in addition to IGF-Ir pathways. Mice deficient in the IGF-II/M6P receptor resulted in larger birth weights and lethality in nearly all

Table 4. Summary of principal IGF family null mutant phenotypes in mice

Gene	Phenotype
IGF-I	60% of normal weight; most die at birth
IGF-II	60% of normal weight; prenatal defects start at day 13.5
IGF-I and IGF-II	30% of normal weight; neonatal lethality
IGF-Ir	45% of normal weight; neonatal lethality
IGF-I and IGF-Ir	45% of normal weight; neonatal lethality
IGF-II and IGF-Ir	30% of normal weight; neonatal lethality
IGF-IIr	Increased birth weight; lethality
IGF-II and IGF-IIr	Normal birth weights

IGF: insulin-like growth factor.
Data from Baker *et al.* (1993).

mutants. If the IGF-II gene of these mice was knocked-out in combination with the IGF-II/M6P receptor, the phenotype was rescued and normal birth weights were observed (for review see Baker *et al.*, 1993; Kaye and Harvey, 1995; Jones and Clemmons, 1995). This result indicated that in the embryo the IGF-II/M6P receptor regulates amounts of IGF-II that may become lethal if increased. Imprinting of the IGF-II gene may represent an additional control important for regulating IGF-II (Rappolee *et al.*, 1992).

IGF Regulation of Preimplantation Development

Several growth factors including IGFs, PDGF, bFGF, TGF-α , and TGF-β , when added exogenously to culture environments stimulate embryo development (for review see Schultz and Heyner, 1993; Schultz *et al.*, 1993; Kaye and Harvey, 1995; Stewart and Cullinan, 1997). These effects on murine preimplantation embryos include increased amino acid uptake, DNA, RNA and protein synthesis, increased embryo cell number and increased frequencies of development to blastocyst stage. The majority of experiments have investigated influences on mouse preimplantation development. Our knowledge of other species is very limited. As our ability to support preimplantation development of non-murine species in simple media improves, it is expected that progress in understanding the physiology of growth factor action on early development in these species will increase sharply.

The anabolic and mitogenic influences of IGF-I and IGF-II on murine preimplantation development are well characterized (see Schultz and Heyner, 1993; Kaye and Harvey, 1995). Treatment of murine blastocysts with IGF-I stimulates increased protein synthesis, inner cell mass proliferation, and increased endocytosis and glucose transport by trophectoderm (Kaye and Harvey, 1995). Effects on early development can be grouped into short and long term and may be mediated by interactions with either the insulin or IGF-Ir. For example, increased protein synthesis can be observed after treatment for 4 h and this response is probably mediated by the insulin-r (Kaye and Harvey, 1995). In contrast, proliferative and morphological responses are probably mediated by the type I-r (Kaye and Harvey, 1995). Recent studies have indicated that IGF-I exerts a beneficial effect on development of early porcine embryos *in vitro* (Xia *et al.*, 1994). The results from bovine studies are not as clear, some authors suggest that IGF-I has little influence on developmental frequencies (Larson *et al.*, 1992), while others report dose responsive influences on development of bovine zygotes to the blastocyst stage (Herrler *et al.*, 1992). It is too early to make definitive conclusions regarding the impact of IGF-I on early development of species such as the cow.

Addition of IGF-II to the culture medium of mouse embryos stimulates increased protein synthesis, cell number and frequencies of development to the blastocyst stage (Rappolee *et al.*, 1992; Schultz and Heyner, 1993; Kaye and Harvey, 1995). These stimulatory effects of IGF-II can be negated by embryo culture in the presence of IGF-II antisense oligodeoxynucleotides (Rappolee *et al.*, 1992). It is unlikely that these IGF-II effects on early development are mediated by interactions with the type 2/M6P receptor. Experiments with mutant IGF-II (modified to interact only with the type 2 and not type 1 receptor) did not result in any stimulatory influences on early development (Rappolee *et al.*, 1992). Furthermore, IGF-II-mediated influences are propagated at the same EC_{50} value as are IGF-I effects implying that both are mediated via the type I IGF-I-r (Kaye and Harvey, 1995). Although the IGF-II mRNA and protein expression patterns have been characterized in early embryos from several other mammalian species (as summarized above), very few additional studies have investigated the physiological effects of IGF-II on early development. We have documented that small pools of bovine blastocysts release greater amounts of IGF-II than IGF-I into culture medium (Winger *et al.*, 1997) and human blastocysts in culture also release IGF-II (Hemmings *et al.*, 1992).

IGFBPs are certain mediators of IGF-I and IGF-II influences on early development. Although mRNAs encoding these genes have been detected during mouse and bovine early development as well as their polypeptides in oviductal fluids from several species (Hahnel and Schultz, 1994; Winger *et al.*, 1997), no studies to date have defined their specific interactions during the first week of development. Studies must be conducted to provide insight into their roles and to elucidate fully the impact of IGF paracrine and autocrine circuits in regulating this early developmental interval.

Significance of IGF Regulation of Early Development

It is clear that growth factors collectively influence a number of events during early development. The preimplantation mammalian embryo develops in an environment that includes all of the necessary gene products (ligands, receptors and binding proteins) required to support the development of 'embryotrophic' maternal and embryonic IGF circuits (Fig. 3). However, phenotypes arising from IGF null mutants indicate that expression of IGF ligand and receptor genes are not essential for preimplantation development. These outcomes may be explained by the autonomous nature of early development. Alternatively, the expression of other growth factor ligand and receptor gene families may regulate their expression in the absence of IGF gene products to provide necessary regulation of early developmental events. It is of concern that the results obtained from rodent species may be simply extended to include all mammalian species. In livestock species it has been possible to influence fetal growth and development (without compromising development to the blastocyst stage) by exposing preimplantation embryos to serum supplemented culture environments (Farin and Farin, 1995; Thompson *et al.*, 1995; Walker *et al.*, 1996). These observations demonstrate that it is possible to affect the preimplantation developmental programme and induce longer term consequences that are not revealed until fetal or postpartum stages. The impact of creating a null mutant may therefore not reveal itself simply by influencing developmental frequencies. Although blastocyst development occurs in all mouse IGF ligand and receptor null mutants, the blastocysts may not be 'normal' and the deleterious influences contributing to abnormal fetal phenotypes may be initiated during the first week of development. The analysis of null mutants should therefore include contrasting gene expression patterns of mutant and wild type preimplantation embryos.

Clearly further experimentation directed at elucidating the functional significance of IGF expression during preimplantation development of other species is required. With the development of effective simple media for supporting early development of bovine, ovine and pig embryos, it is now possible to apply antisense oligodeoxynucleotide approaches to examine consequences of downregulating IGF expression in these species. Since the 'primary goal' of all culture regimens is to support the production of large numbers of healthy blastocysts, it is imperative that the effects of culture on blastocyst quality be defined. Moreover, blastocyst quality can no longer be evaluated solely on the basis of morphology, since there is a clear disparity between morphological appearance of transferred blastocysts and pregnancy outcome. We would propose that evaluation procedures be expanded to include an assessment of the biological/biochemical events occurring within the zygote. Monitoring variations in IGF expression patterns may represent a means for attaining this goal. Continuation of such experimental approaches should ultimately assist in the production of defined conditions for the production of viable bovine preimplantation embryos and also will certainly elucidate the contributions of these growth factor genes during early development.

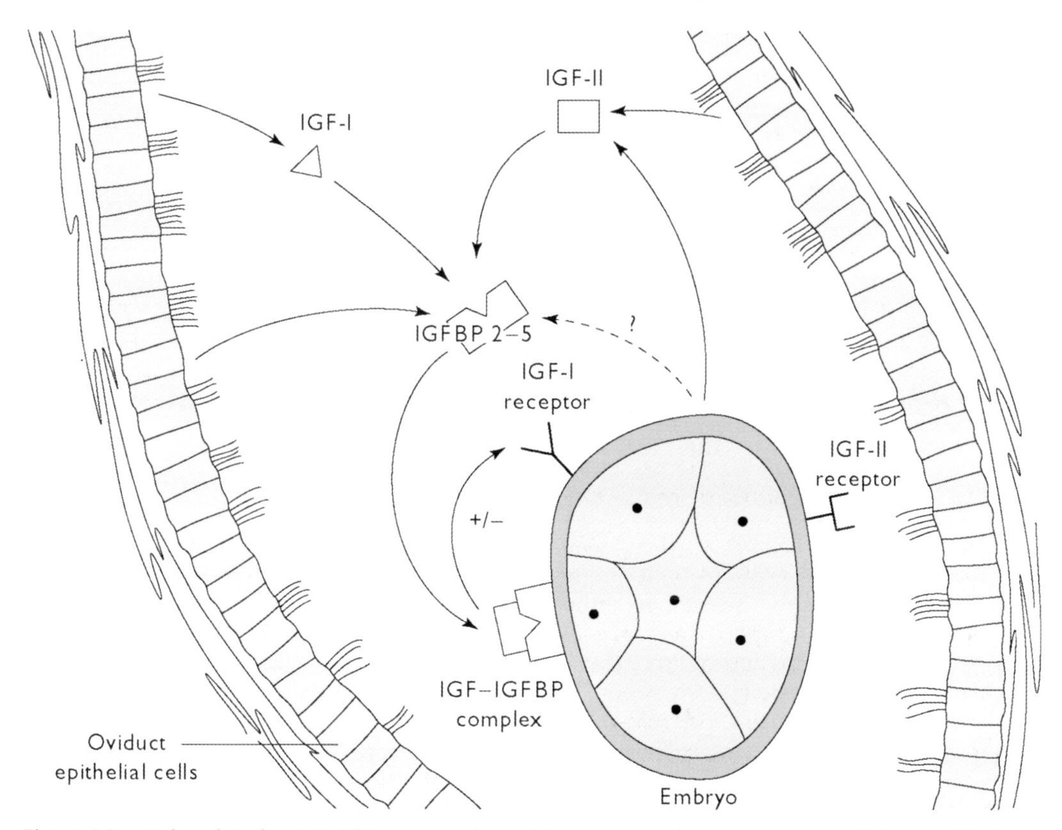

Fig. 3. Maternal and embryonic IGF circuits: detectable amounts of immunoreactive IGF-I and IGF-II are released from primary bovine oviduct cell cultures, and detectable amounts of IGF-II are released from bovine blastocysts. Bovine oviduct primary cultures express transcripts encoding IGFBPs 2–5 and release IGFBPs 2–5 into conditioned media. Bovine zygotes express mRNAs encoding IGFBPs 2–4 through to the blastocyst stage. mRNAs encoding IGFBP-5 were detected in bovine blastocyts. The detection of IGF-I, IGF-II and IGFBPs 2–5 in the culture environment indicates that IGF paracrine and autocrine regulatory circuits are present and may contribute to the events that regulate bovine early development. Future efforts must be directed at measuring the direct influences of these ligands on early development and determining the IGF–IGFBP dynamics that oversee their actions.

The authors are grateful to the ABEL laboratories, University of Guelph, under the direction of Dr Stanley Leibo, for their assistance with the bovine ovary and oviduct collections. Research reported from the authors' laboratories was funded by NSERC and the MRC of Canada. A. J. Watson is also supported by an MRC Scholarship.

References

Baker J, Lie JP, Robertson EJ and Efstratiadis A (1993) Role of insulin-like growth factors in embryonic and postnatal growth *Cell* **75** 73–82

Brown WM, Dziegielewska KM, Foreman RC and Saunders NR (1990) The nucleotide and deduced amino acid sequences of insulin-like growth factor II cDNAs from adult bovine and fetal sheep liver *Nucleic Acids Research* **18** 4614

Carlsson B, Hillensjo T, Nilsson A, Tornell J and Billig H (1993) Expression of insulin-like growth factor I (IGF-I) in the rat fallopian tube: possible autocrine and paracrine action of fallopian tube-derived IGF-I on the fallopian tube and on the preimplantation embryo. *Endocrinology* **133** 2031–2039

Chatot CL, Lewis JL, Torres I and Ziomek CA (1990) Development of 1-cell embryos from different strains of mice in CZB medium *Biology of Reproduction* **42** 432–440

Clemmons DR (1993) IGF binding proteins and their functions *Molecular Reproduction and Development* **35** 368–375

DeChiara TM, Efstratiadis A and Robertson EJ (1991) Paternal imprinting of the mouse insulin-like growth factor II gene *Cell* **64** 849–859

de la Sota RL, Simmen FA, Diaz T and Thatcher WW (1996) Insulin-like growth factor system in bovine first-wave dominant and subordinant follicles *Biology of Reproduction* **55** 803–812

Doherty AS, Temeles GJ and Schultz R (1994) Temporal pattern of IGF-I expression during mouse preimplantation embryogenesis *Molecular Reproduction and Development* **37** 21–26

Farin PW and Farin CE (1995) Transfer of bovine embryos produced *in vivo* or *in vitro:* survival and fetal development *Biology of Reproduction* **52** 676–682

Fotsis T, Murphy C and Gannon F (1990) Nucleotide sequence of the bovine insulin-like growth factor I (IGF-I) and its IGF-IA precursor *Nucleic Acids Research* **18** 676

Gandolfi F and Moor RM (1987) Stimulation of early embryonic development in the sheep by co-culture with oviduct epithelial cells *Journal of Reproduction and Fertility* **81** 23–28

Gardner DK, Lane M, Spitzer A and Batt PA (1994) Enhanced rates of cleavage and development for sheep zygotes cultured to the blastocyst stage *in vitro* in the absence of serum and somatic cells: amino acids, vitamins, and culturing embryos in groups stimulate development *Biology of Reproduction* **50** 390–400

Giudice LC, Dsupin BA, Irwin JC and Eckert RL (1992) Identification of insulin-like growth factor binding proteins in human oviduct *Fertility and Sterility* **57** 294–301

Hahnel A and Schultz GA (1994) Insulin-like growth factor binding proteins are transcribed by preimplantation mouse embryos *Endocrinology* **134** 1956–1959

Hemmings R, Langlais J, Falcone T, Granger L, Miron P and Guyda H (1992) Human embryos produce transforming growth factor activity and insulin like growth factor II *Fertility and Sterility* **58** 101–104

Herrler A, Lucas-Hahn A and Niemann H (1992) Effects of insulin-like growth factor-I in *in vitro* production of bovine embryos *Theriogenology* **37** 1213–1224

Jones JI and Clemmons DR (1995) Insulin-like growth factors and their binding proteins: biological actions *Endocrine Reviews* **16** 3–34

Kane MT, Morgan PM and Coonan C (1997) Peptide growth factors and preimplantation development *Human Reproduction Update* **3** 137–157

Kaye PL and Harvey MB (1995) The roles of growth factors in preimplantation development *Progress in Growth Factor Research* **6** 1–26

Keskintepe L, Burnley CA and Brackett BG (1995) Production of viable bovine blastocyst in defined *in vitro* conditions *Biology of Reproduction* **52** 1410–1417

Larson RC, Ignotz GG and Currie WB (1992) Platelet derived growth factor (PDGF) stimulates development of bovine embryos during the fourth cell cycle *Development* **115** 821–826

Latham KE, Doherty AS, Scott CD and Schultz RM (1994) Igf2r and Igf2 gene expression in androgenetic, gynogenetic, and parthenogenetic preimplantation mouse embryos: absence of regulation by genomic imprinting *Genes and Development* **8** 290–299

LeRoith D, Werner H, Beitner-Johnson D and Roberts CT, Jr (1995) Molecular and cellular aspects of the insulin-like growth factor I receptor *Endocrine Reviews* **16** 143–163

Lesse HJ (1988) The formation and function of oviduct fluid *Journal of Reproduction and Fertility* **82** 843–856

Ling NC, Liu XJ, Malkowski M, Guo YL, Erickson GF and Shimasaki S (1993) Structural and functional studies of insulin-like growth factor binding proteins in the ovary *Growth Regulation* **3** 70–74

Liu JP, Baker J, Perkins AS, Robertson EJ and Efstratiadis A (1993) Mice carrying null mutations of the genes encoding insulin-like growth factor I (Igf-1) and type 1 IGF receptor (Igfr) *Cell* **75** 59–72

Mattson BA, Rosenblum IY, Smith RM and Heyer S (1988) Autoradiographic evidence for stage specific insulin binding to mouse embryo development *Diabetes* **37** 585–590

Morgan DO, Edman JC, Standring DN, Fried VA, Smith MC, Roth RA and Rutter WJ (1987) Insulin-like growth factor II receptor as a multifunctional binding protein *Nature* **326** 300–307

Murphy LJ and Barron DJ (1993) The IGFs and their binding proteins in murine development *Molecular Reproduction and Development* **35** 376–381

O'Neill C (1997) Evidence for the requirement of autocrine growth factors for development of mouse preimplantation embryos *in vitro. Biology of Reproduction* **56** 229–237

Rappolee DA, Sturm KS, Behrendsten O, Schultz GA, Pedersen RA and Werb Z (1992) Insulin-like growth factor II acts through an endogenous growth pathway regulated by imprinting in early mouse embryos *Genes and Development* **6** 939–952

Schmidt A, Einspanier R, Amselgruber W, Sinowatz F and Schams D (1994) Expression of insulin-like growth factor I (IGF-I) in the bovine oviduct during the oestrus cycle *Experimental and Clinical Endocrinology* **102** 364–369

Schultz GA and Heyner S (1993) Growth factors in preimplantation embryos *Oxford Reviews of Reproductive Biology* **15** 43–81

Schultz GA, Hahnel A, Arcellana-Panlilio M, Wang L, Goubau S, Watson AJ and Harvey M (1993) Expression of IGF ligand and receptor genes during preimplantation mammalian development *Molecular Reproduction and Development* **35** 414–420

Sirard MA, Parrish JJ, Ware MI, Leibfried-Rutledge MI and First NL (1988) The culture of bovine oocytes to obtain developmentally competent embryos *Biology of Reproduction* **39** 546–552

Smith RM, Garside WT, Aghayan M, Shi C-Z, Shah N, Jarret L and Heyner S (1993) Mouse preimplantation embryos exhibit receptor-mediated binding and transcytosis of maternal insulin-like growth factor I *Biology of Reproduction* **49** 1–12

Stewart CL and Cullinan EB (1997) Preimplantation development of the mammalian embryo and its regulation by growth factors *Developmental Genetics* **21** 91–101

Summers MC, Bhatnagar PR, Lawitts JA and Biggers JD (1995) Fertilization *in vitro* of mouse ova from inbred and outbred strains: complete preimplantation embryo development in glucose-supplemented KSOM *Biology of Reproduction* **53** 431–437

Tervit HR, Whittingham DG and Rowson LEA (1972) Successful culture *in vitro* of sheep and cattle ova *Journal of Reproduction and Fertility* **30** 493–497

Thompson JG, Gardner DK , Pugh PA, McMillan WH and Tervit HR (1995) Lamb birth weight is affected by culture system utilized during *in vitro* pre-elongation development of ovine embryos *Biology of Reproduction* **53** 1385–1391

Walker SK, Hill JL, Kleemann DO and Nancarrow CD (1996) Development of ovine embryos in synthetic oviductal fluid containing amino acids at oviductal fluid concentrations *Biology of Reproduction* **55** 703–708

Watson AJ, Hogan A, Hahnel A, Wiemer KE and Schultz GA (1992)

Expression of growth factor ligand and receptor genes in the preimplantation bovine embryo *Molecular Reproduction and Development* **31** 87–95

Watson AJ, Watson PH, Arcellana-Panlilio M, Warnes D, Walker SK, Schultz GA, Armstrong DT and Seamark RF (1994) A growth factor phenotype map for ovine preimplantation development *Biology of Reproduction* **50** 725–733

Winger QA, de los Rios P, Han VKM, Armstrong DT, Hill DJ and Watson AJ (1997) Bovine oviductal and embryonic insulin-like growth factor binding proteins: possible regulators of embryotrophic insulin-like growth factor circuits *Biology of Reproduction* **56** 1415–1423

Wiseman DL, Henricks DM, Eberhardt DM and Bridges WC (1992) Identification and content of insulin-like growth factors in porcine oviductal fluid *Biology of Reproduction* **47** 126–132

Xia P, Tekpetey FR and Armstrong DT (1994) Effect of IGF-I on pig oocyte maturation, fertilization, and early embryonic development *in vitro*, and on granulosa and cumulus cell biosynthetic activity *Molecular Reproduction and Development* **38** 373–379

Xia P, Han VKM, Viuff D, Armstrong DT and Watson AJ (1996) Expression of insulin-like growth factors in two bovine oviductal cultures employed for embryo co-culture *Journal of Endocrinology* **148** 41–53

Zhang X, Kidder GM, Watson AJ, Schultz GA and Armstrong DA (1994) Possible roles of insulin and insulin-like growth factors in rat preimplantation development. Investigation of gene expression by reverse transcription–polymerase chain reaction *Journal of Reproduction and Fertility* **100** 375–380

Journal of Reproduction and Fertility Supplement **54**, 317–328

The regulation of interferon-τ production and uterine hormone receptors during early pregnancy

G. E. Mann[1], G. E. Lamming[1], R. S. Robinson[2] and D. C. Wathes[2]

[1]*University of Nottingham, School of Biological Sciences, Division of Animal Physiology, Sutton Bonington, Loughborough, Leics LE12 5RD, UK;* [2]*Department of Veterinary Basic Sciences, Royal Veterinary College, Boltons Park, Hawkshead Road, Potters Bar, Herts EN6 1NB, UK*

During early pregnancy the bovine embryo must produce a protein called interferon τ which inhibits the development of the luteolytic mechanism. Failure to inhibit luteolysis is the major cause of pregnancy loss in cows. The embryo must produce sufficient quantities of interferon τ by about day 16 to prevent luteolysis. Its ability to achieve this is largely dependent on the pattern of maternal progesterone production. A late rise in progesterone after ovulation or poor progesterone secretion during the luteal phase results in the development of poor embryos capable of producing little or no interferon τ at the critical time. The embryo inhibits luteolysis by preventing development of oxytocin receptors on the luminal epithelium of the uterine endometrium and thus oxytocin-induced secretion of $PGF_{2\alpha}$ and by the induction of a prostaglandin synthesis inhibitor within the endometrium. In sheep it has been hypothesised that interferon τ acts to inhibit endometrial oestrogen receptors and thus oestrogen-induced up-regulation of oxytocin receptors. In cows, the embryo inhibits the development of oxytocin receptors and the initiation of luteolysis without causing any change in uterine oestrogen receptors. Thus in the cow, the mechanism by which interferon τ inhibits oxytocin receptor development remains to be determined.

Introduction

Compared with the high level of fertility observed in several species of wild ungulates, pregnancy rates in domestic cattle are currently very low, and appear to be declining. While poor fertility is the result of a number of factors, probably the biggest single cause of poor fertility in cattle is early embryo mortality. For its continued development during early pregnancy, the embryo must prevent the demise of the corpus luteum and thus maintain the secretion of progesterone. To do this it must inhibit the development of the luteolytic mechanism that terminates the luteal phase of the oestrous cycle. Luteolysis results from the release of luteolytic episodes of uterine prostaglandin $F_{2\alpha}$ ($PGF_{2\alpha}$). The majority of current evidence suggests that this occurs in response to the binding of oxytocin to newly developed receptors on the uterine endometrium. The embryo inhibits luteolysis through the secretion of interferon τ, which acts locally within the uterus to inhibit both the development of oxytocin receptors on the endometrium and the secretion of $PGF_{2\alpha}$. Thus the success of early pregnancy in cows, as in other ruminant species, depends upon a fine balance between the development of a maternal luteolytic mechanism and an antiluteolytic embryonic signal. A strong or early signal from the mother for the corpus luteum to regress or a weak or delayed signal from the embryo to prevent the loss of the corpus luteum results in the failure of the pregnancy. Why this balance so often fails has been the focus of much study in recent years.

Hormonal Influences on the Outcome of Early Pregnancy

Many of the mechanisms involved in early pregnancy are influenced by the ovarian steroid hormones, progesterone and oestradiol, and many studies have investigated their roles. A common approach has been to monitor concentrations of hormone in mated cows and then retrospectively analyse these profiles on the basis of whether the animals successfully maintained a pregnancy. It has been established for many years that the concentration of progesterone during early pregnancy has a marked effect on the potential outcome. Lower concentrations of plasma progesterone from about day 12 after mating have been reported in animals in which early pregnancy fails in a number of studies (Lukazewska and Hansel, 1980; Lamming *et al.*, 1989; Mann *et al.*, 1995; Fig. 1). Reanalysis of the data of Mann *et al.* (1995) has revealed that as well as an original difference in milk progesterone between day 12 and day 15, there is also a significantly lower milk progesterone concentration on day 6 in mated not pregnant cows compared with pregnant cows (Fig. 1). A lower concentration of progesterone 6 days after mating in cows with a failed pregnancy was reported in 1971 by Henricks *et al.*, while more recently Lamming and Darwash (1995) reported that a progressive delay in the post-ovulatory rise in progesterone was associated with a marked and progressive reduction in pregnancy rate in mated animals. These studies clearly demonstrate that both a late post-ovulatory rise in progesterone and low luteal phase concentrations of progesterone have a detrimental effect on the outcome of early pregnancy.

Although oestradiol concentrations have not been studied as comprehensively as concentrations of progesterone during early pregnancy, most studies indicate that concentrations of oestradiol do not differ between mated cows in which pregnancy is successful or fails (Lukaszewska and Hansel, 1980; Gyawa and Pope, 1992; Mann *et al.*, 1995). In one study in beef cows by Pritchard *et al.* (1994), a lower pregnancy rate was observed in cows with higher plasma concentrations of oestradiol between day 14 and day 17. However, in this study, luteolysis had begun in some cows and so it is not clear whether the higher concentration of oestradiol was the cause of pregnancy failure or the result of failed embryonic inhibition of luteolysis. Thus current evidence supports the idea that oestradiol does not exert the same degree of influence as progesterone over the outcome of early pregnancy.

Although these studies identify the importance of the concentration of progesterone during early pregnancy, they do not answer the question of how progesterone is exerting its effects. To establish this we must look at the effects of progesterone on the important mechanisms in place during this period. For example, it has now been demonstrated that the concentration of luteal phase progesterone in the cow has a profound influence on the strength of development of the luteolytic signal (Mann and Lamming, 1995a). Furthermore, although oestradiol concentrations do not differ between pregnant and non-pregnant cows, studies have demonstrated an important influence of the concentration of oestradiol in controlling the strength of the luteolytic signal (Mann and Lamming, 1995b).

Production of Interferon τ – The Embryonic Signal

Identification of the anti-luteolytic agent

In 1966, Moor and Rowson established, in sheep, that the presence of an embryo first affected luteal function on day 13 after mating, and Rowson and Moor (1967) found that infusion of homogenates of day 13 and 14 embryos into the uteri of cyclic ewes prolonged luteal life span. This established both the importance of the embryo in the inhibition of luteolysis and the time at which this effect takes place. In cows, Northey and French (1980) demonstrated that removal of the embryo from the uterus on day 15 does not result in a delay in luteolysis, whereas removal on day 17 results in a significant delay. Furthermore, infusion of homogenates of day 17–18 embryos resulted in a delay in luteolysis. Thus it was established that the embryo exerted an anti-luteolytic effect on the cow between day 15 and day 17.

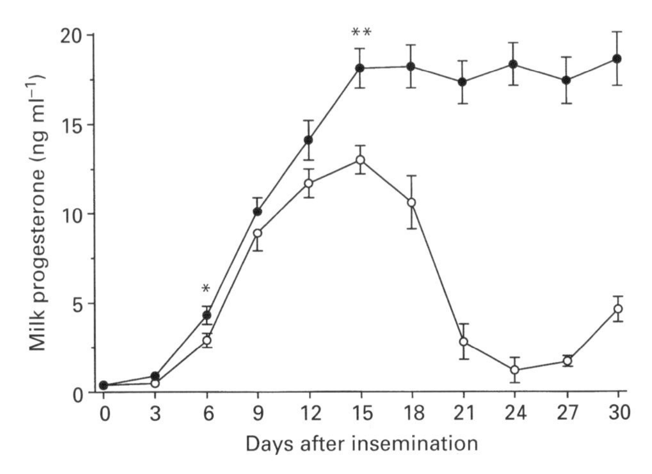

Fig. 1. Mean (± SEM) milk progesterone concentration after insemination in cows that became pregnant (●; n = 28) and cows in which pregnancy failed (○; n = 24). Note the significant differences between pregnant and non-pregnant groups on day 15 ($P < 0.01$) and day 6 ($P < 0.05$). * $P < 0.05$; ** $P < 0.01$. (Based on data from Mann *et al.*, 1995).

Studies using recombinant bovine interferon α, which has approximately 50% amino acid sequence identity with bovine interferon τ, have demonstrated an extension of the duration of the cycle in cattle, but with associated adverse side effects such as hyperthermia, reduction in progesterone secretion and reduced conception rate (Plante *et al.*, 1989, 1991; Barros *et al.*, 1992). Highly purified bTP-1 obtained from culture of day 17–18 conceptuses (Helmer *et al.*, 1989) and more recently recombinant bovine interferon τ (Meyer *et al.*, 1995) have been shown to reduce luteolytic secretion of $PGF_{2\alpha}$ and extend luteal function in the cow, apparently without deleterious side effects.

Timing and control of interferon τ production

Expression of mRNA for interferon τ has been detected as early as day 12 in the cow, is maximal on days 15–16 and continues at least until day 25 (Farin *et al.*, 1990). This expression appears to be limited to the trophectoderm and expression is not apparent in the endoderm or yolk sac (Farin *et al.*, 1990). Despite the appearance of mRNA on day 12, we have found that significant quantities of interferon τ (as measured by antiviral assay) are first detected in uterine flushes between day 14 and day 16, when embryos have begun elongation (Fig. 2)

During the early stages of pregnancy, it is well established that progesterone stimulates the production of the endometrial secretions necessary for embryo development (for review see Geisert *et al.*, 1992). The importance of the pattern of maternal progesterone in producing a suitable uterine environment for the embryo is demonstrated clearly by studies involving asynchronous embryo transfer. In such studies the uterus of a recipient ewe can, by prior treatment with progesterone, be rendered receptive to the transfer of an embryo from a donor ewe at a more advanced stage after mating (Lawson and Cahill, 1983). The effects of this progesterone include increased endometrial protein secretion (Garrett *et al.*, 1988) and increased production of PGE_2 (Vincent *et al.*, 1986).

In cows, maternal concentrations of progesterone have a marked influence on the development of the embryo (Mann *et al.*, 1996) and its ability to produce interferon τ (Mann *et al.*, 1998). We found that cows with a late post-ovulatory increase in progesterone to lower luteal phase concentrations had embryos that, on day 16, exhibited little or no elongation and produced little or no interferon τ.

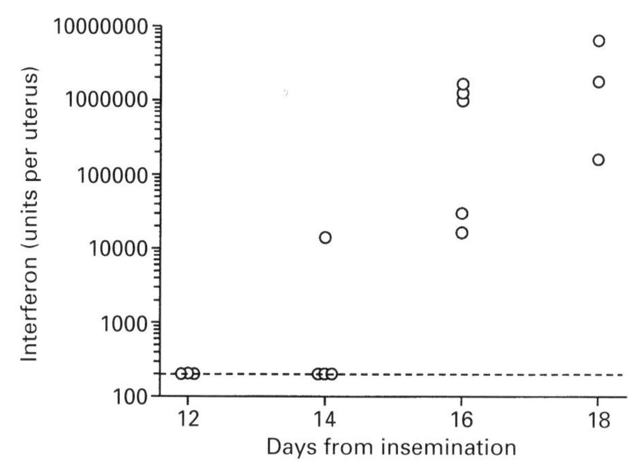

Fig. 2. Total uterine interferon-τ content measured in uterine flushings by antiviral assay in Holstein Friesian cows inseminated at natural oestrus and then slaughtered at various stages of early pregnancy. The dashed line represents the detection limit of the assay system. (GE Mann, unpublished).

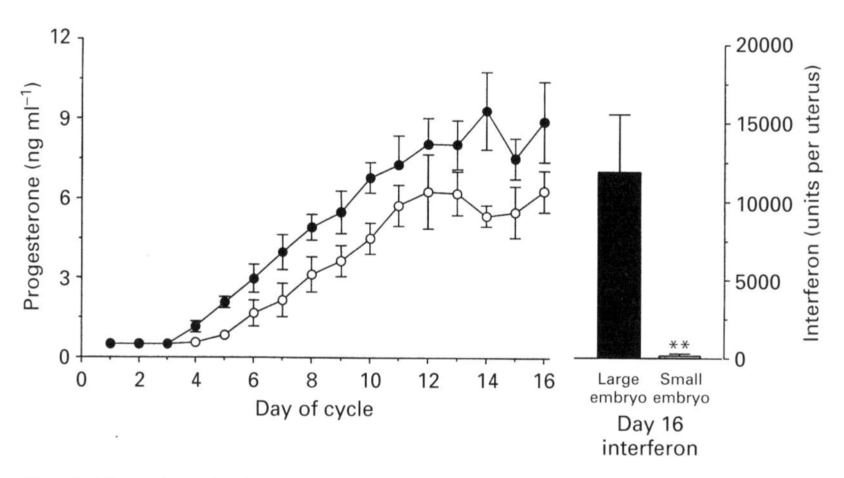

Fig. 3. Mean (± SEM) plasma concentrations of progesterone in inseminated cows slaughtered on day 16 whose uteri were flushed to reveal either a large well-elongated embryo with a high concentration of interferon τ activity (●/bar; $n = 5$) or a small poorly developed embryo (○/bar; $n = 5$). Note both the delayed progesterone rise ($P < 0.01$) and low luteal phase concentrations ($P < 0.05$) leading to low interferon τ production ($P < 0.01$). ** $P < 0.01$. (Based on data from Mann *et al.*, 1996, 1998)

Conversely day 16 embryos of cows with an earlier rise in progesterone to higher luteal phase concentrations were well elongated (> 4 cm) and produced large quantities of interferon-τ (Fig. 3).

A number of other studies have also suggested an important role for early luteal phase progesterone in the control of subsequent embryonic production of interferon-τ. Nephew *et al.* (1991) demonstrated that ewes exhibiting a slightly higher plasma concentration of progesterone on days 2–4 of the luteal phase showed advanced embryo development on day 13 and an associated threefold increase in production of interferon τ. Garrett *et al.* (1988) found that by increasing progesterone from days 2–5 in cows, embryo development on day 14 was advanced significantly (tenfold increase in length of conceptus) and that production of proteins forming the "bovine trophoblast protein complex proposed to be involved with the maintenance of early corpus luteum function" (interferon τ) was present in progesterone-treated cows but not in control cows. Kerbler *et al.* (1997) demonstrated only a slight increase in production of interferon τ by day 18 bovine embryos collected from cows with increased progesterone secretion from day 8 (caused by induction of accessory corpora lutea). These findings suggest that an early increase in progesterone is more important in stimulating embryo development and interferon τ synthesis than are later progesterone concentrations. We have recently demonstrated that progesterone supplementation from day 5 to day 9 gives a significant increase in interferon τ production on day 16 compared with supplementation from day 12 to day 16 (Mann *et al.*, 1998).

Although the maternal progesterone environment appears to have a major influence on the development of the early bovine embryo, and its ability to produce its antiluteolytic interferon τ signal, an understanding of how this is achieved requires an investigation of factors within the uterus. While a wide range of growth factors and cytokines are expressed by the endometrium and the embryo during early pregnancy and are probably involved to various degrees in the control of early embryo development (for review see Martal *et al.*, 1997), little is known about the specific mechanisms controlling early embryo development and production of interferon τ. Some factors that have been demonstrated to affect interferon τ production in sheep specifically include insulin-like growth factors -I and -II (Ko *et al.*, 1991) granulocyte–macrophage-colony stimulating factor (Imakawa *et al.*, 1993) and interleukin 3 (Imakawa *et al.*, 1995). However, further research is required before any detailed control mechanisms can be elucidated.

Development of the Luteolytic Mechanism – The Maternal Signal

Oxytocin receptor development and $PGF_{2\alpha}$ release

In cattle, concentrations of endometrial oxytocin receptors are low or undetectable from about day 6– 8 of the luteal phase to immediately before luteolysis, about day 15–17, when concentrations begin to increase (Meyer *et al.*, 1988; Fuchs *et al.*, 1990; Mann and Lamming, 1994). The pulsatile secretion of luteolytic $PGF_{2\alpha}$ begins on about day 17 and is associated with a small rise in endometrial oxytocin receptor concentration (Mann and Lamming, 1993). By collection of repeated biopsy samples of uterine endometrium, we have found that oxytocin receptors, which are undetectable through much of the luteal phase (< 20 fmol mg^{-1} protein), rise to a concentration of 121 ± 16 fmol mg^{-1} protein when large luteolytic episodes of $PGF_{2\alpha}$ secretion are first observed. The onset of $PGF_{2\alpha}$ secretion is followed by luteolysis within 48 h and oxytocin receptor concentrations continue to increase to maximum concentrations of 500–1000 fmol mg^{-1} protein at oestrus (Fig. 4). Fuchs *et al.* (1990) found that oxytocin receptor concentrations on day 17, the anticipated start of luteolysis were about 10% of the values obtained on day 21 (the anticipated day of oestrus). Furthermore, Mirando *et al.* (1993) found a marked increase in oxytocin-induced $PGF_{2\alpha}$ production in heifers between day 13 and day 16, despite only a slight increase in concentration of endometrial oxytocin receptors; a large increase in oxytocin receptor concentration did not occur until day 19. Thus luteolytic $PGF_{2\alpha}$ release requires only a modest increase in uterine oxytocin receptors. The peak concentrations of oxytocin receptor obtained at oestrus are associated with the fall in progesterone and increase in oestradiol secretion that occur as a result of luteolysis and are not, therefore, the cause of luteolysis.

Receptor localisation studies in ewes have established that this critical first increase in oxytocin receptor activity is restricted to the luminal epithelium; oxytocin receptors appear in the deeper tissues only after luteolysis starts (Wathes and Lamming, 1995). In cows oxytocin receptors first appear on the luminal epithelium of the uterine endometrium (Robinson *et al.*, 1998a). This occurs at the same time as the uterus develops the ability to release $PGF_{2\alpha}$ in response to oxytocin (Mann and Lamming, 1994). This finding demonstrates that in cows, as in ewes, it is the development of oxytocin receptors on the luminal epithelium that is the key event in the development of the

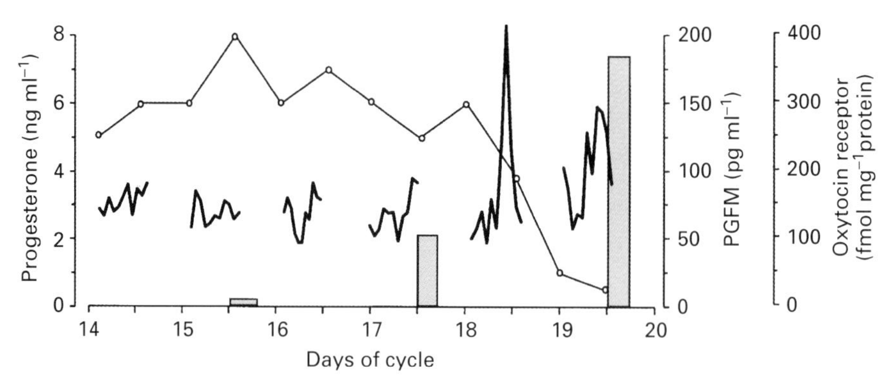

Fig. 4. Relationship between plasma progesterone concentration (○), endometrial oxytocin receptor concentration, measured in repeated endometrial biopsy samples (shaded bars) and plasma 13,14 dihydro-15-keto prostaglandin $F_{2\alpha}$ (PGFM) concentrations (solid lines) during natural luteolysis in a typical Holstein Friesian cow. Note the small increase in oxytocin receptors needed to allow luteolytic $PGF_{2\alpha}$ release and the subsequent further increase in receptor concentration once luteolysis had occurred (Based on data from Mann and Lamming, 1993)

luteolytic mechanism. Further support for this hypothesis, that changes in oxytocin receptor concentrations in the luminal epithelium are the critical factor, comes from the observation that most $PGF_{2\alpha}$ release occurs from epithelium and not from stromal cells (Fortier *et al.*, 1988). Thus the initiation of the luteolytic mechanism requires only a relatively small increase in endometrial oxytocin receptors and it is this initial increase in oxytocin receptor development within the luminal epithelium which the embryo must counteract if it is to prevent luteolysis.

Although many studies have investigated the importance of the oxytocin receptor in luteolysis, Kotwica *et al.* (1997) suggested that treatment of cows with an oxytocin antagonist does not prevent luteolytic $PGF_{2\alpha}$ secretion. These results differ markedly from those in sheep where continuous infusion of a similar oxytocin antagonist completely blocked luteolytic $PGF_{2\alpha}$ secretion (Jenkin, 1992). While it is difficult to ignore the large body of evidence that supports a pivotal role for oxytocin induced $PGF_{2\alpha}$ release in the luteolytic process in cows, the findings of Kotwica *et al.* (1997) do suggest that further evaluation of the mechanisms involved may be required.

The role of steroid hormone receptors

Receptor binding assays demonstrate that progesterone receptors are high at oestrus, increase further to about day 6–8 and then decline to low values by day 10–12 (Meyer *et al.*, 1988). This pattern of expression is largely repeated in receptor localization studies (Boos *et al.*, 1996; Robinson *et al.*, 1998b). Progesterone inhibits the development of oxytocin receptors in ewes (Lau *et al.*, 1992) and it has been postulated that this fall in the number of progesterone receptors is necessary to allow the development of oxytocin receptors needed to initiate the luteolytic mechanism (McCracken *et al.*, 1984). By recreating cyclic changes in hormone concentrations in ovariectomized cows, we have demonstrated that during the luteal phase, progesterone initially inhibits oxytocin receptors (Lamming and Mann, 1995). However, by day 16, despite maintained progesterone concentrations, oxytocin receptors reappeared in the endometrium and their ability to release $PGF_{2\alpha}$ in response to oxytocin returned. Thus the hypothesis that after a suitable period progesterone loses its inhibitory action on the uterus allowing endometrial oxytocin receptor development and $PGF_{2\alpha}$ release appears to be valid for cows as well as ewes.

In sheep, oestradiol receptors are undetectable during the mid-luteal phase owing to inhibition by progesterone. The receptors reappear at about the time of luteolysis and it has been suggested that an increase in oestrogen receptors is needed to allow oestradiol to stimulate the rise in oxytocin receptors needed for luteolysis (McCracken *et al.*, 1984; Spencer *et al.*, 1996). In cattle, receptor binding assays demonstrate peak concentrations of endometrial oestrogen receptor at oestrus and during the first few days of the luteal phase. Concentrations then tend to fall as the luteal phase progresses and an increase is observed as the cow approaches oestrus (Meyer *et al.*, 1988). However, localization studies have shown that while this general pattern occurs in stromal tissues, and to some extent glandular tissues, in the luminal epithelium, the critical region in terms of luteolysis, fluctuating concentrations of oestrogen receptors are present throughout the ovarian cycle (Boos *et al.*, 1996; Robinson *et al.*, 1998b). These fluctuations appear to be correlated with fluctuations in plasma oestradiol (Robinson *et al.*, 1998b). Thus it would appear that during the luteal phase of the cow the oestrogen receptor is not subjected to the same degree of downregulation as appears to be the case in the ewe. The presence of oestrogen receptors on the luminal epithelium during the entire luteal phase demonstrate that in cows, the initial upregulation of oxytocin receptor needed for the initiation of luteolysis is not controlled by the appearance of oestrogen receptors on the endometrium. This differs from current theories in the ewe in which the action of oestrogen on newly upregulated oestrogen receptors has been hypothesized as the trigger for the oxytocin receptor upregulation required for luteolysis (McCracken *et al.*, 1984; Spencer *et al.*, 1996).

The use of pharmacological doses of oestradiol has demonstrated the basic ability of oestradiol to upregulate endometrial oxytocin receptors (Hixon and Flint, 1987; Spencer *et al.*, 1995). However, when postulating a role for upregulation of oxytocin receptors by oestrogen as the mechanism for the initiation of luteolysis, a number of results must be considered. Oxytocin receptor concentrations in the endometrium appear to increase independently of steroid hormones (Lamming and Mann, 1995). Furthermore, in ovariectomized ewes it has been shown that this oxytocin receptor upregulation is not attenuated by the use of passive immunization to neutralize any residual oestradiol that might be stimulating receptor development in ovariectomized animals (Payne *et al.*, 1994). In many studies, no significant increase in peripheral oestradiol concentration has been observed before the initiation of luteolysis. However, Fogwell *et al.* (1985) did report an increase in oestradiol in utero–ovarian venous plasma before luteolysis and it is clear that oestradiol has an effect since luteolysis is prevented by the inhibition of oestradiol (Fogwell *et al.*, 1985).

Uterine Hormone Receptors During Early Pregnancy

A number of studies in both cattle and sheep have demonstrated the general ability of the embryo itself, embryo-derived interferon τ or recombinant interferon τ to inhibit the development of oxytocin receptors on the endometrium. It is now generally accepted that this is the key event in the maintenance of pregnancy in both species. Elsewhere in this review the importance of the initial small rise in oxytocin receptors, localized to the luminal epithelium of the endometrium, in the initiation of luteolysis has been established. The inhibition of this initial rise in oxytocin receptors appears to be the key event in the establishment of pregnancy.

Fuchs *et al.* (1990) found that the presence of a viable conceptus in the uterus completely prevented both the small rise in oxytocin receptors in the uterus between day 14 and day 17, and the much larger rise seen between days 17 and day 21 in cyclic cows. On day 16, a time at which the luteloytic mechanism is beginning to develop, oxytocin receptor mRNA was detectable in the luminal epithelium of over 40% of non-pregnant cows but was undetectable in all cows with an embryo (Robinson *et al.*, 1998a).

In unilaterally pregnant sheep with transected uteri, development of oxytocin receptors is inhibited in the pregnant horn. However, in the non-pregnant horn, deprived of the usual fall in progesterone and rise in oestradiol after luteolysis, development of oxytocin receptors still occurs but is limited to the luminal epithelium (Lamming *et al.*, 1995). Despite the presence of oxytocin receptors in only the luminal epithelium of the non-pregnant horn, large episodes of $PGF_{2\alpha}$ secretion could still be stimulated by the administration of oxytocin (Payne and Lamming, 1994). This finding raises the possibility that changes in receptors on the luminal epithelium may be all that is required for the development of an active luteolytic mechanism. Administration of oestradiol stimulates an increase in oxytocin receptors within 12 h of administration (Hixon and Flint, 1987) and thus within hours of the onset of luteolysis, endometrial hormone receptor changes occur that are the result of the post-luteolytic fall in progesterone and rise in oestradiol. Increases in receptors in regions other than the luminal epithelium, driven by changes in ovarian hormone secretion after luteolysis, are not required for the initiation of luteolysis. However, they may have a role in the final completion of the luteolytic process. As well as the effect of the post-luteolytic increase in oestradiol secretion on uterine hormone receptors, treatment with high concentrations of oestradiol has been shown to stimulate $PGF_{2\alpha}$ secretion within 6 h of administration in cows (Knickerbocker *et al.*, 1986).

We have recently investigated the effects of pregnancy on endometrial oxytocin, oestradiol and progesterone receptors on day 16 in non-pregnant cows and in cows with an embryo present in the uterus (Robinson *et al.*, 1998a). In pregnant cows, there was a significant inhibition of both endometrial oxytocin receptor mRNA concentrations (Fig. 5a and Table 1) and oxytocin-induced secretion of $PGF_{2\alpha}$ (increased secretion in 6/15 cows with an embryo present compared with 14/14 non-pregnant cows). However, despite the inhibitory effect of the embryo on the initiation of a luteolytic mechanism, measurement of both oestradiol receptor mRNA and oestradiol receptor protein revealed no differences between pregnant and non-pregnant cows (Fig. 5b and Table 1). Thus it would appear that in cows, the embryo can inhibit both the initiation of oxytocin receptor and oxytocin-induced secretion of $PGF_{2\alpha}$ without affecting oestradiol receptor concentrations. As with the oestrogen receptor, progesterone receptors were also present at similar concentrations in both pregnant and non-pregnant cows in all regions of the uterus studied (Table 1), supporting the idea that changes in endometrial progesterone receptor concentrations are not involved in the inhibition of luteolysis during early pregnancy.

Maternal Recognition of Pregnancy – Current Theories

The principal role of the embryo during the maternal recognition of pregnancy is to inhibit the development of oxytocin receptors on the endometrium and hence the release of $PGF_{2\alpha}$ that is responsible for the demise of the corpus luteum.

In cattle, as in sheep, the loss of progesterone receptors on the endometrium during the later stages of the luteal phase may prevent continued inhibition by progesterone of the development of oxytocin receptors. One mechanism by which the embryo could inhibit the development of oxytocin receptors on the endometrium could be through the maintenance of progesterone receptors. There is no evidence to suggest that this loss of progesterone receptors is prevented during pregnancy in the sheep (Spencer *et al.*, 1996); similar results were obtained in cows (Robinson *et al.*, 1998a), in which concentrations of progesterone receptor were similar in pregnant and non-pregnant cows at the initiation of the luteolytic mechanism. A study in the rat has, for the first time, demonstrated a direct inhibitory effect of progesterone on the activity of the oxytocin receptor (Grazzini *et al.*, 1998). Whether the embryo induces progesterone to exert a similar role in cows or sheep is not known.

In sheep, it has been postulated that interferon τ prevents the rise in endometrial oestrogen receptors that is thought to precede the rise in oxytocin receptors necessary for the induction of luteolytic release of $PGF_{2\alpha}$. This contention is supported by several studies. Spencer and Bazer (1995) demonstrated that while oestrogen receptor was present on the luminal epithelium on day 13 and 15 in cyclic ewes, it was absent in pregnant ewes at these times. In this study progesterone was lower on day 15 in the cyclic ewes, indicating that there is a rise in oestradiol secretion and consequent oestrogen upregulation of its own receptor in the different hormonal environment of the cyclic group. However, the difference in oestrogen receptor on day 13 does suggest a role for interferon-τ-induced suppression of oestrogen receptor in the establishment of pregnancy in ewes. Furthermore, treatment of ewes with pharmacological concentrations of oestradiol to induce luteolysis induces oestrogen and oxytocin receptors, both of which can be inhibited by the concomitant administration of roIFN τ (Spencer *et al.*, 1995). However, studies in cows indicate that the initial inhibition of oxytocin receptor development on the luminal epithelium and luteolytic uterine $PGF_{2\alpha}$ release occur in the absence of an effect on oestrogen receptor concentrations (Robinson *et al.*, 1998a).

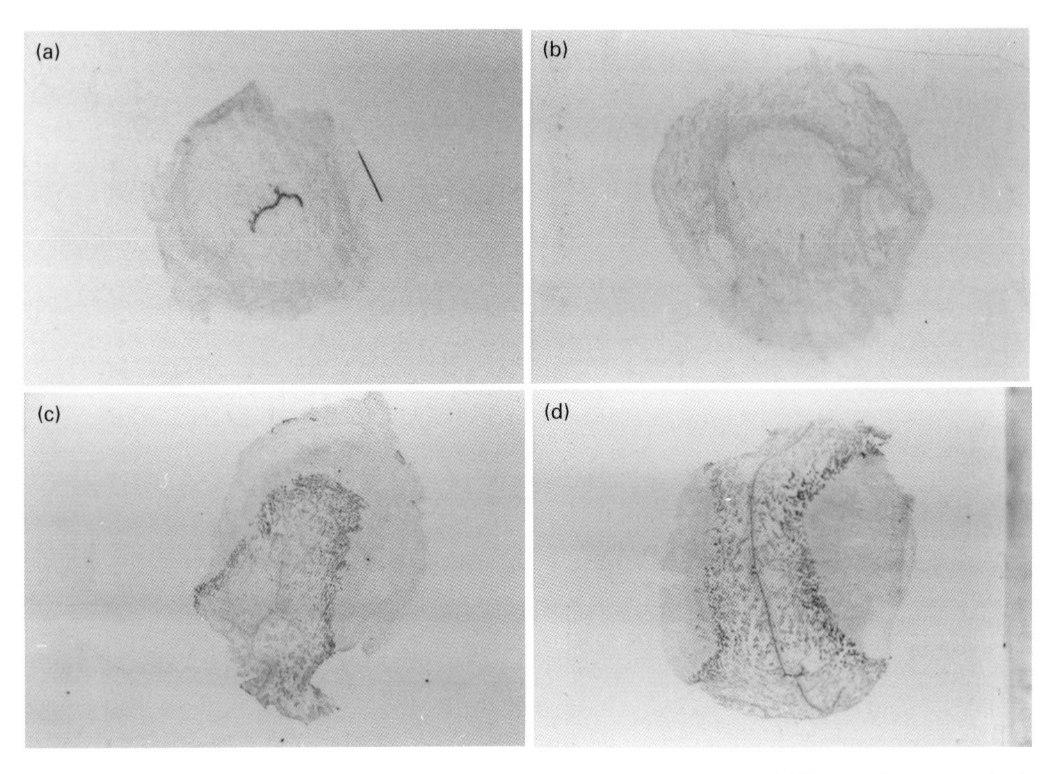

Fig. 5. *In situ* hybridization of day 16 bovine uterus showing the expression of (a) oxytocin receptor in the luminal epithelium of a non-pregnant cow, (b) lack of oxytocin receptor expression in a pregnant cow, (c) expression of oestrogen receptor in a non-pregnant cow and (d) expression of oestrogen receptor in a pregnant cow. Oxytocin receptor mRNA was expressed in the luminal epithelium of 6/14 non-pregnant compared with 0/15 cows with an embryo present. Oestrogen receptor mRNA was expressed at a similar concentration in both the luminal epithelium and the deep glands of both pregnant and non-pregnant cows. Magnification × 5. (Based on Robinson *et al.*, 1998a.)

Table 1. The expression of oxytocin receptor, oestrogen receptor and progesterone receptor mRNA in the uterine endometrium of pregnant (n = 15) and non-pregnant (n = 14) cows on day 16 measured as absorbance units from autoradiographs.

Receptor mRNA	Tissue	Pregnant	Non-pregnant
Oxytocin receptor	Luminal epithelium	nd (< 0.01)	0.04 ± 0.02*
	Stroma	nd (< 0.01)	nd (< 0.01)
Oestradiol receptor	Luminal epithelium	0.06 ± 0.00	0.07 ± 0.01
	Superficial glands	0.05 ± 0.01	0.06 ± 0.01
	Deep glands	0.13 ± 0.01	0.12 ± 0.02
Progesterone receptor	Dense caruncular stroma	0.03 ± 0.01	0.03 ± 0.01
	Caruncular stroma	0.08 ± 0.02	0.06 ± 0.03
	Superficial glands	0.02 ± 0.00	0.02 ± 0.00
	Deep glands	0.03 ± 0.01	0.02 ± 0.00

Values are means ± SEM.
*Expressed in 6/14 non-pregnant compared with 0/15 pregnant cows, $P < 0.05$ (Chi squared test)
Data based on Robinson *et al.* (1998a).

In addition to hypotheses based largely on changes in uterine hormone receptors, it is important to consider other aspects of the control of $PGF_{2\alpha}$ release. One such mechanism in cows involves the stimulation, by the embryo, of an endometrial inhibitor of $PGF_{2\alpha}$ synthesis (Thatcher *et al.*, 1995). In sheep, pregnancy is associated with a reduction in pulsatile release of $PGF_{2\alpha}$ coupled with an increased basal secretion, i.e. a change in the pattern and not the quantity of $PGF_{2\alpha}$ secretion (Payne and Lamming 1994). However, in cattle the attenuation of luteolytic episodes occurs in the absence of an increase in basal secretion of $PGF_{2\alpha}$ (Thatcher *et al.*, 1995). This species difference is clearly supportive of the presence of a direct inhibitory effect of the embryo on $PGF_{2\alpha}$ secretion in cows.

Conclusions

The importance of the maternal progesterone profile in the control of the development of the embryo and its ability to produce the anti-luteolytic interferon τ signal has been established clearly. This is also the case for the role of interferon in the inhibition of endometrial oxytocin receptor upregulation and subsequent oxytocin-induced luteolytic $PGF_{2\alpha}$ secretion and the induction of an inhibitor of $PGF_{2\alpha}$ synthesis. The lack of an effect of the embryo on oestrogen receptor concentrations at the critical initiation of luteolysis demonstrates a species difference between sheep and cows in the mechanism by which the embryo exerts an anti-luteolytic effect and the mechanism by which the embryo prevents oxytocin receptor development in cows remains to be established. In view of the critical role of the luminal epithelium in the luteolytic process, it is this region of the uterine endometrium in which the answers probably lie. However, before a mechanism of embryonic inhibition of oxytocin receptor upregulation can be deduced, the precise cause of the oxytocin receptor upregulation that the embryo must prevent must be determined. In establishing these mechanisms it will be important to discriminate between the specific small changes in the luminal epithelium that initiate luteolysis and the much larger changes in other regions of the endometrium that result from post-luteolytic changes in ovarian hormone secretion.

Many of the studies included in this review were funded by the Ministry of Agriculture Fisheries and Food and the Milk Development Council under the Link Sustainable Livestock Production Programme. This support is gratefully acknowledged.

References

Barros CM, Betts JG, Thatcher WW and Hansen PJ (1992a) Possible mechanisms for reduction of circulating concentrations of progesterone by interferon-$_{\alpha 1}$ in cows: effects on hyperthermia, luteal cells, metabolism of progesterone and secretion of LH *Journal of Endocrinology* **133** 175–182

Boos A, Meyer W, Schwarz R and Grunert E (1996) Immunohistochemical assessment of oestrogen receptor and progesterone receptor distribution in biopsy samples of the bovine endometrium collected throughout the oestrous cycle *Animal Reproduction Science* **44** 11–21

Farin CE, Imakawa K, Hansen TR, McDonnell JJ, Murphy CN, Farin PW and Roberts RM (1990) Expression of trophoblastic interferon genes in sheep and cattle *Biology of Reproduction* **43** 210–218

Fogwell RL, Cowley JL, Wortman JA, Ames NK and Ireland JJ (1985) Luteal function in cows following destruction of ovarian follicles at mid cycle *Theriogenology* **23** 389–398

Fortier MA, Guilbault LA and Grasso F (1988) Specific properties of epithelial and stromal cells from endometrium of cows *Journal of Reproduction and Fertility* **83** 239–248

Fuchs AR, Behrens O, Helmer H, Liu C-H, Barros CM and Fields MJ (1990) Oxytocin and vasopressin receptors in bovine endometrium and myometrium during the estrous cycle and early pregnancy *Endocrinology* **127** 629–636

Garrett JE, Geisert RD, Zavy MT and Morgan GL (1988) Evidence for maternal regulation of early conceptus growth and development in beef cattle *Journal of Reproduction and Fertility* **84** 437–446

Geisert RD, Morgan GL, Short EC and Zavy MT (1992) Endocrine events associated with endometrial function and conceptus development in cattle *Reproduction Fertility and Development* **4** 301–305

Grazzini E, Guillon G, Mouillac B and Zingg HH (1998) Inhibition of oxytocin receptor function by direct binding of progesterone *Nature* **392** 509–512

Gyawa P and Pope GS (1992) Oestradiol-17β in the milk of cows from 6 days before to 14 days after their insemination *British Veterinary Journal* **148** 459–461

Helmer SD, Hansen PJ, Thatcher WW, Johnson JW and Bazer FW (1989) Intrauterine infusion of highly enriched bovine trophoblast protein-1 complex exerts an antiluteolytic effect to extend corpus luteum life span in cyclic cattle *Journal of Reproduction and Fertility* **87** 89–101

Henricks DM, Lamond DR, Hill JR and Dickey JF (1971) Plasma progesterone concentrations before mating and in early pregnancy in the beef heifer *Journal of Animal Science* **33** 450–454

Hixon JE and Flint APF (1987) Effects of a luteolytic dose of oestradiol benzoate on uterine oxytocin receptor concentrations, phosphoinositide turnover and prostaglandin $F_{2\alpha}$ secretion in sheep *Journal of Reproduction and Fertility* **79** 457–467

Imakawa K, Helmer SD, Nephew KP, Meka CSR and Christenson RK (1993) A novel role for GM-CSF: enhancement of pregnancy specific interferon production, ovine trophoblast protein-1 *Endocrinology* **132** 1869–1871

Imakawa K, Tamura K, McGuire WJ, Khan S, Harbison LA, Stanga JP, Helmer SD and Christenson RK (1995) Effect of interleukin-3 on ovine trophoblast interferon during early conceptus development *Endocrine* **3** 511–517

Jenkin G (1992) Oxytocin and prostaglandin interactions in pregnancy and at parturition *Journal of Reproduction and Fertility Supplement* **45** 97–111

Kerbler TL, Buhr MM, Jordan LT, Leslie KE and Walton JS (1997) Relationship between maternal plasma progesterone concentration and interferon-τ synthesis by the conceptus in cattle *Theriogenology* **47** 703–714

Knickerbocker JJ, Thatcher WW, Foster DB, Wolfenson D, Bartol FF and Caton D (1986) Uterine prostaglandin and blood flow responses to estradiol-17β in cyclic cattle *Prostaglandins* **31** 757–775

Ko Y, Lee CY, Ott TL, Davis MA, Simmen RCM, Bazer FW and Simmen FA (1991) Insulin-like growth factors in sheep uterine fluids: concentrations and relationship to ovine trophoblast protein-1 production during early pregnancy *Biology of Reproduction* **45** 135–142

Kotwica J, Skarzynski D, Bogacki M, Merlin P and Starostka B (1997) The use of an oxytocin antagonist to study the function of ovarian oxytocin during luteolysis in cattle *Theriogenology* **48** 1287–1299

Lamming GE and Darwash AO (1995) Effect of inter-luteal interval on subsequent luteal phase length and fertility in post partum dairy cows *Biology of Reproduction* **53** (Supplement 1) Abstract 63

Lamming GE and Mann GE (1995) Control of endometrial oxytocin receptors and prostaglandin $F_{2\alpha}$ responses to oxytocin in ovariectomized cows by progesterone and estradiol *Journal of Reproduction and Fertility* **103** 69–73

Lamming GE, Darwash AO and Back HL (1989) Corpus luteum function in dairy cows and embryo mortality *Journal of Reproduction and Fertility Supplement* **37** 245–252

Lamming GE, Wathes DC, Flint APF, Payne JH, Stevenson KR and Vallet JL (1995) Local action of trophoblast interferon in suppression of the development of oxytocin and oestradiol receptors in ovine endometrium *Journal of Reproduction and Fertility* **105** 165–175

Lau TM, Gow GB and Fairclough RJ (1992) Differential effects of progesterone treatment on the oxytocin-induced prostaglandin $F_{2\alpha}$ response and the levels of endometrial oxytocin receptors in ovariectomised ewes *Biology of Reproduction* **46** 17–22

Lawson RAS and Cahill LP (1983) Modification of embryo maternal relationships in ewes by progesterone treatment early in the oestrous cycle *Journal of Reproduction and Fertility* **67** 473–475

Lukaszewska J and Hansel W (1980) Corpus luteum maintenance during early pregnancy in the cow *Journal of Reproduction and Fertility* **59** 485–493

McCracken JA, Schramm W and Okulicz WC (1984) Hormone receptor control of pulsatile secretion of $PGF_{2\alpha}$ from the uterus and its abrogation in early pregnancy *Animal Reproduction Science* **7** 31–55

Mann GE and Lamming GE (1993) Monitoring endometrial oxytocin receptor development in the cow using a biopsy technique *Journal of Reproduction and Fertility Abstract Series* **11** Abstract 172

Mann GE and Lamming GE (1994) Use of repeated biopsies to monitor endometrial oxytocin receptors in cows *Veterinary Record* **135** 403–405

Mann GE and Lamming GE (1995a) Progesterone inhibition of the development of the luteolytic signal in the cow *Journal of Reproduction and Fertility* **104** 1–5

Mann GE and Lamming GE (1995b) Effect of the level of oestradiol on oxytocin-induced prostaglandin $F_{2\alpha}$ release in the cow *Journal of Endocrinology* **145** 175–180

Mann GE, Lamming GE and Fray MD (1995) Plasma oestradiol during early pregnancy in the cow and the effects of treatment with buserelin *Animal Reproduction Science* **37** 121–131

Mann GE, Mann SJ and Lamming GE (1996) The inter-relationship between the maternal hormone environment and the embryo during the early stages of pregnancy. *Journal of Reproduction and Fertility Abstract Series* **17** Abstract 55

Mann GE, Lamming GE and Fisher PA (1998) Progesterone control of embryonic interferon-τ production during early pregnancy in the cow *Journal of Reproduction and Fertility Abstract Series* **21** Abstract 37

Martal J, Chene N, Camous S, Huynh L, Lantier F, Hermier P, L'Haridon R, Charpigny G, Charlier M and Chaouat G (1997) Recent developments and potentialities for reducing embryo mortality in ruminants: the role of IFN-τ and other cytokines in early pregnancy *Reproduction Fertility and Development* **9** 355–380

Meyer HHD, Mittermeier T and Schams D (1988) Dynamics of oxytocin, estrogen and progestin receptors in the bovine endometrium during the estrous cycle *Acta Endocrinologica* **118** 96–104

Meyer MD, Hansen PJ, Thatcher WW, Drost M, Badinga L, Roberts RM, Li J, Ott TL and Bazer FW (1995) Extension of corpus luteum life span and reduction of uterine secretion of prostaglandin $F_{2\alpha}$ of cows in response to recombinant interferon - τ *Journal of Dairy Science* **78** 1921–1931

Mirando MA, Willard CB and Whiteaker SS (1993) Relationships among endometrial oxytocin receptors, oxytocin stimulated phosphinosotide hydrolysis and prostaglandin $F_{2\alpha}$ secretion *in vitro*, and plasma concentrations of ovarian steroids before and during corpus luteum regression in cyclic heifers *Biology of Reproduction* **48** 874–882

Moor RM and Rowson LEA (1966) The corpus luteum of the sheep: effect of the removal of embryos on luteal function *Journal of Endocrinology* **34** 497–502

Nephew KP, McClure KE, Ott T, Budois DH, Bazer FW and Pope WF (1991) Relationship between variation in conceptus development and differences in estrous cycle duration in ewes *Biology of Reproduction* **44** 536–539

Northey DL and French LR (1980) Effect of embryo removal and intrauterine infusion of embryonic homogenates on the life span of the bovine corpus luteum *Journal of Animal Science* **50** 298–302

Payne JH and Lamming GE (1994) The direct influence of the embryo on uterine $PGF_{2\alpha}$ and PGE_2 production in sheep *Journal of Reproduction and Fertility* **101** 737–741

Payne JH, Mann GE and Lamming GE (1994) Progesterone action in ovariectomised ewes passively immunised against oestradiol *Journal of Reproduction and Fertility Abstract Series* **14** Abstract 70

Plante C, Hansen PJ, Martinod S, Siegenthaler B, Thatcher WW Pollard JW and Leslie MV (1989) Effect of intramuscular administration of interferon-$_{\alpha 1}$ on luteal life span in cattle *Journal of Dairy Science* **72** 1859–1865

Plante C, Thatcher WW and Hansen PJ (1991) Alteration of oestrous cycle length, ovarian function and oxytocin–induced release of prostaglandin $F_{2\alpha}$ by intrauterine and intramuscular administration of recombinant interferon–α to cows *Journal of Reproduction and Fertility* **93** 375–384

Pritchard JY, Schrick FN and Inskeep EK (1994) Relationship of pregnancy rate to peripheral concentrations of progesterone and oestradiol in beef cows *Theriogenology* **42** 247–259

Robinson RS, Mann GE, Lamming GE and Wathes DC (1998a) The effect of pregnancy on the expression of uterine oxytocin, oestrogen and progesterone receptors during early pregnancy in the cow *Journal of Endocrinology* **160** 21–33

Robinson RS, Mann GE, Lamming GE and Wathes DC (1998b) Oxytocin, oestrogen and progesterone receptor mRNA expression in the bovine endometrium throughout the oestrous cycle *Journal of Reproduction and Fertility Abstract Series* **21** Abstract 97

Rowson LEA and Moor RM (1967) The influence of embryonic tissue homogenate infused into the uterus and life-span of the corpus luteum in the sheep *Journal of Reproduction and Fertility* **13** 511–516

Spencer TE and Bazer FW (1995) Temporal and spatial alterations in uterine estrogen receptor and progesterone receptor gene expression during the estrous cycle and early pregnancy in the ewe *Biology of Reproduction* **53** 1527–1543

Spencer TE, Becker WC, George P, Mirando MA, Ogle TF and Bazer FW (1995) Ovine interferon-τ (IFN-τ) inhibits estrogen receptor up-regulation and estrogen-induced luteolysis in cyclic ewes *Endocrinology* **136** 4932–4944

Spencer TE, Ott TL and Bazer FW (1996) τ - interferon: pregnancy recognition signal in ruminants *Proceedings of Experimental Biology and Medicine* **213** 215–229

Thatcher WW, Meyer MD and Danet-Desnoyers G (1995) Maternal recognition of pregnancy *Journal of Reproduction and Fertility Supplement* **49** 15–28

Vincent DL, Meredith S and Inskeep EK (1986) Advancement of uterine secretion of prostaglandin E2 by treatment with progesterone and transfer of asynchronous embryos *Endocrinology* **119** 527–529

Wathes DC and Lamming GE (1995) The oxytocin receptor, luteolysis and the maintenance of pregnancy *Journal of Reproduction and Fertility Supplement* **49** 53–67

Journal of Reproduction and Fertility Supplement **54**, 329–339

Mechanism of action of interferon-tau in the uterus during early pregnancy

T. R. Hansen[1], K. J. Austin[1], D. J. Perry[1], J. K. Pru[1], M. G. Teixeira[2] and G. A. Johnson[1†]

[1]*Department of Animal Science and* [2]*School of Pharmacy, Reproductive Biology Program, University of Wyoming, Laramie, WY 82071, USA*

Early pregnancy is maintained in ruminants through the actions of conceptus-derived interferon (IFN)-tau on the endometrium. IFN-tau alters uterine release of $PGF_{2\alpha}$, which results in rescue of the corpus luteum and continued release of progesterone. The mechanism of action of IFN-tau includes inhibition of oestradiol receptors, consequent reduction in oxytocin receptors, activation of a cyclooxygenase inhibitor, and a shift in the PGs to favour PGE_2 over $PGF_{2\alpha}$. IFN-tau also induces several endometrial proteins that may be critical for survival of the developing embryo. One endometrial protein induced by pregnancy and IFN-tau has been identified as bovine granulocyte chemotactic protein-2 (bGCP-2). This chemotactic cytokine (chemokine) has been used as a marker to delineate IFN-tau from IFN-alpha responses in the endometrium. A second protein, called ubiquitin cross-reactive protein (UCRP), resembles a tandem ubiquitin repeat. UCRP becomes conjugated to cytosolic endometrial proteins in response to IFN-tau and pregnancy. Proteins conjugated to UCRP are either modulated or targeted for processing through the proteasome. The action of IFN-tau is mediated by induction of signal transducer and activator of transcription 1 (STAT-1), STAT-2 and interferon regulatory factor 1 (IRF-1) transcription factors. Induction of these transcription factors, the alpha chemokines and UCRP is the prelude to maternal recognition of pregnancy in ruminants.

Introduction

The mechanism through which interferon-tau (IFN-tau) alters release of the prostaglandins (PGs) during early pregnancy in cows and ewes will be discussed in this review. However, the primary focus will be description of induction of endometrial bovine granulocyte chemotactic protein-2 (bGCP-2) and ubiquitin crossreactive protein (bUCRP) by pregnancy and IFN-tau. Bovine UCRP is induced by IFN-alpha and IFN-tau, whereas bGCP-2 is induced only by IFN-tau. The mechanism of IFN-tau action includes induction of signal transducers and activators of transcription (STAT) and IFN regulatory factor 1(IRF-1) in the bovine endometrium. However, IFN-tau-specific induction of GCP-2 may involve a variant receptor subunit that is associated with the janus kinase (Jak)/STAT/IRF, as well as protein kinase C signal transduction pathways. An overview of function of ubiquitin and human IFN-stimulated gene product 15 (huISG15) is provided as a basis for understanding the function of bUCRP in conjugating to and regulating endometrial proteins during early pregnancy in cows.

Luteolysis

The oxytocin receptor is a seven transmembrane, G-protein-associated receptor that induces inositol triphosphate turnover, cytosolic calcium and protein kinase C (reviewed in Flint *et al.*, 1995). In the

[†]Present address: Department of Animal Science, Texas A & M University, College Station TX 77843.

uterus, these events result in activation of cyclooxygenase 2 (COX-2) which is the inducible form of PG synthase (Asselin *et al.*, 1997a). The resulting increase in $PGF_{2\alpha}$ causes regression of the corpus luteum. Destruction of the corpus luteum by $PGF_{2\alpha}$ results in a decline in circulating progesterone. This decline in progesterone primes the hypothalamus and pituitary to release gonadotrophins that induce ovulation and initiate a new oestrous cycle.

It is currently accepted that the oxytocin receptor becomes uncoupled from associated signal transduction pathways during maternal recognition of pregnancy (reviewed in Bazer *et al.*, 1997). The result is an attenuation of release of $PGF_{2\alpha}$, rescue of the corpus luteum, continued release of progesterone, and uterine support of the developing embryo. The hormone or cytokine implicated in uncoupling the oxytocin receptor during pregnancy in ruminants is conceptus-derived IFN-tau (reviewed in Thatcher *et al.* 1995; Bazer *et al.*, 1997).

Interferon-tau

Interferon-tau, originally called trophoblastin or trophoblast protein-1, was identified as dominant polypeptides that were released by the conceptus on days 16–24 of pregnancy (reviewed in Roberts *et al.*, 1990). Infusion of these trophoblast proteins into uteri of non-pregnant cows caused an extension of the oestrous cycle that was mediated by the attenuation of release of $PGF_{2\alpha}$. Isolation of the cDNA encoding the bovine trophoblast polypeptide and inferred amino acid sequence revealed 45–70% identity with type I IFNs (α, β, ω) (Imakawa *et al.*, 1990). The trophoblast IFNs were subclassified as IFN-tau in 1990 (Roberts *et al.*) because of differences in cDNA sequence (Imakawa *et al.*, 1990), number of genes (Hansen *et al.*, 1991), binding to receptors (Hansen *et al.*, 1989), organization of promoters on genes (Hansen *et al.*, 1991), trophoblast-specific expression (Farin *et al.*, 1990), and distribution of genes across species.

Interferon Receptor and Signal Transduction

Interferon-tau binds to type I IFN receptors that are present in the uterine endometrium (Hansen *et al.*, 1989; Han *et al.*, 1997). The subunits of IFN-alpha receptors have been named IFNAR1 and IFNAR2, and co-expression is necessary for functional high affinity IFN binding and signal transduction (Domanski *et al.*, 1995). The extracellular region of IFNAR1 consists of two distinct 200 amino acid domains (D200).

Janus kinase-1 (JAK-1)and tyrosine kinase-2 (Tyk-2) constitutively associate with IFN receptors. Tyrosine kinase-2 associates with the cytoplasmic domains of both IFNAR1 and IFNAR2 (Colamonici *et al.*, 1994), while Jak-1 associates preferentially with IFNAR2 (Novick *et al.*, 1994). It is believed that these tyrosine kinases directly phosphorylate STAT proteins, although other kinases may be involved (Darnell, 1997). STATs undergo tyrosine phosphorylation, assemble into multimeric complexes, of which IFN-stimulated gene factor 3 (ISGF3) is an example, and translocate to the cell nucleus. ISGF3 is composed of a 48 kDa DNA-binding protein termed p48, STAT-1a (p91), STAT-1b (p84), and STAT-2 (p113) (Kessler *et al.*, 1990). After transport to the nucleus, ISGF3 specifically recognizes the IFN-stimulated response element (ISRE) DNA sequence (*Levy et al.*, 1989). Although a consensus ISRE sequence has been defined, there is considerable sequence heterogeneity. Together, differential activation and combinatorial association of these transcription factors regulate IFN-stimulated genes (ISGs).

IFN-tau and the Oxytocin Receptor

Regulation of the oxytocin receptor is an essential event in the mechanism of action of IFN-tau. In pregnant ewes, it is likely that IFN-tau attenuates release of $PGF_{2\alpha}$ by inhibiting formation of oxytocin receptors (Vallet *et al.*, 1990). Progesterone, oestrogen, oxytocin, and their respective

```
                 10        20         30         40         50         60         70    75
bGCP-2 VAAVVRELR CVCLTTTPG IHPKTVSDLQ VIAAGPQCSK VEVIATLKNG REVCLDPEAP LIKKIVQKIL DSGKN
8K-9      (100%)                  TVSDLQ VIAAGPQ
8K-13     ( 92%)                                                      EVCLxPEAP LIK

mMIP-2    AVVASEL RCQCLKTLPG IDLKNIQSLS VTPPGPHCAQ TEVIATLKGG QKVCLDPEAP LVQKIIQKIL NKGKAN
8K-10     (88%)                              TPPGPHsgQ TEVIATL
8K-12     (67%)      eCLeTLeG IhLK
```

Fig. 1. Amino acid sequence of peptides derived from the purified uterine 8 kDa protein and identity with the alpha chemokines: bovine granulocyte chemotactic protein 2 (bGCP-2) and murine macrophage inflammatory protein 2 (mMIP-2). The 8 kDa protein was purified and digested to yield peptides (8K-9, 8K-13, 8K-10 and 8K-12) that were partially sequenced. Peptide amino acids and identities (in parentheses) are shown underneath amino acid sequences for bGCP-2 and mMIP-2. Amino acid sequences have been reported previously for the 8K-9 and 8K-13 peptides (Teixeira *et al.*, 1997).

receptors are involved in this process. The theory is that progesterone blocks oestrogen-induced oxytocin receptor synthesis immediately before luteolysis (McCracken *et al.*, 1984). Interferon-tau continues to inhibit oestrogen receptors and oxytocin receptors during early pregnancy, resulting in attenuation of pulsatile $PGF_{2\alpha}$ release. In sheep, IFN-tau suppresses oestrogen receptor gene expression in the endometrial epithelium (Spencer and Bazer, 1996), which in turn inhibits oxytocin receptor formation and oxytocin-induced signal transduction through inositol triphosphate turnover (Mirando *et al.*, 1990; Ott *et al.*, 1992). These effects on the oxytocin receptor, coupled with activation of a PG synthase inhibitor in cows (Thatcher *et al.*, 1995), result in attenuation of luteolytic $PGF_{2\alpha}$ pulses. In addition to inhibiting $PGF_{2\alpha}$ release, IFN-tau also might stimulate release of PGE_2 at the expense of $PGF_{2\alpha}$ in cows (Asselin *et al.*, 1997b). This concept is intriguing in the context of the luteotrophic effects of PGE_2 and in regard to how synthesis of PGs might be regulated by IFN-tau during early pregnancy. IFN-tau might induce degradation or activation of an enzyme downstream from COX-2 (i.e., PGE_2-9-ketoreductase) that favours synthesis of PGE_2 rather than $PGF_{2\alpha}$ in endometrial epithelial cells. It also might activate transcription factors that inhibit, or inhibit transcription factors that activate, the oestrogen receptor gene.

IFN-tau and Uterine Proteins

Uterine chemokines

IFN-tau induces 8 kDa uterine proteins during early pregnancy in cows (Rueda *et al.*, 1993). Amino acid sequencing of four internal peptides derived from the 8 kDa uterine proteins revealed identity with the alpha chemokine family: 92–100% identity with bovine bGCP-2 (Teixeira *et al.*, 1997), and 67–88% identity with murine (m) macrophage inflammatory protein-2 (mMIP-2; Fig. 1). Antiserum against bGCP-2 peptide was generated and used in western blot studies to describe release of bGCP-2 by the endometrium during early pregnancy and in response to conceptus-derived IFN-tau (Fig. 2). Bovine GCP-2 was not released by endometrium representing the oestrous cycle or in response to the closely related IFN-alpha unless high concentrations were used (Fig. 2b; Staggs *et al.*, 1998). This unique induction of bGCP-2 prompted the hypothesis that it could be used as an index of IFN-tau-specific signal transduction. Another interesting aspect was that phorbol ester mimicked the effects of IFN-tau in inducing the release of bGCP-2 by cultured bovine primary endometrial cells (BEND cells) (Staggs *et al.*, 1998). It was concluded from these experiments that IFN-tau signal transduction was complex in the bovine endometrium and might involve transcription factors other than or ancillary to those described for the Jak/STAT pathway. Type I IFNs compete for binding to a common cell surface receptor. The AR1 subunit of the IFN receptor undergoes ligand-dependent tyrosine phosphorylation. It is possible that AR1 undergoes a specific conformational change in response to IFN-tau to allow phosphorylation of transcription factors that are unique to IFN-tau-specific signal transduction.

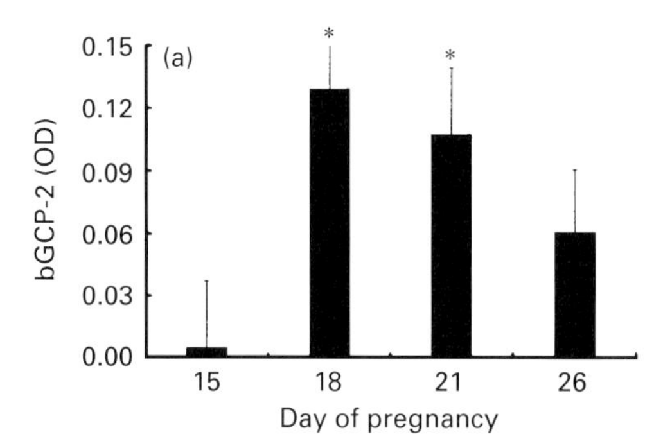

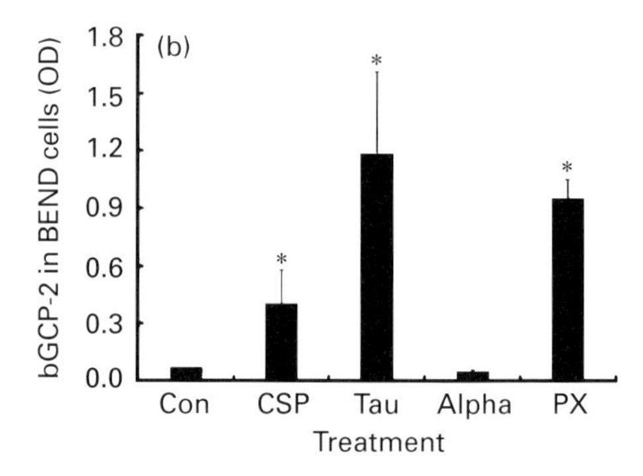

Fig. 2. Release of bovine granulocyte chemotactic protein 2 (bGCP-2) in response to pregnancy (a) and conceptus secretory protein (CSP; 200 μg ml^{-1}), IFN-tau (Tau; 25 nmol l^{-1}), IFN-alpha (Alpha; 25 nmol l^{-1}) and phorbol ester (PX; 100 ng ml^{-1}), Con: control (b). In (a), explants were collected from pregnant cows on days 15, 18, 21, or 26 of pregnancy (n = 3 cows per day) and cultured for 24 h as described by Austin *et al.* (1996a). In (b), bovine endometrial (BEND) cells (n = 3 replicates) were cultured for 24 h with respective treatments as described by Staggs *et al.* (1998). Bovine GCP-2 was detected using one-dimensional PAGE and western blot analysis (anti-boGCP-2 peptide antiserum; Teixeira *et al.*, 1997). Western blots were scanned using densitometry. *Optical density (OD) means were significantly different ($P < 0.05$) when compared with day 15 of pregnancy (a) or controls (b). Data in (b) were redrawn from Staggs *et al.* (1998) with permission from Biology of Reproduction.

Several uterine proteins are induced by IFN-tau, but only a few of these proteins have been identified. For example, the interferon-induced protein Mx (Charleston and Stewart, 1993), 2′, 5′ oligoadenylate synthase (Short *et al.*, 1991), and β_2-microglobin (Vallet *et al.*, 1991) are induced by IFN-tau. However, the specific function of these proteins during early pregnancy remains elusive. The chemokines, although specific markers for IFN-tau action, also belong to this growing category of identified proteins with no known specific function during early pregnancy.

Pregnancy is dependent upon a receptive uterus that allows attachment and invasion of the conceptus while preventing immunological rejection. Chemokines are potent chemo-attractants for cells of the immune system and have been implicated in cell adhesion, inflammatory, and angiogenic processes (reviewed in Oppenheim *et al.*, 1991). The chemokines may attract the conceptus or cells of the immune system to implantation sites. They also may direct an immunostimulatory phenotype that, coupled with scavenger effects (macrophage, neutrophils), controls inflammation and adhesion associated with limited invasion of the maternal caruncles by the conceptus. Available recombinant bovine uterine chemokines would facilitate experiments designed to examine these hypotheses.

Ubiquitin crossreactive protein

A 17 kDa uterine protein was identified through examining proteins released into medium of cultured endometrial explants from day 18 pregnant cows (Naivar *et al.* 1995). This protein was released by endometrial explants in response to both recombinant (r)bIFN-tau and rbIFN-alpha. Because the 17 kDa protein was similar in size to huISG15 (also called huUCRP), and immunoreacted with antiserum against ubiquitin, it was called bUCRP (Austin *et al.*, 1996a). Antiserum against ubiquitin was used to demonstrate that bUCRP was released by the endometrium at times coincident with IFN-tau release from the conceptus. Bovine UCRP was first detected in the medium of cultured endometrial explants representing day 15 of pregnancy. Release of bUCRP increased on day 18 and remained high through day 26 of pregnancy. Recombinant bIFN-tau induced release of bUCRP by endometrial explants representing day 12 of the oestrous cycle in a dose-dependent manner. Likewise, bUCRP was found in significant amounts in uterine flushings from day 18 pregnant cows.

IFN induces release of ISG15 by mouse Ehrlich ascites tumour cells (Farrell *et al.*, 1979). In addition, ISG15 is released by several cell lines in response to IFN and during the acquisition of an antiviral state. Human ISG15 is released in response to type I IFNs and it induces release of IFN-gamma by T- and B-lymphocytes (Recht *et al.*, 1991). Through inducing release of IFN-gamma from T-cells, ISG15 may augment natural killer cell proliferation and activate monocytes and macrophages.

Endometrial UCRP may be the bovine counterpart to huISG15. Bovine UCRP is released by the endometrium in response to IFN-tau during early pregnancy. Whether bUCRP induces IFN-gamma remains to be determined. The activation of natural killer cells by bUCRP in the endometrium is intriguing, as these effects might be considered detrimental to the developing conceptus (Robertson *et al.*, 1994). However, an extensive description of cell-mediated and humoral cytokines, and consequences of action in the bovine endometrium during early pregnancy is lacking. Bovine UCRP has limited amino acid sequence identity with ubiquitin and huISG15, but retains the C-terminal amino acids, Arg-Gly-Gly (RGG) that have been implicated in the first step in covalent conjugation to cytosolic proteins (Austin *et al.*, 1996b). Because bUCRP retains functional amino acids of ubiquitin and is released from endometrial explants in response to IFN-tau during early pregnancy, it probably has cytosolic and extracellular roles that are consistent with the maintenance of pregnancy.

Ubiquitination. Proteolysis is an important intracellular regulatory mechanism for controlling many biological processes. In eukaryotes, ubiquitin-dependent proteolysis is essential for protein turnover. Ubiquitin is a highly conserved 76 amino acid polypeptide that is involved in a wide array of biological systems including cell cycle regulation, DNA repair, receptor modification, signal transduction, antigen presentation, the stress response, and both proteasomal and non-proteasomal protein degradation (Finley and Chau, 1991). Conjugation of ubiquitin to proteins can result in either polyubiquitination or mono-ubiquitination. Polyubiquitination is believed to target proteins for rapid degradation (Finley and Chau, 1991): for example, it may be involved in the process of luteal regression (Murdoch *et al.*, 1996). Carboxyl terminal amino acids on ubiquitin, via the RGG sequence, link to primary amines on targeted proteins (Hershko and Ciechanover, 1992). Packed multi-ubiquitin chains are then added through isopeptide bonds between carboxyl terminals and Lys-48 residues on successive ubiquitin monomers. This quaternary structure is unique and aids in directing the complex to the proteasome where the targeted protein is degraded and ubiquitin monomers are released and recycled (Finley and Chau, 1991).

Mono-ubiquitinated proteins, such as actin, histones, microtubules and the growth hormone receptor are longer-lived and probably undergo ubiquitin protein modification as well as modest rates of targeted degradation (reviewed in Finley and Chau, 1991). A case in point is the transcription factor nuclear factor kappa B (NF-κB), for which the proteasome proteolytic pathway is required for proper activation (Palombella *et al.*, 1994). This transcription factor regulates a variety of genes involved in immune system function such as those encoding immunoglobulin kappa light chains, interleukin 2 (IL-2) receptor alpha chain, class I major histocompatibility complex (MHC), and a number of cytokines that include IL-2, IL-6, granulocyte–macrophage colony-stimulating factor (GM-CSF), and IFN-beta. NF-κB also has been implicated in the expression of cell adhesion genes that encode selectin, intercellular adhesion molecule 1 and viral cellular adhesion molecule 1 (Palombella *et al.*, 1994).

Unique primary structure of bUCRP. Human ISG15 has 29–31% amino acid sequence identity with two tandem ubiquitin monomers. Inferred amino acid sequence of the bUCRP cDNA and gene have 68% identity with huISG15 and 31% identity with a tandem ubiquitin repeat (Austin *et al.*, 1996b). Only one major bUCRP gene exists (Perry *et al.*, 1997). Thus, it appears likely that bUCRP is the counterpart to huISG15. However, there are significant differences in primary structure between huISG15 and bUCRP. The huISG15 C-terminal amino acids are removed via post-translational processing to yield a 15 kDa protein that ends in RGG (Feltham *et al.*, 1989). The bUCRP cDNA encodes a 17 kDa protein in which translation stops immediately after RGG amino acids (Austin *et al.*, 1996b). The coding region of the bUCRP gene is identical to that reported for the cDNA. Functional residues of ubiquitin that are retained in bUCRP include the C-terminal RGG residues and Arg-72 of ubiquitin that have been retained as Arg-150 in bUCRP. This residue is involved with binding to ubiquitin conjugating enzyme (E1), and is required for ordered addition of substrate in a complex ligation pathway (Burch and Haas, 1994). Other ubiquitin residues implicated in binding to targeted proteins (i.e., Arg54), degradation through the 26 S proteasome (i.e. His68) (Ecker *et al.*, 1987), and polymerization of monomers (Lys48) (Finley and Chau, 1991) are absent in bUCRP.

Unique cytosolic role of bUCRP. It was hypothesized that bUCRP has a cytosolic role in the modification of uterine proteins during early pregnancy. In a manner similar to ubiquitin, bUCRP may become conjugated to cytosolic endometrial proteins. Because bUCRP lacked residues known to form ubiquitin polymers and to target proteins for degradation, it was suspected that the array of proteins conjugated to bUCRP, and the destiny of these proteins might be different from those conjugated to ubiquitin.

Antiserum was developed against a bUCRP peptide that immunoreacted with bUCRP, but did not immunoreact with ubiquitin on western blots. This antibody also detected bUCRP when conjugated to an array of endometrial proteins (Johnson *et al.*, 1998). These complexes with bUCRP were called conjugates and represented proteins coupled to a single bUCRP molecule (monomer), or to polymers of bUCRP. Co-incubation of antibody against bUCRP with antigenic bUCRP peptide blocked detection of bUCRP and its conjugates. Pre-absorption of bUCRP antibody with ubiquitin had no effect on detection of bUCRP or bUCRP conjugates. Thus, detection of bUCRP and its conjugates by the anti-bUCRP antibody was immunospecific.

Ubiquitin and its conjugates are abundant in the bovine endometrium and are not regulated by IFN-tau (Johnson *et al.*, 1998). Because bUCRP and its conjugates were induced by pregnancy and IFN-tau (Fig. 3a), it was concluded that proteins conjugated to bUCRP were distinct from those conjugated to ubiquitin. The induction of bUCRP and its conjugates in endometrial explants required exposure to rbIFN-tau for 12 h. This appearance of bUCRP and its conjugates was consistent with the time required for transcription, translation, and conjugation of bUCRP to cytosolic proteins.

Phosphorylated STATs become ubiquitinated and degraded within minutes after treatment with IFN-gamma (Kim and Maniatis, 1996). This process is so fast that use of proteasome inhibitors was required to observe an accumulation of ubiquitinated STAT-1. Interestingly, proteins conjugated to bUCRP continued to accumulate through 48 h after treatment with rbIFN-tau (Johnson *et al.*, 1998). This accumulation of proteins that were conjugated to bUCRP might reflect increased protein modulation or stabilization as opposed to degradation. It was inferred from these experiments that regulation or degradation of endometrial proteins by bUCRP was a component of establishing early pregnancy in ruminants. For example, bUCRP might ligate to and modify, or initiate proteasomal degradation of cytosolic uterine proteins that are involved with the release of $PGF_{2\alpha}$.

IFN-tau may induce conjugation of bUCRP to the oxytocin receptor so that it becomes unable to bind to oxytocin during early pregnancy in cows (Fig. 4). Alternatively, bUCRP may conjugate to the oestrogen receptor, or to transcription factors that induce transcription of the oestrogen receptor gene and target these proteins to degradation. Likewise, enzymes involved with the synthesis of PGs may be modified by bUCRP such that $PGF_{2\alpha}$ is attenuated and PGE_2 is stimulated. In addition, IFN-tau may dictate the selective proteolysis or modulation of uterine cytokines by bUCRP to accommodate the changing immune demands of pregnancy. The identification of proteins that are conjugated to bUCRP during early pregnancy and in response to IFN-tau would address these hypothesized roles of bUCRP.

Induction of bUCRP mRNA in the endometrium by IFN-tau. Another major difference between huISG15 and bUCRP is the apparent widespread expression of the huISG15 gene in several cell lines and tissues (Lowe *et al.*, 1995). Northern blot analysis revealed that expression of the bUCRP gene was

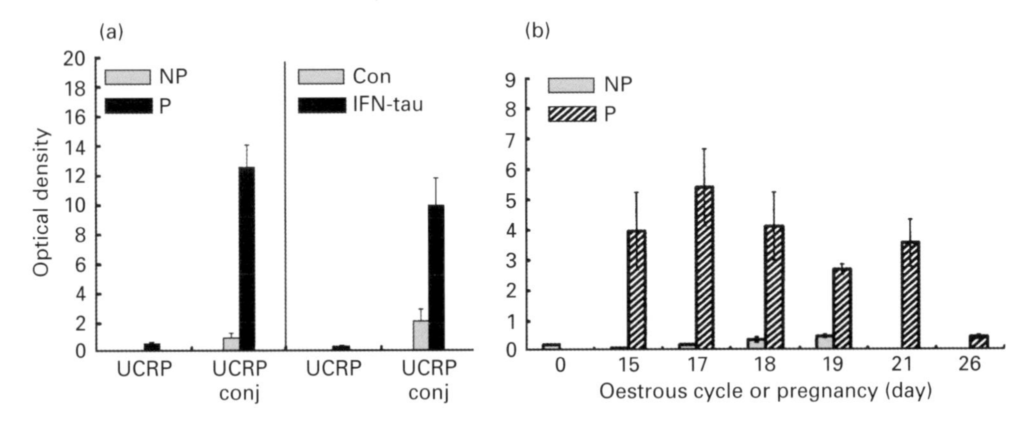

Fig. 3. Induction of bovine ubiquitin crossreactive protein (bUCRP) in endometrial explants. (a) Induction of bUCRP and its conjugates by pregnancy and IFN-tau. Explants from day 18 pregnant (P; n = 4) or non-pregnant (NP, n = 4) cows were cultured for 24 h. Cytosolic proteins were separated using one-dimensional PAGE and analysed for the presence of bUCRP and its conjugates using western blotting as described by Johnson *et al.* (1998). Endometrial explants from day 14 non-pregnant cows (n = 3) were cultured in the absence (Con) or presence (25 nmol l^{-1}) of rbIFN-tau for 24 h. Bovine UCRP and its conjugates were detected as described in Johnson *et al.* (1998). Data in (a) are adapted from Johnson *et al.* (1998) with permission from Biology of Reproduction. (b) Endometrium was collected from non-pregnant and pregnant cows on the days specified (n = 4 cows on each day). RNA was isolated, northern blotted, hybridized with radiolabelled bUCRP cDNA and quantitated as described in Hansen *et al.* (1997); (b) is reprinted from Hansen *et al.* (1997) with permission from The Endocrine Society. *Optical density means were significantly different ($P < 0.05$) from NP or Con.

restricted to endometria from pregnant cows or from endometria from non-pregnant cows that had been treated *in vitro* with rbIFN-tau (Hansen *et al.*, 1997).

Bovine UCRP mRNA was not detected using northern blot analysis in endometrium from non-pregnant cows (days 15, 17, 18 or 19) or in spleen, kidney, liver, corpus luteum or muscle (Hansen *et al.*, 1997). The bUCRP cDNA did not hybridize to ubiquitin mRNAs. Both major ubiquitin transcripts UbB (1.2 kb) and UbC (2.5 kb) were present in all tissues examined and were not regulated by IFN-tau. Bovine UCRP mRNA was detected in endometria from pregnant cows by day 15, reached highest values by day 17, remained high on days 18, 19 and 21, and then declined to amounts that were not detectable on day 26 (Fig. 3b). This pattern of transcription closely paralleled detection of UCRP protein in cultured endometrial explants from cows on the same days of pregnancy (Austin *et al.*, 1996a). The bUCRP mRNA transcript (~700 bp) can be induced in bovine endometrial cells (i.e. BEND cells) derived from non-pregnant cows when cultured with 25 nmol rbIFN-tau l^{-1} (Hansen *et al.*, 1997). Although the role of huISG15 may be general to many cell types, transcription of the bUCRP gene appears to be a specific uterine response to conceptus-derived IFN-tau during early pregnancy in cows.

The bovine UCRP gene. The huISG15 promoter contains a functional tandem ISRE at position –39 relative to the mRNA cap site (Reich *et al.*, 1987), but little is known about sequences that are upstream of this tandem ISRE. The TATA box (TATTAAA) in the bUCRP gene is located at position –31 from the mRNA cap site (Perry *et al.*, 1997). Both genes contain a single intron after the initiation codon, followed by coding, termination, 3′ non-coding, and polyadenylation sequences. The bUCRP gene promoter is similar to the huISG15 gene promoter in the placement of a conserved tandem ISRE at position –90. However, three additional putative ISREs are present in the bUCRP gene at positions –123, –332 and –525.

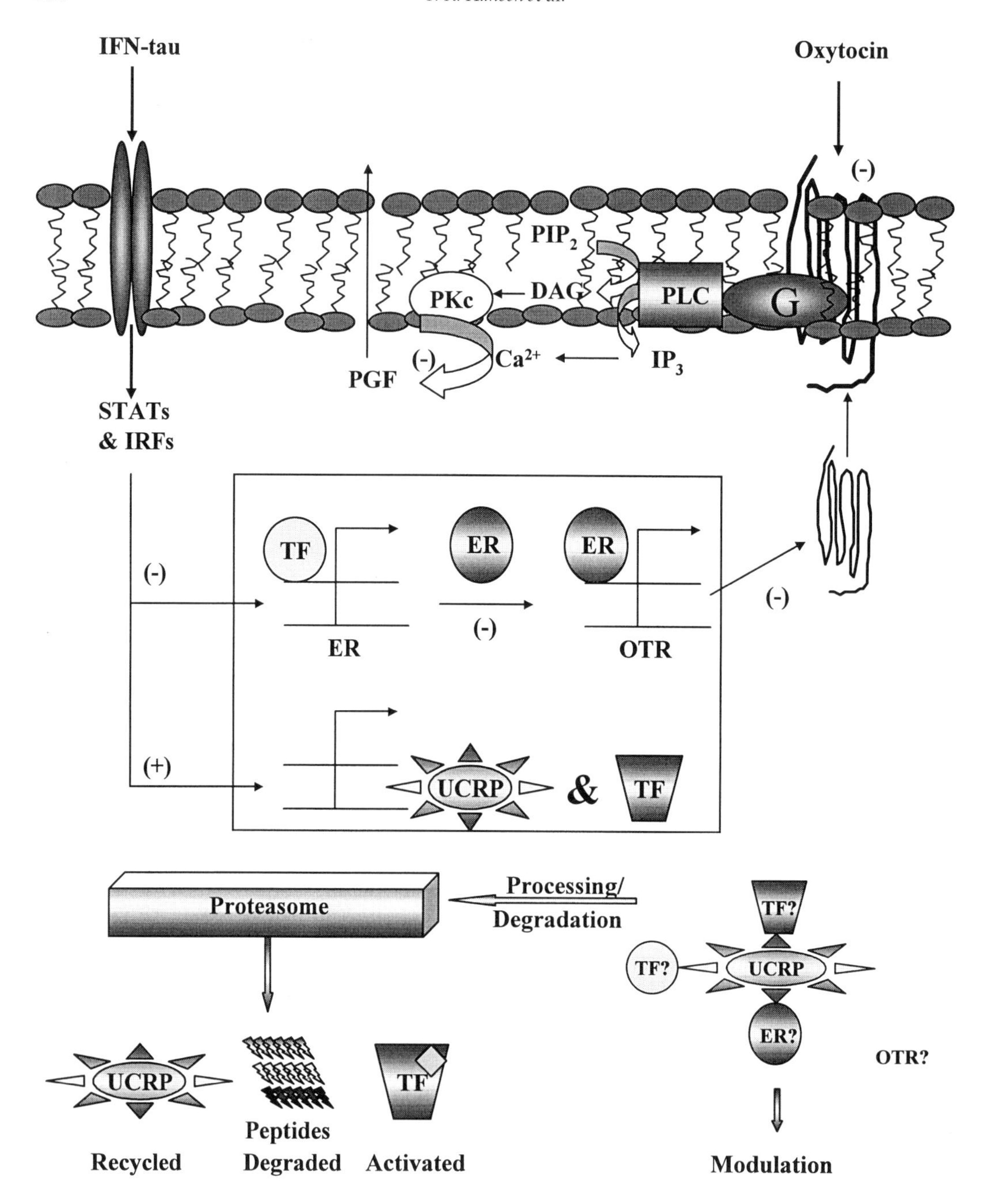

Fig. 4. Hypothetical mechanism of IFN-tau action in bovine uterine endometrium. IFN-tau binds to an endometrial receptor and induces phosphorylation of signal transducers and activators of transcription (STATs). STATs induce transcription and subsequent phosphorylation of interferon regulatory proteins (IRFs). The STATs and IRFs may have direct actions through inhibiting the oestradiol receptor (ER), and potentially, the oxytocin receptor (OTR) genes. Disruption of oxytocin receptor gene transcription results in reduced binding to oxytocin. Consequently, G proteins (G), phospholipase C (PLC), second messengers associated with activation of protein kinase C (PKc), and increased calcium become uncoupled resulting in diminished release of $PGF_{2\alpha}$. In addition to inhibiting transcription of the oestradiol receptor gene, IFN-tau, through inducing the STATs and IRF-1 also induces transcription of bovine ubiquitin crossreactive protein (bUCRP) and potentially transcription factor (TF)

Interferon receptors exhibit preference for ligand and sort cytosolic responses through tyrosine kinase phosphorylation of at least six STAT transcription factors. Nuclear proteins induced by rbIFN-tau in BEND cells interact with bUCRP ISRE DNA in mobility shift assay (Perry *et al.*, 1998). At least four slower migrating bands were induced by rbIFN-tau. IFN-tau induced appearance of STAT-1, STAT-2 and IRF-1 in nuclear proteins extracted from BEND cells (Perry *et al.*, 1998; Binelli *et al.*, 1998). The temporal relationship between appearance of the STATs (0.5 h) and IRF-1 (2 h) in BEND cells was consistent with induction of these nuclear proteins in other cells by other type I IFNs. The STATs are activated within minutes of IFN treatment. Induction of phosphorylation of cytoplasmic STATs initiates translocation to the nucleus, formation of transcriptional complexes and induction of secondary transcription factors called IRFs. The bUCRP gene may be induced by STATs and IRF-1. The intriguing and more complex issue raised by these experiments is how the STATs and IRF-1 regulate other genes in the endometrium during early pregnancy. Genes involved with the conjugation of bUCRP to targeted cytosolic proteins may be induced by these transcription factors. Likewise, genes encoding enzymes that shunt arachidonic acid to $PGF_{2\alpha}$ may be inhibited by these transcription factors.

Conclusions

The study of uterine responses to the conceptus during early pregnancy resulted in identification of several uterine proteins that may prepare the uterus for the implanting embryo. One critical event that must be circumvented or altered is uterine release of $PGF_{2\alpha}$ (Fig. 4). Inhibition of transcription of the oestrogen receptor gene is one critical action of IFN-tau in the modulation of $PGF_{2\alpha}$ release (Spencer and Bazer, 1996). Exactly how IFN-tau regulates the oestrogen receptor gene remains to be determined. It may involve direct interaction of the STATs or IRFs (Bazer *et al.*, 1997; Spencer et al., 1998) with negative acting response elements. Alternatively, it may involve removal of positive *trans*-acting transcription factors through conjugation to bUCRP and degradation through the proteasome. Likewise, enzymes involved in the synthesis of $PGF_{2\alpha}$ could be degraded through conjugation to bUCRP, whereas enzymes involved in the synthesis of PGE_2 could be activated or stabilized.

Alpha chemokines have only recently been identified in the uterine endometrium (Teixeira *et al.*, 1997). The function of these chemotactic cytokines in the uterus is unknown, but may involve preparing the endometrium for implantation or changing the immune environment to an environment that nurtures survival of the embryo. Several years ago it was suggested that the IFN-tau receptor was complex based on curvilinear Scatchard plot and cross-linking experiments (Hansen *et al.* 1989). Recent experiments, using cultured endometrial explants and BEND cells, revealed that the uterine endometrium can distinguish between type I IFNs (Staggs *et al.*, 1998). Phorbol ester and IFN-tau, but not the closely related IFN-alpha, induced release of bGCP-2. However, the IFN receptor cloned from ovine endometrium appears to be identical in nucleotide sequence to the IFN-alpha receptor (Han *et al.*, 1997). Because this IFN receptor was cloned using IFN-alpha receptor sequences, it may not be surprising that the receptor identified in ovine and bovine endometrium was identical. It remains unresolved whether there is a unique IFN-tau receptor subunit in the endometrium. Likewise it remains unknown if the IFNAR might change conformation after binding IFN-tau to expose a cytosolic domain that activates IFN-tau-specific STATs or IRFs. IFN-tau signal transduction appears to involve the STATs, and IRF-1. Future experiments using the bUCRP gene promoter and potentially the bGCP-2 gene promoter may delineate exactly which transcription factors are induced in response to IFN-tau. Certainly

---

genes. Bovine UCRP becomes conjugated to cytosolic proteins and either modulates these proteins or targets them to the proteasome where they are either degraded into peptides or activated. An example of activation is provided by the transcription factor, NF-kappa B that must be processed through the proteasome prior to regulating transcription of genes (Palombella *et al.*, 1994). DAG: diacyl glycerol; PIP_2: phosphoinositol-4,5 diphosphate.

development of the BEND cell line will facilitate these experiments as well as those planned to identify endometrial proteins that become conjugated to bUCRP in response to pregnancy and IFN-tau.

*This research was supported by grants from the National Institutes of Health (# HD 32475), United States Department of Agriculture (# 97-35203-4808), and University of Wyoming Agricultural Competitive Grants Program.

References

Asselin E, Drolet P and Fortier MA (1997a) Cellular mechanisms involved during oxytocin-induced prostaglandin $F_{2\alpha}$ production in endometrial epithelial cells *in vitro*: role of cyclooxygenase-2 *Endocrinology* **138** 4798–4805

Asselin E, Bazer FW and Fortier MA (1997b) Recombinant ovine and bovine interferons tau regulate prostaglandin production and oxytocin response in cultured bovine endometrial cells *Biology of Reproduction* **56** 402–408

Austin KJ, Ward SK, Teixeira MG, Dean VC, Moore DW and Hansen TR (1996a) Ubiquitin cross-reactive protein is released by the bovine uterus in response to interferon during early pregnancy *Biology of Reproduction* **54** 600–606

Austin KJ, Pru JK and Hansen TR (1996b) Complementary deoxyribonucleic acid sequence encoding bovine ubiquitin cross-reactive protein: a comparison with ubiquitin and a 15-kDa ubiquitin homolog *Endocrine* **5** 191–197

Bazer FW, Spencer TE and Ott TL (1997) Interferon tau: a novel pregnancy recognition signal *American Journal of Reproductive Immunology* **37** 412–420

Binelli M, Diaz T, Hansen TR and Thatcher WW (1998) Bovine interferon-tau induces phosphorylation and nuclear translocation of STAT-1, -2, and -3 in endometrial epithelial cells *Biology of Reproduction* (Supplement 1) 214 (Abstract)

Burch TJ and Haas AL (1994) Site-directed mutagenesis of ubiquitin. Differential roles for arginine in the interaction with ubiquitin-activating enzyme *Biochemistry* **33** 7300–7308

Charleston B and Stewart HJ (1993) An interferon-induced Mx protein: cDNA sequence and high-level expression in the endometrium of pregnant sheep *Gene* **137** 327–331

Colamonici O, Yan H, Domanski P, Handa R, Smalley D, Mullersman J, Witte M, Krishnan K and Krolewski J (1994) Direct binding to and tyrosine phosphorylation of the alpha subunit of the type 1 interferon receptor by p135tyk2 tyrosine kinase *Molecular and Cellular Biology* **14** 8133–8142

Darnell JE, Jr (1997) STATs and gene regulation *Science* **277** 1630–1635

Domanski P, Witte M, Kellum M, Rubinstein M, Hackett R, Pitha P and Colamonici OR (1995) Cloning and expression of a long form of the interferon alpha beta receptor that is required for signaling *Journal of Biological Chemistry* **270** 21606–21611

Ecker DJ, Butt TR, Marsh J, Sternberg EJ, Margolis N, Monia BP, Jonnalagadda S, Khan MI, Weber PL, Mueller L and Crooke ST (1987) Gene synthesis, expression, structures, and functional activities of site-specific mutants of ubiquitin *Journal of Biological Chemistry* **262** 14213–14221

Farin CE, Imakawa K, Hansen TR, McDonnell JJ, Murphy CN, Farin PW and Roberts RM (1990) Expression of trophoblastic interferon genes in sheep and cattle *Biology of Reproduction* **43** 210–218

Farrell PJ, Broeze RJ and Lengyel P (1979) Accumulation of an mRNA and protein in interferon-treated Ehrlich ascites tumour cells *Nature* **279** 523–525

Feltham N, Hillman M, Jr, Cordova B, Fahey D, Larsen B, Blomstrom D and Knight E, Jr (1989) A 15-kD interferon-induced protein and its 17-kDa precursor: expression in *Escherichia coli*, purification, and characterization *Journal of Interferon Research* **9** 493–507

Finley D and Chau V (1991) Ubiquitination *Annual Review of Cell Biology* **7** 25–69

Flint APF, Riley PR, Stewart HJ and Abayasekara DRE (1995) The sheep endometrial oxytocin receptor *Advances in Experimental Medicine and Biology* **395** 281–294

Han CS, Mathialagan N, Klemann SW and Roberts RM (1997) Molecular cloning of ovine and bovine type I interferon receptor subunits from uteri, and endometrial expression of messenger ribonucleic acid for ovine receptors during the estrous cycle and pregnancy *Endocrinology* **138** 4757–4767

Hansen TR, Kazemi M, Keisler DH, Malathy P-V, Imakawa K and Roberts RM (1989) Complex binding of the embryonic interferon, ovine trophoblast protein-1 to endometrial receptors *Journal of Interferon Research* **9** 215–225

Hansen TR, Leaman DW, Cross JC, Mathialagan N, Bixby JA and Roberts RM (1991) The genes for the trophoblast interferons and the related interferon αII possess distinct 5′-promoter and 3′-flanking sequences *Journal of Biological Chemistry* **266** 3060–3066

Hansen TR, Austin KJ and Johnson GA (1997) Transient ubiquitin cross-reactive protein gene expression in bovine endometrium *Endocrinology* **138** 5079–5082

Hershko A and Ciechanover A (1992) The ubiquitin system for protein degradation *Annual Review of Biochemistry* **61** 761–807

Imakawa K, Hansen TR, Malathy P-V, Anthony RV, Polites HG, Marotti KR and Roberts RM (1990) Molecular cloning and characterization of complementary deoxyribonucleic acids corresponding to bovine trophoblast protein-1: a comparison with ovine trophoblast protein-1 and bovine interferon-α_{II} *Molecular Endocrinology* **3** 127–139

Johnson GA, Austin KJ, Van Kirk EA and Hansen TR (1998) Pregnancy and interferon-tau induce conjugation of bovine ubiquitin cross-reactive protein to cytosolic uterine proteins *Biology of Reproduction* **58** 898–904

Kessler DS, Veals SB, Fu XY and Levy DE (1990) IFN-alpha regulates nuclear translocation and DNA-binding affinity of ISGF3, a multimeric transcriptional activator *Genes and Development* **4** 1753–1765

Kim TK and Maniatis T (1996) Regulation of interferon-γ-activated STAT1 by the ubiquitin–proteasome pathway *Science* **273** 1717–1719

Levy DE, Kessler DS, Pine R and Darnell JE (1989) Cytoplasmic activation of ISGF3, the positive regulator of interferon-alpha-stimulated transcription, reconstituted *in vitro*. *Genes and Development* **3** 1362–1371

Lowe J, McDermott H, Loeb K, Landon M, Haas AL and Mayer RJ (1995) Immunohistochemical localization of ubiquitin cross-reactive protein in human tissues *Journal of Pathology* **177** 163–169

McCracken JA, Schramm W and Okulicz WC (1984) Hormone receptor control of pulsatile secretion of PGF2α from the ovine uterus during luteolysis and its abrogation in early pregnancy *Animal Reproduction Science* **7** 31–55

Mirando MA, Ott TL, Vallet JL, Davis M and Bazer FW (1990) Oxytocin-stimulated inositol phosphate turnover in endometrium of ewes is influenced by stage of the estrous cycle, pregnancy, and intrauterine infusion of ovine conceptus secretory proteins *Biology of Reproduction* **42** 98–105

Murdoch WJ, Austin KJ and Hansen TR (1996) Polyubiquitin up-regulation in corpora lutea of prostaglandin treated ewes *Endocrinology* **137** 4526–4529

Naivar KA, Ward SK, Austin KJ, Moore DW and Hansen TR (1995) Secretion of bovine uterine proteins in response to type-1 interferons *Biology of Reproduction* **52** 848–854

Novick D, Cohen B and Rubenstein M (1994) The human interferon alpha/beta receptor: characterization and molecular cloning *Cell* **77** 391–400

Oppenheim JJ, Zachariae COC, Mukaida N and Matsushima K (1991) Properties of the novel proinflammatory supergene "intercrine" cytokine family *Annual Review of Immunology* **9** 617–648

Ott TL, Mirando MA, Davis MA and Bazer FW (1992) Effects of ovine conceptus secretory proteins and progesterone on oxytocin-stimulated endometrial production of prostaglandin and turnover of inositol phosphate in ovariectomized ewes *Journal of Reproduction and Fertility* **95** 19–29

Palombella VJ, Rando OJ, Goldberg AL and Maniatis T (1994) The ubiquitin-proteasome pathway is required for processing the NF-kappa B1 precursor protein and the activation of NF-kappa B *Cell* **78** 773–785

Perry DJ, Austin KJ and Hansen TR (1997) The uterine bovine ubiquitin cross-reactive protein gene: a comparison with human interferon stimulated gene 15 *Biology of Reproduction* **56** (Supplement 1) 128 (Abstract 181)

Perry DJ, Austin KJ and Hansen TR (1998) Transcription factors that are involved with the mechanism of interferon-tau action in bovine endometrial cells *The Endocrine Society Program Abstracts of the 80th Annual Meeting* p222

Recht M, Borden EC and Knight E, Jr (1991) A human 15-kDa IFN-induced protein induces the secretion of IFN-γ *Journal of Immunology* **147** 2617–2623

Reich N, Evans B, Levy D, Fahley D, Knight E, Jr and Darnell JE, Jr (1987) Interferon-induced transcription of a gene encoding a 15-kDa protein depends on an upstream enhancer element *Proceedings of the National Academy of Sciences USA* **84** 6394–6398

Roberts RM, Cross JC and Leaman DW (1990) Interferons as hormones of pregnancy *Endocrine Reviews* **13** 432–452

Robertson SA, Seamark RF, Guilbert LJ and Wegmann TG (1994) The role of cytokines in gestation *Critical Reviews in Immunology* **14** 239–292

Rueda BR, Naivar KA, George EM, Austin KJ, Francis H and Hansen TR (1993) Recombinant interferon-τ regulates secretion of two bovine endometrial proteins *Journal of Interferon Research* **13** 303–309

Short EC, Geisert RD, Helmer SD, Zavy MT and Fulton RW (1991) Expression of antiviral activity and induction of 2′, 5′-oligoadenylate synthetase by conceptus secretory proteins enriched in bovine trophoblast protein-1 *Biology of Reproduction* **44** 261–268

Spencer TE and Bazer FW (1996) Ovine interferon tau suppresses transcription of the estrogen receptor and oxytocin receptor genes in ovine endometrium *Endocrinology* **137** 1144–1147

Spencer TE, Ott TL and Bazer FW (1998) Expression of interferon regulatory factors one and two in the ovine endometrium: effects of pregnancy and ovine interferon tau *Biology of Reproduction* **58** 1154–1162

Staggs KL, Austin KJ, Johnson GA, Teixeira MG, Talbott CT, Dooley VA and Hansen TR (1998) Complex induction of bovine uterine proteins by interferon-tau *Biology of Reproduction* **59** 293–297

Teixeira MG, Austin KJ, Perry DJ, Dooley VD, Johnson GA, Francis BR and Hansen TR (1997) Bovine granulocyte chemotactic protein-2 is secreted by the endometrium in response to interferon-tau (IFN-τ) *Endocrine* **6** 31–37

Thatcher WW, Meyer MD and Danet-Desnoyers G (1995) Maternal recognition of pregnancy *Journal of Reproduction and Fertility Supplement* **49** 15–28

Vallet JL, Lamming GE and Batten M (1990) Control of endometrial oxytocin receptor and uterine response to oxytocin by progesterone and oestradiol in the ewe *Journal of Reproduction and Fertility* **90** 625–634

Vallet JL, Barker PJ, Lamming GE, Skinner N and Huskisson NS (1991) A low molecular weight endometrial secretory protein which is increased by ovine trophoblast protein-1 is β2-microglobulin-like protein *Journal of Endocrinology* **130** R1–R4

LOCAL CELLULAR AND TISSUE COMMUNICATION

Chair
D. Schams

Journal of Reproduction and Fertility Supplement **54**, 343–352

Roles of extracellular matrix in follicular development

R. J. Rodgers, I. L. van Wezel, H. F. Irving-Rodgers, T. C. Lavranos, C. M. Irvine and M. Krupa

Department of Medicine, Flinders University of South Australia, Bedford Park, South Australia 5042, Australia

The cellular biology and changes in the extracellular matrix of ovarian follicles during their development are reviewed. During growth of the bovine ovarian follicle the follicular basal lamina doubles 19 times in surface area. It changes in composition, having collagen IV α1–26 and laminin α1, β2 and γ1 at the primordial stage, and collagen IV α1 and α2, reduced amounts of α3–α5, and a higher content of laminin α1, β2 and γ1 at the antral stage. In atretic antral follicles laminin α2 was also detected. The follicular epithelium also changes from one layer to many layers during follicular growth. It is clear that not all granulosal cells have equal potential to divide, and we have evidence that the granulosal cells arise from a population of stem cells. This finding has important ramifications and supports the concept that different follicular growth factors can act on different subsets of granulosal cells. In antral follicles, the replication of cells occurs in the middle layers of the membrana granulosa, with older granulosal cells towards the antrum and towards the basal lamina. The basal cells in the membrana granulosa have also been observed to vary in shape between follicles. In smaller antral follicles, they were either columnar or rounded, and in follicles > 5 mm the cells were all rounded. The reasons for these changes in matrix and cell shapes are discussed in relation to follicular development.

Introduction

The mammalian ovary produces mature oocytes capable of being fertilized and sustaining embryonic development. It also produces steroid hormones which primarily communicate to the other organs the degree of progress in the development of the oocyte, both before and after ovulation. These functions are accomplished by follicles and corpora lutea. Follicles contain an epithelioid layer, the membrana granulosa, surrounded by a basal lamina which in turn separates it from the surrounding stroma (preantral follicles) or the theca interna (antral follicles). At ovulation, the epithelioid granulosal cells redifferentiate into the mesenchymal luteal cells.

Compared with other organs, very little is known about the cellular biology of developing follicles. Even some of the most basic cellular biology has not been described for this structure. The aim of this review is to focus on some of our new ideas and discoveries that relate particularly to the extracellular matrix, including the follicular basal lamina. The development of the membrana granulosa is also considered, and comparisons to other epithelia are made where appropriate.

Extracellular Matrix

The extracellular matrix, more aptly termed intercellular matrix, has many different roles, including effects on cell behaviour, such as migration, division, differentiation, cell death and cell anchorage.

The extracellular matrix can play a role in the fluid dynamics of a tissue, either providing osmotic forces, or filtering material from solutions as they pass through the matrix. They can provide mechanical support for tissues, either rigid or elastic. In addition, nutritional materials and hormones and other extracellular signals are often required to traverse the extracellular matrix to reach target cells. The extracellular matrix can bind growth factors, either directly or indirectly via the specific binding proteins for the growth factors, ensuring that they act locally. In essence the extracellular matrix defines and provides microenvironments, enabling cells to specialize. Not surprisingly extracellular matrix is generally a diverse mixture of components.

There are a number of different compartments and extracellular matrices in follicles. These include the follicular basal lamina, follicular fluid, zona pellucida, membrana granulosa, cumulus, and either theca interna and theca externa in larger antral follicles, or the stroma in the smaller primordial and preantral follicles. These matrices have been studied to various degrees. The study of the zona pellucida is well advanced and has been reviewed many times; the production of hyaluronin by cumulus cells has also been well studied. Follicular fluid has been analysed for its composition of glycosaminoglycans (GAGs: Yanagishita *et al.*, 1979; Bellin and Ax, 1984; Grimek, 1984) and the biochemical production of the GAGs by granulosal cells has been studied by Yanagishita and colleagues. However, very little attention has been paid to the proteoglycans from which the GAGs were derived, or their physiological roles. The mesenchymal area surrounding the membrana granulosa has not been the focus of systematic study, which is unfortunate as the phenotype of the cells changes as follicular development occurs from stroma to theca.

Basal Lamina

Basal laminae are specialized sheets of extracellular matrix that separate epithelial cell layers from underlying mesenchyme in organs throughout the body including the ovary. They influence epithelial cell migration, proliferation and differentiation, and can selectively retard the passage of molecules from one side of a basal lamina to the other. Basal laminae are a lattice-type network of collagen IV intertwined with a network of laminin. This structure is stabilized by the binding of entactin to the collagen and laminin, and by low-affinity interactions between collagen IV and laminin (Yurchenco and Schittny, 1990; Paulsson, 1992). Fibronectin, heparan sulphate proteoglycans (HSPGs) and other molecules are associated with the collagen IV–laminin backbone. Importantly, basal laminae in different regions of the body differ in the ratio of all these components. Furthermore, each 'component' is in effect a class of several components. Thus each collagen IV molecule is composed of three α chains, but six different types of α chain have been discovered to date. Potentially, any combination of these might be present (Hay, 1991; Zhou *et al.*, 1994). Similarly, each laminin molecule is composed of one α (A in the old nomenclature), one β (B1 in the old nomenclature) and one γ (B2 in the old nomenclature) chain (Burgeson *et al.*, 1994), yet five different α chains, three β chains and two γ chains have been discovered. There are at least twenty different isoforms of fibronectin, due to alternative splicing of mRNA. It is considered that the unique composition of each basal lamina contributes to its specific functional properties (Engvall, 1993).

Numerous studies *in vitro* have shown that cell morphology was altered according to the type of extracellular matrix component on which the cells were cultured (Watt, 1986). Thus alterations to the composition of the basal lamina may affect the fate of the associated cells. The composition of basal laminae also affects their ability to filter materials selectively. For example, in the normal neonatal mouse, laminin $\beta 1$ is replaced by $\beta 2$ in the kidney glomerular basement membrane as the kidney develops. However, mice with a null mutation in the laminin $\beta 2$ gene continued with $\beta 1$, but then failed to retard the passage of plasma proteins across the glomerular basement membrane, despite the membrane being structurally intact. These mice died of proteinurea within one month of birth (Noakes *et al.*, 1995).

In ovarian follicles there are many different basal laminae including the follicular basal lamina. The vasculature of the theca has a subendothelial basal lamina, and the smooth muscle cells of the arterioles have basal laminae.

Follicular Basal Lamina

In the ovary, the membrana granulosa of each ovarian follicle is enveloped by a follicular basal lamina, which separates it from the surrounding stromal elements in primordial follicles (van Wezel and Rodgers, 1996) or from the theca in antral follicles (Gosden *et al.*, 1988; Luck, 1994). The follicular basal lamina is believed to play a role in influencing granulosal cell proliferation and differentiation (Amsterdam *et al.*, 1989; Richardson *et al.*, 1992; Luck, 1994). In addition, in healthy follicles it excludes capillaries, white blood cells and nerve processes from the granulosal compartment until ovulation, at which time the basal lamina is degraded.

The follicular basal lamina probably has a role in retarding entry of larger molecular weight plasma proteins and molecules (for example low density lipoproteins, LDL) into the follicular antrum (Andersen *et al.*, 1976). Conversely if the flow of material in the other direction is similarly retarded, the follicular basal lamina may trap large molecules (for example some proteoglycans) that are synthesized by granulosal cells and oocytes in the follicular fluid. The molecular mass cut off is calculated to be 100–850 kDa based upon comparisons of the composition of follicular fluid with that of plasma (Anderson *et al.*, 1976). It is not known whether the size of material that can cross the follicular basal lamina changes during the course of follicular development, particularly before and after antrum formation.

The molecular weight cut off of the follicular basal lamina is considerably larger than most growth factors. It is probably for this reason that many authors have assumed that growth factors can readily move from the thecal layer to the membrana granulosa, and vice versa. However, the follicular basal lamina could be a barrier to the movement of growth factors from theca interna to membrana granulosa, and vice versa. This is due to the nature of growth factors (for example fibroblast growth factor 2), or their binding proteins (for example follistatin, insulin-like growth factor binding factor 5), to bind avidly to extracellular matrix components, especially heparan sulphate proteoglycans. It is also possible that the follicular basal lamina serves as a reservoir of attached growth factors. Thus, in contrast to the wealth of literature on the production of growth factors and the expression of their receptors, and the notion that thecal and granulosal cells signal each other via growth factors, there is scant evidence that this can actually occur.

We have estimated that the surface area of the bovine follicle doubles nineteen times during follicular development, implying that continuous remodelling of the follicular basal lamina occurs (van Wezel and Rodgers, 1996). We hypothesized that the composition of the follicular basal lamina is altered during follicular development, particularly at the time when follicular fluid accumulates to form an antrum, and during follicular atresia. Immunolocalization studies have demonstrated the presence of collagen IV (Bagavandoss *et al.*, 1983; Kaneko *et al.*, 1984; Palotie *et al.*, 1984), laminin (Wordinger *et al.*, 1983; Bagavandoss *et al.*, 1983; Palotie *et al.*, 1984; Leu *et al.*, 1986; Christiane *et al.*, 1988; Yoshinaga-Hirabayashi *et al.*, 1990; Leardkamolkarn and Abrahamson, 1992; Fröjdman *et al.*, 1995), and fibronectin (Bagavandoss *et al.*, 1983; Yoshimura *et al.*, 1991; Figueiredo *et al.*, 1995) in the follicular basal laminae of antral follicles. However, none of these studies has differentiated between the different isoforms of any of these components, except for one study that compared the localisation of $\alpha 1$ versus $\beta 1$–$\gamma 1$ laminin in the fetal mouse ovary (Fröjdman *et al.*, 1995). Other studies using western and northern blotting identified the expression of a few of the subtypes of collagen and laminin (Zhao and Luck, 1995; Iivanainen *et al.*, 1995) but did not specifically localize these components to the follicular basal lamina; this is important as there are other basal laminae in follicles, such as those associated with blood vessels.

In recent studies of the bovine follicle (Fig. 1), we found that laminin $\alpha 1$ chain was present in the follicular basal lamina at all stages of follicular development, while laminin $\alpha 2$ chain was present only in atretic antral follicles and a few healthy antral follicles and absent from primordial and growing preantral follicles (van Wezel *et al.*, 1998). The laminin $\beta 2$ chain was present in the follicular basal lamina of follicles of all stages, but the laminin $\beta 1$ chain was detected only in the basal lamina of large preantral follicles. Staining for laminin $\alpha 1$, $\beta 2$ or $\gamma 1$ chains appeared to increase in intensity from the preantral to healthy antral stages, and the basal lamina of atretic antral follicles appeared thicker but stained less intensely than that of healthy antral follicles. The basal laminae of the thecal

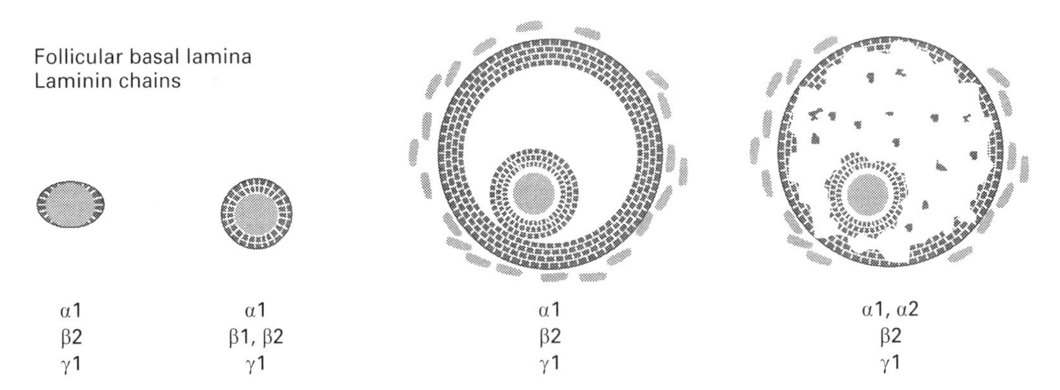

Fig. 1. The laminin chains identified in the follicular basal lamina of bovine follicles (summarized from van Wezel *et al.*, 1998). Follicles from left to right represent primordial, preantral, healthy antral and atretic antral follicles.

vasculature stained positively for laminin β1 and β2 chains, whereas the laminin γ1 chain localized more generally throughout the theca in areas that do not have a recognized conventional basal lamina. The follicular basal lamina of almost all primordial and preantral follicles was postive for all of the type IV collagen chains α1–α6. However, only a proportion of antral follicles had a basal lamina immunopositive for the type IV collagen chains α3, α4 or α5. In addition to staining the follicular basal lamina, α1α2 was also present in the theca and not associated with a structural basal lamina (Rodgers *et al.*, 1998).

Examination of the bovine follicular basal lamina by electron microscope has shown that it is a single layer closely associated with the granulosal cells in primordial follicles (van Wezel and Rodgers, 1996). The follicular basal lamina of a number of larger follicles is composed of many layers of basal lamina material forming a branching network (Rodgers *et al.*, 1995). This finding is consistent with the notion that the basal lamina is shed and replaced by a newly synthesized basal lamina closer to the granulosal cells as the follicle grows. These observations and that of different laminin and collagen type IV isoforms at different stages of follicular development (see above) allow us to suggest that the follicular basal lamina is continually remodelled during follicular development. On the basis of the knowledge of other basal laminae we suggest that these changes in the follicular basal lamina are related key changes during follicular development. These changes could include the formation of follicular fluid or the migration, proliferation or differentiation of the granulosal cells.

We have also found that larger antral follicles either have a single layer of basal lamina, like that of primordial follicles, or have multilayers that form a branching network. The latter have vesicles attached to the basal lamina material not immediately adjacent to the cell surface (Rodgers *et al.*, 1995). Vesicles have been observed rarely in other basal laminae but have been observed in the camel kidney (Safer and Katchburian, 1991). The significance of these vesicles, or of the two phenotypes of follicular basal lamina, is not clear yet.

The cellular origin of the follicular basal lamina is a contentious issue (see van Wezel and Rodgers, 1996). Concerning the laminin component of the follicular basal lamina, the γ1 chain but not β1 chain was expressed by granulosal cells as detected by northern blot analysis, and this is consistent with the present study localizing the γ1 but not β1 laminin chain to the follicular basal lamina of antral follicles. Furthermore, a previous immunoelectron study in the rat ovary (Leardkamolkarn and Abrahamson, 1992) localized laminin to Call-Exner bodies, which are similar in ultrastructure to basal lamina and have been observed within the membrana granulosa of follicles *in vivo* (cow: van Wezel *et al.*, 1999a; rabbit: Gosden *et al.*, 1988). Laminin was also localized intracellularly in both granulosal and thecal cells, but this latter observation may represent degradation rather than synthesis. We have also shown that granulosal cells cultured under anchorage-independent conditions produce a basal lamina, collagen type IV (Rodgers *et al.*, 1995) and fibronectin (Rodgers *et al.*, 1996). All of these studies suggest that the granulosal cells are capable of secreting many of the components of a basal lamina. The contribution of the theca in larger follicles, or indeed the stroma surrounding preantral follicles, to the production of the follicular basal lamina is not known. In other systems, it is predominantly epithelial rather than stromal cells that synthesize basal lamina components, although in some tissues both cell types make a contribution (Timpl and Dziadek, 1986). The thecal compartment of antral follicles has been shown by northern blot analyses to express laminin β1 and γ1 chains (Zhao and Luck, 1995). Our recent observations showing laminin β1 in the thecal vasculature and undetectable in the follicular basal lamina of antral follicles (van Wezel *et al.*, 1998) indicate that the thecal expression of β1 is not necessarily a contribution to the follicular basal lamina. Similarly, laminin γ1 was found widely distributed in the theca and, although it is present in the follicular basal lamina, expression in the theca cannot be considered as proof that the theca contributes laminin γ1 to the follicular basal lamina. However, it remains possible that *in vivo*, the follicular basal lamina requires a contribution from both the granulosal cells and the stromal or thecal cells.

'Thecal Matrix'

Immunostaining for laminin γ1chain or EHS (α1β1γ1) laminin has been observed in bovine follicles throughout the theca interna (van Wezel *et al.*, 1998). This staining was extracellular and in areas where there are no conventional basal laminae, such as those of blood vessels. We have also observed this pattern of staining using antibodies to some types of collagen IV (Rodgers *et al.*, 1998). In other species (rat: Bagavandoss *et al.*, 1983; Leardkamolkarn and Abrahamson, 1992; mouse: Fröjdman *et al.*, 1995) and in the interstitial tissue of developing gonads (Fröjdman *et al.*, 1989, 1992a,b, 1993, 1995; Smith and MacKay, 1991), similar immunostaining patterns have been obtained using antibodies to EHS laminin. At the level of the electron microscope, fragments of basal lamina-like, electron-dense material have been observed in the theca interna in sheep (O'Shea *et al.*, 1978), rats (Leardkamolkarn and Abrahamson, 1992) and cows (see Fig. 1 in Rodgers *et al.*, 1986). It is likely, though not proven, that collagen IV and some of the laminin chains are present in these fragments of basal lamina-like material in the theca. The origins and functions of this 'thecal matrix' are not known. It is possible that they are involved in the vascularization and the formation and expansion of the theca that occurs during follicular development.

Follicular Epithelium

The membrana granulosa is multilayered, and there is a persistent coordinated interaction between the granulosal cells and the oocyte and theca (Buccione *et al.*, 1990). The fate of the granulosal cells is either redifferentiation at the time of ovulation to form a mesenchymal cell type, the luteal cells, or death of the granulosal cells and destruction of the membrana granulosa in follicular atresia. Before luteinization, the granulosal cells are regarded as an epithelial or epithelioid cell type (see Hirshfield, 1991).

Quite a lot is known about other epithelia, such as the epidermis of skin and the luminal epithelium of the gut. Both of these epithelia lie on a basal lamina and are highly structured. Subpopulations of cells are located in specific regions of the epithelia. There is a population of stem cells and, at a distance from these, a population of well differentiated cells (Stenn, 1983). In the gut crypt, there is also a separate population of more rapidly dividing cells between the stem cells and differentiated cells (Potten and Loeffler, 1990). The membrana granulosa of ovarian follicles also lies on a basal lamina (van Wezel *et al.*, 1998) and there is some indication that granulosal cells in different regions of the membrana granulosa have different shapes (Marion *et al.*, 1968) and biochemical properties (Amsterdam *et al.*, 1975; Bortolussi *et al.*, 1977; Zoller and Weisz, 1978, 1979; Dunaif *et al.*, 1982; Zlotkin *et al.*, 1986; Tabarowski and Szoltys, 1987; Salustri *et al.*, 1992).

In contrast to the other epithelial cell types, very little has been discovered about the membrana

Fig. 2. The structure of the membrana granulosa observed in bovine small (< 5 mm) and large (> 5 mm) antral follicles. Some smaller follicles have a zone of columnar cells adjacent to the follicular basal lamina, a zone of rounded cells closer to the antrum, and a zone of flattened granulosal cells immediately adjacent to the antrum; other cells have rounded basal cells. Larger follicles have rounded basal granulosal cells.

granulosa as an epithelium. Its overall structure has not been reported, particularly as a function of follicular development, nor has the location of the dividing cells and the location of differentiated cells, if there are any. The membrana granulosa is also more complex than other epithelia for a number of reasons. First, it expands from a single to a multi-layered epithelium as the follicle grows. In the transition from pre- to post-antral follicles, the amount of fluid traversing the epithelium changes considerably. The epithelium expands laterally with time as the follicle grows. Total destruction of the epithelium occurs as the follicle becomes atretic, and death of the granulosal cells is one of the first indicators of follicular atresia. Thus, not only is the follicular epithelium poorly understood, it is considerably complex. Recently, we have undertaken to study the cellular biology of the membrana granulosa.

Changes occur in the membrana granulosa and its environment during follicular growth. These changes include the formation of an antrum, a net increase in the surface area of the follicle (19 doublings to form an 18 mm bovine follicle, calculated from van Wezel and Rodgers, 1996), and a net increase in the number of granulosal cells (estimated at 21 doublings, if 40×10^6 granulosal cells were present as reported by McNatty *et al.*, 1984 – see van Wezel and Rodgers, 1996). The net increase in the number of cells is related to the extent of cell division and loss of cells via apoptosis or terminal differentiation (van Wezel *et al.*, 1999b). The net increases in both cell numbers and surface area have important ramifications for the number of layers of cells in the membrana granulosa. For example, if the cell numbers double 21 times and the surface area doubles 19 times (see van Wezel and Rodgers, 1996), it is predicted that cell layers in the membrana granulosa would increase from one layer, as in primordial follicles, to four $[(21–19)^2]$ layers. In a recent study, we found considerable variation in the numbers of layers per follicle (van Wezel *et al.*, 1999c), and this is consistent with a reported variation in the number of granulosal cells obtained from follicles of the same size (McNatty *et al.*, 1979). On the basis of these observations and the variation in basal cell shapes (columnar or rounded) in the smaller follicles (≤ 5 mm) (see below), we suggest that the rate of granulosal cell proliferation and maturation is not tightly or coordinately regulated with the timing or rate of antrum formation or to follicular expansion. Thus, follicle size is not necessarily a good indication of follicle maturation.

We have found three structurally distinct zones in the membrana granulosa of many small bovine antral follicles: a zone of columnar cells adjacent to the follicular basal lamina, a zone of rounded cells closer to the antrum, and a zone of flattened granulosal cells immediately adjacent to the antrum (Fig 2). The presence of columnar and rounded cells has been described in a range of species, while the layer of flattened cells closest to the antrum was mentioned in the literature thirty years ago (Marion *et al.*, 1968) but has been ignored in more recent literature. In a recent study (van Wezel *et al.*, 1999c), we have demonstrated that the structure of the membrana granulosa varies between follicles: distinct layers of columnar, rounded and flattened cells were only found in small follicles (≤ 5 mm; ovulation is from follicles ≥ 10 mm; Staigmiller and England, 1982), and larger follicles contained predominantly rounded cells throughout the membrana granulosa (Fig. 2). It is therefore possible that there is a developmental progression from columnar to rounded cells in the basal zone of the follicle as the antral follicle enlarges.

The issue of whether cells in different regions of the membrana granulosa (basal to antral) are at different stages of differentiation is complex. Cells tend to differentiate in G0 and by far the greatest proportion of non-dividing cells was found in the most basal and most antral layers of bovine antral follicles (see below). Therefore, it would appear that cells replicate in the middle zones of the membrana granulosa in antral follicles and then migrate in either of two opposite directions: towards the antrum or towards the basal lamina. As they migrate and spend longer in G0, it is probable that their phenotype and hence pattern of gene expression change. The nature of the differentiation has still to be addressed. Certainly many authors recognize expression of cytochrome P450 aromatase and oestradiol production as markers of differentiation, but there may be other specialized functions that granulosal cells undertake, particularly at the time of antrum formation.

It has been suggested that cells in the basal zone are more differentiated than are the antrally situated cells of the membrana granulosa (see Amsterdam, 1987). These suggestions arose from numerous studies localizing LH receptors (Amsterdam and Rotmensch, 1975, Bortolussi *et al.*, 1977), steroidogenic enzymes (Zoller and Weisz, 1978; 1979; Zlotkin *et al.*, 1986; Tabarowski and Szoltys, 1987), or housekeeping enzymes (Zoller and Weisz, 1979; Zoller and Enelow, 1983) in the follicle of a number of species. The amounts of these enzymes and receptors were apparently higher (usually about double) in the basal zone than in the antral regions of the membrana granulosa. Unfortunately many of these studies did not measure the receptors/enzymes on a per cell basis, but rather the level of staining per unit area of the membrana granulosa. In bovine antral follicles we have observed that columnar basal cells were more compact than the rounded cells found in the middle or antral zones. This difference in the level of 'cellularity' between basal and antral zones has not previously been considered, and it may be a cause of the apparent reduced staining for receptors or enzymes of the antral zone observed in many studies. We therefore consider that a reanalysis of this area is warranted.

Clearly the issues of differentiation of cells in the membrana granulosa are complex and still unresolved. However, we propose that the membrana granulosa is more similar to other epithelia than previously recognized. The shape of the granulosal cells in different zones is comparable to skin epidermis, where cells in the 'basal layer' are columnar, cells in the 'spinous layer' (in the suprabasal position) are rounded, and the cells that are furthest from the epithelial basal lamina in the 'granular', 'transitional', and 'horny' layers are increasingly flattened (Stenn, 1983). Furthermore, we propose that there is a spatial progression in the membrana granulosa from stem cells (see below) near the basal lamina to terminally differentiated cells near the antrum which are more flattened and slough from the membrana granulosa into the lumen (van Wezel *et al.*, 1999b). This is also similar to skin, where the stem cells are located in the basal layer and the terminally differentiated keratinocytes are furthest from the basal lamina and are desquamated. However, whereas cell proliferation in the epidermis is limited to cells in the basal layer (Stenn, 1983), cell division in the membrana granulosa is more common in the middle regions (see below). This could be similar to the luminal epithelium of gut, in which the stem cells and the rapidly dividing cells are spatially distinct (Potten and Loeffler, 1990). One important difference between the membrana granulosa and the skin epidermis is that the membrana granulosa expands laterally as the follicular antrum grows. Thus in addition to cellular movement in basal–antral directions, movement in a sideways direction could also occur. The degree of this will depend upon the rate of cellular replication versus the rate of follicular antrum expansion.

Granulosal Stem Cells

In some tissues, cell division is carried out exclusively by stem cells. True or even committed stem cells are totipotent or pluripotent, and they can divide *in vivo* without contact inhibition or *in vitro* without the need for anchorage. We have argued that granulosal cells divide *in vivo* without contact inhibition, at least during the preantral stages when the cells are in very close physical contact with each other. In a recent study of antral follicles, most of the dividing cells were found in the middle layers of the membrana granulosa (van Wezel *et al.*, 1999c). Thus, *in vivo*, granulosal cells can divide and are not inhibited from doing so by contact with adjacent cells. *In vitro* we have demonstrated that a proportion of granulosal cells can divide in soft agar or methycellulose solution, neither of which provide anchorage (Lavranos and Rodgers, 1994; Lavranos *et al.*, 1994, 1996; Rodgers *et al.*, 1995). On the basis of these observations, we have postulated that there is a population of granulosal stem cells (Lavranos and Rodgers, 1994; Lavranos *et al.*, 1994, 1996; Rodgers *et al.*, 1995).

It is not known where the granulosal stem cells reside in the membrana granulosa. Cell division is not always carried out by stem cells alone. In the luminal epithelium of the gut crypt, a population of cells that are spatially distinct from the stem cell subpopulation has been described; these are more differentiated and more rapidly dividing. In a recent study of the membrana granulosa of bovine antral follicles, most of the mitotic figures were observed in the middle regions, although such figures were present throughout the membrana granulosa and theoretically any of these could be the stem cell subpopulation (van Wezel *et al.*, 1999c). However, on the basis of our observation that 19% of colonies grown from bovine granulosal cells under anchorage-independent conditions produced a basal lamina like material (Rodgers *et al.*, 1996), we suggest that *in vivo* at least some granulosal cells with properties of stem cells are located close to the follicular basal lamina.

Conclusion

Most studies on 'folliculogenesis' have focused on the hormones and growth factors involved. However, in many other organ systems, structural studies similar to our recent investigations of the membrana granulosa were undertaken decades ago. We have also discussed the follicular basal lamina and thecal matrix. Clearly there are many important and informative studies to be done in both of these areas, and of course there are many other areas of discovery within the follicle. We have also highlighted areas where we consider the literature to be too superficial in its assumptions on how follicles function. In conclusion, the study of the cellular biology of the ovarian follicle is not as advanced as that of many other organs but promises to hold many exciting discoveries.

The authors would like to acknowledge the support of the National Health and Medical Research Council of Australia, Flinders University and Flinders Medical Research Foundation.

References

Amsterdam A and Rotmensch S (1987) Structure–function relationships during granulosa cell differentiation *Endocrine Reviews* **8** 309–337

Amsterdam A, Koch Y, Lieberman ME and Lindner HR (1975) Distribution of binding sites for human chorionic gonadotropin in the preovulatory follicle of the rat *Journal of Cell Biology* **67** 894–900

Amsterdam A, Rotmensch S, Furman A, Venter EA and Vlodavsky I (1989) Synergistic effect of human chorionic gonadotrophin and extracellular matrix on *in vitro* differentiation of human granulsoa cells: progesterone production and gap junction formation *Endocrinology* **124** 1956–1964

Andersen MM, Krøll J, Byskov AG and Faber M (1976) Protein composition in the fluid of individual bovine follicles *Journal of Reproduction and Fertility* **48** 109–118

Bagavandoss P, Midgley AR, Jr and Wicha M (1983) Developmental changes in the ovarian follicular basal lamina detected by immunofluorescence and electron microscopy *Journal of Histochemistry and Cytochemistry* **31** 633–640

Bellin ME and Ax RL (1984) Chondroitin sulfate: an indicator of atresia in bovine follicles *Endocrinology* **114** 428–434

Bortolussi M, Marini G and Dal Lago A (1977) Autoradiographic study of the distribution of LH (HCG) receptors in the ovary of untreated and gonadotrophin-primed immature rats *Cell and Tissue Research* **183** 329–342

Buccione R, Schroeder AC and Eppig JJ (1990) Interactions between somatic cells and germ cells throughout mammalian oogenesis *Biology of Reproduction* **43** 543–547

Burgeson RE, Chiquet M, Deutzmann R, Ekblom P, Engel J, Kleinman H, Martin GR, Meneguzzi G, Paulsson M, Sanes J, Timpl R, Tryggvason K, Yamada Y and Yurchenco PD (1994) A new nomenclature for the laminins *Matrix Biology* **14** 209–211

Christiane Y, Demoulin A, Gillain D and Leroy F (1988) Laminin and type III procollagen peptide in human preovulatory follicular fluid *Fertility and Sterility* **50** 48–51

Dunaif AE, Zimmerman EA, Friesen HG and Frantz AG (1982) Intracellular localization of prolactin receptor and prolactin in the rat ovary by immunocytochemistry *Endocrinology* **110** 1465–1471

Engvall E (1993) Laminin variants: why, where and when? *Kidney International* **43** 2–6

Figueiredo JR, Hulshof SCJ, Thiry M, Van Den Hurk R, Bevers MM, Nusgens B and Beckers JF (1995) Extracellular matrix proteins and basement membrane: their identification in bovine ovaries and significance for the attachment of cultured preantral follicles *Theriogenology* **43** 845–858

Fröjdman K, Paranko J, Kuopio T and Pelliniemi LJ (1989) Structural proteins in sexual differentiation of embryonic gonads *International Journal of Developmental Biology* **33**, 99–103

Fröjdman K, Malmi R and Pelliniemi LJ (1992a) Lectin-binding carbohydrates in sexual differentiation of rat male and female gonads *Histochemistry* **97** 469–477

Fröjdman K, Paranko J, Virtanen I and Pelliniemi LJ (1992b) Intermediate filaments and epithelial differentiation of male rat embryonic gonad *Differentiation* **50** 113–123

Fröjdman K, Paranko J, Virtanen I and Pelliniemi LJ (1993) Intermediate filament proteins and epithelial differentiation in the embryonic ovary of the rat *Differentiation* **55** 47–55

Fröjdman K, Ekblom P, Sorokin L, Yagi A and Pelliniemi J (1995) Differential distribution of laminin chains in the development and sex differentiation of mouse internal genitalia *International Journal of Developmental Biology* **39** 335–344

Gosden RG, Hunter RHF, Telfer E, Torrance C and Brown N (1988) Physiological factors underlying the formation of ovarian follicular fluid *Journal of Reproduction and Fertility* **82** 813–825

Grimek HJ (1984) Characteristics of proteoglycans isolated from small and large bovine ovarian follicles *Biology of Reproduction* **30** 397–409

Hay ED (1991) *Cell Biology of Extracellular Matrix* (2nd Edn) Plenum Press, New York

Hirshfield AN (1991) Development of follicles in the mammalian ovary *International Review of Cytology* **124** 43–101

Iivanainen A, Sainio K, Sariola H and Tryggvason KB (1995) Primary structure and expression of a novel human laminin alpha-4 chain *FEBS Letters* **365** 183–188

Kaneko Y, Hirakawa S, Momose K and Konomi H (1984) Immunochemical localization of Type I, III, IV and V collagens in the normal and polycystic ovarian capsules *Acta Obstetrica Gynaecologica Japonica* **36** 2473–2474 (Abstract)

Lavranos TC and Rodgers RJ (1994) An assay of tritiated thymidine incorporation into DNA by cells cultured under anchorage-independent conditions *Analytical Biochemistry* **223** 325–327

Lavranos TC, Rodgers HF, Bertoncello I and Rodgers RJ (1994) Anchorage-independent culture of bovine granulosa cells: the effects of basic fibroblast growth factor and dibutyryl cAMP on cell division and differentiation *Experimental Cell Research* **211** 245–251

Lavranos TC, O'Leary PC and Rodgers RJ (1996) Effects of insulin-like growth factors and binding protein-1 on bovine granulosa cell division in anchorage-independent culture *Journal of Reproduction and Fertility* **106** 221–228

Leardkamolkarn V and Abrahamson DR (1992) Immunoelectron microscopic localization of laminin in rat ovarian follicles *Anatomical Record* **233** 41–52

Leu FJ, Engvall E and Damjanov I (1996) Heterogeneity of basement membranes of the genitourinary tract revealed by sequential immunofluorescence staining with monoclonal antibodies to laminin *Journal of Histochemistry and Cytochemistry* **34** 483–489

Luck MR (1994) The gonadal extracellular matrix *Oxford Reviews in Reproductive Biology* **16** 33–85

McNatty KP, Smith DM, Makris A, Osathanondh R and Ryan KJ (1979) The microenvironment of the human antral follicle: interrelationships among the steroid levels in antral fluid, the population of granulosa cells, and the status of the oocyte *in vivo* and *in vitro*. *Journal of Clinical Endocrinology and Metabolism* **49** 851–860

McNatty KP, Heath DA, Lun S, Fannin JM, McDiarmid JM and Henderson KM (1984) Steroidogenesis by bovine theca interna in an *in vitro* perifusion system. *Biology of Reproduction* **30** 159–170

Marion GB, Gier HT and Choudary JB (1968) Micromorphology of the bovine ovarian follicular system. *Journal of Animal Science* **27** 451–465

Noakes PG, Gautam M, Mudd J, Sanes JR and Merlie JP (1995) The renal glomerulus of mice lacking s-laminin/laminin β2: nephrosis despite molecular compensation by laminin β1 *Nature Genetics* **10** 400–406

O'Shea JD, Cran DG, Hay MF and Moor RM (1978) Ultrastructure of the theca interna of ovarian follicles in sheep *Cell and Tissue Research* **187** 473–478

Palotie A, Peltonen L, Foidart J-M and Rajaniemi H (1984) Immunohistochemical localization of basement membrane components and interstitial collagen types in preovulatory rat ovarian follicles *Collagen Related Research* **4** 279–287

Paulsson M (1992) Basement membrane proteins: structure, assembly, and cellular interactions *Critical Reviews in Biochemistry and Molecular Biology* **27** 93–127

Potten CS and Loeffler M (1990) Stem cells: attributes, cycles, spirals, uncertainties and pitfalls: lessons for and from the crypt *Development* **110** 1001–1019

Richardson MC, Davies DW, Watson RH, Dunsford ML, Inman CB and Masson GM (1992) Cultured human granulosa cells as a model for corpus luteum function: relative roles of gonadotrophin and low density lipoprotein studied under defined culture conditions *Human Reproduction* **7** 12–18

Rodgers HF, Lavranos TC, Vella CA and Rodgers RJ (1995) Basal lamina and other extracellular matrix produced by bovine granulosa cells in anchorage-independent culture *Cell Tissue Research* **282** 463–471

Rodgers RJ, Rodgers HF, Hall PF, Waterman MR and Simpson ER (1986) Immunolocalization of cholesterol side-chain-cleavage cytochrome P-450 and 17α-hydroxylase cytochrome P-450 in bovine ovarian follicles *Journal of Reproduction and Fertility* **78** 627–638

Rodgers RJ, Vella CA, Rodgers HF, Scott K and Lavranos TC (1996) Production of extracellular matrix, fibronectin and steroidogenic enzymes, and growth of bovine granulosa cells in anchorage-independent culture. *Reproduction Fertility and Development* **8** 249–257

Rodgers HF, Irvine CM, van Wezel IL, Lavranos TC, Luck MR, Sao Y, Ninomiya Y and Rodgers RJ (1998) Distribution of the α1 to α6 chains of Type IV collagen in bovine follicles *Biology of Reproduction* **59** 1334–1341

Safer AM and Katchburian I (1991) Unusual membrane-bound bodies in the basal lamina of the uriniferous tubules of the camel *Camelus dromedarius*. Freeze-fracture and ultrathin-section study *Acta Anatomica* **140** 156–162

Salustri A, Yanagishita M, Underhill CB, Laurent TC and Hascall VC (1992) Localization and synthesis of hyaluronic acid in the cumulus cells and mural granulosa cells of the preovulatory follicle *Developmental Biology* **151** 541–551

Smith C and MacKay S (1991) Morphological development and fate of the mouse mesonephros *Journal of Anatomy* **174** 171–184

Staigmiller RB and England BG (1982) Folliculogenesis in the bovine *Theriogenology* **17** 43–52

Stenn KS (1983) The Skin. In *Histology: Cell and Tissue Biology* (2nd Edn) pp569–606 Ed. L Weiss. Elsevier Science, New York

Tabarowski Z and Szoltys M (1987) Histochemical localization of Δ^5-3β-HSDH activity in preovulatory rat follicles *Folia Histochemistry Cytobiology* **25** 149–154

Timpl R and Dziadek M (1986) Structure, development and molecular pathology of basement membranes *International Reviews in Experimental Pathology* **29** 1–112

van Wezel IL and Rodgers RJ (1996) Morphological characterization of bovine primordial follicles and their environment *in vivo*. *Biology of Reproduction* **55** 1003–1011

van Wezel IL, Rodgers HF and Rodgers RJ (1998) Differential localisation of laminin chains in bovine follicles *Journal of Reproduction and Fertility* **112** 267–278

Watt FM (1986) The extracellular matrix and cell shape *Trends in Biochemistry Science* **11** 482–585

van Wezel IL, Rodgers HF, Saod Y, Ninomiya Y and Rodgers RJ (1999a) Ultrastructure and composition of Call-Exner bodies in bovine follicles *Cell Tissue Research* (in press)

van Wezel IL, Dbarmarajan AM, Lavranos TC and Rodgers RJ (1999b) Evidence for alternative pathways of granulosa cell death in healthy and slightly atretic bovine antral follicles *Endocrinology* **140** (in press)

van Wezel IL, Krupa M and Rodgers RJ (1999c) Development of the membrana granulosa of bovine antral follicles: structure, location of mitosis and pyknosis, and immunolocalization of involucrin and vimentin *Reproduction Fertility and Development* (in press)

Wordinger RJ, Rudick VL and Rudick MJ (1983) Immunohistochemical localization of laminin within the mouse ovary *Journal Experimental Zoology* **228** 141–143

Yoshimura Y, Okamoto T and Tamura T (1991) Localization of fibronectin in the bovine ovary *Animal Science Technology* **62** 529–532

Yanagishita M, Rodbard V and Hascall VC (1979) Isolation and characterization of proteoglycans from porcine follicular fluid *Journal Biological Chemistry* **254** 911–920

Yoshinaga-Hirabayashi T, Ishimura K, Fujita H, Kitawaki J and Osawa Y (1990) Immunocytochemical localization of aromatase in immature rat ovaries treated with PMSG and hCG, and in pregnant rat ovaries *Histochemistry* **93** 223–228

Yurchenco PD and Schittny JC (1990) Molecular architecture of basement membranes *FASEB Journal* **4** 1577–1590

Zhao Y and Luck MR (1995) Gene expression and protein distribution of collagen, fibronectin and laminin in bovine follicles and corpora lutea *Journal of Reproduction and Fertility* **104** 115–123

Zhou J, Ding M, Zhao Z and Reeders ST (1994) Complete primary structure of the sixth chain of human basement membrane collagen, α6(IV) *Journal Biological Chemistry* **269** 13193–13199

Zlotkin T, Farkash Y and Orly J (1986) Cell-specific expression of immunoreactive cholesterol side-chain cleavage cytochrome P-450 during follicular development in the rat ovary *Endocrinology* **119** 2809–2820

Zoller LC and Enelow R (1983) A quantitative histochemical study of lactate dehydrogenase and succinate dehydrogenase activities in the membrana granulosa of the ovulatory follicle in the rat *Histochemical Journal* **15** 1055–1064

Zoller LC and Weisz J (1978) Identification of cytochrome P-450, and its distribution in the membrana granulosa of the preovulatory follicle, using quantitative cytochemistry *Endocrinology* **103** 310–313

Zoller LC and Weisz J (1979) A quantitative cytochemical study of glucose-6-phosphate dehydrogenase and Δ^5-3β-hydroxysteroid dehydrogenase activity in the membrana granulosa of the ovulable type of follicle in the rat *Histochemistry* **62** 125–135

Journal of Reproduction and Fertility Supplement **54**, 353–358

Plasmin–tumour necrosis factor interaction in the ovulatory process

W. J. Murdoch

Department of Animal Science, University of Wyoming, Laramie, WY 82071, USA

Collagen breakdown and apoptotic cell death within the apex of the preovulatory ovine follicle are hallmarks of impending ovarian rupture. An integrative mechanism is proposed whereby gonadotrophic stimulation of urokinase-type plasminogen activator secretion by the follicular-contiguous ovarian surface epithelium elicits a localized increase in tissue plasmin, which activates collagenolysis and tumour necrosis factor α-induced cell death within the formative ovulatory stigma.

Introduction

Regulatory mechanisms of ovulatory follicular rupture have been a subject of investigation for more than a century (see the comprehensive overview by Espey and Lipner, 1994); notwithstanding, essential ovarian pathways remain uncertain. A role for proteolytic enzymes in the degradation of connective tissue elements of the ovarian wall was apparent from the outset. Numerous studies have since implicated the plasminogen activator/plasmin and collagenolytic systems in the mechanism of ovulation. That programmed cell death (apoptosis) occurs within the developmental site of ovulation is a new discovery. A prospective mediator of ovulatory ovarian apoptosis is tumour necrosis factor α (TNFα). The objective of this overview is to summarize recent evidence based largely on work in sheep denoting an interaction between plasmin and TNFα in ovulation.

Experimental Paradigm

Mature western-range ewes were penned daily with vasectomized rams and observed for oestrous behaviour. The first day of oestrus was considered day 0 of the oestrous cycle. Animals were treated with prostaglandin $F_{2\alpha}$ ($PGF_{2\alpha}$) on day 14 to synchronize luteal regression. A synthetic agonist of GnRH was administered 36 h after $PGF_{2\alpha}$ to generate a preovulatory surge of gonadotrophins (natural surges commence at approximately 40 h). The follicle of greatest visible diameter within the pair of ovaries will ovulate approximately 24 h after injection of GnRH and form a normal corpus luteum (Roberts *et al.*, 1985). A translucent ovulatory stigma develops within 2 h of follicular rupture (Murdoch, 1985).

Plasmin Upregulation at the Ovarian Surface–Preovulatory Follicular Interface

Plasmin (fibrinolysin) is a pleiotrophic serine protease that is derived from the zymogen plasminogen by enzymatic activation. Two forms of plasminogen activator have been characterized in vertebrates – urokinase (uPA) and tissue (tPA) types. Catalytic uPA can exist either as high (about 50 kDa) or low (about 30 kDa) molecular mass variants. The major tPA has a molecular mass of approximately 70 kDa and has a strong affinity (unlike uPA) for fibrin (Danø *et al.*, 1985). Secretion of plasminogen activators by thecal and granulosal cells of gonadotrophin-stimulated follicles has been established; both uPA and tPA apparently contribute to preovulatory ovarian plasmin biosynthesis in rodents (Tsafriri and Reich, 1991; Hägglund *et al.*, 1996).

An auxiliary source of plasminogen activator is the ovarian surface epithelium. Accordingly, an increase in plasmin within the apical hemisphere of preovulatory ovine follicles (plus conjoined tunica albuginea) at 12 h after GnRH administration was attributed to secretion of low molecular mass uPA by ovarian surface epithelial cells (tPA was undetectable). When ovarian surface epithelium was surgically removed at 8 h after GnRH treatment, the follicular rise in uPA was negated and ovarian rupture was inhibited. Furthermore, ovulation was suppressed by intrafollicular injections of uPA (but not tPA) antibodies at 8 h (Colgin and Murdoch, 1997) or α_2-antiplasmin at 16 h (Murdoch, 1998a) after GnRH. Plasminogen activators were also increased preferentially within the apices of preovulatory porcine (Smokovitis *et al.*, 1988) and rat (Peng *et al.*, 1993) follicles and intrabursal administration of inhibitors of the plasminogen activator/plasmin system decreased ovulation rates in rats (Tsafriri and Reich, 1991). Interestingly, in certain species (for example horse and armadillo) ovulation is restricted to a discrete ovarian depression (fossa) covered by prototypical (coelomic mesothelial-derived) surface epithelium (Mossman and Duke, 1973).

Local controlling mechanisms of uPA secretion by sheep ovarian surface epithelial cells are equivocal. Receptors for gonadotrophins have been detected on ovarian surface cells (Godwin *et al.*, 1993). In fact, isolated sheep ovarian surface epithelial cells secrete uPA in response to LH (D. C. Colgin and W. J. Murdoch, unpublished observation). It is conceivable that cells in close proximity to the preovulatory follicle are readily exposed to surge gonadotrophin concentrations because of a marked acute increase (4–12 h after GnRH) in permeability of the thecal vascular wreath (Halterman and Murdoch, 1986; Cavender and Murdoch, 1988).

Collagenolytic Activation

It appears that plasmin activates latent collagenases (Danø *et al.*, 1985) which degrade the matrices of connective tissue of the follicular theca and tunica albuginea, thereby weakening the ovarian wall (Tsafriri and Reich, 1991). The general consensus of functional studies on plasminogen activators is that uPA regulates tissue degradation, while tPA is involved in thrombolysis (Danø *et al.*, 1985; Hart and Rehemtulla, 1988).

In preovulatory ovine follicles there is a close correlation between apical plasmin accumulation (Colgin and Murdoch, 1997) and the onset of collagen dissolution (Murdoch and McCormick, 1992). Explants of follicular wall released hydroxyproline-containing peptides (degraded collagen) upon exposure to plasmin and injection of α_2-antiplasmin into preovulatory follicles inhibited collagenase bioactivity of tissue extracts (Murdoch, 1998a). Morphological observations indicate that preovulatory connective tissue breakdown begins at the ovarian surface and advances inward toward the follicular wall (Bjersing and Cajander, 1975; Talbot *et al.*, 1987).

Induction of Cell Death by TNFα

Marked alterations in organ morphology are often associated with a programmed process of active physiological cell death or apoptosis. Early-stage apoptosis is distinguished by calcium influx, endonuclease activation, internucleosomal DNA fragmentation, and nuclear pyknosis. Apoptotic cells shrink and lose contact with their neighbours and supporting basement membranes. Residual bodies typically are resorbed by adjacent epithelial cells or resident macrophages. Cells undergoing apoptosis may disappear completely within a few hours (Ellis *et al.*, 1991; Schwartzman and Cidlowski, 1993).

Direct *in situ* fluorescence detection of digoxigenin end-labelled genomic DNA was used as a marker of nuclear apoptosis within preovulatory ovine follicles and surrounding ovarian tissues. As the time of ovulation approached (16–24 h after GnRH), there was a progressive increase in apoptotic cells within the ovarian surface epithelium, tunica albuginea and apical follicular wall. At the avascular site of impending rupture, follicles were almost devoid of ovarian surface and granulosal epithelia (dispersion of granulosa within the basal region of preovulatory follicles was not associated with apoptosis). Sloughing of ovarian surface epithelial cells occurred first, followed by cell losses within the tunica albuginea and follicular wall (Murdoch, 1994, 1995a,b). Thus, discrete physicochemical interactions between preovulatory follicles and the ovarian surface are evidently a prelude to programmed cell deletion and ovulation.

Our initial inclination was that prostaglandins were somehow involved in the biomechanics of apoptotic cell death within the formative ovulation papilla of sheep follicles. That prostaglandins are produced by follicular and ovarian surface cells during the preovulatory period (Murdoch *et al.*, 1991, 1993) and (at high doses *in vitro*) can provoke ovarian apoptotic cell death was established (Ackerman and Murdoch, 1993). It was therefore predicted that indomethacin, an inhibitor of prostaglandin biosynthesis and ovulation (Murdoch *et al.*, 1993), would protect apical ovarian cells from programmed death. The anti-ovulatory potencies of two systemic doses of indomethacin (200 or 800 mg given 8 h after GnRH) were tested. Ovulation did not occur after administration of 800 mg indomethacin but was not inhibited by 200 mg indomethacin. Both doses of the drug suppressed follicular prostaglandin production below pre-gonadotrophin values. Fragmentation of DNA was averted among ovarian surface epithelial and granulosal cells recovered from the apical dome of follicles (16 h after GnRH) of ewes given 800 mg indomethacin, whereas apoptosis ensued after 200 mg indomethacin. Intracellular calcium accretion detected by fluorescence of fura-2 was increased in ovarian cells of animals destined to ovulate (200 mg indomethacin) in comparison to (safeguarded) cells of anovulatory ewes (800 mg indomethacin) (Murdoch, 1996a). These observations provided circumstantial evidence that apical ovarian cell degeneration by calcium-mediated apoptosis is a determinant of follicular instability and rupture, but that these events are unrelated to the gonadotrophin-induced increase in prostanoid production characteristic of preovulatory follicles.

We then investigated the close relationship between plasmin upregulation and TNFα secretion within the apex of preovulatory follicles (Murdoch *et al.*, 1997). Tumour necrosis factor α is expressed as a 26 kDa integral transmembrane precursor molecule which upon proteolytic cleavage yields a 17 kDa extracellular domain subunit. Mature (cytotoxic) TNFα is a noncovalent trimer. Common cell types known to produce TNFα are leucocytes, smooth muscle, fibroblasts, and endothelium. Plasma membrane receptors for TNFα (R55, R75) are present on almost all nucleated cells (Vilcek and Lee, 1991; Vandenabeele *et al.*, 1995), including cells of the mammalian ovary (Terranova, 1997). It is now apparent that, in addition to its ability to induce lytic cell death (haemorrhagic necrosis), TNFα can transduce an apoptotic signal that results in programmed cell death (Larrick and Wright, 1990; Haanen and Vermes, 1995; Steller, 1995).

Tumour necrosis factor α was localized by indirect immunofluorescence microscopy to thecal endothelial cells of preovulatory ovine follicles (Murdoch *et al.*, 1997). Immunostaining of endothelial cells within the follicular apex declined abruptly with the approach of ovulation (cells within the counterpart basal wall were unaffected); it appeared that TNFα had been released into the progenitor site of ovulation. Intrafollicular injection (10 h after GnRH) of TNFα antiserum circumvented ovarian DNA fragmentation and blocked ovulation in ewes. Moreover, TNFα (at physiologically relevant concentrations) induced ovarian cell apoptosis *in vitro* (Murdoch *et al.*, 1997) and ovulation rates were enhanced by addition of TNFα to perfusates of rat ovaries (Brännström *et al.*, 1995).

The biomechanics of TNFα expression and release from resident ovarian cells have not been elucidated. It is unlikely that a stimulatory effect of gonadotrophin on TNFα secretion is direct, but rather is mediated by other agents in response to hormonal stimulation (Terranova, 1997). Tumour necrosis factor α is a candidate substrate for serine protease (perhaps plasmin) attack (Scuderi, 1989; Perona and Craik, 1995). Intrafollicular α_2-antiplasmin averted preovulatory apical TNFα-mediated cellular DNA fragmentation and plasmin-stimulated bioactive TNFα release from follicular explants (Murdoch, 1998a). In recent studies (W. J. Murdoch and E. A. Van Kirk, unpublished), cleavage by plasmin of TNFα exodomain from its membrane anchor on thecal endothelial cells appeared to be responsible for programming apoptotic death among ovarian cells within a limited diffusion radius. At high tissue concentrations, TNFα also initiates microvascular coagulation associated with necrotic cell death (Larrick and Wright, 1990) and inflammatory tissue damage symptomatic of the ovulatory process (Espey, 1980). Vascular lesions typical of haemorrhagic necrosis are observed

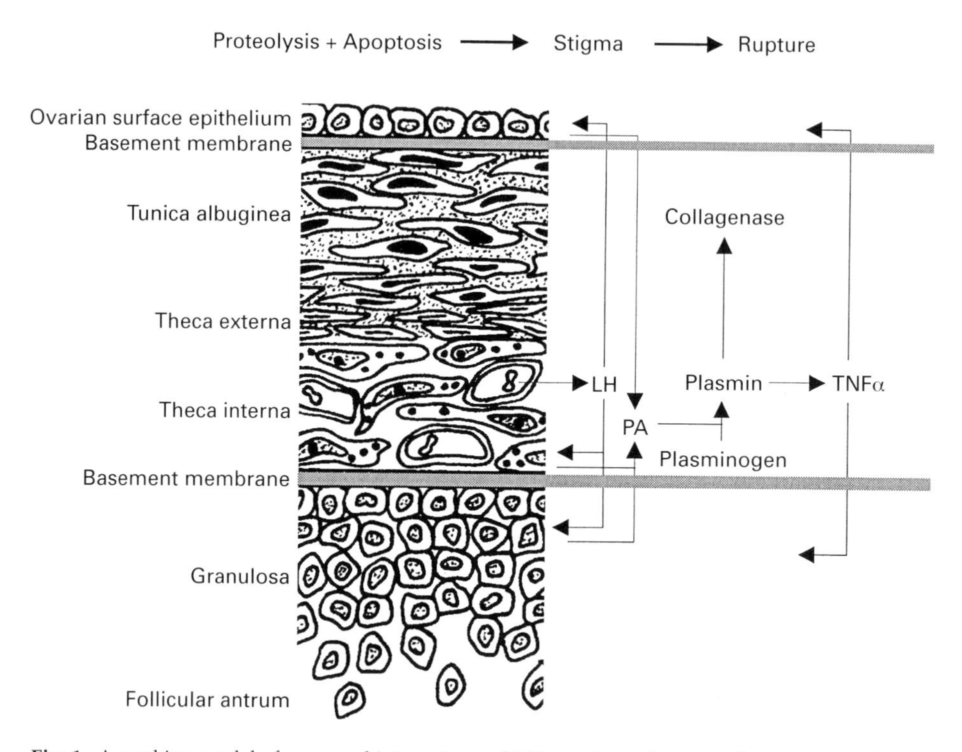

Fig. 1. A working model of proposed interactions of LH, ovarian cell types, plasminogen activator (PA)/plasmin, and tumour necrosis factor α (TNFα) in the breakdown of the apical follicular wall during ovulation in sheep: vascular transudate containing gonadotrophin is delivered to receptor-bearing cells (i.e. granulosa, theca interna, surface epithelium) of the ovarian wall, thereby stimulating secretion of plasminogen activator; interstitial plasminogen is converted to plasmin, which activates latent collagenases and cleaves TNFα from its endothelial mooring; collagenases disrupt the fibril network of the theca and tunica albuginea and promote disintegration of the basement membranes supporting the ovarian and granulosal epithelia; TNFα induces apoptosis; collagenolysis and cellular effacement dictate stigma development and follicular rupture.

within the immediate area surrounding the stigma of the preovulatory ovine follicle (Cavender and Murdoch, 1988; Murdoch and Cavender, 1989). A lack of blood flow (ischaemia) into the ovulation papilla, leading to oxygen deprivation and toxic metabolite accumulation, would predictably potentiate cell death. Experiments are underway to determine whether the anti-ovulatory effect of indomethacin is related to a (prostaglandin-independent) abrogation of TNFα action.

Finally, mechanical forces may combine with connective tissue degradation and cell elimination to assure tissue thinning and follicular rupture. Retraction of the basal theca due to contractility (Martin and Talbot, 1981) would theoretically cause the opposing wall to recede from the ovarian surface.

Ovulation and Wound Repair

Ovulation creates a wound along the ovarian surface that is repaired during the ensuing luteal phase. Inauspiciously, some cells at the margins of ruptured follicles that endure the ovulatory (TNFα) insult contain fragmented DNA (Ackerman and Murdoch, 1993; Murdoch, 1994, 1995a, 1998b). Damage to DNA that is uncorrected could be problematic if propagated; indeed, most ovarian cancers apparently originate by malignant clonal transformation of a surface epithelial cell traumatized at ovulation (Hamilton, 1992; Godwin *et al.*, 1993; Murdoch, 1996b). Sublethal damage to DNA that is inflicted upon ovarian surface cells can evidently be reconciled by repair enzymes induced on a localized basis by progesterone of luteal origin (Murdoch, 1998b).

Conclusion

On the basis of the cited investigations, plasmin has an intermediary role in the proteolytic and cell death mechanisms of follicular stigma formation and ovulatory ovarian rupture in sheep. A role for plasmin in the stimulation of collagenases is well known. However, a pivotal function of plasmin in the bioactivation of TNFα is novel. A synopsis of putative interactions of gonadotrophin, apical ovarian tissues, plasminogen activator/plasmin, and TNFα in the ovulatory process of the sheep is depicted in Fig. 1.

References

Ackerman RC and Murdoch WJ (1993) Prostaglandin-induced apoptosis of ovarian surface epithelial cells *Prostaglandins* **45** 473–483

Bjersing L and Cajander S (1975) Ovulation and the role of the ovarian surface epithelium *Experientia* **31** 605–608

Brännström M, Bonello N, Wang LJ and Norman RJ (1995) Effects of tumour necrosis factor α on ovulation in the rat ovary *Reproduction Fertility and Development* **7** 67–73

Cavender JL and Murdoch WJ (1988) Morphological studies of the microcirculatory system of periovulatory ovine follicles *Biology of Reproduction* **39** 989–997

Colgin DC and Murdoch WJ (1997) Evidence for a role of the ovarian surface epithelium in the ovulatory mechanism of the sheep: secretion of urokinase-type plasminogen activator *Animal Reproduction Science* **47** 197–204

Danø K, Andreasen PA, Grøndahl-Hansen J, Kristensen P, Nielsen LS and Skriver L (1985) Plasminogen activators, tissue degradation, and cancer *Advances in Cancer Research* **44** 139–266

Ellis RE, Yuan J and Horvitz HR (1991) Mechanisms and functions of cell death *Annual Review of Cell Biology* **7** 663–698

Espey LL (1980) Ovulation as an inflammatory reaction – a hypothesis *Biology of Reproduction* **22** 73–106

Espey LL and Lipner H (1994) Ovulation. In *The Physiology of Reproduction* pp 725–780 Eds E Knobil and JD Neill. Raven Press, New York

Godwin AK, Testa JR and Hamilton TC (1993) The biology of ovarian cancer development *Cancer* **71** 530–536

Haanen C and Vermes I (1995) Apoptosis and inflammation *Mediators of Inflammation* **4** 5–15

Hägglund A-C, Ny A, Liu K and Ny T (1996) Coordinated and cell-specific induction of both physiological plasminogen activators creates functionally redundant mechanisms for plasmin formation during ovulation *Endocrinology* **137** 5671–5677

Halterman SD and Murdoch WJ (1986) Ovarian function in ewes treated with antihistamines *Endocrinology* **119** 2417–2421

Hamilton TC (1992) Ovarian cancer, biology *Current Problems in Cancer* **16** 5–57

Hart DA and Rehemtulla A (1988) Plasminogen activators and their inhibitors: regulators of extracellular proteolysis and cell function *Comparative Biochemistry and Physiology* **90B** 691–708

Larrick JW and Wright SC (1990) Cytotoxic mechanism of tumor necrosis factor-α *FASEB Journal* **4** 3215–3223

Martin GG and Talbot P (1981) The role of follicular smooth muscle cells in hamster ovulation *Journal of Experimental Zoology* **216** 469–482

Mossman HW and Duke KL (1973) *Comparative Morphology of the Mammalian Ovary* University of Wisconsin Press, Madison

Murdoch WJ (1985) Follicular determinants of ovulation in the ewe *Domestic Animal Endocrinology* **2** 105–121

Murdoch WJ (1994) Ovarian surface epithelium during ovulatory and anovulatory ovine estrous cycles *Anatomical Record* **240** 322–326

Murdoch WJ (1995a) Programmed cell death in preovulatory ovine follicles *Biology of Reproduction* **53** 8–12

Murdoch WJ (1995b) Endothelial cell death in preovulatory ovine follicles: possible implication in the biomechanics of rupture *Journal of Reproduction and Fertility* **105** 161–164

Murdoch WJ (1996a) Differential effects of indomethacin on the sheep ovary: prostaglandin biosynthesis, intracellular calcium, apoptosis, and ovulation *Prostaglandins* **52** 497–506

Murdoch WJ (1996b) Ovarian surface epithelium, ovulation, and carcinogenesis *Biological Reviews* **71** 529–543

Murdoch WJ (1998a) Regulation of collagenolysis and cell death by plasmin within the formative stigma of preovulatory ovine follicles *Journal of Reproduction and Fertility* **113** 331–336

Murdoch WJ (1998b) Perturbation of sheep ovarian surface epithelial cells by ovulation: evidence for roles of progesterone and poly(ADP-ribose) polymerase in the restoration of DNA integrity *Journal of Endocrinology* **156** 503–508

Murdoch WJ and Cavender JL (1989) Effect of indomethacin on the vascular architecture of preovulatory ovine follicles: possible implication in the luteinized unruptured follicle syndrome *Fertility and Sterility* **51** 153–155

Murdoch WJ and McCormick RJ (1992) Enhanced degradation of collagen within apical vs. basal wall of ovulatory ovine follicle *American Journal of Physiology* **263** E221–E225

Murdoch WJ, Slaughter RG and Ji TH (1991) In situ hybridization analysis of ovarian prostaglandin endoperoxide synthase mRNA throughout the periovulatory period of the ewe *Domestic Animal Endocrinology* **8** 457–459

Murdoch WJ, Hansen TR and McPherson LA (1993) Role of eicosanoids in vertebrate ovulation *Prostaglandins* **46** 85–115

Murdoch WJ, Colgin DC and Ellis JA (1997) Role of tumor necrosis factor-α in the ovulatory mechanism of ewes *Journal of Animal Science* **75** 1601–1605

Peng X-R, Hsueh AJW and Ny T (1993) Transient and cell-specific expression of tissue-type plasminogen activator and plasminogen-activator-inhibitor type 1 results in controlled and directed proteolysis during gonadotropin-induced ovulation *European Journal of Biochemistry* **214** 147–156

Perona JJ and Craik CS (1995) Structural specificity in the serine proteases *Protein Science* **4** 337–360

Roberts AJ, Dunn TG and Murdoch WJ (1985) Induction of ovulation in proestrous ewes: identification of the ovulatory follicle and functional status of the corpus luteum *Domestic Animal Endocrinology* **2** 207–210

Schwartzman RA and Cidlowski JA (1993) Apoptosis: the biochemistry and molecular biology of programmed cell death *Endocrine Reviews* **14** 133–151

Scuderi P (1989) Suppression of human leukocyte tumor necrosis factor secretion by the serine protease inhibitor *p*-toluenesulfonyl-L-arginine methyl ester *Journal of Immunology* **143** 168–173

Smokovitis A, Kokolis N and Alexaki-Tzivanidou E (1988) The plasminogen activator activity is markedly increased mainly at the area of the rupture of the follicular wall at the time of ovulation *Animal Reproduction Science* **16** 285–294

Steller H (1995) Mechanisms and genes of cellular suicide *Science* **267** 1445–1449

Talbot P, Martin GG and Ashby H (1987) Formation of the rupture site in preovulatory hamster and mouse follicles: loss of surface epithelium *Gamete Research* **17** 287–302

Terranova PF (1997) Potential roles of tumor necrosis factor-α in follicular development, ovulation, and the life span of the corpus luteum *Domestic Animal Endocrinology* **14** 1–15

Tsafriri A and Reich R (1991) Plasminogen activators in the preovulatory follicle: role in ovulation. In *Plasminogen Activators: From Cloning To Therapy* pp 81–93 Eds R Abbate, T Barni and A Tsafriri. Raven Press, New York

Vandenabeele P, Declercq W, Beyaert R and Fiers W (1995) Two tumour necrosis factor receptors: structure and function *Trends in Cell Biology* **5** 392–399

Vilcek J and Lee TH (1991) Tumor necrosis factor *Journal of Biological Chemistry* **266** 7313–7316

Journal of Reproduction and Fertility Supplement **54**, 359–365

Growth factors and extracellular matrix proteins in interactions of cumulus–oocyte complex, spermatozoa and oviduct

R. Einspanier[1], C. Gabler[1], B. Bieser[1], A. Einspanier[2], B. Berisha[1], M. Kosmann[1], K. Wollenhaupt[3] and D. Schams[1]

[1]*Institute of Physiology, TU Munich, D-85350 Freising, Germany;* [2]*German Primate Center, D-37077 Göttingen, Germany; and* [3]*FBN, D-18196 Dummerstorf, Germany*

The expression and localization of selected growth factor systems and extracellular matrix (ECM) components that may influence oocyte maturation and fertilization within the mammalian oviduct are reported. Fibroblast growth factor (FGF) and vascular endothelial growth factor (VEGF) systems could be detected by use of RT–PCR, RNase protection assay (RPA) and immunohistochemistry in bovine follicles, bovine cumulus–oocyte complexes (COC) and bovine and marmoset oviducts. Two different subtypes of the FGF receptor (FGFR-1 and -2) were identified in distinct cell types, indicating a functional difference. A complete epidermal growth factor (EGF) system was found in the porcine, but not in the bovine, oviduct. There were additional differences between bovine and primate oviducts: FGF-1/2 and FGFR were increased in the marmoset around ovulation, in contrast to an increase in FGF-1 in the cow. Immunohistochemistry revealed accumulation and storage of FGF and VEGF on the surface of the epithelium, possibly due to their binding property on heparan-glycoproteins. Other ECM components, matrix metalloproteinase 1 (MMP-1) and tissue inhibitor of metalloproteinase 1 (TIMP-1), were found to be modulated in the ovarian follicle, COC and oviduct during the cycle. An oviduct-mediated depletion of sperm surface proteins (BSP1–3) was discovered as well as a sperm-induced novel oviductal mRNA related to an anti-oxidant protein family. Associated systems of growth factors and ECM components can be suggested as paracrine or autocrine mediators during fertilization in a species-, cycle- and tissue-dependent manner.

The Oviduct as the Site of Fertilization and Early Embryonic Development

The central events of fertilization take place in the environment of the mammalian oviduct, which appears to be controlled by gonadotrophins, sex steroids or both. Such hormonal actions may be sustained by a paracrine action mediated by growth factors or extracellular matrix components (ECM) as in the development of mammalian ovarian follicle (Monniaux *et al.*, 1997). These multifunctional growth factors are known to both stimulate and inhibit cell proliferation and differentiation (Sporn and Roberts, 1988) and therefore represent an important class of substances in regulating cell–cell interactions. Known as the organ of limited storage and final capacitation of the sperm (Lefebvre *et al.*, 1995), the oviduct has attracted much scientific interest. The nature of oestrogen-dependent glycoproteins has been analysed for humans, primates and ruminants (Donnelly *et al.*, 1991; Arias *et al.*, 1994; DeSouza and Murray 1995; Sendai *et al.*, 1995) and indicates that they bind to the oocyte or spermatozoa.

Embryotrophic factors and mitogens, such as growth factors, are present in variable amounts in the oviduct (Gandolfi *et al.*, 1989) and Ellington (1991) provided a detailed review on this important reproductive organ. This review deals with the expression of bioactive proteins in bovine ovarian

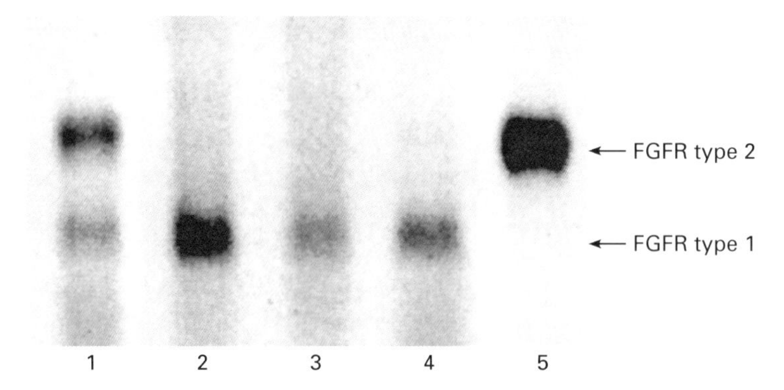

Fig. 1. Detection of fibroblast growth factor receptor (FGFR) subtype 1 and 2 in bovine ovarian thecal tissue (1), granulosal cells before LH (2), granulosal cells after LH (3), matured cumulus–oocyte complexes (COC) (4) and oviductal epithelium (5). Uniform RT–PCR products were separated by intercalating agarose electrophoresis depending on sequence differences.

follicles, cumulus–oocyte complexes (COC) and mammalian oviducts in relation to their developmental state. Furthermore, the initial interactions between the oviduct and spermatozoa are described.

Protein Expression in Ovarian and Oviductal Cells

Growth factor systems

Within the ovarian follicle, at the site of oocyte maturation, a complex network of different growth factors is present, modulating the autocrine and paracrine effects observed under hormonal influence. We elucidated some changes during the maturation of bovine follicles by detecting distinct gene products of the fibroblast growth factor (FGF) and vascular endothelial growth factor (VEGF) system. In thecal tissue, the overall expression of FGF-1, FGF-2, and FGFR was found to be relatively high when compared with that of granulosal cells (Einspanier *et al.*, 1997). FGF-2 mRNA was found in considerable amounts only after the LH surge in granulosal cells *in vivo*. The same effect could be observed in cumulus cells during *in vitro* maturation (IVM) of COC which was supported by detection of immunoreactive FGF-2 protein. The complete VEGF system was found in bovine follicles: the ligand and both receptors (flk, flt) were identified in thecal and granulosal cells and, moreover, VEGF protein was increased in follicular fluid just before ovulation. This observation was in agreement with a large increase in VEGF mRNA synthesis found in granulosal cells *in vivo* after LH. In contrast, cumulus cells surrounding the oocyte contained only low concentration of VEGF transcript after IVM. A second difference between granulosal and cumulus cells is a probable lack of the flk-type receptor in cumulus cells. After applying a gel-retardation technique (Plath *et al.*, 1997) two different types of FGFR were found using a simple electrophoretic separation of specific PCR products. A sole FGFR type 1 was found in granulosal and cumulus cells, the type 2 only in oviductal epithelium, but a mixture of both in thecal tissue (Fig. 1). A functional difference between these receptor types could support their distinct tasks.

The ideal environment for fertilization and early embryonic development is found in the oviduct. Our search for specific growth factors revealed the following details: all components of the FGF and VEGF systems are expressed within the bovine and primate oviduct epithelium. Specifically, a cycle-dependent regulation of the FGF-1 was detected by measuring significantly higher protein concentrations in bovine oviductal flushings during ovulation (5.3 ± 0.5 ng ml^{-1})

compared with the luteal phase (3.5 ± 0.5 ng ml^{-1}). In parallel, there was an increase in FGF-1 transcripts before ovulation in oviductal cells and for the whole cycle the FGF-2 and FGFR expression appeared to stay constant (Gabler *et al.*, 1997a). VEGF was enriched just before ovulation, as was its flt-type receptor, but expression of the flk-receptor was unchanged during the cycle, indicating an independent regulation of both receptor types within the bovine oviduct. In contrast, when these growth factors were investigated in the marmoset monkey, highest mRNA contents were found during the ovulatory period for all growth factors and their receptors. Likewise, FGF-1, FGF-2 and VEGF immunoreactive proteins were mainly found on the epithelial surface of both the bovine and primate oviduct and at the same location as heparan sulfate molecules. This phenomenon will be discussed in the ECM section below. In the cow, we could not find a complete epidermal growth factor (EGF) system, which is known to represent a prominent growth factor in other mammals (Adachi *et al.*, 1995; Wollenhaupt *et al.*, 1997). In contrast, the porcine endometrium and oviduct contain mRNA encoding both ligands (EGF and TGFα) and the receptor (EGFR). Furthermore, a biologically active endometrial or oviductal EGFR could be demonstrated by ligand binding and EGF-dependent phosphorylation of the receptor protein (Wollenhaupt *et al.*, 1998).

ECM components

Collagenolytic activity has been shown to be an important process during gonadotrophin-mediated follicular rupture in the rabbit (Tadakuma *et al.*, 1993). According to this, MMP-1 expression was below the detection limit in bovine thecal and granulosal cells before the LH surge, but thereafter increased expression in both cell types indicates that they have a role at the time of ovulation. The inhibitor of MMP-1, TIMP-1, reacted in a similar way indicating a fine-tuned proteolytic cascade within the follicle. When *in vitro* matured COC were sequentially analysed for the expression of MMP-1 and TIMP-1 using zymographic electrophoresis, two proteolytically active enzymes of 55 and 67 kDa were detected, and one of these proteins may be the latent MMP-1 protein (Lauer and Einspanier, 1996a). In parallel, using RT–PCR, we detected increasing MMP-1 transcripts during IVM from 6 h. A similar increase of specific mRNA could be observed for TIMP-1 after 6 h of maturation, with a relatively stable high expression up to 24 h of IVM. Parallel cumulus expansion suggested some influences of these proteolytic ECM components.

Heparan sulfate-like molecules could be detected by immunohistology on the surface of bovine and marmoset oviduct epithelial cell layer, as shown by Gabler *et al.* (1997a, 1998). On this basis, it is possible that heparin-binding proteins like FGF and VEGF could be easily fixed on such structures, changing their biological availability. A longer half-life of these growth factors could result from this process, mediating local effects. In this context, the ECM protease MMP-1 showed increased activity and expression during the luteal phase within the bovine oviduct, possibly contributing to a cellular turnover or reassembly. At the same time, there was a decrease of its inhibitor TIMP-1, but during ovulation a rise of transcript concentrations could be seen (Fig. 2), possibly preventing a matrix degradation during ovulation. However, a second embryo-stimulating effect of TIMP-1 in the oviduct, as suggested earlier by Satoh *et al.* (1994), cannot be excluded.

Interactions Between Oviduct and Spermatozoa

Sperm capacitation takes place not only in the cervix but also continues during passage through the oviduct. Therefore, final maturation of spermatozoa may occur in transit. Furthermore, storage of viable spermatozoa was observed in the isthmus of the oviduct (Hunter, 1995). After exposure of bovine spermatozoa to cultured oviductal epithelial cells or oviductal fluid, we found that three specific proteins were liberated in a time-dependent manner from sperm surfaces. N-terminal characterization of these proteins revealed that they were known bovine seminal proteins BSP1, -2 and -3 (Lauer *et al.*, 1995). Such proteins have high affinities for heparin and can easily react with heparan molecules on the oviductal surface. Furthermore, BSPs can be removed by addition of heparin *in vitro*, which may explain why heparin has been commonly introduced to bovine sperm

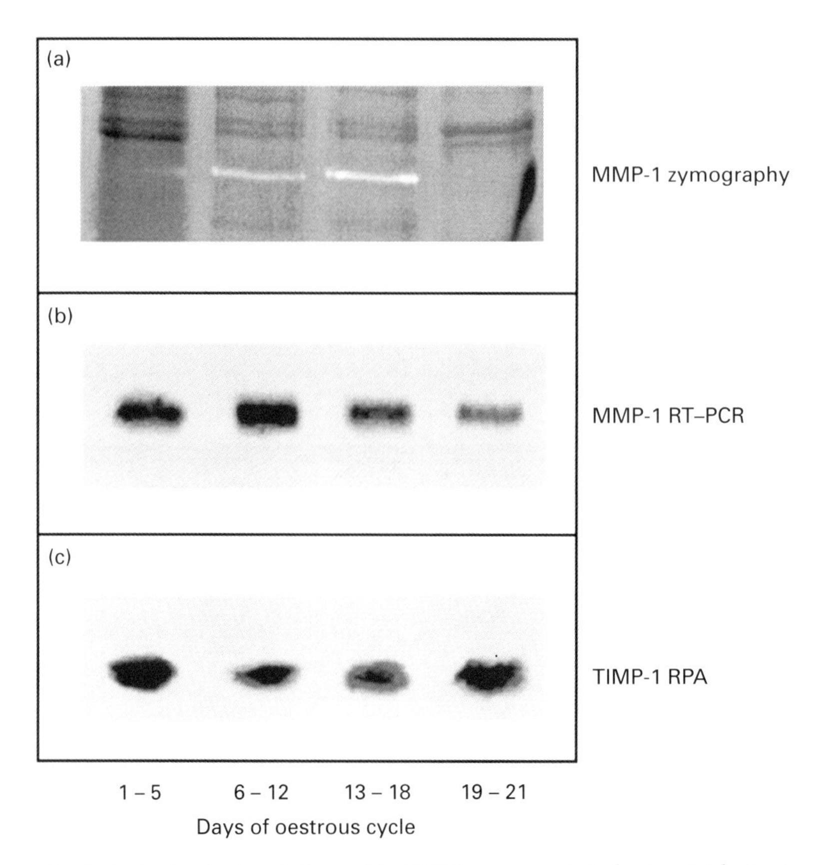

Fig. 2. Analysis of extracellular matrix (ECM) components in bovine oviducts at different stages of the oestrous cycle: presence of matrix metalloproteinase 1 (MMP-1)-like enzymatic activity detected by zymography (a; one of two experiments). Specific mRNAs in epithelial cells were shown by RT–PCR for MMP-1 (b; 37 cycles, one of three experiments) and by RNase protection assay for tissue inhibitor of metalloproteinase 1 (TIMP-1) (c; 30 μg total RNA, one of six experiments) as indicated.

capacitation media. *In vivo* related heparin analogues are found within the oviduct fluid in variable amounts depending on the stage of the ovarian cycle (Lauer and Einspanier, 1996b).

A new random PCR technique (randomly amplified RNA (RAP)-PCR) was used to study new transcripts within the oviductal cell that may be triggered by sperm contact. Such a sperm-induced mRNA of about 440 bases was isolated from cultured oviduct epithelial cells. After cDNA sequencing, the transcript was identified as a new bovine mRNA fragment (EMBL-Ac.nr: Z86040) encoding a protein related to a large anti-oxidant protein family. The deduced amino acid structure of the S5-clone possesses a very high homology (96%) to a known human protein (ULA06) affiliated to the AhpC/TSA (alkyl-hydroperoxide-reductase/thiol-specific antioxidant) family (Fig. 3). All members of this protein family show properties associated with oxygen-regulating enzyme systems that may be important contributors of embryonic viability; comparable redox systems within the bovine oviduct were reported to be necessary by Harvey and co-workers (1995).

```
S5     ..LQLTAEKRVATPVDWKNGDSVMVLPTIPEEEAKKLFPKGVFTKELPSGKKYLRYTPQP-C
         ++++++++++++++++ ++++++++++++++++++++++++++++++++++++++++
ULA06  ..LQLTAEKRVATPVDWKDGDSVMVLPTIPEEEAKKLFPKGVFTKELPSGKKYLRYTPQP-C
```

Fig. 3. Sequence comparison of the new sperm-induced oviductal protein S5 and a known human protein (ULA06) related to a reductase/antioxidant (AhpC/TSA) protein family. The C-terminal part is shown. Homologous amino acids are marked (+).

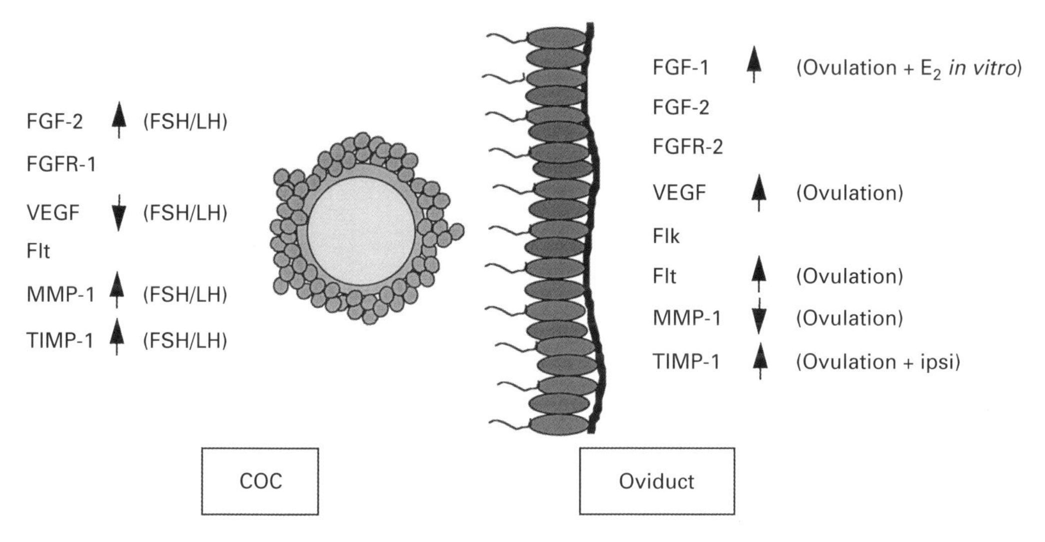

Fig. 4. Specific gene expression is identified in bovine cumulus–oocyte complexes (COC) and oviduct epithelium. Obvious regulation is increased (↑) or decreased (↓) amounts of mRNA after the FSH/LH surge, during ovulation, oestradiol-mediated (E_2) or in the ipsilateral oviduct (ipsi). FGF-1, fibroblast growth factor 1; FGF-2, fibroblast growth factor 2; FGFR-1, fibroblast growth factor receptor 1; FGFR-2, fibroblast growth factor 2; Flt/Flk, vascular endothelial growth factor (VEGF) receptors; MMP-1, matrix metalloproteinase 1; TIMP-1, tissue inhibitor of metalloproteinase 1.

Conclusions

The FGF and VEGF growth factor systems were detected and the ECM components MMP-1 and TIMP-1 were expressed in a specific manner within ovarian follicles, the COC, and the oviduct. Species-specific and cycle-dependent variations in expression of growth factors were found especially correlated with the cell type (Gabler *et al.*, 1997b). Specific distributions of different FGFR types indicate an important role of the receptor structure that is necessary for transducing intracellular signals (Bifkalvi *et al.*, 1997). All known biological effects of the potent mitogenic FGFs are related to cellular differentiation, angiogenesis, wound healing and tumour growth (Gospodarowicz *et al.*, 1987), whereas endothelial cells may be the target of VEGF to promote angiogenesis and permeability (Ferrara *et al.*, 1992). Differentiation and nutritional support could be partly regulated by these growth factors within the ruminant and primate oviduct.

Furthermore, increased TIMP-1 expression could support an embryotrophic effect within the bovine oviduct (Gabler and Einspanier 1998). As summarized in Fig. 4, within the bovine oviduct many possible paracrine interactions between the COC and oviduct can be proposed. In addition, ECM components are directly connected to important reproductive responses by either liberating biologically active proteins from epithelial surfaces or changing sperm-surface structures (Einspanier *et al.*, 1996). However, the selection and dominance of ovarian follicles may be influenced by growth factors as proposed for the FGF- and VEGF-systems (Kamhuber *et al.*, 1997; Schams *et al.*, 1997). The proposed paracrine cross-talk involving different growth factor and ECM systems leads to a complex network within the oviduct enabling alternating reaction cascades leading to an interesting fuzzy-logic approach (Horseman *et al.*, 1997). Our results support such a concept of paracrine and autocrine interactions containing different growth factor and ECM systems with respect to a species-specific appearance in the mammalian oviduct. This enables a limited insight into the naturally occurring conditions for supporting the young embryo, and in the future the biological relevance of the proteins described will be further characterized.

We would like to thank S. Rode, K. Fuhrmann and A. Marten for technical help. This work was supported by grants of the DFG (Ei 296/4-2; Ei 296/6-1; Wo 663/1-1).

References

Adachi K, Kurachi H, Adachi H, Imai T, Sakata M, Homma H, Higashiguchi O, Yamamoto T and Miyake A (1995) Menstrual cycle specific expression of epidermal growth factor receptors in human fallopian tube epithelium *Journal of Endocrinology* **147** 553–563

Arias EB, Verhage HG and Jaffe RC (1994) Complementary deoxyribonucleic acid cloning and molecular characterization of an estrogen-dependent human oviductal glycoprotein *Biology of Reproduction* **51** 685–694

Bifkalvi A, Klein S, Pintucci G and Rifkin DB (1997) Biological roles of fibroblast growth factor-2 *Endocrine Reviews* **18** 26–44

DeSouza MM and Murray MK (1995) An estrogen-dependent secretory protein, which shares identity with chitinases, is expressed in a temporally and regionally specific manner in the sheep oviduct at the time of fertilization and embryo development *Endocrinology* **136** 2485–2496

Donnelly KM, Fazleabas and Verhage HG (1991) Cloning of a recombinant complementary DNA to a baboon estradiol-dependent oviduct-specific glycoprotein *Molecular Endocrinology* **5** 356–364

Einspanier R, Gabler C and Lauer B (1996) Expression of urokinase plasminogen activator (uPA) is correlated with soluble heparin-analogues and basic fibroblast growth factor (bFGF) in bovine oviducts during oestrous cycle *Journal of Reproduction and Fertility Abstract Series* **17** Abstract 26

Einspanier R, Lauer B, Gabler C, Kamhuber M and Schams D (1997) Egg–cumulus–oviduct interactions and fertilization. In *The Fate of the Male Germ Cell Advances in Experimental Medicine and Biology* pp 279–289 Ed. R Ivell and Holstein. Plenum Press.

Ellington JE (1991) The bovine oviduct and its role in reproduction: a review of the literature *Cornell Veterinarian* **81** 313–328

Ferrara N, Houck K, Jakeman L and Leung DW (1992) Molecular and biological properties of the VEGF family of proteins *Endocrine Reviews* **13** 18–32

Gabler C and Einspanier R (1998) Increased expression of tissue inhibitor of metalloproteinase 1 (TIMP-1) during ovulation indicates its remarkable biological action in the bovine oviduct *Experimental and Clinical Endocrinology and Diabetes* **106** 8 (Abstract)

Gabler C, Lauer B, Einspanier A, Schams D and Einspanier R (1997a) Detection of mRNA and immunoreactive proteins for acidic and basic fibroblast growth factor and expression of the fibroblast growth factor receptor in the bovine oviduct *Journal of Reproduction and Fertility* **109** 213–221

Gabler C, Einspanier A and Einspanier R (1997b) Species-specific differences of growth factor expression in the female reproductive tract: FGF- and IGF-systems in primate versus bovine oviduct *Reproduction in Domestic Animals* **32** 7 (Abstract)

Gabler C, Plath-Gabler A, Einspanier A and Einspanier R (1998) The localization and expression of insulin-like growth factors, fibroblast growth factors, transforming growth factor alpha and their receptors indicate an auto-/paracrine influence in the oviduct of the common marmoset monkey (*Call jacch*) *Biology of Reproduction* **58** 1451–1457

Gandolfi F, Brevini TAL and Moor RM (1989) Effects of oviduct environment on embryonic development *Journal of Reproduction and Fertility* **38** 107–115

Gospodarowicz D, Ferrara N, Schweigerer L and Neufeld G (1987) Structural characterization and biological functions of FGF *Endocrine Review* **8** 95–114

Harvey MB, Arcellana-Panlilio MY, Zhang X, Schultz GA and Watson AJ (1995) Expression of genes encoding antioxidant enzymes in pre-implantation mouse and cow embryos and primary bovine oviduct cultures employed for embryo co-culture *Biology of Reproduction* **53** 532–540

Horseman N, Engle S and Ralescu A (1997) The logic of signalling from the cell surface to the nucleus *Trends in Endocrinology and Metabolism* **8** 123–129

Hunter RH (1995) Human sperm reservoirs and Fallopian tube function: a role for the intra-mural portion? *Acta Obstetrica et Gynecologia Scandinavia* **74** 677–681

Kamhuber M, Amselgruber W, Einspanier R and Schams D (1997) Possible role of growth factors for follicle selection during the bovine oestrous cycle *Reproduction in Domestic Animals* **32** 8 (Abstract)

Lauer B and Einspanier R (1996a) Cumulus expansion at the molecular level: expression of extracellular matrix components *Archives of Animal Breeding, Dummerstorf* **39** 51 (Abstract)

Lauer B and Einspanier R (1996b) Are distinct levels of soluble heparin-analogues responsible for different alterations of sperm in the bovine oviduct? *Experimental and Clinical Endocrinology and Diabetes* **104** 131 (Abstract)

Lauer B, Wollenhaupt K and Einspanier R (1995) Detection of a new estradiol-mediated bovine oviduct protein and sperm–oviduct interactions using a tissue-perfusion system *Journal of Reproduction and Fertility Abstract Series* **15** Abstract 94

Levebvre R, Chenoweth PJ, Drost M, LeClear CT, MacCubbin M, Dutton JT and Suarez SS (1995) Characterization of the oviductal sperm reservoir in cattle *Biology of Reproduction* **53** 1066–1074

Monniaux D, Huet C, Besnard N, Clement F, Bosc M, Pisselet C, Monget P and Mariana JC (1997) Follicular growth and ovarian dynamics in mammals *Journal of Reproduction and Fertility* **51** 3–23

Plath A, Krause I and Einspanier R (1997) Species identification in dairy products by three different DNA-based techniques *Zeitschrift Lebensmittel Untersuchung Forschung A* **205** 437–441

Satoh T, Kobayashi K, Yamashita S, Kikuchi M, Sendai Y and Hoshi H (1994) Tissue inhibitor of metalloproteinases (TIMP-1) produced by granulosa and oviduct cells enhances *in vitro* development of bovine embryo *Biology of Reproduction* **50** 835–844

Schams D, Kamhuber M and Einspanier R (1997) Expression changes of growth factors, receptors and P450-aromatase during recruitment, selection and dominance of bovine follicles *Biology of Reproduction* **56** 169 (Abstract)

Sendai Y, Komiya H, Suzuki K, Onuma T, Kikuchi M, Hoshi H and Araki Y (1995) Molecular cloning and characterization of a mouse oviduct-specific glycoprotein *Biology of Reproduction* **53** 285–294

Sporn MB and Roberts AB (1988) Peptide growth factors are multifunctional *Nature* **332** 217–219

Tadakuma H, Okamura H, Kitaoka M, Iyama K and Usuku G (1993) Association of immunolocalization of matrix metalloproteinase 1 with ovulation in hCG-treated rabbit ovary *Journal of Reproduction and Fertility* **98** 503–508

Wollenhaupt K, Tiemann U, Einspanier R, Schneider F, Kanitz W and Brüssow KP (1997) Characterization of the epidermal growth factor receptor (EGF-R) in the pig oviduct and endometrium *Journal of Reproduction and Fertility* **111** 173–181

Wollenhaupt K, Gabler C, Einspanier R and Schneider F (1998) Regulation of the EGF/EGF-R-system in the endometrium of early pregnant pigs *Reproduction in Domestic Animals* **5** 154 (Abstract)

Journal of Reproduction and Fertility Supplement **54**, 367–381

Regulation of ovarian extracellular matrix remodelling by metalloproteinases and their tissue inhibitors: effects on follicular development, ovulation and luteal function

M. F. Smith[1], E. W. McIntush[2], W. A. Ricke[1], F. N. Kojima[1] and G. W. Smith[3]

[1]*Department of Animal Sciences, University of Missouri-Columbia, Columbia, MO 65211, USA;* [2]*Animal Reproduction and Biotechnology Laboratory, Colorado State University, Ft Collins, CO 80523, USA; and* [3]*Department of Animal Sciences, Michigan State University, East Lansing, MI 48824, USA*

In most organs, remodelling of tissues after morphogenesis is minimal; however, normal ovarian function depends upon cyclical remodelling of the extracellular matrix (ECM). The ECM has a profound effect on cellular functions and probably plays an important role in the processes of follicular development and atresia, ovulation, and development, maintenance and regression of corpora lutea. Matrix metalloproteinases (MMPs; collagenases, gelatinases, stromelysins and membrane-type MMPs) cleave specific components of the ECM and are inhibited by tissue inhibitors of metalloproteinases (TIMPs). MMPs have been detected at all stages of follicular development and probably modulate follicular expansion or atresia within the ovarian stroma. In addition, increased MMP activity appears to be required for ovulation since follicular rupture occurred in the absence of plasminogen activator activity and inhibitors of MMPs blocked follicular rupture. Development and luteolysis of the corpus luteum are accompanied by extensive remodelling of the ECM. Differentiation and regression of luteal cells are associated with construction and degradation of ECM, respectively. There is increasing evidence that ECM components enhance luteinization; whereas loss of ECM results in luteal cell death. Ovine large luteal cells may be the primary type of cell responsible for controlling the extent of remodelling of luteal ECM since they produce TIMP-1, TIMP-2 and plasminogen activator inhibitor 1. The ratio of active MMPs to TIMPs may be important in maintaining an ECM microenvironment conducive to the differentiation of follicular-derived cells into luteal cells, and maintenance of the phenotype of luteal cells.

Introduction

In female mammals, glandular systems including the cyclic endometrium, mammary gland, ovarian follicle and corpus luteum undergo growth, maturation and involution at various stages in the reproductive cycle or lifespan of the animal. Remodelling of the extracellular matrix (ECM) is required for the dynamic tissue reorganization characteristic of these tissues. The ECM consists of proteinaceous and nonproteinaceous molecules that provide the tissue-specific, extracellular architecture to which cells attach. Furthermore, interaction of cellular receptors (integrins) with proteins of the ECM can regulate cellular structure, second messenger generation and gene expression. Selective synthesis and degradation of proteinaceous components of the ECM are essential for follicular growth, ovulation, luteal formation and luteolysis (reviewed by McIntush and Smith, 1998). Two families of enzymes that regulate remodelling of ECM are the matrix metalloproteinase (MMP) and the plasminogen activator/plasmin families. The focus of this review is on the role of MMPs and their inhibitors (TIMPs) in remodelling of ovarian ECM and subsequent effects on follicular and luteal function. Emphasis has been given to ruminant species whenever possible.

The Extracellular Matrix and Cellular Function

Within follicles and corpora lutea, cells are exposed to various ECM ligands that bind to integrins (receptors for ECM) as components of ECM are degraded and replaced (Luck, 1994). Proteinaceous and nonproteinaceous components of the ECM vary between and within tissues and, thereby, provide specialised microenvironments for specific cells. The primary proteinaceous components of the ECM have been reviewed elsewhere (Luck, 1994). Components of the ECM actively modulate the function of cells through integrins, which serve as a class of receptor on the cellular surface. Integrins are heterodimers containing α- and β-subunits. With 14 α-subunits and 8 β-subunits that associate in different combinations, integrins can form at least 20 distinct receptors with different ligands. Binding of integrins to their respective ligands modulates the generation of second messengers by hormones, growth factors and cytokines. In addition, interactions between ECM and integrins activate a variety of intracellular signalling molecules including serine–threonine, tyrosine, and lipid kinases and phospholipases. Consequently, the regulation of ECM degradation by MMPs and their natural inhibitors (tissue inhibitors of metalloproteinases; TIMPs) can have a profound influence on the cellular microenvironment and, thereby, modulate the function of follicular and luteal cells.

Matrix Metalloproteinases

Matrix metalloproteinases are zinc- and calcium-dependent enzymes that include collagenases, gelatinases, stromelysins and the membrane bound metalloproteinases (Nagase, 1997). These enzymes share various biochemical properties and are largely responsible for degrading proteinaceous components of the ECM. The family of MMPs currently includes at least 17 members that have different specificities (Table 1); however, new members will undoubtedly be discovered in the future. Activity of MMPs is highly regulated and subject to control at several different points (Fig. 1; Kleiner and Stetler-Stevenson, 1993). Two important points of control include activation of latent enzymes and association with tissue-derived inhibitors (TIMPs; see below). Most MMPs are secreted as proenzymes and are activated by proteolytic cleavage of an N-terminal peptide. Mast cell proteinases (Suzuki *et al.*, 1995), serine proteinases (plasmin and kallikriens; Espey, 1992) and other MMPs including the membrane-bound MMPs (Sato *et al.*, 1994) activate latent MMPs in the extracellular milieu. However, certain MMPs, such as stromelysin-3 and the membrane type-1 metalloproteinase are activated intracellularly. Although most MMPs are secreted, four membrane bound MMPs, which preferentially activate gelatinase A, have been described (Sato *et al.*, 1994; Takino *et al.*, 1995; Will and Hinzmann, 1995). Furthermore, gelatinase A was localized to the cellular surface by association with an integrin (Brooks *et al.*, 1996), which implies another mechanism for temporal and spatial regulation of remodelling of ECM.

Tissue Inhibitors of Metalloproteinases

The TIMPs and the liver-derived, serum-borne α_2 macroglobulin control activity of MMPs in the extracellular microenvironment. Within tissues, a family of TIMPs, including TIMP-1, TIMP-2, TIMP-3 and TIMP-4, regulate activity of MMPs (Greene *et al.*, 1996; Salamonsen 1996; Table 2). All TIMPs share twelve cysteine residues, which are considered to form six disulfide bonds; however,

Table 1. Matrix metalloproteinase (MMP) family of mammalian extracellular matrix proteinases

Family	Enzyme	MMP No.	Matrix substrates of functions
Collagenase	Interstitial collagenase	MMP-1	Collagens I, II, III, VII and X
	Neutrophil collagenase	MMP-8	Collagens I, II and III
	Collagenase 3	MMP-13	Collagens I, II and III
	Collagenase 4 (*Xenopus*)	MMP-18	Collagen I
Gelatinases	Gelatinase A	MMP-2	Gelatins, collagens IV, V, VII, X and XI, fibronectin, laminin
	Gelatinase B	MMP-9	Gelatins, collagens IV, V, XIV, fibronectin
Stromelysins	Stromelysin 1	MMP-3	Gelatins, fibronectin, laminin, collagens III, IV, IX and X, vitronectin: activates proMMP-1
	Stromelysin 2	MMP-10	Fibronectin, collagen IV
	Enamelysin	MMP-20	Amelogenin
Membrane-type MMPs	MT1-MMP	MMP-14	Collagens I, II, III, fibronectin, laminin, vitronectin: activates proMMP-2 and proMMP-13
	MT2-MMP	MMP-15	Gelatin, fibronectin, laminin: activates proMMP-2
	MT3-MMP	MMP-16	Activates proMMP-2
	MT4-MMP	MMP-17	Not known
Others	Matrilysin	MMP-7	Fibronectin, laminin, gelatins, collagen IV
	Stromelysin 3	MMP-11	Weak activity on fibronectin, laminin, collagen IV, gelatins
	Metalloelastase	MMP-12	Elastin
	(Unnamed)	MMP-19	Not known

TIMPs differ in relative molecular size, degree of glycosylation and *in vitro* or *in vivo* expression (Greene *et al.*, 1996; Salamonsen 1996; Table 2). Although they are separate gene products, each TIMP can inhibit most members of the MMP family by noncovalently binding with a 1:1 stoichiometry and high affinity ($K_i < 10^{-9}$ mol l^{-1}; Willenbrock and Murphy, 1994).

Remodelling of the ECM depends upon the ratio of active MMPs to TIMPs. Degradation of the ECM may occur when the ratio favours MMP activity, whereas deposition of ECM may occur when the ratio favours TIMPs. There is a considerable body of literature on the activation of latent MMPs; however, information regarding post-secretory mechanisms for regulating TIMPs is lacking. In particular, few mechanisms have been elucidated for liberating active MMPs from complexes with TIMPs or for the destruction of TIMPs. Potential physiological mechanisms for regulating TIMP-1 include: (1) degradation of TIMP-1 by proteinases (Itoh and Nagase, 1995); (2) degradation of TIMP-1 by peroxynitrite (a product of superoxide radicals and hydrogen peroxide; Frears *et al.*, 1996); and (3) liberation of active MMPs from complexes of TIMP-1 by endothelial stimulating angiogenesis factor (ESAF; McLaughlin *et al.*, 1991). The potential relevance of the preceding mechanisms for inactivating TIMP-1 to luteal function will be described in a subsequent section.

TIMPs are multifunctional molecules that stimulate proliferation of various types of cell (TIMP-1 and -2; Edwards *et al.*, 1996), and promote steroidogenesis (TIMP-1; Boujrad *et al.*, 1995) in addition

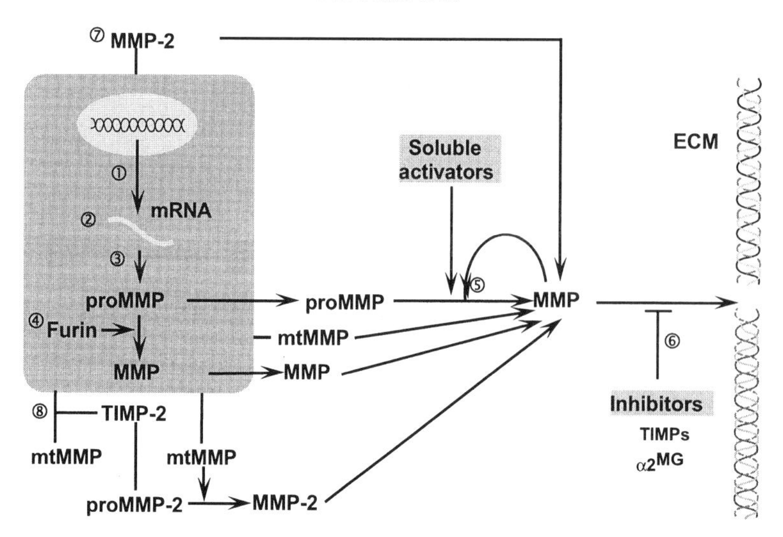

Fig. 1. A schematic representation of the pathways for matrix metalloproteinase (MMP) production, activation and inhibition. Production of MMPs can be regulated at the level of gene activation/transcription (1), mRNA stability (2) and mRNA translation of latent proenzyme (proMMP; 3). Activation of proMMP can occur via intracellular processing by furin (4) or extracellular processing by soluble activators (5; mast cell proteases and serine proteases (plasmin and kallikriens)). Proteolytically active MMPs (MMP) and membrane bound MMP (mtMMP) can also activate proMMP (5). MMPs degrade the proteinaceous components of the extracellular matrix (ECM) and the degree of ECM degradation is primarily regulated by locally produced inhibitors (6) such as tissue inhibitor of metalloproteinases (TIMPs-1, -2, -3 and -4) and possibly the serum derived inhibitor (6) α_2 macroglobulin (α_2MG). ECM degradation can be localized to the cell surface by the association of MMP-2 with an integrin (7). Another mechanism (8) for concentrating MMP activity near the cellular surface involves formation of a tri-molecular (mt-MMP–TIMP-2–MMP-2) complex that activates MMP-2 (progelatinase A). (Redrawn with permission from McIntush and Smith, 1998.)

to inhibiting MMPs. Boujrad and coworkers (1995) identified TIMP-1 as the active component of a TIMP-1–procathepsin L complex that stimulated production of progesterone by various types of steroidogenic cell (including luteinized granulosal cells). Cathepsin L is a lysosomal cysteine proteinase that is involved in intracellular protein metabolism and is found in numerous tissues. However, cathepsin L can be secreted and can cleave a variety of proteins including proteins of the ECM.

Follicular Growth and Atresia

Follicular expansion

Growth of bovine follicles from the primordial to the preovulatory stage is characterized by an approximately 360 000-fold increase in surface area as the follicular basement membrane expands within the limits of the ovarian stroma. Modification of the adjacent ovarian stroma is necessary for a growing follicle to reach a place on the surface of the ovary where the oocyte can be released at ovulation; therefore, remodelling of ECM occurs at all stages of follicular development. Remodelling of ECM also occurs during recruitment of a vascular supply within the thecal layer. The MMPs

Table 2. Tissue inhibitor of metalloproteinase (TIMP) family

Inhibitor	Relative molecular weight	Glycosylation	Extracellular location
TIMP-1	28 000	Glycosylated	Soluble in ECM and body fluids
TIMP-2	21 000	Not glycosylated	Soluble in ECM and body fluids
TIMP-3	24 000	Not glycosylated	Bound to ECM
TIMP-4	22 000	nd	nd

nd = not determined.
ECM: extracellular matrix.
Reproduced with permission from McIntush and Smith (1998).

and TIMPs probably regulate remodelling of ECM accompanying follicular expansion and angiogenesis. In rats, collagenase 3 was localized to thecal cells and interstitial tissue, but not to granulosal cells of antral follicles. Maximal expression of the enzyme occurred during pro-oestrus (Balbin *et al.*, 1996). Hence, collagenase 3 may be involved in remodelling of the ovarian stroma during follicular development to the preovulatory stage. Interstitial collagenase is also present within thecal and granulosal cells of rabbit follicles at all stages of development (Tadakuma *et al.*, 1993).

Bioavailability of growth factors

Follicular development requires significant proliferation of thecal and granulosal cells. The role of growth factors in control of ovarian cell proliferation has been examined extensively. Often growth factors are secreted constitutively and sequestered in the extracellular matrix in an inactive form or in association with specific binding proteins where they can subsequently be liberated by proteolysis of the ECM. Therefore, MMPs probably play a key role in regulating the availability of growth factors and their activities within developing follicles. For example, MMPs can liberate fibroblast growth factor molecules bound to heparin sulfate proteoglycans in the extracellular matrix and hence increase growth factor availability to nearby cells. Other growth factors and growth factor-binding proteins that are capable of binding to components of the ECM include transforming growth factor β, platelet derived growth factor, hepatocyte growth factor, heparin binding-epidermal growth factor, and insulin-like growth factor binding protein 3 (IGFBP-3).

Proteolytic degradation of the insulin-like growth factor binding proteins is rapidly emerging as a physiologically important mechanism for regulating availability of insulin-like growth factor (IGF) within follicles. In general, follicular growth is characterized by a decrease in concentrations of specific IGF-binding proteins in follicular fluid, presumably resulting in increased bioavailability of IGF to healthy follicles. The reduced concentration of binding protein is believed to result, at least in part, from proteolytic degradation. MMPs have been shown to degrade IGFBP-3 in other systems (Fowlkes *et al.*, 1994). However, the biochemical properties of proteinases responsible for degradation of IGF-binding proteins in follicular fluid are inconsistent with those of MMPs (Besnard *et al.*, 1996).

Atresia

Increased expression of specific activities of MMPs may facilitate the process of atresia. MMP activity is linked to involution in numerous tissues, such as the postpartum uterus and the mammary gland. Gelatinolytic activity corresponding to gelatinase A and B is increased within follicular fluid of atretic ovine follicles collected after hypophysectomy (Huet *et al.*, 1997). The increased activity of these enzymes is probably required for the breakdown of the basement membrane characteristic of later stages of atresia. The precise contribution of MMPs and their inhibitors to follicular growth and atresia remains to be determined.

Ovulation

The preovulatory gonadotrophin surge initiates a complex cascade of events resulting in follicular rupture, release of an oocyte, and luteinization of the remaining follicular cells. A growing body of evidence indicates that proteolytic degradation of the ECM at the follicular apex is the rate-limiting step in the ovulatory process. The apex region of the preovulatory follicular wall is composed of granulosal cells, basement membrane, theca interna, theca externa, tunica albuginea, and the surface epithelium with its underlying basement membrane. Distinct morphological changes that are indicative of degradation of ECM occur after exposure to the surge of gonadotrophins (Espey and Lipner, 1994).

Although a role for proteolytic enzymes in the process of ovulation was proposed in 1916, the identification and subsequent characterization of enzymes required for follicular rupture is incomplete. The serine proteinases, tissue plasminogen, activator (tPA) and urokinase plasminogen activator (uPA), convert the inactive zymogen, plasminogen into its active form plasmin, which has broad substrate specificity. Recent studies in mice with targeted mutations in the tPA and uPA genes have shed important insight on the regulation of the ovulatory process and on the potential requirement of MMPs.

Mice with a homozygous null mutation in either the tPA or the uPA gene displayed normal rates of ovulation. However, mice carrying homozygous null mutations in both genes exhibited a 26% reduction in rates of ovulation (Leonardsson *et al.*, 1995). These authors suggested that the plasminogen activator–plasmin enzyme family contributes to proteolysis of the ECM accompanying ovulation, although ovulation still occurs, albeit at a reduced rate, in mice lacking both plasminogen activators. Thus, the activity of this family of enzymes is not solely responsible for the marked proteolysis that occurs before ovulation, as they cannot cleave collagenous components of the ECM. The most important role of the plasminogen activator–plasmin system during the periovulatory period may be to increase the rate of activation of the latent proenzyme form of interstitial collagenase.

A growing body of evidence, primarily in rodents, indicates that MMPs help mediate follicular rupture. Administration of synthetic inhibitors of MMPs disrupted ovulation in rats (Brannstrom *et al.*, 1988). However, the requirement of individual MMPs for the ovulatory process has not been determined. The ovarian capsule, theca externa, and tunica albuginea are rich in type I and III collagens. Mice with a targeted mutation in the type I collagen gene, resulting in a molecule that is resistant to collagenase, display markedly reduced fertility. The reduction in fertility is potentially due to an impairment of the ovulatory process (Liu *et al.*, 1995). Collagenolytic activity, as measured by liberation of ^{3}H hydroxyproline or cleavage of tritiated type I collagen, is increased within rat ovaries (Reich *et al.*, 1985; Curry *et al.*, 1986) and sheep follicles (Murdoch and McCormick, 1992) after the preovulatory LH surge. In sheep, activity is substantially higher within the follicular apex than in the base; however, the mechanisms that accounted for spatial differences in collagenolytic activity within preovulatory follicles are not understood.

Collagenases

Collagenases, including interstitial collagenase and collagenase-3, may be responsible for the initial degradation and unwinding of the triple helical fibres of collagen within the follicular apex before ovulation. Expression of interstitial collagenase during the periovulatory period has been examined in rodents and rabbits. Messenger RNA expression of interstitial collagenase was increased twenty-five fold within rat follicles after exposure to an ovulatory dose of hCG and the mRNA was present within both isolated granulosal cells and the remaining ovarian tissue (Reich *et al.*, 1991). Immunohistochemical analysis of rabbit follicles indicated that this enzyme was present within the thecal and granulosal layers and increased within thecal and granulosal cells after follicular rupture (Tadakuma *et al.*, 1993). Although collagenase 3 is highly expressed by thecal cells/stroma of rat antral follicles (Balbin *et al.*, 1996), regulation of its expression by the preovulatory surge of gonadotrophins has not been demonstrated. The role of collagenase 3 in the ovulatory process is undefined.

Gelatinases

The gelatinases (gelatinase A and gelatinase B) are most noted for their ability to cleave the denatured helix of collagen (gelatin) and type IV collagen, a major component of basement membranes. It has been postulated that during the ovulatory process these enzymes play a key role in facilitating breakdown of the basement membrane and further hydrolysis of the denatured fibrils of collagen after their initial cleavage by collagenase(s).

Gelatinase A: Messenger RNA expression (Reich *et al.*, 1991) and enzymatic activity (Curry *et al.*, 1992) of gelatinase A were increased within rat ovaries after exposure to the LH surge and the mRNA was present within the residual ovarian tissue but not within isolated granulosal cells (Reich *et al.*, 1991). Immunization of ewes against the N-terminal peptide of the 43 kDa subunit of α-N inhibin resulted in reduced concentration of gelatinase A in follicular fluid and an impairment of the ovulatory process (Russell *et al.*, 1995). Immunized animals displayed reduced oviductal recovery of oocytes and abnormal corpora lutea in which the normal tissue remodelling process was partially disrupted. Mice with a targeted mutation in the gelatinase A gene have been generated. Surprisingly, a preliminary report indicated that disruption of the gelatinase A gene did not reduce fertility (Itoh *et al.*, 1997). Elucidation of the role of gelatinase A in the ovulatory process will require further investigation.

Gelatinase B: The role of gelatinase B in the ovulatory process is unclear. Like gelatinase A, gelatinase B can cleave type IV collagen and may play a role in basement membrane breakdown. Evidence in rats indicates that interleukin-1β, an LH regulated-putative paracrine mediator of the ovulatory process, can regulate expression of gelatinase B by preovulatory follicles. Expression of gelatinase B is increased by treatment of whole ovarian dispersates or enriched theca, but not granulosa cells with interleukin 1β (Hurwitz *et al.*, 1993). However, activity of gelatinase B was low or undetectable by gelatin zymographic analysis of rat ovarian extracts (Curry *et al.*, 1992) and ovine follicular fluid (Russell *et al.*, 1995) collected during the periovulatory period. Mice null for the gelatinase B gene were fertile (Vu *et al.*, 1998).

Membrane-bound MMPs

Members of the newly discovered family of membrane-bound metalloproteinases, such as the membrane type 1 metalloproteinase, may also be involved in ovulation. Membrane type-1 metalloproteinase was expressed within bovine follicles and corpora lutea (G. W. Smith *et al.*, unpublished). This enzyme can activate progelatinase A through formation of a tri-molecular complex with TIMP-2 (Strongin *et al.*, 1995). During activation, TIMP-2 first binds to the membrane type 1 metalloproteinase. Next, TIMP-2 binds progelatinase A through its carboxy-terminal hemopexin domain. Binding to TIMP-2 localizes progelatinase A at the cellular surface in immediate proximity to the membrane type 1 metalloproteinase where the propiece of progelatinase A can be cleaved. In addition to activation of progelatinase A, the membrane type-1 metalloproteinase can hydrolyse type I and III collagen, fibronectin, laminin and proteoglycans (Nagase, 1997). Therefore, it may participate in degradation of ECM during ovulation. Espey and Lipner (1994) postulated that proteinases localized on the surface of cells promote dissociation of cells and breakdown of the surrounding fibres of collagen within the theca externa and tunica albuginea during ovulation.

Tissue inhibitors of metalloproteinases

Regulation of MMPs is important for tissue homeostasis. As discussed previously, the activity of MMPs in the extracellular milieu is regulated through production of specific inhibitors. The MMPs and their inhibitors are often secreted in parallel; i.e. agents that stimulate expression of MMPs also increase expression of the inhibitors (Murphy *et al.*, 1985). Therefore, net proteolysis during ovulation may be regulated by the ratio of enzyme to inhibitor.

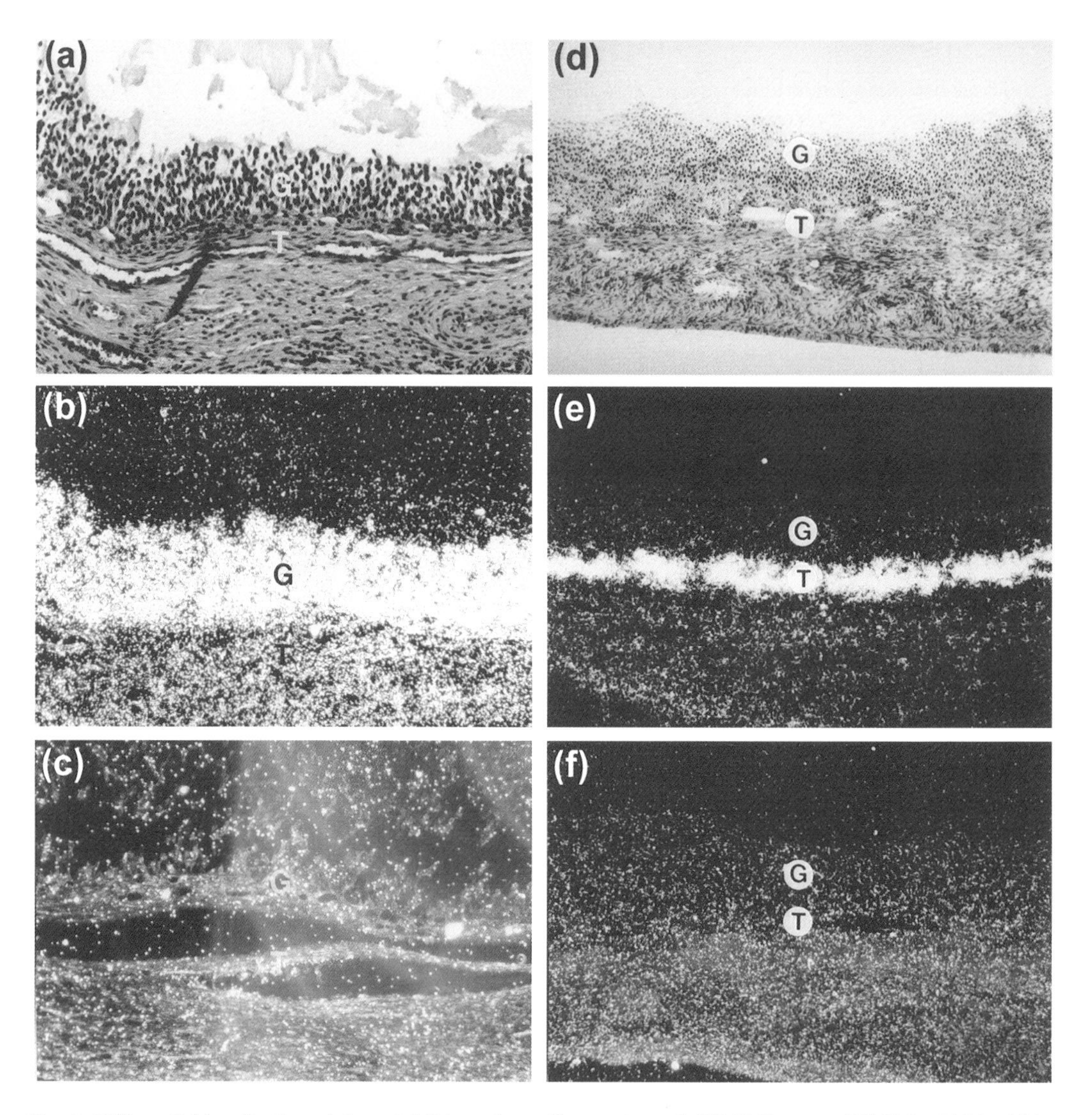

Fig. 2. Differential localization of tissue inhibitor of metalloproteinase 1 (TIMP-1) versus TIMP-2 mRNA within ovine follicles collected 12 h after the LH surge (post-surge). (a,d) Bright field images of sections of post-surge ovine follicles hybridized with ^{35}S antisense TIMP-1 (a) or TIMP-2 (d) cRNA probes. (b,e) Dark field images of sections of post-surge ovine follicles hybridized with ^{35}S antisense TIMP-1 (b) or TIMP-2 (e) cRNA probes. (b) TIMP-1 mRNA was localized to the granulosal layer of ovine post-surge follicles, while (e) TIMP-2 mRNA was localized specifically to the thecal layer. (c,f) Specific hybridization was not detected when sections were hybridized with sense (negative control) TIMP-1 (c) or TIMP-2 (f) cRNA probes. G (granulosal cell layer) T (thecal cell layer). (Reproduced with permission from Smith *et al.*, 1994; 1995; © The Endocrine Society.)

We determined the effect of the preovulatory LH surge on expression of two members of the TIMP family (TIMP-1 and -2) within ovine follicles (Fig. 2). Expression of TIMP-1 mRNA (Smith *et al.*, 1994a) and follicular fluid concentration of TIMP-1 (McIntush *et al.*, 1997) were increased approximately 10-fold within ovine follicles after a preovulatory LH surge. The granulosal cells were the primary source of the increased expression of TIMP-1 (Smith *et al.*, 1994a, Fig. 2, and McIntush *et al.*, 1996). In contrast, expression of TIMP-2 was constitutive within ovine follicles collected at similar time points after the preovulatory LH surge, and TIMP-2 was localized to the thecal layer (Fig. 2;

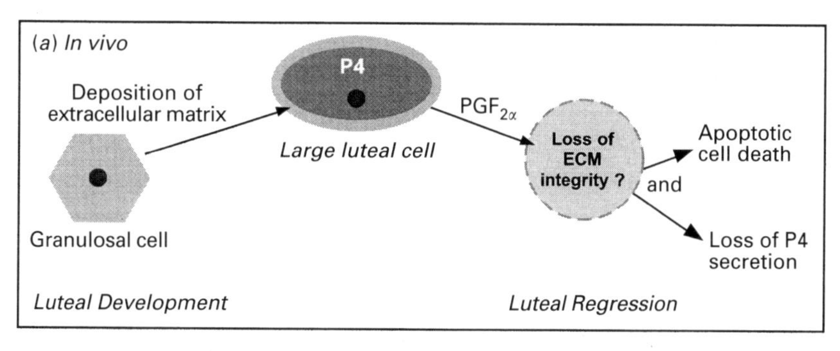

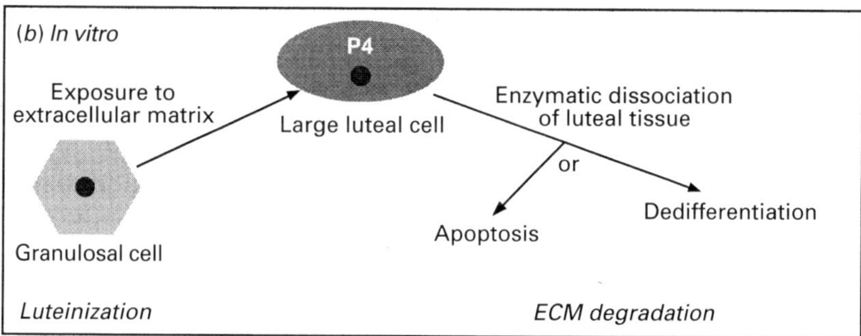

Fig. 3. Relationship between extracellular matrix (ECM) and luteal cell function *in vivo* and *in vitro*. (a) *In vivo*. Relationship of luteal development and regression to changes in ECM. Differentiation of follicular cells into large luteal cells is associated with net accumulation of ECM during luteal development. Likewise, reduction in progesterone (P4) secretion and apoptotic cell death is associated with loss of ECM integrity. (b) *In vitro*. Granulosal cell luteinization is enhanced by exposure to basal lamina components (laminin and fibronectin). In addition, enzymatic dissociation of luteal tissue leads to apoptosis or dedifferentiation of luteal cells. (Reproduced with permission from McIntush and Smith, 1998.)

Smith *et al.*, 1995). The distinct localization and temporal expression of TIMP-1 versus TIMP-2 within ovine follicles indicates complementary yet distinct roles for each inhibitor during the periovulatory period. TIMP-1 probably regulates the extent of proteolysis within the granulosal layer during the ovulatory process. In contrast, TIMP-2 within the thecal layer may enhance proteolysis through localization of progelatinase A at the surface of cells with the membrane type 1-metalloproteinase.

Corpus luteum

The transition of a preovulatory follicle into a corpus luteum is a complex process involving mechanisms similar to wound healing and tumour formation. Corpus luteum development and luteolysis are accompanied by extensive remodelling of ECM, which can modulate specific cellular processes including mitosis, migration, differentiation, apoptosis and gene expression. Presumably, MMPs and their inhibitors play an important role in remodelling of luteal tissue. Indeed, expression of MMP genes and activity of MMPs in corpora lutea have been demonstrated. Likewise, luteal tissue expresses inhibitors of these proteinases ((TIMP-1, TIMP-2, TIMP-3) and plasminogen activator inhibitor-1 (PAI-1)), presumably to control the extent of breakdown of ECM. Although many reproductive and non-reproductive tissues produce TIMP-1, the production of TIMP-1 mRNA (Hampton *et al.*, 1995) and protein (McIntush and Smith, 1997) by ovine corpora lutea was 30 to 3 000

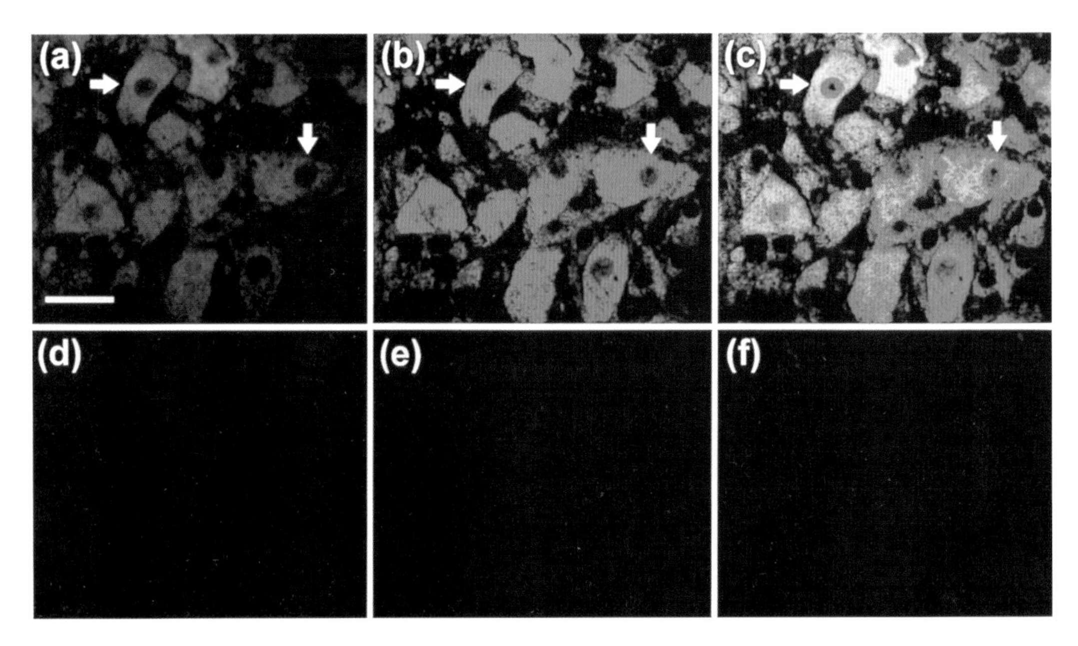

Fig. 4. (a–c) Confocal micrographs illustrating portions of ovine corpora lutea collected on day 10 after oestrus (day 0 = oestrus). Tissue inhibitor of metalloproteinase 1 TIMP-1 (a) and oxytocin (b) were colocalized (c) within day 10 large luteal cells (arrows indicate examples of cells in which TIMP-1 and oxytocin were colocalized). Control tissues treated with second antibody alone (d–f) did not show immunofluorescence. Scale bar represents 25 μm. (Reproduced with permission from McIntush *et al.*, 1996.)

fold greater than that of other tissues. On the basis of reports of growth- (Edwards *et al.*, 1996) and steroidogenesis-promoting (Boujrad *et al.*, 1995) activities of TIMP-1, it may be that this protein modulates several processes within corpora lutea.

There are similarities between luteal development–luteolysis *in vivo* and luteal cell differentiation/dedifferentiation or death of luteal cells *in vitro*. A model for the role of the ECM during differentiation and regression of luteal cells *in vivo* is shown (Fig. 3a). Evidence derived from studies *in vitro* indicates that ECM components enhance luteinization of follicular cells and that loss of ECM results in death or loss of differentiated phenotype of luteal cells (Fig. 3b). Aspects of this model were derived from studies of ovine and rat corpora lutea. Specific details may not apply to all species, but similar principles may apply. Experimental evidence to support this model are provided in subsequent sections.

Luteal development

Luteinization. The process of luteinization involves morphological and biochemical changes that follicular cells undergo during transformation into the steroidogenic cells of a corpus luteum. In sheep and cattle, corpora lutea contain two types of steroidogenic cell (large luteal cells (LLC), presumably derived from follicular granulosal cells and small luteal cells (SLC), presumably derived from follicular thecal cells (reviewed by Smith *et al.*, 1994b)). These cells differ biochemically, physiologically and morphologically (Farin *et al.*, 1986). Ovine large luteal cells appear to be the primary type of cell responsible for controlling the extent of remodelling of luteal ECM as they produced TIMP-1 (Fig. 4; Smith *et al.*, 1994a; McIntush *et al.*, 1996), TIMP-2 (Smith *et al.*, 1995; McIntush *et al.*, 1996), and plasminogen activator inhibitor 1 (G. W. Smith and M. F. Smith, unpublished).

Components of the ECM have an important effect on the differentiation of a variety of types of

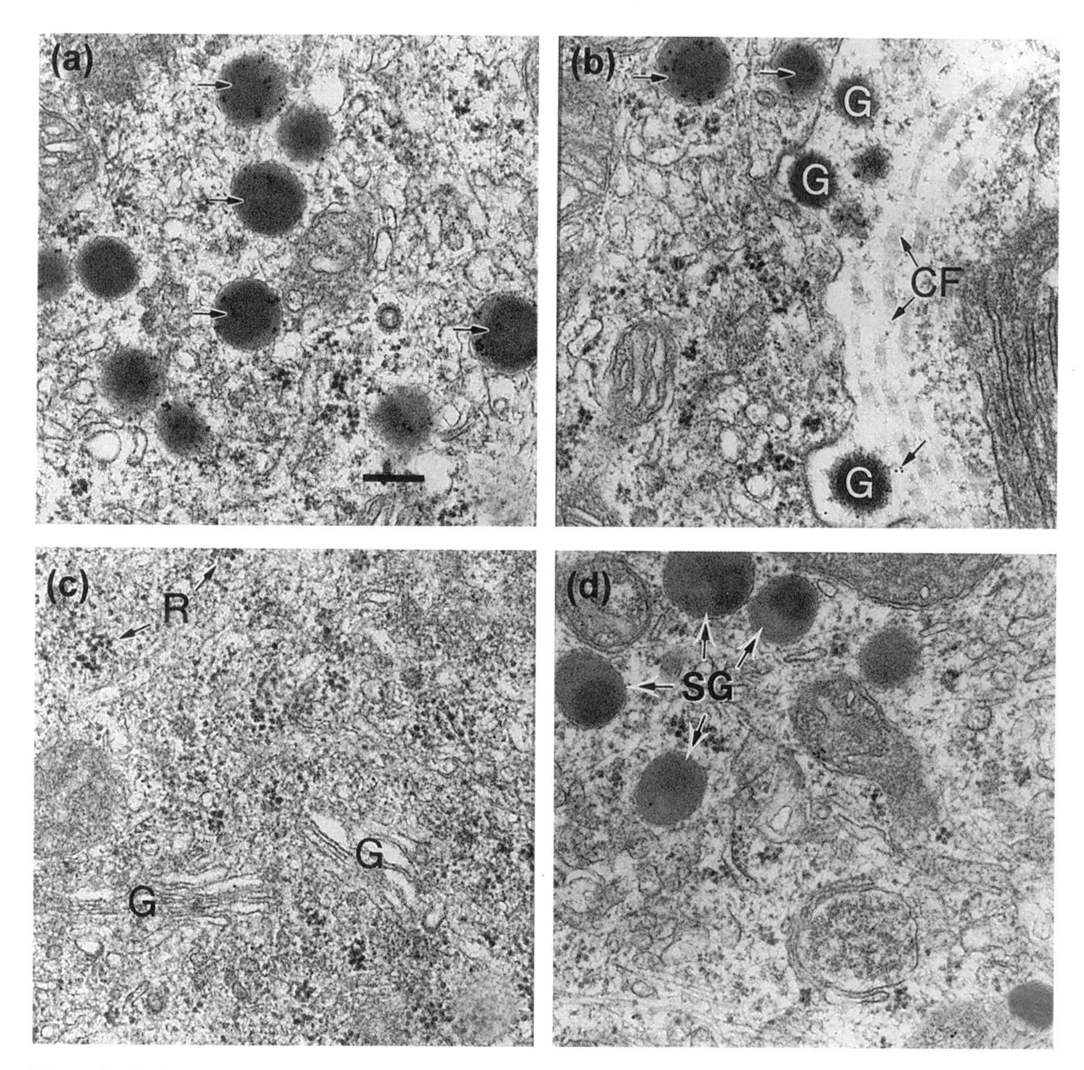

Fig. 5. (a–d) Electron micrographs illustrating portions of luteal cells from ovine corpora lutea collected on day 10 after oestrus (day 0 = oestrus). Scale bar represents 200 nm. (a) Colloidal gold particles indicative of positive immunostaining for tissue inhibitor of metalloproteinase 1 (TIMP-1) were restricted to secretory granules (arrows) dispersed throughout the cytoplasm of large luteal cells. (b) The contents of secretory granules in the cytoplasm of large luteal cells (arrows) consistently stained positive for TIMP-1. After release of secretory granules (G) into the extracellular spaces, particles of colloidal gold were frequently observed in association with granule contents undergoing dissolution. Collagen fibrils (CF) were present in the extracellular matrix. (c) Portion of a small luteal cell located in close proximity to the large luteal cell shown in (a). Aggregates of ribosomes (R) and Golgi cisternae (G) are evident. Note the absence of colloidal gold particles. (d) Negative control. A group of secretory granules (SG) in the cytoplasm of a large luteal cell in a section that was incubated in preimmune serum rather than the primary antiserum (M17W4). In negative control sections, background (nonspecific) staining associated with secretory granules was minimal (compared with positive staining secretory granules in (a)). (Reproduced with permission from McIntush *et al.*, 1996.)

cell, including granulosal cells. Aten and coworkers (1995) demonstrated that fibronectin or laminin (two components of the basal lamina) promote the differentiation of rat granulosal cells into luteal cells, while an antibody to the integrin β_1 subunit inhibits differentiation of granulosal cells. Interaction of luteal cells with the ECM probably changes during the luteal phase as components of the matrix are degraded and replaced coincident with cell proliferation and migration. Changing contacts between cells and the ECM may even be necessary for differentiation of follicular cells into luteal cells, as is the case for differentiation of other types of cell. Differentiation of granulosal cells into large luteal cells is likely to require sequential exposure to different environments of ECM.

Approximately 80% of the progesterone secreted *in vivo* by ovine corpora lutea is believed to come from LLC (Niswender *et al.*, 1985). The factor(s) responsible for regulating secretion of progesterone by LLC are unknown. Recently, a TIMP-1–procathepsin L complex was shown to stimulate synthesis of progesterone by several steroidogenic cell types including luteinized granulosal cells (Boujrad *et al.*, 1995). TIMP-1 may be an important stimulator of synthesis of progesterone in luteinizing granulosal cells and LLC.

Cellular proliferation/migration. Luteal development has been characterized by cellular proliferation equal to that of rapidly growing tumours (Jablonka-Shariff *et al.*, 1993). Such a rapid rate of proliferation is not surprising since corpora lutea increase in size 10- to 20-fold over a few days (Jablonka-Shariff *et al.*, 1993) and develop into one of the most vascular tissues known. Many cells migrate at this time, as the distinctly compartmentalized tissue of the follicle makes the transition into a corpus luteum consisting of a heterogeneous population of cells. Many of the migrating and proliferating cells are endothelial cells (Jablonka-Shariff *et al.*, 1993). TIMP-1 and -2 promote growth of a variety of types of cell, including fibroblasts and endothelial cells. The mechanism by which TIMPs stimulates cellular proliferation is unclear but may involve membrane receptors (Edwards *et al.*, 1996). Alternatively, MMPs and TIMPs may influence proliferation of cells by regulating the bioavailability of growth factors (see Follicular Growth and Atresia section).

Angiogenesis. During formation of the corpus luteum, the follicular wall containing an avascular granulosal layer undergoes a transition to become one of the most vascular tissues in the body. Two important stages of angiogenesis include breakdown of the basement membrane and migration of endothelial cells. It is likely that MMPs play a significant role in both stages of angiogenesis. Interestingly, endothelial cells within melanomas were found to contain active gelatinase A associated with integrin $\alpha_v\beta_3$ (Brooks *et al.*, 1996). A naturally occurring, noncatalytic fragment of gelatinase A competes with gelatinase A for binding to the $\alpha_v\beta_3$ integrin and may function to limit the extent of migration and invasion of endothelial cells during angiogenesis (Brooks *et al.*, 1998). The activity of gelatinase A localized at the cell surface, as well as that of the secreted and membrane-bound MMPs may facilitate migration of endothelial cells within the developing corpus luteum.

Tissue inhibitor of metalloproteinases 1 is expressed abundantly during ovine luteal development (Smith *et al.*, 1994a) and TIMP-1 has also been shown to inhibit angiogenesis. These results suggest an apparent paradox since TIMP-1 inhibited neovascularization and migration of cells (Khokha, 1994). A possible explanation is that expression of TIMP-1 mRNA is not indicative of the concentration of TIMP-1 in the extracellular milieu. Although TIMP-1 mRNA per unit of DNA decreased (Freudenstein *et al.*, 1990; Smith *et al.*, 1996) from early to mid-luteal phase, TIMP-1 protein per milligram of luteal tissue increased approximately threefold from early to the mid-luteal phase (McIntush and Smith, 1997). The secretory granules to which TIMP-1 was localized (McIntush *et al.*, 1996) fit the morphological criteria for dense core secretory granules (Fig. 5), which are often associated with a regulated pathway of protein secretion. Thus, increasing TIMP-1 protein, decreasing TIMP-1 mRNA, and localization of TIMP-1 to secretory granules are consistent with the hypothesis that, although some TIMP-1 may be secreted, TIMP-1 accumulates within secretory granules of LLC. This possibility helps resolve the paradox of high TIMP-1 expression in the midst of extensive neovascularization.

Luteolysis

Regression of the corpus luteum is marked by loss of adhesion of cells to matrix (Nett *et al.*, 1976), loss of capacity to synthesize progesterone (Niswender and Nett, 1994), and apoptosis (Juengel *et al.*, 1993). Furthermore, changes in luteal gene expression in response to $PGF_{2\alpha}$ (the natural luteolysin in numerous species) are similar to changes in gene expression induced by enzymatic dissociation of luteal tissue (G. D. Niswender, personal communication). Ovine LLC are the cells that possess receptors for and respond to $PGF_{2\alpha}$. If degradation of matrix is an important component of the physiological response to $PGF_{2\alpha}$, vesting LLC with both maintenance of the ECM of luteal tissue and responsiveness to the luteolysin ($PGF_{2\alpha}$) appears logical. In such a scenario, response to $PGF_{2\alpha}$ may include removal of substances involved in preservation of the ECM. Indeed, TIMP-1 mRNA (Duncan *et al.*, 1996) and protein (McIntush and Smith, 1997) were lower within hours after administration of $PGF_{2\alpha}$. Furthermore, dissolution of microtubules within LLC was noted soon (2 h) after administration of $PGF_{2\alpha}$ (Murdoch, 1996). These data may indicate that contacts between LLC and ECM were disrupted during luteal regression. Thus, maintenance of corpora lutea appears dependent upon establishment and maintenance of proper contact with ECM. Luteal tissue may produce TIMP-1, TIMP-2, TIMP-3 and PAI-1 to preserve the integrity of components of the ECM upon which luteal cells appear dependent.

If inhibitors of ECM proteinases are important for luteal maintenance, destruction of inhibitors may be important for luteal regression. Inactivation of TIMP-1 could be accomplished by phenomena that have been implicated in luteolysis. Leukocytes, which appear to be involved in luteolysis (Murdoch, 1987), could inactivate TIMP-1 by proteolytic degradation (Itoh and Nagase, 1995). In addition, highly reactive peroxynitrite ($ONOO^-$; generated by the reaction of superoxide (O_2^-) with nitric oxide (NO)), which inactivated TIMP-1 (Frears *et al.*, 1996) may be generated by luteal tissue during the generation of oxygen radicals associated with luteolysis. Combining data on development, maintenance and regression of mammary and luteal tissues provides the rationale for the hypothesis that remodelling of the ECM plays a fundamental role in luteal function. However, confirmation of this hypothesis awaits further investigation.

Conclusions

The extracellular matrix is integrally involved in the function of tissues. In adult mammals, few tissues undergo extensive remodelling of the ECM. In fact, remodelling of the ECM in adult tissues is often linked to pathologies such as osteoarthritis and tumour cell metastasis. However, the ovary is one tissue that undergoes remodelling as a continuous, physiological phenomenon. Information on the ability of the ECM to direct the proliferation, differentiation and function of cells implies that breakdown and regeneration of the ECM plays more than a permissive role in normal ovarian function. The ECM probably plays an active part in directing the processes of follicular development and atresia, ovulation, and development, maintenance and regression of corpora lutea. Matrix metalloproteinases appear to be key enzymes that determine which molecules of the ECM are degraded. The extent of degradation of ECM molecules by MMPs is controlled to a large extent by availability of TIMPs. To date, most studies have focused upon correlating expression of MMPs and TIMPs with various stages of the reproductive cycle. From these studies, many key molecules have been identified. However, a better understanding of the role of individual MMPs and TIMPs in ovarian function awaits additional studies that characterize the temporal and spatial expression of these proteins more thoroughly. Subsequently, creative manipulative studies must be undertaken to determine the complex interactions of these molecules if a clear understanding of their function is to be obtained.

This manuscript is a contribution of the Missouri Agricultural Experiment Station. Journal Series Number 12,770. This work was partially funded by USDA Grant, USDA CSRS 92-37203-7950. The Michigan Agricultural Experiment Station is gratefully acknowledged for their support (GWS).

References

Aten RF, Kolodecik TR and Berhman HR (1995) A cell adhesion receptor antiserum abolishes, whereas laminin and fibronectin promote, luteinization of cultured rat granulosa cells *Endocrinology* **136** 1753–1758

Balbin M, Fueyo A, Lopez JM, Diez-Itza I, Velasco G and Lopez-Otin C (1996) Expression of collagenase-3 in the rat ovary during the ovulatory process *Journal of Endocrinology* **149** 405–415

Besnard N, Pisselet C, Zapf J, Hornebeck W, Monniaux D and Monget P (1996) Proteolytic activity is involved in changes in intrafollicular insulin-like growth factor-binding protein levels during growth and atresia of ovine ovarian follicles *Endocrinology* **137** 1599–1607

Boujrad N, Ogwuegbu SO, Garnier M, Lee C, Martin BM and Papadopoulos V (1995) Identification of a stimulator of steroid hormone synthesis isolated from testis *Science* **268** 1609–1612

Brannstrom M, Woessner JF, Jr, Koos RD, Sear CH and LeMaire WJ (1988) Inhibitors of mammalian tissue collagenase and metalloproteinases suppress ovulation in the perfused rat ovary *Endocrinology* **122** 1715–1721

Brooks PC, Stromblad S, Sanders LC, von Schalscha TL, Aimes RT, Stetler-Stevenson WG, Quigley JP and Cheresh DA (1996) Localization of matrix metalloproteinase MMP-2 to the surface of invasive cells by interaction with integrin $\alpha_V \beta_3$ *Cell* **85** 683–693

Brooks PC, Siletti S, von Schalscha TL, Friedlander M and Cheresh D (1998) Disruption of angiogenesis by PEX, a noncatalytic metalloproteinase fragment with integrin binding activity *Cell* **92** 391–400

Curry TE, Jr, Clark MR, Dean DD, Woessner JF, Jr and LeMaire WJ (1986) The preovulatory increase in ovarian collagenase activity in the rat is independent of prostaglandin production *Endocrinology* **118** 1823–1828

Curry TE, Jr, Mann JS, Huang MH and Keeble SC (1992) Gelatinase and proteoglycanase activity during the periovulatory period in the rat *Biology of Reproduction* **46** 256–264

Duncan WC, Illingworth PJ and Fraser HM (1996) Expression of tissue inhibitor of metalloproteinases-1 in the primate ovary during induced luteal regression *Journal of Endocrinology* **151** 203–213

Edwards DR, Beaudry PP, Laing TD, Kowal V, Leco KJ, Leco PA and Lim SA (1996) The roles of tissue inhibitors of metalloproteinases in tissue remodelling and cell growth. *International Journal of Obesity* **20** 9–15

Espey LL (1992) Ovulation as an inflammatory process. In *Local Regulation of Ovarian Function* pp 183–200 Eds N Sjoberg, L Hamberger, P Janson, C Owman and H Coelingh-Bennink. Parthenon Publishing Group, Park Ridge

Espey LL and Lipner H (1994) Ovulation. In *The Physiology of Reproduction* pp 725–780 Eds E Knobil and JD Neill. Raven Press, New York

Farin CE, Moeller CL, Sawyer HR, Gamboni F and Niswender GD (1986) Morphometric analysis of cell types in the ovine corpus luteum throughout the estrous cycle *Biology of Reproduction* **35** 1299–1308

Fowlkes JL, Enghild JJ, Suzuki K and Nagase H (1994) Matrix metalloproteinases degrade insulin-like growth factor-binding protein-3 in dermal fibroblast culture *Journal of Biological Chemistry* **269** 25742–25746

Frears ER, Zhang Z, Blake DR, O'Connell JP and Winyard PG (1996) Inactivation of tissue inhibitor of metalloproteinase-1 by peroxynitrite *FEBS Letters* **381** 21–24

Freudenstein J, Wagner S, Luck MR, Einspanier R and Scheit KH (1990) mRNA of bovine tissue inhibitor of metalloproteinases: sequence and expression in bovine ovarian tissue *Biochemical and Biophysical Research Communications* **171** 250–256

Greene J, Wang M, Liu YE, Raymond LA, Rosen C and Shi YE (1996) Molecular cloning and characterization of human tissue inhibitor of metalloproteinase 4 *Journal of Biological Chemistry* **271** 30375–30380

Hampton AL, Butt AR, Riley SC and Salamonsen LA (1995) Tissue inhibitor of metalloproteinases in endometrium of ovariectomized steroid-treated ewes and during the estrous cycle and early pregnancy *Biology of Reproduction* **53** 302–311

Huet C, Monget P, Pisselet C and Monniaux D (1997) Changes in extracellular matrix components and steroidogenic enzymes during growth and atresia of antral ovarian follicles in the sheep *Biology of Reproduction* **56** 1025–1034

Hurwitz A, Dushnik M, Solomon H, Ben-Chetrit A, Finci-Yeheskel Z, Milwidsky A, Mayer M, Adashi EY and Yagel S (1993) Cytokine-mediated regulation of rat ovarian function: interleukin-1 stimulates the accumulation of a 92-kilodalton gelatinase *Endocrinology* **132** 2709–2714

Itoh Y and Nagase H (1995) Preferential inactivation of tissue inhibitor of metalloproteinases-1 that is bound to the precursor of matrix metalloproteinase 9 (progelatinase B) by human neutrophil elastase *Journal of Biological Chemistry* **270** 16518–16521

Itoh T, Ikeda T, Gomi H, Nakao S, Suzuki T and Itohara S (1997) Unaltered secretion of beta-amyloid percursor protein in gelatinase A (matrix metalloproteinase 2)-deficient mice *Journal of Biological Chemistry* **272** 22389–22392

Jablonka-Shariff A, Grazul-Bilska AT, Redmer DA and Reynolds LP (1993) Growth and cellular proliferation of ovine corpora lutea throughout the estrous cycle *Endocrinology* **133** 1871–1879

Juengel JL, Garverick HA, Johnson AL, Youngquist RS and Smith MF (1993) Apoptosis during luteal regression in cattle *Endocrinology* **132** 249–254

Khokha R (1994) Suppression of tumorigenic and metastatic abilities of murine B16-F10 melanoma cells *in vivo* by overexpression of the tissue inhibitor of metalloproteinases-1 *Journal of the National Cancer Institute* **86** 299–304

Kleiner DE and Stetler-Stevenson WG (1993) Structural biochemistry and activation of the matrix metalloproteinases *Current Opinion in Cell Biology* **5** 891–897

Leonardsson G, Peng XR, Liu K, Nordstrom L, Carmeliet P, Mulligan R, Collen D and Ny T (1995) Ovulation efficiency is reduced in mice that lack plasminogen activator gene function: functional redundancy among physiological plasminogen activators *Proceedings of the National Academy of Sciences USA* **92** 12445–12450

Liu X, Wu H, Byrne M, Jeffrey J, Krane S and Jaenisch R (1995) A targeted mutation at the known collagenase cleavage site in mouse type I collagen impairs tissue remodelling *Journal of Cell Biology* **130** 227–237

Luck MR (1994) The gonadal extracellular matrix *Oxford Reviews of Reproductive Biology* **16** 34–85

McIntush EM and Smith MF (1997) Concentration of tissue inhibitor of metalloproteinases (TIMP-1) in ovine follicular and luteal tissues *Biology of Reproduction* **56** (Supplement 1) 123

McIntush EM and Smith MF (1998) Matrix metalloproteinases and tissue inhibitors of metalloproteinases in ovarian function *Reviews in Reproduction* **3** 23–30

McIntush EW, Pletz JD, Smith GW, Long DK, Sawyer HR and Smith MF (1996) Immunolocalization of tissue inhibitor of metalloproteinase-1 within ovine periovulatory follicular and luteal tissues *Biology of Reproduction* **54** 871–878

McIntush EW, Keisler DH and Smith MF (1997) Concentration of tissue inhibitor of metalloproteinases (TIMP)-1 in ovine follicular fluid and serum *Journal of Animal Science* **75** 3255–3261

McLaughlin B, Cawston T and Weiss JB (1991) Activation of the matrix metalloproteinase inhibitor complex by a low molecular weight angiogenic factor *Biochimica et Biophysica Acta* **1073** 295–298

Murdoch WJ (1987) Treatment of sheep with prostaglandin $F_2\alpha$ enhances production of a luteal chemoattractant for eosinophils *American Journal of Reproductive Immunology* **15** 52–56

Murdoch WJ (1996) Microtubular dynamics in granulosa cells of periovulatory follicles and granulosa-derived (large) lutein cells of sheep: relationships to the steroidogenic folliculo-luteal shift and functional luteolysis *Biology of Reproduction* **54** 1135–1140

Murdoch WJ and McCormick RJ (1992) Enhanced degradation of collagen within apical vs. basal wall of ovulatory ovine follicle *American Journal of Physiology* **263** E221–225

Murphy G, Reynolds JJ and Werb Z (1985) Biosynthesis of tissue inhibitor of metalloproteinases by human fibroblasts in culture: stimulation by 12-*0*-tetradecanoylphorbol 13-acetate and interleukin 1 in parallel with collagenase *Journal of Biological Chemistry* **260** 3079–3083

Nagase H (1997) Activation mechanisms of matrix metalloproteinases *Biological Chemistry* **378** 151–160

Nett TM, McClellan MC and Niswender GD (1976) Effects of prostaglandins on the ovarian corpus luteum: blood flow, secretion of progesterone and morphology *Biology of Reproduction* **15** 66–78

Niswender GD and Nett TM (1994) The corpus luteum and its control in infraprimate species. In *The Physiology of Reproduction* pp 781–816 Ed. E Knobil and JD Neill. 2nd Edn. Raven Press, New York

Niswender GD, Schwall RH, Fitz TA, Farin CE and Saywer HR (1985) Regulation of luteal function in domestic ruminants: new concepts *Recent Progress in Hormone Research* **41** 101–151

Reich R, Tsafriri A and Mechanic GL (1985) The involvement of collagenolysis in ovulation in the rat *Endocrinology* **116** 522–527

Reich R, Daphna Iken D, Chun SY, Popliker M, Slager R, Adelmann Grill BC and Tsafriri A (1991) Preovulatory changes in ovarian expression of collagenases and tissue metalloproteinase inhibitor messenger ribonucleic acid: role of eicosanoids *Endocrinology* **129** 1869–1875

Russell DL, Salamonsen LA and Findlay JK (1995) Immunization against the N-terminal peptide of the inhibin alpha 43-subunit (alpha N) disrupts tissue remodelling and the increase in matrix metalloproteinase-2 during ovulation *Endocrinology* **136** 3657–3664

Salamonsen LA (1996) Matrix metalloproteinases and their tissue inhibitors in endocrinology *Trends in Endocrinology and Metabolism* **7** 28–34

Sato H, Takino T, Okada Y, Cao J, Shinagawa A, Yamamoto E and Seiki M (1994) A matrix metalloproteinase expressed on the surface of invasive tumor cells *Nature* **370** 61–65

Smith GW, Goetz TL, Anthony RV and Smith MF (1994a) Molecular cloning of an ovine ovarian tissue inhibitor of metalloproteinases: ontogeny of messenger ribonucleic acid expression and *in situ* localization within preovulatory follicles and luteal tissue *Endocrinology* **134** 344–352

Smith MF, McIntush EW and Smith GW (1994b) Mechanisms associated with corpus luteum development *Journal of Animal Science* **72** 1857–1872

Smith GW, McCrone S, Petersen SL and Smith MF (1995) Expression of messenger ribonucleic acid encoding tissue inhibitor of metalloproteinases-2 within ovine follicles and corpora lutea *Endocrinology* **136** 570–576

Smith GW, Juengel JL, McIntush EW, Youngquist RS, Garverick HA and Smith MF (1996) Ontogenies of messenger RNA encoding tissue inhibitor of metalloproteinases 1 and 2 within bovine periovulatory follicles and luteal tissue *Domestic Animal Endocrinology* **13** 151–160

Strongin AY, Collier I, Bannikov G, Marmer BL, Grant GA and Goldberg GI (1995) Mechanism of cell surface activation of 72-kDa type IV collagenase: isolation of the activated form of the membrane metalloproteinase *Journal of Biological Chemistry* **270** 5331–5338

Suzuki K, Lees M, Newlands G, Nagase H and Woolley DE (1995) Activation of precursors for matrix metalloproteinases 1 (interstitial collagenase) and 3 (stromelysin) by rat mast-cell proteinases I and II *Biochemical Journal* **305** 301–306

Tadakuma H, Okamura H, Kitaoka M, Iyama K and Usuku G (1993) Association of immunolocalization of matrix metalloproteinase-1 with ovulation in hCG-treated rabbit ovary *Journal of Reproduction and Fertility* **98** 503–508

Takino T, Sato H, Shinagawa A and Seiki M (1995) Identification of the second membrane-type matrix metalloproteinase (MT-MMP-2) gene from a human placenta cDNA library: MT-MMP form a unique membrane type sub-class in the MMP family *Journal of Biological Chemistry* **270** 23013–23020

Vu TH, Shipley JM, Bergers G, Gergers JE, Helms JA, Hanahan D, Shapiro SD, Senior RM and Werb Z (1998) MMP-9/gelatinase B is a key regulator of growth plate angiogenesis and apoptosis of hypertrophic chondrocytes *Cell* **93** 411–422

Will H and Hinzmann B (1995) cDNA sequence and mRNA tissue distribution of a novel human matrix metalloproteinase with a potential transmembrane segment *European Journal of Biochemistry* **231** 602–608

Willenbrock F and Murphy G (1994) Structure–function relationships in the tissue inhibitors of metalloproteinases *American Journal of Respiration and Critical Medicine* **150** 165–170

NUTRITION AND METABOLIC SIGNALLING

Chair
R. P. Wettemann

Journal of Reproduction and Fertility Supplement **54**, 385–399

Nutrition and fetal growth: paradoxical effects in the overnourished adolescent sheep

J. M. Wallace, D. A. Bourke and R. P. Aitken

The Rowett Research Institute, Greenburn Road, Bucksburn, Aberdeen, UK

Inappropriate maternal nutrient intake at key developmental timepoints during ovine pregnancy has a profound influence on the outcome of pregnancy and aspects of postnatal productivity. However, the responses to alterations in maternal nutrition in adult sheep are often highly variable and inconsistent between studies. The growing adolescent sheep provides a new, robust and nutritionally sensitive paradigm with which to study the causes, consequences and reversibility of prenatal growth restriction. Overnourishing the adolescent dam to promote rapid maternal growth throughout pregnancy results in a major restriction in placental mass, and leads to a significant decrease in birthweight relative to moderately fed, normally growing adolescents of equivalent gynaecological age. Maternal insulin and IGF-I concentrations are increased from an early stage of gestation in overnourished adolescent dams and these hormones ensure that the anabolic drive required to promote maternal tissue synthesis is initiated at a time when the nutrient requirements of the gravid uterus are low. The major restriction in fetal growth in rapidly growing dams occurs irrespective of high concentrations of essential nutrients in the maternal circulation and suggests that the small size or altered metabolic and transport capacity of the placenta is the primary constraint to fetal growth. The decrease in placental weight in the overnourished animals reflects a significant reduction in both fetal cotyledon number and mean cotyledon weight. The role of nutritionally mediated alterations in progesterone and the components of the IGF system in this early pregnancy placental phenomenon are being investigated. Nutritional switch-over studies have demonstrated that reducing maternal nutrient intake at the end of the first third of pregnancy can stimulate placental growth and enhance pregnancy outcome, but increasing nutrient intake at this time has a deleterious effect on placental development and fetal growth.

Introduction

Alterations in the prenatal growth process influence both size and viability at birth. Low birthweight lambs have a relatively large surface area per unit weight, reduced insulation and inadequate lipid reserves, and depending on climatic conditions perinatally, have an increased risk of either hypothermia or dehydration and heat stress (Alexander, 1974). Low birthweight lambs are unable to compete with larger siblings for colostrum and hence are vulnerable to both starvation and infection during neonatal life. For those animals that survive the rigours of the early neonatal period, the long-term consequences of prenatal growth restriction include reduced postnatal and skeletal growth, altered carcass composition and failure to attain mature body size (reviewed by Bell, 1992). In addition, there is evidence that undernutrition during fetal life reduces lifetime reproductive capability (Williams, 1984; Gunn *et al.*, 1995) and the quality and quantity of wool production (Kelly *et al.*, 1996). At the other end of the scale, lamb or calf birthweights above the breed norm are associated with an increased incidence of distocia-related neonatal mortality, which has significant welfare and economic implications (Walker *et al.*, 1992).

As 80% of fetal growth occurs during the final third of pregnancy, it is not surprising that most early studies on the effects of nutrition on fetal growth and development concentrated on this period. However, there is an increasing body of evidence that the prenatal growth trajectory is sensitive to maternal dietary intake throughout pregnancy including the earliest stages of embryonic life. This is not unexpected if one considers pregnancy as a continuum from ovulation, fertilization, embryo differentiation, implantation, placental and fetal growth and that pregnancy outcome is dependent on each of the preceding events. As extremes in food intake are a common feature of sheep production systems, this review will concentrate on how alterations in maternal nutrition, at various developmental times during the 145 day gestation period, influence ovine prenatal growth. It will also detail recent studies using a novel adolescent sheep paradigm that has proved to be highly sensitive to maternal dietary intake and which is being used to determine the nutritionally mediated mechanisms underlying prenatal growth restriction and the consequences for the developing fetus.

Periovulatory and Early Pregnancy Nutrition

Exposure of the early cleavage stage ruminant embryo to a short period of *in vitro* culture (1–7 days) is associated with fetal oversize and is linked with a longer duration of gestation and high neonatal mortality (Walker *et al.*, 1992; Farin *et al.*, 1994). Culture of sheep embryos for 5 days in synthetic oviduct fluid medium (SOF) containing a human serum supplement before transfer to recipients resulted in lamb birthweights that were significantly higher than those in control spontaneously ovulating dams (4.2±0.2 versus 3.4±0.2 kg, $P < 0.01$, Thompson *et al.*, 1994). A recent, more stringently controlled study, in which *in vivo* derived sheep embryos were compared with those cultured in a granulosal cell co-culture system for 5 days, demonstrated that a major enhancement of fetal growth was evident as early as day 61 of pregnancy (74.2 versus 84.5 g respectively, $P < 0.01$) and was independent of placental mass (Sinclair *et al.*, 1997). The increase in weight observed in the fetuses that were derived from co-culture was associated with increased or precocious primary muscle fibre hypertrophy and secondary fibre hyperplasia (Maxfield *et al.*, 1997). The mechanisms whereby different *in vitro* culture systems influence the growth trajectory of the ruminant embryo are unknown, but could be largely nutritional in origin. Ammonium generated from amino acids in culture inhibits development and cleavage of the sheep blastocyst (Gardner *et al.*, 1994). *In vivo*, high dietary intakes of rumen degradable nitrogen increase plasma urea and ammonia concentrations, and are associated with increased embryo mortality (McEvoy *et al.*, 1997). However, in the latter study, exposure to ammonium appeared to upregulate protein metabolism in some embryos, and fetal growth was enhanced among those that survived autotransfer. It is conceivable that embryos in culture are being subjected to an inappropriate, nutrient-depleted environment, leading to changes in gene expression, which may be either lethal for the developing embryo or result in an alteration in the growth trajectory. Indirect evidence to support this hypothesis comes from studies using F9 embryonal carcinoma cells that display a differentiation pattern analogous to that of the developing embryo. In this rapidly dividing cell line, amino acid deficiency altered expression of a variety of growth-arrest genes (Fleming *et al.*, 1998). However, it is unknown whether the enhanced *in utero* growth of *in vitro* derived embryos reflects a permanent resetting of the growth trajectory, or whether these embryos could be influenced by manipulation of recipient dam nutrition during later stages of prenatal life.

In animals expressing their natural ovulation rate, maternal dietary intake premating and body composition at mating have both been shown to influence embryo survival (reviewed by Rhind, 1992). Short-term increases in maternal dietary intake premating may also influence early embryo development in superovulated animals. High dietary intakes that suppressed progesterone concentrations during the period of follicular recruitment, oocyte maturation and ovulation resulted in embryos that were developmentally retarded both on recovery 4 days after insemination and after 72 h in culture (McEvoy *et al.*, 1995). It remains to be established whether these embryos would have survived and developed appropriately if returned to a synchronous recipient uterus. Although there are reports that clearly demonstrate that a low plane of nutrition during early pregnancy is detrimental to embryo survival (Edey, 1966), the weight of evidence suggests that overfeeding during the early postmating period compromises establishment of pregnancy. In sheep, high intakes equivalent to twice maintenance rations for only 12 or 14 days commencing on day 2 after mating significantly reduced pregnancy rate (Parr *et al.*, 1987) and embryo survival (Cumming *et al.*, 1975). Maternal feed intakes and peripheral progesterone concentrations after mating are inversely related (Williams and Cumming, 1982) and the 20% reduction in pregnancy rate measured in high intake ewes by Parr *et al.* (1987) was reversed and enhanced relative to controls by progesterone supplementation on days 8–14 after mating. Administration of epostane, an inhibitor of 3β hydroxysteroid dehydrogenase, to reduce the concentration of progesterone for windows of 48 h between days 9 and 13 after mating revealed that pregnancy rates were lower when progesterone concentrations were temporarily reduced during days 11 and 12 (Parr, 1992). The inhibition of luteolysis and the maintenance of adequate progesterone secretion by the corpus luteum is central to the maternal recognition of pregnancy, and progesterone plays a major role in controlling maternal secretion of nutrients, immunosuppressive agents and enzymes required for successful embryo development at this time. It has been suggested that the mechanisms underlying nutritionally induced differences in pregnancy rate and embryo survival may operate via progesterone-dependent modifications of protein synthesis by either conceptus or endometrium (Robinson, 1990). Direct evidence to support this hypothesis is lacking. In a highly controlled study at the Rowett Research Institute, we collected embryos from ewes fed maintenance rations and transferred them in singleton on day 3 of the ovarian cycle to ewes receiving a high (2 × maintenance) or low (0.7 × maintenance) plane of nutrition from day 0. Progesterone concentrations were significantly reduced throughout the luteal phase in high intake ewes, but survival and growth of the conceptus at day 16 of pregnancy were equivalent between groups. Furthermore, the ability of the uterine endometrium and conceptus tissues to synthesis and secrete *de novo* proteins *in vitro*, and the conceptus secretion of antiluteolytic interferon tau were independent of maternal intake (Wallace *et al.*, 1994).

Studies examining the influence of nutrition during early pregnancy on the prenatal growth trajectory beyond the preimplantation stage generally indicate that food restriction is detrimental to fetal growth. Singleton fetuses of ewes fed a 0.5 × maintenance ration from mating until day 35 of gestation, and therafter fed normally, weighed less at day 35 of gestation. This difference was still evident at day 90 of gestation but did not persist until birth (Parr *et al.*, 1986). A more severe feed restriction (0.15 × maintenance) from mating until day 60 of gestation reduced both lamb birthweight and viability by 15 and 36%, respectively (Vincent *et al.*, 1985).

Mid-pregnancy Nutrition

Previous reviews have highlighted the central role of placental size in the determination of lamb birthweight (Mellor, 1983; Bell, 1984). Maximum proliferative growth of the ovine placenta occurs between days 50 and 60 of gestation (Ehrhardt and Bell, 1995) and placental weight peaks prior to the end of the second trimester of gestation; thus, mid-pregnancy is potentially a critical time when alterations in maternal nutrition may influence placental growth and hence pregnancy outcome. Kelly (1992) reviewed 16 sets of studies that examined the role of mid-pregnancy nutrition on placental growth and found that only two of these failed to demonstrate a significant effect of nutrient intake on placental size. At that time the consensus was that a high plane of nutrition from approximately day 40 to 100 of gestation significantly increased placental weight (nine studies). However, in three studies, ewes on a low plane of nutrition during mid-pregnancy had significantly heavier placentas. This observation has recently been reinforced by Heasman *et al.* (1998) who suggested that it is the fetal component of the placenta that is enhanced following such dietary restriction. There is evidence of an interaction between maternal liveweight together with body condition score (an indicator of body fat status) at mating and mid-pregnancy nutrition on the growth of the placenta or fetus. Russel *et al.* (1981) varied feeding levels between days 30 and 98 of gestation in primiparous ewes that were heavy or light at mating and in good or poor body condition, respectively. In ewes that were heavy at mating, feed restriction during mid-pregnancy increased lamb birthweight by 17%, while in ewes that were light at mating a low nutrient intake was associated with a 13% decrease in birthweight. Similar effects have been documented by De Barro *et al.* (1992) and appear to be mediated by alterations in placental growth and morphology.

Nutrition in Late Pregnancy

Fetal growth in late pregnancy is clearly dependent on the placental component, particularly if growth of the placenta has been compromised during early or mid-pregnancy. However, where placental growth is normal, it is largely assumed that the availability of essential nutrients in the maternal circulation can directly influence umbilical nutrient uptake and hence fetal growth (Bell, 1984).

Early studies reviewed by Hammond (1944) revealed that in singleton-bearing ewes fed a restricted diet (liveweight gain 0.5 kg during the last 60 days of pregnancy) versus well nourished animals (gaining 17.7 kg during the same period), lamb birthweights were similar. In contrast, in twin-bearing animals experiencing the same nutritional treatments, combined lamb birthweight from ewes on the restricted diet was reduced by 33%. It is now generally accepted that failure to increase nutrient intake in the final third of gestation in line with the increasing needs of the rapidly growing fetal load results in a reduction in birthweight. Robinson (1983) reviewed these studies and highlighted the wide variations in the amount of metabolizable energy (ME) required to elicit a given reduction in birthweight, and suggested that this may be due to differences in genotype, maternal reserves and the ME content of the different diets used. These factors, in addition to variable numbers of fetuses carried, differences in climate, housing and exercise, potentially confound interpretation of the effects of maternal nutrition on placental and fetal growth at all the developmental times discussed in this review.

It has been suggested that birthweight is a poor estimation of fetal growth rate, in that a variety of different growth patterns may result in the same final size (Harding and Johnston, 1995). The classic studies of David Mellor, which predate the use of ultrasonography, used a crown–rump length or girth measuring device to measure fetal growth on a daily basis throughout late gestation and provide a valuable insight into the sensitivity of the fetus to both short- and long-term alterations in maternal nutrition. When ewes that had been well fed were severely underfed during the last third of gestation, fetal growth rate decreased markedly within 3 days and resulted in a 20–30% decrease in birthweight at term (Mellor and Matheson, 1979). When fetuses were exposed to short periods of severe feed restriction of between 7 and 16 days, fetal growth rate generally increased when feed levels were restored. However, no recovery in growth rate occurred if the period of feed restriction was extended to 21 days (Mellor and Murray, 1982). Comparison of severe feed restriction in late pregnancy versus moderate restriction throughout mid- and late pregnancy produced similar reductions in late pregnancy growth rates and fetal weights. The former presumably occurs via a direct effect of nutrient availability in the maternal circulation and the latter via compromised placental growth (Mellor, 1983). More recent data suggest that the response of the fetus to a 10 day period of undernutrition during late pregnancy (days 105–115) may depend on its prior growth rate. Thus, the growth rate of fetuses from ewes that were well nourished for 2 months before until 1 month after mating was more perturbed during a late pregnancy nutritional insult than that of fetuses from ewes that were moderately undernourished during the same periconceptual period and were already growing at a slower rate (Harding and Johnston, 1995).

In late gestation, the adverse effect of feed restriction on birthweight is most pronounced when accompanied by low protein intakes (Robinson, 1983). In this review of data, it was calculated that for ewes receiving ME intakes in late pregnancy that only meet the maintenance requirement of the maternal body, the inclusion of an undegraded protein supplement during the final 3 weeks of pregnancy would increase the birthweight of twin lambs by up to 25%. In field conditions, the effects of supplementary protein in late pregnancy on lamb birthweight have been equivocal, but a recent study has reported a beneficial effect on lamb survival to weaning (Hinch *et al.*, 1996). This is probably due to a stimulatory effect of protein supplementation on colostrum production, as demonstrated by Robinson (1987).

Nutrition and the Pregnant Adolescent

In the literature reviewed in the preceding sections, it has clearly been demonstrated that inappropriate maternal nutrition at various developmental times during pregnancy can have a profound influence on pregnancy outcome and postnatal performance in an agriculturally important species. However, the responses to alterations in maternal nutrition have not always been consistent, even when studies have been performed by the same group of researchers using a single genotype. Thus, there is a clear requirement for the development of more rigorously controlled models to investigate the nutritionally mediated causes and consequences of alterations in the prenatal growth process. Human adolescent mothers have a risk of delivering low birthweight and premature infants who exhibit high mortality rates within the first year of life (McAnarney, 1987). The risk of adverse pregnancy outcome is poorly understood but has been attributed to the growth or nutritional status of the mother at the time of conception, or her young gynaecological age (Fraser *et al.*, 1995). At the Rowett Research Institute, we initially developed the sheep model outlined below to resolve this issue and to investigate nutrient partitioning between the maternal and fetal compartments during adolescent pregnancy (Wallace *et al.*, 1996, 1997a). However, it soon became evident that this experimental paradigm has both clinical and agricultural relevance that extends beyond the particular problems of the growing adolescent.

Experimental Model

Embryo recovery and transfer procedures are used to establish singleton pregnancies in peripubertal adolescent sheep. This removes the confounding influence of variation in fetal number and by using a single sire and a small number of adult donors maximizes the homeogeneity of the resulting fetuses. The adolescent recipients are of equivalent age, liveweight and body condition score at the time of embryo transfer and, in addition, care is taken to randomize for ovulation rate and embryo source. The adolescent recipient dams are housed in individual pens throughout pregnancy to facilitate precise nutritional management. Immediately after embryo transfer on day 4 of the oestrous cycle, the recipient dams are offered either a high or moderate quantity of a complete diet to promote rapid or normal maternal growth, respectively. Typically, these two dietary manipulations result in a liveweight gain of 200–300 g per day compared with 50–75 g per day during the first 100 days of gestation and significant differences in liveweight and body condition score are easily detectable by the end of the first third of pregnancy. After day 100 of gestation the feed intake of the normally growing group is adjusted each week to maintain body condition score and to meet the increasing nutrient demand of the gravid uterus.

Adolescent Pregnancy Outcome

The key features of pregnancy outcome for adolescent dams offered a high or moderate nutrient intake throughout their entire pregnancy and delivering live young at term are summarized in Table 1. Overnourishing dams to promote rapid maternal growth throughout pregnancy results in a major restriction in total placental mass (36%) and leads to a significant decrease in birthweight (33%) relative to normally growing adolescents (Wallace *et al.*, 1996, 1997b,c and J. M. Wallace, D. A. Bourke and R. P. Aitken, unpublished data). This somewhat paradoxical effect of overfeeding is consistent and completely repeatable between studies. Total placental mass and lamb birthweight are highly correlated in these adolescent animals (Fig. 1) but the placental : fetal weight ratio is not altered significantly by nutritional treatment. For those ewes delivering live young, high nutrient intakes are associated with a shorter duration of gestation and, irrespective of treatment group, duration of gestation is positively correlated with both placental weight ($r = 0.461$, $P < 0.001$) and lamb birthweight ($r = 0.572$, $P < 0.001$). Maternal liveweight gains during the first 95 days of gestation were 281±8.1 and 72±2.9 g per day for the high compared with the moderate intake

Table 1. Duration of gestation, lamb birthweight, placental parameters and colostrum production in normally growing (moderate nutrient intake) versus rapidly growing (high nutrient intake) adolescent dams delivering live young.

	Normal maternal growth Moderate nutrient intake	Rapid maternal growth High nutrient intake	Significance
Number of adolescents	34	37	
Duration of gestation (days)	145.1±0.50	142.4±0.42	***
(range)	(142–158)	(135–147)	
Lamb birthweight (g)	4758±154	3175±123	***
(range)	(2950–7050)	(1910–4560)	
Total placental weight (g)	470±18.6	299±14.2	***
(range)	(245–759)	(134–553)	
Number of fetal cotyledons	97±2.9	78±2.6	***
(range)	(54–127)	(44–107)	
Total fetal cotyledon weight (g)	130±6.3	64±2.9	***
(range)	(61–223)	23–98)	
Mean fetal cotyledon weight per placenta (g)	1.4±0.06	0.8±0.04	***
(range)	(0.63–2.04)	(0.28–1.45)	
Fetal : placental weight ratio	10.3±0.27	11.2±0.38	n.s.
Colostrum yield at parturition (g)	375±35.1	128±19.5	***

Values are means ± SEM. Data summarized from Wallace *et al.*, 1996, 1997a,b and J. M. Wallace, D. A. Bourke and R.P. Aitken, unpublished observations.
*** $P < 0.001$.

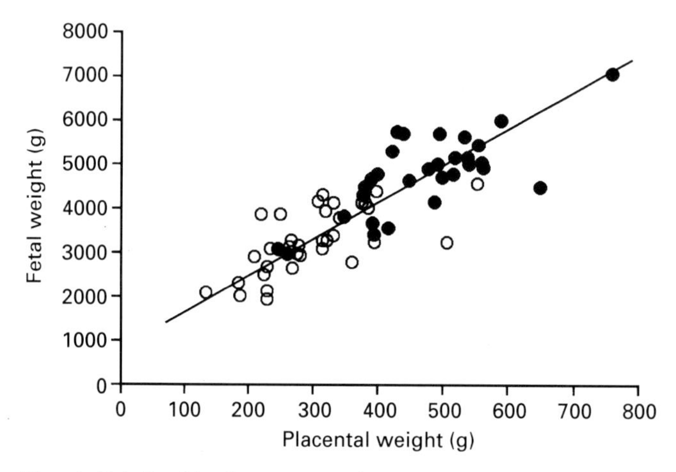

Fig. 1. Relationship between total placental weight and lamb birthweight in adolescent dams delivering singleton fetuses and offered a high (○) or moderate (●) nutrient intake throughout pregnancy to promote rapid or normal maternal growth, respectively.

dams, respectively. Irrespective of treatment group, placental weight ($r = -0.593$, $P < 0.001$) and lamb birthweight ($r = -0.609$, $P < 0.001$) were correlated negatively with liveweight gain. As the adolescent dams are all the same age, these results strongly suggest that maternal nutritional status rather than gynaecological immaturity predisposes the adolescent to poor pregnancy outcome.

Table 2. Relative differences in maternal concentrations of metabolic and placental hormones in adolescent dams offered a high or moderate nutrient intake throughout pregnancy. Arrows indicate when hormone concentrations (measured in blood samples collected one or three times a week) in high intake dams relative to moderate intake dams began to diverge significantly, and, where appropriate, the stage and direction of changes during gestation

	Stage of gestation		
	Early	Mid-	Late
Metabolic hormones			
Insulin	↑	↑	↑
IGF-I	↗	↑	↑
Tri-iodothyronine	→	↗	↑
Thyroxine	→	→	↑
GH*	–	↓	↓
Placental hormones			
Progesterone	↓	↓	↓
Pregnancy-specific protein B	↓	↓	↓

↑ : hormone concentrations in high intake dams significantly higher than in moderate intake dams.
→ : hormone concentrations equivalent in high and moderate intake dams.
↓ : hormone concentrations in high intake dams significantly lower then in moderate intake dams.
Summarized from Wallace *et al.*, 1997a and b.
*GH concentrations determined at 15 min intervals for 8 h on day 68 and day 122 of gestation.

Putative Endocrine Nutrient Partitioning Agents

Maternal hormones such as insulin, insulin-like growth factor I (IGF-I), growth hormone and thyroid hormones do not cross the placenta in physiologically significant quantities (Brown and Thorburn, 1989), but may indirectly regulate nutrient partitioning between the maternal and fetal compartments via secondary changes in maternal or placental metabolism, utero–placental blood flow or placental growth and nutrient transport functions. Similarly, placental hormones such as progesterone, oestrogen and placental lactogen are secreted largely into the maternal circulation and are considered to modify maternal physiology to the advantage of the growing conceptus. Many of these putative endocrine regulators of gestational nutrient partitioning are highly nutritionally sensitive and, in contrast to the numerous studies outlined above in which the dam and her fetus are both undernourished, the adolescent model involves overnutrition of the dam and placentally mediated restriction of nutrient supply to the fetus. We have examined the circulating maternal concentrations of a number of putative endocrine partitioning agents in relation to placental and fetal growth in this paradoxical model (Wallace *et al.*, 1997a,b) and the results are summarized in Table 2. High nutrient intakes are associated with high maternal insulin and, to a lesser extent, IGF-I concentrations from an early stage of gestation. These nutritionally sensitive hormones may ensure that the anabolic drive required to promote maternal tissue synthesis is initiated during the first third of pregnancy, at a time when the nutrient requirements for both placental and fetal growth are low. The potential for continuing maternal growth during pregnancy is high in the overnourished adolescent sheep and the major drive to maternal tissue deposition, particularly of adipose tissue, is maintained throughout gestation in spite of the gradually increasing nutrient demands of the gravid uterus. During normal pregnancy in the adult sheep, the amount of glucose available to non-uterine maternal tissues is reduced as pregnancy advances (Oddy *et al.*, 1985). It has been suggested that insulin and placental lactogen play key roles in mediating these metabolic changes. In adult ewes maternal insulin concentrations normally decrease during the final third of pregnancy and are inversely related to concentrations of placental lactogen (Vernon *et al.*, 1981). It appears that placental lactogen may have a causative role because removal of the hormone during a short-term infusion with a specific antibody leads to a significant rise in maternal insulin concentrations (Waters *et al.,* 1985). In contrast, in the overnourished adolescent, insulin concentrations continue to rise during the final third of pregnancy and may reflect low secretion of placental lactogen by the growth-restricted placenta. This, in turn, appears to result in increased glucose utilisation by the maternal tissues and continued lipid accumulation. Although we have not yet measured placental lactogen in these animals, we have clearly demonstrated that maternal peripheral concentrations of progesterone and of the pregnancy-specific protein B produced by the binucleate cells of the placenta were low throughout gestation in the overnourished animals and were associated positively with placental mass, particularly during the second half of gestation (Wallace *et al.,* 1997a).

Placental Growth

The unexpected but major restriction in fetal growth, observed in rapidly growing mothers, occurs irrespective of high levels of essential nutrients available in the maternal circulation. High intake dams are relatively hyperglycaemic throughout gestation, and an attenuated glucose response to an exogenous insulin challenge during mid- and late pregnancy implies a degree of insulin resistance (Wallace *et al.,* 1997b). Small size, altered metabolic and transport capacity of the placenta, or both factors may be the primary constraint to fetal growth in overnourished adolescent sheep.

The sheep has an epitheliochorialcotyledonary placenta and number and size of the individual cotyledons determine the available area for nutrient exchange between the maternal and fetal systems. The decrease in placental weight observed at term in the overnourished animals reflects a significant reduction in both the number of fetal cotyledons per placenta and mean fetal cotyledon weight (Table 1). These studies are the first to demonstrate consistently that maternal nutrition can influence the number of maternal caruncles used by the developing trophoblast. We are beginning to examine the putative nutritionally mediated endocrine mechanisms underlying this early pregnancy placental phenomenon.

Exogenous progesterone administration during the first 3 days of pregnancy enhanced fetal growth at day 74 of gestation by 11% (Kleemann *et al.,* 1994). Although placental data were not reported, these researchers observed that progesterone influenced blastocyst differentiation in favour of the trophectoderm cells and stimulated earlier trophoblast elongation compared with control animals (Hartwich *et al.,* 1995). Since maternal dietary intakes are related inversely to concentrations of progesterone in peripheral plasma in the adolescent dams (Table 2), it seemed probable that suboptimal progesterone in overnourished dams could compromise growth of the differentiating conceptus, resulting in fewer uterine caruncles being occupied. In a preliminary study of progesterone supplementation, we have investigated this hypothesis and the results are presented in Table 3. Progesterone concentrations were lower in high intake versus moderate intake dams. Daily administration of physiological doses of progesterone to a third group of high intake dams from day 5 to day 55 of gestation restored moderate circulating progesterone concentrations throughout the first third of pregnancy. Thereafter, peripheral progesterone concentrations were similar in the high intake and high intake plus progesterone groups and significantly lower than in the moderate intake group. At term, mean fetal weight in the high intake plus progesterone group was intermediate between the high and moderate intake groups but this increase in birthweight was not mediated by significant alterations in any of the gross placental measurements. These preliminary results on a very small number of animals do not preclude the possibility of more subtle effects of exogenous progesterone on placental morphology or nutrient exchange capacity, but progesterone may be influencing the inner cell mass directly.

The IGF system may play a pivotal role in the regulation of early placental growth in that the ovine placenta contains type 1 receptors for IGFs throughout gestation (Lacroix *et al.,* 1995; Reynolds *et al.,* 1997a). These receptors could be a target for locally produced or circulating IGFs from the maternal or fetal circulation and hence may have a role in the proliferative growth or metabolic

Table 3. Influence of progesterone supplementation in high intake ewes during the first third of pregnancy on peripheral progesterone concentrations and pregnancy outcome

	Maternal nutrient intakes		
	Moderate	High	High + progesterone (Day 5–55 of gestation)
Number of adolescents	7	6	7
*Progesterone concentrations			
(ng ml^{-1}) first trimester	8.7±0.68	6.6±0.91	9.5±1.05
second trimester	11.8±0.78	6.5±1.46	6.3±0.63
third trimester	20.9±1.60	7.1±1.17	8.6±1.10
Lamb birthweight (g)	5164±150	3118±363	4150±389
Total placental weight (g)	498±18.9	313±65.9	318±41.5
Fetal cotyledon number	85±6.9	83±10.4	76±12.6
Total fetal cotyledon weight (g)	136±12.1	61±8.9	76±10.2

Values are means ±SEM. J. M. Wallace, D. A. Bourke and R. P. Aitken, unpublished.
*Based on overall mean individual progesterone concentrations measured in blood samples collected three times a week.

activity of the developing placenta. In the overnourished adolescent, maternal IGF-I concentrations are high while placental growth is restricted. Preliminary data indicate that utero–placental IGF-I receptor expression at the end of the second third of pregnancy is low in these animals (Reynolds *et al.*, 1997b). Prolonged exposure to high maternal IGF-I concentrations may downregulate placental IGF-I receptor expression and hence placental growth. Alternatively, if placentally derived IGF-I plays a major autocrine role in placental growth, attenuated IGF-I secretion by the growth-restricted placenta *per se* may be a secondary consequence of reduced nutrient availability at the utero–placental level.

We intend to extend these studies on the nutritionally mediated mechanisms underlying early placental growth to investigate the role of maternal nutrition on the utero–placental secretion and expression of the angiogenic growth factors and how these relate to blood vessel formation and utero–placental blood flow.

Nutritionally Sensitive Windows of Placental Growth and their Reversibility

Although the number of cotyledonary attachment sites occupied by the developing trophoblast is fixed by day 30 of gestation (Barcroft and Kennedy, 1939), the proliferative growth of the placenta continues until the end of the second third of pregnancy. A recent study has examined whether placental growth and hence pregnancy outcome can be altered by switching adolescent dams from an anabolic to a catabolic state, and vice versa, at the end of the first third of pregnancy (Wallace *et al.*, 1997c). After embryo transfer adolescent dams were offered a high (H, $n = 33$) or moderate (M, $n = 32$) nutrient intake to promote rapid or normal maternal growth as described previously. At day 50 of gestation half the ewes had their dietary intakes switched to yield HH, MM, HM and MH treatments (Fig. 2). After day 100, feed intake of the MM and HM group was adjusted weekly to maintain body condition score during the final trimester. A subset of ewes ($n = 4$ or 5 per group) were slaughtered at day 104 of gestation, while the remaining ewes were allowed to deliver spontaneously. For ewes maintaining pregnancies to term, total placental mass (258±33 versus 457±75 g), fetal cotyledon number (79±7.8 versus 103±3.4), mean cotyledon weight (0.7±0.06 versus 1.3±0.19 g) and lamb birthweight (3.03±0.34 versus 4.94±0.57 kg) were lower ($P < 0.02$) in HH than in MM groups as reported above for earlier studies. Abruptly decreasing maternal dietary intake at the end of the first third of the pregnancy (HM) resulted in a moderate but non-significant increase

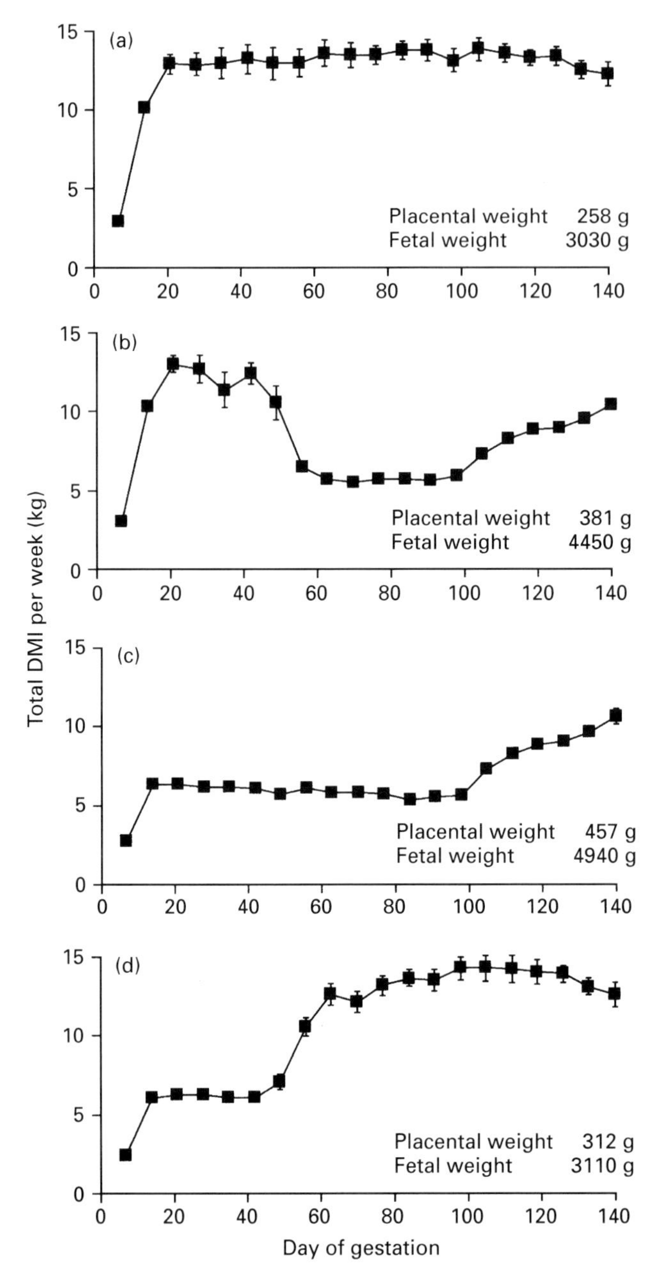

Fig. 2. Weekly dry matter intakes (DMI) throughout gestation in relation to pregnancy outcome in adolescent dams delivering singleton fetuses. After embryo transfer ewes were initially offered a high (H) or moderate (M) nutrient intake to promote rapid or normal maternal growth. At day 50 of gestation half the ewes in each treatment had their dietary intakes switched to yield (a) HH (n = 7), (b) HM (n = 9), (c) MM (n = 6) and (d) MH (n = 8) treatments. After day 100, feed intake of the MM and HM groups was adjusted weekly to maintain body condition score during the final third of pregnancy.

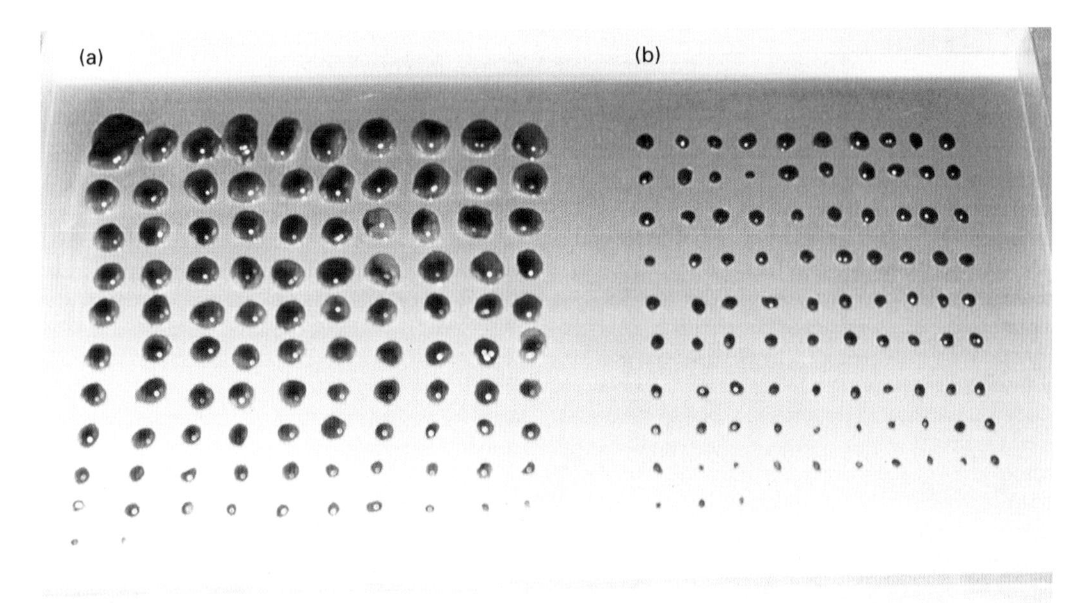

Fig. 3. Individual fetal cotyledons dissected on day 104 of gestation from (a) an adolescent ewe offered a moderate intake for the first two thirds of gestation and (b) an adolescent ewe offered a moderate intake for the first third followed by a high intake for the second third of gestation.

in the number of fetal cotyledons (18%) and a major stimulation in individual cotyledon weight (1.1±0.13 g, $P < 0.05$) compared with the HH group. Consequently, lamb birthweights in the HM group (4.45±0.34 kg) were equivalent to the MM group. In contrast, ewes initially on moderate intakes had a normal number of cotyledons (109±3.9) and when their intakes were increased markedly at the end of the first third of pregnancy (MH), the growth of the individual cotyledons was severely reduced (0.6±0.07 g, $P < 0.01$) relative to the MM group (Fig. 3). Hence total placental mass was lower (312±22 g) and lead to a major restriction in fetal growth (3.11±0.24 kg, $P < 0.02$). Dissection of the placenta at day 104 of gestation revealed that moderate nutrient intakes during mid-gestation preferentially stimulate growth of the fetal rather than the maternal component of the placenta; total fetal cotyledon weight in MM and HM compared with HH and MH groups was 196±18.1 versus 104±19.8 g, $P < 0.02$, while maternal cotyledon weight was 157±20.5 versus 120±19.8 g, respectively. At this stage of gestation, the fetus weighs less than a third of its potential birthweight and mean fetal weights in the various nutritional treatment groups were not significantly different. However, a positive relationship between placental size and fetal growth was already evident since, irrespective of treatment group, fetal weight ($r = 0.650$) and the weight of the fetal liver ($r = 0.760$), spleen ($r = 0.784$), kidneys ($r = 0.521$) and intestines ($r = 0.599$) were significantly correlated with total cotyledon mass. Thus, in this highly controlled nutritionally sensitive paradigm, reducing maternal dietary intake at the end of the first third of the pregnancy can stimulate placental growth and enhance pregnancy outcome, but increasing dietary intake at this time has a deleterious effect on placental development and fetal growth. These nutritional switch-over studies have obvious implications for both agricultural and clinical practice. In humans, intrauterine growth restriction is rarely diagnosed before the final third of the pregnancy and is associated primarily with poor placental growth. It remains to be established in this overnourished sheep model whether abruptly reducing maternal nutrient intake at the end of the second third of pregnancy can alter the metabolic and transport functions of a stunted placenta to redirect essential nutrients to the fetus.

Table 4. Conformation and absolute fetal organ weights of singleton fetuses derived from dams that were offered a high or moderate nutrient intake from day 4 after oestrus until slaughter on day 128 of gestation:

	Normal Maternal Growth Moderate Nutrient Intake	Rapid Maternal Growth High Nutrient Intake	Significance
Number of dams/fetuses	7	7	
Total cotyledon weight[a] (g)	448±23.5	219±34.7	***
Fetal weight (g)	4190±87	2645±385	**
Biparietal head diameter (mm)	67.9±0.46	58.9±2.57	**
Umbilical girth (cm)	34.8±0.63	27.0±1.81	**
Crown–rump length (cm)	49.4±0.70	41.5±2.72	*
Brain (g)	40.8±1.24	34.5±1.61	**
Lungs (g)	156.8±7.54	95.8±15.8	**
Heart (g)	33.2±1.01	22.7±3.57	*
Liver (g)	128.7±3.86	75.0±11.8	***
Kidneys (g)	28.1±0.81	19.5±2.28	**
Empty gut (g)	136.6±4.62	94.7±14.38	*
Pancreas (g)	4.03±0.148	2.39±0.344	***
Spleen (g)	7.35±0.338	4.82±1.017	*
Adrenals (g)	0.49±0.023	0.41±0.034	ns
Thyroid (g)	0.89±0.067	0.58±0.104	*
Gastrocnemius muscle (g)	8.11±0.344	5.45±0.816	*
Empty carcass (g)	2655±61.8	1624±254.6	**

Values are means ±SEM. J. M. Wallace, D. A. Bourke and R. P. Aitken, unpublished data.
[a] Combined fetal cotyledon and maternal caruncle weight.
*** $P < 0.001$, ** $P < 0.01$, * $P < 0.05$; ns, not significant.

Consequences of a Disrupted Growth Trajectory

We are examining the consequences of a disrupted placental growth trajectory on fetal organ growth, structure and function. An initial study compared fetuses at day 128 of gestation from adolescent dams that were overnourished or moderately fed throughout pregnancy and had growth restricted and normal placentae, respectively (Table 4). All variables of fetal conformation and absolute fetal organ weights, with the exception of the adrenals, were significantly lower in the fetuses from overnourished dams. However, relative organ and tissue weights expressed as g per kg fetal body or fetal empty carcass weight were not influenced significantly by maternal nutrient intake (data not shown), with the exception of the gut. The implication is that growth restriction in these fetuses is largely symmetrical as indicated by the allometric plots for three of the major fetal organs (Fig. 4). Analysis of single blood samples collected immediately before these fetuses were killed reveals that concentrations of insulin, IGF-I and glucose were significantly attenuated in growth-restricted fetuses from overnourished dams. In addition, plasma urea was high, indicating that the fetuses may be relatively catabolic compared with the normally growing fetuses from the moderately fed dams. Detailed examination of the effects of maternal nutrition on the functional development of the major fetal organ systems in late gestation fetuses has been initiated. Of key economic interest to all agricultural sectors is the concept that aspects of our adult reproductive potential may be programmed during prenatal life. Thus, it is potentially significant that we have detected reduced expression of LH and FSH-β mRNAs in the fetal pituitary of severely growth restricted versus normally growing late gestation fetuses. Moreover, we have observed alterations in gonadal germ cell number and stage of differentiation at this developmental time, which may be GnRH dependent, and which could underlie reduced postnatal reproductive capability (P. Da. Silva, N. Brooks, S. M. Rhind and J. M. Wallace, unpublished data).

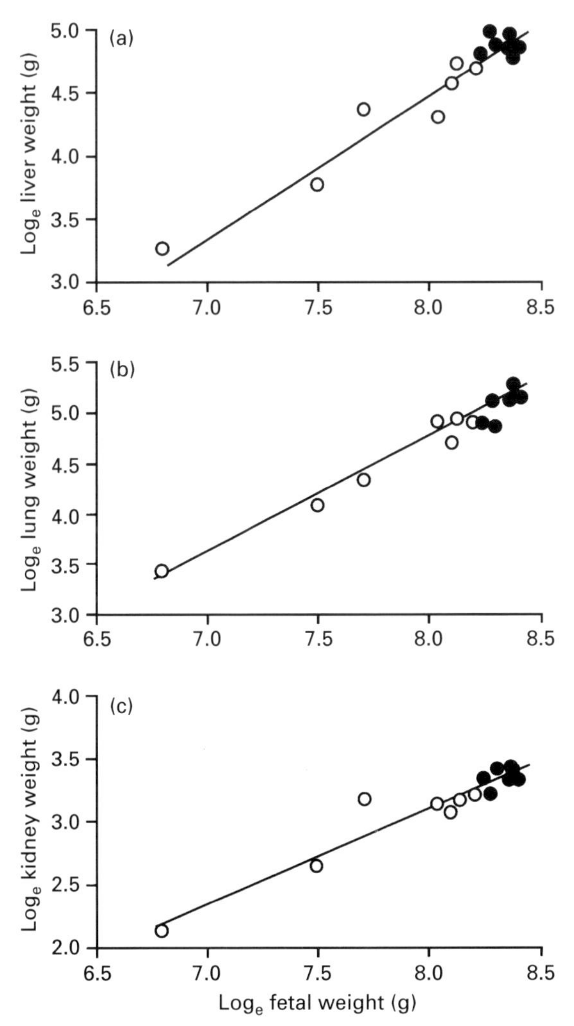

Fig. 4. Allometric plots of (a) liver, (b) lungs and (c) kidneys of fetuses from dams offered a high (○) or moderate (●) nutrient intake throughout pregnancy and killed on day 128.

Conclusion

We have demonstrated that inadequate placental growth is the primary limitation to fetal growth in the overnourished adolescent sheep. Consequently, the growing adolescent sheep provides a new and non-invasive paradigm to investigate the putative nutritionally mediated endocrine and paracrine mechanisms underlying early placental growth. The restriction in fetal growth in overnourished dams occurs in spite of the ready availability of nutrients in the maternal system and future studies will determine placental uptake, metabolism and transfer of nutrients by the growth-restricted placenta to the developing fetus. Initial studies indicate that this form of placentally mediated fetal growth restriction programmes a legacy of structural and functional defects that may have a major impact on postnatal health, growth and development. As such, this paradoxical adolescent sheep model is of major agricultural and clinical relevance and may play a key role in the development of more informed nutritional strategies to alleviate or prevent intrauterine growth retardation.

The authors thank P. Da Silva and M. Cruickshank for technical assistance and gratefully acknowledge the continued financial support of the Scottish Office Agriculture Environment and Fisheries Department.

References

Alexander G (1974) Birth weight of lambs: influences and consequences. In *Size at Birth* pp 215–239. Eds K Elliot and J Knight. Associated Scientific Publishers, Amsterdam

Barcroft J and Kennedy JA (1939) The distribution of blood flow between the fetus and placenta in sheep *Journal of Physiology* **95** 173–186

Bell AW (1984) Factors controlling placental and foetal growth and their effects on future production. In *Reproduction in Sheep* pp 144–152. Eds DR Lindsay and PT Pearce. Australian Academy of Science, Canberra

Bell AW (1992) Foetal growth and its influence on postnatal growth and development. In *The Control of Fat and Lean Deposition* pp 111–127. Eds KN Boormann, PJ Buttery and DB Lindsay. Butterworth-Heinemann Ltd, Oxford

Brown CA and Thorburn GD (1989) Endocrine control of fetal growth *Biology of the Neonate* **55** 331–346

Cumming IA, Blockey MADeB, Winfield CG, Parr RA and Williams AH (1975) A study of the relationships of breed, time of mating, level of nutrition, live weight, body condition, and face cover to embryo survival in ewes *Journal of Agricultural Science* **84** 559–565

De Barro TM, Owens JA, Earl CR and Robinson JS (1992) Nutrition during early pregnancy interacts with mating weight to affect placental growth *Proceedings of the Australian Society of Reproductive Biology* **35** 70 (Abstract)

Edey TN (1966) Nutritional stress and pre-implantation embryonic mortality in Merino sheep *Journal of Agricultural Science* **67** 287–293

Ehrhardt RA and Bell AW (1995) Growth and metabolism of the ovine placenta during mid-gestation *Placenta* **16** 727–741

Farin PW, Farin CE and Yang L (1994) *In vitro* production of bovine embryos is associated with altered fetal development *Theriogenology* **41** 193 (Abstract)

Fleming JV, Hay SM, Harries N and Rees WD (1998) Effects of nutrient deprivation and differentiation on the expression of growth-arrest genes (*gas and gadd)* in F9 embryonal carcinoma cells *Biochemical Journal* **330** 573–579

Fraser AM, Brockert JE and Ward RH (1995) Association of young maternal age with adverse reproductive outcomes *The New England Journal of Medicine* **332** 1113–1117

Gardner DK, Lane M, Spitzer A and Batt PA (1994) Enhanced rates of cleavage and development for sheep zygotes cultured to the blastocyst stage *in vitro* in the absence of serum and somatic cells: amino acids, vitamins and culturing embryos in groups stimulate development *Biology of Reproduction* **50** 390–400

Gunn RG, Sim DA and Hunter EA (1995) Effects of nutrition *in utero* and in early life on the subsequent lifetime reproductive performance of Scottish Blackface ewes in two management systems *Animal Science* **60** 223–230

Hammond J (1944) Physiological factors affecting birthweight *Proceedings of The Nutrition Society* **2** 8–14

Harding JE and Johnston BM (1995) Nutrition and fetal growth *Reproduction Fertility and Development* **7** 539–547

Hartwich KM, Walker SK, Owens JA and Seamark RF (1995) Progesterone supplementation in the ewe alters cell allocation to the inner cell mass *Proceedings of the Australian Society of Medical Research* **26** 128 (Abstract)

Heasman L, Clarke L, Firth K, Stephenson T and Symonds ME (1998) Influence of restricted maternal nutrition in early to mid gestation on placental and fetal development at term in sheep *Paediatric Research* (in press)

Hinch GN, Lynch JJ, Nolan JV, Leng RA, Bindon BM and Piper LR (1996) Supplementation of high fecundity Border Leicester × Merino ewes with a high protein feed: its effect on lamb survival *Australian Journal of Experimental Agriculture* **36** 129–136

Kelly RW (1992) Nutrition and placental development *Proceedings of The Nutrition Society of Australia* **17** 203–211

Kelly RW, Macleod I, Hynd P and Greeff J (1996) Nutrition during fetal life alters annual wool production and quality in young Merino sheep *Australian Journal of Experimental Agriculture* **36** 259–267

Kleemann DO, Walker SK and Seamark RF (1994) Enhanced fetal growth in sheep administered progesterone during the first three days of pregnancy *Journal of Reproduction and Fertility* **102** 411–417

Lacroix MC, Servely JL and Kann G (1995) IGF-I and IGF-II receptors in the sheep placenta: evolution during the course of pregnancy *Journal of Endocrinology* **144** 179–191

McAnarney ER (1987) Young maternal age and adverse neonatal outcome *American Journal of Diseases of Children* **141** 1053–1059

McEvoy TG, Robinson JJ, Aitken RP, Findlay PA, Palmer RM and Robertson IS (1995) Dietary-induced suppression of pre-ovulatory progesterone concentrations in superovulated ewes impairs the subsequent *in vivo* and *in vitro* development of their ova *Animal Reproduction Science* **39** 89–107

McEvoy TG, Robinson JJ, Aitken RP, Findlay PA and Robertson IS (1997) Dietary excesses of urea influence the viability and metabolism of preimplantation sheep embryos and may affect fetal growth among survivors *Animal Reproduction Science* **47** 71–90

Maxfield EK, Sinclair KD, Dolman DF, Staines ME and Maltin CA (1997) *In vitro* culture of sheep embryos increases weight, primary fibre size and secondary to primary fibre ratio in fetal muscle at day 61 of gestation *Theriogenology* **47** 376 (Abstract)

Mellor DJ (1983) Nutritional and placental determinants of foetal growth rate in sheep and consequences for the newborn lamb *British Veterinary Journal* **139** 307–324

Mellor DJ and Matheson IC (1979) Daily changes in the curved crown–rump length of individual sheep fetuses during the last 60 days of pregnancy and effects of different levels of maternal nutrition *Quarterly Journal of Experimental Physiology* **64** 119–131

Mellor DJ and Murray L (1982) Effects on the rate of increase in fetal girth of refeeding ewes after short periods of severe undernutrition during late pregnancy *Research in Veterinary Science* **32** 377–382

Oddy VH, Gooden JM, Hough GM, Teleni E and Annison EF (1985) Partitioning of nutrients in Merino ewes II. Glucose utilisation by skeletal muscle, the pregnant uterus and the lactating mammary gland in relation to whole body glucose utilisation *Australian Journal of Biological Science* **38** 95–108

Parr RA (1992) Nutrition–progesterone interactions during early pregnancy in sheep *Reproduction Fertility and Development* **4** 297–300

Parr RA, Williams AH, Cambell IP, Witcome GF and Roberts AM (1986) Low nutrition of ewes in early pregnancy and the residual effect on the offspring *Journal of Agricultural Science* **106** 81–87

Parr RA, Davis IF, Fairclough RJ and Miles MA (1987) Overfeeding during early pregnancy reduces peripheral progesterone concentration and pregnancy rate in sheep *Journal of Reproduction and Fertility* **80** 317–320

Reynolds TS, Stevenson KR and Wathes DC (1997a) Pregnancy-specific alterations in the expression of the insulin-like growth factor system during early placental development in the ewe *Endocrinology* **138** 886–897

Reynolds TS, Wathes DC, Aitken RP and Wallace JM (1997b) Effect of maternal nutrition on components of the insulin-like growth factor (IGF) system and placental growth *Journal of Endocrinology* **152** (Supplement) p248

Rhind SM (1992) Nutrition: its effects on reproductive performance and its hormonal control in female sheep and goats. In *Progress in Sheep and Goat Research* pp 25–52 Ed. AW Speedy. CAB International, Oxford

Robinson JJ (1983) Nutrition of the pregnant ewe. In *Sheep Production* pp 111–131 Ed. W Haresign. Butterworths, London

Robinson JJ (1987) Energy and protein requirements of the ewe. In *Recent Advances in Animal Nutrition* pp187–204 Eds W Haresign and DJA Cole. Butterworths, London

Robinson JJ (1990) Nutrition in the reproduction of farm animals *Nutrition Research Reviews* **3** 253–276

Russel AJF, Foot JZ and White IR (1981) The effect of weight at mating and of nutrition during mid-pregnancy on the birthweight of lambs from primiparous ewes *Journal of Agricultural Science* **97** 723–729

Sinclair KD, Maxfield EK, Robinson JJ, Maltin CA, McEvoy TG, Dunne LD, Young LE and Broadbent PJ (1997) Culture of sheep zygotes can alter fetal growth and development *Theriogenology* **47** 380 (Abstract)

Thompson JG, Gardner DK, Pugh PA, McMillan WH and Tervit HR (1994) Lamb birthweight following transfer is affected by the culture system used for pre-elongation development of embryos *Journal of Reproduction and Fertility Abstract Series* **13** Abstract 25

Vernon RG, Clegg RA and Flint DJ (1981) Metabolism of sheep adipose tissue during pregnancy and lactation *Biochemistry Journal* **200** 307–314

Vincent IC, Williams HLI and Hill R (1985) The influence of a low-nutrient intake after mating on gestation and perinatal survival of lambs *British Veterinary Journal* **141** 611–617

Walker SK, Heard TM and Seamark RF (1992) *In vitro* culture of sheep embryos without co-culture: successes and perspectives *Theriogenology* **37** 111–126

Wallace JM, Aitken RP and Cheyne MA (1994) Effect of post-ovulation nutritional status in ewes on early conceptus survival and growth *in vivo* and luteotrophic protein secretion *in vitro. Reproduction, Fertility and Development* **6** 253–259

Wallace JM, Aitken RP and Cheyne MA (1996) Nutrient partitioning and fetal growth in rapidly growing adolescent ewes *Journal of Reproduction and Fertility* **107** 183–190

Wallace JM, Aitken RP, Cheyne MA and Humblot P (1997a) Pregnancy-specific protein B and progesterone concentrations in relation to nutritional regimen, placental mass and pregnancy outcome in growing adolescent ewes carrying singleton fetuses *Journal of Reproduction and Fertility* **109** 53–58

Wallace JM, Da Silva P, Aitken RP and Cruickshank MA (1997b) Maternal endocrine status in relation to pregnancy outcome in rapidly growing adolescent sheep *Journal of Endocrinology* **155** 359–368

Wallace JM, Bourke DA, Aitken RP and Cruickshank MA (1997c) Effect of switching maternal nutrient intake at the end of the first trimester on placental development and fetal growth in adolescent sheep carrying singleton fetuses *Journal of Reproduction and Fertility Abstract Series* **19** Abstract 21

Waters MJ, Oddy VH, McCloghry CE, Gluckmann PD, Duplock R, Owens PC and Brinsmead MW (1985) An examination of the proposed roles of placental lactogen in the ewe by means of antibody neutralization *Journal of Endocrinology* **106** 377–386

Williams AH (1984) Long-term effects of nutrition of ewe lambs in the neonatal period. In *Reproduction in Sheep* pp 272–273. Eds DR Lindsay and PT Pearce. Australian Academy of Science, Canberra

Williams AH and Cumming IA (1982) Inverse relationship between concentration of progesterone and nutrition in ewes *Journal of Agricultural Science* **98** 517–522

Journal of Reproduction and Fertility Supplement **54**, 401–410

Placental transport of nutrients and its implications for fetal growth

A. W. Bell[1], W. W. Hay, Jr[2] and R. A. Ehrhardt[1]

[1]*Department of Animal Science, Cornell University, Ithaca, NY 14853; and* [2]*Department of Pediatrics, Division of Perinatal Medicine, University of Colorado School of Medicine, 4200 E. 9th Ave, Denver, CO 80262, USA*

Placental growth during early and mid-pregnancy has a powerful, constraining influence on fetal growth during late pregnancy. Studies involving surgical and environmental reduction of placental size in sheep have shown an associated reduction in capacity to transport oxygen, glucose and amino acids. Oxygen transport is limited by placental blood flow but transport of glucose and amino acids is determined by the abundance and activity of specific transport proteins. Glucose transporters include the GLUT1 and GLUT3 isoforms previously identified in brain and other tissues; systems for active transport of amino acids have been inferred but not characterized. Placental metabolism of glucose and amino acids has major effects both on the quantity of carbon and nitrogen delivered to the fetus, and on the composition of substrates involved. For example, the uteroplacental tissues consume more than 60% of uterine glucose uptake during late pregnancy, and the placenta substantially modifies the pattern of amino acids delivered to fetal blood. The placenta also participates in the array of metabolic adaptations of maternal and conceptus tissues to altered maternal nutrient supply. Placental capacity for glucose transport in moderately undernourished ewes is upregulated, partly by increased expression of the GLUT3 transport protein. During more severe glucose deprivation, placental transfer and fetal uptake of glucose are constrained in proportion with maternal supply, leading to fetal growth retardation.

Introduction

The placenta is a unique organ of pregnancy of higher animals, including domestic ruminants. Its various, highly specialized functions include exchange of nutrients and excreta between mother and fetus, endocrine regulation of numerous pregnancy-specific physiological and metabolic adaptations in fetal and maternal tissues, and immunological protection of the conceptus from its maternal host. This review will address only the nutrient transport functions of the placenta.

Domestic ruminants, principally sheep, have provided most of the experimental evidence regarding placental nutrient transfer and the importance of placental influence on prenatal growth (Alexander, 1974; Battaglia and Meschia, 1988; Ferrell, 1989; Robinson *et al.*, 1995). Unless stated otherwise, examples in this review will be confined to studies on sheep, particularly those focussed on the integration of placental function and fetal response. Major themes include relations between placental growth and capacity for nutrient transport, and placental adaptations to changing fetal nutrient requirements during gestational development and altered maternal nutrition.

Placental Influences on Fetal Growth

Growth patterns of conceptus tissues

In ruminants as in most other mammals, the major phase of placental growth occurs during the first half of gestation, substantially preceding that of fetal growth during later gestation. In the sheep, polycotyledonary, epitheliochorial placentation is fully established by about 30 days after conception, and the number of placentomes attached to each fetus is fixed at or soon after this time. Rapid hyperplastic growth then occurs until about day 55, before declining to minimal rates by mid-pregnancy at approximately 75 days (Ehrhardt and Bell, 1995). Placental mass (that is, total mass of placentomes) declines appreciably between mid-pregnancy and term, due to tissue dehydration, associated with loss of hyaluronic acid and related glycosaminoglycans, and extensive tissue remodelling. This pattern contrasts somewhat with that in cows, in which modest placental growth, confined to the maternal (caruncular) component, continues into the third trimester (A. W. Bell and R. A. Ehrhardt, unpublished).

Effect of placental size on fetal growth

During the latter half of pregnancy, positive correlations between fetal and placental weights become progressively stronger, such that within a few weeks of term, variation in placental weight accounts for more than 80% of variation in fetal weight (Stegeman, 1974). Such statistical associations have been used to imply, but do not necessarily prove, placental cause and fetal effect. Persuasive evidence that placental weight is indeed a powerful determinant of fetal growth during late gestation was first obtained by the deceptively elegant carunclectomy experiments of Alexander (see Alexander, 1974). In other studies, chronic heat stress, sufficient to cause persistent hyperthermia in pregnant ewes, caused a profound reduction of placental weight that was followed by fetal growth restriction (Alexander, 1974; Vatnick *et al.*, 1991; McCrabb *et al.*, 1993). Premating carunclectomy and maternal heat stress result in similar patterns of association between fetal and placental weights near term (Bell and Ehrhardt, 1998), suggesting that they may provide comparable models of placental insufficiency during late pregnancy.

Effects of maternal nutrition on placental growth are usually more variable and subtle than those caused by physical ablation or heat stress. Undernutrition of ewes during early and mid-pregnancy has caused conflicting positive (Faichney and White, 1987; McCrabb *et al.*, 1992) and negative (McCrabb *et al.*, 1992; Clarke *et al.*, 1998) effects on placental growth. Variation in body condition during early pregnancy may partly explain this confusion, in that fatter ewes are more likely to respond to underfeeding with a compensatory increase in placental growth, whereas the opposite occurs in lean ewes (Bell and Ehrhardt, 1998). In contrast, recent, novel studies have shown that overfeeding and rapid maternal growth of primiparous ewes during early–mid- pregnancy causes profound reductions in placental and fetal weights at term (Wallace *et al.*, 1996; Table 1). Early indications are that the fetal growth retardation in this model of adolescent pregnancy is a consequence of placental insufficiency that is due to a primary failure of cotyledonary growth (J. M. Wallace, personal communication).

Functional correlates of placental size

In both carunclectomized and heat-treated ewes, reduction in placental size is highly correlated with decreases in several important determinants and indices of placental transport, and with consequent changes in fetal metabolic characteristics during late gestation (see reviews by Bell, 1987; Owens *et al.*, 1989; Robinson *et al.*, 1995; Bell and Ehrhardt, 1998). These include reductions in uterine and umbilical blood flows, consistent with reduced placental clearance of highly diffusible, flow-limited materials such as antipyrine or ethanol, and metabolic consequences such as reduced placental oxygen uptake and transport, and development of fetal hypoxaemia.

Placental capacity for glucose transport also was reduced substantially, as were uteroplacental glucose consumption rate and fetal glycaemia in carunclectomized (Owens *et al.*, 1989) and heat-treated ewes (Bell, 1987; Thureen *et al.*, 1992). At least part of the absolute reduction in glucose transport capacity is presumed to be due to a reduction in exchange surface area of the trophoblastic membrane, as also shown in carunclectomized ewes (Robinson *et al.*, 1995). In previously heat-

Table 1. Fetal weight and placental variables at term in normally grown and rapidly grown adolescent ewes

Variable	Normally grown	Rapidly grown	Significance of difference (P)
Number of ewes	11	8	
Duration of gestation (days)	143 ± 0.3	140 ± 0.9	< 0.01
Fetal weight (kg)	4.34 ± 0.27	2.74 ± 0.25	< 0.001
Placental weight (g)	438 ± 44.6	263 ± 16.8	< 0.01
Number of placentomes	90 ± 7.5	74 ± 5.4	ns

Values are means ± SEM. ns: not significant
Data from Wallace *et al.* (1996)

treated (Thureen *et al.*, 1992), but not in carunclectomized (Owens *et al.*, 1989) ewes, placental weight-specific glucose transport capacity also was reduced. This implies that chronic heat stress, which reduces average weight but not total number of placentomes, additionally reduces number or activity of specific glucose transport proteins at maternal or fetal exchange surfaces. In contrast, carunclectomy, which reduces placentome number but may stimulate a compensatory increase in average weight of individual placentomes, caused a modest increase in the placental weight-specific clearance of the nonmetabolizable glucose analogue, 3-*O*-methyl glucose (Owens *et al.*, 1989). This implies that glucose transporter expression was preserved or increased in the remaining placentomes.

Placental insufficiency in heat-treated ewes also extends to impaired capacity for amino acid transport, including major reductions in placental uptake and fetal transfer of leucine, and in the normally extensive placental catabolism of this branched-chain amino acid (Ross *et al.*, 1996). The molecular basis for the reduction in placental weight-specific transfer of leucine, and perhaps, other essential amino acids, is unknown. Presumably there is decreased abundance of specific transporter proteins, especially those responsible for active transport and concentration of amino acids in trophoblast cells (Hay, 1998).

Conceptus Requirements and Placental Transport of Macronutrients

Glucose

Glucose is a principal energy substrate for fetal and placental metabolism in ruminants (Battaglia and Meschia, 1988; Ferrell, 1989; Bell, 1993). For example, in the well-nourished, late-pregnant ewe, glucose accounts for approximately 60% of the net uptake of carbon by the gravid uterus, as calculated from the data of Carver and Hay (1995) for non-nitrogenous substrates, and of Chung *et al.* (1998) for amino acids. Under these favourable conditions fetal glucose requirements are met entirely by placental transport and fetal uptake of glucose from the umbilical circulation. Oxidation of glucose, directly and via its fetoplacental metabolite, lactate, then accounts for about 60% of fetal ATP synthesis (see Hay, 1995).

Analysis of the kinetics of placental glucose transport *in vivo* has confirmed that in sheep and other species, this process is achieved by facilitated diffusion (see Hay, 1995). We have shown that the predominant glucose transporter protein isoforms in sheep placenta are GLUT1 and GLUT3, and that mRNA and protein abundance of these transporters, especially those of GLUT3, increase from mid- to late pregnancy (Ehrhardt and Bell, 1997; Table 2). This appears to account for much of the five-fold increase in glucose transport capacity of the ovine placenta *in vivo* over this period (Molina *et al.*, 1991). Our failure to detect placental expression of the insulin-responsive isoform, GLUT4, is entirely consistent with the lack of a direct effect of maternal or fetal insulinaemia on uteroplacental uptake and placental transport of glucose *in vivo* in the pregnant ewe (see Bell, 1993; Hay, 1995).

Table 2. Developmental changes in expression of GLUT 1 and GLUT 3 protein and RNA in the sheep placenta

Variable	Day of pregnancy		
	75	110	140
Protein abundance			
GLUT 1	1.0 ± 0.06[a]	2.1 ± 0.19[b]	2.4 ± 0.15[b]
GLUT 3	1.0 ± 0.04[a]	1.9 ± 0.09[b]	2.9 ± 0.09[c]
mRNA abundance			
GLUT 1	1.0 ± 0.07[a]	1.6 ± 0.19[b]	1.8 ± 0.21[b]
GLUT 3	1.0 ± 0.05[a]	2.3 ± 0.5[b]	4.0 ± 0.17[c]

Values are means ± SEM in arbitrary, densitometric units expressed relative to day 75 of pregnancy.
[a,b,c] Values with different superscripts within rows are significantly different ($P < 0.05$).
Adapted from Ehrhardt and Bell (1997)

Other possible influences on placental glucose transport include uterine and umbilical blood flow and placental glucose metabolism. Consistent with its diffusion-limited transport mechanism, the placental delivery of glucose to the umbilical circulation is not responsive to physiological variations in uterine blood flow (Wilkening *et al.*, 1985).

Amino acids

Fetal requirements for amino acids are determined by rates of tissue growth and protein deposition that change with gestational age, and by fetal energy demands that result in extensive catabolism of amino acids throughout the latter half of gestation, even in well-nourished animals. Fractional rates of fetal tissue protein synthesis decline from approximately 25% per day at mid-gestation to < 10% per day near term (Kennaugh *et al.*, 1987), concomitant with a decline in fractional rate of protein deposition from 12% per day to approximately 4% per day (Van Veen *et al.*, 1987; Bell *et al.*, 1989). Throughout this period, the anabolic use of amino acids is accompanied by extensive oxidative deamination and fetal ureagenesis, sufficient to support 30–35% of fetal energy requirements (Faichney and White, 1987). Thus, placental transport of amino acid nitrogen into the umbilical circulation is considerably greater than that required for fetal synthetic purposes in sheep (Hay, 1998) and cows (Ferrell, 1989). The gestational decline in relative rates of umbilical uptake of amino acids (Bell *et al.*, 1989) is consistent with accompanying declines in fractional rates of fetal protein synthesis (Kennaugh *et al.*, 1987) and oxygen consumption (Bell *et al.*, 1987).

Most amino acids taken up by the placenta are transported against a fetal–maternal concentration gradient by energy-dependent mechanisms that have been elaborated in various mammalian tissues. These mechanisms, including identity and characterization of specific transporters, have not been studied in ruminant placentae. However, it is assumed that in ruminant as in human placental microvesicles, for example, there are at least ten sodium-dependent and sodium-independent transporter systems that have different levels of activity at different placental membrane surfaces. Hay (1998) has recently summarized the specific amino acids transported by each system, conditions favouring or inhibiting or affected by each system, and location (maternal or fetal trophoblast membrane) for each system.

Known mechanisms of placental amino acid transport imply diffusion-limited rather than flow-limited clearance, and, therefore, insulation against moderate fluctuations in placental blood flow. More severe restriction of placental perfusion, as can occur during exercise or acute heat stress (Bell, 1987), may indirectly affect amino acid transport through negative effects on placental energetics and ion gradients, as discussed by Hay (1998).

Fatty acids

Acetate and other derivatives of rumen fermentation, such as 3-hydroxybutyrate, are relatively abundant in maternal blood and are important energy sources for maternal tissues of ruminant animals (Bell, 1993). However, these short-chain fatty acids and keto acids are poorly transported by the ruminant placenta and make relatively minor contributions to fetal energy requirements in sheep and cattle (Battaglia and Meschia, 1988; Bell, 1993). The capacity for placental transport of long-chain, nonesterified fatty acids (NEFA), which are the primary vehicle for plasma delivery to other tissues of fatty acids mobilized from adipose tissue stores, is also extremely limited in sheep (Elphick *et al.*, 1979) and, presumably, other ruminants. Regarding the adequacy of fetal supplies of the C18 essential fatty acids, Noble *et al.* (1985) have identified active systems for desaturation and chain-elongation of linoleic and linolenic acids in the sheep placenta. In addition, the placenta takes up and metabolizes esterified lipids from maternal plasma (Hay, 1996), which, in ruminants, are richer than plasma NEFA in linoleic and linolenic acids. Thus, placental metabolism ensures an adequate fetal supply of the longer-chain ω6 and ω3 metabolites of the C18 polyunsaturated fatty acids, which are the forms ultimately required by tissues.

Impact of Placental Metabolism on Maternal–Fetal Nutrient Transfer

Oxygen consumption

The vital role of the placenta in transporting nutrients from the maternal to the fetal bloodstream, as well as functions such as peptide synthesis and maintenance of ion gradients, have a disproportionately high metabolic cost. This greatly affects the partitioning of nutrients within the gravid uterus, as well as adding substantially to the nutrient demands of pregnancy on the dam. In the late-pregnant ewe, the aggregate weight of placentomes is less than 15% that of the attached fetus. However, the weight-specific metabolic rate of the placenta is so great that the uteroplacental tissues (placentomes, endometrium, myometrium) consume 40–50% of oxygen taken up by the uterus in ewes (Bell and Ehrhardt, 1998) and cows (Ferrell, 1989). Estimates based on measurement of uteroplacental oxygen consumption *in vivo* and of placental oxygen consumption *in vitro* suggest that neither absolute nor dry weight-specific rates of placental energy expenditure change appreciably between mid- and late gestation (Vatnick and Bell, 1992). During mid-gestation, when fetal demands are small, much of this energy presumably is used to support active placental growth, whereas in late gestation the high rate of placental ATP synthesis must be related to functional demands.

Glucose metabolism

Uteroplacental consumption accounts for 60–70% of uterine net uptake of glucose during late pregnancy in ewes (Hay, 1995) and cows (Ferrell, 1989). Glucose uptake by the entire conceptus is determined directly by the maternal arterial glucose concentration, and glucose transport to the fetus is dependent directly on the maternal–fetal concentration gradient. The transplacental glucose concentration gradient, in turn, is directly related to both placental and fetal glucose consumption. Partitioning of the uterine glucose supply into placental and fetal rates of glucose consumption, however, is dependent on the fetal glucose concentration. For example, as fetal glucose concentration decreases relative to that of the mother, increasing the maternal–fetal gradient, glucose transport to the fetus increases at the expense of placental glucose consumption (Hay, 1995).

We have recently examined the metabolic fate of glucose consumed by the ovine placenta (Aldoretta *et al.*, 1994). Rapid metabolism to lactate (about 35%), fructose (about 4%), and CO_2 (about 17%) accounted for about 56% of uteroplacental glucose consumption in late-pregnant ewes with low or high maternal plasma glucose concentrations (Table 3). The metabolic fate of the remaining approximately 44% of glucose consumed is not known and requires investigation. Glucose oxidation accounted for 23–34% of uteroplacental oxygen consumption, depending on maternal glycaemia.

Table 3. Effect of glucose supply on uteroplacental glucose metabolism in ewes during late pregnancy

Variable	Glucose supply	
	Low	High
Plasma glucose concentration (mmol l^{-1})	2.23 ± 0.13	4.93 ± 0.29
Uteroplacental metabolic rates (mmol min^{-1})		
Glucose consumption	164 ± 22	313 ± 39
Lactate production	109 ± 12	182 ± 21
Fructose production	3.9 ± 1.1	7.0 ± 2.5
Glucose oxidation	26.7 ± 3.5	42.7 ± 7.5

Values are means ± SEM (n = 8).
Data from Aldoretta *et al.* (1994)

Oxidizable substrates that might contribute to the remaining 66–77% of uteroplacental respiration include ketones (Carver and Hay, 1995) and acetate, at least in caruncular tissues (Bell, 1993), certain amino acids, and carbon derived from the turnover of carbohydrate and lipid stores in placental tissues. The significance of placental synthesis of lactate and fructose for fetal metabolism is reviewed elsewhere (Battaglia and Meschia, 1988; Bell, 1993; Hay, 1995). In short, umbilical uptake and fetal oxidation of these glucose-derived substrates are estimated to contribute up to 20% of fetal energy requirements, additional to the 40–50% contributed by the direct oxidation of glucose.

Amino acid metabolism

Placental metabolism substantially affects both the quantity and composition of amino acids delivered to umbilical venous blood. The turnover rate of placental constitutive proteins is very rapid but net deposition of protein is negligible during the latter half of ovine pregnancy when placental dry weight is essentially static (Ehrhardt and Bell, 1995). Nevertheless, net consumption by uteroplacental tissues of glutamate, serine, and the branched chain amino acids is appreciable (Liechty *et al.*, 1991; Chung *et al.*, 1998), implying significant catabolism or transamination of these acids. An additional, small fraction of this net loss of amino acids will be in the form of secreted peptides. Placental net catabolism was estimated recently to account for 24% of uterine uptake of amino acid nitrogen in well-fed, late-pregnant ewes (Chung *et al.*, 1998).

The ovine placenta has very little enzymatic capacity for urea synthesis but produces considerable amounts of ammonia, much of which is released into maternal and, to a lesser extent, fetal circulations (see Hay, 1998). This is consistent with reports of extensive placental deamination of branched chain amino acids to their respective keto acids, which are released into fetal and maternal bloodstreams (Smeaton *et al.*, 1989; Loy *et al.*, 1990), and with rapid rates of glutamate oxidation in the placenta (Moores *et al.*, 1994). Transamination of branched chain amino acids accounts for some of the net glutamate acquisition by the placenta, the remainder of which is taken up from the umbilical circulation (Moores *et al.*, 1994). That which is not quickly oxidized combines with ammonia to synthesize glutamine, which is then released back into the umbilical bloodstream (Chung *et al.*, 1998). Some of this glutamine is converted back to glutamate by the fetal liver, which produces most of the glutamate consumed by the placenta (Vaughn *et al.*, 1995). This establishes a glutamate–glutamine shuttle which promotes placental oxidation of glutamate and fetal hepatic utilization of the amide group of glutamine.

Another example of the influence of placental metabolism on the pattern of amino acids delivered to the fetus is the almost quantitative conversion of serine, mostly taken up from maternal blood, to glycine by the placenta (Chung *et al.*, 1998). This reconciles earlier observations of major discrepancies between negligible net uptake of glycine by the uterus and substantial net release of

Table 4. Maternal weight change, fetal weight, and indices of placental glucose transport at day 135 of pregnancy in ditocous ewes fed 100% (Fed) or 60% (Underfed) of predicted energy requirements for the preceding 14 days

Variable	Fed	Underfed	PSE	Significance of difference (*P*)
Δ maternal weight (kg)	5.3	–2.7	1.0	< 0.001
Fetal weight (kg)	3.58	3.46	0.16	ns
Plasma glucose (mmol l^{-1})				
Maternal	3.72	2.84	0.09	< 0.001
Fetal	0.57	0.49	0.03	< 0.05
Maternal–fetal gradient	3.15	2.33	0.03	< 0.001
Placental 3MG clearance (ml min^{-1} kg^{-1} placental weight)	117	176	7	< 0.001
CB sites (pmol mg^{-1} protein)	105	126	3	< 0.01
GLUT protein (arbitrary units)[a]				
GLUT 1	1.00	0.83	0.06	ns
GLUT 3	1.00	1.19	0.04	< 0.05

Values are means (n = 5).
PSE: pooled standard error; ns: not significant; 3MG: 3–*O*-methyl glucose; CB: cytochalasin B binding sites.
[a]Expresssed relative to Fed group.
Adapted from Ehrhardt (1997)

this amino acid into the umbilical circulation (see Hay, 1998). In addition to ensuring a supply of the most abundant amino acid in fetal blood, this process is important for placental purine synthesis via the donation of the side-chain β-carbon atom of serine to form methylenetetrahydrofolate.

Placental Adaptations to Altered Supplies of Energy and Nitrogen

Maternal nutrition is often uncertain and variable, especially in extensively managed ruminant herds and flocks, with potential effects on fetal nutrient supply, growth and well-being, particularly during late pregnancy. Domestic ruminants, like other mammals, have developed various adaptive mechanisms to ameliorate the direct impact on fetal growth and development of all but the most serious nutritional vagaries. Some of these involve altered responses of maternal tissues such as liver, adipose tissue, and muscle to insulin and, possibly, other regulatory hormones. The net result is substitution of NEFA for glucose as energy sources in maternal insulin-responsive tissues, and increased availability of glucose for insulin-independent uptake by the placenta (Bell, 1993; Bell and Bauman, 1997). Fetal metabolic adaptations become necessary when maternal and placental responses fail to maintain a fully adequate fetal glucose supply (Bell, 1993; Hay, 1996). These include induction of hepatic gluconeogenesis, increased reliance on amino acids as a primary energy source, and, inevitably, reduced rates of protein synthesis and growth.

Glucose transport and metabolism

We recently tested the idea that the placenta is more than a passive beneficiary of maternal metabolic adaptations designed to support fetal glucose uptake in the face of a fluctuating maternal energy supply. Various indices of placental glucose transport capacity were compared in well-fed and moderately undernourished, ditocous ewes during late pregnancy (Ehrhardt, 1997). Results are summarized in Table 4. Restriction of maternal energy intake to 60% of predicted requirements for 2 weeks caused moderate maternal and fetal hypoglycaemia and a 26% decrease in maternal–fetal glucose concentration gradient. In these ewes, placental glucose transport capacity, assessed *in vivo* by measurement of clearance of the nonmetabolizable analogue, 3-*O*-methyl glucose, was increased 50% over values in well-fed ewes. Estimation of placental glucose transporter abundance *in vitro* by binding of cytochalasin B, and by the concentration of the GLUT3 transport protein as measured by Western blotting, were each increased by about 20%; concentration of the other major placental glucose transporter isoform, GLUT1, was unchanged. The effectiveness of these adaptations was indicated by unimpaired fetal growth in the underfed ewes (Table 4).

During more severe maternal undernutrition or starvation for several days, the ability to repartition maternal glucose in favour of the conceptus becomes limited and uterine and umbilical net uptake of glucose dwindle directly with the decline in maternal glucose supply (Bell, 1993; Hay, 1995). Under these conditions, the development of profound fetal hypoglycaemia helps to sustain the maternal–fetal gradient in glucose concentration by restricting the reverse transfer of glucose to the placenta, and reducing placental glucose consumption (Hay, 1995). More specific manipulation of maternal and fetal glycaemia by prolonged maternal infusion with insulin has shown that the decline in fetal glucose concentration is not proportional to that of the mother. This tends to decrease the maternal–fetal glucose concentration gradient, protecting placental glucose consumption at the expense of the fetus. In response, fetal glucose needs are diminished by a reduction in fetal growth rate (Carver and Hay, 1995).

Amino acid transport and metabolism

Fasting ewes for 5 days during late pregnancy had relatively little effect on placental delivery of amino acids to the fetus despite significant reductions in maternal plasma concentrations of many amino acids (Lemons and Schreiner, 1983). This suggests that during short term energy or protein deprivation, placental mechanisms for active transport of amino acids are unimpaired and may even be upregulated. Under similar fasting conditions, the uteroplacental deamination of branched chain amino acids appeared to be increased, as judged from a threefold increase of the efflux of α-ketoisocaproate into uterine and umbilical circulations (Liechty *et al.*, 1991). This finding suggests that increased protein catabolism and amino acid oxidation may partly substitute for the likely reduction in placental glucose oxidation under these conditions.

Placental responses to more prolonged restriction of energy or protein have not been investigated. However, in ewes fed adequate energy but insufficient protein during the last month of pregnancy, fetal growth and protein deposition over this period were reduced by 18% (McNeill *et al.*, 1997). This finding implies that neither maternal mobilization of labile protein stores nor putative adaptations in placental capacity for amino acid transport were sufficient to offset a 50% reduction in maternal supply of absorbed amino acids.

Conclusions

In healthy, well-managed ruminants, the placenta exerts an appropriate constraint on fetal growth during late gestation. When environmental factors, including maternal nutrition, retard placental development, associated reductions in nutrient transport capacity can lead to fetal growth retardation. Normal regulation and environmental modulation of early placental growth appear to be critically important in this regard, but are poorly understood. Recent observations of placental stunting in overfed adolescent ewes (Wallace *et al.*, 1996) offer an intriguing model for further study. The specific aspect(s) of placental nutrient transport that are normally most limiting have not been conclusively defined. However, it is revealing that even in very well-fed, monotocous ewes, direct fetal infusion with glucose during the last month of gestation caused an 18% increase in birth weight of lambs (Stevens *et al.*, 1990). Identification of the specific proteins responsible for placental glucose transport, together with early evidence for their molecular regulation, offer possibilities for more fundamental studies of gene expression. The complexity of placental transport systems for amino acids has so far defied detailed investigation in ruminants. The unique importance of amino acids as substrates for both fetal growth and oxidative metabolism demands serious study of the means by which their placental transfer is regulated.

References

Aldoretta PW, Gresores A and Hay WW, Jr (1994) Effect of glucose supply on ovine uteroplacental glucose utilization, oxidation, and lactate production *Proceedings of the Society for Gynecological Investigation* p 138 (Abstract 0104)

Alexander G (1974) Birth weight of lambs: influences and consequences. In *Size at Birth* pp 215–245 Eds K Elliot and J Knight. Elsevier, Amsterdam

Battaglia FC and Meschia G (1988) Fetal nutrition *Annual Review of Nutrition* **8** 43–61

Bell AW (1987) Consequences of severe heat stress for fetal development. In *Heat Stress: Physical Exertion and Environment* pp 313–333 Eds JRS Hales and DAB Richards. Elsevier, Amsterdam

Bell AW (1993) Pregnancy and fetal metabolism. In *Quantitative Aspects of Ruminant Digestion and Metabolism* pp 405–431 Eds JM Forbes and J France. CAB International, Wallingford

Bell AW and Bauman DE (1997) Adaptations of glucose metabolism during pregnancy and lactation *Journal of Mammary Gland Biology and Neoplasia* **2** 265–278

Bell AW and Ehrhardt RA (1998) Placental regulation of nutrient partitioning during pregnancy. In *Nutrition and Reproduction* pp 229–254 Eds GA Bray, W Hansel and DH Ryan. Louisiana State University Press, Baton Rouge.

Bell AW, Battaglia FC and Meschia G (1987) Relation between metabolic rate and body size in the ovine fetus *Journal of Nutrition* **117** 1181–1186

Bell AW, Kennaugh JM, Battaglia FC and Meschia G (1989) Uptake of amino acids and ammonia at mid-gestation by the fetal lamb *Quarterly Journal of Experimental Physiology* **74** 635–643

Carver TD and Hay WW, Jr (1995) Uteroplacental carbon substrate metabolism and O_2 consumption after long-term hypoglycemia in pregnant sheep *American Journal of Physiology* **269** E299–E308

Chung M, Teng C, Timmerman M, Meschia G and Battaglia FC (1998) Production and utilization of amino acids by ovine placenta *in vivo*. *American Journal of Physiology* **274** E13–E22

Clarke L, Heasman L, Juniper DT and Symonds ME (1998) Maternal nutrition in early–mid gestation and placental size in sheep *British Journal of Nutrition* **79** 359–364

Ehrhardt RA (1997) *Regulation of Glucose Transport and Partitioning by the Placenta in the Second Half of Pregnancy in Sheep*. PhD Dissertation, Cornell University, Ithaca NY

Ehrhardt RA and Bell AW (1995) Growth and metabolism of the ovine placenta during mid-gestation *Placenta* **16** 727–741

Ehrhardt RA and Bell AW (1997) Developmental increases in glucose transporter concentration in the sheep placenta *American Journal of Physiology* **273** R1132–R1141

Elphick MC, Hull D and Broughton Pipkin F (1979) The transfer of fatty acids across the sheep placenta *Journal of Developmental Physiology* **1** 31–45

Faichney GJ and White GA (1987) Effects of maternal nutritional status on fetal and placental growth and on fetal urea synthesis in sheep *Australian Journal of Biological Sciences* **40** 365–377

Ferrell CL (1989) Placental regulation of fetal growth. In *Animal Growth Regulation* pp 1–19 Eds DR Campion, GJ Hausman and RJ Martin. Plenum Press, New York

Hay WW, Jr (1995) Regulation of placental metabolism by glucose supply *Reproduction, Fertility and Development* **7** 365–375

Hay WW, Jr (1996) Nutrition and development of the fetus: carbohydrate and lipid metabolism. In *Nutrition in Pediatrics* pp 364–378 Eds WA Walker and JB Watkins. BC Decker, Hamilton

Hay WW, Jr (1998) Fetal requirements and placental transfer of nitrogenous compounds. In *Fetal and Neonatal Physiology* (2nd Edn) pp 619–634 Eds RA Polin and WW Fox. Saunders, Philadelphia

Kennaugh JM, Bell AW, Teng C, Meschia G and Battaglia FC (1987) Ontogenetic changes in the rates of protein synthesis and leucine oxidation during fetal life *Pediatric Research* **22** 688–692

Lemons JA and Schreiner RL (1983) Amino acid in the ovine fetus *American Journal of Physiology* **244** E459–E466

Liechty EA, Kelley J and Lemons JA (1991) Effect of fasting on uteroplacental amino acid metabolism in the pregnant sheep *Biology of the Neonate* **60** 207–214

Loy GL, Quick AN, Jr, Hay WW, Jr, Meschia G, Battaglia FC and Fennessey FC (1990) Fetoplacental deamination and decarboxylation of leucine *American Journal of Physiology* **259** E492–E497

McCrabb GJ, Egan AR and Hosking BJ (1992) Maternal undernutrition during mid-pregnancy in sheep: variable effects on placental growth *Journal of Agricultural Science, Cambridge* **118** 127–132

McCrabb GJ, McDonald BJ and Hennoste LM (1993) Heat stress during mid-pregnancy in sheep and the consequences for placental and fetal growth *Journal of Agricultural Science, Cambridge* **120** 265–271

McNeill DM, Slepetis R, Ehrhardt RA, Smith DM and Bell AW (1997) Protein requirements of sheep in late pregnancy: partitioning of nitrogen between gravid uterus and maternal tissues *Journal of Animal Science* **75** 809–816

Molina RD, Mechia G, Battaglia FC and Hay WW, Jr (1991) Gestational maturation of placental glucose transfer capacity in sheep *American Journal of Physiology* **261** R697–R704

Moores RR, Jr, Vaughn PR, Battaglia FC, Fennessey PV, Wilkening RB and Meschia G (1994) Glutamate metabolism in the fetus and placenta of late-gestation sheep *American Journal of Physiology* **267** R89–R96

Noble RC, Shand, JH and Christie WW (1985) Synthesis of C20 and C22 polyunsaturated fatty acids by the placenta of the sheep *Biology of the Neonate* **47** 333–338

Owens JA, Owens PC and Robinson JS (1989) Experimental growth retardation: metabolic and endocrine consequences. In *Research in Perinatal Medicine VIII. Advances in Fetal Physiology* pp 263–286 Eds PD Gluckman, BM Johnston and PW Nathanielsz. Perinatology Press, Ithaca

Robinson J, Chidzanja S, Kind K, Lok F, Owens P and Owens J (1995) Placental control of fetal growth *Reproduction, Fertility and Development* **7** 333–344

Ross JC, Fennessey PV, Wilkening RB, Battaglia FC and Meschia G

(1996) Placental transport and fetal utilization of leucine in a model of fetal growth retardation *American Journal of Physiology* **270** E491–E503

Schneider H (1991) Placental transport function *Reproduction, Fertility and Development* **3** 345–353

Smeaton TC, Owens JA, Kind KL and Robinson JS (1989 The placenta releases branched-chain keto acids into the umbilical and uterine circulations in the pregnant sheep *Journal of Developmental Physiology* **12** 95–99

Stegeman JHG (1974) Placental development in the sheep and its relation to fetal development *Bijdragen tot de Dierkunde* **44** 1–72

Stevens D, Alexander G and Bell AW (1990) Effect of prolonged glucose infusion into fetal sheep on body growth, fat deposition and gestation length *Journal of Developmental Physiology* **13** 277–281

Thureen P, Trembler KA, Meschia G, Makowski EL and Wilkening RB (1992) Placental glucose transport in heat-induced fetal growth retardation *American Journal of Physiology* **263** R578– R585

van Veen LCP, Teng C, Hay WW, Jr, Meschia G and Battaglia FC (1987) Leucine disposal and oxidation rates in the fetal lamb *Metabolism* **36** 48–53

Vatnick I and Bell AW (1992) Ontogeny of hepatic and placental growth and metabolism in sheep *American Journal of Physiology* **263** R619–R623

Vatnick I, Ignotz G, McBride BW and Bell AW (1991) Effect of heat stress on ovine placental growth in early pregnancy *Journal of Developmental Physiology* **16** 163–166

Vaughn PR, Lobo C, Battaglia FC, Fennessey PV, Wilkening RB and Meschia G (1995) Glutamine–glutamate exchange between placenta and fetal liver *American Journal of Physiology* **268** E705–E711

Wallace JM, Aitken RP and Cheyne MA (1996) Nutrient partitioning and fetal growth in rapidly growing adolescent ewes *Journal of Reproduction and Fertility* **107** 183–190

Wilkening RB, Battaglia FC and Meschia G (1985) The relationship of umbilical glucose uptake to uterine blood flow *Journal of Developmental Physiology* **7** 313–331

Journal of Reproduction and Fertility Supplement **54**, 411–424

Effects of energy balance on follicular development and first ovulation in postpartum dairy cows

S. W. Beam[1] and W. R. Butler[2]

[1]*Dept of Animal Science, University of California, Davis CA 95616, USA; and* [2]*Dept of Animal Science, Cornell University, Ithaca, NY 14853, USA*

As milk production has increased during the past four decades, conception rates in lactating cows have declined. Although reduced reproductive performance has been associated with high milk yields, measures of postpartum ovarian activity have been more closely related to energy balance. The relationship between daily energy balance and postpartum reproductive activity is confirmed by longer intervals to first ovulation in cows with greater body condition loss. Patterns in daily energy balance, such as improvement from nadir, have been correlated with enhanced follicular function and a shorter interval to first ovulation. Such observations are consistent with increased LH pulse frequency following the energy balance nadir in lactating dairy cows. Evidence indicates a primarily hypothalamic locus for the modulation of LH secretion during negative energy balance. Formation of follicular waves after parturition begins synchronously in response to increased FSH in the first week postpartum, and is typically not a limiting factor in reproductive recrudescence. Altered follicular responsiveness to gonadotrophic support through changes in metabolic hormones such as insulin-like growth factor I (IGF-I) and insulin may contribute to impaired function of dominant follicles early postpartum. Positive relationships between changes in energy balance, peripheral IGF-I and function of dominant follicles support the identification of IGF-I and the day of the energy balance nadir as metabolic modulators of postpartum ovarian activity in dairy cows.

Introduction

The reproductive process in the ruminant female entails a complex series of physiological events that may be impacted at many levels to influence overall fertility. Although a myriad of regulatory mechanisms remain to be elucidated, certain fundamental factors mediating reproductive activity have become apparent. One of the earliest factors to be recognized is the often profound influence of nutritional status. In the USA, apparent metabolic influences on reproductive performance have been reflected in declining conception rates (66% to 40% since 1951) coincident with increasing yearly milk production per cow (218% increase since 1950; National Agricultural Statistics Service, USDA). Conception rates after artificial insemination (AI) in nonlactating heifers have remained at 70–80% during this period (Butler and Smith, 1989), indicating that there is no direct genetic trend against fertility. Understanding the metabolic constraints on postparturient ovarian function is an important step towards the development of management practices that positively influence postpartum reproductive efficiency. The focus of this review is on current knowledge of the resumption of cyclic ovarian activity after parturition in high producing dairy cows, with an emphasis on patterns of follicular development and their relationship to energy balance and metabolic hormones. This review is not intended to be comprehensive for all nutritional interactions. Current reviews describe the effects of supplemental fat (Staples *et al.*, 1998) or the level of dietary protein (Butler, 1998) on postpartum reproduction in dairy cows.

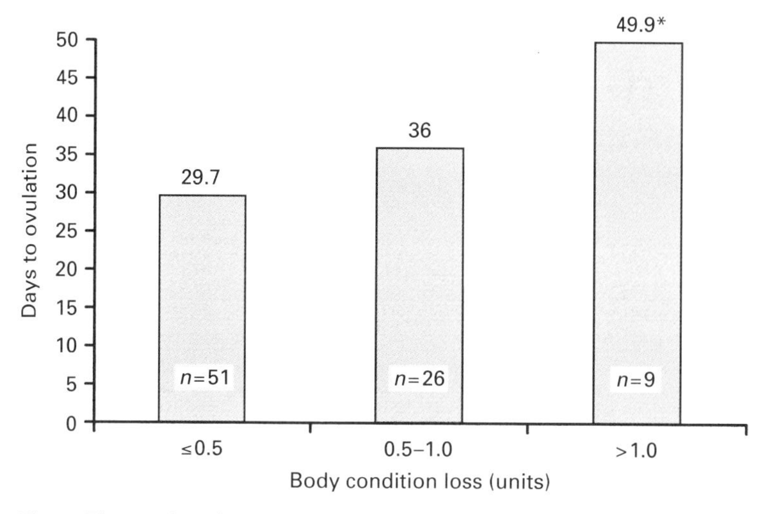

Fig. 1. The number of days to first ovulation in dairy cows with different body condition score loss during the first 30 days postpartum. Scores of 1 = emaciated and 5 = obese. Pooled data from Beam and Butler (1997, 1998). * Significantly different from other two means ($P < 0.05$).

Energy Balance and Postpartum Reproduction: General Relationships

In the context of the postpartum dairy cow, energy balance is the difference between the dietary intake of utilizable energy and the expenditure of energy for body maintenance and milk synthesis. Owing to a lower rate of increase in feed intake compared with that of milk production, high producing dairy cows typically experience a variable period of negative energy balance during early lactation that is characterized by the loss of body weight and mobilization of body fat stores. Negative energy balance may persist for 10–12 weeks of lactation (Bauman and Currie, 1980), and the level of energy deficit is often related directly to milk yield (Butler *et al.*, 1981). High milk yields have in turn been associated with lower reproductive efficiency. Data summarized by Nebel and McGilliard (1993) from over 4550 herds of Holstein cows have shown an inverse relationship between first-service conception rate and per cow yearly milk yield (52% for cows producing 6300–6800 kg per year versus 38% for cows producing > 10 400 kg per year). However, despite the apparent negative relationship between conception rate and milk production, measures of reproductive performance, in several studies, have been related more closely to energy balance than to milk yield (Butler and Smith, 1989). Staples *et al.* (1990) reported lower milk production, lower feed intake and more negative energy balance in anoestrous dairy cows compared with cows that returned to cyclic ovarian activity before day 60 postpartum. Likewise, multiparous dairy cows ovulating after day 42 postpartum consumed less feed, but had similar fat corrected milk production, compared with cows ovulating before day 42 postpartum (Lucy *et al.*, 1992a). Furthermore, variation in energy balance among cows within a herd is due largely to variation in energy intake ($r = 0.73$) as compared with milk yield ($r = -0.25$; Villa-Godoy *et al.*, 1988), and increased utilization of body energy reserves for milk production has been associated with longer intervals to cyclic ovarian activity (Staples *et al.*, 1990). Effects of a given lactation on reproduction are dependent upon concomitant patterns of feed energy intake and subsequent energy availability to support proper postparturient function of the hypothalamic–pituitary–ovarian (HPO) axis. The inverse relationship between energy balance and postpartum ovarian function is represented most clearly by changes in body condition; cows losing more body condition during the first month postpartum experience longer intervals to first ovulation (Fig. 1).

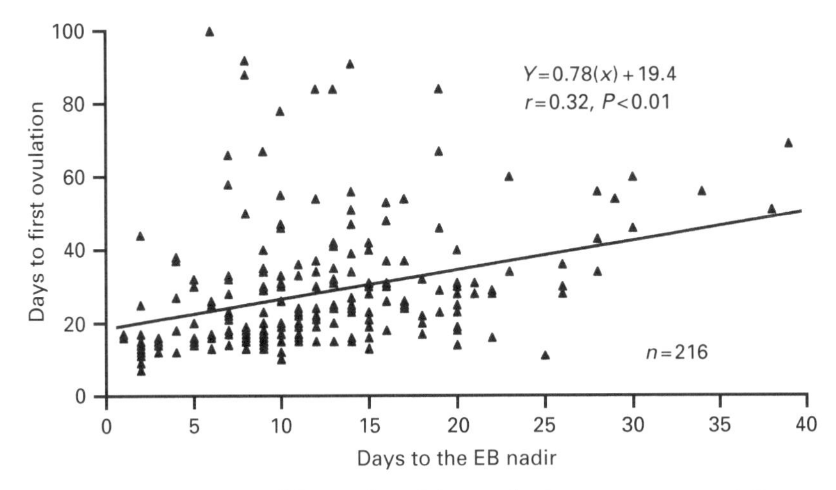

Fig. 2. Linear regression of the number of days to the first postpartum ovulation on the number of days to the energy balance (EB) nadir in dairy cows. Data include observations from five separate studies: Canfield *et al.* (1990); Canfield and Butler (1990, 1991); Beam and Butler (1997, 1998).

Energy balance and days to first ovulation

Because the number of oestrous cycles preceding AI has been shown to influence conception rate (Thatcher and Wilcox, 1973; Lucy *et al.*, 1992a), the duration of the postpartum anovulatory interval can serve as a measure of potential energy balance mediated effects on reproductive performance. Although effects of energy balance on the number of days to first ovulation have been observed inconsistently, both the level of negative energy balance (Butler *et al.*, 1981; Staples *et al.*, 1990) and changes in energy balance over time (Canfield and Butler, 1990; Canfield *et al.*, 1990; Beam and Butler, 1997) have been implicated in the timing of the first ovulatory event after parturition. In modern dairy cows, the average day of first ovulation is approximately 25–30 days postpartum with a typical range between 17 and 42 days (Butler and Smith, 1989). Thus, energy balance during the first three to four weeks postpartum has been correlated with the interval to first ovulation (Lucy *et al.*, 1991; Beam and Butler, 1998). Average energy balance over longer periods (6 to 12 weeks) often has no significant relationship to the resumption of ovarian cycles (Staples *et al.*, 1990; Beam and Butler, 1998). In addition to the influence of the level of negative energy balance during the early weeks postpartum, recovery of daily energy balance from its most negative value (nadir) appears to provide an important signal for initiation of cyclic ovarian activity. The number of days to the energy balance nadir is positively correlated with the number of days to first ovulation (Fig. 2; Canfield and Butler, 1990; Canfield *et al.*, 1990; Beam and Butler, 1997). Evaluation of relationships between energy balance profiles and the day of first ovulation, although informative, does not reveal the nature of ovarian follicular activity within the anovulatory period. Examination of follicular development leading to the first ovulation postpartum has produced important insights into factors limiting reproductive recrudescence during negative energy balance.

Postpartum Follicular Development

The term 'follicular dynamics' has been defined by Lucy *et al.* (1992b) as the continual growth and regression of antral follicles leading to preovulatory follicle development. In cattle, this process is characterized by the formation of follicular waves; two or three waves typically occur during the normal course of the oestrous cycle (Pierson and Ginther, 1984). Although follicular dynamics during the oestrous cycle have been extensively examined, there is less information on follicular development following parturition and before the first postpartum ovulation, particularly in modern high producing dairy cows. The use of transrectal ultrasonography to monitor follicular development has added greatly to the understanding of postpartum ovarian activity. Savio *et al.* (1990) examined follicular development by ultrasonography in dairy cows from day 5 postpartum and reported a variable period until the detection of a dominant (> 9 mm) follicle (range 5–39 days). Follicular development before dominant follicle formation was characterized by growth and regression of small (< 5 mm) and medium (5–9 mm) sized follicles; the first dominant follicle either ovulated (14/19 cows) or became cystic (4/19 cows) in all but one cow. Ultrasonography in dairy cows beginning on day 14 postpartum revealed first ovulation in eight of ten cows before day 25 postpartum; one cow developed a follicular cyst and one cow displayed multiple waves of anovulatory follicle development until first ovulation at day 55 (Rajamahendran and Taylor, 1990). More recent studies using ultrasonography revealed the regular formation of follicular waves, including the development of dominant follicles, throughout most of pregnancy in dairy heifers (Ginther *et al.*, 1996), and the rather synchronous initiation of a new follicular wave 5–7 days postpartum in multiparous dairy cows (Beam and Butler, 1997).

During the oestrous cycle in cattle, new waves of follicular development are preceded by peak increases in mean circulating concentrations of FSH that are believed to be responsible for initiating the growth of each new cohort of ovarian follicles (Adams *et al.*, 1993). This observation was extended to follicular waves occurring throughout most of pregnancy in heifers by Ginther *et al.* (1996), in which dominant follicles developed at 6–8 day intervals in association with peaks in plasma FSH. However, the size of dominant follicles observed by Ginther *et al.* (1996) decreased linearly during months 4–9 of gestation, and no follicles > 6 mm in diameter were detected during the last 3 weeks before parturition. This reduced follicular development during the final weeks of gestation was associated with altered FSH secretion patterns, including a longer interpeak interval, and was postulated to result from increased concentrations of gestational oestrogens (oestradiol and oestrone). This finding is in agreement with a report by Beam and Butler (1997) in which mean daily plasma FSH concentrations increased to peak values on days 4–5 postpartum after gestational oestradiol concentrations had declined. The FSH peak was followed immediately by the initiation of a new follicular wave and development of the first dominant follicle postpartum. This new period of follicular dominance occurred in all cows examined despite average negative energy balance of –7.5 Mcal day^{-1} during the first 3 weeks postpartum. It appears that initiation of follicular waves in the early postpartum cow is unperturbed by negative energy balance and occurs in response to the re-establishment of periodic FSH surges synchronized by the end of gestation. In milked dairy cows experiencing negative energy balance, LH, but not FSH, appears to be deficient after the first week postpartum (Lamming *et al.*, 1981; Beam and Butler, 1997). Therefore, FSH appears to be insensitive to metabolic energy status and signals the development of ovarian follicular waves at regular intervals following parturition in dairy cows. Because formation of a dominant follicle does not appear to be a limiting factor in the resumption of ovarian cycles postpartum, the function (that is steroidogenic capability) and fate of dominant follicles during negative energy balance becomes an important focus of investigation.

Patterns of dominant follicle fate

Although most dairy cows appear to develop dominant follicles during the second week postpartum, three patterns of follicular development based on the fate of the first-wave dominant follicle have been described (Savio *et al.*, 1990; Rajamahendran and Taylor, 1990; Beam and Butler, 1997): (1) ovulation of a dominant follicle during the first follicular wave after parturition; (2) development of a first-wave anovulatory dominant follicle followed by additional waves of follicular development before first ovulation; or (3) development of a first-wave dominant follicle that becomes cystic (Fig. 3). Pattern 1 (ovulatory) and pattern 3 (cystic) are characterized by development of oestrogen-active dominant follicles, whereas pattern 2 (anovulatory) is characterized by growth of dominant follicles that produce low peripheral concentrations of

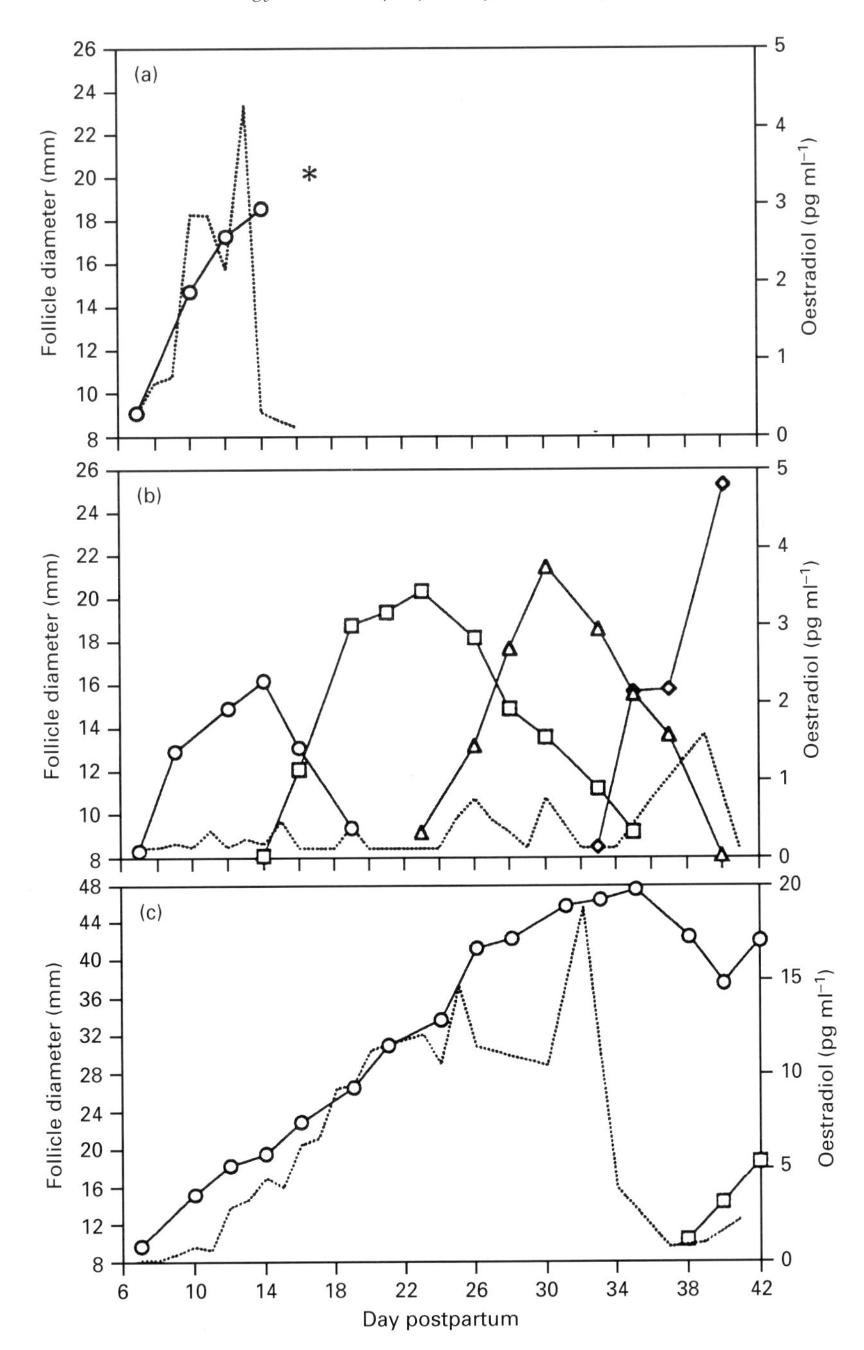

Fig. 3. Patterns of dominant follicle growth and oestradiol concentrations in three representative postpartum dairy cows that (a) ovulated the dominant follicle of the first follicular wave postpartum, (b) experienced multiple waves of anovulatory dominant follicle development and low plasma oestradiol, and (c) developed an oestrogen-active follicular cyst. *Denotes ovulation. Reproduced with permission from Beam and Butler (1997).

oestradiol and become atretic (Beam and Butler, 1997). The fate of the first-wave dominant follicle has a significant impact on the postpartum anovulatory interval (Beam and Butler, 1997); regression of the first-wave dominant follicle or formation of a follicular cyst results in a similarly prolonged interval to first ovulation (51 and 48 days, respectively) compared with that of cows that ovulate their first dominant follicle postpartum (20 days). Combined data from two studies examining postpartum follicular development in multiparous dairy cows ($n = 87$) show that 46% of cows ovulated the first dominant follicle postpartum; 31% experienced at least two waves of dominant follicle development before first ovulation, and the remaining 23% became cystic, with all but one cyst forming from the first follicular wave postpartum (Beam and Butler, 1997, 1998). The average duration of the anovulatory period in cows with anovulatory (noncystic) first-wave dominant follicles was 40 days, indicating that in most cows early ovulation failure is followed by additional waves of anovulatory follicle development before first ovulation is achieved. These data emphasize the importance of investigating the mechanisms regulating the emergence and function of the first dominant ovarian follicle after parturition, including possible metabolic differences between cows with different follicular fates and the effects of energy balance and specific metabolic hormones on follicular dynamics.

The effects of energetic stress on the HPO axis have been examined primarily at the hypothalamus and anterior pituitary, and the loss of pulsatile LH secretion has been shown to result from prolonged inadequate intake of dietary energy in postpartum beef cows (Perry *et al.*, 1991). Energy restriction in postpartum beef cows does not alter pituitary GnRH receptor density (Moss *et al.*, 1985). However, the results of studies examining pituitary responsiveness to GnRH in the postpartum cow remain equivocal: dietary energy restrictions have both decreased (Rutter and Randel, 1984) and increased (Whisnant *et al.*, 1985) responsiveness to GnRH.

The re-establishment of a pulsatile LH secretion pattern conducive to preovulatory follicular development and function is recognized as a key event in the return of ovarian cyclicity by the postpartum dairy cow experiencing negative energy balance (Lamming *et al.*, 1981; Malven, 1984; Canfield and Butler, 1991). As shown in Fig. 4, the frequency of LH pulses is significantly lower during the first follicular wave postpartum (days 8–12) in cows that develop an anovulatory dominant follicle compared with cows that develop an ovulatory dominant follicle (Beam and Butler, 1994). Peters *et al.* (1985) treated dairy cows between days 3 and 8 postpartum with 2.5 μg GnRH intravenously at intervals of 2 h for 48 h, and reported the induction of an episodic pattern of LH release and a sustained rise in plasma oestradiol. However, a preovulatory-type LH surge was exhibited in only one of nine animals. When the same GnRH treatment protocol was applied to anovulatory dairy cows later than day 10 postpartum, the accelerated pulsatile LH pattern was followed by a preovulatory-type LH surge, ovulation and normal luteal function (Lamming *et al.*, 1982). Thus, endocrine responses in the postpartum cow are highly dependent on the stage postpartum. Differences in the severity of either induced or spontaneous negative energy balance is also likely to be an important variable. Overall, it appears that postparturient recovery of pituitary LH content and responsiveness to GnRH is complete by day 10 postpartum in dairy cows (Lamming *et al.*, 1982; Malven, 1984), and available evidence across species suggests a predominantly hypothalamic locus for the primary effect of decreased energy intake (Schillo, 1992). In the early postpartum dairy cow the reduced activity of the GnRH pulse generator is expressed as reduced pulsatile LH support of follicular steroidogenesis necessary for the induction of an LH surge and ovulation. However, a seemingly low LH pulse frequency (2 pulses per 6 h) is apparently adequate to sustain the morphological development of dominant ovarian follicles by the second week postpartum. This observation is consistent with the growth and differentiation of competent dominant follicles during the mid-luteal phase of the oestrous cycle in cattle when LH pulse frequency is low (Rahe *et al.*, 1980; Driancourt *et al.*, 1991).

Searching for a Signal

Metabolic and hormonal cues

It is reasonable to predict that factors signalling the HPO axis of an energy deficit are part of the metabolic and endocrine milieu characteristic of negative energy balance and early lactation.

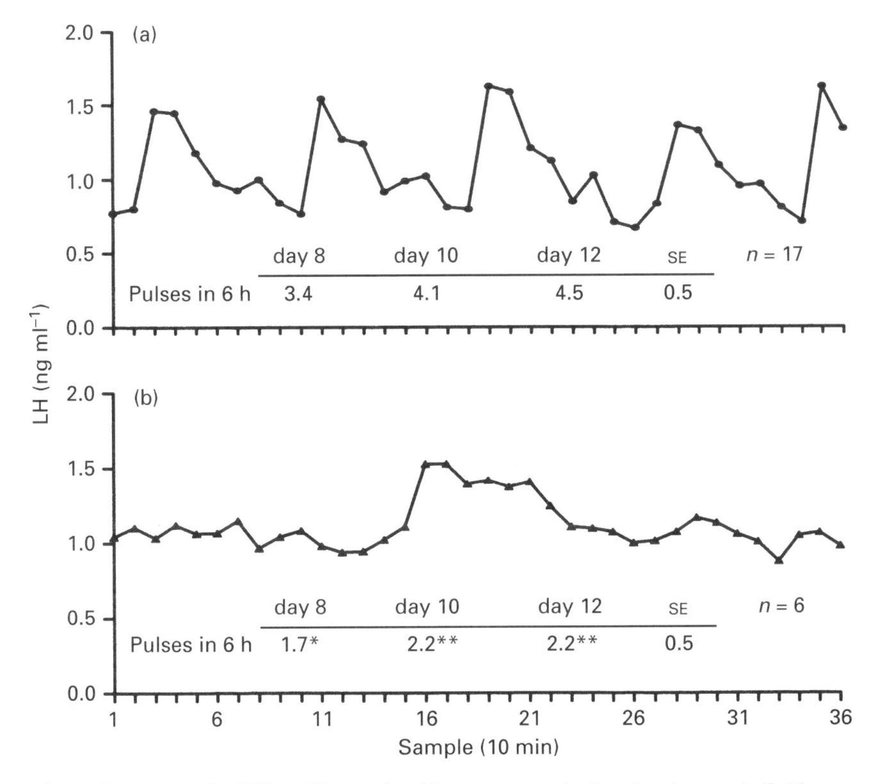

Fig. 4. Representative LH profiles on day 12 postpartum during development of either an ovulatory (a) or anovulatory (b) dominant follicle in the first follicular wave postpartum. Mean number of LH pulses per 6 h on day 8, 10 and 12 postpartum for each group is shown as an inset. Pulse frequency was significantly different (** $P < 0.01$, * $P < 0.05$) within day between groups. Data from Beam and Butler (1994).

Therefore, studies investigating potential metabolic signals for ovarian activity have focused primarily on blood metabolites and metabolic hormones known to fluctuate during altered metabolic states. Increased plasma non-esterified fatty acids (NEFA) have been suggested as a potential cue of energy balance status on initiation of first ovulation (Canfield and Butler, 1990; Canfield and Butler, 1991). However, plasma NEFA and glucose are similar during the first 2–3 weeks postpartum in dairy cows developing either an ovulatory or anovulatory first-wave dominant follicle postpartum (Beam and Butler, 1997, 1998). It is generally accepted that any mechanism coupling metabolic status with ovarian function will ultimately involve a hormonal component. Metabolic data for postpartum dairy cows from a previous study (Beam and Butler, 1997), including measures of energy balance and the metabolic hormones insulin and GH during the first 2 weeks postpartum, were analysed by step-up logistic regression analysis in which ovulation or anovulation of the first postpartum dominant follicle served as the binary dependent variable (Table 1). Logistic regression analysis of these variables produced a significant model that correctly classified 86.5% of the experimental animals (32 of 37 cows) as having either an ovulatory or anovulatory first-wave dominant follicle. The plasma insulin:GH ratio during week 1 postpartum and the day of the energy balance nadir served as the most significant contributors to the model. The insulin:GH ratio was increased due to both higher insulin and lower GH concentrations during only the first week postpartum in cows possessing dominant follicles that ovulated (Beam and Butler, 1997). These results indicate that hormonal differences in the immediate postpartum period may influence follicular function during the first follicular wave after parturition. Insulin has been shown

Table 1. Step-up logistic regression analysis of metabolic variables during the first 2 weeks postpartum in lactating dairy cows with an ovulatory or anovulatory dominant follicle[a,b]

Variable	Chi-square $\beta = 0$	Probability $\beta = 0$	Last R-Square
Insulin:GH ratio Week 1	4.95	0.0261	0.1339
Day of EB Nadir	4.88	0.0271	0.1324
Mean insulin Week 1	3.85	0.0496	0.1075
Change in EB Day 1–14	3.74	0.0530	0.1047

[a]Model r-square = 0.4552, $P<.0001$, model Chi-square = 26.73; ovulation or anovulation of the first-wave dominant follicle postpartum served as the binary dependent variable. The model correctly classified 17/19 ovulators and 15/18 nonovulators (86.5% correct).
[b]Analysis of data from Beam and Butler (1997). Analysis variables for the first 2 weeks postpartum included plasma insulin, GH, insulin-like growth factor I (IGF-I), insulin:GH ratio, net energy intake, mean energy balance (EB), day of the energy balance nadir and change in energy balance.

to stimulate follicular cells *in vitro* in a variety of species including cattle (Spicer *et al.* 1993) and small increases postpartum could have important effects during the very early stages of follicular development. Furthermore, an increased insulin:GH ratio following parturition may be conducive to greater hepatic IGF-I production (McGuire *et al.*, 1995), resulting in increased amounts of this growth factor earlier postpartum. Indeed, in cows developing oestrogen-active, ovulatory dominant follicles during the first follicular wave postpartum, circulating insulin-like growth factor I (IGF-I) is significantly higher during the first 2 weeks postpartum compared with that in cows developing oestrogen-inactive dominant follicles that regress (Beam and Butler, 1997, 1998). Concentrations of circulating IGF-I are in turn positively correlated with IGF-I concentrations in follicular fluid of large bovine follicles (Echternkamp *et al.*, 1990). Similar to insulin, IGF-I is known to affect follicular cell function *in vitro*; stimulation of steroidogenesis and proliferation in both thecal (Spicer and Stewart, 1996) and granulosal cells (Spicer *et al.*, 1993) are well documented. In bovine thecal cells, IGF-I increases the number of LH-binding sites and enhances LH-induced production of progesterone and androstenedione *in vitro* (Spicer and Stewart, 1996; Stewart *et al.*, 1996). Both oestradiol and FSH increase the number of IGF-I receptors in granulosal cells, which have been shown to be greater in number in large bovine follicles compared with small follicles (Spicer *et al.*, 1994), and may thus form a self-amplifying system of IGF-I stimulation in the growing and differentiating dominant follicle postpartum. In dairy cows, there is an apparent relationship between the steroidogenic activity of the first dominant follicle postpartum and circulating concentrations of IGF-I (Fig. 5).

Patterns of energy balance

The recovery of pulsatile LH secretion patterns leading to the enhancement of follicular function has been associated with changing patterns of energy balance over time in the postpartum dairy cow. In a study of energy balance and dietary lipid effects on postpartum follicular development in high producing dairy cows (Beam and Butler, 1997), mean daily energy balance did not have a significant relationship with follicular function. However, improvement in energy balance from its most negative level (nadir) enhanced follicular competence, because dominant follicles during the first follicular wave postpartum exhibited greater steroidogenic output and ovulatory success when development occurred after the day of the energy balance nadir (75% ovulated) compared with before the day of the energy balance nadir (24% ovulated). Furthermore, in cows that experienced consecutive waves of follicular development before first ovulation (type 2 of Fig. 3), dominant

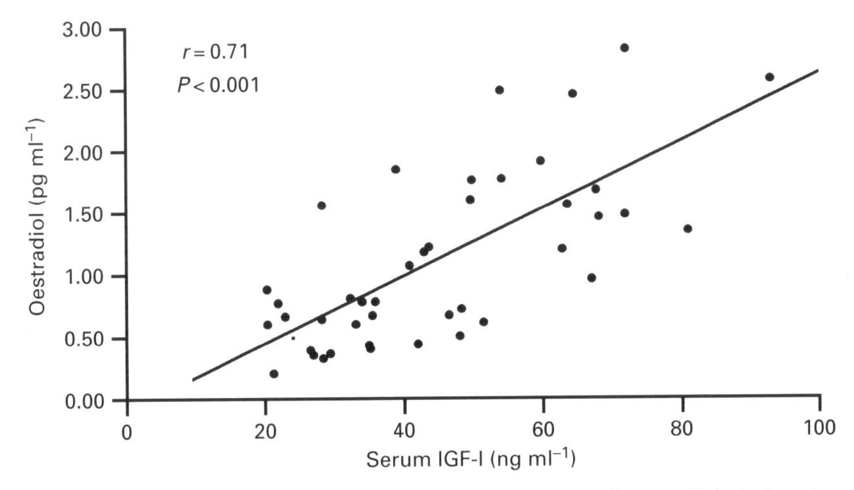

Fig. 5. Linear regression of mean plasma concentrations of oestradiol during the development of the first dominant follicle postpartum (day 8–14) on mean serum concentrations of insulin-like growth factor I (IGF-I) during days 1–21 postpartum in dairy cows ($n = 42$). Analysis of data from Beam and Butler (1998).

follicle diameter and plasma oestradiol concentrations were increased after the energy balance nadir compared with the follicular wave occurring before the energy balance nadir (Fig. 6). Examination of the pooled data from two recent studies by our laboratory (Beam and Butler, 1997; 1998) shows a significant difference ($P < 0.01$) in the day of the energy balance nadir between cows that developed either an ovulatory (6.9 ± 1.3 days, $n = 40$) or anovulatory (15.5 ± 1.6 days, $n = 27$) dominant follicle during the first follicular wave postpartum. Enhancement of follicular steroidogenesis and diameter after the energy balance nadir is consistent with effects of increased LH pulse frequency on follicular function (Glencross, 1987), and with previously reported increases in LH pulse frequency following the day of the energy balance nadir (Canfield and Butler, 1991). The specific signal after the energy balance nadir leading to increased LH pulse frequency remains unknown, but does not involve endogenous opioid peptides because naloxone administration to postpartum dairy cows does not affect any variable of LH secretion (Canfield and Butler, 1991).

Energy Balance and Postpartum Follicular Dynamics

Few studies have examined directly the relationship between energy balance and follicular dynamics during the early postpartum period in dairy cows. Lucy *et al.* (1991) examined follicular development in dairy cows by ultrasonography and reported an effect of energy balance on different populations of ovarian follicles postpartum. The number of class 1 (3–5 mm) and class 2 (6–9 mm) follicles decreased, and the number of class 3 (10–15 mm) follicles increased with more positive energy balance before day 25 postpartum. The authors suggested that as cows improve in energy balance, the movement of smaller follicles into larger size classes is enhanced. Comparisons between lactating and nonlactating cows have also shown differences in follicular recruitment, with nonlactating cows having a greater number of class 1 (3–5 mm), class 2 (6–9 mm) and class 3 (10–15 mm) follicles during the first follicular wave of a synchronized oestrous cycle (De La Sota *et al.*, 1993). During the first follicular wave postpartum (days 8–14) in cows receiving three levels of dietary fat, the number of class 1 and class 2 follicles was not correlated with energy balance during either the first or second week postpartum, regardless of diet (Beam and Butler, 1997). Likewise, the number of class 1 and class 2 follicles on day 8 postpartum was not correlated with energy balance

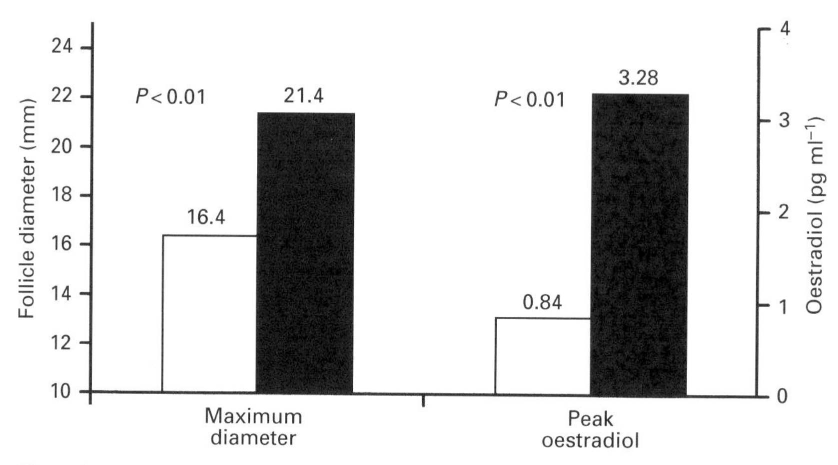

Fig. 6. Mean maximum diameter of dominant follicle and peak oestradiol concentration during consecutive follicular waves occurring immediately before (□) and after (■) the energy balance nadir in postpartum dairy cows (n = 11). Both maximum follicle diameter and peak oestradiol concentrations were greater in follicular waves developing after the energy balance nadir (Beam and Butler, 1997).

during the first week postpartum in dairy cows fed prilled fatty-acids at 0 or 2.9% of ration DM (Beam and Butler, 1998). Furthermore, the interaction between energy balance and follicle size classes reported by Lucy *et al.* (1991) was not significant when cows receiving calcium salts of long-chain fatty acids were removed from the analysis, and no effect of energy balance on follicle populations was noted after day 25 postpartum. A clear relationship between energy balance and follicular recruitment remains to be demonstrated during the very early stages of lactation.

The development of a dominant follicle postpartum in dairy cows is tolerant to periods of energy deficiency as demonstrated by the selection and growth of a follicle over 15 mm in diameter during the second week postpartum despite negative energy balance (Beam and Butler, 1997, 1998). However, results of several studies indicate that the ultimate diameter and oestrogen production of dominant follicles are influenced by metabolic factors. In prepubertal heifers (Bergfeld *et al.*, 1994), postpartum suckled beef cows (Perry *et al.*, 1991) and cyclic lactating dairy cows (Lucy *et al.*, 1992c), growth of dominant follicles is reduced during dietary energy restriction. As noted previously, dominant follicle diameter and plasma oestradiol increased after energy balance improved from its most negative level in early postpartum dairy cows (Beam and Butler, 1997). Comparisons between lactating and nonlactating dairy cows have revealed differences in dominant follicle development, and smaller dominant follicles are observed in nonlactating compared with lactating cows during the first follicular wave of a synchronized oestrous cycle (De La Sota *et al.*, 1993). In the same study, plasma oestradiol concentrations during the preovulatory period were several-fold higher in nonlactating compared with lactating cows. In a study conducted by Beam and Butler (1994), cows were either not milked (DRY), milked twice per day (2 ×) or three times per day (3 ×) following parturition. This resulted in different energy balance and body weight loss among groups during the first 4 weeks postpartum. Although peak plasma oestradiol was similar among groups, maximum diameter of the dominant preovulatory follicle from the first follicular wave postpartum was larger in both 3 × (23.9 ± 1.2 mm) and 2 × cows (21.0 ± 0.9 mm) compared with DRY cows (16.5 ± 0.9 mm). Therefore, lactational status or large differences in energy balance do not prevent the formation of follicular waves but apparently alter the growth and ultimate diameter of dominant follicles. Collectively, results of studies comparing lactating and nonlactating dairy cows indicate that dominant follicles of lactating cows in negative energy balance have a lower oestradiol output per unit of gross follicular size compared with nonlactating cows in positive energy balance. Differences in metabolic hormones such as insulin or IGF-I could be involved in the large difference in

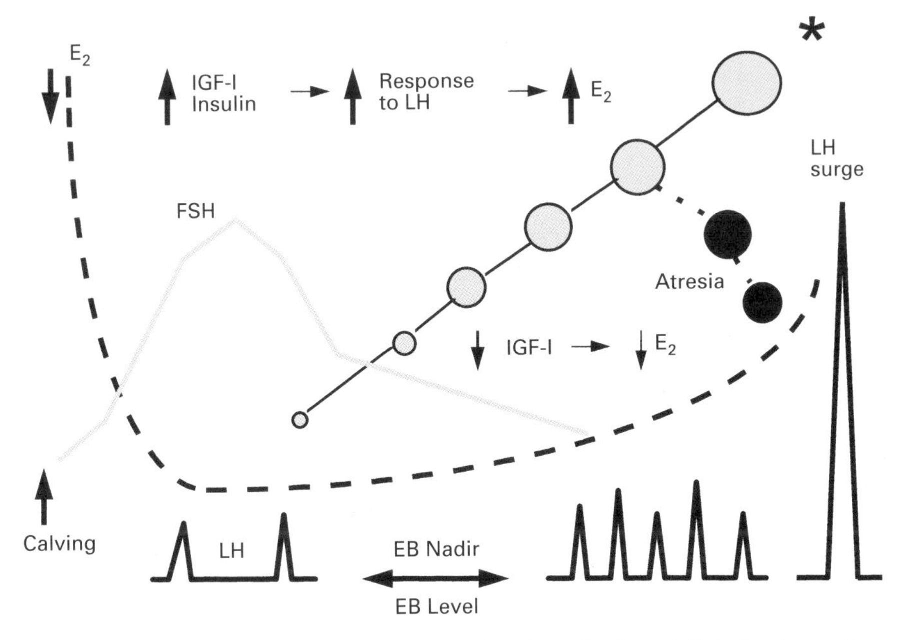

Fig. 7. Schematic representation of a basic model describing dominant follicle development (circles) and function in relation to changing metabolic and reproductive hormones, and energy balance (EB), during the first follicular wave postpartum in dairy cows. The first-wave follicle either ovulates (*) or undergoes atresia (dark circles). LH pulse frequency is modulated by the day of the EB nadir and, to a lesser extent, the level of EB. The large upward arrows indicate increased insulin-like growth factor I (IGF-I) and insulin leading to improved responsiveness to LH and greater oestradiol (E_2) production by the dominant follicle.

preovulatory follicular diameter between lactating and nonlactating dairy cows. For example, insulin treatment increased follicular diameter and concentrations of oestradiol in follicular fluid of superovulated cows (Simpson *et al.*, 1994). Other studies have implicated a role of the GH:IGF-I axis in follicular growth and turnover in cyclic cattle. In lactating dairy cows (days 60–100 postpartum) bovine somatotrophin (bST) increased the rate of dominant follicle turnover, preovulatory follicular growth and oestrogen secretion during a synchronized oestrous cycle compared with cows treated with saline (De La Sota *et al.*, 1993). Treatment of lactating dairy cows with recombinant bST during follicular waves of the oestrous cycle has also increased the size of the subordinate (or second largest) ovarian follicle (Lucy *et al.*, 1993). Moreover, an association has been reported between twinning rate in cattle and concentrations of both peripheral and follicular fluid IGF-I (Echternkamp *et al.*, 1990). Perhaps the most convincing evidence for the role of peripheral IGF-I in follicular function *in vivo* is data collected during the oestrous cycle in cattle that have a growth hormone receptor deficiency (GHRD; Chase *et al.*, 1998). In these cattle, GH is chronically increased and IGF-I is low in the peripheral circulation. The dominant follicle of the first follicular wave of the oestrous cycle stops growing at approximately 9 mm in diameter, when expression of LH receptors on granulosal cells is known to begin and to contribute to final maturation of the dominant follicle (Xu *et al.*, 1995). According to the authors, the timing of follicular regression in cows with GHRD was an indication of a failure in gonadotrophic support, and it was speculated that low circulating IGF-I led to inadequate LH receptor function or expression within granulosal cells of the first-wave dominant follicle. A similar scenario may be occurring in the early postpartum cow experiencing negative energy balance. As noted previously, serum IGF-I is low in dairy cows that develop anovulatory dominant follicles during the first follicular wave postpartum, and differences in IGF-I between cows with either anovulatory or ovulatory dominant follicles precede the observed differences in follicular oestradiol production (Beam and Butler, 1998). Changes in systemic concentrations of IGF-I and IGF-binding proteins affect their concentrations in follicular fluid and follicular development in heifers (Cohick *et al.*, 1996), and IGF-I is known to contribute to induction of LH receptors in granulosal cells and to increase the sensitivity of follicular cells to LH stimulation (Spicer and Echternkamp, 1995). Lucy *et al.* (1992c) have shown a positive relationship between the ratio of oestrogen to progesterone in follicular fluid and plasma IGF-I in lactating Holstein cows. In the early postpartum cow, low concentrations of circulating IGF-I may contribute to reduced follicular responsiveness to a given level of gonadotrophic support, low oestradiol synthesis, and anovulation of dominant ovarian follicles.

A Working Model

On the basis of the preceding discussion, the following model is proposed for postpartum follicular development leading to first ovulation in lactating dairy cows experiencing negative energy balance. Major aspects of the model are represented schematically in Fig. 7. The release of negative feedback effects on FSH secretion by the clearance of high gestational oestrogens allows for an increase in mean plasma concentrations of FSH during days 3–7 postpartum. The increased FSH initiates a wave of follicular development during days 6–8 postpartum characterized by the recruitment of a pool of small (< 9 mm) follicles, from which usually one follicle is selected to become dominant. The dominant follicle emerges and continues to grow throughout the second week postpartum. Both the initiation of the follicular wave, including follicular recruitment, and emergence of a dominant follicle are largely insensitive to the severity of early postpartum negative energy balance. The steroidogenic activity, as measured in peripheral plasma, and ovulatory competence of this first dominant follicle are related to when postpartum energy balance begins to improve from its most negative level and the associated increase in LH pulse frequency. The level or severity of negative energy balance has a lesser modulating role in LH pulse frequency, but may influence the overall hormonal milieu including concentrations of circulating insulin and IGF-I conducive to enhanced follicular cell responsiveness to the ongoing pulsatile LH secretion pattern. If the day of the energy balance nadir occurs before, or soon after, the emergence of the dominant follicle, a concurrent increase in LH pulse frequency, responding to metabolic signals at the hypothalamus, would promote differentiation of the dominant follicle and increased oestradiol production leading to the induction of an LH surge and first ovulation. A delayed energy balance nadir, even in situations of relatively modest energy deficits, would probably result in low LH pulse frequency, low production of oestradiol, ovulation failure and atresia of the dominant follicle. Reduced concentrations of IGF-I, and perhaps insulin, contribute to reduced follicular responsiveness to gonadotrophic stimulation and to low oestradiol production by the dominant follicle, exacerbating the effects of low pulsatile LH secretion. The fate of the first-wave dominant follicle affects the duration of the postpartum anovulatory interval, with either the development of an anovulatory or cystic dominant follicle resulting in a greater number of days to first ovulation. Prolonged anovulatory intervals in turn impact negatively on first-service conception rates and overall reproductive performance.

Conclusion

The metabolic demands on modern dairy cattle impact negatively on the physiology regulating reproductive organs postpartum. In addition to the level of negative energy balance, changes in energy balance over time, such as improvement from the energy balance nadir, appear to be modulating postpartum ovarian activity. Together with patterns of LH secretion, evidence is strong for a role of the GH:IGF-I system in promoting follicular function postpartum and for IGF-I serving as a metabolic modulator of ovarian activity during negative energy balance. Challenges for the future include: greater delineation of the role of circulating IGF-I, as well as insulin, in follicular function during negative energy balance; the determination of the importance of ovulatory follicle size to subsequent fertility in lactating cows; and the development of nutritional strategies to increase postpartum energy intake and foster improvements in energy balance with attendant increase in circulating metabolic hormones beneficial to final follicular differentiation and ovulation at an early stage postpartum.

References

Adams GP, Matteri RL, Kastelic RP, Ko JCH and Ginther OJ (1993) Association between surges of follicle-stimulating hormone and the emergence of follicular waves in heifers *Journal of Reproduction and Fertility* **94** 177–188

Bauman DE and Currie WB (1980) Partitioning of nutrients during pregnancy and lactation: a review of mechanisms involving homeostasis and homeorhesis *Journal of Dairy Science* **63** 1514–1529

Beam SW and Butler WR (1994) Ovulatory follicle development during the first follicular wave postpartum in cows differing in energy balance *Journal of Animal Science* **72** (Supplement 1) 77

Beam SW and Butler WR (1997) Energy balance and ovarian follicle development prior to the first ovulation postpartum in dairy cows receiving three levels of dietary fat *Biology of Reproduction* **56** 133–142

Beam SW and Butler WR (1998) Energy balance, metabolic hormones, and early postpartum follicular development in dairy cows fed prilled lipid *Journal of Dairy Science* **81** 121–131

Bergfeld EGM, Kojima FN, Cupp AS, Wehrman, ME, Peters KE, Garcia-Winder M and Kinder JE (1994) Ovarian follicular development in prepubertal heifers is influenced by level of dietary energy intake *Biology of Reproduction* **51** 1051–1057

Butler WR (1998) Effect of protein nutrition on ovarian and uterine physiology in dairy cattle *Journal of Dairy Science* **81** 2533–2539

Butler WR and Smith RD (1989) Interrelationships between energy balance and postpartum reproductive function in dairy cattle *Journal of Dairy Science* **72** 767–783

Butler WR, Everett RW and Coppock CE (1981) The relationships between energy balance, milk production and ovulation in postpartum Holstein cows *Journal of Animal Science* **53** 742–748

Canfield RW and Butler WR (1990) Energy balance and pulsatile luteinizing hormone secretion in early postpartum dairy cows *Domestic Animal Endocrinology* **7** 323–330

Canfield RW and Butler WR (1991) Energy balance, first ovulation and the effects of naloxone on LH secretion in early postpartum dairy cows *Journal of Animal Science* **69** 740–746

Canfield RW, Sniffen CJ and Butler WR (1990) Effects of excess degradable protein on postpartum reproduction and energy balance in dairy cattle *Journal of Dairy Science* **73** 2342–2349

Chase CC, Jr, Kirby CJ, Hammond AC, Olson TA and Lucy MC (1998) Patterns of ovarian growth and development in cattle with a growth hormone receptor deficiency *Journal of Animal Science* **76** 212–219

Cohick WS, Armstrong JD, Whitacre MD, Lucy MC, Harvey RW and Campbell RM (1996) Ovarian expression of insulin-like growth factor-I (IGF-I), IGF binding proteins, and growth hormone (GH) receptor in heifers actively immunized against GH-releasing factor *Endocrinology* **137** 1670–1677

De La Sota RL, Lucy MC, Staples CR and Thatcher WW (1993) Effects of recombinant bovine somatotropin (sometribove) on ovarian function in lactating and nonlactating dairy cows *Journal of Dairy Science* **76** 1002–1013

Driancourt MA, Thatcher WW, Terqui M and Andrieu D (1991) Dynamics of ovarian follicular development in cattle during the estrous cycle, early pregnancy and in response to PMSG *Domestic Animal Endocrinology* **8** 209–221

Echternkamp SE, Spicer LJ, Gregory KE, Canning SF and Hammond JM (1990) Concentrations of insulin-like growth factor-I in blood and ovarian follicular fluid of cattle selected for twins *Biology of Reproduction* **43** 8–14

Ginther OJ, Kot K, Kulick LJ, Martin S and Wiltbank MC (1996) Relationships between FSH and ovarian follicular waves during the last six months of pregnancy in cattle *Journal of Reproduction and Fertility* **108** 271–279

Glencross RG (1987) Effect of pulsatile infusion of gonadotrophin-releasing hormone on plasma oestradiol-17β concentrations and follicle development during naturally and artificially maintained high levels of plasma progesterone in heifers *Journal of Endocrinology* **112** 77–85

Lamming GE, Wathes DC and Peters AR (1981) Endocrine patterns of the post-partum cow *Journal of Reproduction and Fertility Supplement* **30** 155–170

Lamming GE, Peters AR, Riley GM and Fisher W (1982) Endocrine regulation of postpartum function *Current Topics in Veterinary Medicine and Animal Science* **20** 148–172

Lucy MC, Staples CR, Michel FM and Thatcher WW (1991) Energy balance and size and number of ovarian follicles detected by ultrasonography in early postpartum dairy cows *Journal of Dairy Science* **74** 473–482

Lucy MC, Staples CR, Thatcher WW, Erickson PS, Cleale RM, Firkins JL, Clark JH, Murphy MR and Brodie BO (1992a) Influence of diet composition, dry matter intake, milk production and energy balance on time of postpartum ovulation and fertility in dairy cows *Animal Production* **54** 323–331

Lucy MC, Savio JD, Badinga L, De La Sota RL and Thatcher WW (1992b) Factors that affect ovarian follicular dynamics in cattle *Journal of Animal Science* **70** 3615–3626

Lucy MC, Beck J, Staples CR, Head HH, De La Sota RL and Thatcher WW (1992c) Follicular dynamics, plasma metabolites, hormones and insulin-like growth factor I (IGF-I) in lactating cows with positive or negative energy balance during the preovulatory period *Reproduction Nutrition and Development* **32** 331–341

Lucy MC, De La Sota RL, Staples CR and Thatcher WW (1993) Ovarian follicular populations in lactating dairy cows treated with recombinant bovine somatotropin (sometribove) or saline and fed diets differing in fat content and energy *Journal of Dairy Science* **76** 1014–1027

McGuire MA, Bauman DE, Dwyer DA and Cohick WS (1995)

Nutritional modulation of the somatotropin/insulin-like growth factor system: response to feed deprivation in lactating cows *Journal of Nutrition* **125** 493–502

Malven PV (1984) Pathophysiology of the puerperium: definition of the problem *Proceedings of the 10th International Congress on Animal Reproduction and Artificial Insemination* **4**(III) 1–8

Moss GE, Parfet JR, Marvin CA, Allrich RD and Diekman MA (1985) Pituitary concentrations of gonadotropins and receptors for GnRH in suckled beef cows at various intervals after calving *Journal of Animal Science* **60** 285

National Agricultural Statistics Service, United States Department of Agriculture (1997) *Dairy Facts* p 5 Department of Agriculture, Trade and Consumer Protection Madison, WI

Nebel RL and McGilliard ML (1993) Interactions of high milk yield and reproductive performance in dairy cows *Journal of Dairy Science* **76** 3257–3268

Perry RC, Corah LR, Cochran RC, Beal WE, Stevenson JS, Minton JE, Simms DD and Brethour JR (1991) Influence of dietary energy on follicular development, serum gonadotropins, and first postpartum ovulation in suckled beef cows *Journal of Animal Science* **69** 3762–3773

Peters AR, Pimentel MG and Lamming GE (1985) Hormone responses to exogenous GnRH pulses in post-partum dairy cows *Journal of Reproduction and Fertility* **75** 557–565

Pierson RA and Ginther OJ (1984) Ultrasonography of the bovine ovary *Theriogenology* **21** 495–504

Rahe CH, Owens RE, Fleeger JL, Newton HJ and Harms PG (1980) Patterns of plasma luteinizing hormone in the cyclic cow: dependence upon the period of the cycle *Endocrinology* **107** 498–503

Rajamahendran R and Taylor C (1990) Characterization of ovarian activity in postpartum dairy cows using ultrasound imaging and progesterone profiles *Animal Reproduction Science* **22** 171–180

Rutter LM and Randel RD (1984) Postpartum nutrient intake and body condition: effect on pituitary function and onset of oestrus in beef cattle *Journal of Animal Science* **58** 265–274

Savio JD, Boland MP, Hynes N and Roche JF (1990) Resumption of follicular activity in the early postpartum period of dairy cows *Journal of Reproduction and Fertility* **88** 569–579

Schillo KK (1992) Effects of dietary energy on control of luteinizing hormone secretion in cattle and sheep *Journal of Animal Science* **70** 1271–1282

Simpson RB, Chase CC, Jr, Spicer LJ, Vernon RK, Hammond AC and Rae DO (1994) Effect of exogenous insulin on plasma and follicular insulin-like growth factor I, insulin-like growth factor binding protein activity, follicular oestradiol and progesterone, and follicular growth in superovulated angus and brahman cows *Journal of Reproduction and Fertility* **102** 483–492

Spicer LJ and Echternkamp SE (1995) The ovarian insulin and insulin-like growth factor system with an emphasis on domestic animals *Domestic Animal Endocrinology* **12** 223–245

Spicer LJ and Stewart RE (1996) Interactions among basic fibroblast growth factor, epidermal growth factor, insulin, and insulin-like growth factor-I (IGF-I) on cell numbers and steroidogenesis of bovine thecal cells: role of IGF-I receptors *Biology of Reproduction* **54** 255–263

Spicer LJ, Alpizar E and Echternkamp SE (1993) Effects of insulin, insulin-like growth factor I, and gonadotropins on bovine granulosa cell proliferation, progesterone production, estradiol production, and(or) insulin-like growth factor I production *in vitro*. *Journal of Animal Science* **71** 1232–1241

Spicer LJ, Alpizar E and Vernon RK (1994) Insulin-like growth factor-I receptors in ovarian granulosa cells: effect of follicle size and hormones *Molecular and Cellular Endocrinology* **102** 69–76

Staples CR, Thatcher WW and Clark JH (1990) Relationship between ovarian activity and energy status during the early postpartum period of high producing dairy cows *Journal of Dairy Science* **73** 938–947

Staples CR, Burke JM and Thatcher WW (1998) Influence of supplemental fats on reproductive tissues and performance of lactating cows *Journal of Dairy Science* **81** 856–871

Stewart RE, Spicer LJ, Hamilton TD, Keefer BE, Dawson LJ, Morgan GL and Echternkamp SE (1996) Levels of insulin-like growth factor (IGF) binding proteins, luteinizing hormone and IGF-I receptors, and steroids in dominant follicles during the first follicular wave in cattle exhibiting regular estrous cycles *Endocrinology* **137** 2842–2850

Thatcher WW and Wilcox CJ (1973) Postpartum estrus as an indicator of reproductive status in the dairy cow *Journal of Dairy Science* **56** 608–610

Villa-Godoy A, Hughes TL, Emery RS, Chapin LT and Fogwell RL (1988) Association between energy balance and luteal function in lactating dairy cows *Journal of Dairy Science* **71** 1063–1072

Whisnant CS, Kiser TE, Thompson FN and Hall JB (1985) Effect of nutrition on the LH response to calf removal and GnRH *Theriogenology* **24** 565–573

Xu ZZ, Garverick HA, Smith GW, Smith MF, Hamilton SA and Youngquist RS (1995) Expression of follicle-stimulating hormone and luteinizing hormone receptor messenger ribonucleic acids in bovine follicles during the first follicular wave *Biology of Reproduction* **53** 951–957

Journal of Reproduction and Fertility Supplement **54**, 425–435

The role of leptin in nutritional status and reproductive function

D. H. Keisler, J. A. Daniel and C. D. Morrison

Department of Animal Science, University of Missouri, Columbia, MO 65211, USA

Infertility associated with suboptimal nutrition is a major concern among livestock producers. Undernourished prepubertal animals will not enter puberty until they are well fed; similarly, adult, normally cyclic females will stop cycling when faced with extreme undernutrition. Work in our laboratory has focused on how body fat (or adiposity) of an animal can communicate to the brain and regulate reproductive competence. In 1994, the discovery in rodents of the obese (*ob*) gene product leptin, secreted as a hormone from adipocytes, provided a unique opportunity to understand and hence regulate whole body compositional changes. There is now evidence that similar mechanisms are functioning in livestock species in which food intake, body composition, and reproductive performance are of considerable economic importance. Leptin has been reported to be a potent regulator of food intake and reproduction in rodents. There is evidence indicating that at least some of the effects of leptin occur through receptor-mediated regulation of the hypothalamic protein neuropeptide Y (NPY). NPY is a potent stimulator of food intake, is present at high concentrations in feed-restricted cattle and ewes, and is an inhibitor of LH secretion in these livestock species. In our investigations in sheep, we have cloned a partial cDNA corresponding to the ovine long-form leptin receptor, presumably the only fully active form, and have localized the long-form leptin receptor in the ventromedial and arcuate nuclei of the hypothalamus. Leptin receptor mRNA expression was colocalized with NPY mRNA-containing cell bodies in those regions. We have also determined that hypothalamic leptin receptor expression is greater in feed-restricted ewes than in well-fed ewes. These observations provide a foundation for future investigations into the nutritional modulators of reproduction in livestock.

Introduction

The ancient sculpture known as *Venus of Willendorf* (Fig. 1) illustrates man's long known association between nutritional status and fertility. What is not known, however, is the mechanism by which the nutritional status of an animal regulates reproductive processes. In 1953, Kennedy acknowledged the association between nutrition and reproduction and proposed the 'lipostat' or 'setpoint' theory, which asserted that the reproductive performance of an animal was related positively to the animal's body-fat mass. Kennedy's work was followed by many studies which included the classical ventromedial hypothalamic (VMH) lesioning studies by Hervey (1958). Hervey concluded that VMH lesions ablated the satiety centre and resulted in mice that exhibited hyperphagia, obesity and infertility.

In addition, the parabiotic-mice trials of Coleman *et al.* (reviewed in 1978) were extraordinarily insightful. These researchers worked with two strains of obese mice, notably the *ob/ob* mice (identified as the first genetic anomaly of its kind; Ingalis *et al.*, 1950), and the *db/db* (or diabetic) mice. The consequence of parabiosing *ob/ob* and *db/db* mice with each other or with wild-type mice are illustrated in Fig. 2. Mating of wild-type strains of mice with *ob/ob* or *db/db* mice resulted in offspring expressing simple Mendelian probabilities for the recessive traits. As a result of the lesioning and

Fig. 1. *Venus of Willendorf* – a symbol of the Mother and Fertility Goddess. (ca. ~25 000 years old).

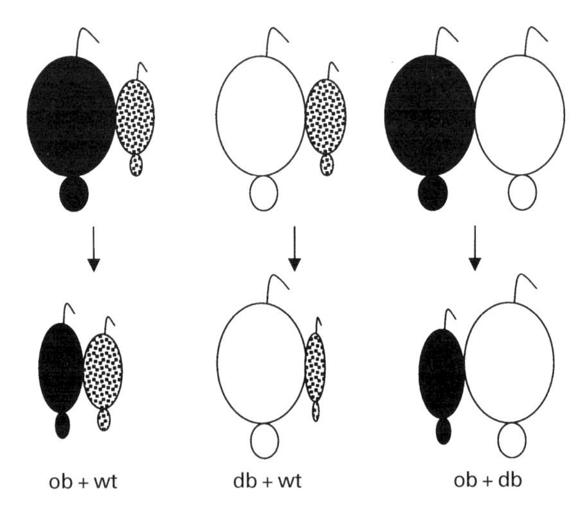

Fig. 2. Consequence of parabiosing *ob/ob* and *db/db* mice together or with wild-type mice.

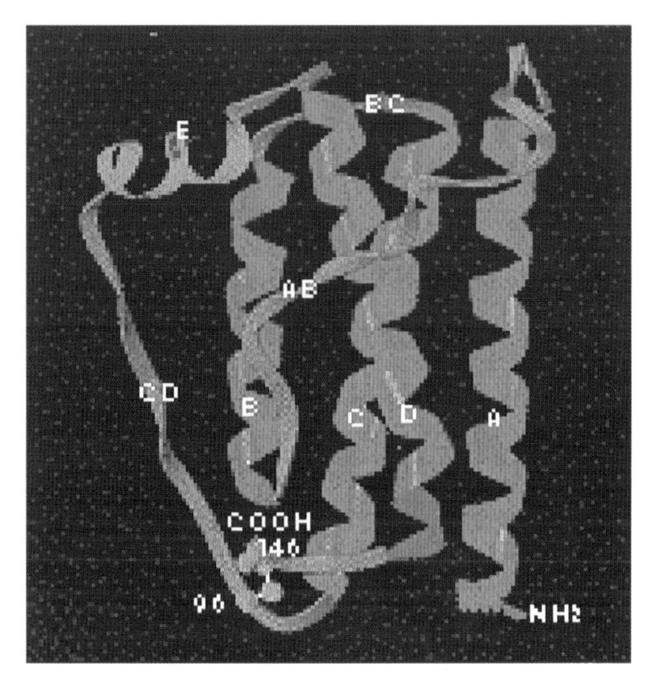

Fig. 3. Three-dimensional structural model of leptin.

parabiotic trials, it was concluded that blood-borne factors were responsible for communicating the animal's body composition to its brain. Elucidation (at least in part) of that 'body-to-brain' signal required more than two decades of additional work.

In 1994, Friedman and co-workers (Zhang *et al.*, 1994) used positional cloning techniques to identify the 167 amino acid protein product of the *ob* gene which was named leptin (derived from the Greek term 'leptos' meaning 'thin'). Subsequent nuclear magnetic resonance analysis of a crystalline form of leptin (E100; Zhang *et al.*, 1997) revealed that it was present as a quadra-helical protein (Fig. 3). Structurally, leptin contains a single disulfide bond which links cysteines within the C and D helices and which has proven critical to the biological activity of leptin.

Originally, Friedman and colleagues (Zhang *et al.*, 1994) reported that leptin was secreted exclusively by adipocytes; however, the list of tissues expressing leptin now includes the placental trophoblast, the mammary epithelium in primates and rodents, and avian liver. Unique to each of these tissues are their lipogenic–lipolytic capabilities. The importance of the lipogenic–lipolytic mechanisms in these tissues and particularly in the liver of oviparous and ovoviviparous animals (to include fish, birds, reptiles, select mammals (spp. Monotremata) and others), likely portends future reports of species-specific detection of leptin expression, particularly in liver or mammary epithelial tissue.

Concomitant with the identification of leptin, leptin receptors were cloned and are now known to occur in at least five variably spliced forms (*OB*-Ra-e; Li *et al.*, 1998). The high structural similarity of leptin receptors to cytokine receptors has led to numerous reports referring to leptin as a member of the cytokine family. When leptin binds to its receptor, receptor dimerization occurs; this is also characteristic of members of the growth hormone–cytokine family of receptors (Devos *et al.*, 1997; Girard, 1997; Liu *et al.*, 1997; Nakashima *et al.*, 1997). However, only one of the leptin receptors, referred to as *Ob*-Rb, or the 'long-form' of the leptin receptor is known to both span the cellular membrane and possess a 302 amino acid cytoplasmic domain. The G-protein-like signal transduction mechanism of the long-form of the leptin receptor is mediated most likely via

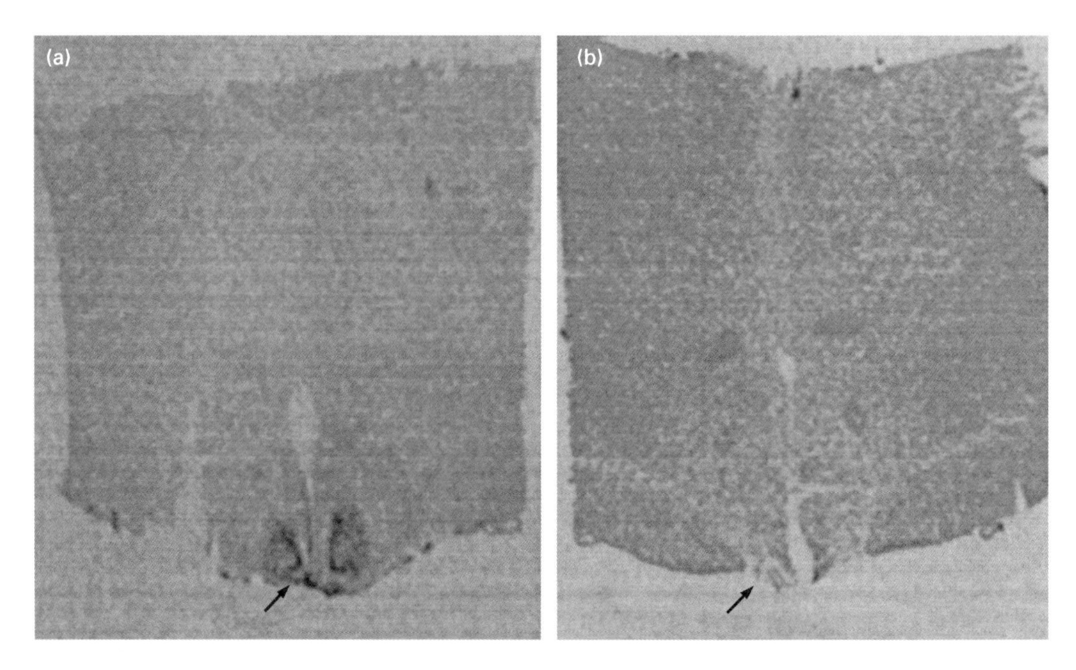

Fig. 4. Leptin receptor expression (arrow) in the ventromedial hypothalamus of feed-restricted (a) versus well-fed (b) ewes.

activation of janus kinase (JAK)-2/signal transducer and activator of transcription (STAT)-3, -5 or-6 pathways (Bjorbaek *et al.*, 1997). At least three of the short-forms of the leptin receptors (*Ob*-Ra, c, and d) also possess cytoplasmic domains. However, their cytoplasmic domains are truncated to 30–40 amino acids. Initially the short-form leptin receptors were believed not to be involved in signal transduction, but to mediate transmembrane movement or clearance of leptin. However, recent investigations (Murakami *et al.*, 1997; Yamashita *et al.*, 1998) have provided evidence that the short forms of the leptin receptors (*Ob*-Ra, c, and d) may indeed possess functional signal transduction capabilities (ex. via JAK/mitogen-activated protein kinase (MAPK) or phosphotidyl inositol [PI]-3 kinase activation). The leptin receptor denoted as Ob-Re is completely devoid of a cytoplasmic domain and is believed to occur in several circulating forms, possibly as soluble binding proteins (Gavrilova *et al.*, 1997; Lollmann *et al.*, 1997; Li et *al.*, 1998).

Initially leptin receptors were cloned from the choroid plexus (Tartaglia *et al.*, 1995). The significance of the choroid plexus is that it is one of the circumventricular organs within the brain where the brain is said to be 'permeable' to the exchange of substances with the peripheral blood supply. Location of leptin receptors in these neuroanatomical tissues may be a potential point of control in the regulation of the 'body-to-brain' signal. Indeed, there is evidence to support the hypothesis that the leptin transport process in the circumventricular organs is specific and saturable (Banks *et al.*, 1996; Caro *et al.*, 1996; Schwartz *et al.*, 1996; Diamond *et al.*, 1997; Corp *et al.*, 1998; Karonen *et al.*, 1998). In addition to the choroid plexus, leptin receptor expression has been localized in the brain to the ventromedial hypothalamus (Fig. 4), arcuate nucleus, hippocampus, thalamus, piriform cortex and anterior pituitary, and peripherally in tissues including adipocytes, liver, pancreas, fetal cartilage–bone, hair follicles, ovary, testis, uterus, heart, skeletal muscle, lung, lymph nodes, thyroid, adrenals, kidney, spleen and prostate gland (Hoggard *et al.*, 1997; Mendiola *et al.*, 1997; Zamorano *et al.*, 1997). The function, if any, of leptin in many of these tissues has yet to be determined.

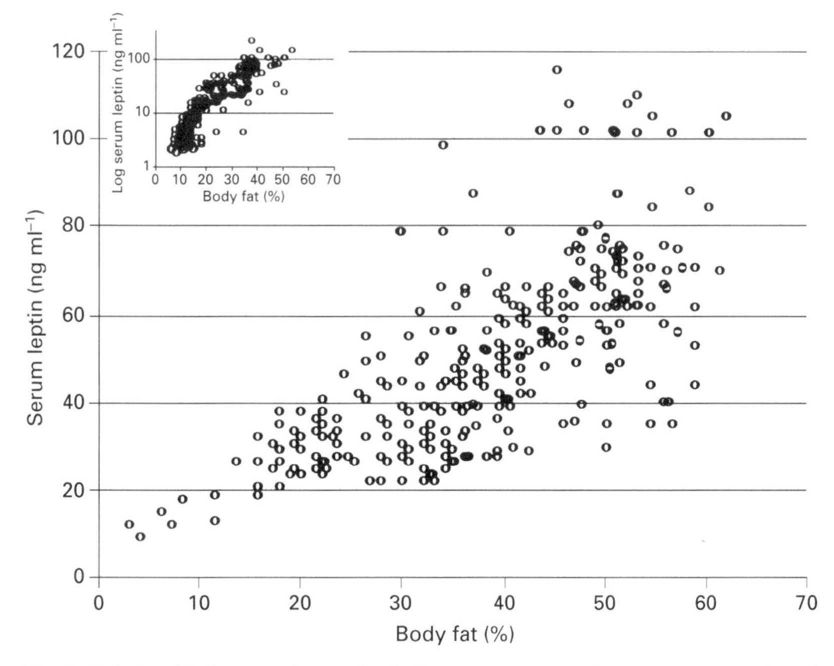

Fig. 5. Relationship between human body fat mass content and serum concentration of leptin.

Role of Leptin in Nutritional Status

All systems that perform work require fuel to function and physiological processes are not exceptions. However, unlike mechanical engines, animals possess the ability to integrate what they know about their environment and physiological status to anticipate their need for additional energy and to partition the energy among the systems in need. In order for an animal to anticipate its need for fuel and partition it appropriately, some variable or combination of variables (that is, inputs to the system) must reflect the well-being or status of each of the components of the system and the amount of fuel contained within the animal. Consequently, the fuel itself may serve as a dynamic indicator of the well-being or status of the animal by providing instantaneous yet discrete (yes versus no) assessments of status.

The fuel is important in determining whether the system functions or not, but how robust the system operates must be determined by some indicator(s) of the quality or quantity of fuel present within the animal. The issue of fuel quality implies that there are several sources of fuel which differ in energetic or satiating capabilities; and this is known to be true. For example, the brain is heavily dependent on glucose, its preferred energy substrate, but the brain can also function (within limitations) on reserves of ketones (Owen *et al.*, 1967). Another example is that fatty acids can vary greatly in carbon-length, structural branching, and proportion of hydrogen saturation, all of which can influence the insulinogenic response within the animal. Consequently, although the quality of fuel is a mechanism by which physiological responses can be affected, it has so far received limited attention but will probably be an important area of study in the future.

Assessments of the quantity of fuel present within an animal has been long sought and has resulted in the development of a variety of electromechanical approaches, including body-condition scores, skin fold thickness, dual-energy X-ray absorptiometry (DEXA), K^{40} counting, and ultrasonography, all of which are focused, in essence, on estimating fat mass. The discovery of leptin and subsequent development of methods to assess it in primates and rodents has provided

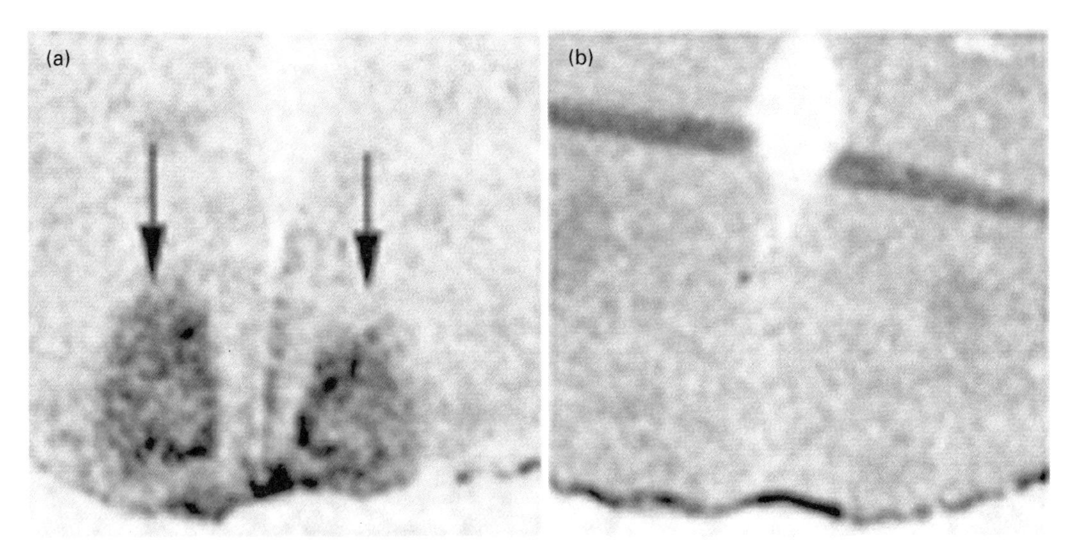

Fig. 6. NPY mRNA expression (arrows) in the ventromedial hypothalamus of feed-restricted (a) versus well-fed (b) ewes.

investigators with a relatively simple and accurate indicator of body-fat mass in these species (Fig. 5; Maffei *et al.*, 1995; Campfield *et al.*, 1996; Blum *et al.*, 1997; Perry *et al.*, 1997; Shimizu *et al.*, 1997; Langendonk *et al.*, 1998). Unfortunately, this relatively easily applied 'tool' has resulted in a large number of reports of replicated efforts and clinical associations between fat-mass and a variety of weakly linked conditions. The few quality studies reported so far have allowed us to focus on the relationship between fat-mass and physiological function and shape understanding of the regulatory processes involved in body 'fat and function'.

Similar tests for assessing leptin in livestock species have been developed with opportunities to: (1) monitor more accurately the 'fuel reserves' that an animal possesses in order to facilitate management of the animal and (2) potentially provide meat producers, processors, and consumers with an objective quantifier of meat quality.

As a result of years of focus on fundamental hypothesis driven research, it has been established that when animals lose body fuel reserves or fat mass, as in starvation conditions, satiety centres within the brain stimulate appetite (Hoebel, 1997; Hirschberg, 1998). Neurochemically, this is mediated, at least in part, if not predominately, by an increase in brain content of neuropeptide Y (NPY; Fig. 6; Tomaszuk *et al.*, 1996; Kalra, 1997; Yu *et al.*, 1997; Xu *et al.*, 1998). It is also important to note here that NPY is not being considered as the sole mediator of these processes, as there must be other collateral mechanisms to ensure the redundancy and thus preservation of this axis of communication. However, the significance of the focus on NPY is that there is evidence to support a pivotal role for NPY in regulating both nutritional status and reproductive function.

Neuropeptide Y is a 37 amino acid peptide, which is known to be one of the most potent stimulators of appetite in a broad range of species. But how secretion of NPY is regulated, or more specifically which somatic signals arise outside of the brain that are capable of communicating quantity or quality of body fuel reserves to the NPY neurones are not known. The answer may be that when many of the traditional indicators of nutritional status increase, especially insulin, secretion of NPY decreases (Kalra, 1997). In addition, there are direct effects of leptin on NPY neurones, as leptin receptors have been colocalized with NPY neurones in rodents (Mercer *et al.*, 1996) and sheep (Keisler *et al.*, unpublished observations; see Fig. 7). Consequently, at least one axis of the 'body-to-brain' signalling pathways is direct. The functionality of this pathway has been confirmed via intracerebroventricular infusions of leptin into both well-fed and feed-restricted ewes (Henry *et al.*, 1998; Morrison *et al.*, 1998). The mean (± one standard error) of feed-intake profiles of

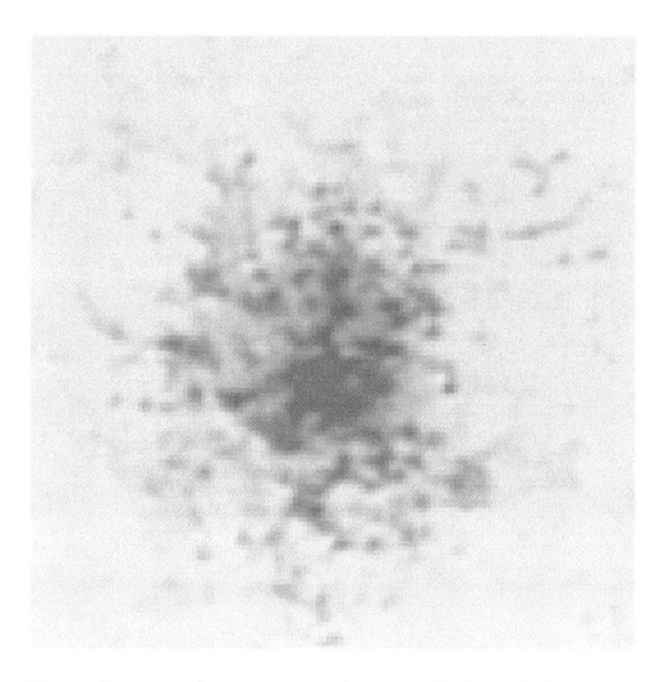

Fig. 7. 40 × magnification grey scale image of a hypothalamic NPY neurone (NPY mRNA in grey) exhibiting colocalization with leptin receptor mRNA (dark silver grains).

well-fed and feed-restricted ewes receiving vehicle or recombinant ovine leptin is shown in Fig. 8. Recombinant ovine leptin was infused in a linearly increasing dose from 0 $\mu g\ kg^{-1}\ h^{-1}$ on day 0 to a maximum dose of 1.25 $\mu g\ kg^{-1}\ h^{-1}$ on day 8. Well-fed ewes receiving leptin began to decrease intake of feed on day 4 and ceased eating by day 7 of infusion. However, feed-restricted ewes receiving leptin began to limit their intake only while receiving maximum amounts of leptin. These observations may reflect the ability of an animal to balance its presumptive desire for food against its constitutive demand for food. In well-fed ewes, leptin suppressed the animals' desire for food, while fat reserves within the animal supplied its demand for food. In contrast, the feed-restricted ewes were almost devoid of body fat and, therefore, their constitutive demand for food (for survival) superseded signals (exogenous leptin) suppressing their presumptive disinterest in eating. In essence, the message that communicated the quantity (or possibly quality) of available body fuel reserves (that is, leptin) did not concur with the flagrant absence of fuel; consequently, the thin ewes continued to eat to survive.

Role of Leptin in Reproductive Function

In 1996, Barash and coworkers revealed evidence implicating a role for leptin in the reproductive axis. They reported that treatment of *ob/ob* mice with leptin increased reproductive organ weight and serum concentrations of gonadotrophins. They thus declared leptin as a 'metabolic signal to the reproductive system'. In 1997, Chehab and coworkers substantiated the involvement of leptin in the reproductive axis by reporting that treatment of wild-type mice with leptin advanced the onset of sexual maturation by 9 days. Similarly, Carro *et al.* (1997) reported that administration of leptin antiserum to ovariectomized rats led to a marked decrease in secretion of LH. The validity of the observations of Chehab *et al.* (1997) and perhaps the observations of Carro *et al.* (1997) have now

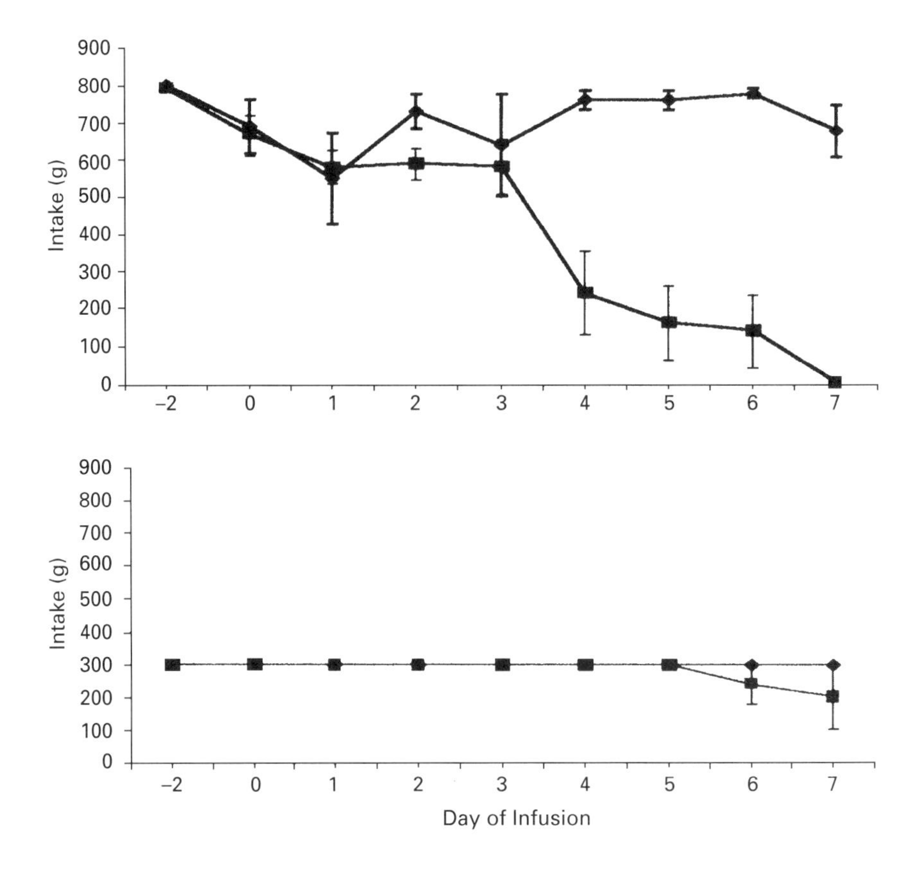

Fig. 8. Daily feed-intake of well-fed and feed-restricted ewes receiving intracerebroventricular infusions of vehicle (◆) or recombinant ovine leptin (■).

been challenged by the findings of Gruaz *et al.* (1998), who suggested that the action of leptin on the reproductive axis was confounded by its ability to decrease food intake and thus it is difficult to discriminate between these actions.

In addition, it is known that serum concentrations of leptin in primates and rodents: (1) are positively correlated with body weight and body mass index, (2) increase before puberty in females and to a lesser extent in males, (3) are typically three to four times greater in females than in males, and (4) are inversely related to serum concentrations of testosterone. It is also known that serum concentrations of leptin: (1) increase significantly during early pregnancy before any major changes occur in maternal body fat mass; (2) are significantly lower in newborns with intrauterine growth retardation than in newborns with normal intrauterine growth profiles and (3) are positively correlated with birth weight. In ruminants (and livestock in general), acceptable methods for assessing serum concentrations of leptin are only now being developed and information is not available so far. Furthermore, only recently has a large-scale method for the preparation of biologically active recombinant ruminant (ovine) leptin been described (Gertler *et al.*, 1998).

Irrespective of these impediments, Henry *et al.* (1998) and Morrison *et al.* (1998) infused recombinant human and ovine leptin (respectively) into the cerebroventricles of well-fed or well-fed and feed-restricted ewes, respectively. Independently, both groups of investigators reported that well-fed ewes reduced their food intake in response to leptin. This observation is consistent with the report by Henry *et al.* (1998) that hypothalamic expression of NPY (a potent orexigenic protein) was reduced in well-fed ewes infused with leptin. In contrast, neither Henry *et al.* (1998) nor Morrison *et al.* (1998) observed any discernible change in secretion of gonadotrophins in the well-fed ewes in response to leptin. It is possible that (1) gonadotrophin secretion in the well-fed ewes was already maximal, (2) systems governing feed intake were more responsive than systems governing reproductive responses, or (3) a further reduction in the already low hypothalamic content of NPY in the well-fed ewes was inconsequential to mechanisms governing secretion of gonadotrophins. In feed-restricted ewes (Morrison *et al.*, 1998), cerebroventricular infusion of leptin began to reduce food intake only at maximum amounts of leptin, but again failed to alter serum concentrations of gonadotrophins. We suggest that in the feed-restricted ewes collateral mechanisms or signals (such as insulin) provide redundancy, and thus ensure the signal for food intake is maintained for the preservation of the animal. The presence of several pathways thus permits a form of checks and balances of the system. With regards to the reproductive axis, perhaps gonadotrophin secretion in the feed-restricted ewes was so inhibited that any change in NPY was not of sufficient magnitude or duration to affect an increase in gonadotrophins or collateral consequences of the feed-restricted condition prohibited an increase in gonadotrophins. Regardless of the precise mechanism of action, these observations are consistent with the hypothesis that the effects of leptin on the reproductive axis may be mediated largely, if not exclusively, by central inhibition of NPY. As stated earlier, leptin receptors have been colocalized on NPY neurones (see Fig. 7) implicating a direct body-to-brain mechanism for communicating nutritional status to the reproductive axis.

Conclusion

Although there is significant evidence implicating the role of leptin in communicating nutritional status to the reproductive axis, it is not only highly unlikely but also perilous for an animal to rely on a single mechanism to mediate these body-to-brain communications. This is not to imply that the role of leptin in this process is minimal, but rather to place the role of leptin in perspective relative to several alternative mechanisms previously examined and yet to be examined. We are only at the beginning of understanding the complexity of these 'body-to-brain' communications and their interactions and the contribution of each of the potential mediators of these processes on the local, peripheral and central tissues that constitute the animal.

Contribution from the Missouri Agric. Exp. Sta. Journal Series.

References

Banks WA, Kastin AJ, Huang WT, Jaspan JB and Maness LM (1996) Leptin enters the brain by a saturable system independent of insulin *Peptides* **17** 305–311

Barash IA, Cheung CC, Weigle DS, Ren HP, Kabigting EB, Kuijper JL, Clifton DK and Steiner RA (1996) Leptin is a metabolic signal to the reproductive system *Endocrinology* **137** 3144–3147

Bjorbaek C, Uotani S, da Silva B and Flier JS (1997) Divergent signaling capacities of the long and short isoforms of the leptin receptor *Journal of Biological Chemistry* **272** 32686–32695

Blum WF, Englaro P, Hanitsch S, Juul A, Hertel NT, Muller J, Skakkebaek NE, Heiman ML, Birkett M, Attanasio AM, Kiess W and Rascher W (1997) Plasma leptin levels in healthy children and adolescents: dependence on body mass index, body fat mass, gender, pubertal stage, and testosterone *Journal of Clinical Endocrinology and Metabolism* **82** 2904–2910

Campfield LA, Smith FJ and Burn P (1996) The ob protein (leptin) pathway – a link between adipose tissue mass and central neural networks *Hormone and Metabolic Research* **28** 619–632

Caro JF, Kolaczynski JW, Nyce MR, Ohannesian JP, Opentanova I, Goldman WH, Lynn RB, Zhang PL, Sinha MK and Considine RV (1996) Decreased cerebrospinal-fluid/serum leptin ratio in obesity – a possible mechanism for leptin resistance *Lancet* **348** 159–161

Carro E, Pinilla L, Seoane LM, Considine RV, Aguilar E, Casanueva FF and Dieguez C (1997) Influence of endogenous leptin tone on the estrous cycle and luteinizing hormone pulsatility in female rats *Neuroendocrinology* **66** 375–377

Chehab FF, Mounzih K, Lu RH and Lim ME (1997) Early onset of reproductive function in normal female mice treated with leptin *Science* **275** 88–90

Coleman DL (1978) Obese and diabetes: two mutant genes causing diabetes–obesity syndromes in mice *Diabetologia* **14** 141–148

Corp ES, Conze DB, Smith F and Campfield LA (1998) Regional localization of specific [^{125}I]leptin binding sites in rat forebrain *Brain Research* **789** 40–47

Devos R, Guisez Y, Van der Heyden J, White DW, Kalai M, Fountoulakis M and Plaetinck G (1997) Ligand-independent dimerization of the extracellular domain of the leptin receptor and determination of the stoichiometry of leptin binding *Journal of Biological Chemistry* **272** 18304–18310

Diamond FB, Eichler DC, Duckett G, Jorgensen EV, Shulman D and Root AW (1997) Demonstration of a leptin binding factor in human serum *Biochemical and Biophysical Research Communications* **233** 818–822

Gavrilova O, Barr V, Marcus-Samuels B and Reitman M (1997) Hyperleptinemia of pregnancy associated with the appearance of a circulating form of the leptin receptor *Journal of Biological Chemistry* **272** 30546–30551

Gertler A, Simmons J and Keisler DH (1998) Large-scale preparation of biologically active recombinant ovine obese protein (leptin) *FEBS Letters* **422** 137–140

Girard J (1997) Is leptin the link between obesity and insulin resistance? *Diabetes and Metabolism* **23** (Supplement 3) 16–24

Gruaz NM, Lalaoui M, Pierroz DD, Raposinho PD, Blum WF and Aubert ML (1998) Failure of leptin administration to advance sexual maturation in the female rat *80th Annual Meeting of the Endocrine Society*, New Orleans Abstract P3-668

Henry B, Goding J, Alexander W, Tilbrook A, Canny B and Clarke I (1998) High doses of leptin can reduce food intake in sheep whilst not affecting the secretion of pituitary hormones *80th Annual meeting of the Endocrine Society*, New Orleans Abstract OR38-1

Hervey GR (1958) The effects of lesions in the hypothalamus in parabiotic rats *Journal of Physiology* **145** 336–352

Hirschberg AL (1998) Hormonal regulation of appetite and food intake *Annals of Medicine* **30** 7–20

Hoebel BG (1997) Neuroscience and appetitive behavior research – 25 years *Appetite* **29** 119–133

Hoggard N, Hunter L, Duncan JS, Williams LM, Trayhurn P and Mercer JG (1997) Leptin and leptin receptor mRNA and protein expression in the murine fetus and placenta *Proceedings of the National Academy of Sciences of the USA* **94** 11073–11078

Ingalis AM, Dickie MM and Snell GD (1950) Obese, a new mutation in the mouse *Journal of Heredity* **41** 317–318

Kalra SP (1997) Appetite and body weight regulation – is it all in the brain *Neuron* **19** 227–230

Karonen SL, Koistinen HA, Nikkinen P and Koivisto VA (1998) Is brain uptake of leptin *in vivo* saturable and reduced by fasting? *European Journal of Nuclear Medicine* **25** 607–612

Kennedy GC (1953) The role of depot fat in the hypothalamic control of food intake in the rat *Proceedings of the Royal Society of Medicine* **140** 578–592

Langendonk JG, Pijl H, Toornvliet AC, Burggraaf J, Frolich M, Schoemaker RC, Doornbos J, Cohen AF and Meinders AE (1998) Circadian rhythm of plasma leptin levels in upper and lower body obese women: influence of body fat distribution and weight loss *Journal of Clinical Endocrinology and Metabolism* **83** 1706–1712

Li C, Ioffe E, Fidahusein N, Connolly E and Friedman JM (1998) Absence of soluble leptin receptor in plasma from dbpas/dbpas and other db/db mice *Journal of Biological Chemistry* **273** 10078–10082

Liu CL, Liu XJ, Barry G, Ling N, Maki RA and Desouza EB (1997) Expression and characterization of a putative high affinity human soluble leptin receptor *Endocrinology* **138** 3548–3554

Lollmann B, Gruninger S, Stricker-Krongrad A and Chiesi M (1997) Detection and quantification of the leptin receptor splice variants Ob-Ra, b, and, e in different mouse tissues *Biochemical and Biophysical Research Communications* **238** 648–652

Maffei M, Halaas J, Ravussin E, Pratley RE, Lee GH, Zhang Y, Fei H, Kim S, Lallone R, Ranganathan S, Kern PA and Friedman JM (1995) Leptin levels in human and rodent – measurement of plasma leptin and ob RNA in obese and weight-reduced subjects *Nature Medicine* **1** 1155–1161

Mendiola J, Janzen M, Cruz M and Louis CF (1997) Cloning and tissue distribution of leptin mRNA in the pig *Animal Biotechnology* **8** 227–236

Mercer JG, Hoggard N, Williams LM, Lawrence CB, Hannah LT, Morgan PJ and Trayhurn P (1996) Coexpression of leptin receptor and prepreneuropeptide Y mRNA in arcuate nucleus of mouse hypothalamus *Journal of Neuroendocrinology* **8** 733–735

Morrison CD, Daniel JA Holmberg B, Bolden OU and Keisler DH (1998) Effects of lateral cerebroventricular infusion of leptin on ewe lambs *Journal of Animal Science* **76** 225

Murakami T, Yamashita T, Iida M, Kuwajima M and Shima K (1997) A short form of leptin receptor performs signal transduction *Biochemical and Biophysical Research Communications* **231** 26–29

Nakashima K, Narazaki M and Taga T (1997) Leptin receptor (ob-r) oligomerizes with itself but not with its closely related cytokine signal transducer gp130 *FEBS Letters* **403** 79–82

Owen OE, Morgan AP, Kemp HG, Sullivan JM, Herrera MG and Cahill GF (1967) Brain metabolism during fasting *Journal of Clinical Investigation* **46** 1589–1595

Perry HM, Morley JE, Horowitz M, Kaiser FE, Miller DK and Wittert G (1997) Body composition and age in african-american and caucasian women – relationship to plasma leptin levels *Metabolism Clinical and Experimental* **46** 1399–1405

Schwartz MW, Peskind E, Raskind M, Boyko EJ and Porte D (1996) Cerebrospinal fluid leptin levels – relationship to plasma levels and to adiposity in humans *Nature Medicine* **2** 589–593

Shimizu H, Shimomura Y, Hayashi R, Ohtani K, Sato N, Futawatari T and Mori M (1997) Serum leptin concentration is associated with total body fat mass, but not abdominal fat distribution *International Journal of Obesity and Related Metabolic Disorders* **21** 536–541

Tartaglia LA, Dembski M, Weng X, Deng NH, Culpepper J, Devos R, Richards GJ, Campfield LA, Clark FT, Deeds J, Muir C, Sanker S, Moriarty A, Moore KJ, Smutko JS, Mays GG, Woolf EA, Monroe CA and Tepper RI (1995) Identification and expression cloning of a leptin receptor, ob-r *Cell* **83** 1263–1271

Tomaszuk A, Simpson C and Williams G (1996) Neuropeptide Y, the hypothalamus and the regulation of energy homeostasis *Hormone Research* **46** 53–58

Yamashita T, Murakami T, Otani S, Kuwajima M and Shima K (1998) Leptin receptor signal transduction: ObRa and ObRb of fa type *Biochemical and Biophysical Research Communications* **246** 752–759

Yu WH, Kimura M, Walczewska A, Karanth S and McCann SM (1997) Role of leptin in hypothalamic–pituitary function *Proceedings of the National Academy of Sciences USA* **94** 1023–1028

Xu B, Dube MG, Kalra PS, Farmerie WG, Kaibara A, Moldawer LL, Martin D and Kalra SP (1998) Anorectic effects of the cytokine, ciliary neurotropic factor, are mediated by hypothalamic neuropeptide Y: comparison with leptin *Endocrinology* **139** 466–473

Zamorano PL, Mahesh VB, Desevilla LM, Chorich LP, Bhat GK and Brann DW (1997) Expression and localization of the leptin receptor in endocrine and neuroendocrine tissues of the rat *Neuroendocrinology* **65** 223–228

Zhang Y, Proenca R, Maffei M, Barone M, Leopold L and Friedman JL (1994) Positional cloning of the mouse obese gene and its human homologue *Nature* **372** 425–432

Zhang FM, Basinski MB, Beals JM, Briggs SL, Churgay LM, Clawson DK, Dimarchi RD, Furman TC, Hale JE, Hsiung HM, Schoner BE, Smith DP, Zhang XY, Wery JP and Schevitz RW (1997) Crystal structure of the obese protein leptin-E100 *Nature* **387** 206–209

REPRODUCTIVE TECHNOLOGY

Chair
S. J. Dieleman

Journal of Reproduction and Fertility Supplement 54, 439–448

Activation of primordial follicles *in vitro*

J. E. Fortune*, S. Kito† and D. D. Byrd‡

Department and Section of Physiology, College of Veterinary Medicine, Cornell University, Ithaca, NY 14853, USA

The resting pool of primordial follicles in mammalian ovaries is a potential resource for the genetic manipulation of domestic animals, the preservation of endangered species, and the amelioration of some forms of infertility in humans. Exploitation of this large reservoir of follicles depends on the development of methods for activating primordial follicles to begin growth *in vitro* and of methods for sustaining follicular growth to the stage at which oocytes are capable of meiotic maturation, fertilization and development to live young. It has been shown that primordial follicles of rodents, cattle and primates can initiate growth *in vitro*, even in serum-free medium. The signals that cause primordial follicles to leave the resting pool or remain quiescent are unknown. However, of interest is the observation that in cultures of whole rodent ovaries an apparently normal number of follicles leaves the resting pool and begins to grow, whereas in cultures of isolated bovine or primate ovarian cortex almost all primordial follicles activate and develop into primary follicles. This finding suggests that non-cortical portions of the ovary may regulate the flow of follicles from the resting reservoir. In cattle, it has been difficult to sustain follicular growth beyond the primary stage and the development of methods for doing so are critical for achievement of the practical goal of use of the primordial pool for embryo production. However, the development of murine follicles *in vitro* from the primordial stage through oocyte maturation and fertilization, and the birth of one pup, provides encouragement for efforts to achieve similar results in large mammals.

Introduction

Mammalian ovaries contain a reservoir of non-growing primordial follicles. Primordial follicles are formed when primary oocytes become invested with a layer of flattened pre-granulosal cells. These follicles constitute a pool from which follicles will be drawn gradually to begin growth, starting soon after follicle formation and continuing throughout the reproductive life span. What causes follicles to leave (or remain in) the resting pool and how a continuous 'trickle' of exiting follicles is achieved are unknown. The regulation of follicular quiescence versus growth is currently perhaps the most intriguing question in the area of regulation of ovarian follicular development. Although some progress has been made and will be reviewed in this manuscript, the most fundamental questions remain to be answered.

Two types of mammalian species have been used as models to address questions of the initiation of follicular growth – rodents and larger mammals, especially cattle and primates. In rats and mice follicle formation occurs at a specific time, one or two days after birth, depending on the species or strain. After follicle formation, follicles begin immediately to leave the resting pool and the first cohort reaches the antral stage after about two weeks (Hirshfield, 1991). There are several important advantages to rodents as models for exploring the mechanisms that govern the initiation of follicular growth, a process also referred to as follicle activation. The ovaries of newborn animals are small and

*Correspondence
†Current address: NRIS, Division of Education and Scientific Services, 4-9-1 Anagawa, Inage-ku, Chiba 263-8555, Japan
‡Current address: Cornell University Medical College, 445 E. 69th St, New York, NY 10021 USA

soft, and they can be cultured as whole ovaries, thus maintaining relationships among ovarian components, or they can be dissociated with enzymes to allow retrieval and culture of individual follicles (see for example Eppig and O'Brien, 1996). In addition the presence of only primordial follicles in newborn rodents allows analysis, during the first few weeks of life, of a synchronous cohort, the first follicles that leave the resting reservoir and initiate growth.

In larger mammals, such as ruminants and primates, the formation of primordial follicles occurs during fetal life over a much more protracted period, compared with that in rodents, of weeks or months (Henricson and Rajakoski, 1959; van Wagenen and Simpson, 1965; Russe, 1983). Some follicles begin to leave the resting pool before others have been formed, so that in these species follicular formation and the initiation of growth are occurring simultaneously within the same ovary. This offers the advantage that fetal ovaries, which may be more readily available and more economical and which have large numbers of primordial follicles, can be used to study follicle growth initiation. However, there are two distinct disadvantages to ruminants and primates as animal models for studies of activation of primordial follicles. First, since some follicles leave the resting pool before other primordial follicles have been formed, there is no easily identifiable time during development when the ovary contains only primordial follicles in large numbers. Second, the stroma of ruminant and primate ovaries is much denser and tougher than the stroma of rodent ovaries, making enzymatic dissociation or mechanical dissection of component parts difficult. This difficulty can be partially obviated by the use of fetal ovaries, which are much softer. However, bovine oocytes are more sensitive to enzymatic treatments that are tolerated well by rodent oocytes and can be easily damaged by enzymatic dissociation of ovarian tissue (Wandji *et al.*, 1996a).

Despite the difficulties of studying the regulation of follicle activation in large mammals, they have been used as models because of the practical benefits that could result from greater knowledge of the signals that regulate follicular growth and differentiation. The resting pool of primordial follicles is a potential resource that could be tapped to increase the reproductive potential of valuable domestic animals, members of endangered species, and women with fertility problems. The similarity of bovine and primate ovaries makes cattle excellent models for humans, as well as important models in themselves because of their agricultural importance. The ultimate goal of studies on the activation of primordial follicles *in vitro* is to develop conditions that will sustain follicular development to the stage where the oocyte is capable of meiotic maturation, fertilization and normal development.

Initiation of Follicular Growth *In Vitro*

The initiation of follicular growth has been achieved *in vitro* for several species. Blandau *et al.* (1965) cultured fetal mouse ovaries in serum-containing medium and reported that some oocytes grew in culture, specifically those that were surrounded by one or more layers of somatic cells. Eppig and O'Brien (1996) cultured ovaries obtained from newborn mice, which contain only newly formed primordial follicles, in medium containing 10% serum for 8 days. During this interval, follicular development *in vitro* was qualitatively similar to development *in vivo*, in that some primordial follicles initiated growth and developed to the secondary stage. More remarkably, these authors then isolated growing preantral follicles from the ovaries after 8 days in culture and grew them to the stage of oocyte competence for meiotic maturation, fertilization and embryonic development. One live pup was produced after embryo transfer. These experiments showed that murine primordial follicles can initiate growth *in vitro*, at least in the presence of 10% serum.

Several years ago, our laboratory began experiments to determine whether bovine primordial follicles could be activated *in vitro* in serum-free medium. Since the large size of bovine ovaries precludes whole-organ cultures, we isolated small pieces of ovarian cortex (about 0.5 mm × 0.5 mm × 0.3 mm). Because primordial follicles are located in the cortical region of the ovary, cortical pieces are rich in primordial follicles. Since follicle formation in cattle begins in mid-gestation and fetal ovaries are much softer and easier to dissect than adult ovaries, fetuses in the third trimester of gestation were used. Pieces of cortex were cultured on transwell membrane inserts in Waymouth MB

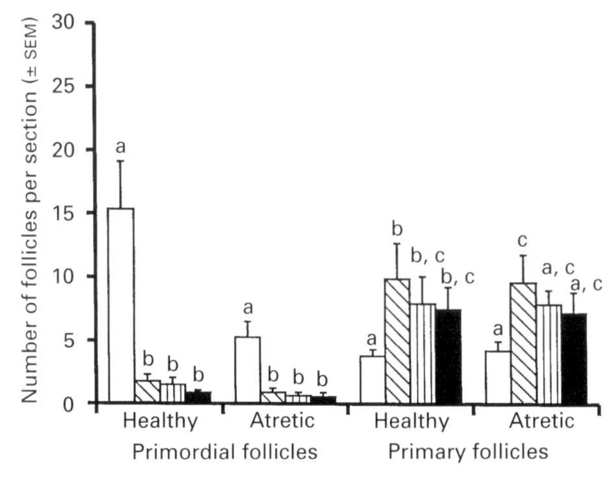

Fig. 1. Numbers of healthy and atretic primordial and primary follicles (mean per histological section ±SEM, n = 4 fetuses, with 59–71 sections examined per fetus) in fetal bovine ovarian cortex after 0 (□), 2(▧), 4 (▥) or 7 (■) days in culture. Within each group of four bars, bars with no common superscript are significantly different (a,b $P < 0.01$; a,c $P < 0.05$). (Reprinted from Wandji *et al.*, 1996b with permission.)

752/1 medium containing antibiotics and ITS+ (insulin, transferrin, selenium, BSA and linoleic acid) for up to 7 days (Wandji *et al.*, 1996b). At the initiation of culture (day 0), the cortical pieces contained mostly primordial follicles (Fig. 1), characterized by an oocyte surrounded by a single layer of flattened granulosal cells. As early as the second day of culture the number of primordial follicles had declined markedly, whereas the number of primary follicles, characterized by a single layer of cuboidal granulosal cells, had increased (Wandji *et al.*, 1996b). In addition, the diameter of primary follicles and their oocytes increased gradually throughout the 7 day culture (Fig. 2; Wandji *et al.*, 1996b). These results indicate that bovine primordial follicles can activate *in vitro* in serum-free medium and differentiate into primary follicles. This indicates, as the experiments with rodents had implied, that the activation of primordial follicles does not depend on specific endocrine signals. The experiments with cattle further support the contention that signals from non-cortical components of the ovary are not needed to stimulate initiation of follicle growth.

What is puzzling about the results presented in Fig. 1 is that such a large percentage of the primordial follicles in pieces of fetal bovine cortex initiated growth; almost all of them became activated *in vitro*. This mass exodus of follicles from the resting stage was not due to the fetal origin of the cultured cortical pieces, since Braw-Tal and Yossefi (1997) reported a similar loss of follicles from the primordial pool and an increase in primary follicles after 2 day cultures of ovarian cortical pieces from adult cattle (Table 1). Hence, something about the conditions of culture obviated the mechanisms that normally 'tell' each follicle when its 'turn' has come to leave the resting pool. The serum-free culture of ovarian cortical pieces provides a system that may be used to explore some of the factors that regulate follicle activation. Some potential regulators will be discussed below. The mass movement of follicles out of the resting pool and their development and growth as primary follicles under these experimental conditions is not unique to cattle. We have used identical methods to isolate and culture cortical pieces from ovaries from baboon fetuses and obtained similar results (Wandji *et al.*, 1997). In addition the results of Hovatta *et al.* (1997) indicate that primordial follicles may activate in cultured slices of human ovarian tissue.

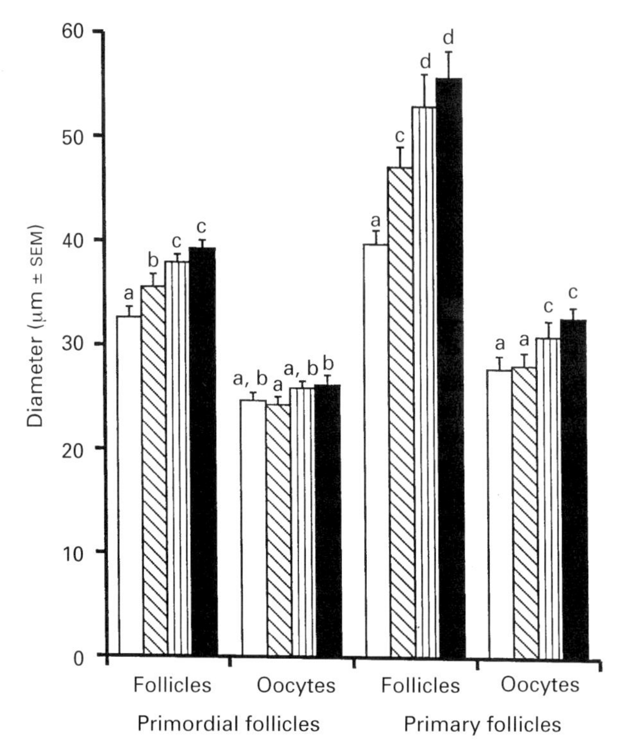

Fig. 2. Mean diameter (μm ± SEM) of healthy primordial and primary follicles and oocytes in pieces of fetal bovine ovarian cortex after 0 (□), 2 (▧), 4 (▥) or 7 (■) days in culture ($n = 4$ fetuses, with 107–204 primordial and 231–346 primary follicles/oocytes measured per fetus). Within each group of four bars, bars with no common superscript are significantly different (a,c,d $P < 0.01$; a,b $P < 0.05$). (Reprinted from Wandji *et al.*, 1996b, with permission.)

What Regulates the Activation of Primordial Follicles?

As discussed above, experiments to date with rodent, bovine and primate ovaries show that primordial follicles can activate in whole ovaries or ovarian cortical pieces maintained *in vitro* even, in the case of experiments with cattle and baboons, in serum-free medium. Thus, gonadotrophins or other blood-borne factors do not appear to be necessary for initiation of follicle growth. Of interest is the observation that in cultures of whole rodent ovaries only some of the primordial follicles initiate growth *in vitro*, as occurs *in vivo* (Eppig and O'Brien, 1996), whereas in isolated pieces of bovine or baboon ovarian cortex the vast majority of primordial follicles is activated (Wandji *et al.*, 1996b; Braw-Tal and Yossefi, 1997; Wandji *et al.*, 1997). It is therefore possible that the more central, medullary portion of the ovary regulates the flow of primordial follicles into the pool of growing preantral follicles by secreting an inhibitory factor(s) that keeps most primordial follicles quiescent. Alternatively, the artificial reduction in the size of the pool of primordial follicles in the cortical pieces may remove some inhibitory factor(s) that normally emanates from that compartment of the ovary. There is evidence that a higher percentage of primordial follicles becomes activated if the size of the resting pool is reduced (Krarup *et al.*, 1969; Hirshfield, 1994). Another possibility is that the conditions *in vitro* for cultured bovine and baboon cortical pieces are richer in some way(s) than their situation *in vivo*. For example, the ovarian cortex is known to be poorly vascularized (Guraya,

Table 1. Effect of culture and FSH (100 ng ml^{-1}) on development of bovine follicles *in vitro*

Day of culture (*n*)	FSH	Primordial follicles (% of total)	Primary or transitory follicles (% of total)	Preantral follicles (% of total)	Follicle diameter (mean ± SEM, μm)	Oocyte diameter (mean ± SEM, μm)	Number of granulosal cells*
0	–	50^{a}	17^{a}	2	37.94 ± 1.54^{a}	28.50 ± 0.49^{a}	9.81 ± 1.48^{a}
(69)		(72.0)	(25.6)	(2.9)			
2	–	6^{b}	50^{b}	2	50.83 ± 2.16^{b}	27.91 ± 0.52^{a}	11.47 ± 1.22^{a}
(58)		(10.3)	(86.2)	(3.4)			
2	+	3^{b}	53^{b}	1	50.77 ± 2.67^{b}	27.91 ± 0.53^{a}	10.84 ± 1.85^{a}
(57)		(5.3)	(93.0)	(1.7)			

Within columns, values with different superscripts are significantly different ($P < 0.05$).
n: number of non-atretic follicles examined.
*Largest cross-section of the follicle.
(Reprinted with permission from Braw-Tal and Yossefi, 1997).

1985; van Wezel and Rodgers, 1996), so primordial follicles may have better access to nutrients *in vitro* and/or a higher oxygen concentration than *in vivo*.

The only specific factor that has thus far been linked to the activation of primordial follicles is kit ligand (also called stem cell factor or steel factor). Kit ligand is produced by granulosal cells, whereas primordial germ cells, oocytes and theca cells express the receptor for kit ligand, c-kit (Manova *et al.*, 1993; Motro and Bernstein, 1993). Yoshida *et al.* (1997) injected mice with a function-blocking antibody to c-kit at various times during the first two weeks of life and concluded that kit ligand is needed for the activation of primordial follicles, but not for their formation. Parrott and Skinner (1997) reported that addition of kit ligand to cultures of rat ovaries induced primordial follicles to begin development, whereas an antibody that blocks the function of c-kit (the receptor for kit ligand) blocked spontaneous activation of primordial follicles. Although workers in our laboratory did not observe any effects of kit ligand on the activation of primordial follicles or subsequent growth of primary follicles (Wandji and Fortune, unpublished), in our experimental model (isolated cortex) almost all of the primordial follicles activate spontaneously. Hence further studies on the potential role of kit ligand and c-kit in the initiation of follicle growth will be of interest.

Can Follicular Growth be Maintained *In Vitro*?

The previous section summarized current limited knowledge of the signals that regulate the movement of primordial follicles into the growing pool. Clearly this question is of interest if we are to understand this critical first step in follicular growth and differentiation. However, determining factors and conditions that will maintain the growth of follicles once they have been activated *in vitro* is also of interest. In cultures of bovine or baboon ovarian cortex, growth beyond the primary stage was rare (Wandji *et al.*, 1996b, 1997), in contrast to the development of secondary follicles in organ cultures of whole newborn mouse or rat ovaries (Eppig and O'Brien, 1996; Mayerhofer *et al.*, 1997). Sustained follicular growth after activation of primordial follicles in ovarian cultures from larger mammals is a necessary step if the pool of primordial follicles is to provide oocytes that can be fertilized. The sections below discuss factors or conditions which may enhance the development of small preantral follicles *in vitro*.

Culture media

Recently we have conducted experiments to determine whether media containing fetal bovine serum (FBS) or a combination of FBS and ITS+ in various proportions would support the growth of

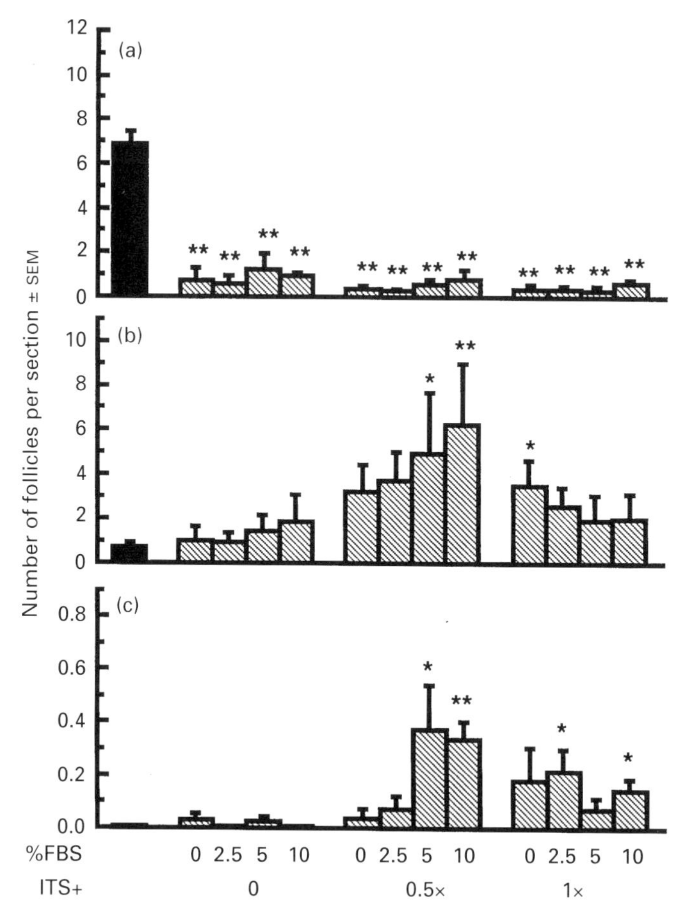

Fig. 3. Effects of culture in different media on numbers of primordial (a), primary (b) and secondary (c) follicles in pieces of fetal bovine ovarian cortex ($n = 4$ fetuses). Cortical pieces were fixed immediately after isolation (day 0 control; ■) or after 10 days of culture (▧) in 0, 2.5, 5, or 10% fetal bovine serum (FBS) in the presence or absence of full strength (1 ×) or half-strength (0.5 ×) ITS+ (insulin, transferrin, selenium, BSA and linoleic acid). Asterisks indicate significant differences from the day 0 control (*, $P < 0.05$; **, $P < 0.01$).

secondary follicles after activation of primordial follicles. Cortical pieces from four bovine fetuses were cultured for 10 days in Waymouth MB 752/1 medium containing 0, 2, 5 or 10% FBS in the presence or absence of half-strength ITS+ (0.5 × the normal concentration) or full-strength ITS+ (1 ×) and then subjected to histological morphometry and statistical analysis by methods described by Wandji *et al.* (1996b). The combination of 0.5 × ITS+ and 5% or 10% FBS was most effective, of the media tested, at supporting the growth of follicles to the primary and secondary stages (Fig.3). Surprisingly, Waymouth medium plus 10% FBS, which supports normal follicular development in newborn mouse ovaries (Eppig and O'Brien, 1996) and activates baboon oocytes without activating their granulosal cells (Wandji *et al.*, 1997), provided a very poor environment for the activation and growth of bovine follicles. These results indicate that the type of culture medium can markedly affect the growth of follicles after activation of primordial follicles and that the optimal medium conditions may vary from species to species.

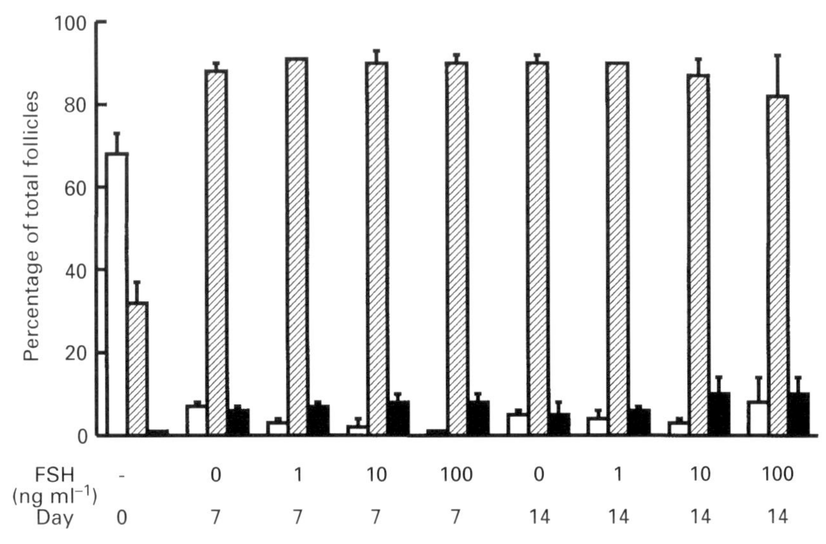

Fig. 4. Lack of effect of FSH on cultures of fetal bovine ovarian cortex. Bars indicate percentages (±SEM) of healthy follicles that were primordial (□), primary (▨), or early secondary (■) stage in freshly isolated tissue or after culture with graded doses of FSH (0, 1, 10 or 100 ng ml^{-1}) for 7 or 14 days (n = 4 cortical pieces, 2 from each of 2 fetuses; with 472–1232 follicles examined per treatment). (Reprinted from Fortune *et al.*, 1998 with permission.)

FSH

The development of preantral follicles in hypophysectomized mammals (Dufour *et al.*, 1979; Hirshfield, 1985) indicates that gonadotrophins are not absolutely required for follicular development until the antral stage. However, since hypophysectomy reduces the number of growing preantral follicles in sheep (Dufour *et al.*, 1979) and since bovine ovarian follicles bind FSH (Wandji *et al.*, 1992a) and ovine follicles express messenger RNA for the FSH receptor (Tisdall *et al.*, 1995) beginning with the primary stage, it is possible that FSH could facilitate the growth of bovine follicles *in vitro*, after the activation of primordial follicles. Braw-Tal and Yossefi (1997) cultured bovine cortical pieces from adult ovaries for 2 days in medium containing FSH (100 ng ml^{-1}, NIDDK oFSH-17) and found no effect on the distribution of follicles among the primordial, primary and preantral size classes (Table 1). To determine whether a longer period of exposure to FSH would induce newly activated follicles in cortical pieces from fetal bovine ovaries to develop to the secondary stage, we cultured pieces for 7 or 14 days with graded doses of FSH (0, 1, 10 or 100 ng ml^{-1}; NIDDK oFSH-17). No dose of FSH had a significant effect on the distribution of follicles among the primordial, primary and secondary size classes after either 7 or 14 days of culture (Fig. 4; Fortune *et al.*, 1998). These results are consistent with the suggestion of Wandji *et al.* (1992b) that the FSH receptors in fetal ovaries may not be linked to the adenylate cyclase second messenger system.

Mayerhofer *et al.* (1997) reported that treating neonatal rat ovaries, which contain only primordial follicles, with vasoactive intestinal peptide (VIP) or other agents that increase cAMP induced messenger RNA (mRNA) for FSH receptors. In addition, a short pretreatment with these agents induced the ovaries to become responsive to subsequent treatment with exogenous FSH in terms of cAMP secretion and in terms of follicular growth, which proceeded rapidly to the secondary stage in some follicles. To determine whether the lack of response of bovine cortical pieces to FSH that we (Fortune *et al.*, 1998) and Braw-Tal and Yossefi (1997) had observed could be reversed by previous exposure of cortical pieces to agents that increase cyclic AMP, we replicated the

Table 2. Cyclic AMP secretion (pg h^{-1} per culture ±SEM) by pieces of bovine ovarian cortex cultured for 32 h (n = 6 cultures, 2 from each of 3 fetuses)[a]

A. Effects of VIP or forskolin 0–8 h of culture		B. Effects of FSH (300 ng ml^{-1}) 8–32 h of culture	
Treatment	cAMP secretion	Treatment	cAMP secretion
Control	32 ± 15[b]	Control	22 ± 4[b]
	31 ± 9[b]	+ FSH	34 ± 9[b]
VIP	391 ± 42[c]	Control	180 ± 14[d]
(10 μmol l^{-1})	416 ± 81[c]	+ FSH	195 ± 14[d]
Forskolin	753 ± 71[c,d]	Control	124 ± 15[d]
(40 μmol l^{-1})	792 ± 108[d]	+ FSH	164 ± 38[d]

[a] Each culture contained four pieces of freshly isolated, fetal ovarian cortex (approximately 0.5 mm × 0.5 mm × 0.3 mm) on a membrane insert in 350 μl Waymouth MB containing ITS+ (insulin, transferrin, selenium, and BSA); medium also contained isobutyl methylxanthine, IBMX (0.5 mmol l^{-1}) to inhibit metabolism of cAMP.
[b,c,d] Within columns means with no common superscript are significantly different. b versus c or c versus d, $P < 0.05$; b versus d, $P < 0.01$.

experiment of Mayerhofer *et al.* (1997) by treating bovine cortical pieces with VIP or forskolin for 8 h, followed by treatment with FSH for 24 h (300 ng ml^{-1}; NIDDK oFSH-17). Media were collected and measured for cAMP by radioimmunoassay. Both VIP and forskolin increased the secretion of cAMP during the first 8 h of culture (Table 2). In addition, they continued to exert a 'carry-over' stimulatory effect during the next 24 h. However, FSH did not increase the secretion of cAMP whether or not tissue had been pretreated with VIP or forskolin. These results suggest that, at least under the conditions used, a short exposure to increased cAMP is not sufficient to induce functional FSH receptors and a cAMP response to FSH in bovine ovarian cortex, in contrast to the results for neonatal rat ovaries (Mayerhofer *et al.*, 1997). Therefore, it appears that the addition of FSH to cultures of bovine primary follicles grown *in vitro* is not useful. However, FSH does have effects on cultures of larger preantral follicles from mice (Eppig and O'Brien, 1996; Cortvrindt *et al.*, 1997). Therefore, if methods can be devised for growing bovine follicles to later preantral stages, FSH could then be tested for facilitation of further growth *in vitro* of larger preantral follicles.

Other factors: growth factors (GDF-9, bFGF), activin, WT1

Several hormones and growth factors have been implicated indirectly in early follicular development. It would be of interest to determine whether one or more of these agonists could stimulate the growth of follicles activated *in vitro* to the early secondary stage and beyond. Growth differentiation factor 9 (GDF-9), a member of the transforming growth factor-β superfamily, is of particular interest since mRNA for GDF-9 is found only in oocytes from the primary follicle stage through ovulation. In mice homozygous for a GDF-9 'knockout', follicles were activated but did not proceed beyond the primary stage (Dong *et al.*, 1996), indicating that production of GDF-9 by the oocyte is critical for follicular development after the primary stage. Basic fibroblast growth factor (bFGF) is also of interest. Van Wezel *et al.* (1995) immunolocalized bFGF to bovine primordial and primary oocytes and suggested a role for this growth factor in stimulating granulosal cell proliferation. This hypothesis is consistent with experiments that showed that the binding of radiolabelled bFGF to bovine follicle cells is highest in the preantral stages, including primary follicles (Wandji *et al.*, 1992c) and that bFGF stimulates thymidine incorporation into bovine granulosal cells of preantral follicles *in vitro* (Wandji *et al.*, 1996a).

Activin A markedly stimulated the growth of preantral follicles (100–120 μm in diameter) and synergized with FSH when follicles were obtained from immature mice, but not adult mice (Yokota *et al.*, 1997). Tisdall *et al.* (1995) detected mRNA for β_B inhibin (needed for synthesis of activin B) as early as the primary stage of ovine follicular development. Finally Wilms' tumour gene, WT1, is a gene deleted in some Wilms' tumours that codes for a transcription factor; its mRNA is expressed strongly in the early stages of rat preantral follicular development (Hsu *et al.*, 1995). What role(s) activin or WT1 may play in early follicular development remains to be elucidated.

Conclusions

Primordial follicles of rodents, cattle and primates can be activated *in vitro* to begin growth. In the experiments conducted thus far, the culture of intact whole ovaries results in the activation of an approximately normal number of primordial follicles and some of them grow to the multilayered secondary stage within a few days to a week of culture (Eppig and O'Brien, 1996; Mayerhofer *et al.*, 1997). In contrast, in isolated pieces of ovarian cortex from cattle and primates most primordial follicles initiate growth, but few follicles proceed to the secondary stage (Wandji *et al.*, 1996b, 1997; Fortune *et al.*, 1998). These findings raise interesting questions about what regulates the activation of primordial follicles *in vivo*. Although preliminary evidence implicates kit ligand as a stimulator of primordial follicle activation (Parrott and Skinner, 1997), the results also suggest that an inhibitor(s) is involved. The development of conditions *in vitro* that will allow the development of ruminant and primate follicles activated *in vitro* to a size where they might be isolated for further culture is essential to achieve the goal of producing embryos from the reservoir of primordial follicles. A number of laboratories have cultured larger preantral follicles (for review see van den Hurk *et al.*, 1997) and their experience will be helpful in developing culture strategies for large preantral follicles grown *in vitro*, if methods can be devised for getting them to that stage. Thus far one live mouse has been produced from a primordial follicle (Eppig and O'Brien, 1996). This shows that it is possible to use the primordial pool as a source of oocytes to produce embryos, but the difficulties of doing so will be much greater in ruminants and primates because of the larger size and longer developmental period of their oocytes. Some progress has been made towards the goal, but much more remains to be done. However, recent reports of follicle growth *in vitro* or production of live offspring after cryopreservation of rodent or human ovarian tissue (Carroll and Gosden, 1993; Hovatta *et al.*, 1997; Sztein *et al.*, 1998) indicate that the 'banking' of frozen ovarian tissue from valuable domestic animals, endangered species or women scheduled for radiation or chemotherapy, coupled with the ability to produce embryos from primordial follicles, would provide a powerful method for enhancing fertility in these groups.

The contributions of Drs Wandji, Voss and Srsen, Ms Hansen, Ms Bartholomew and Mr Murphy, to the experiments presented in Figs 1, 2 and 3 and Table 2 are gratefully acknowledged. The unpublished data presented herein were generated through the support of the NIH (HD-35168 to J. E. Fortune). The NIDDK generously provided the oFSH and Taylor Packing Inc. (Wyalusing, PA) donated the fetal bovine ovaries used in those studies.

References

Blandau RJ, Warrick E and Rumery RE (1965) *In vitro* cultivation of fetal mouse ovaries *Fertility and Sterility* **16** 705–715

Braw-Tal R and Yossefi S (1997) Studies *in vivo* and *in vitro* on the initiation of follicle growth in the bovine ovary *Journal of Reproduction and Fertility* **109** 165–171

Carroll J and Gosden RG (1993) Transplantation of frozen–thawed mouse primordial follicles *Human Reproduction* **8** 1163–1167

Cortvrindt R, Smitz J and Van Steirteghem AC (1997) Assessment of the need for follicle stimulating hormone in early preantral mouse follicle culture *in vitro*. *Human Reproduction* **12** 759–768

Dong J, Albertini DF, Nishimori K, Kumar TR, Lu N and Matzuk MM (1996) Growth differentiation factor-9 is required during early ovarian folliculogenesis *Nature* **383** 531–535

Dufour J, Cahill LP and Mauleon P (1979) Short- and long-term effects of hypophysectomy and unilateral ovariectomy on ovarian follicular populations in sheep *Journal of Reproduction and Fertility* **57** 301–309

Eppig JJ and O'Brien MJ (1996) Development *in vitro* of mouse oocytes from primordial follicles *Biology of Reproduction* **54** 197–207

Fortune JE, Kito S, Wandji S-A and Srsen V (1998) Activation of bovine and baboon primordial follicles *in vitro*. *Theriogenology* **49** 441–449

Guraya SS (1985) Primordial follicle. In *Biology of Ovarian Follicles in Mammals* pp 3–14 Springer-Verlag, New York

Henricson B and Rajakoski E (1959) Studies of oocytogenesis in cattle *Cornell Veterinarian* **49** 494–503

Hirshfield AN (1985) Comparison of granulosa cell proliferation in small follicles of hypophysectomized, prepubertal, and mature rats *Biology of Reproduction* **32** 979–987

Hirshfield AN (1991) Development of follicles in the mammalian ovary *International Review of Cytology* **124** 43–101

Hirshfield AN (1994) Relationship between the supply of primordial follicles and the onset of follicular growth in rats *Biology of Reproduction* **50** 421–428

Hovatta O, Silye R, Abir R, Krausz T and Winston RML (1997) Extracellular matrix improves survival of both stored and fresh human primordial and primary ovarian follicles in long-term culture *Human Reproduction* **12** 1032–1036

Hsu SY, Kubo M, Chun S-Y, Haluska FG, Housman DE and Hsueh AJW (1995) Wilms' tumor protein WT1 as an ovarian transcription factor: decreases in expression during follicle development and repression of inhibin-α gene promoter *Molecular Endocrinology* **9** 1356–1366

Krarup T, Pedersen T and Faber M (1969) Regulation of oocyte growth in the mouse ovary *Nature* **224** 187–188

Manova K, Huang EJ, Angeles M, De Leon V, Sanchez S, Pronovost SM, Besmer P and Bachvarova RF (1993) The expression pattern of the c-*kit* ligand in gonads of mice supports a role for the c-*kit* receptor in oocyte growth and in proliferation of spermatogonia *Developmental Biology* **157** 85–99

Mayerhofer A, Dissen GA, Costa ME and Ojeda SR (1997) A role for neurotransmitters in early follicular development: induction of functional follicle-stimulating hormone receptors in newly formed follicles of the rat ovary *Endocrinology* **138** 3320–3329

Motro B and Bernstein A (1993) Dynamic changes in ovarian c-*kit* and *Steel* expression during the estrous reproductive cycle *Developmental Dynamics* **197** 69–79

Parrott JA and Skinner MK (1997) Theca cell–granulosa cell interactions that induce primordial follicle development and promote folliculogenesis *Biology of Reproduction* **56,** Supplement 1 125

Russe I (1983) Oogenesis in cattle and sheep *Bibliotheca Anatomica* **24** 77–92

Sztein J, Sweet H, Farley J and Mobraaten L (1998) Cryopreservation and orthotopic transplantation of mouse ovaries: new approach in gamete banking *Biology of Reproduction* **58** 1071–1074

Tisdall DJ, Smith P, Leeuwenberg B and McNatty KP (1995) FSH-receptor, b_B inhibin subunit, follistatin, β_A and α inhibin subunits and IGF-I genes are expressed sequentially in ovine granulosa cells during early follicular development *Journal of Reproduction and Fertility Abstract Series* **15** Abstract 28

van den Hurk R, Bevers MM and Beckers JF (1997) *In-vivo* and *in-vitro* development of preantral follicles *Theriogenology* **47** 73–82

van Wagenen G and Simpson ME (1965) *Embryology of the Ovary and Testis: Homo sapiens and Macaca mulatta* Yale University Press, New Haven

van Wezel IL and Rodgers RJ (1996) Morphological characterization of bovine primordial follicles and their environment *in vivo. Biology of Reproduction* **55** 1003–1011

van Wezel IL, Umapathysivam K, Tilley WD and Rodgers RJ (1995) Immunohistochemical localization of basic fibroblast growth factor in bovine ovarian follicles *Molecular and Cellular Endocrinology* **115** 133–140

Wandji S-A, Pelletier G and Sirard M-A (1992a) Ontogeny and cellular localization of ^{125}I-labeled insulin-like growth factor-I, ^{125}I-labeled follicle-stimulating hormone, and ^{125}I-labeled human chorionic gonadotropin binding sites in ovaries from bovine fetuses and neonatal calves *Biology of Reproduction* **47** 814–822

Wandji S-A, Fortier MA and Sirard M-A (1992b) Differential response to gonadotropins and prostaglandin E_2 in ovarian tissue during prenatal and postnatal development in cattle *Biology of Reproduction* **46** 1034–1041

Wandji S-A, Pelletier G and Sirard M-A (1992c) Ontogeny and cellular localization of ^{125}I-labeled basic fibroblast growth factor and ^{125}I-labeled epidermal growth factor binding sites in ovaries from bovine fetuses and neonatal calves *Biology of Reproduction* **47** 807–813

Wandji S-A, Eppig JJ and Fortune JE (1996a) FSH and growth factors affect the growth and endocrine function *in vitro* of granulosa cells of bovine preantral follicles *Theriogenology* **45** 817–832

Wandji S-A, Srsen V, Voss AK, Eppig JJ and Fortune JE (1996b) Initiation *in vitro* of growth of bovine primordial follicles *Biology of Reproduction* **55** 942–948

Wandji S-A, Srsen V, Nathanielsz PW, Eppig JJ and Fortune JE (1997) Initiation of growth of baboon primordial follicles *in vitro. Human Reproduction* **12** 1993–2001

Yokota H, Yamada K, Liu X, Kobayashi J, Abe Y, Mizunuma H and Ibuki Y (1997) Paradoxical action of activin A on folliculogenesis in immature and adult mice *Endocrinology* **138** 4572–4576

Yoshida H, Takakura N, Kataoka H, Kunisada T, Okamura H and Nishikawa S-I (1997) Stepwise requirement of *c*-kit tyrosine kinase in mouse ovarian follicle development *Developmental Biology* **184** 122–137

Journal of Reproduction and Fertility Supplement **54**, 449–460

Aspects of follicular and oocyte maturation that affect the developmental potential of embryos

P. Mermillod, B. Oussaid and Y. Cognié

Institut National de la Recherche Agronomique, Station de Physiologie de la Reproduction des Mammifères Domestiques, 37380 Nouzilly, France

The ability to mature, be fertilized and finally to develop into a viable embryo is acquired gradually by the oocyte during progressive differentiation throughout folliculogenesis. This process starts with oocyte growth during the first steps of follicular development. As the oocyte is close to its final size, other modifications occur, less spectacular but at least as important in determining the resulting ability of the oocyte to accomplish its reproductive purpose (developmental competence). These modifications, referred to as 'oocyte capacitation', are probably influenced by the follicle. The proportion of developmentally competent oocytes increases with follicular size. However, the relationship between follicular growth and oocyte competence is not very strict, since a given oocyte may acquire its competence at any stage of follicular growth and since some examples of functional disjunction between follicular size and oocyte competence are described. Follicular atresia may impair the acquisition of oocyte competence, as evidenced by the parallel study of follicular characteristics and of the developmental potential of their oocytes treated individually through *in vitro* maturation, fertilization and development. However, when atresia is experimentally induced in large preovulatory follicles, oocytes remain competent, indicating that once competence is acquired, it is no longer sensitive to atresia. Oocyte maturation represents only the end of this long and progressive process and validates the preparation of the oocyte by conferring its final developmental ability. As evidenced by recent cloning experiments, the cytoplasmic aspects of oocyte maturation are crucial for the acquisition of developmental competence. This cytoplasmic maturation may be activated *in vitro* by the use of complex media supplement (serum, follicular fluid) but the use of defined media for maturation allowed the identification of some active factors (such as epidermal growth factor, growth hormone, inhibin and activin). The study of some differential models of oocyte competence (follicular size and atresia, Booroola gene, prepubertal oocytes) will provide a better understanding of oocyte capacitation and maturation, and allow the improvement of *in vitro* methods for oocyte maturation, which represent the most limiting step of *in vitro* production of embryos in large mammals.

Introduction

Despite the vast amount of work during the last twenty years aimed at improving *in vitro* production (IVP) of embryos in domestic species, the percentages of *in vitro* matured (IVM), *in vitro* fertilized (IVF) oocytes reaching the blastocyst stage after *in vitro* development (IVD) still reach a plateau at 30–40%. *In vitro* development has been the subject of many studies in recent years and many advances have been made. The use of oviductal cells in coculture first allowed the species-specific block of development to be overcome and produced the first success in IVP. Semi-defined and fully defined embryo culture systems have been devised and used with equal or better success rates than coculture (see Thompson, 1996 for review). After much investigation of embryo culture systems, it now seems more likely that oocyte maturation is the more limiting step in the *in vitro* production of embryos. The second limitation of the use of IVP concerns embryo viability after transfer, especially with frozen–thawed embryos. It appears that the origin of this limitation is in the system used for embryo development (reviews by Massip *et al.*, 1995; Thompson, 1997).

There are differences between oocytes in their ability to be fertilized and to develop to the blastocyst stage in any IVF–IVD system. This ability has been referred to as oocyte quality or oocyte developmental competence. The competence of a mature oocyte depends on at least two factors: (1) oocyte capacitation, that is the preparation of the oocyte during folliculogenesis and especially during the later phases of follicular growth (Hyttel *et al.*, 1997), and (2) the morphological and biochemical modifications of oocytes taking place during maturation, after the LH surge or after removal of oocytes from the inhibitory effects of the follicle. These two aspects of oocyte quality do not have the same consequences for IVP. It is possible to adapt IVM conditions to improve the quality of oocyte maturation and several advances have been made during the last decade. However, improving oocyte capacitation is more difficult, because oocytes resume meiosis *in vitro* and capacitation ceases as soon as they are removed from the follicle. The follicular environment is probably one of the most important parameters regulating capacitation. Until now, no well established culture system for germinal vesicle stage oocytes has been available for large domestic animals. Consequently, the improvement of the quality of oocytes used for IVM depends upon the manipulation of the physiology of the whole animal to increase the number of follicles containing competent oocytes. Much work has been devoted to the isolation and *in vitro* development of small preantral follicles and some success have been achieved in rodent species (review by Eppig *et al.*, 1996). In larger animals it has been shown that a large number of preantral follicles can be obtained from ovaries (Nuttinck *et al.*, 1993). However, a better understanding of oocyte differentiation and interactions between oocyte and follicular cells is required before efficient systems can be set up to produce fully grown and competent oocytes from preantral or small antral follicles, even though some encouraging results have been obtained by Harada *et al.* (1997).

The first step in the acquisition of developmental competence is oocyte growth at the start of follicular development. This important differentiation process will not be considered here, but good reviews of mechanisms and regulations involved include those by Eppig *et al.* (1996), Gosden *et al.* (1997) and Hyttel *et al.* (1997).

In this review follicular factors that affect the capacitation and maturation of oocytes and their resulting ability to develop into viable embryos are discussed.

Oocyte Capacitation

Acquisition of oocyte developmental competence occurs continuously throughout folliculogenesis (Fig. 1). This acquisition can be divided in three separate stages defined by particular physiological events: (1) oocyte growth, which takes place mainly during the beginning of follicle emergence (primary and secondary preantral follicles); (2) oocyte capacitation, starting at the end of oocyte growth in antral (tertiary) follicles; and (3) oocyte maturation, starting after the LH surge in preovulatory follicles or after removal of the oocyte from the follicular environment which inhibits meiotic resumption.

Among a batch of oocytes collected from small to medium size follicles of cattle, sheep or goat ovaries at an abbatoir, more than 90% will reach metaphase II of meiosis after IVM; more than 70% will be successfully fertilized (two pronuclei after IVF) and cleave; but only a third will develop to the morula–blastocyst stage after IVD (Fig. 2). This finding indicates that among a batch of oocytes, a large number are unable to develop normally beyond the first cleavage. This is a common observation in different species and in different IVP systems (see review by Mermillod *et al.*, 1996). The explanation for this phenomenon becomes apparent when it is considered that oocyte developmental competence is continuously increasing during folliculogenesis and is also regulated by other characteristics of follicular physiology: oocytes collected from the heterogeneous follicular population of ovaries from an abattoir are of different developmental ability.

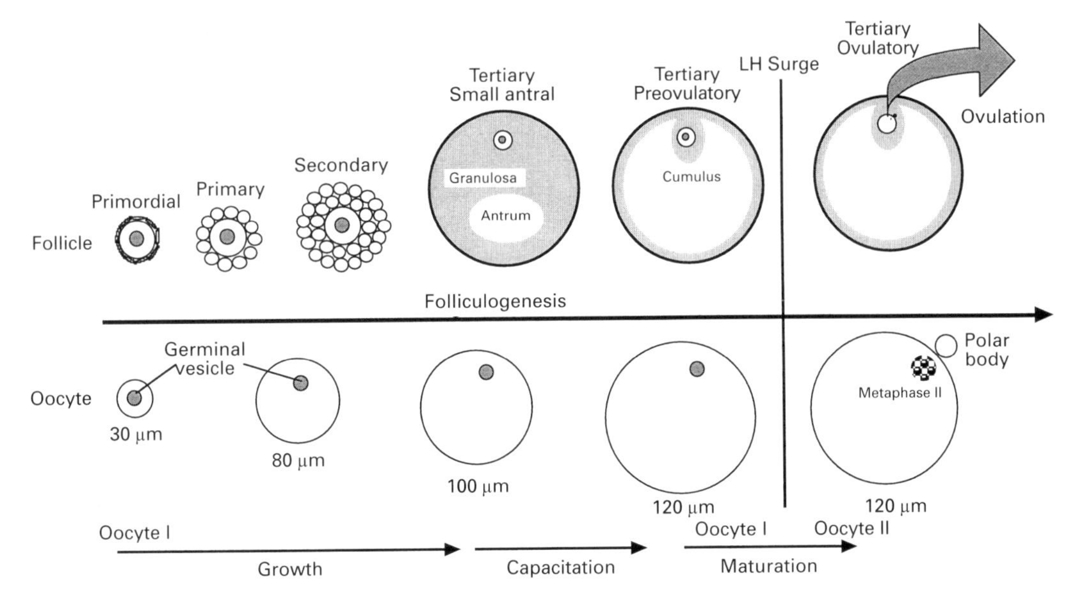

Fig. 1. Schematic representation of oocyte growth, capacitation and maturation along the axis of folliculogenesis. Oocyte growth is initiated as soon as the follicle enters the pool of growing follicles and is almost completed upon antrum formation even if some increase in size is observed up to ovulation. Other modifications take place in an antral follicle during oocyte capacitation, conferring developmental competence to the oocyte. Maturation, after the LH surge or removal of oocyte from the inhibitory environment of the follicle, is necessary to prepare the haploid chromosome complement of oocyte II and to express the developmental potential acquired during capacitation.

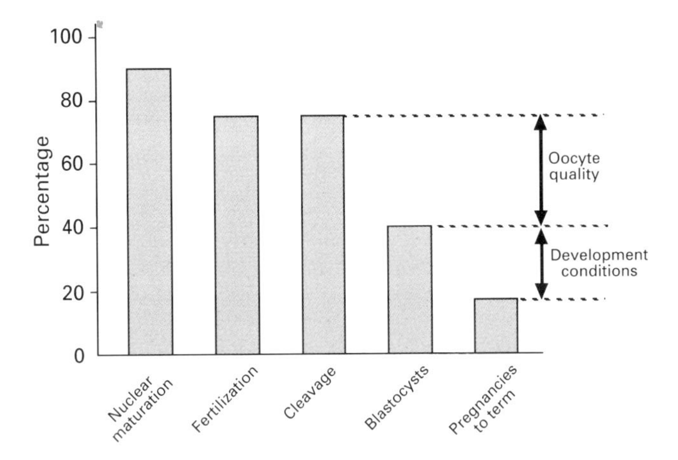

Fig. 2. Successive steps of embryo production *in vitro*. Meiotic maturation occurs at very high rates as do fertilization and cleavage of early embryos in three domestic ruminant species (cattle, sheep and goat). The first decrease in success is observed at the blastocyst stage and may be attributed to the low oocyte capacitation or suboptimal conditions used for IVM. A second decrease is observed in the pregnancy rate obtained after transfer of *in vitro* produced embryos (particularly after a cryopreservation step) that could be attributed to the conditions used for embryo development.

Follicle size

The first indication of the presence of oocyte capacitation was obtained by evaluating the developmental potential of immature oocytes collected from follicles of known size (cattle: Pavlok *et al.*, 1992; Lonergan *et al.*, 1994a; goat: Crozet *et al.*, 1995; and sheep: Cognié *et al.* 1998). When follicles are dissected and classified according to their size (size classes are different between species), the oocytes harvested from larger follicles provide better development results than those from smaller follicles. Consequently, large follicles contain a higher proportion of developmentally competent oocytes. Since 'blastocyst rate' is measured at the end of IVP and some blastocysts are observed even when oocytes are harvested from smaller follicles, it is more likely that oocyte capacitation is occurring with different kinetics between follicles. If we consider the smaller follicles, only a few oocytes appear competent (the faster oocytes), whereas in larger follicles, even more slowly developing oocytes have had time to complete capacitation.

Thus, the question is: what determines the kinetics of oocyte capacitation during folliculogenesis? Capacitation could be regulated by the oocyte itself or by follicular characteristics other than size. In this view, it is of interest to note that for an equal follicular size, oocytes from peripheral follicles (follicles visible at the surface of the ovary) are larger and have a higher meiotic and developmental competence than oocytes from follicles localized in the ovarian cortex (Arlotto *et al.*, 1996). This result indicates that the timing of capacitation of an oocyte could differ in accordance with the location of its follicle.

Another example of functional disjunction between follicular growth and acquisition of meiotic and developmental ability by the oocyte is found in ewes bearing the Booroola fecundity gene (Bindon, 1984). It has been shown that in heterozygous ewes (Fec^B Fec^+), the rate of development of oocytes from a given follicular size class is superior to the rate observed for oocytes obtained from the same size class in females of wild (Fec^+ Fec^+) genotype (Cognié *et al.*, 1998). This difference in the timing of acquisition of competence is reflected earlier since, even in preantral follicles, oocytes from ewes bearing the Booroola gene are larger than oocytes from wildtype animals at the same follicular stage. As a consequence, a single gene may be able to regulate the kinetics of the capacitation of the oocyte inside the growing follicle. The identification of this gene will improve understanding of oocyte capacitation.

Considering these facts, it is clear that some important but unknown events leading to acquisition of developmental competence take place in oocytes at variable stages of follicular development. Oocytes that do not complete this preparation are not competent to develop even if they are already able to resume meiosis and to be fertilized. The morphological and biochemical basis of this capacitation remains to be determined; however, it probably involves the transcription of some genes important for survival or for the regulation of gene expression in early embryos. The reconstitution *in vitro* of this capacitation step will require the control of meiotic inhibition to allow the oocyte to complete transcription of necessary genes before chromosome condensation. Knowledge of the mechanisms of meiotic inhibition is not yet sufficient for the establishment of culture systems for germinal vesicle stage oocytes even though some encouraging results have been obtained (Lonergan *et al.*, 1997; Van Tol *et al.*, 1996).

To summarize, it appears that the proportion of competent oocytes increases with follicular size. Oocyte capacitation (that is, the ability to accomplish the cytoplasmic aspects of maturation successfully) may be acquired by the oocyte at any time during growth of antral follicles. The mechanisms and the follicular regulation of capacitation are unknown.

Follicular atresia

Follicular degeneration, or atresia, is the more probable destiny of any given follicle. A high proportion of follicles present at the surface of ovaries are at various stages of atresia (review by Monniaux *et al.*, 1997). Consequently, it is important to know to what extent atresia could influence the acquisition of developmental competence by the oocyte and to determine whether oocytes from atretic follicles could be rescued during *in vitro* culture when they would have been lost normally during normal degeneration processes *in vivo* (review by Sirard and Blondin, 1996).

The main problem encountered in the study of the effect of atresia on oocyte developmental competence is that no macroscopic criteria are available for selecting and classifying follicles according to their health status. The criteria usually used for the determination of atresia are biochemical: steroid content of the follicular fluid (Grime and Ireland, 1986), or pattern of expression of insulin-like growth factor binding proteins (IGFBP, Monget *et al.*, 1993). Atresia may also be evaluated by the rate of mitosis–pycnosis or apoptosis in granulosa cells (review by Monniaux *et al.*, 1997). All these methods are time consuming and thus do not allow classification and pooling of oocytes before IVM according to the stage of atresia of the follicles they come from. Consequently, the parallel study of atresia and oocyte competence requires the use of IVM, IVF, IVD methods designed for individual oocytes, whereas oocytes are usually treated in groups and cooperation among oocytes or embryos has been reported (Ferry *et al.*, 1994; Blondin and Sirard, 1995). However, it appeared that IVP techniques could be successfully adapted to individual oocytes or embryos (Carolan *et al.*, 1996a) and some results in the comparative study of follicular characteristics, morphology of cumulus–oocyte complexes and oocyte developmental competence using IVM, IVF and IVD systems designed for individual oocytes have been obtained recently in cattle. These results are slightly controversial. For example, in one study (Hazeleger *et al.*, 1995), follicles containing highly competent oocytes were characterized by lower progesterone concentration in follicular fluid, whereas oestradiol was not affected. In a more recent study (Driancourt *et al.*, 1998), the wall of follicles containing oocytes able to develop to the blastocyst stage had a higher aromatase activity compared with follicles containing oocytes unable to develop beyond the 8–16-cell stage (Fig. 3), although the steroid content of follicular fluid was not affected. These authors also reported a higher concentration of inhibin (α subunit) in the fluid of follicles containing developmentally competent oocytes. This finding is of interest considering the positive effect of inhibin–activin on oocyte maturation (see below) and considering that the concentration of inhibin α subunit is strongly reduced during follicular regression caused by atresia (Guilbault *et al.*, 1993). In conclusion, the clear relationship between follicular atresia and acquisition of developmental competence by an oocyte is not fully established, although some data indicate that competent oocytes are more frequently found in physiologically active follicles. This is not surprising as it is known that developmental competence may be acquired by oocytes at any time during follicular development. The occurrence of atresia before that time could deprive the slower oocytes of any chance of gaining this competence (Fig. 4).

Once an oocyte has reached its full developmental competence, it appears less dependent upon follicular health status. To test this hypothesis, we maintained preovulatory follicles for increasing periods after FSH superovulation of synchronized heifers treated with the GnRH antagonist Antarelix (Gift from Europeptides, Dr Deghenghi, Argenteuil, France) to inhibit the ovulatory surge of LH (Oussaid, Lonergan and Mermillod unpublished data). FSH treatment consisted of a total of 24 mg pFSH (kindly provided by J-F Beckers, University of Liège) injected twice a day for 4 days in a regimen of decreasing doses. Animals received 1 ml Prosolvin (Intervet, Boxmeer) on the last day of FSH treatment. Heifers were divided among three treatments : (1) control group, killed 24 h after the last FSH injection ($n = 4$); (2) group 36 h, killed 36 h after the control group and injected with Antarelix every 12 h (1.6 mg) starting 24 h after the last FSH injection ($n = 4$); and (3) group 60 h, killed 60 h after the control group and injected with Antarelix in the same way ($n = 4$). Ovaries were collected at slaughter and all follicles larger than 5 mm present at their surface were dissected free of surrounding tissues and opened with a scalpel. The oocyte was retrieved; a sample of follicular fluid was collected and centrifuged for measurements of steroids (radioimmunoassay after extraction, Thibier and Saumande, 1975) and some granulosa cells were smeared on to a microscope slide for evaluation of cell viability (mitosis and pycnosis frequencies after Feulgen staining). Oocytes were treated individually through IVM, IVF, IVD protocols previously described for individual oocytes (Carolan *et al.*, 1996a) and the developmental stage reached by each oocyte was scored as well as the number of cells in the resulting embryo (Hoechst fluorescent staining after ethanol fixation). The maintenance of preovulatory follicles was detrimental to follicular health, inducing a high rate of atresia as evidenced by the pycnotic index of granulosal cells (less than 10% of follicles showing pycnotic bodies in the control group compared with 58 and 90% in the 36 and 60 h groups,

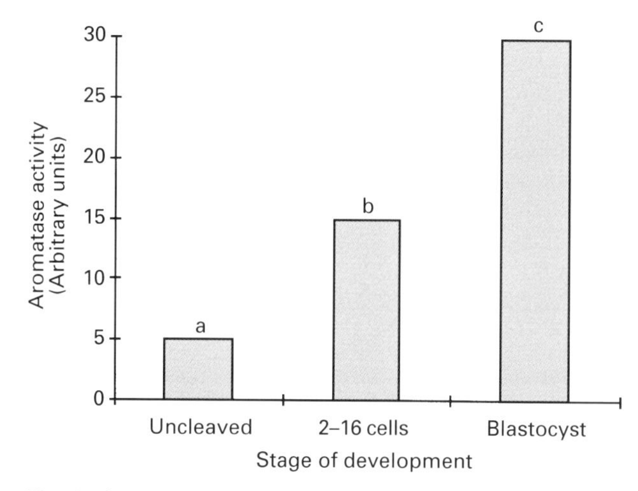

Fig. 3. Aromatase activity in the wall of follicles containing oocytes that are unable to cleave, blocked at early cleavage stage or developing to the blastocyst stage in individual conditions (MA Driancourt, B Thuel, P Mermillod and P Lonergan, unpublished). [a,b,c] Columns with different letters are significantly different ($P < 0.05$, Chi square).

respectively) and a decrease in the concentrations of oestradiol and progesterone. This result was observed after 36 h and increased after 60 h (Table 1). However, the developmental potential of the oocytes collected from these advanced atretic follicles was not affected and the quality of the resulting embryos was not affected, as indicated by the blastocyst rate and number of cells (Table 2).

To summarize these observations, it appears that atresia occurring in small–medium size follicles may be detrimental to the developmental potential of the oocytes, at least when physiological characteristics of the follicle are affected (for example aromatase activity). In large preovulatory follicles, where most oocytes are already fully competent, atresia no longer affects competence.

Oocyte Maturation

The preovulatory LH surge or removal of the oocyte from its follicular inhibitory environment triggers the resumption of meiosis, the most visible feature of oocyte maturation. The meiotic process confers on the secondary oocyte its final haploid DNA complement. Meiosis and its molecular regulation have been studied and described extensively (for reviews see Wassarman and Albertini, 1994; Downs, 1996; Taieb *et al.*, 1997). Beyond these nuclear aspects of oocyte maturation, cytoplasmic events also occur and seem important for the fertilization and development ability of the oocytes (Eppig, 1996). These aspects have been termed cytoplasmic maturation as opposed to the meiotic events called nuclear maturation. Some ultrastructural and biochemical features of cytoplasmic maturation have been described (Wassarman and Albertini, 1994) but the events determining the final quality of the mature oocyte remain to be identified. Recent cloning experiments using fully differentiated donor nuclei have underlined the power of oocyte cytoplasm in reprogramming the nucleus for complete development (Campbell and Wilmut, 1997). Other cloning experiments also indicated the importance of cytoplasmic maturation in determining developmental potential. When enucleated oocytes of different origins are used as recipients for the same batch of embryonic nuclei, they develop at different rates. This has been shown by comparing enucleated oocytes from adult and prepubertal cattle (Mermillod *et al.*, 1998) and from different classes of follicular size (Kubota and Yang, 1998).

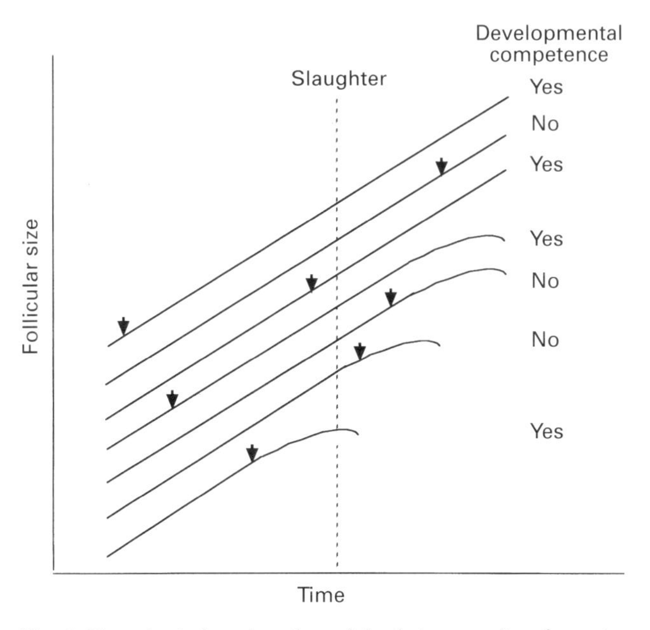

Fig. 4. Hypothetical explanation of the heterogeneity of oocytes collected from a particular ovary. Oocyte developmental competence can appear at any stage of antral follicle development, thus competent oocytes are recovered more frequently from large follicles. As a consequence, some oocytes from large follicles are still incompetent and some oocytes from small follicles are already competent. Atresia develops at any stage of follicular growth. When it appears in large follicles, the oocyte is often already competent and remains competent during the first steps of follicle regression. When atresia occurs in smaller follicles, the oocyte is rarely already competent and hence capacitation is stopped before competence is acquired. Arrows represent acquisition of competence and atresia is represented by inflexion of the curve.

Nuclear maturation occurs spontaneously *in vitro* in most oocytes and large numbers of mature (metaphase II) oocytes can be obtained in very simple culture conditions (Lonergan *et al.*, 1994b). Even oocytes from small follicles are able to mature properly, and most of them show normal haploid metaphase II after IVM (90%, follicles of 1–3 mm in diameter in cattle, Ectors *et al.*, 1995). Cytoplasmic maturation, on the other hand, is dependent on conditions used for IVM and could be stimulated by addition of growth factors to the culture medium. *In vivo*, follicular environment could regulate cytoplasmic maturation as evidenced by experiments of Moor *et al.* (1996), showing that the first 6 h spent by the oocyte inside the LH-stimulated follicle in sheep are important for its developmental competence.

Follicular fluid used as a supplement in maturation medium is a potent activator of oocyte maturation, at least in the cytoplasm. This positive effect of follicular fluid seems to be independent of follicular size. However, it seems to be influenced by the stage of atresia (sheep: Cognié *et al.*, 1995; cattle: Carolan *et al.*, 1996b) as well as by the stage of follicular development (growing or regressing follicles, Sirard *et al.*, 1995). Follicular fluid seems thus to contain maturation-activating factors from early stages of follicular growth. The hormonal environment may act by regulating the effect of such factors on the cumulus–oocyte complexes. The molecular effector(s) mediating this positive influence of the follicular environment are not yet fully elucidated. However, immunohistochemical studies of the ovary as well as *in vitro* maturation experiments have allowed the identification of some factors in follicles (ligand and receptor) and of their effect on *in vitro* maturation.

Table 1. Concentrations of oestradiol and progesterone (ng ml^{-1} ± SD) in follicular fluids of superovulated heifers slaughtered 24 h after the last FSH injection (control), 36 h later with injection of 1.6 mg Antarelix every 12 h (36 h) or 60 h later with Antarelix in the same way (60 h)

	Treatment		
	Control (n = 54)	36 h (n = 64)	60 h (n = 52)
Oestradiol	788.3 ± 39.3^{a}	17.3 ± 1.2^{b}	4.0 ± 0.5^{c}
Progesterone	736.7 ± 46.2^{a}	54.8 ± 3.5^{b}	54.3 ± 3.5^{b}

^{a,b,c}Values in the same row with different superscripts are significantly different ($P < 0.05$, t test).
n = number of follicles obtained from the four heifers in each group.

Table 2. Development of bovine oocytes after individual IVM/IVF/IVC of oocytes recovered from superovulated heifers killed 24 h after the last FSH injection (control) or maintained for a further 36 or 60 h with administration of Antarelix to inhibit the LH surge before slaughter

			Development 8 days pi		
Treatment	n	Cleavage at 72 h pi n (%)	Morula–blastocysts(%)	Blastocysts (%)	Cell number x ± SEM
Control	54	46 (85)	22 (41)^{a}	22 (41)	95 ± 8
36 h	64	55 (86)	39 (61)^{b}	36 (56)	93 ± 5
60 h	52	46 (88)	30 (58)^{ab}	30 (58)	79 ± 4

^{a,b}Significantly different ($P < 0.05$, Chi square); pi, post insemination.

Gonadotrophins (FSH and LH) are widely used as supplements in maturation media and their action on cumulus expansion, meiotic resumption and cytoplasmic maturation has been shown in different systems (reviewed by Gordon, 1994). However, effects of gonadotrophins *in vitro* remain controversial in cattle, possibly due to the diversity of the preparations and of the concentrations used by different workers (reviewed by Bevers *et al.*, 1997). It is still not clear whether gonadotrophins have direct effects on cumulus and oocyte, or potentiate the action of growth factors present in follicular fluid (review by Gandolfi, 1996).

The LH peak triggers a change in the steroid content of follicular fluid. Consequently, the meiotic process occurs in a changing steroid environment with a progressive decrease in oestradiol concentration and increase in progesterone. However, the possible effect of steroids on meiotic resumption or on cytoplasmic maturation remains controversial in domestic mammals (review by Bevers *et al.*, 1997), although oestradiol is frequently added to maturation media. Oestradiol has been shown to increase the developmental potential of maturing human oocytes without affecting meiotic progression (Tesarik and Mendoza, 1995). In this case, oestradiol acts directly at the plasma membrane (non-genomic effect) by allowing influx of calcium ions, initiating several calcium oscillations in the ooplasm.

Some recent data support the possible intervention of growth factors, hormones and intraovarian peptides (review by Bevers *et al.*, 1997). The effect of epidermal growth factor has been investigated in numerous species including some domestic ruminants. A role for EGF in both cytoplasmic and nuclear oocyte maturation as well as in development of the preimplantation embryo has been identified in cattle (Lorenzo *et al.*, 1995; Lonergan *et al.*, 1996). The addition of 10 ng EGF ml^{-1} to M199 medium resulted in increased rates of development that were similar to the addition of 10% fetal calf serum. The presence of EGF in follicular fluid as well as intrafollicular

EGF-binding sites have been reported in numerous species and the expression of this factor and of its receptor seems to be under the control of both gonadotrophins and steroids (see Lonergan *et al.*, 1996 for references). Taken together, these observations indicate a central function for EGF or related molecules in the mediation of intrafollicular regulation of oocyte capacitation and maturation.

The insulin-like growth factors (IGF-I and IGF-II) and related proteins (IGF-binding proteins and IGFBP-specific proteases) are suspected to play a major role in intra-ovarian regulation of follicular growth and differentiation (review by Monniaux *et al.*, 1997). The action of IGFs on oocyte maturation has been widely investigated but remains controversial and IGF activity seems to be dependent on the presence of other factors. Growth hormone (GH) is known to improve the superovulatory response in cattle. GH is thought to exert its action by regulating expression of IGFs in various organs including the ovary. GH has recently been shown to accelerate oocyte maturation in cattle and to increase the developmental potential of mature oocytes (Izadyar *et al.*, 1996). However, this effect is not mediated by IGFs since the effect of GH can be obtained in the presence of anti-IGF-I antibody. Furthermore, bovine oocytes as well as cumulus cells express GH receptor and, consequently, are able to respond directly to GH stimulation (Izadyar *et al.*, 1997). A positive effect of the inhibin–activin family of ovarian glycoproteins on oocyte developmental competence has been reported recently in cattle (Stock *et al.*, 1997). Human recombinant inhibin A and activin A added to defined IVM medium enhanced the development of cleaved bovine embryos to the blastocyst stage alone (10 ng ml^{-1}) or in combination (1 ng ml^{-1} of each) without affecting the cleavage rate. Thus inhibin and activin appear to increase the quality of oocyte cytoplasmic maturation. This effect could be mediated through an action on meiotic resumption and progression since nuclear maturation was advanced by the addition of activin, inhibin or both factors, whereas the final metaphase II rate was not affected. A possible link between the kinetics of progression of meiotic events and the resulting developmental ability of the oocytes has already been reported in cattle (Dominko and First, 1997), indicating that these two events (nuclear and cytoplasmic maturation) are strongly correlated and might be regulated by common factors such as the activin–inhibin system. Inhibin and activin may act directly on oocytes, because their action does not involve cumulus expansion (Stock *et al.*, 1997) and because they are able to stimulate the cytoplasmic maturation of denuded oocytes (Silva and Knight, 1998). In addition, the expression of the activin receptor has been detected by RT–PCR in oocytes as well as in cumulus cells (Izadyar *et al.*, 1998). Inhibin and activin are present in follicular fluid and their concentration is increased by the time of meiotic resumption in pig follicles (Miller *et al.*, 1991), indicating that they might be involved in the regulation of meiosis.

To summarize, oocyte maturation includes nuclear (meiotic resumption and progression from prophase I to metaphase II) and cytoplasmic events. Nuclear maturation is spontaneous when the follicular inhibitory signal is suppressed (LH surge or *in vitro* culture). However, some oocytes need stimulation to complete meiosis. More oocytes are dependent upon external signalling for completion of cytoplasmic maturation. EGF seems to be one of the most potent factors stimulating both aspects of maturation, but other factors (GH, activin, inhibin) contained in follicular fluid may also regulate the different aspects of oocyte maturation.

Criteria of Oocyte Quality

Some of the events related to oocyte capacitation and maturation are known. However, we still lack knowledge of real markers strictly linked to the final developmental competence of the oocyte, except the observation of their development after IVF. Such markers would be helpful in investigating the conditions of *in vitro* maturation or to test treatments on animals to increase the occurrence of follicles containing competent oocytes. It would therefore be of interest to determine some morphological or biochemical characteristics of highly competent oocytes, of their surrounding cells and of follicles containing such oocytes.

The first visible characteristic that might be linked to the developmental competence is the morphology of the cumulus–oocyte complex (COC) obtained after follicular aspiration. Most of the IVP laboratories use the same morphological evaluation and selection of the COCs entering IVM based on the appearance of the cumulus (several layers, continuous, compact) and of the ooplasm (Leibfried and First, 1979). However, this morphological evaluation is not sufficient to predict the developmental potential of a given oocyte and the identification of markers more strictly linked to this potential would be helpful.

The discovery of such markers of oocyte quality requires some experimental models providing highly and poorly competent oocytes in a reproducible way. Some follicular characteristics such as follicular diameter or stage of atresia could provide comparative models of developmental competence as could some genotype particularities in sheep, such as the Booroola gene (see above). In addition, oocytes originated from prepubertal animals provide negative models of developmental competence in cattle (Revel *et al.*, 1995; Khatir *et al.*, 1996), sheep (O'Brien *et al.*, 1997) and goat (Martino *et al.*, 1994). These oocytes can resume meiosis, reach metaphase II, be fertilized and cleave in the same proportion as their adult counterparts, but their ability to develop to the blastocyst stage is strongly reduced. The same discrepancy was observed when nuclear transfer was used instead of IVF in cattle, underlining the cytoplasmic origin of the deficiency of prepubertal oocytes (Mermillod *et al.*, 1998).

The differential analysis of oocytes from these models with different approaches (such as morphology, ultrastructure, protein synthesis and phosphorylation and expression of receptors) has already provided some clues toward a better comprehension of oocyte quality and will certainly allow the identification of some markers of oocyte quality that will help to improve *in vitro* embryo production in future.

Conclusion

In conclusion, it is clear that oocyte quality is decisive for the success of *in vitro* production of embryos from domestic ruminants. Oocyte quality is the consequence of both oocyte capacitation during the course of folliculogenesis and oocyte maturation after the LH surge *in vivo* or after *in vitro* culture during IVM. An appropriate follicular environment is essential to facilitate these two aspects of oocyte differentiation.

References

Arlotto T, Schwartz J-L, First NL and Leibfried-Rutledge ML (1996) Aspects of follicle and oocyte stage that affect *in vitro* maturation and development of bovine oocytes *Theriogenology* **45** 943–956

Bevers MM, Dieleman SJ, van den Hurk R and Izadyar F (1997) Regulation and modulation of oocyte maturation in the bovine *Theriogenology* **47** 13–22

Bindon BM (1984) Reproductive biology of the Booroola Merino sheep *Australian Journal of Biological Science* **37** 163–189

Blondin P and Sirard MA (1995) Oocyte and follicular morphology as determining characteristics for developmental competence in bovine oocytes *Molecular Reproduction and Development* **41** 54–62

Campbell KHS and Wilmut I (1997) Totipotency and multipotentiality of cultured cells: applications and progress *Theriogenology* **47** 63–72

Carolan C, Lonergan P, Khatir H and Mermillod P (1996a) The *in vitro* production of bovine embryos using individual oocytes *Molecular Reproduction and Development* **45** 145–154

Carolan C, Lonergan P, Monget P, Monniaux D and Mermillod P (1996b) Effect of follicle size and quality on the ability of follicular fluid to support cytoplasmic maturation of bovine oocytes *Molecular Reproduction and Development* **43** 477–483

Cognié Y, Poulin N, Pisselet C and Monniaux D (1995) Effect of atresia on the ability of follicular fluid to support cytoplasmic maturation of sheep oocytes *in vitro*. *Theriogenology* **43** 188 (Abstract)

Cognié Y, Benoit F, Poulin N, Khatir H and Driancourt MA (1998) Effect of follicle size and of the FecB booroola gene on oocyte function in sheep *Journal of Reproduction and Fertility* **112** 379–386

Crozet N, Ahmed-Ali M and Dubos MP (1995) Developmental competence of goat oocytes from follicles of different size categories following maturation, fertilization and culture *in vitro*. *Journal of Reproduction and Fertility* **103** 293–298

Dominko T and First NL (1997) Timing of meiotic progression in bovine oocytes and its effect on early embryo development *Molecular Reproduction and Development* **47** 456–467

Downs SM (1996) Regulation of meiotic arrest and resumption in mammalian oocytes. In *The Ovary: Regulation, Dysfunction And Treatment* pp 141–148 Ed. M Filicori and C. Flamigni. Elsevier, Amsterdam

Driancourt MA, Thuel B, Mermillod P and Lonergan P (1998) Relationship between oocyte quality (measured after IVM, IVF and IVC of individual oocytes) and follicle function in cattle *Theriogenology* **49** 345 (Abstract)

Ectors FJ, Koulischer L, Jamar M, Herens C, Verloes A, Remy B and Beckers JF (1995) Cytogenetic study of bovine oocytes matured *in vitro*. *Theriogenology* **44** 445–450

Eppig JJ (1996) Coordination of nuclear and cytoplasmic oocyte maturation in eutherian mammals *Reproduction Fertility and Development* **8** 485–489

Eppig JJ, O'Brien M and Wiggleworth K (1996) Mammalian oocyte growth and development *in vitro*. *Molecular Reproduction and Development* **44** 260–273

Ferry L, Mermillod P, Massip A and Dessy F (1994) Bovine embryos cultured in serum-poor conditioned medium need cooperation to reach the blastocyst stage *Theriogenology* **42** 445–453

Gandolfi F (1996) Intra-ovarian regulation of oocyte developmental competence in cattle *Zygote* **4** 323–326

Gordon I (1994) *Laboratory Production of Cattle Embryos* CAB International, Oxon

Gosden R, Krapez J and Briggs D (1997) Growth and development of the mammalian oocyte *BioEssays* **19** 875–882

Grime RW and Ireland JJ (1986) Relationship of macroscopic appearance of the surface of bovine ovarian follicles, concentrations of steroids in follicular fluid, and maturation of oocytes *in vitro*. *Biology of Reproduction* **35** 725–732

Guilbault LA, Rouillier P, Matton P, Glencross RG, Beard AJ and Knight PG (1993) Relationships between the level of atresia and inhibin contents (alpha subunit and ab dimer) in morphologically dominant follicles during their growing and regressing phases of development in cattle *Biology of Reproduction* **48** 268–276

Harada M, Miyano T, Matsumura K, Osaki S, Miyake M and Kato S (1997) Bovine oocytes from early antral follicles grow to meiotic competence *in vitro*: effect of FSH and hypoxanthine *Theriogenology* **78** 743–755

Hazeleger NL, Hill DJ, Stubbings RB and Walton JS (1995) Relationship of morphology and follicular fluid environment of bovine oocytes to their developmental potential *in vitro*. *Theriogenology* **43** 509–522

Hyttel P, Fair T, Callesen H and Greve T (1997) Oocyte growth, capacitation and final maturation in cattle *Theriogenology* **47** 23–32

Izadyar F, Colenbrander B and Bevers MM (1996) *In vitro* maturation of bovine oocytes in the presence of growth hormone accelerates nuclear maturation and promotes subsequent embryonic development *Molecular Reproduction and Development* **45** 372–377

Izadyar F, Van Tol HTA, Colenbrander B and Bevers MM (1997) Stimulatory effect of growth hormone on *in vitro* maturation of bovine oocytes is exerted through cumulus cells and not mediated by IGF-I *Molecular Reproduction and Development* **47** 175–180

Izadyar F, Colenbrander B and Bevers MM (1998) Stimulatory effect of growth hormone on *in vitro* maturation of bovine oocytes is exerted through the cyclic adenosine 3′,5′-monophosphate signaling pathway *Biology of Reproduction* **57** 1484–1489

Khatir H, Lonergan P, Carolan C and Mermillod P (1996) The prepubertal oocytes as a negative model in the study of bovine oocyte developmental competence acquisition *Molecular Reproduction and Development* **45** 231–239

Kubota C and Yang X (1998) Cytoplasmic incompetence results in poor development of bovine oocytes derived from small follicles *Theriogenology* **47** 183 (Abstract)

Leibfried L and First NL (1979) Characterization of bovine follicular oocytes and their ability to mature *in vitro*. *Journal of Animal Science* **48** 76–86

Lonergan P, Monaghan P, Rizos D, Boland MP and Gordon I (1994a) Effect of follicle size on bovine oocyte quality and developmental competence following maturation, fertilization, and culture *in vitro*. *Molecular Reproduction and Development* **37** 48–53

Lonergan P, Carolan C and Mermillod P (1994b) Development of bovine embryos *in vitro* following oocyte maturation under defined conditions *Reproduction, Nutrition and Development* **34** 329–339

Lonergan P, Carolan C, Van Langendonckt A, Donnay I, Khatir H and Mermillod P (1996) Role of epidermal growth factor in bovine oocyte maturation and preimplantation embryo development *Biology of Reproduction* **54** 1412–1421

Lonergan P, Khatir H, Carolan C and Mermillod P (1997) Bovine blastocyst production *in vitro* following inhibition of oocyte meiotic resumption for 24 h *Journal of Reproduction and Fertility* **109** 355–365

Lorenzo PL, Illera MJ, Illera JC and Illera M (1995) Role of EGF, IGF-I, sera and cumulus cells on maturation *in vitro* of bovine oocytes *Theriogenology* **44** 109–118

Martino A, Mogas T, Palomo MJ and Paramio MT (1994) Meiotic competence of prepubertal goat oocytes *Theriogenology* **41** 969–980

Massip A, Mermillod P and Dinnyes A (1995) Morphology and biochemistry of *in-vitro* produced bovine embryos: implications for their cryopreservation *Human Reproduction* **10** 3004–3011

Mermillod P, Lonergan P, Carolan C, Khatir H, Poulin N and Cognié Y (1996) Oocyte maturation in domestic ruminants *Contraception Fertilité Sexualité* **24** 552–558

Mermillod P, Peynot N, Lonergan P, Khatir H, Driancourt MA, Renard JP and Heyman Y (1998) Developmental potential of oocytes collected from 8–15 day old unstimulated or FSH treated calves *Theriogenology* **47** 294 (Abstract)

Miller KF, Xie S and Pope W (1991) Immunoreactive inhibin in follicular fluid is related to meiotic stage of the oocyte during final maturation of the porcine follicle *Molecular Reproduction and Development* **28** 35–39

Monget P, Monniaux D, Pisselet C and Durand P (1993) Changes in insulin-like growth factor-I (IGF-I), IGF-II, and their binding proteins during growth and atresia of ovine ovarian follicles *Endocrinology* **132** 1438–1446

Monniaux D, Huet C, Besnard N, Clément F, Bosc M, Pisselet C, Monget P and Mariana JC (1997) Follicular growth and ovarian dynamics in mammals *Journal of Reproduction and Fertility Supplement* **51** 3–23

Moor RM, Lee C, Dai YF and Fulka J, Jr (1996) Antral follicles confer developmental competence on oocytes *Zygote* **4** 289–293

Nuttinck F, Mermillod P, Massip A and Dessy F (1993) Characterization of *in vitro* growth of bovine preantral ovarian follicles: a preliminary study *Theriogenology* **39** 811–821

O'Brien JK, Catt SL, Ireland KA, Maxwell WMC and Evans G (1997) *In vitro* and *in vivo* developmental capacity of oocytes from prepubertal and adult sheep *Theriogenology* **47** 1433–1443

Pavlok A, Lucas-Hahn A and Niemann H (1992) Fertilization and developmental competence of bovine oocytes derived from different categories of antral follicles *Molecular Reproduction and Development* **31** 63–67

Revel F, Mermillod P, Peynot N, Renard JP and Heyman Y (1995) Comparison of developmental ability of oocytes from prepubertal calves and adult cows *Journal of Reproduction and Fertility* **103** 115–120

Silva CC and Knight PG (1998) Modulatory actions of activin-A and follistatin on the developmental competence of *in vitro*-matured bovine oocytes *Biology of Reproduction* **58** 558–565

Sirard MA and Blondin P (1996) Oocyte maturation and IVF in cattle *Animal Reproduction Science* **42** 417–426

Sirard MA, Roy F, Mermillod P and Guilbault LA (1995) The origin of follicular fluid added to the media during bovine IVM influences embryonic development *Theriogenology* **44** 85–94

Stock AE, Woodruff TK and Smith LC (1997) Effects of inhibin A and activin A during *in vitro* maturation of bovine oocytes in hormone- and serum-free medium *Biology of Reproduction* **56** 1559–1564

Taieb F, Thibier C and Jessus C (1997) On cyclins, oocytes, and eggs *Molecular Reproduction and Development* **48** 397–411

Tesarik J and Mendoza C (1995) Nongenomic effects of 17 beta-estradiol on maturing human oocytes: relationship to oocyte developmental potential *Journal of Clinical Endocrinology and Metabolism* **80** 1438–1443

Thibier M and Saumande J (1975) Oestradiol-17ß, progesterone and 17alpha-hydroxyprogesterone concentration in jugular venous plasma in cows prior to and during oestrus *Journal of Steroid Biochemistry* **6** 1433–1437

Thompson JG (1996) Defining the requirements for bovine embryo culture *Theriogenology* **45** 27–40

Thompson JG (1997) Comparison between *in vivo*-derived and *in vitro*-produced pre-elongation embryos from domestic ruminants *Reproduction Fertility and Development* **9** 341–354

Van Tol HTA, Van Eijk MJT, Mummery CL, Van Den Hurk R and Bevers MM (1996) Influence of FSH and hCG on the resumption of meiosis of bovine oocytes surrounded by cumulus cells connected to membrana granulosa *Molecular Reproduction and Development* **45** 218–224

Wassarman PM and Albertini DF (1994) The mammalian ovum. In *Physiology of Reproduction* pp A79–A122 Ed. E Knobil and JD Neill. Raven Press, New York

Journal of Reproduction and Fertility Supplement **54**, 461–475

Development of serum-free culture systems for the ruminant embryo and subsequent assessment of embryo viability

D. K. Gardner

Colorado Center for Reproductive Medicine, 799 East Hampden Ave, Suite 300, Englewood, CO 80110, USA

The mammalian embryo undergoes considerable changes in its physiology and energy metabolism as it proceeds from the zygote to the blastocyst stage. Complete development of the mammalian zygote *in vitro* was restricted to a few strains of mice and their F1 hybrids for many years, as the ruminant embryo arrested development at the 8- to 16-cell stage. The introduction of co-culture of ruminant embryos with somatic cells in the mid-1980s helped to alleviate this *in vitro* induced arrest. However, such culture systems required the use of complex tissue culture media and serum. Serum has subsequently been shown to induce several abnormalities during embryo development in culture and has been associated with the production of offspring with significantly greater birth weights than normal, leading to both difficulties in pregnancy management and an unacceptable frequency of neonatal death. Resurgence of interest in mammalian embryo physiology has culminated in the formulation of defined embryo culture media, capable of supporting a high percentage of viable blastocyst development *in vitro*. Optimum embryo development in culture has been shown to take place not in one, but two or more media, each designed to cater for the changing requirements and metabolism of the embryo as it develops. The development of viability assays to identify those embryos with the highest developmental potential will further increase the efficiency of embryo transfer procedures. Assays based upon nutrient uptake and subsequent utilization make promising candidates.

Introduction

Before attempting to culture any cell type, be it embryonic or somatic, it is important to consider the physiology of the cell to establish its nutrient requirements. The mammalian embryo therefore poses an intriguing problem in that its physiology changes during the preimplantation period. Subsequently, nutrient requirements by the embryo also change with successive stages of development. In parallel with such changes in nutrient requirements of the embryo, the environment to which the embryo is exposed in the female reproductive tract differs as the embryo progresses through the oviduct to the uterus. Data are emerging that show that embryo development in culture and maintenance of viability is facilitated not by a single culture medium, but rather by the use of sequential media that provide for the different requirements of the developing embryo. The ability to maintain the embryos of sheep and cattle in culture without compromising developmental potential will greatly expedite the development and introduction of procedures such as transgenesis and cloning.

Nutrient Requirements and Metabolism

Changes in physiology as the embryo develops are reflected in different nutrient requirements and energy metabolism. The significance of energy metabolism during embryo development cannot be understated, as impairment of metabolism is associated with developmental delay or even arrest (Gardner, 1998a). Indeed embryo development will not occur in the absence of exogenous substrates. Successful early embryo development is therefore dependent upon the ability of the embryo to generate energy through the appropriate metabolic pathway(s) at specific times during the preimplantation period (Gardner, 1998a). The ruminant embryo before compaction exhibits low levels of oxidative metabolism and oxygen consumption, whereas the later stages (i.e. after compaction) exhibit both high levels of glycolysis and high oxygen consumption (Thompson *et al.*, 1996). The ruminant embryo has a limited capacity to utilise glucose until after compaction (Tiffin *et al.*, 1991; Rieger *et al.*, 1992; Thompson *et al.*, 1992a; Gardner *et al.*, 1993). Up to this stage of development, plasma concentrations of glucose can be detrimental to early mammalian embryo development in specific culture conditions (Thompson *et al.*, 1992a). Accordingly, the early embryo utilises pyruvate and lactate alone or in combination with amino acids (Leese, 1991; Gardner and Lane, 1993a) to generate the required energy. The relatively low level of metabolism in the pre-compacted embryo reflects the quiescent state of the oocyte, which remains energetically relatively dormant within the ovary. As the oocyte and zygote have relatively low levels of biosynthesis before embryonic genome activation and expression, there will be a high ATP:ADP ratio within the blastomeres, which will in turn allosterically reduce the flux through the glycolytic pathway. As the embryo becomes increasingly transcriptionally active and protein synthesis increases, and as the blastocoel is formed through the action of the basolateral ATPases, the ATP:ADP ratio will fall and an increased glycolytic flux will become possible (Gardner, 1998a). Indeed, glucose metabolism by both sheep and cattle embryos increases with development, with the highest rates of utilization occurring at the blastocyst stage (Tiffin *et al.*, 1991; Rieger *et al.*, 1992; Thompson *et al.*, 1992a; Gardner *et al*,. 1993).

In the blastocyst of primates and rodents, the high levels of glucose conversion to lactate even in the presence of adequate concentrations of oxygen for oxidative metabolism (termed 'aerobic glycolysis'), have been interpreted as the adaptation of the embryo to its imminent invasion of the endometrium. As the ruminant blastocyst does not attach to the endometrium for several days after entry into the uterus, alternative explanations for this type of metabolism are required. It is plausible that the oxygen tension within the ruminant uterus is very low, and so glycolysis represents an adaptation of the blastocyst to such conditions. Alternatively, a high level of aerobic glycolysis is a common characteristic of rapidly dividing cells and tumours (Rieger, 1992; Gardner, 1998b). As well as being used to generate energy for blastocoel expansion and mitosis, glucose is therefore required for the synthesis of triacylglycerols and phospholipids and as a precursor for complex sugars of mucopolysaccharides and glycoproteins. Glucose metabolized by the pentose phosphate pathway (PPP) generates ribose moieties required for nucleic acid synthesis and the NADPH required for the biosynthesis of lipids and other complex molecules. NADPH is also required for the reduction of intracellular glutathione, an important antioxidant for the embryo (Rieger, 1992). The production of nucleic acids is probably an important biosynthetic role of glucose in the blastocyst. As the ruminant blastocyst undergoes considerable expansion prior to attachment, it is feasible that high glucose flux through glycolysis will ensure that there is sufficient substrate available for biosynthetic pathways, such as DNA replication, RNA transcription and synthesis of new membranes, at the required times during cellular proliferation. This suggests that there may be times within the cell cycle during which the PPP is more active than others.

The mammalian embryo possesses a considerable degree of plasticity and can adapt to the absence of one substrate by increasing its utilization of others (Gardner 1998a). Furthermore, inadequate culture conditions *in vitro* induce serious metabolic aberrations in the embryo, whereby the pattern of energy metabolism is significantly different from that of embryos developed *in vivo*. The more unphysiological the culture conditions are, the greater the metabolic stress placed on the embryo, which in turn is associated with loss of developmental competence (Gardner, 1998a).

Culture of the Ruminant Embryo

Tervit *et al.* (1972) demonstrated that by formulating a simple culture medium (Synthetic Oviduct Fluid, SOF) based upon the concentrations of ions and carbohydrates present in sheep oviduct fluid

(Restall and Wales, 1966), it was possible to observe development of the early sheep and cow embryo to the morula or blastocyst stage in culture. However, this approach was not widely adopted. Rather the embryos of sheep and cattle were routinely transferred to the ligated oviduct of either a rabbit or sheep. Such an approach was deemed more physiological. In an attempt to reproduce conditions similar to those *in vivo*, Gandolfi and Moor (1987) used a monolayer of epithelial cells taken from the sheep oviduct to support the development of sheep embryos. When such a co-culture approach was used 42% of the embryos reached the blastocyst stage. Following on from this work, the beneficial effects of co-culture were reported for the goat and cow (Sakkas *et al.*, 1989; Wiemer *et al.*, 1991). However, the media used in co-culture systems were designed to maintain specific somatic cell lines in culture and were not designed for the requirements of the embryo. Furthermore, such co-culture systems required the presence of serum to keep the somatic cells viable. Two points arise directly from this: first, the media used in co-culture systems do not themselves support embryo development per se, and secondly the use of whole serum has serious consequences for subsequent fetal development (see below). In spite of this, co-culture has been adopted widely as a routine culture system, and the resultant fetal oversize and neonatal deaths have become accepted costs of this approach (Behboodi *et al.*, 1995).

Interestingly, the oxygen concentration within the lumen of the female reproductive tract (3–8%; Mastroianni and Jones, 1965; Fischer and Bavister, 1993) is significantly lower than that present in air (20%). In line with these physiological data, development of ruminant embryos in culture is significantly increased when the oxygen concentration is between 5 and 10% compared with 20%. This was the first factor identified that allowed embryo development in a simple medium such as SOF (Tervit *et al.*, 1972; Thompson *et al.*, 1990; Batt *et al.*, 1991). Indeed it is plausible that reducing the oxygen tension in the vicinity of the embryo is one mechanism by which somatic cells can confer benefit to the embryo in a co-culture system (Edwards *et al.*, 1997).

There appears to have been a north–south divide regarding the choice of culture system for the ruminant embryo. In Australia and New Zealand, unlike the United States and Europe, co-culture has not been the culture system of choice. Rather the SOF medium has been used extensively. Thompson *et al.* (1992b) and Walker *et al.* (1992) both reported that the supplementation of SOF with human serum could support high rates of blastocyst development (80 to 90%) from the culture of 1- and 2-cell sheep embryos for 5 days *in vitro*. Furthermore, such blastocysts were shown to have a high viability with 70–80% of embryos transferred surviving to term. However, similar to the use of co-culture systems, a 37% post-natal mortality rate was observed, most likely due to the high birth weight of the dead lambs. Recent data on the sheep embryo have shown that it is the serum in the culture media that adversely affects embryo development in a number of ways: precocious blastocoel formation (Walker *et al.*, 1992; Thompson *et al.*, 1995), sequestration of lipid (Thompson *et al.*, 1995), abnormal mitochondrial ultrastructure (Thompson *et al.*, 1995) and perturbations in metabolism (Gardner *et al.*, 1994). Indeed, embryos cultured in the presence of serum have very different morphologies from those cultured in its absence (Fig. 1). However, it was the finding that the presence of serum in the culture system was associated with abnormally large offspring in sheep that is of greatest significance (Thompson *et al.*, 1995). The mechanism by which serum induces these aberrations remains to be resolved. However, the role of growth factors in serum in inducing altered patterns of development, and in the overexpression of growth factor genes cannot be overlooked.

Removal of Serum, Moving Towards Defined Culture Media

The key to the successful replacement of serum in ruminant embryo culture systems was the supplementation of the SOF medium with BSA (fatty-acid free) and amino acids (SOFaa). Oviduct and uterine fluids of mammals are characterized by high concentrations of free amino acids (Miller and Schultz, 1987; Moses *et al.*, 1997). Specifically, the amino acids alanine, aspartate, glutamate, glycine, serine and taurine are present at high concentrations. The fact that both oocytes and embryos possess specific transport systems for amino acids, readily take up amino acids from the surrounding culture medium, and maintain an endogenous pool of amino acids indicates that

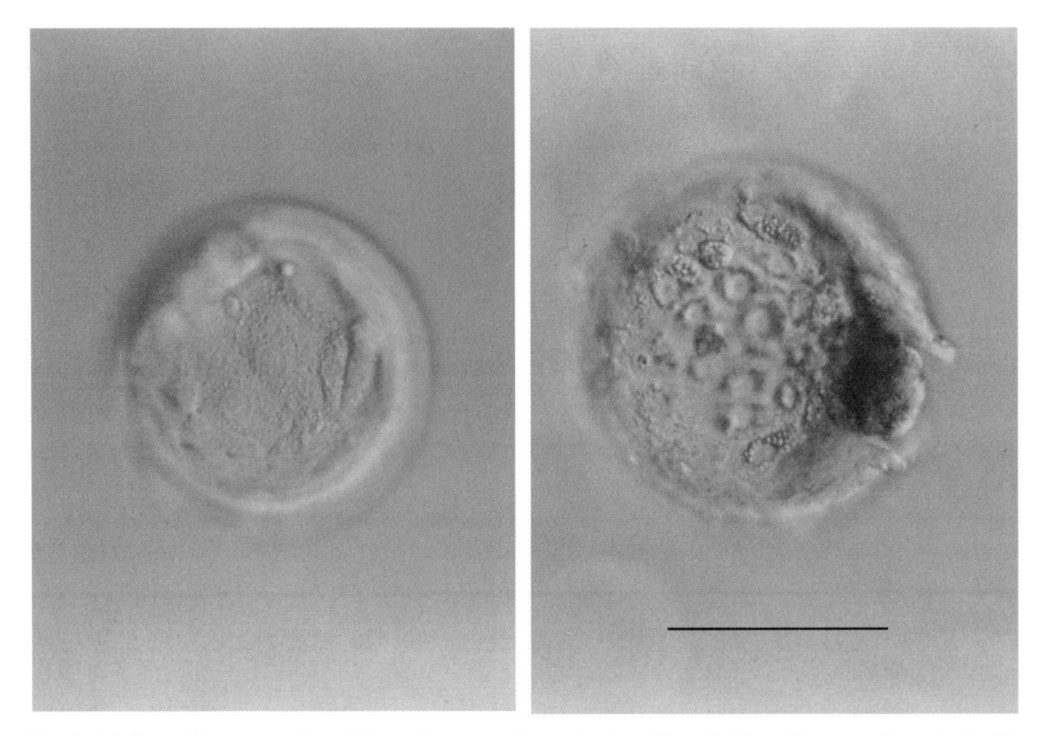

Fig. 1. (a) Sheep blastocyst cultured from the zygote in synthetic oviduct fluid medium supplemented with all 20 of Eagle's amino acids (SOFaa + 8 mg BSA ml^{-1}). Note the translucent appearance of the trophectoderm. (b) Sheep blastocyst (sibling of that shown in Fig 1a) cultured from the zygote in SOF medium supplemented with 20% human serum. The trophectoderm exhibits many vesicular inclusions, which have been shown to be lipid through osmium staining. Scale bar represents 100 μm. From Gardner (1994) with permission.

amino acids have a physiological role in the preimplantation period of mammalian embryo development (Gardner and Lane, 1993b). The inclusion of Eagle's amino acids in SOF medium significantly increased the development of sheep and cattle zygotes to the blastocyst stage in culture (Gardner, 1994; Gardner *et al.*, 1994). In a study on sheep zygotes fertilized *in vivo*, 95% of blastocyst formation on day 6 of culture was obtained, compared with 67% when the SOF medium lacked amino acids and was supplemented with 20% human serum. Furthermore, 79% of the blastocysts hatched after culture in SOFaa medium. The mean number of cells of such blastocysts (173 ± 6) was equivalent to control blastocysts developed *in vivo* (160 ± 9), in contrast to the number of cells of blastocysts cultured in SOF without amino acids (75 ± 7). More importantly, the viability of blastocysts cultured with amino acids was equivalent to that of embryos developed *in vivo* (Gardner *et al.*, 1994). When this culture system was used for the development of IVM/IVF cattle embryos, about 50% of fertilized oocytes reached the blastocyst stage after just 6 days of culture, day 7 of development (Gardner, 1994; Edwards *et al.*, 1997). Furthermore, the calving rate of these blastocysts was 55% (Gardner, 1994). The typical morphology of cattle blastocysts derived from the *in vitro* maturation and *in vitro* fertilization of oocytes, followed by culture for 6 days in SOFaa medium, is shown (Fig. 2). Similarly, Rosenkrans and First (1994) observed that the addition of the amino acids present in either Eagle's minimum essential medium or those in Eagle's basal medium to the culture medium CR1 increased development of cattle embryos to the blastocyst stage in culture. Liu and Foote (1995) using the culture medium KSOM (Lawitts and Biggers, 1992) found that Eagle's non-essential amino acids with essential amino acids (at half the concentration used by Eagle) gave greatest cow blastocyst development. The composition of these different media have been reviewed

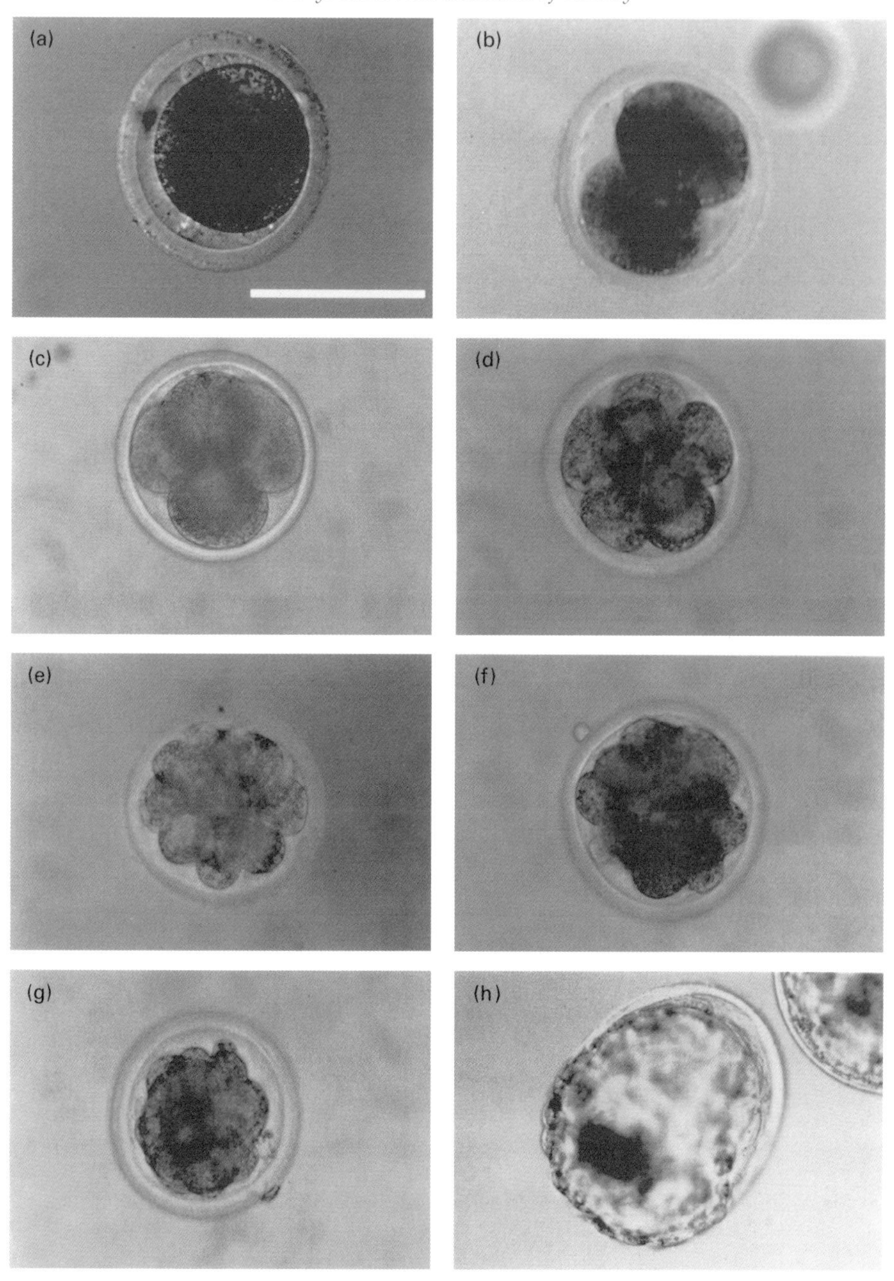

Fig. 2. Temporal development of cattle embryos in culture. Embryos were grown in synthetic oviduct fluid medium supplemented with all 20 of Eagle's amino acids (SOFaa + 8 mg BSA ml^{-1}). (a) Denuded fertilized oocyte 20 h post insemination (PI). Scale bar represents 100 μm. (b) Two-cell embryo, at 32 h PI. (c) Four-cell embryo at 44 h PI. (d) Eight-cell embryo at 68 h PI. (e) 16-cell embryo at 92 h PI. (f) 16- to 32-cell embryo undergoing the initial phase of compaction at about 116 h PI. (g) Morula at 140 h PI. (h) Expanded hatching blastocyst at 168 h PI (6 days of culture from the zygote). Note the inner cell mass in the bottom left of the blastocoel. Reproduced from Gardner (1998b) with permission.

elsewhere (Thompson, 1996; Gardner 1998b), and their ability to support embryo development is shown (Table 1).

The mammalian embryo appears to undergo a switch in its amino acid requirement as development proceeds. In both cattle and mice, the pre-compacted embryo development is faster in the presence of Eagle's non-essential amino acids and glutamine (Gardner and Lane, 1993b; Steeves and Gardner, 1997) (Fig. 3). Beneficial effects of this group of amino acids may well be attributable to their protective role in maintaining cellular function. Amino acids can act as osmolytes to protect against ionic stress (Lawitts and Biggers, 1992), and as regulators of intracellular pH and metabolism (Gardner and Lane, 1997). After compaction, embryo development is enhanced by the inclusion of all 20 amino acids. It has been shown in the mouse that while the non-essential amino acids stimulate the trophectoderm and glutamine stimulates hatching, the inclusion of essential amino acids causes an increase in the number of inner cell mass cells (Lane and Gardner, 1997b).

A cautionary note when dealing with amino acids in embryo culture media is that they are labile in solution at 37°C, the result of which is the release of ammonium into the medium (Gardner and Lane, 1993b). It is therefore very important that this toxicity is alleviated for optimum embryo growth to occur by renewing the culture medium every 48 h (mouse and human) or 72 h (sheep and cattle).

For defining culture media completely, alternatives to BSA, such as PVA (Pinyopummintr and Bavister, 1991; Liu and Foote, 1995; Keskintepe and Brackett, 1996) and hyaluronate (Gardner *et al.*, 1997a; Kano *et al.*, 1998), have also been used in the extended culture of embryos. However, embryos cultured in the presence of PVA do differ biochemically from those cultured in the presence of BSA (Thompson *et al.*, 1998). The suitability of different macromolecules therefore awaits extensive field trials.

Embryo Grouping

It has been demonstrated in several species that the culture of embryos in reduced volumes of medium or in groups significantly increases blastocyst development (Paria and Dey, 1990; Lane and Gardner, 1992) as well as increasing the number of blastocyst cells (Paria and Dey, 1990; Lane and Gardner, 1992; Gardner *et al.*, 1994) (Fig. 4a). More importantly, however, culturing embryos in reduced volumes increases subsequent viability after transfer (Lane and Gardner, 1992). It has been proposed that the beneficial effects of growing embryos in small volumes and groups is due to the production of autocrine/paracrine factor(s) by the embryos that stimulate their own development or that of surrounding embryos. Therefore, culture in large volumes results in a dilution of the factor so that it becomes ineffectual (Gardner, 1994). Furthermore, decreasing the incubation volume:embryo ratio specifically stimulates the development of the inner cell mass (ICM). Both mouse and cow blastocysts cultured in a reduced incubation volume:embryo ratio had significantly more ICM cells than those cultured in large volumes, whereas the number of trophectoderm cells was unaffected (Gardner *et al.*, 1997c; Ahern and Gardner, 1998) (Fig. 4b). These findings explain the increased viability of embryos cultured in reduced volumes or groups (Lane and Gardner, 1992). It is plausible that this effect is manifest through a specific embryo derived factor(s). Possible candidates for such autocrine/paracrine factor(s) include platelet-activating factor and insulin-like growth factor II (O'Neill, 1997).

Sequential Embryo Culture Media

The changes in carbohydrate metabolism and amino acid requirement during development support the application of sequential culture media. *In vivo* carbohydrate concentrations in the fluid of the female tract change with the day of the oestrous or menstrual cycle (Nichol *et al.*, 1992; Gardner *et al.*, 1996a). Sequential media have been shown to improve embryo development and viability markedly in both mice (Gardner and Lane, 1996; Lane and Gardner, 1997) and humans (Gardner *et al.*, 1998). Results show that sequential media also confer benefit to both sheep (Steeves and Gardner, 1997)

Table 1. Development of *in vitro* produced cattle embryos in different culture system

Culture system	Reference	% Blastocyst from cleaved oocytes	Blastocyst cell number	Number of days in culture
Co-culture with bovine oviduct epithelial cells and serum	Wiemer *et al.* (1991)	47	nd	8
HECM	Pinyopummintr and Bavister (1991)	10	nd	8
SOFaa with BSA	Gardner (1994)	45	123	6
	Edwards *et al.* (1997)	64	97	6
SOFaa with PVA	Keskintepe *et al.* (1995)	50	nd	6
KSOM with PVA First 48 h in KSOM with no amino acids followed by culture in KSOM with amino acids for 6 days	Liu and Foote (1995)	41	nd	8
mHECM-3 with lactate, 11 amino acids and PVA for 2 days followed by culture in TCM 199 + 10% bovine serum for 6 days	Pinyopummintr and Bavister (1996)	41	nd	8

HECM: hamster embryo culture medium; SOFaa: synthetic oviduct fluid with amino acids; KSOM: simplex optimized medium; mHECM3: modified hamster embryo culture medium version 3; PVA: polyvinyl alcohol. From Gardner (1998) with permission.

and cow (Gardner *et al.*, 1997b) embryos in culture. In the case of the cow embryo, this can be demonstrated clearly by the addition of EDTA to the culture medium. The inclusion of EDTA in the culture medium for the first 72 h prevents 'metabolic transformation' (Gardner, 1998a), by suppressing glycolytic activity. However, should EDTA be present for the second 72 h of culture to the blastocyst stage, it impairs embryo development, specifically reducing inner cell mass development (Fig. 5). The mechanism by which EDTA acts has been discussed elsewhere (Gardner and Lane, 1997; Gardner, 1998a). In summary, culture conditions, which support optimal development of the zygote, do not support good development and differentiation of the blastocyst. Conversely, those conditions, which favour blastocyst development and differentiation, are detrimental to the zygote (Gardner and Lane, 1997). Ultimately, with the use of perfusion culture (Thompson, 1996), it will be possible to expose the embryo to a number of nutrient gradients in culture, while at the same time being able to remove toxins such as ammonium that are generated in a static system.

Quality Control

For the successful development and use of defined embryo culture media, it is paramount that the laboratory has a most rigorous quality control system. In a co-culture system both the somatic cells and serum undoubtedly help to remove any toxins present, thereby conferring a greater degree of tolerance to the system. Until defined media become commercially available, it is essential to run bioassays on each new component of the culture system. The most suitable bioassay for any media is the cell type that is being grown. However, in the case of domestic animal embryos this is not always feasible or economical. Therefore, an alternative is the use of a mouse embryo bioassay (Gardner and Lane, 1993a; 1997). In such a bioassay each medium component can be screened for toxicity by

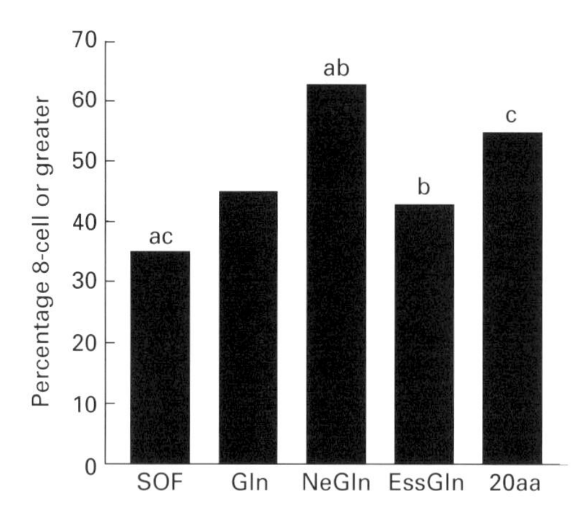

Fig. 3. Development of *in vitro* matured and *in vitro* fertilized bovine embryos. SOF (synthetic oviduct fluid with 8 mg BSA ml^{-1}); Gln (synthetic oviduct fluid with 8 mg BSA ml^{-1} + 1 mmol glutamine l^{-1}); NeGln (synthetic oviduct fluid with 8 mg BSA ml^{-1} + 1 mmol glutamine l^{-1} + non-essential amino acids); EssGln (synthetic oviduct fluid with 8 mg BSA ml^{-1} + 1 mmol glutamine l^{-1} + essential amino acids); 20 aa (synthetic oviduct fluid with 8 mg BSA ml^{-1} + 1 mmol glutamine l^{-1} + non-essential + essential amino acids). Values with the same letter are significantly different; $P < 0.05$. Data from Steeves and Gardner (1997).

culturing mouse zygotes in protein-free medium for 4 days, by which time more than 80% of the embryos should have reached the expanded blastocyst stage.

Assessment of Embryo Viability

Assessment of embryo viability in culture is rather subjective: gross embryo morphology is used as the most common method for selecting embryos for transfer. The ability to identify the most viable embryos from within a given cohort should increase the overall success of assisted reproductive procedures. There are several comprehensive reviews on the suitability of viability tests for embryos (Rieger, 1984; Gardner and Leese, 1993; Overstrom, 1996). Here, those methods most likely to identify viable embryos before transfer are discussed.

Nutrient uptake and energy metabolism

In light of the significance of embryo metabolism to the developing embryo in culture, it is worth considering the potential of quantitating metabolism to assess embryo viability before transfer. Renard *et al.* (1980) observed that day 10 cattle blastocysts that had a glucose uptake higher than 5 μg h^{-1} developed better both in culture and *in vivo* after transfer than blastocysts with a glucose uptake below this value. However, due to the insensitivity of the spectrophotometric method used they were not able to quantify glucose uptake by earlier stage embryos. Rieger (1984) showed that morphologically normal day 7 cattle blastocysts took up significantly more radiolabelled glucose than did degenerating blastocysts, and proposed that embryonic metabolism may be a suitable

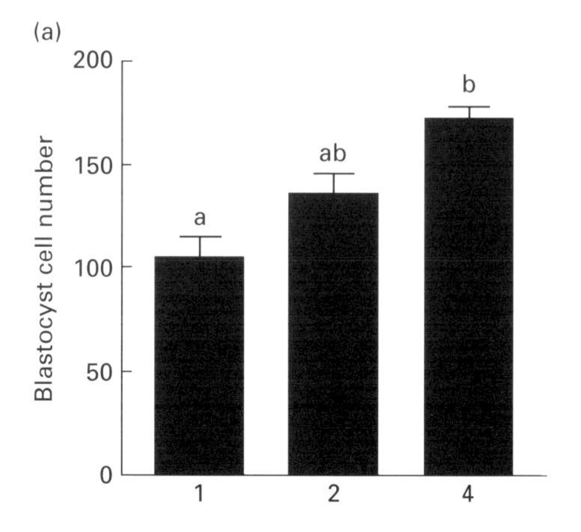

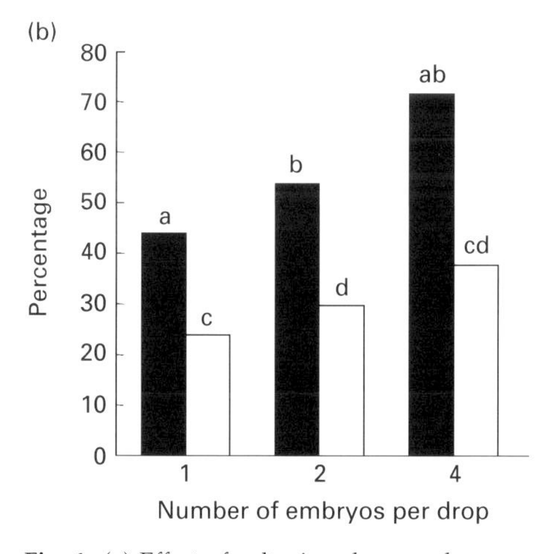

Fig. 4. (a) Effect of culturing sheep embryos in groups. Sheep zygotes were cultured individually, or in groups of two or four, in 20 µl of synthetic oviduct fluid with 8 mg BSA ml^{-1} + 20 amino acids (SOFaa). Values with the same letter are significantly different; a, $P < 0.05$; b, $P < 0.01$. Data from Gardner *et al.* (1994). (b) Effect of culturing bovine embryos in groups. Percentage blastocyst development (closed bars) and percentage of inner cell mass/total number of cells (open bars). Bovine embryos were cultured in groups of 50 in 500 µl of medium SOFaa at 39°C, in 7% O_2, 5% CO_2 and 88% N_2 for 72 h. After this time embryos with more than 8 cells were transferred to 50 µl drops of SOFaa for a further 72 h of culture. Embryos were cultured singly or in groups of two or four. Values with the same letters are significantly different; $P < 0.05$. Data from Ahern and Gardner (1998).

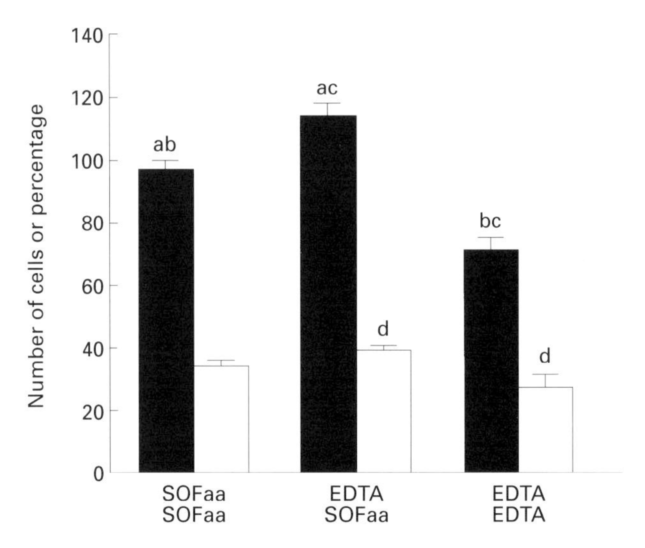

Fig. 5. Number of bovine blastocyst cells (solid bars) and percentage of inner cell mass/total number of cells (open bars). Bovine embryos, obtained from *in vitro* maturation and *in vitro* fertilization, were cultured for the first 72 h in either SOFaa (synthetic oviduct fluid with 8 mg BSA ml^{-1} + 20 amino acids) or EDTA (SOFaa + 100 μmol EDTA l^{-1}). After 72 h, embryos at about the 8- to 16-cell stage were transferred to fresh medium. Values with the same letter are significantly different; $P < 0.01$. Data from Gardner *et al.* (1997b).

method for assessing viability before transfer. With the application of non-invasive microfluorescence, it became possible to quantify glucose uptake by individual day 4 mouse blastocysts before transfer to recipient females (Gardner and Leese, 1987). The embryos that went to term were found to have a significantly higher glucose uptake in culture than embryos that failed to develop after transfer. Unfortunately these studies were retrospective and therefore could not demonstrate conclusively whether it was possible to identify viable embryos before transfer using metabolic criteria.

However, a study on day 7 cattle blastocysts before and after cryopreservation showed that it was possible to identify blastocysts capable of re-expansion in the hours immediately after thawing. Blastocysts that survived the freeze–thaw procedure had a significantly higher glucose uptake and lactate production than those embryos that did not re-expand and subsequently died (Gardner *et al.*, 1996b; Fig. 6). The significance of this study is that there was no overlap in the distribution of glucose uptake by the viable and non-viable embryos. Thus it may indeed be possible to use metabolic criteria for prospective selection of viable embryos. Lane and Gardner (1996) therefore performed a prospective trial in which day 5 mouse blastocysts were classified as either viable or non-viable according to their glycolytic activity. Glucose consumption and lactate production were measured in individual blastocysts of equivalent morphology and the same diameter. It was observed that although the blastocysts had the same appearance, there was a great difference between metabolic profile of blastocysts, confirming that morphology is a poor criterion upon which to base embryo selection (Gardner and Lane, 1997; Lane and Gardner, 1997b). A hypothesis was established in which blastocysts with glycolytic activity similar to that of an *in vivo* developed blastocyst, that is a high rate of glucose consumption but low lactate production, were deemed to be the most viable

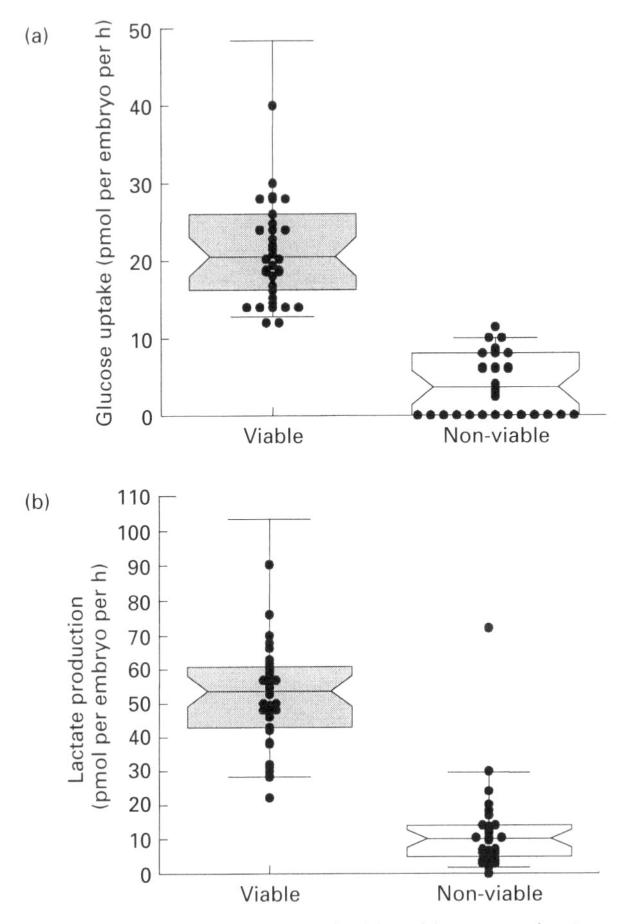

Fig. 6. Box plots of glucose uptake (a) and lactate production (b) by individual bovine blastocysts after thawing. Blastocysts were classified retrospectively as either 'viable' or 'non-viable' based upon their ability, or otherwise, to re-expand within 14 h after thawing. Data from Gardner *et al.* (1996b).

embryos. In contrast, blastocysts with a low glucose uptake and a high lactate production were deemed to be of low viability. Subsequently, metabolism of individual blastocysts was assessed before transfer. The control for this study was blastocysts transferred on the basis of morphology alone. Fetal development per blastocyst transferred in the control group was 20%, while fetal development of blastocysts classified as viable before the transfer was 80%. In contrast, fetal development of blastocysts classified as having low viability was just 6%. Furthermore, when the rate of glucose uptake was analysed retrospectively, it was found that embryos classified as viable had a significantly higher glucose uptake than embryos classified as non-viable. So from this study it was evident that both the rate and fate of nutrient utilization appear important.

Although abnormally high levels of glycolysis indicate a loss of viability in the mouse blastocyst, it is not clear whether glycolysis is the most suitable marker for the ruminant embryo based on the observations that sheep and cattle embryos have a much higher glycolytic rate than the mouse (Rieger *et al.*, 1992; Thompson *et al.*, 1992a; Gardner *et al.*, 1993). Furthermore, assessment of metabolism in the mouse blastocyst is performed in a matter of hours before implantation, whereas

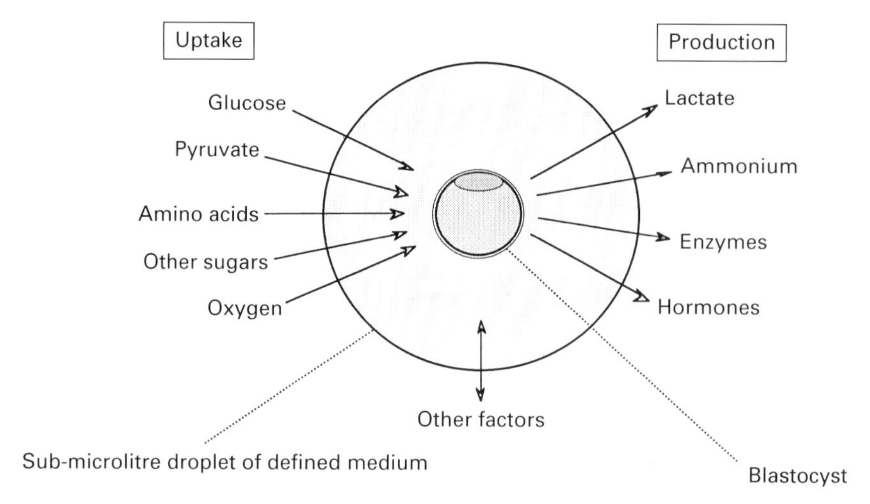

Fig. 7. Non-invasive assessment of blastocyst viability. Blastocysts are incubated individually in a known volume, for example 0.5 µl, of defined medium. Serial nanolitre samples can then be taken and analysed for carbohydrates, amino acids, ammonium, oxygen and enzymes. The concomitant measurement of glucose consumption and lactate production can give an indirect measure of glycolytic activity. The concomitant measurement of amino acid consumption and ammonium production can give an indirect measure of amino acid utilization. The release of enzymes, such as lactate dehydrogenase, into the surrounding culture medium reflects impairment in membrane integrity, and as such may be useful in assessing freezing damage. From Gardner and Schoolcraft (1998) with permission.

in sheep and cows, the blastocyst is routinely assessed and transferred at about day 7, several days before implantation occurs. It is most likely that due to the plasticity of the embryo, a single marker of viability will have limited potential. It is therefore proposed that several markers be used in parallel to assess development potential before transfer (see below and Fig. 7).

Oxygen uptake

With the development of both multichannel embryo microrespiration systems (Overstrom, 1996) and fluorometric assays (Thompson *et al.*, 1996), it is now possible to quantify oxygen consumption by individual and small groups of cattle embryos non-invasively. Such procedures will be invaluable in assessing mammalian embryo metabolism and respiration. Whether oxygen consumption by embryos is correlated with developmental potential has yet to be determined. It is envisaged that this approach will be most applicable in conjunction with the assessment of nutrient uptake, such as pyruvate or amino acids, to determine the rate and fate of the nutrient consumed.

Enzyme leakage

A marker of plasma membrane integrity, and therefore cellular viability, is the leakage of cytosolic enzymes into the surrounding medium. Johnson *et al.* (1991) used such an approach to assess the developmental capacity of day 6.5 to 7.5 cattle embryos produced *in vivo* after culture for 24 h. There was a significant inverse relationship between the appearance of the enzyme lactate dehydrogenase (LDH) in the surrounding culture medium and embryo development, that is embryos that failed to develop released significantly more LDH into the medium. Whether this approach can be used prospectively to select viable embryos has yet to be determined. However, such an approach may be significant in the comparison of different cryopreservation procedures, specifically their induction of membrane damage.

Hormone and growth factor production

Hernandez-Ledmezma *et al.* (1993) have shown that the production of trophoblast interferon by cattle blastocysts was determined by both the developmental stage and quality of the embryo. Such an approach is by default more specific than general metabolism, and as such may be useful in determining appropriate levels of genome activation, which in turn may reflect subsequent viability.

Conclusions

The physiology of the mammalian embryo undergoes considerable changes during the preimplantation period. Concomitantly, the embryo exhibits changes in its nutrient requirements and metabolism. The application of sequential media or perfusion culture enables the changing requirements of the embryos to be met. The use of such gradient culture systems appears to induce less stress on the embryo, thereby maintaining high levels of viability. As culture of the *in vivo* fertilized embryo gives rise to more blastocysts than the culture of embryos derived from *in vitro* maturation and *in vitro* fertilization (Thompson *et al.*, 1995), it may be appropriate to focus more on maturation and fertilization conditions, both of which have a significant impact on subsequent embryo development (De Matos *et al.*, 1995; Earl *et al.*, 1997).

First and foremost I would like to thank Paul Batt for his endless enthusiasm and encouragement. Without Paul our ruminant endeavours would never have succeeded. Sincerest thanks to Lisa Maclellan, Tracey Steeves and Tim Ahern for their dedication to the cause, and to Drs Jeremy Thompson and Michelle Lane for their comments on this manuscript and for keeping the fire burning.

References

Ahern TJ and Gardner DK (1998) Culturing bovine embryos in groups stimulates blastocyst development and cell allocation to the inner cell mass *Theriogenology* **49** 194 (Abstract)

Batt PA, Gardner DK and Cameron AWN (1991) Oxygen concentration and protein source affect the development of preimplantation goat embryos *in vitro*. *Reproduction Fertility and Development* **3** 601–607

Behboodi E, Anderson GB, BonDurant RH, Cargill SL, Kreuscher BR, Medrano JF and Murray JD (1995) Birth of large calves that developed from *in vitro*-derived bovine embryos *Theriogenology* **44** 227–232

De Matos DG, Furnus CC, Moses DF and Baldassarre H (1995) Effect of cysteamine on glutathione level and developmental capacity of bovine oocytes matured *in vitro*. *Molecular Reproduction and Development* **42** 432–436

Earl CR, Kelly J, Rowe J and Armstrong DT (1997) Glutathione treatment of bovine sperm enhances *in vitro* blastocyst production rates *Theriogenology* **47** 255 (Abstract)

Edwards LE, Batt PA, Gandolfi F and Gardner DK (1997) Modifications made to culture medium by bovine oviduct epithelial cells: changes to carbohydrates stimulate bovine embryo development *Molecular Reproduction and Development* **46** 146–154

Fischer B and Bavister BD (1993) Oxygen tension in the oviduct and uterus of rhesus monkeys, hamsters and rabbits *Journal of Reproduction and Fertility* **99** 673–679

Gandolfi F and Moor RM (1987) Stimulation of early embryonic development in the sheep by coculture with oviduct cells *Journal of Reproduction and Fertility* **81** 23–28

Gardner DK (1994) Culture of mammalian embryos in the absence of serum and somatic cells *Cell Biology International* **18** 1163–1179

Gardner DK (1998a) Changes in requirements and utilization of nutrients during mammalian preimplantation embryo development and their significance in embryo culture *Theriogenology* **49** 83–102

Gardner DK (1998b) Embryo development and culture techniques. In *Animal Breeding: Technology for the 21st Century* pp13–46 Ed. J Clark. Harwood Academic Publishers, London

Gardner DK and Lane M (1993a) Embryo culture systems. In *Handbook of In Vitro Fertilization* pp 85–114 Eds A Trounson and DK Gardner. CRC Press, Boca Raton

Gardner DK and Lane M (1993b) Amino acids and ammonium regulate the development of mouse embryos in culture *Biology of Reproduction* **4** 377–385

Gardner DK and Lane M (1996) Alleviation of the "2-cell block" and development to the blastocyst of CF1 mouse embryos: role of amino acids, EDTA and physical factors *Human Reproduction* **11** 2703–2712

Gardner DK and Lane M (1997) Culture and selection of viable blastocysts: a feasible proposition for human IVF? *Human Reproduction Update* **3** 367–382

Gardner DK and Leese HJ (1987) Assessment of embryo viability prior to transfer by the non-invasive measurement of glucose uptake *Journal of Experimental Zoology* **242** 103–105

Gardner DK and Leese HJ (1993) Assessment of embryo metabolism and viability. In *Handbook of In Vitro Fertilization* pp 195–211 Eds A Trounson and DK Gardner. CRC Press, Boca Raton

Gardner DK and Schoolcraft WB (1998) Elimination of high order multiple gestations by blastocyst culture and transfer. In *Female Infertility Therapy: Current Practice* pp 267–274 Eds Z Shoham, C Howles and H Jacobs. Martin Dunnitz, London

Gardner DK, Lane M and Batt PA (1993) The uptake and metabolism of pyruvate and glucose by individual pre-attachment sheep embryos developed *in vivo*. *Molecular Reproduction and Development* **36** 313–319

Gardner DK, Lane M, Spitzer A and Batt PA (1994) Enhanced rates of cleavage and development for sheep zygotes cultured to the blastocyst stage *in vitro* in the absence of serum and somatic cells: amino acids, vitamins and culturing embryos in groups stimulate development *Biology of Reproduction* **50** 390–400

Gardner DK, Lane M, Calderon I and Leeton J (1996a) The environment of the human embryo *in vivo*: analysis of oviduct and uterine fluids during the menstrual cycle and metabolism of cumulus cells *Fertility and Sterility* **65** 349–353

Gardner DK, Pawelczynski M and Trounson A (1996b) Nutrient uptake and utilization can be used to select viable day 7 bovine blastocysts after cryopreservation *Molecular Reproduction and Development* **44** 472–475

Gardner DK, Lane M and Rodriguez-Martinez H (1997a) Fetal development after transfer is increased by replacing protein with the glycosaminoglycan hyaluronate for embryo culture *Human Reproduction* **12** Abstract Book 1, O-215 (Abstract)

Gardner DK, Lane MW and Lane M (1997b) Bovine blastocyst cell number is increased by culture with EDTA for the 72 hours of development from the zygote *Theriogenology* **47** 278 (Abstract)

Gardner DK, Lane MW and Lane M (1997c) Development of the inner cell mass in mouse blastocysts is stimulated by reducing the embryo:incubation volume ratio *Human Reproduction* **12** Abstract Book 1 P–132

Gardner DK, Vella P, Lane M, Wagely L, Schlenker T and Schoolcraft WB (1998) Culture and transfer of human blastocysts increases implantation rates and reduces the need for multiple embryo transfers *Fertility and Sterility* **69** 84–88

Hernandez-Ledmezma JJ, Mathialagan N, Villanueva C, Sikes JH and Roberts RM (1993) Expression of bovine trophoblast interferons by *in vitro*-derived blastocysts is correlated with their morphological quality and stage of development *Molecular Reproduction and Development* **36** 1–6

Johnson SK, Jordan JE, Dean RG and Page RD (1991) The quantification of bovine embryo viability using a bioluminescent assay for lactate dehydrogenase *Theriogenology* **35** 425–433

Kano K, Miyano T and Kato S (1998) Effects of glycosaminoglycans on the development of *in vitro* matured and fertilized porcine oocytes to the blastocyst stage *in vitro*. *Biology of Reproduction* **58** 1226–1232

Keskintepe L and Brackett BG (1996) *In vitro* developmental competence of *in vitro*-matured bovine oocytes fertilized and cultured in completely defined media *Biology of Reproduction* **55** 333–339

Keskintepe L, Burnely CA and Brackett BG (1995) Production of viable bovine blastocysts in defined *in vitro* conditions *Biology of Reproduction* **52** 1410–1427

Lane M and Gardner DK (1992) Effect of incubation volume and embryo density on the development and viability of mouse embryos *in vitro*. *Human Reproduction* **7** 558–562

Lane M and Gardner DK (1996) Prospective selection of viable mouse embryos prior to transfer using metabolic rate *Human Reproduction* **11** 1975–1978

Lane M and Gardner DK (1997) Differential regulation of mouse embryo development and viability by amino acids *Journal of Reproduction and Fertility* **109** 153–164

Lawitts JA and Biggers JD (1992) Joint effects of sodium chloride, glutamine, and glucose in mouse preimplantation embryo culture media *Molecular Reproduction and Development* **31** 189–194

Leese HJ (1991) Metabolism of the preimplantation mammalian embryo. In *Oxford Reviews of Reproductive Biology* **13** 35–72

Liu Z and Foote RH (1995) Effects of amino acids on the development of *in-vitro* matured/*in-vitro* fertilized bovine embryos in a simple protein-free medium *Human Reproduction* **11** 2985–2991

Mastroianni L, Jr and Jones R (1965) Oxygen tension within rabbit fallopian tube *Journal of Reproduction and Fertility* **9** 99–102

Miller JGO and Schultz GA (1987) Amino acid content of preimplantation rabbit embryos and fluids of the reproductive tract *Biology of Reproduction* **36** 125–129

Moses DF, Matkovic M, Cabrera Fisher E and Martinez AG (1997) Amino acid contents of sheep oviductal and uterine fluids *Theriogenology* **47** 336 (Abstract)

Nichol R, Hunter RHF, Gardner DK, Leese HJ and Cooke GM (1992) Concentration of energy substrates in porcine oviduct fluid and blood plasma during the peri-ovulatory period *Journal of Reproduction and Fertility* **96** 699–707

O'Neill C (1997) Evidence for the requirement of autocrine growth factors for the development of mouse pre-implantation embryos *in vitro*. *Biology of Reproduction* **56** 229–237

Overstrom EW (1996) *In vitro* assessment of embryo viability *Theriogenology* **45** 3–16

Paria PC and Dey SK (1990) Preimplantation embryo development *in vitro*: cooperative interactions among embryos and role of growth factors *Proceedings of the National Academy of Sciences, USA* **87** 3756–3760

Pinyopummintr T and Bavister BD (1991) *In vitro* matured/*in vitro* fertilized bovine oocytes can develop into morulae/blastocysts in chemically defined, protein-free culture media *Biology of Reproduction* **45** 736–742

Pinyopummintr T and Bavister BD (1996) Energy substrate requirements for *in vitro* development of early cleavage-stage bovine embryos *Molecular Reproduction and Development* **44** 193–199

Renard JP, Philippon A and Menezo Y (1980) *In vitro* uptake of glucose by bovine blastocysts *Journal of Reproduction and Fertility* **58** 161–164

Restall BJ and Wales RG (1966) The Fallopian tube of the sheep III. The chemical composition of the fluid from the Fallopian tube *Australian Journal of Biological Science* **19** 687–698

Rieger D (1984) The measurement of metabolic activity as an approach to evaluating viability and diagnosing sex in early embryos *Theriogenology* **21** 138–149

Rieger D (1992) Relationship between energy metabolism and development of the early embryo *Theriogenology* **37** 75–93

Rieger D, Loskutoff NM and Betteridge KJ (1992) Developmentally related changes in the metabolism of glucose and glutamine by cattle embryos produced and co-cultured *in vitro*. *Journal of Reproduction and Fertility* **95** 585–595

Rosenkrans CF, Jr and First NL (1994) Effect of free amino acids and vitamins on cleavage and developmental rate of bovine zygotes *in vitro*. *Journal of Animal Science* **72** 434–437

Sakkas D, Batt PA and Cameron WN (1989) Development of preimplantation goat (*Capra hircus*) embryos *in vivo* and *in vitro*. *Journal of Reproduction and Fertility* **87** 359–365

Steeves TE and Gardner DK (1997) Temporal effects of amino acids on bovine embryo development in culture *Biology of Reproduction* **57** (Supplement 1) 25 (Abstract)

Tervit HR, Whittingham DG and Rowson LEA (1972) Successful culture *in vitro* of sheep and cattle ova *Journal of Reproduction and Fertility* **30** 493–497

Thompson JG (1996) Defining the requirements for bovine embryo culture *Theriogenology* **45** 27–40

Thompson JGE, Simpson AC, Pugh PA, Donnelly PE and Tervit HR (1990) Effect of oxygen concentration on *in-vitro* development of preimplantation sheep and cattle embryos *Journal of Reproduction and Fertility* **89** 573–578

Thompson JG, Simpson AC, Pugh PA and Tervit HR (1992a) Requirement for glucose during *in vitro* culture of sheep preimplantation embryos *Molecular Reproduction and Development* **31** 253–257

Thompson JG, Simpson AC, Pugh PA and Tervit HR (1992b) *In vitro* development of early sheep embryos is superior in medium supplemented with human serum compared with sheep serum or human serum albumin *Animal Reproduction Science* **29** 61–68

Thompson JG, Gardner DK, Pugh PA, McMillan J and Tervit RH (1995) Lamb birth weight following transfer is affected by the culture system used for pre-elongation development of embryos *Biology of Reproduction* **53** 1385–1391

Thompson JG, Partridge RJ, Houghton FD, Cox CI and Leese HJ (1996) Oxygen uptake and carbohydrate metabolism by *in vitro* derived bovine embryos *Journal of Reproduction and Fertility* **106** 299–306

Thompson JG, Sherman ANM, Allen NW, McGowan LT and Tervit HR (1998) Total protein content and protein synthesis within pre-elongation stage bovine embryos *Molecular Reproduction and Development* **50** 139–145

Tiffin GJ, Rieger D, Betteridge KJ, Yadav BR and King WA (1991) Glucose and glutamine metabolism in pre-attachment cattle embryos in relation to sex and stage of development *Journal of Reproduction and Fertility* **93** 125–132

Walker SK, Heard TM and Seamark RF (1992) *In vitro* culture of sheep embryos without co-culture: success and perspectives *Theriogenology* **37** 111–126

Wiemer KE, Watson AJ, Polanski V, McKena AI, Fick GH and Schultz GA (1991) Effects of maturation and co-culture treatments on the development capacity of early bovine embryos *Molecular Reproduction and Development* **30** 330–338

Journal of Reproduction and Fertility Supplement **54**, 477–487

Sexing mammalian spermatozoa and embryos – state of the art

G. E. Seidel, Jr

Animal Reproduction and Biotechnology Laboratory, Colorado State University, Fort Collins, CO 80523, USA

Methods for sexing preimplantation embryos range from karyotyping to recording speed of development *in vitro*. The only method used routinely on a commercial scale is to biopsy embryos and amplify Y-chromosome-specific DNA using the polymerase chain reaction. This method is effective for more than 90% of embryos and is > 95% accurate. Within males, spermatozoa are essentially identical phenotypically due to: (1) connection of spermatogenic cells by intercellular bridges, (2) transcriptional inactivation of sex chromosomes during meiosis and spermiogenesis, (3) severe limitation of all gene expression during the later stages of spermiogenesis, and (4) coating all spermatozoa with common macromolecules during and after spermiogenesis. One consequence is that no convincing phenotypic difference has been detected between X- and Y-chromosome-bearing spermatozoa. The only consistently successful, nondestructive approach to sexing spermatozoa is to quantify DNA in spermatozoa using a fluorescing DNA-binding dye followed by flow cytometry and cell sorting. X-chromosome-bearing ruminant spermatozoa have about 4% more DNA compared with Y-chromosome-bearing spermatozoa; accuracy of sorting can exceed 90% routinely, and sorting rates currently exceed 10^3 live spermatozoa of each sex chromosome composition s^{-1}. Hundreds of apparently normal offspring from a number of species have been produced from sexed semen, some via intrauterine artificial insemination.

Introduction

The most sought-after reproductive biotechnology for humans and animals has been a means to obtain offspring of a pre-determined sex. Potential applications are numerous. For example, in humans, pre-conception selection of the sex of offspring to balance family gender probably would be used widely. Although there are ethical concerns, enabling parents to select the sex of their children would lead to smaller families, ameliorating social ills resulting from overpopulation; in the short term, if the procedure were inexpensive, it would probably be a very effective, non-coercive method for limiting population growth. It certainly would be less objectionable on moral grounds than some current procedures, including selective and non-selective abortion. This technique also can be used to produce girls to avoid X-linked genetic diseases (Levinson *et al.*, 1995).

Applications for farm animals range from the obvious, for example bulls do not produce milk, to the subtle, for example there is less calving difficulty with female calves, due to their smaller size (Burfening *et al.*, 1978). Such technology would be used widely for companion animals, and might even make the difference to whether an endangered species could be rescued from extinction. The focus of this review is on methodology, so applications will not be discussed beyond having given this context for justification. Most applications of sexing concern cattle, so much of the scientific literature concerns this species.

There are several methods for accurate sexing of embryos, but only one method for sexing semen that does not render spermatozoa infertile (Johnson *et al.*, 1989) and can be applied reliably across species. There are many reviews of sexing procedures (including: Kiddy and Hafs, 1971; Betteridge *et al.*, 1981; Amann and Seidel, 1982; Seidel, 1988; Amann, 1989; Bondioli, 1992; Johnson, 1994; Johnson, 1995; Cran and Johnson, 1996). Emphasis in this review will be on recent work.

From a theoretical standpoint, the phenotypic sex of an animal that will be produced by a spermatozoon or embryo can be determined most accurately at the gene level. Even at this level, there will be occasional errors due to mutations. Although presence or absence of the Sry gene (Koopman *et al.*, 1991) is the primary determinant of phenotypic sex, a cascade of additional genetic events is required to produce the specified sexual structures. This cascade, which involves genes on autosomes and sex chromosomes, is poorly understood in mammals (Bogan and Page, 1994).

Sex-chromosomal composition is another level at which sex potentially can be detected; functionally, this is the level at which nature operates. Several genes that appear unrelated to sex determination are transmitted incidentally with the sex-determining arm (non-pseudoautosomal region) of the Y chromosome (for example Greenfield *et al.*, 1996). There is debate on whether determining sex-chromosomal composition is a genetic or phenotypic trait. This is relevant, for example, when measurement of total DNA content is the method of identifying an X or Y chromosome in a spermatozoon.

The most studied and used methods for attempting sexing are phenotypic, and range from measuring enzyme activity of embryos (Williams, 1986) to determining parameters of sperm motion (Penfold *et al.*, 1998). Sexing fetuses by skilled technicians using ultrasound approaches 99% accuracy by 2 months of gestation in humans, cattle and horses (Curran and Ginther, 1991). Although this special case concerns recording rather than selecting, many find information on fetal sex valuable in for example experimentation, merchandizing, management and planning. Occasionally this approach is used together with abortion of fetuses of the undesired gender to control the sex of offspring produced.

Sexing Preimplantation Embryos

The many procedures for analysing the sex of preimplantation embryos are summarized in Table 1. Results, of course, are obtained too late to choose the sex except by discarding the sex that is less valuable to the client. This is not merely a trite observation. For example, many embryo transfer practitioners fail to use or promote procedures for sexing embryos because their income from a particular donor animal decreases considerably if half of the embryos are discarded.

The only method of sexing embryos in routine commercial use is to biopsy embryos, usually with tools connected to a micromanipulator, followed by amplification of Y-chromosome-specific DNA in the sample using the polymerase chain reaction (PCR) and appropriate primers (Bondioli *et al.*, 1989; Herr and Reed, 1991; Thibier and Nibart, 1995). Obtaining the biopsy is art and science. Ideally, only a few cells are removed without damaging the embryo and without contamination by accessory spermatozoa attached to or embedded in the zona pellucida. In practice, about ten cells are generally excised. There are several species-specific aspects to biopsy; for example, compromising the capsule of the equine blastocyst can decrease pregnancy rates.

There are two options for selecting Y-chromosome-specific DNA: single copy sequences such as Sry, or sequences with no known function that are repeated hundreds to thousands of times. Mammalian Y chromosomes abound with such species-specific repeats; they have the advantage of having been amplified even before any PCR procedures. Unfortunately, different sequences must be identified for each individual species.

Once potential Y-chromosome-specific DNA has been amplified by PCR, the most thorough identification procedure would be gel electrophoresis followed by Southern blotting. This is rarely done in practice. The most common procedure is to observe the gel for the presence or absence of the correct size band of DNA, as revealed by ethidium bromide intercalated into the DNA (visualized with UV light). Often a positive control for amplified DNA is engineered into the system: for example, an additional set of PCR primers for each sample can be incorporated that is specific to autosomes or the X chromosome (Thibier and Nibart, 1995). This procedure confirms that DNA from

Table 1. Procedures for sexing preimplantation embryos

Method	Sexable (%)	Accuracy	Reference	Comments
Biopsy; sex chromatin	40–50	> 95%	Gardner and Edwards, 1968	First method
Biopsy; karyotype	30–70	> 95%	Hare *et al.*, 1976	First definitive method
Compaction response to antibodies to male-specific antigen	90–95	~75%	Utsumi *et al.*, 1984	Subjective
Male embryos develop more rapidly *in vitro*	60–70	~ 70%	Tsunoda *et al.*, 1985	Depends on culture system
Gene dosage of G6PD	60–70	~ 65%	Williams, 1986	Narrow window; subjective
Biopsy; Y-specific *in situ* hybridization	80–90	> 90%	West *et al.*, 1987	Tedious
Immunological detection of male-specific antigen	80–85	~ 80%	Anderson, 1987	No biopsy required; subjective
Biopsy; Y-specific DNA, electrophoresis	90–95	> 95%	Bondioli *et al.*, 1989	Used commercially
Biopsy; Y-specific DNA, no electrophoresis	90–95	~ 95%	Bredbacka *et al.*, 1995	Being refined

the biopsy was in fact placed into the assay vessel and that the entire system was functioning. Without such a control, system failure cannot be distinguished from a female embryo. Note that great care is required for these procedures because of the extreme sensitivity of PCR; just one contaminating cell or contaminating DNA from serum can lead to an incorrect diagnosis. An example of precautions taken to avoid contamination is to have all reagents in a single container throughout the process, with the containers prepared in advance in a controlled environment, and with quality-control sampling before use.

An improvement to the technique just described eliminates the need for gel electrophoresis, simplifying the sexing procedure and shortening the time required (Bredbacka et al., 1995). In this case, PCR primers are made that bind to multiple sites along a highly repeated (~60 000 ×) bovine sequence of Y-chromosome-specific DNA. Concentrations of PCR primers and nucleotides and other PCR conditions are such that large quantities of DNA are produced by PCR if this sequence is present.

The endpoint of this assay is simply to observe the PCR reaction mix, which contains ethidium bromide, under UV light. Tubes containing the Y-specific sequence will have large quantities of DNA, and thus are pink; those without Y-specific sequence just have DNA from the biopsy and have barely any colour. Numerous unpublished improvements have been made to the original paper by Bredbacka *et al.* (1995); a control to distinguish system failure from a female embryo has not yet been published for this method.

All methods of sexing embryos have limitations. Those based on PCR of Y-chromosome-specific sequences are very accurate, but require biopsy of embryos. The biopsy process is time consuming, and an occasional embryo is lost or destroyed in the process; however, there is little effect on pregnancy rates as practised in cattle, with the possible exception of cryopreserved, biopsied embryos.

The methods involving antibodies to male-specific antigen (sometimes termed H-Y antigen, perhaps incorrectly) are subjective, and reagents are not commercially available. A major drawback of the method is that the molecule being detected is unknown, although it appears to be conserved across species. Moreover, it is easily detected only between the eight-cell and early blastocyst stages (Anderson, 1987), probably because the gene is not expressed in early stages, and the gene product is masked or internalized after the early blastocyst stage. The molecule seems to be a very weak antigen, requiring various immunological tricks to raise usable antibodies (Anderson, 1987). A candidate gene is the H-Y antigen (Wang *et al.*, 1995); this molecule differs in only two amino acids from the female antigen in humans (but see Greenfield *et al.*, 1996), which would explain why male-specific antibodies are so difficult to produce. Possibly a renewed effort in this area based on antibodies to a synthetic sequence would lead to a reasonably reliable procedure for sexing embryos without the need for biopsy.

The other methods listed in Table 1 have limitations of subjectivity, impractical timing, low accuracy, and/or require too much time to be practical. Investing resources for improved methods may be risky from a commercial standpoint because reliable methods of sexing spermatozoa might make sexing embryos obsolete. However, it is quite possible that genotyping biopsies of embryos for alleles of economic interest will become a major animal breeding tool, in which case confirming sex may simply become important ancillary information, even with sexed semen.

Sexing Spermatozoa

Theoretical considerations

From an evolutionary perspective, there is a clear theoretical advantage to a 50:50 sex ratio (Fisher, 1930). In most cases, species with a 50:50 sex ratio will out-compete those in the same ecological niche that do not have a 50:50 sex ratio, although the optimal sex ratio produced by individual wildlife parents may vary because of a variety of environmental and sociological factors (reviewed by Gosling, 1986). In mammals, numbers of X- and Y-chromosome-bearing spermatozoa produced are equal as a consequence of the way that chromosomes segregate at meiosis. One problem remains, that is to ensure that X- and Y-chromosome-bearing spermatozoa have an equal chance of fertilizing an oocyte.

X- and Y-chromosome-bearing spermatozoa are identical phenotypically and this ensures a 50:50 sex ratio; if the spermatozoa are not different from each other, each spermatozoon will have an equal chance of fertilizing the ovum. With rare exceptions such as the T/t complex in mice (Silver, 1985), this principle applies to all genes, not just to those determining sex; that is, the phenotype of the spermatozoa does not vary with the haploid genotype. Thus, within each meiotic event, the four spermatozoa produced from a primary spermatocyte are fairly homogeneous phenotypically.

Note that if this were not true, there would frequently be a distortion from a 50:50 ratio of transmitting alleles. Conceivably, a heterozygous male for black/red colour would not produce half black and half red offspring when mated to red females if there were chance pleiotropic effects of these alleles on sperm function; many genes affect sperm function. Thus, although there are great differences between males in sperm phenotype, including fertilizing potential, velocity of swimming, and ability to compete (demonstrated convincingly and repeatedly in heterospermic insemination experiments; for example Nelson *et al.*, 1975; Dziuk, 1996), within males, alleles usually are transmitted randomly. There is not complete success in achieving a 50:50 allele distribution within males; for example, in most mammals a slight excess of males is produced consistently due to mechanisms that remain unexplained. The distorted ratios of transmitting the autosomal T/t alleles represent another failure (Silver, 1985). However, the exceptions are very rare and the distortions generally slight. In addition it should be remembered that X-chromosome-bearing spermatozoa are male, not female, tissue.

Four major mechanisms ensure that spermatozoa within a male are phenotypically equivalent. The first is sex-chromosome specific. During post-meiotic manufacture of spermatozoa, a process that takes weeks to accomplish, the sex chromosomes are in a heterochromatic sex vesicle, and there is little expression of genes on these chromosomes (Meistrich, 1982; Schmid, 1985). Both X and Y chromosomes are inactive. Because many genes on the X chromosome are essential for cells to live, inactivation of the X chromosome requires one of two compensatory ploys: expression of substitute autosomal genes active only during spermatogenesis, or production of required gene products by the Sertoli cells as they carry out their nurse cell functions (Bellvé, 1982).

The second mechanism to enhance phenotypic equivalence of spermatozoa is that developing spermatids are interconnected by cytoplasmic bridges (deKretser and Kerr, 1988). Thus, there is cytoplasmic continuity between X- and Y-chromosome-bearing spermatids. Little is known about the kinetics of information exchange through these cytoplasmic bridges, but some exchange is inevitable (Bellvé, 1982), including mRNA (Braun *et al.*, 1989). Maintenance of these bridges through meiotic cell divisions is shown in Fig. 1.

The third mechanism is limitation of post-meiotic gene expression, particularly in the later stages of sperm formation. Thus, most of the mRNA guiding transformation of a spherical spermatid into a spermatozoon is synthesized in the large primary spermatocyte, which has the diploid (actually duplicated diploid) genome expressing genes, and retained for later use (Hecht, 1993). The primary spermatocyte proceeds through the two meiotic cell divisions, leading to four very similar spermatids, including similar mRNA content (Fig. 1). Although mRNA from some genes continues to be transcribed in the haploid spermatids (which, however, are connected to each other by cytoplasmic bridges), it is at a relatively low level, slowing to near zero as the chromatin in the spermatid nucleus condenses (Bellvé, 1982; Hecht, 1993).

The fourth mechanism of making sperm phenotypically equivalent is to coat their surfaces with secretions of fluids from Sertoli cells and cells of the rete testis, efferent ducts, epididymis, and accessory sex glands (for example Iusem *et al.*, 1989; Amann *et al.*, 1993). Many molecules from these fluids have relatively strong binding characteristics.

To summarize this section, nature goes to great lengths to make spermatozoa within a male uniform phenotypically. Mechanisms include inactivation of sex chromosomes early in the meiotic process, maintaining cytoplasmic continuity among the developing sperm cells, turning off most gene expression once haploid cells are formed, and coating the resulting spermatozoa with numerous molecules. Thus, it would be expected that spermatozoa within a male are similar in size, cell surface properties, and other features. Undoubtedly, within each male there are subtle, mostly intracellular phenotypic differences between spermatozoa with different sex chromosomes, but whether these will be sufficient for practical use is unknown (Howes *et al.*, 1997). The repeated failures in separating X- and Y-chromosome-bearing spermatozoa (Amann and Seidel, 1982; Howes *et al.*, 1997) attest to success in producing a 50:50 sex ratio by making phenotypically identical spermatozoa. However, there is one substantial phenotypic difference between X- and Y-chromosome-bearing spermatozoa: there is more DNA in X than in Y chromosomes.

Principles of separating X- and Y-chromosome-bearing spermatozoa based on DNA content

Rapid measurement of DNA in single cells became possible with flow cytometry techniques. This was first accomplished for spermatozoa in the early 1980s, and was made even more useful by adding a sorting mechanism to the detection system. This work has been reviewed in detail several times recently (Johnson, 1994, 1995; Cran and Johnson, 1996).

Before 1997, X- and Y-chromosome-bearing spermatozoa from most mammals were analysed and sorted with a standard flow cytometer/cell sorter at rates of approximately 100 spermatozoa s^{-1} of each sex with about 90% accuracy. The principle is to incubate spermatozoa with the cell membrane-permeable, fluorescing dye Hoechst 33342, which binds quantitatively to AT-rich regions of DNA. Since X-chromosome-bearing spermatozoa have more DNA (about 3.8% more in cattle) than do Y-chromosome-bearing spermatozoa (Johnson, 1992), when stained they emit a stronger signal when excited by light at the correct wavelength, usually from a laser. The emission signal is amplified, analysed, and a determination made of whether the spermatozoon probably is X-chromosome bearing, Y-chromosome bearing, or ambiguous to the detection system. The majority of spermatozoa are classified as ambiguous, not because of intermediate DNA content, but because of imprecision in the staining and detection procedures.

One difficulty is that the fluorescence signal emitted is highly dependent on the orientation of

Primary spermatocytes

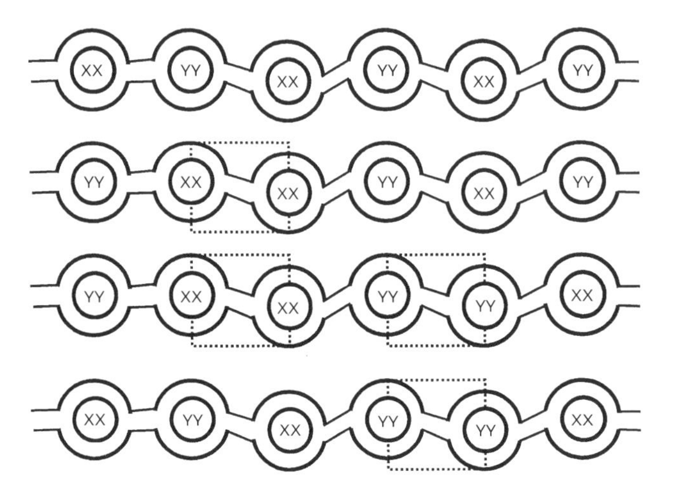

The four possible configurations resulting from divisions of three primary spermatocytes

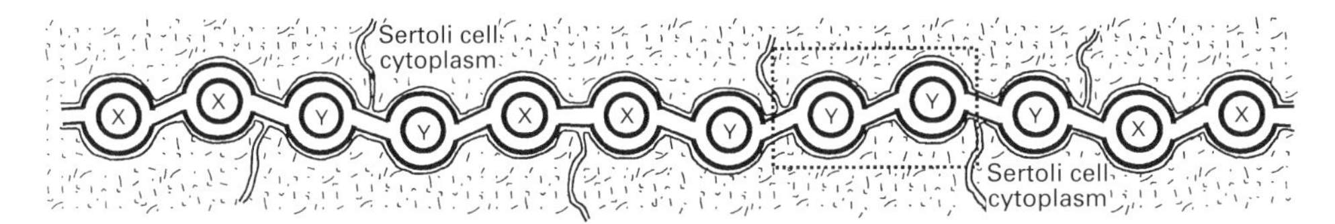

Resulting spermatids (illustrated for last configuration only) showing continuum of Sertoli cells that further affect homogeneity of resulting sperm cells

The boxes indicate the spermatids that will form that are connected to 2X or 2Y spermatids

Fig. 1. Segregation of sex chromosomes during meiosis. Depending on the species, up to 128 primary spermatocytes are interconnected by intercellular bridges due to incomplete cell division. Cell death may result in broken bridges and reduce the numbers of germ cells connected. Because centromeres do not divide during the first meiotic division, each secondary spermatocyte will be either of XX or YY sex chromosome composition, never XY. Possibilities of sex chromosome distributions during the first meiotic cell divisions of three interconnected primary spermatocytes are shown. It is not known whether the four possibilities illustrated occur with equal probability. Whatever the configuration, each secondary spermatocyte is connected to one XX and one YY secondary spermatocyte.

If the four configurations occur at random, excluding ends of sets, when a secondary spermatocyte divides to form spermatids, three situations would occur with the probabilities indicated in parentheses:

(1) A given spermatid connected to one X- and one Y-chromosome-bearing spermatid (75%).
(2) An X-chromosome-bearing spermatid connected to 2 X-chromosome-bearing spermatids (12.5%).
(3) A Y-chromosome-bearing spermatid connected to 2 Y-chromosome-bearing spermatids (12.5%).

Since the sex chromosomes are inactive transcriptionally, the asymmetry illustrated is of no phenotypic consequence. The course that nature has taken, shutting down transcription of sex the spermatozoa, because mammalian spermatozoa are asymmetrical. In practice, a second detector is used to determine orientation, and only those spermatozoa oriented correctly are analysed. Orientation can be regulated in several ways, for example by presenting spermatozoa to the quantifying detector in a ribbon rather than a cylinder of fluid (Gledhill *et al.*, 1982); the paddle-shaped ruminant spermatozoa tend to orient with their flat plane in the plane of the ribbon, although this is by no means a perfect relationship.

The sexing process is more productive if dead or deteriorating spermatozoa, which frequently constitute more than 10% of the ejaculate, depending on the species and the individual male, are not collected. This is accomplished by adding a molecule to the medium containing spermatozoa that is excluded from the sperm cytoplasm by a healthy plasma membrane, but that readily diffuses into the cytoplasm of spermatozoa with compromised cell membranes. This molecule quenches fluorescence of Hoechst 33342 that is bound to DNA, and thus marks dead spermatozoa. Propidium iodide and some food-colouring dyes are examples of such molecules (Johnson *et al.*, 1998).

What is new in flow cytometry/cell sorting of spermatozoa for DNA content?

The number of droplets encasing spermatozoa produced by flow cytometers per unit time increases as input pressure of the fluid stream increases. A new generation of flow cytometers has recently been developed that operate at higher pressure than earlier models, resulting in more rapid throughput and droplet production rates approaching 10^5 s^{-1}. Some of these also have improved electronics; a few microseconds difference in accomplishing each of the various steps of signal detection and processing can make a great difference in performance of such instruments. We have been using one such instrument routinely to obtain over 600 live spermatozoa s^{-1} of each sex (Seidel *et al.*, 1998). Another improvement is in nozzle design (Johnson *et al.*, 1998). One principle exploited is to move the detection system closer to the nozzle where there are oriented spermatozoa so that a higher percentage of spermatozoa remain oriented with their flat side toward the quantifying detector. Johnson *et al.* (1998) reported sorting rates between 1 000 and 1 500 spermatozoa s^{-1} of each sex at about 90% accuracy using this approach.

Further refinements of flow cytometers and cell sorters for sexing spermatozoa are inevitable. New detection systems (Sharpe *et al.*, 1997) might completely circumvent the problem of orientation of asymmetrical cells. Slit-scan procedures may solve the problem of overlap in distributions of signal strength of X- and Y-chromosome-bearing spermatozoa that occurs even when spermatozoa are in ideal orientation (Rens *et al.*, 1996).

There are theoretical limits to the number of spermatozoa that can flow past a detector per unit of time without obfuscating the signals of the individual spermatozoa. Similarly, a drop exiting the system may contain two spermatozoa with opposite sex-chromosomes; such drops are discarded. Perhaps such limitations will set a theoretical upper limit to sorting in the range of $5–10 \times 10^3$ live spermatozoa s^{-1} of each sex with flow cytometers. It is possible to resort to parallel processing, with individual components serving several streams, akin to one fuel pump for a multi-cylinder internal combustion engine. Ideally, systems for sexing spermatozoa will evolve that do not require evaluation of spermatozoa one at a time. Up to the present time, no such system has proven reliable.

Special problems and benefits of sorting spermatozoa with flow cytometers/cell sorters

Incubating spermatozoa with Hoechst 33342, which has a high affinity for AT-rich regions of DNA, exposing them to laser light, pumping them through fine tubing at high pressure so that they exit a small orifice at about 100 km h^{-1}, and storing them at ambient temperature for hours as numbers accumulate are processes that are not conducive to sperm health. We have evaluated sorted

---

chromosomes and limiting transcription of autosomes post-meiotically circumvents potential problems of transferring excess genetic information from X or Y chromosomes to some spermatids via intercellular bridges. Note that all germ cells are embedded in Sertoli cell cytoplasm, another mechanism for making sperm cell membranes identical (illustrated for spermatids only).

spermatozoa using many different procedures and invariably find that spermatozoa are damaged to some extent by the sexing procedures described. However, it is unclear which aspects of the process cause damage. To what extent such damage can be decreased by improved techniques or overcome by compensatory numbers of spermatozoa or other means is unknown. However, it is clear that pregnancy rates are not compromised greatly with sorted spermatozoa, even though they have sustained some damage (Seidel *et al.*, 1998).

A phenomenon documented by McNutt and Johnson (1996) is that the first cell cycle is delayed in rabbit embryos fertilized by spermatozoa treated with Hoechst 33342. The mechanism involved is unknown, but could be interference of the dye molecules as DNA is replicated or transcribed. This might be responsible for lowered pregnancy rates with sexed spermatozoa. Fortunately, no increased incidence of abnormalities has been noted in the hundreds of offspring of various species produced by sexed semen (Cran and Johnson, 1996). However, rigorous tests of normality using hundreds of offspring in controlled trials involving multiple generations remain to be done.

All is not negative. There are theoretical benefits to using spermatozoa for insemination that are processed by the methods described. For example, dead spermatozoa as determined by dye exclusion are eliminated. In addition, some aneuploidies, including diploid spermatozoa, are eliminated via gating of fluorescence signals during the sorting process. Finally, it is possible that *in vitro* capacitation of flow-sorted spermatozoa will be facilitated.

Artificial insemination with sexed semen

Although impressive improvements have been made in flow cytometer/cell sorter performance, even the most optimistic projections result in insufficient spermatozoa throughput to use sexed semen with procedures for routine artificial insemination. Current sperm sorting rates are more than adequate, however, for procedures such as *in vitro* fertilization (Cran *et al.*, 1993) and intra-cytoplasmic sperm injection.

In some species, laparoscopic insemination with low numbers of spermatozoa may be feasible (for example, sheep; Cran *et al.*, 1997). In other species, artificial insemination procedures can probably be modified to achieve acceptable pregnancy rates with a smaller number of spermatozoa than are used routinely. For example, we have achieved reasonable pregnancy rates in heifers using only 5×10^5 total (unsexed) spermatozoa per cryopreserved insemination dose (Seidel *et al.*, 1996). In these and related studies with unfrozen semen (Seidel *et al.*, 1997), we inseminated spermatozoa in 0.1–0.2 ml into the uterine horns using atraumatic, side-opening sheaths. Which, if any, of these modifications to conventional insemination procedures contributes to efficacy was not studied, and results may not apply to all bulls. Den Daas *et al.* (1998) found that the decrease in fertility as numbers of spermatozoa frozen per insemination dose were lowered varied markedly among bulls; for some bulls fertility was normal at the lowest numbers of spermatozoa tested, about 2×10^6 spermatozoa per insemination into the uterine body.

In two recent studies, we combined procedures for inseminating low doses of semen with the flow sorting procedures described earlier. In the first study (Seidel *et al.*, 1997), a conventional flow cytometer providing about 100 spermatozoa s^{-1} of each sex was used. The bulls, flow cytometer, and cattle to be inseminated were located great distances from each other, so coordination of procedures was imperfect. Success with this liquid, sexed semen varied among replicates from 0/38 to 9/20 pregnant to term; 14 of the 17 calves produced in this study were of the predicted sex. In this preliminary trial, there was a strong suggestion of bull-to-bull variation in pregnancy rates with low doses of sexed semen. In addition, the pregnancy rate declined markedly if the interval from sorting to insemination exceeded 12 h.

The second study (Seidel *et al.*, 1998) was logistically less demanding. A high performance flow cytometer and cell sorter provided 500–600 live spermatozoa s^{-1} of each sex, and the cattle were only 200 km distant from the instrument. The pregnancy rate to term after insemination of doses of 3×10^5 live (total) spermatozoa selected to have X chromosomes was 42% ($n = 45$); for the liquid semen control of 3×10^5 motile spermatozoa per insemination, the pregnancy rate was 54% ($n = 28$); and for the frozen semen control of 15.6×10^6 motile spermatozoa per dose after thawing, it was 52% ($n = 29$;

Seidel *et al.*, 1998). These data were balanced across three bulls and two inseminators. Eighteen of 19 calves born from sexed semen were females. Only one sexed and one control pregnancy detected by ultrasonography at 31–34 days after insemination did not develop to term, indicating that pregnancy wastage was minimal after 1 month of gestation. Although fertility of this sexed semen was only about 80% of controls, these results are promising. Early data from a field trial in progress (Schenk *et al.*, unpublished) indicate that pregnancy rates after artificial insemination of heifers with sexed, frozen semen with twice the number of spermatozoa (10^6 per dose) to compensate for damage due to cryopreservation were virtually identical to those using sexed, unfrozen semen (5×10^5 spermatozoa per dose).

Future prospects for applying sexed semen technology

There are many studies that convincingly demonstrate that it is possible to change the sex ratio at conception to approximately 90% of the desired sex by sorting spermatozoa on the basis of DNA content using flow cytometer and cell sorter technology (for example Johnson *et al.*, 1989; Cran *et al.*, 1993; Seidel *et al.*, 1998). Even with sorting rates of 1×10^3–1.5×10^3 live spermatozoa s^{-1} of each sex, this remains primarily a research technique, with possible niche applications for *in vitro* fertilization and artificial insemination of very valuable breeding animals.

Flow cytometer and cell sorter equipment is very expensive, and requires skilled personnel for operation. To date, general purpose research instruments have been used. Possibly flow cytometers and cell sorters specifically designed for sorting spermatozoa would be simpler to operate, less expensive, and more efficacious. For serving mass agricultural markets, it would seem that instruments sorting a minimum of 10–20 insemination doses of semen h^{-1} of each sex would be required.

Even with optimistic assumptions, sorting spermatozoa one at a time as they flow past a detector is not an appealing approach for agricultural applications. This is a difficult process to scale up. Perhaps other approaches will be developed. Many patents have been issued for sexing semen (summarized by Ericsson and Glass, 1982), although the majority are of little value. Before the early 1980s, there was no reliable method for determining whether a particular technique to sex spermatozoa was efficacious, even if spermatozoa were killed, other than inseminating many animals for each assay. The expense and time required for obtaining such information for practical purposes precluded developing and improving sperm sexing methodology. Flow cytometry provides a robust tool to test, within minutes, other approaches to sexing semen. To date, all alternative approaches to sexing semen examined have failed this quality control test (Gledhill *et al.*, 1982; Johnson, 1988).

With flow cytometry or molecular approaches such as fluorescence-in situ hybridization for assays and our rapidly increasing knowledge of molecular and cellular biology of spermatozoa, more practical methods of sexing semen might become available eventually. However, for the foreseeable future, the only reliable method for sexing spermatozoa while maintaining fertility is with a flow cytometer and cell sorter.

Conclusions

Thousands of bovine preimplantation embryos are sexed annually on a commercial basis with > 95% accuracy and with little decrease in pregnancy rates, by subjecting an embryonic biopsy to the polymerase chain reaction. Other methods of sexing embryos, while accurate to various degrees, remain limited to research applications.

There is only one efficacious method of sexing spermatozoa that does not kill them – quantifying DNA by flow cytometry and cell sorting after exposing spermatozoa to a fluorescing DNA-binding dye. Sorting techniques are improving rapidly; currently it is possible to sort more than 2×10^3 live spermatozoa per second, half enriched for each sex chromosome. Sorted spermatozoa are damaged to some extent with current procedures, but hundreds of apparently normal offspring have been produced. Sexing spermatozoa is still a research technique, but commercialization for artificial insemination may occur in the near future.

Drs Rupert Amann and David Cran critically read the manuscript and provided helpful comments. Research from our laboratory has been supported in part by XY, Inc., Fort Collins, Colorado, the Colorado State University Experiment Station, the Colorado Advanced Technology Institute, Genex, Inc., and the National Association of Animal Breeders.

References

Amann RP (1989) Treatment of sperm to predetermine sex *Theriogenology* **31** 49–60

Amann RP and Seidel GE, Jr (1982) *Prospects For Sexing Mammalian Sperm.* Colorado Associated University Press, Boulder, CO

Amann RP, Hammerstedt RH and Veeramachaneni DNR (1993) The epididymis and sperm maturation: a perspective *Reproduction, Fertility and Development* **5** 361–381

Anderson GB (1987) Identification of embryonic sex by detection of H-Y antigen *Theriogenology* **27** 81–97

Bellvé AR (1982) Biogenesis of the mammalian spermatozoan. In *Prospects for Sexing Mammalian Sperm* pp 69–102 Ed. RP Amann and GE Seidel, Jr. Colorado Associated University Press, Boulder, CO

Betteridge KJ, Hare WCD and Singh EL (1981) Approaches to sex selection in farm animals. In *New Technologies in Animal Breeding* pp 109–125 Ed. BG Bracket, GE Seidel, Jr and SM Seidel. Academic Press, New York

Bogan JS and Page DC (1994) Ovary? Testis? A mammalian dilemma *Cell* **76** 603–607

Bondioli K (1992) Embryo sexing: a review of current techniques and their potential for commercial application in livestock production *Journal of Animal Science* **70** (Supplement 2) 19–29

Bondioli KR, Ellis SB, Prior JH, Williams MW and Harpold MM (1989) The use of male-specific chromosomal DNA fragments to determine the sex of preimplantation bovine embryos *Theriogenology* **31** 95–104

Braun RE, Behringer RR, Peschon JJ, Brinster RL and Palmiter RD (1989) Genetically haploid spermatids are phenotypically diploid *Nature* **337** 373–376

Bredbacka P, Kankaanpää A and Peippo J (1995) PCR-sexing of bovine embryos: a simplified protocol *Theriogenology* **40** 167–176

Burfening PJ, Kress DD, Friedrich RL and Vaniman DD (1978) Phenotypic and genetic relationships between calving ease gestation length, birth weight and pre-weaning growth *Journal of Animal Science* **47** 595–600

Cran DG and Johnson LA (1996) The predetermination of embryonic sex using flow cytometrically separated X and Y spermatozoa *Human Reproduction Update* **2** 355–363

Cran DG, Johnson LA, Miller NGA, Cochrane D and Polge C (1993) Production of calves following separation of X and Y chromosome bearing sperm and *in vitro* fertilization *Veterinary Record* **132** 40–41

Cran DG, McKelvey WAC, King ME, Dolman DF, McEvoy TG, Broadbent PJ and Robinson JJ (1997) Production of lambs by low dose intrauterine insemination with flow cytometrically sorted and unsorted semen *Theriogenology* **47** 267 (Abstract)

Curran S and Ginther OJ (1991) Ultrasonic determination of fetal gender in horses and cattle under farm conditions *Theriogenology* **36** 809–814

deKretser DM and Kerr JB (1988) The cytology of the testis. In *The Physiology of Reproduction.* Vol. 1. pp 837–932 Ed. E Knobil and JD Neill. Raven Press, New York

den Daas JHG, De Jong G, Lansbergen LMTE and Van Wagtendonk-De Leeuw AM (1998) The relationship between the number of spermatozoa inseminated and the reproductive efficiency of individual dairy bulls *Journal of Dairy Science* **81** 1714–1723

Dziuk PJ (1996) Factors that influence the proportion of offspring sired by a male following heterospermic insemination *Animal Reproduction Science* **43** 65–88

Ericsson RJ and Glass RH (1982) Functional differences between sperm bearing the X or Y chromosome. In *Prospects for Sexing Mammalian Sperm* pp 201–211 Ed. RP Amann and GE Seidel, Jr. Colorado Associated University Press, Boulder, CO

Fisher RA (1930) *The Genetical Theory of Natural Selection.* Oxford University Press, Oxford

Gardner RL and Edwards RG (1968) Control of the sex ratio at full term in the rabbit by transferring sexed blastocysts *Nature* **218** 346–348

Gledhill BL, Pinkel D, Garner DL and Van Dilla MA (1982) Identifying X- and Y-chromosome-bearing sperm by DNA content: retrospective perspectives and prospective opinions. In *Prospects for Sexing Mammalian Sperm* pp 177–191 Ed. RP Amann and GE Seidel, Jr. Colorado Associated University Press, Boulder, CO

Gosling LM (1986) Selective abortion of entire litters in the Coypu: adaptive control of offspring production in relation to quality and sex *American Naturalist* **127** 772–795

Greenfield H, Scott D, Pennisi D, Ehrman I, Ellis P, Cooper L, Simpson E and Koopman P (1996) An H-YDb epitope is encoded by a novel Y chromosome gene *Nature Genetics* **4** 474–478

Hare WCD, Mitchell D, Betteridge KJ, Eaglesome MD and Randall GCB (1976) Sexing two-week old bovine embryos by chromosomal analysis prior to surgical transfer: preliminary methods and results *Theriogenology* **5** 243–253

Hecht NB (1993) Gene expression during male germ cell development. In *Cell and Molecular Biology of the Testes* pp 400–432 Ed. C Desjardins and L Ewing. Oxford University Press, Oxford

Herr CM and Reed KC (1991) Micromanipulation of bovine embryos for sex determination *Theriogenology* **35** 45–54

Howes EA, Miller NGA, Dolby C, Hutchings A, Butcher GW and Jones R (1997) A search for sex-specific antigens on bovine spermatozoa using immunological and biochemical techniques to compare the protein profiles of X and Y chromosome-bearing sperm populations separated by fluorescence-activated cell sorting *Journal of Reproduction and Fertility* **110** 195–204

Iusem ND, Pineiro L, Blaquier JA and Belocopitow E (1989) Identification of a major secretory glycoprotein from rat epididymis: interaction with spermatozoa *Biology of Reproduction* **40** 307–316

Johnson LA (1988) Flow cytometric determination of sperm sex ratio in semen purportedly enriched for X- and Y-bearing sperm *Theriogenology* **29** 265 (Abstract)

Johnson LA (1991) Sex preselection in swine: altered sex ratios in offspring following surgical insemination of flow sorted X and Y chromosome bearing sperm *Reproduction in Domestic Animals* **26** 309–314

Johnson LA (1992) Gender preselection in domestic animals using flow cytometrically sorted sperm *Journal of Animal Science* **70** (Supplement 1) 8–18

Johnson LA (1994) Isolation of X- and Y-bearing sperm for sex preselection *Oxford Reviews of Reproductive Biology* **16** 303–326

Johnson LA (1995) Sex preselection by flow cytometric separation of X and Y chromosome-bearing sperm based on DNA difference: a review *Reproduction, Fertility and Development* **7** 893–903

Johnson LA, Flook JP and Hawk HW (1989) Sex preselection in rabbits: live births from X and Y sperm separated by DNA and cell sorting *Biology of Reproduction* **41** 199–203

Johnson LA, Welch GR, Rens W and Dobrinsky JR (1998) Enhanced flow cytometric sorting of mammalian X and Y sperm: high speed sorting and orienting nozzle for artificial insemination *Theriogenology* **49** 361 (Abstract)

Kiddy CA and Hafs HD, Eds (1971) *Sex Ratio at Birth – Prospects for Control*. American Society for Animal Science, Savoy, IL

Koopman P, Gubbay J, Vivian N, Goodfellow P and Lovell-Badge R (1991) Male development of chromosomally female mice transgenic for SRY *Nature* **351** 117–121

Levinson G, Keyvanfar K, Wu JC, Fugger EF, Fields RA, Harton GL, Palmer FT, Sisson ME, Starr KM, Dennison-Lagos L, Calvo L, Sherins RJ, Bick D, Schulman JD and Black SH (1995) DNA-based X-enriched sperm separation as an adjunct to preimplantation genetic testing for the prevention of X-linked disease *Molecular Human Reproduction* **10** 979–982

McNutt TL and Johnson LA (1996) Flow cytometric sorting of sperm: influence on fertilization and embryo/fetal development in the rabbit *Molecular Reproduction and Development* **43** 261–267

Meistrich ML (1982) Events of meiosis, spermiogenesis and post-meiotic gene expression related to haploid gene expression. In *Prospects for Sexing Mammalian Sperm* pp 103–106 Ed. RP Amann and GE Seidel, Jr. Colorado Associated University Press, Boulder, CO

Nelson LD, Pickett BW and Seidel GE, Jr (1975) Effect of heterospermic insemination on fertility of cattle *Journal of Animal Science* **40** 1124–1129

Penfold LM, Holt C, Holt WV, Welch GR, Cran DG and Johnson LA (1998) Comparative motility of X and Y chromosome-bearing bovine sperm separated on the basis of DNA content by flow sorting *Molecular Reproduction and Development* **50** 323–327

Rens W, Welch GR, Houck DW, van Oven CH and Johnson LA (1996) Slit-scan flow cytometry for consistent high resolution DNA analysis of X- and Y-chromosome bearing sperm *Cytometry* **25** 191–199

Schmid M (1985) Arrangement of the Y chromosome in interphase and metaphase cells. In *The Y Chromosome, Part A: Basic Characteristics of the Y Chromosome* pp 17–61. AR Liss, New York

Seidel GE, Jr (1988) Sexing mammalian sperm and embryos *Proceedings of the 11th International Congress on Animal Reproduction and Artificial Insemination* (Dublin) **5** 136–144

Seidel GE, Jr, Allen CH, Brink Z, Holland MD and Cattell M-B (1996) Insemination of heifers with very low numbers of frozen spermatozoa *Journal of Animal Science* **74** (Supplement 1) 235 (Abstract)

Seidel GE, Jr, Allen CH, Johnson LA, Holland MD, Brink Z, Welch GR, Graham JK and Cattell M-B (1997) Uterine horn insemination of heifers with very low numbers of non-frozen and sexed spermatozoa *Theriogenology* **48** 1255–1264

Seidel GE, Jr, Herickhoff LA, Schenk JL, Doyle SP and Green RD (1998) Artificial insemination of heifers with cooled, unfrozen sexed semen *Theriogenology* **49** 365 (Abstract)

Sharpe JC, Schaare PN and Kunnemeyer R (1997) Radially symmetric excitation and collection optics for flow cytometric sorting of asymmetrical cells *Cytometry* **29** 363–370

Silver LM (1985) Mouse t haplotypes *Annual Review of Genetics* **19** 179–208

Thibier M and Nibart M (1995) The sexing of bovine embryos in the field *Theriogenology* **43** 71–80

Tsunoda Y, Tokunaga T and Sugie T (1985) Altered sex ratio of live young after transfer of fast- and slow-developing mouse embryos *Gamete Research* **12** 301–304

Utsumi K, Satoh E and Yuhara M (1984) Sexing of mammalian embryos exposed to H-Y antisera *Proceedings of the 10th International Congress on Animal Reproduction and Artificial Insemination* (Urbana) **2** 234

Wang W, Meadows LR, den Hahn JMM, Sherman NE, et al. (1995) Human H-Y: a male specific histocompatibility antigen derived from the SMCY protein *Science* **269** 1588–1590

West JD, Godsen JR, Angell RR, Hastie ND, Thatcher SS, Glasier AF and Baird DT (1987) Sexing the human pre-embryo by DNA–DNA *in situ* hybridization *Lancet* **1** 1345–1347

Williams TJ (1986) A technique for sexing mouse embryos by a visual colorimetric assay of the X-linked enzyme glucose 6-phosphate dehydrogenase *Theriogenology* **25** 733–739

Journal of Reproduction and Fertility Supplement **54**, 489–497

Nuclear transfer from somatic cells: applications in farm animal species

W. H. Eyestone[1] and K. H. S. Campbell[2]

PPL Therapeutics, [1]1700 Kraft Drive, Blacksburg, VA and [2]Roslin Institute, Midlothian EH25 9PP, UK

The reconstruction of mammalian embryos by transfer of a blastomere nucleus to an enucleated oocyte or zygote allows for the production of genetically identical individuals. This has advantages for research (that is, as biological controls) and commercial applications (that is, multiplication of genetically valuable livestock). However, the number of offspring that can be produced from a single embryo is limited both by the number of blastomeres (embryos at the 32–64-cell stage are the most widely used in farm animal species) and the limited efficiency of the nuclear transfer procedure. The ability to produce live offspring by nuclear transfer from cells that can be propagated and maintained in culture offers many advantages, including the production of many identical offspring over an extended period (since cultured cells can be frozen and stored indefinitely) and the ability to modify genetically or to select populations of cells of specific genotypes or phenotypes before embryo reconstruction. This objective has been achieved with the production of lambs using nuclei from cultured cells established from embryonic, fetal and adult material. In addition, lambs transgenic for human factor IX have been produced from fetal fibroblasts transfected and selected in culture.

Introduction

Embryo reconstruction by nuclear transfer involves the transfer of a single nucleus to an unfertilized oocyte or zygote from which the native genetic material has been removed. In farm animals, unfertilized oocytes arrested at metaphase of the second meiotic division (MII) as recipient cells have become the 'recipient' cell of choice. The nuclear 'donor' cell used for reconstruction depends upon the application of the technology. For embryo multiplication, blastomeres are used as nuclear donors (Prather *et al.*, 1987). More recently, live lambs were produced using cultured cells derived from embryonic, fetal and adult tissues as nuclear donors (Campbell *et al.*, 1995, 1996a; Wilmut *et al.*, 1997). This approach has recently been applied to cattle and viable offspring have been reported by several laboratories, including ours (Christmann, Chen, Polejaeva, Campbell and Eyestone, unpublished), using cultured fetal cells as nuclear donors (Cibelli *et al.*, 1998).

The development of embryos reconstructed by nuclear transfer is dependent on many factors. These include the quality of the recipient oocytes, the cell cycle stage of both the donor cell and the recipient cytoplasm, the method of oocyte culture and activation, and the developmental stage or differentiated state of the donor nucleus. More recently it has been suggested that the state of chromatin in the donor nucleus at the time of transfer may affect development of reconstructed embryos (Campbell and Wilmut, 1997).

Methods of Embryo Reconstruction

Historically, two types of recipient cell have been used for nuclear transfer: oocytes arrested at MII or pronuclear zygotes. In mice, enucleated, two-cell stage blastomeres have been used as recipients

(Tsunoda *et al.*, 1987). In farm species, development does not occur when pronuclear zygotes are used, except when pronuclei are exchanged between zygotes (cattle: Robl *et al.*, 1987; pig: Prather *et al.*, 1989); thus, MII-arrested oocytes have become the recipient of choice, since development to live offspring has been obtained using this type of recipient.

Sources of recipient oocytes

Oocytes for use as recipient cells in embryo reconstruction can be obtained after *in vivo* maturation from mature, unovulated follicles, or by flushing ovulated oocytes from the oviducts (Prather *et al.*, 1987; Prather *et al.*, 1989; Campbell *et al.*, 1994). Alternatively, oocytes recovered at slaughter in cattle (Barnes *et al.*, 1993), sheep (Pugh *et al.*, 1991) and pigs (Hirao *et al.*, 1994) and matured *in vitro* have also been used as recipients. In cattle, immature oocytes may also be obtained by aspirating ovarian follicles *in vivo* in cattle (for review, see Bols *et al.*, 1994).

Enucleation of recipient oocytes

The term enucleation is used to describe the removal of the genetic material from the recipient cell. Oocytes arrested at MII do not contain a nucleus; rather, the chromatin is condensed as chromosomes arranged on the meiotic spindle. In farm animal oocytes, the MII chromosomes, or metaphase plate, are not visible under the light microscope. However, they are generally situated subjacent to the first polar body, which can be used as a convenient landmark for locating the metaphase plate. Enucleation is accomplished by piercing the zona pellucida with a glass pipette (15–20 μm in diameter), and then placing the pipette tip over the polar body (and thus the metaphase plate) and withdrawing a membrane-enclosed portion of cytoplasm containing the meiotic chromosomes by applying gentle suction. In some cases, oocytes may be treated with the microtubule inhibitor cytochalasin B to render the plasma membrane elastic to facilitate enucleation and reduce mechanical stress and damage to the oocyte as a result of the enucleation procedure. Enucleation is confirmed by staining the karyoplast with a DNA-specific fluorochrome (i.e. Hoescht 3332) either following aspiration (Westhusin *et al.*, 1990) or during the aspiration procedure as is used routinely in our laboratory (Campbell *et al.*, 1993a).

Embryo reconstruction

After enucleation, the genetic material from the donor cell (karyoplast) must be introduced into the enucleated oocyte (cytoplast). In general this has been achieved by fusion, although direct injection techniques have been used successfully by some workers (Collas and Barnes, 1994; Ritchie and Campbell, 1995). Fusion is induced by a number of agents including Sendai virus (Graham, 1969), polyethylene glycol (PEG; Kanka *et al.*, 1991) or application of a DC electric current (Willadsen, 1986). In farm animal species, electrofusion is the most commonly used method. Use of Sendai virus is efficient in mice, although its effects are variable in other species, for example sheep (Willadsen, 1986). The use of PEG requires its fast and efficient removal after fusion because of its toxicity.

Activation of the reconstructed embryos

After introducing the donor genetic material, the reconstructed embryo has to initiate embryo development. Normally, development is initiated by an activation event induced by the spermatozoa at fertilization. Fertilization stimulates a series of intracellular calcium peaks that appear to be necessary and sufficient for activation of development. In the absence of fertilization, treatments must be applied that mimic these events to induce development. During recent years many treatments have been reported to cause oocyte activation. Such treatments have included electrical stimulation using either a single DC pulse (pig: Prochazka *et al.*, 1992; cattle: Kono *et al.*,

1989; for review see Robl *et al.*, 1992) or multiple electrical stimuli coinciding with the reported calcium peaks following fertilization (rabbit: Ozil, 1990; cattle: Collas *et al.*, 1993a), various chemical treatments including phorbol ester (mouse: Cuthbertson and Cobbold, 1985), calcium ionophores (i.e. A23187: Ware *et al.*, 1989; Aoyagi, 1992), ionomycin (cattle: Susko Parrish *et al.*, 1994), components of various second messenger systems (inositol Tris-phosphate, mouse: Jones *et al.*, 1995a; cattle: White and Yue, 1996), ethanol (cattle: Nagai, 1987) and strontium chloride (mouse: O'Neill *et al.*, 1991).

The rate and frequency of oocyte activation are dependent upon the age of the oocyte after the onset of maturation. As oocyte age increases, activation can occur spontaneously as a result of changes in temperature or other manipulations. In addition, pronuclear formation occurs more rapidly (cattle: Ware *et al.*, 1989; Presicce and Yang, 1994). In contrast, activation of 'young' oocytes has proved more difficult. In cattle and pigs, treatment of young MII oocytes with an activation stimulus combined with inhibitors of protein synthesis (that is, cycloheximide or puromycin) can overcome this block, for example in cattle (Presicce and Yang, 1994) and pigs (Nussbaum and Prather, 1995). More recently, soluble sperm factors have been extracted that can induce activation after they are injected into mouse oocytes (Parrington *et al.*, 1996; for review see Swann, 1996).

The role of activation in subsequent development is being elucidated slowly. Calcium oscillations continue throughout the first cell cycle in the mouse and are associated with mitotic division (Kono *et al.*, 1996; Jones *et al.*, 1995b). Activation of mouse oocytes with strontium chloride and continued exposure until after the first cleavage division have been shown to increase the size of measurable calcium oscillations at first mitosis to that observed in fertilized zygotes (Kono *et al.*, 1996; Jones *et al.*, 1995a). In addition, the number of cells at the blastocyst stage also increases (Jones *et al.*, 1996) indicating that the efficiency of activation may have far-reaching developmental effects.

Culture of reconstructed embryos

After reconstruction, the embryo must be cultured to a stage at which it can be transferred to a synchronized recipient animal for development to term, generally at the morula or blastocyst stage. Briefly, two options are available: culture *in vitro* or culture *in vivo*. There are many systems for the culture of embryos *in vitro* and research in this area has resulted in an increase in the frequency and the quality of development (for review see Campbell and Wilmut, 1994). Traditionally, reconstructed embryos from cattle and sheep have been cultured in ligated oviducts of temporary recipient ewes (Willadsen, 1986). Owing to the hole made in the zona pellucida during manipulation, this method requires the encapsulation of each embryo in agar. The function of the agar 'chip' is two-fold: first, it holds the embryo in the zona pellucida and second, it prevents attack of the embryo by macrophages within the oviduct.

Nuclear–cytoplasmic interactions in reconstructed embryos

The successful development of reconstituted embryos is dependent upon a large number of factors. In fertilized zygotes early events are controlled by maternally inherited RNAs and proteins, until the transcription is initiated in blastomere nuclei during early cleavage stages. In reconstituted embryos both the cytoplasm and the transferred nucleus must be able to recapitulate these events. Changes in nuclear structure, chromatin structure and gene activity have been reported; these will be discussed in relation to the differentiated state of the donor nucleus. However, for these so called 'reprogramming' events to be successful the chromatin must remain free of damage and the embryo must maintain normal ploidy.

Nuclear events during the first cell cycle

Interactions between cytoplasmic factors within the cytoplast and the cell cycle stage of the donor nucleus at the time of fusion are crucial to the avoidance of DNA damage and the

Table 1. Effects of various cell cycle stage combinations of donor and recipient cells on chromatin and ploidy of reconstructed mammalian embryos during the first cell cycle

Cell cycle stage of recipient	MPF activity	Stage of ploidy (cell cycle stage of donor)	Effect on nucleus	Effect on chromatin	DNA synthesis	Ploidy daughter cells
MII	High	2C[a] (G0/G1)	NEBD[c]	SCs[d]	+	2C
MII	High	4C[b] (G2)	NEBD	DCs[e]	+	4C
MII	High	2–4C (S)	NEBD	PULVERISED	+	? 2–4C[f]
G1/S	Low	2C (G0/G1)	No NEBD		+	2C
G1/S	Low	4C (G2)	No NEBD		+	2C
G1/S	Low	2–4C (S)	No NEBD		–	2C

MPF, maturation/mitosis/meiosis promoting factor; [a]2C = diploid; [b]4C = tetraploid; [c]Nucelar envelope breakdown; [d]single chromatids; [e]double chromatids
[f]?2–4C = unknown ploidy

maintenance of correct ploidy (for review see Campbell *et al.*, 1996b). Early development is characterized by a series of reduction divisions and no net growth occurs. For simplicity the major events of the first cell cycle can be described as those that concern the nucleus relative to DNA replication during which the genetic material is duplicated and this is followed by mitosis and cleavage when the duplicated material is equally segregated to the two daughter cells. The onset of both meiotic and mitotic divisions is controlled by a cytoplasmic activity factor termed MPF (maturation/mitosis/meiosis promoting factor). MII oocytes arrest at metaphase of the second meiotic division and contain high MPF activity. Upon fertilization or activation, MPF activity declines to basal values until the G2 phase of the cycle when increasing MPF activity induces entry to first mitosis. When nuclei are fused to MII oocytes, MPF activity induces the transferred nucleus to enter a mitotic division precociously, characterized by nuclear envelope breakdown (NEBD) and chromatin condensation. The effects of this premature entry to mitosis, or premature chromosome condensation (PCC) as it has been termed, on the transferred nucleus are dependent upon its cell cycle phase. Nuclei in S-phase undergo large amounts of DNA damage; in contrast, nuclei that are before (2C) or after (4C) S-phase form single or double chromatids, respectively, and appear to avoid DNA damage. NEBD also results in DNA synthesis occurring in all nuclei regardless of their cell cycle stage; thus only diploid nuclei will avoid DNA damage and retain correct ploidy. In contrast, if embryos are reconstructed after the decline of MPF activity, NEBD or PCC do not occur, DNA synthesis is controlled by the cell cycle stage of the donor nucleus and correct ploidy is maintained (see Table 1).

The occurrence of NEBD and PCC may not be related solely to the activity of MPF. In cattle oocytes MPF activity, when measured biochemically, declines within 2–3 h after activation (Campbell *et al.*, 1993a,b; Collas *et al.*, 1993b). However, in embryos reconstructed at different times after activation and examined one hour after application of the fusion pulse, NEBD of the transferred nucleus is observed for approximately 9 h after activation (Campbell *et al.*, 1993a). This discrepancy may be the result of further cytoplasmic factor/s, one such activity is MAP kinase. This cytoplasmic kinase which becomes activated early during the first cell cycle of murine zygotes is incompatible with pronuclear formation (for review see Whittaker, 1996).

Cell cycle co-ordination and development

As detailed earlier, inappropriate choice of donor and recipient cell cycle stages can result in chromosomal damage and aneuploidy during the first cell cycle. Two approaches can be used to avoid these problems; first, diploid nuclei can be transferred to MII oocytes and secondly G0/G1, S and G2-phase nuclei can be transferred after the decline of cytoplasmic activities that induce NEBD (for review see Campbell *et al.*, 1996b). The former of these approaches requires the synchronization or selection of diploid nuclei.

Embryonic blastomeres as nuclear donors

In early embryos, at any one time, most nuclei are in S-phase. Although synchronization of embryonic blastomeres has been successful in mice (Otaegui *et al.*, 1994), similar methods have proved unreliable in farm animal species (authors' unpublished observations). The use of pre-activated oocytes as cytoplast recipients for unsynchronized blastomeres results in a significant increase in the frequency of development to the blastocyst stage in both sheep (Campbell *et al.*, 1994) and cattle (Stice *et al.*, 1994). As an alternative to preactivation, treatment of MI-enucleated cattle oocytes with 6-dimethyl amino purine results in the oocyte arresting in an unactivated state with low MPF (Susko-Parrish *et al.*, 1994). This situation mimics the use of preactivated oocytes as cytoplast recipients in that the transferred nuclei do not undergo NEBD and upon subsequent activation DNA replication is controlled by the cell cycle phase of the karyoplast, thus maintaining ploidy.

The ability to synchronize the blastomeres of early murine embryos has allowed more detailed studies of cytoplast/ karyoplast cell cycle combinations. Recent reports in mice have demonstrated the use of mitotically arrested cells as donors of genetic material (Kwon and Kono, 1996). Briefly, blastomeres from nocodazole-treated embryos arrested in mitosis are transferred to enucleated MII oocytes. Extrusion of a polar body is inhibited by treatment with cytochalasin B, which results in the formation of two (pro) nuclei. One of these nuclei is then transferred to an enucleated, one-cell embryo. Six identical pups have been produced from a single four-cell embryo using this method, and development has also been obtained from 20-cell embryos (Kono, personal communication). In a further study, Otaegui (personal communication) suggests that during the late G2 and early G1 phase of the donor cell cycle, the chromatin can re-direct development more effectively. Together these two studies indicate that during late G2, there is an M and early G1 permissive state which allows nuclear reprogramming. One possible explanation for these observations is that, during these cell cycle phases, certain factors are released from the chromatin, thus allowing access to oocyte-derived factors. This hypothesis is supported by two lines of evidence; first during mitosis transcription factors become displaced from the chromatin (Schermoen and O'Farrel, 1991; Martinez-Balbas *et al.*, 1995) and second, recent evidence indicates that live offspring can be produced from cultured, differentiated cell lines induced to exit the growth cycle and enter a state of quiescence (see below).

Other cell types as nuclear donors

One of the aims of developmental biologists and biotechnologists is to produce offspring from cell populations that can be maintained in culture. For nuclear transfer the availability of a cultured cell line would facilitate cell cycle synchronization of the donor nucleus and allow optimization of cell cycle co-ordination in the reconstituted embryo. In frogs, development to adult was obtained following the transplantation of intestinal epithelial cell nuclei into enucleated oocytes (Gurdon, 1962a,b). In mammals a number of cell types including embryonic stem cells in the mouse and primordial germ cells may be totipotent (for review see Wilmut *et al.*, 1992); however, as yet development to term has not been demonstrated, i.e. cattle PGCs (Moens *et al.*, 1996), murine ES cells (Modlinski *et al.*, 1996). Recently we reported development to term of ovine embryos reconstructed using cell populations derived from embryo, fetal and adult tissues as nuclear donors (Campbell *et al.*, 1995, 1996b; Wilmut *et al.*, 1997). In these experiments the cells were arrested in a G0 or quiescent state by the reduction of serum concentrations in the growth medium. Quiescent cells exit the growth cycle during the G1 phase and arrest with a diploid DNA content. The role of quiescent donor nuclei in the success of these studies may be related to a number of factors. First, a stable population of diploid cells allows the co-ordination of donor and recipient cell cycle; second, when cells enter quiescence a number of changes occur: these include a reduction in transcription, a reduction in translation, and active degradation of mRNA and chromatin condensation (Whitfield *et al.*, 1985). These changes may render both the cytoplasm and the chromatin more compatible with the cytoplast and facilitate a greater reprogramming of the donor chromatin by maternally derived cytoplasmic factors as discussed in the previous section. The role of quiescence in changing chromatin structure and the ability of donor nuclei to re-control development after nuclear transfer requires further studies; however at PPL this technology has been transferred to cattle and a live calf produced (Chen *et al.*, unpublished).

Techniques for and uses of Genetic Modification in Farm Animal Species

The aims of any scheme for genetic modification or selection are to obtain stable desirable phenotypes that transmit the required traits through the germ line. In farm animal species we were until recently limited to the technique of pronuclear injection; however, nuclear transfer from cultured cell populations provides an alternative route to genetic modification.

Pronuclear injection

The addition of genetic material or production of a transgenic animal can be achieved by the injection of the required gene into the pronucleus of a zygote. Although this technique has been applied successfully in a number of species including mice, rabbits, pigs, sheep, goats and cattle (for review see Wall 1996), there are a number of disadvantages. Integration does not always occur during the first cell cycle resulting in the production of mosaic embryos (Burdon and Wall, 1992). Integration occurs at random within the genome resulting in variable expression of the gene product (see Wall, 1996). At present only simple gene additions may be performed. The selection of transgenic embryos before their transfer is hampered by mosaicism (Rusconi, 1991). The production of the required phenotype coupled to germ line transmission may require the generation of several transgenic lines. Multiplication of the required phenotype or its dissemination into the population is restricted by breeding programmes.

Nuclear transfer

The production of animals from cells that can be maintained in culture offers a number of advantages over the technique of pronuclear injection. First, the cells to be used as nuclear donors can be sexed, genetically modified and selected in culture before their use for nuclear transfer. The resultant animal is produced from a single nucleus of the desired genotype; therefore, mosaics will not be produced and the genetic modification should be transferred to the offspring. As all of the cells in the animal contain the transgene, then dependent upon its site of integration and tissue specific promotor, transgene expression should be obtained in the tissue of interest. The use of cultured cells will allow multiple genetic modifications and also facilitate precise genetic modifications which are presently not possible. For instance, specific genes may be removed (knocked out), replaced (knocked in) or specific chromosomal regions modified. The combination of cell culture and nuclear transfer allows multiple genetic modifications to be carried out by isolating new cell populations to act as nuclear donors from any of the embryos, fetuses or adult animals created. Thus, modifications not possible within the limited lifespan of a primary cell population may be facilitated by its rejuvenation via nuclear transfer.

Genetically modified, selected, clonally derived cells may be stored until expression data are obtained from the resultant animals. An 'instant' flock or herd of animals may then be produced by nuclear transfer, thus reducing the period required by natural breeding for production purposes.

Applications of nuclear transfer

In the short term, the major applications of such technology are most likely to be in the biopharmaceutical industry. The ability to produce animals from cultured cell populations will allow the precise genetic modification of these cells before embryo reconstruction. Such applications will not only include the addition of genes previously demonstrated using pronuclear injection, but also the targeting of such genes to precise regions of the genome, the removal or replacement of genes by targeting technology (i.e. removal of the PrP gene for scrapie resistance or removal of the bovine serum albumin gene and its replacement with the human homologue for therapeutic use) or the precise modification of endogenous genes (i.e. down- or upregulation of endogenous genes by changes in promoter sequences). The combination of these techniques will have applications in the production of pharmaceutical proteins, production of animal models of human disease and the modification of animals for organ transplantation.

In the longer term the combination of such technologies with the information obtained from genome mapping projects may allow the transfer of large segments of DNA coding for specific animal production traits. These may include genes or chromosomal segments coding for disease resistance, meat quality, growth, reproductive, nutritional or behavioural traits.

Incorporation into traditional breeding regimens

The application of this technology will allow the production of specific selected or modified progenitor animals for the dissemination of required traits into the population as a whole. In addition embryo multiplication by nuclear transfer will allow multiplication of embryos or animals that contain the genetic modification or exhibit the required traits in order to hasten the rate of dissemination into the population.

Other uses of nuclear transfer from somatic cell populations

As with the Amazonian rainforest, unknown genetic resources are also being lost from farm animal species, for example rare breeds of cattle in parts of Africa. At present preservation of these unknown resources is limited to the storage of semen or early embryos. Unfortunately, owing to many factors this approach has proved difficult. The demonstration that cultures of somatic cells isolated from adult animals (Wilmut *et al.*, 1997) can be used successfully in nuclear transfer offers an alternative route to the storage of genetic material.

Conclusion

The use of cultured somatic cells for nuclear transfer holds great potential for the cloning of large numbers of identical animals for agricultural applications. In addition, the relative ease by which cultured somatic cells may be rendered transgenic coupled with use of such cells for nuclear transfer offers a new and more efficient procedure to generate transgenic animals for both agricultural and industrial applications. The possibility of applying gene targeting techniques may allow for the generation of transgenic animals in which specific genes have been silenced, or 'knocked out', and replaced by other genes (knocked in). For example, Yom and Bremmel (1993) suggested that genes coding for the major proteins in cow's milk could be knocked out and replaced by their human counterparts. Cows modified in this fashion would produce milk containing a human milk protein profile. Such milk would presumably be hypoallergenic and more nutritious for human infants and thus be more suitable for infant formula manufacture than normal cow's milk. Gene targeting could also be applied to generate transgenic swine suitable for donating organs to human patients in need of transplant therapy. Use of gene targeting strategies should prevent organ rejection in recipient patients by 'knocking out' expression of proteins that trigger the immune rejection response.

Numerous technical hurdles must be overcome before somatic cell nuclear transfer can be considered routine. The overall efficiency of the process is quite low; many reconstructed embryos have been required so far to generate a few offspring (Wilmut *et al.*, 1997; Schnieke *et al.*, 1997; Cibelli *et al.*, 1998; Christmann, Chen, Polejaeva, Campbell and Eyestone, unpublished). Losses of potential offspring occur throughout the process, from early embryo development, during gestation and even perinatally. Such losses are presumably due to incomplete reprogramming of differentiated nuclei. A more thorough understanding of the reprogramming mechanism could lead to the development of better methods of inducing reprogramming during nuclear transfer. For transgenic work, especially if it involves gene targeting, cultured cells must be maintained in culture through numerous population doublings to allow for transfection and selection of genetically modified cells and may suffer karyotypic abnormalities or other mutations during extended culture. Improved methods of cell culture and more efficient gene targeting and selection methods could reduce the incidence of mutations induced during culture and thus lead to more robust cell populations for nuclear transfer.

References

Aoyagi Y (1992) Artificial activation of bovine oocytes matured *in vitro* by electric shock or exposure to ionophore A23187 *Theriogenology* **37** 180–188

Barnes F, Endebrock M, Looney C, Powell R, Westhusin M and Bondioli K (1993) Embryo cloning in cattle: the use of *in vitro* matured oocytes *Journal of Reproduction and Fertility* **97** 317–320

Bols PEJ, Van Soom A and De Kruif A (1994) Ovum pick-up in the cow, a review *Vlaams Diergeneeskundig Tijdschrift* **63** 101–108

Burdon TG and Wall RJ (1992) Fate of microinjected genes in preimplantation mouse embryos *Molecular Reproduction and Development* **33** 436–442

Campbell K and Wilmut I (1994) Recent advances on *in vitro* culture and cloning of ungulate embryos. In *Proceedings of the 5th World Congress on Genetics Applied to Livestock Production* University of Guelph, Guelph, Ontario, Canada pp 7–12

Campbell KHS and Wilmut I (1997) Totipotency or multipotentiallity of cultured cells: applications and progress *Theriogenology* **47** 63–72

Campbell KHS, Ritchie WA and Wilmut I (1993a) Disappearance of maturation promoting factor and the formation of pronuclei in electrically activated *in-vitro* matured bovine oocytes *Theriogenology* **39** 190–199

Campbell KHS, Ritchie WA and Wilmut I (1993b) Nuclear–cytoplasmic interactions during the first cell cycle of nuclear transfer reconstructed bovine embryos: implications for deoxyribonucleic acid replication and development *Biology of Reproduction* **49** 933–942

Campbell KHS, Loi P, Cappai P and Wilmut I (1994) Improved development to blastocyst of ovine nuclear transfer embryos reconstructed during the presumptive S-phase of enucleated activated oocytes *Biology of Reproduction* **50** 1385–1393

Campbell KHS, McWhir J, Ritchie W and Wilmut I (1995) Production of live lambs following nuclear transfer of cultured embryonic disc cells *Theriogenology* **43** 181 (Abstract)

Campbell KHS, McWhir J, Ritchie WA and Wilmut I (1996a) Sheep cloned by nuclear transfer from a cultured cell line *Nature* **380** 64–66

Campbell KHS, Loi P, Otaegui PJ and Wilmut I (1996b) Cell cycle co-ordination in embryo cloning by nuclear transfer *Reviews of Reproduction* **1** 40–46

Cibelli JB, Stice SL, Golueke PJ, Kane JE, Jerry J, Blackwell C, Abel Ponce de Leon F and Robl JM (1998) Cloned transgenic calves produced from non transgenic fetal fibroblasts *Science* **280** 1256–1258

Collas P and Barnes FL (1994) Nuclear transplantation by microinjection of inner cell mass and granulosa cell nuclei *Molecular Reproduction and Development* **38** 264–267

Collas P, Fissore R, Robl JM, Sullivan EJ and Barnes FL (1993a) Electrically induced calcium elevation, activation, and parthenogenetic development of bovine oocytes *Molecular Reproduction and Development* **34** 212–223

Collas P, Sullivan EJ and Barnes FL (1993b) Histone H1 kinase activity in bovine oocytes following calcium stimulation *Molecular Reproduction and Development* **34** 224–231

Cuthbertson KSR and Cobbold PH (1985) Phorbol ester and sperm activates mouse oocytes by inducing sustained oscillations in cell Ca^{2+} *Nature* **316** 541–542

Graham CF (1969) The fusion of cells with one- and two-cell mouse embryos. In *Wistar Institute Symposium Monograph* **9** pp 19–10

Gurdon JB (1962a) The developmental capacity of nuclei taken from intestinal epithelium cells of feeding tadpoles *Journal of Embryology and Experimental Morphology* **10** 622–640

Gurdon JB (1962b) Adult frogs from the nuclei of single somatic cells *Developmental Biology* **4** 256–273

Hirao Y, Nagai T, Kubo M, Miyano T, Miyake M and Kato S (1994) *In vitro* growth and maturation of pig oocytes *Journal of Reproduction and Fertility* **100** 333–339

Jones KT, Carroll J and Whittingham DG (1995a) Ionomycin, thapsigargin, ryanodine, and sperm induced Ca^{2+} release increase during meiotic maturation of mouse oocytes *Journal of Biological Chemistry* **270** 6671–6677

Jones KT, Carroll J, Merriman JA, Whittingham DG and Kono T (1995b) Repetitive sperm-induced Ca^{2+} transients in mouse oocytes are cell cycle dependent *Development* **132** 915–923

Jones KT, Bos-Mikich A and Whittingham DG (1996) Ca^{2+} oscillations during exit from meiosis and first mitotic division influence inner cell mass and trophectoderm number in mouse blastocysts *Journal of Reproduction and Fertility* **17** 5(Abstract)

Kanka J, Fulka J, Jr, Fulka J and Petr J (1991) Nuclear transplantation in bovine embryo: fine structural and autoradiographic studies *Molecular Reproduction and Development* **29** 110–116

Kono T, Iwasaki S and Nakahara T (1989) Parthenogenetic activation by electric stimulus of bovine oocytes matured *in vitro*. *Theriogenology* **33** 569–576

Kono T, Jones KT, Bos-Mikich A, Whittingham DG and Carroll J (1996) A cell cycle-associated change in Ca^{2+} releasing activity leads to the generation of Ca^{2+} transients in mouse embryos during the first mitotic division *Journal of Cell Biology* **152** 915–923

Kwon OY and Kono T (1996) Production of sextuplet mice by transferring metaphase nuclei from 4-cell embryos *Journal of Reproduction and Fertility* **17** 30 (Abstract)

Martinez-Balbas MA, Dey A, Rabindran SK, Ozato K and Wu C (1995) Displacement of sequence specific transcription factors from mitotic chromatin *Cell* **83** 29–38

Modlinski JA, Reed MA, Wagner TE and Karasiewicz J (1996)

Embryonic stem cells: developmental capabilities and their possible use in mammalian embryo cloning. In *Animal Reproduction: Research and Practice* pp 437–446 Ed. GM Stone and G Evans. Elsevier, Amsterdam

Moens A, Chesne P, Delhaise F, Delval F, Ectors F, Dessy F and Renard JP (1996) Assessment of nuclear totipotency of fetal bovine diploid germ cells by nuclear transfer *Therigenology* **46** 871–880

Nagai T (1987) Parthenogenetic activation of cattle follicular oocytes *in vitro* with ethanol *Gamete Research* **16** 243–249

Nussbaum DJ and Prather RS (1995) Differential effects of protein synthesis inhibitors on porcine oocyte activation *Molecular Reproduction and Development* **41** 70–75

O'Neill GT, Rolfe LR and Kaufman MH (1991) Developmental potential and chromosome constitution of strontium-induced mouse parthenogenones *Molecular Reproduction and Development* **30** 214–219

Otaegui PJ, O' Neill GT, Campbell KHS and Wilmut I (1994) Transfer of nuclei from 8-cell stage mouse embryos following use of nocodazole to control the cell cycle *Molecular Reproduction and Development* **39** 147–152

Ozil JP (1990) The parthenogenetic development of rabbit oocytes after repetitive pulsatile electrical stimulation *Development* **109** 117–127

Parrington J, Swann K, Shevchenko VI, Sesay AK and Lai FA (1996) Calcium oscillations in mammalian eggs triggered by a soluble sperm protein *Nature* **379** 364–368

Prather RS, Barnes FL, Sims MM, Robl JM, Eyestone WH and First NL (1987) Nuclear transfer in the bovine embryo: assessment of donor nuclei and recipient oocyte *Biology of Reproduction* **37** 859–866

Prather RS, Sims MM and First NL (1989) Nuclear transplantation in early pig embryos *Biology of Reproduction* **41** 414–418

Presicce GA and Yang X (1994) Nuclear dynamics of parthenogenesis of bovine oocytes matured *in vitro* for 20 and 40 hours and activated with combined ethanol and cycloheximide treatment *Molecular Reproduction and Development* **37** 61–68

Prochazka R, Kanka J, Sutovsky P, Fulka J and Motlik J (1992) Development of pronuclei in pig oocytes activated by a single electric pulse *Journal of Reproduction and Fertility* **96** 725–734

Pugh PA, Fukui Y, Tervit HR and Thompson JG (1991) Developmental ability of *in vitro* matured sheep oocytes collected during the nonbreeding season and fertilized *in vitro* with frozen ram semen *Theriogenology* **36** 771–778

Ritchie WA and Campbell KHS (1995) Intracytoplasmic nuclear injection as an alternative to cell fusion for the production of bovine embryos by nuclear transfer *Journal of Reproduction and Fertility* **15** 60 (Abstract)

Robl JM, Prather R, Barnes F, Eyestone W, Northey D, Gilligan B and First NL (1987) Nuclear transplantation in bovine embryos *Journal of Animal Science* **64** 642–647

Robl JM, Fissore R, Collas P and Duby RT (1992) Cell fusion and oocyte activation. In *Symposium on Cloning Mammals by Nuclear Transplantation* pp 24–27. Ed. GE Seidel, Jr. Colorado State University, Fort Collins

Rusconi S (1991) Transgenic regulation in laboratory animals *Experientia* **47** 866–877

Schermoen AW and O'Farrell PH (1991) Progression of the cell cycle through mitosis leads to abortion of nascent transcripts *Cell* **67** 303–310

Schnieke A E, Kind AJ, Ritchie WA, Mycock K, Scott AR, Ritchie M, Wilmut I, Colman A and Campbell KHS (1997) Sheep transgenic for human factor IX produced by transfer of nuclei from transfected fetal fibroblasts *Science* **278** 2130–2133

Stice SL, Keefer CL and Matthews L (1994) Bovine nuclear transfer embryos: oocyte activation prior to blastomere fusion *Molecular Reproduction and Development* **38** 61–68

Susko-Parrish JL, Leibfried-Rutledge ML, Northey D.L, Schutzkus V and First NL (1994) Inhibition of protein kinases after an induced calcium transient causes transition of bovine oocytes to embryonic cycles without meiotic completion *Developmental Biology* **166** 729–739

Swann K (1996) Soluble sperm factors and Ca^{2+} release in eggs at fertilization *Reviews of Reproduction* **1** 33–39

Tsunoda Y, Yasui T, Shioda Y, Nakamura K, Uchida T and Sugie T (1987) Full-term development of mouse blastomere nuclei transplanted into enucleated two-cell embryos *Journal of Experimental Zoology* **242** 140–147

Wall RJ (1996) Transgenic livestock: progress and prospects for the future *Theriogenology* **45** 57–68

Ware CB, Barnes FL, Maiki-Laurila M and First NL (1989) Age dependence of bovine oocyte activation *Gamete Research* **22** 265–275

Westhusin ME, Levanduski MJ, Scarborough R, Looney CR and Bondioli KR (1990) Utilization of fluorescent staining to identify enucleated demi-oocytes for utilization in bovine nuclear transfer *Biology of Reproduction* (Supplement) **42** 176–170

White KL and Yue C (1996) Intracellular receptors and agents that induce activation in bovine oocytes *Theriogenology* **45** 91–100

Whitfield JF, Boynton, AL, Rixon AL and Youdale T (1985) The control of cell proliferation by calcium, Ca^{2+}–calmodulin and cyclic AMP. In *Control of Animal Cell Proliferation* Vol. 1 pp 331–365 Eds AL Boynton and Leffert HL. Academic Press, London

Whittaker M (1996) Control of meiotic arrest *Reviews of Reproduction* **1** 127–135

Willadsen SM (1986) Nuclear transplantation in sheep embryos *Nature* **320** 63–65

Wilmut I, Campbell KHS and O'Neill GT (1992) Sources of totipotent nuclei including embryonic stem cells. In *Symposium on Cloning Mammals by Nuclear Transfer* pp 8–16 Ed. GE Seidel, Jr. Colorado State University, Fort Collins

Wilmut I, Schnieke AE, McWhir J, Kind AJ and Campbell KHS (1997) Viable offspring derived from fetal and adult mammalian cells *Nature* **385** 810–813

Yom H-C and Bremmel RD (1993) Genetic engineering of milk composition: modification of milk components in lactating transgenic animals *American Journal of Clinical Nutrition* 58 (Supplement) 3060S

ABSTRACTS

Journal of Reproduction and Fertility Supplement **54**, 501–523

Abstracts

Influence of early post-partum ovarian activity on the re-establishment of pregnancy in multiparous and primiparous dairy cattle. M. C. A. Smith and J. M. Wallace, *Rowett Research Institute, Bucksburn, Aberdeen AB21 9SB, UK*

Multiparous ($n = 87$) and primiparous ($n = 60$) Holstein/Friesian dairy cattle from a single herd were monitored throughout the post-partum period to investigate the influence of early ovarian activity on subsequent fertility. Milk progesterone concentrations were determined twice a week from calving until at least 100 days post-partum (PP), and cattle were inseminated from 42 days PP. Early ovarian activity was given a pre-determined value of ≤ 21 days PP. Multiparous cows showing ovarian activity before 21 days PP ($n = 41$, first ovarian activity 16.8 ± 0.39 days PP) compared with cows showing ovarian activity after 21 days PP ($n = 46$, first ovarian activity 37.8 ± 2.09 days PP) had a longer calving to conception interval (CCI, 138.6 ± 10.37 versus 107.2 ± 7.41 days; $P < 0.01$), required more services per cow (2.7 ± 0.27 versus 2.1 ± 0.17; $P < 0.05$) and had a lower pregnancy rate to all inseminations (29.3% versus 47.6%; $P < 0.05$). They also required a higher rate of exogenous fertility treatment (31.7% versus 10.9%; $P < 0.05$) and were more likely to be culled for failure to conceive (27.0% versus 7.1%; $P < 0.05$). A higher proportion of multiparous cows that commenced ovarian activity before 21 days PP failed to conceive by 100 days PP compared with those that commenced ovarian activity after 21 days PP (65.9% versus 41.3%; $P < 0.05$), and by 150 days PP the difference between groups was greater (51.2% versus 21.7%; $P < 0.01$). Early ovarian activity was associated with a higher incidence of persistent corpora lutea, sub-optimal progesterone concentrations and extended inter-luteal intervals compared with those cows that commenced ovarian activity after 21 days PP (31.7% versus 15.2%; $P < 0.1$). In contrast, in primiparous cows, no significant relationships were detected between the onset of ovarian activity before ($n = 32$) or after ($n = 28$) day 21 PP with CCI (123.8 ± 11.09 versus 133.3 ± 13.68 days), number of services per cow (2.3 ± 0.22 versus 2.3 ± 0.25) and pregnancy rates to all inseminations (37.5% versus 40.0%). A higher proportion of primiparous cows that commenced ovarian activity before 21 days PP had persistent corpora lutea, sub-optimal progesterone concentrations or extended inter-luteal intervals compared with those that commenced ovarian activity after 21 days PP (25.0% versus 0%; $P < 0.01$). In conclusion, early resumption of ovarian activity after calving has a detrimental effect on the re-establishment of pregnancy in multiparous but not primiparous dairy cows.

Expression of growth and differentiation factor-9 (gdf-9) mRNA in domestic ruminants. K. J. Bodensteiner, C. M. Clay and H. R. Sawyer, *Animal Reproduction and Biotechnology Laboratory, Department of Physiology, Colorado State University, Fort Collins, CO 80523, USA*

Recently a novel member of the TGF-β superfamily termed growth and differentiation factor-9 (gdf-9) was shown to be essential for normal follicular development beyond the primary follicle stage in mice (Dong *et al.*, 1996). This novel growth factor has been shown to be expressed in human and mouse ovaries and appears to be localized to oocytes at all stages of follicular growth, except primordial, in neonatal and adult mice (McGrath *et al.*, 1995). In the present study, expression of gdf-9 in ovaries of domestic ruminants was examined. Genomic fragments of bovine and ovine gdf-9 were isolated by PCR with primers directed against homologous sequences of the human and mouse gdf-9 genes. PCR was performed using 100 ng genomic DNA and resulted in the amplification of a 277 bp product for both species. The bovine and ovine predicted amino acid identities were 98% homologous to each other and approximately 91% and 84% homologous to human and mouse gdf-9, respectively. Distribution and relative abundance of gdf-9 mRNA was examined using *in situ* hybridization. Frozen tissue sections of bovine and ovine ovaries were cut at thicknesses of 15–20 μm and were incubated with ^{35}S-UTP labelled bovine or ovine gdf-9 antisense and sense (negative control) cRNA probes. Slides were stringently washed, dipped and exposed to emulsion for 32 days at 4°C. After development, slides were stained with haematoxylin and eosin and examined using both bright-field and dark-field optics. Specific hybridization using the antisense probe was restricted to oocytes and was present from the primordial follicle stage onwards. In addition, levels of gdf-9 expression appeared to correspond to the stage of follicular development, with increasing levels of expression as follicles increased in size. The oocyte specific expression of gdf-9 in domestic ruminants correlates well with data from mice. However, expression of gdf-9 in primordial follicles and increasing levels of expression with increasing follicle size have not been reported previously. The pattern of gdf-9 expression in domestic ruminants is consistent with a possible role for gdf-9 in the initiation and maintenance of folliculogenesis in these species.

Bioassay of ovine gonadotrophin preparations using ovine follicular cells cultured in serum free medium. B. K. Campbell and D. T. Baird, *Department of Obstetrics and Gynaecology, University of Edinburgh, Centre for Reproductive Biology, 37 Chalmers Street, Edinburgh EH3 9EW, UK*

The recent development of serum-free culture systems for sheep granulosal (Campbell *et al.*, 1996 *Journal of Reproduction and Fertility* **106** 7–16) and thecal cells (Campbell *et al.*, *Journal of Reproduction and Fertility*, in press) that allow *in vitro* differentiation of cells from small follicles in response to physiological doses of gonadotrophins raises the possibility of using these culture systems as *in vitro* bioassays. In the present study, the abilities of NIADDK-oFSH-16 (bioactivity 20 U mg^{-1} determined by Steelman–Pohley Bioassay, where one unit equals 1 mg NIH-FSH-S1), NIDDK-oFSH-20 (175 U mg^{-1} or 4453 IU mg^{-1} when calibrated to WHO second IRP-HMG) and Ovagen™ (20 U mg^{-1}; ICP, NZ) to stimulate oestradiol production by granulosal cells from small follicles (2–3 mm in diameter) during long term culture were determined. Similarly, the abilities of NIADDK-oLH-26 (bioactivity 2.3 U mg^{-1} by ovarian ascorbic acid depletion bioassay where one unit equals 1 mg NIH-LH-S1) and Ovagen™ (LH activity unspecified) to stimulate androstenedione production by thecal cells from small follicles during long term culture were determined. When corrected for their stated biopotencies, NIADDK-oFSH-16 and NIDDK-oFSH-20 gave identical FSH dose–response curves with an ED_{50} value of 100 μU ml^{-1} (5 ng oFSH-16 ml^{-1}) and an ED_{max} value of 500 μU ml^{-1}. However, Ovagen™ was more potent (3–10 times depending on the culture; $P < 0.01$) and consistently stimulated more oestradiol (20–80%; $P < 0.05$) than at the ED_{max} value determined for the other FSH preparations. These discrepancies could not be accounted for fully by variation in FSH content as RIA showed that Ovagen™ contained only 50% more FSH than oFSH-16 and oFSH-20. In contrast, thecal cell culture showed that Ovagen™ was 0.01 times as potent as oLH-26 in stimulating androstenedione production. In conclusion, despite large differences in purity, oFSH-16 and NIDDK-oFSH-20 are equivalent preparations, whereas Ovagen™, despite very low levels of LH activity, displays higher levels of bioactivity in this granulosal cell culture system than the FSH preparations it is calibrated against.

Supported by MRC Program Grant G8929853.

Effects of oestradiol benzoate and progesterone on patterns of follicle waves during induced luteolysis in cattle. C. R. Burke, S. R. Morgan, B. A. Clark and F. M. Rhodes, *Dairying Research Corporation Ltd, Private Bag 3123, Hamilton, New Zealand*

Fertility is reduced after progestin synchronization in cattle unless ovarian follicular development is controlled during treatment. This study tested the ability of oestradiol benzoate (OEB) and progesterone from an intravaginal implant (CIDR) to promote follicle wave turnover during the occurrence of induced luteolysis. Oestrus was synchronized in 24 lactating Friesian cows. From day 7 (oestrus = day 0), the ovaries of each cow were examined each day by ultrasonography, until subsequent ovulation. On day 13, every animal received an injection of 500 μg cloprostenol i.m. to induce luteolysis, and one of these treatments: (i) 1 mg OEB i.m. ($n = 7$); (ii) a CIDR device ($n = 6$); (iii) 1 mg OEB and a CIDR device ($n = 5$); or (iv) 2 mg OEB and a CIDR device ($n = 5$). The CIDR devices were removed from the vagina after 6 days. Plasma concentrations of progesterone and oestradiol-17β were determined using commercial RIA kits. The second dominant follicle emerged on day 9.5 ± 0.3 (mean ± SEM) and luteolysis was induced successfully in every cow. Plasma progesterone concentrations were maintained between 1.8 and 4.1 (± 0.2) in animals from groups with a CIDR device. Basal oestradiol-17β concentrations of 1.2 ± 0.2 pg ml^{-1} were elevated to 5.8 ± 0.6 pg ml^{-1} and 11.1 ± 1.7 pg ml^{-1} at 24 h after 1 and 2 mg OEB, respectively. Oestrous cycles comprising two follicle waves predominated when either OEB or the CIDR were administered alone, compared with mostly three wave cycles when used in combination ($P < 0.01$). The day of ovulation was 15.4 ± 0.3, 21.2 ± 0.2 and 22.5 ± 0.2 for OEB only, CIDR only and 1 or 2 mg OEB plus CIDR cows, respectively ($P < 0.01$). The interval from day of emergence to ovulation of the ovulatory follicle was less ($P < 0.01$) in cows treated with OEB only (6.0 ± 0.3 days) or 1 or 2 mg OEB plus a CIDR (7.2 ± 0.6 days), compared with those treated with a CIDR only (11.0 ± 0.8 days). These results show that insertion of a CIDR implant facilitates OEB-induced follicle wave turnover during the occurrence of luteolysis in most cows.

Regulation of relaxin-like factor (RLF) mRNA expression in bovine thecal cell cultures. R. A. D. Bathgate, N. Moniac, B. Bartlick and R. Ivell, *Institute for Hormone and Fertility Research, University of Hamburg, Grandweg 64, 22529 Hamburg, Germany*

The relaxin-like factor (RLF) was characterized initially as the Leydig insulin-like peptide, as it was predominantly expressed in porcine Leydig cells. Recently, we have shown that the RLF gene is highly expressed in the bovine ovary in follicular thecal cells and in the corpus luteum of the cycle and pregnancy. As the highest expression of RLF mRNA in bovine ovaries is in thecal cells, we have established a thecal cell culture system to study the expression and regulation of RLF mRNA. Using Northern blots, RLF mRNA appears to be downregulated after 4 days in culture under fetal calf serum (FCS)-stimulated or FCS-free conditions. However, after 6 days in FCS-free conditions together with insulin or IGF-I alone, or in combination with LH, RLF mRNA is again highly expressed. The expression of RLF mRNA under these different treatment conditions is differential in both size of the transcript and timing of expression. Thecal cells under insulin or IGF-I stimulation express a larger RLF mRNA transcript, which is expressed from days 6 to 12 in culture, whereas expression under LH stimulation is maximal at 6 days and decreases thereafter. Using a combination of 3′ RACE PCR and RNase H treatment of the RNA, we were able to show that the differences in size of the RLF mRNA are due to differences in the length of the poly A tail. Cells under LH/insulin stimulation are luteinized, as indicated by the continued and increased expression of both P450 SCC mRNA and progesterone production. Therefore, the expression of the RLF mRNA in these luteinizing cell cultures appears to parallel the expression of RLF mRNA in thecal/lutein cells of the early cycle *in vivo*. Neither progesterone nor P450 SCC mRNA expression is increased by insulin or IGF-I alone, indicating that such cells are not luteinizing. Our studies indicate that RLF mRNA expression may be an interesting and important marker for distinguishing bovine thecal cell differentiation.

Research supported by the Deutsche Forschungsgemeinschaft (Iv7/1-3).

Long-term follicular dynamics and biochemical characteristics of dominant follicles in dairy cows subjected to heat-stress. J. D. Ambrose[1], A. Guzeloglu[2], M. J. Thatcher[2], T. Kassa[2], T. Diaz[2] and W. W. Thatcher[2], [1]*Alberta Agriculture, Food and Rural Development, Edmonton, Canada; and* [2]*Department of Dairy and Poultry Sciences, University of Florida, Gainesville 32611, USA*

The objective of this study was to determine the quality of successive dominant follicles after induced heat stress. Non-lactating, cyclic Holstein cows were allocated randomly to heat stress ($n = 8$) or control ($n = 8$) groups. Cows received 100 μg GnRH i.m. on day 0, a CIDR-B device was placed intravaginally on day 4 and was removed on day 7 at $PGF_{2\alpha}$ (25 mg) administration. Dominant follicles and follicles > 5 mm in diameter were aspirated on day 8, and 100 μg GnRH was injected after aspiration, to initiate a new follicular wave. Thus, a dominant follicle was aspirated every 8 days (one 'follicular cycle') for 10 cycles. After one cycle (day 8), heat stress cows were placed in environmental chambers for 7 days (8 h per day at 43.3°C and 16 h per day at 22°C for 4 days and 8 h per day at 43.3°C and 16 h per day at 32.2°C for 3 days; 40% relative humidity) and maintained thereafter in an outdoor lot with control cows at ambient temperature (22°C). Rectal temperature increased significantly ($P < 0.01$) in heat stressed cows compared with control cows (39.28 ± 0.01 versus 38.78 ± 0.01°C), with a maximum of 40.1°C on the last day. Heat stress appeared to cause a temporal decrease in follicular dominance based upon: an increase in the number of class 1 (≤ 5 mm in diameter) follicles on day 4 for cycles 2–6 ($P < 0.05$); the number of class 2 (6–9 mm in diameter) follicles on day 4 declined in cycles after heat stress ($P < 0.04$); and the number of class 3 (≥ 10 mm in diameter) follicles on days 7 and 8 increased for cycles 2 and 3 ($P < 0.02$). The concentrations of oestradiol (1563.8 ± 181.9 versus 1500.7 ± 170.5 ng ml^{-1}) and progesterone (43.2 ± 6.2 versus 59.7 ± 5.95 ng ml^{-1}), and the ratio of oestradiol to progesterone in follicular fluid (39.0 ± 3.9 versus 33.3 ± 3.7) did not differ between control and heat stressed cows, respectively. IGF-2 ligand blots were run on follicular fluid samples ($n = 104$) from four heat stressed and four control cows. There was a prominence of IGFBP-3 in 100 of 104 follicular fluid samples; 3.8% of follicular fluid ($n = 4$) had low molecular weight IGFBPs indicative of a poor quality dominant follicle. Heat stress induced a temporal decrease in follicular dominance, but GnRH-induced follicle cycles resulted in development of healthy preovulatory follicles in both groups.

Two or three waves of ovarian follicle development during the oestrous cycle in sheep. A. C. O. Evans[1], P. Duffy[2] and M. P. Boland[1], [1]*Department of Animal Science and Production, University College Dublin, Ireland; and* [2]*Lyons Research Farm, University College Dublin, Ireland*

The aim of this study was to characterize the pattern of development of ovarian follicles in maiden cyclic lambs. The ovaries of nine Texel and 11 Suffolk-Cross lambs (40–45 kg) were examined each day using transrectal ultrasonography (7.5 MHz rigid transducer) for one or two oestrous cycles, respectively. Identified follicles were defined as those that grew to ≥ 4 mm in diameter and were present at ≥ 3 mm in diameter for ≥ 3 days. Data were analysed by repeated measures ANOVA and compared among days using Fisher's (protected) LSD test. There were no differences between breeds ($P > 0.05$). Two ($n = 10$ cycles), three ($n = 20$ cycles) or four ($n = 1$ cycle) periods (waves) of growth of identified follicles occurred during individual cycles, with oestrous cycle lengths of 15.6 ± 1.6, 16.1 ± 1.1 and 17 days, respectively. The number of identified follicles emerging (identified retrospectively as 2 or 3 mm in diameter) in animals with two waves of follicle growth was greatest ($P < 0.05$) on days 1 (0.7 follicles emerging per ewe per day) and 9 (0.7), and least on days 0 (0.1), 5 (0) and from days 13 to 16 (all 0). The number of follicles emerging in animals with three waves of follicle growth was greatest ($P < 0.05$) on days 3 (0.9), 7 (0.6) and 12 (0.7), and least on days 0 (0), 4 (0.1), 10 (0.2) and from days 14 to 16 (all 0). In animals with two or three waves of follicle growth, the numbers of small (2 and 3 mm diameters), medium (MF, 4 and 5 mm diameters) and large (LF, ≥ 6 mm in diameter) follicles differed ($P < 0.001$) among days. In animals with two waves of follicle growth, numbers of medium and large follicles were greatest ($P < 0.05$) on days 5 and 12 (1.3 and 1.8 MF, respectively) and days 8 and 16 (0.5 and 1.1 LF, respectively), and least on days 0, 8 and 16 (0.1, 0.5 and 0 MF, respectively) and days 3 and 12 (0 and 0.1 LF, respectively). In animals with three waves of follicle growth, the numbers of medium and large follicles were greatest ($P < 0.05$) on days 5, 8 and 13 (1.3, 1.5 and 1.7 MF, respectively) and 6 and 16 (0.3 and 0.9 LF, respectively), and least on days 1 and 16 (0.2 and 0.5 MF, respectively) and days 2 and 13 (0 and 0.2 LF, respectively). In conclusion, ovarian follicle growth in ewe lambs occurred in two or three organised waves during which there was synchronous emergence of follicles associated with fluctuations in numbers of follicles in size classes.

Follicular fluid steroids and inhibin-A determined *in vivo* during loss of dominance of the first dominant follicle in beef heifers with two or three dominant follicles during the oestrous cycle.
M. Mihm[1], P. G. Knight[2] and J. F. Roche[1], [1]*Faculty of Veterinary Medicine, University College Dublin, Dublin, Ireland; and* [2]*School of Animal and Microbial Sciences, University of Reading, Reading RG6 6AJ, UK*

The timing of the second transient FSH rise and associated emergence of the second follicle wave is variable in cattle and dependent on the number of dominant follicles selected during the cycle. The aims of this study were: (i) to validate a model that allows *in vivo* sampling of follicular fluid from the first dominant follicle without affecting its subsequent fate; and (ii) to determine whether onset of atresia in the first dominant follicle (i.e. reduction in follicular fluid oestradiol concentrations and increased concentrations of follicular fluid inhibin-A) is delayed in heifers with two dominant follicles per cycle. On day 8 of the cycle (day 0 = oestrus) crossbred beef heifers underwent midline laparotomy under general anaesthesia with (follicular fluid aspiration; $n = 7$) or without (sham aspiration; $n = 6$) aspiration of 20 µl of follicular fluid from the first dominant follicle, using a transstromal approach. Blood samples were taken from heifers at 12 h intervals and ultrasound examinations of ovaries were performed each day from day 0 to subsequent ovulation. Three follicular fluid aspiration and four sham aspiration heifers were bled and scanned in the previous cycle (controls). The interovulatory interval and the ratio of heifers with two or three dominant follicles per cycle (controls 2:5; sham-aspiration 1:5; follicular fluid aspiration 4:3) were not ($P > 0.05$) influenced by surgery or aspiration of follicular fluid, but the maximum size of the first dominant follicle was reduced ($P < 0.05$) in follicular fluid aspiration heifers compared with controls. No other characteristics of the first, second or third follicle wave were altered by surgery or aspiration of follicular fluid. The day of maximum FSH concentrations during the second transient rise (day 10.4 ± 0.4 versus 9.3 ± 0.2) and the day of emergence of the second wave (day 11.3 ± 0.3 versus 10.0 ± 0.3) were delayed ($P < 0.05$) in heifers with two dominant follicles per cycle compared with heifers with three dominant follicles per cycle. In the first dominant follicle on day 8, mean follicular fluid oestradiol concentrations were lower, and mean progesterone and inhibin-A concentrations were higher, in heifers with two compared with heifers with three dominant follicles per cycle. However, these differences were not significant ($P > 0.09$). In conclusion: (i) individual first dominant follicle function can be monitored by measurements of follicular fluid in addition to ultrasound determination of subsequent dominant follicle turnover; and (ii) the delay in the timing of events leading to the second dominant follicle in cycles with only two dominant follicles may not be due to maintained health of the first dominant follicle.

Matrix metalloproteinase (MMP) and tissue inhibitor of metalloproteinases (TIMP-1 and -2) activities in bovine follicular fluid collected at different stages of development.
F. N. Kojima[1], W. A. Ricke[1], R. N. Funston[2], J. E. Kinder[3] and M. F. Smith[1], [1]*Department of Animal Science, University of Missouri, Columbia, MO 65211, USA;* [2]*Chadron State College, Chadron, NE 69337, USA; and* [3]*Department of Animal Science, University of Nebraska, Lincoln, NE 68583, USA*

Folliculogenesis requires proliferation of follicular cells as well as expansion of the basement membrane within the confines of the ovarian stroma. Furthermore, proteolytic degradation of the extracellular matrix (ECM) may affect the cellular microenvironment of ovarian follicles by altering ECM–cell contacts and (or) altering sequestered growth factor availability and activity. Matrix metalloproteinases and their inhibitors (TIMPs) are candidates for modulating basement membrane expansion and ECM remodelling within ovarian follicles. Therefore, the objective of this study was to characterize matrix metalloproteinase (MMP) and TIMP activities in bovine follicular fluid during different stages of follicular development and atresia. After synchronization of oestrus, growth of bovine follicles was monitored by real-time ultrasonography and cows were ovariectomized at different stages of follicular development. Follicles were classified into four groups ($n = 3$–5 per group) based on ultrasonography and follicular fluid concentrations of oestradiol-17β and progesterone: growing small follicles (GS; 5–10 mm in diameter); growing large follicles (GL;

> 10 mm in diameter); atretic small follicles (AS); and atretic large (AL) follicles. In addition, preovulatory follicles were collected at 0, 8.5 or 20.5 h after injection of GnRH 48 h after $PGF_{2\alpha}$-induced luteolysis. Gelatin zymography detected gelatinolytic activities and the relative molecular sizes (Mr) of the gelatinases were 72 000 and 92 000, which correspond to the Mr of progelatinase A (proMMP-2) and progelatinase B (proMMP-9), respectively. Gelatinase activities were inhibited in the presence of an MMP inhibitor. Reverse zymography detected MMP inhibitor activities at Mr 30 000 and 22 000, which correspond to the Mr of TIMP-1 and –2, respectively. Atretic follicles had greater ($P < 0.05$) proMMP-2 and TIMP-1 activities than growing follicles. TIMP-2 activity was greater ($P < 0.05$) in AS follicles compared with all other follicular groups. In preovulatory follicles, proMMP-2 and TIMP-1 and -2 activities were increased ($P < 0.05$) by 20.5 h after GnRH injection compared with follicles collected at 0 h. Increased proMMP-2 activity in atretic and preovulatory follicles may indicate that proMMP-2 is associated with both follicular demise and rupture. A parallel increase in TIMP activity in atretic and preovulatory follicles indicates the importance of regulating MMP activity.

Bull exposure increases ovarian cyclicity in Brahman heifers. P. Bastidas, J. Ruiz, M. Manzo and O. Silva, *Instituto de Reproducción Animal e Inseminación Artificial, Facultad de Ciencias Veterinarias. Universidad Central de Venezuela, Maracay, Aragua, Venezuela*

The objective of this study was to determine whether stimulation of pre-pubertal Brahman heifers by mature bulls would alter the proportion of heifers exhibiting ovarian activity. Forty Brahman heifers at 13 months of age, weighing 285 kg, were allotted randomly to two groups: group 1 ($n = 20$) was exposed to a vasectomized bull equipped with a chinball marker; and group 2 ($n = 20$) remained isolated from bulls. Ultrasound examination of the ovaries was done each week for 34 weeks. Blood samples were collected from all heifers on day 0 (first day of trial) and at weekly intervals until the end of the experiment. The numbers of small (3–6 mm diameter) and large (7–10 mm diameter) follicles were determined. The pre-pubertal status of heifers was confirmed by the absence of detectable luteal tissue by ultrasound evaluation and by plasma progesterone concentrations (< 1 ng ml^{-1}). The percentage of heifers showing cyclic activity was 70% (14/20) in group 1 and 55% (11/20) in group 2. Exposure to the mature bull had a significant effect on cumulative plasma progesterone concentrations ($P < 0.004$) and on the number of small ($P < 0.001$) and large ($P < 0.007$) follicles, during the 34 week period. Progesterone concentrations (mean ± SEM) were 7.6 ± 1.8 and 4.7 ± 1.9 ng ml^{-1} for groups 1 and 2, respectively. The number of small and large follicles accumulated during the experimental 34 week period was greater in group 1 (4.6 ± 1.2 and 2.1 ± 0.8, respectively) than in group 2 (2.4 ± 0.9 and 1.4 ± 0.8, respectively). In conclusion, the presence of mature bulls is effective in promoting ovarian activity of pre-pubertal Brahman heifers without affecting age at puberty.

Development of the dominant follicle is suppressed in postpartum dairy cows induced experimentally to produce maximum milk yield. J. G. Gong[1], C. H. Knight[2], D. N. Logue[3], W. M. Crawshaw[3] and R. Webb[4], [1]*Roslin Institute, Roslin, Midlothian EH25 9PS, UK;* [2]*Hannah Research Institute, Ayr KA6 5HL, UK;* [3]*SAC, Auchincruive, Ayr KA6 5HW, UK; and* [4]*School of Biological Sciences, University of Nottingham, Sutton Bonington, Loughborough LE12 5RD, UK*

Reproductive performance in modern dairy cows has been shown to be declining in association with increased milk yields, although the underlying mechanisms are not well understood. In this study, ovarian follicular development was examined in cows induced to produce maximum milk output around the peak of lactation. Twenty-four cows which had been selected for genetic merit for milk-yield (12 high and 12 low) were managed as in normal commercial practice and milked twice each day after calving. From week 6 of lactation, half of the cows (six high and six low) were induced to increase milk output using a staggered multiple stimuli approach: milking four times each day, recombinant bovine somatotropin (Monsanto) and thyroxine (Sigma) treatment. The oestrous cycles of all cows were then synchronized using PRID. The ovaries were scanned and plasma samples collected each day for two weeks. There were no differences in any of the responses measured between high and low genetic merit cows and therefore the results are pooled into two groups: control and treated. In the control group, plasma progesterone concentrations in 11 of 12 animals increased after synchronized oestrus, reaching a maximum of 6.0 ± 0.7 ng ml^{-1} at the end of the experiment. In contrast, progesterone concentrations in all 12 treated animals remained below the assay detection limit (0.3 ng ml^{-1}) throughout the experiment. After removal of the PRID, a dominant follicle developed and ovulated in 11 of 12 control cows, followed by a new follicular wave with the selection of a dominant follicle that reached a maximum size of 14.7 ± 0.4 mm diameter. However, the largest follicles in the treated cows only reached a mean diameter of 7.7 ± 0.2 mm. Basal plasma FSH (control, 0.76 ± 0.06; treated, 0.75 ± 0.07 ng ml^{-1}) and LH (control, 1.33 ± 0.12; treated, 1.29 ± 0.16 ng ml^{-1}) concentrations were not different between the two groups. In conclusion, this study has demonstrated that development of dominant follicles in dairy cows can be completely suppressed by injections of somatotropin or thyroxine or by milking the cows four times per day to dramatically increase milk output, irrespective of the genetic merit of the animals.

Nitric oxide is involved in the regulation of LH secretion in heifer calves and may mediate the stimulatory effects of *N*-methyl-D,L-aspartic acid on LH release. A. Honaramooz, S. J. Cook, A. P. Beard, P. M. Bartlewski and N. C. Rawlings, *Department of Veterinary Physiological Sciences, Western College of Veterinary Medicine, 52 Campus Drive, University of Saskatchewan, Saskatoon, Saskatchewan S7N 5B4, Canada*

We have shown that *N*-methyl-D,L-aspartic acid (NMA, an excitatory amino acid agonist) induces an immediate LH release in prepubertal heifer calves. Nitric oxide has emerged as an important regulator of LH release in rats. This study was designed to test the role of nitric oxide in the regulation of LH secretion, as well as the possible mediation of the effects of NMA on LH release in heifer calves by nitric oxide. Four groups of five prepubertal heifers (33 weeks old) received one of these treatments: (i) *N*-G-nitro-L-arginine methyl ester (L-NAME, a nitric oxide synthase inhibitor; one dose of 35 mg kg^{-1}, i.v.); (ii) NMA (one dose of 4.7 mg kg^{-1}, i.v.); (iii) combined dose of NMA and L-NAME (same doses as previously); and (iv) vehicle (saline, i.v.). All heifers were also challenged with a bolus injection of GnRH (10 ng kg^{-1}, i.v.). Blood samples were collected through jugular catheters every 15 min for 10 h. L-NAME was injected after the first blood sample, NMA after 2 h and GnRH after 6 h of blood sampling. Before injection of L-NAME or NMA, mean serum concentrations of LH did not differ among groups ($P > 0.05$). Administration of L-NAME alone suppressed spontaneous pulses of LH ($P < 0.04$), when compared with the control animals. Three animals (out of five) responded to NMA in both the NMA and combined groups. The amplitude of the induced LH pulses in heifers in the NMA group was higher than in heifers in the combined group (2.47 ± 0.7 versus 0.52±0.3 ng ml^{-1}, respectively; mean ± SEM; $P = 0.057$). One animal in the NMA group had an LH surge (LH concentrations increased for 4 h). GnRH challenge induced an LH pulse in all animals in all groups; however, heifers in the L-NAME group had a higher LH pulse amplitude compared with the controls (2.55 ± 0.3 versus 1.07 ± 0.2 ng ml^{-1}, respectively; mean ± SEM; $P < 0.002$). The number of LH pulses after GnRH treatment was higher in the NMA group than in the L-NAME group ($P < 0.03$). In conclusion, nitric oxide is involved in the regulation of LH secretion and at least partly mediates the effects of NMA on LH release.

Opioidergic and dopaminergic neuronal control of LH secretion in heifer calves. A. Honaramooz, R. K. Chandolia, A. P. Beard and N. C. Rawlings, *Department of Veterinary Physiological Sciences, Western College of Veterinary Medicine, 52 Campus Drive, University of Saskatchewan, Saskatoon, Saskatchewan S7N 5B4, Canada*

Endogenous opioids inhibit LH release in the early post-natal period in heifers; however, it is not clear if this effect is direct or by way of other neuronal systems. To test if opioidergic and dopaminergic neuronal systems interact in the control of LH secretion, four groups of five heifer calves were bled every 15 min for 10 h and received one of the following treatments, at 4, 14, 24, 36 and 48 weeks of age: (i) naloxone (NAL, opioid antagonist; 1 mg kg^{-1}, i. v.) every hour for 10 h (up to 24 weeks of age) or as a single injection (at 36 and 48 weeks of age); (ii) sulpiride (SULP, a D2-dopaminergic antagonist; one dose of 0.59 mg kg^{-1}, s.c.); (iii) naloxone and sulpiride (NAL+SULP combined); and (iv) vehicle. Treatments began after the first blood sample was taken. An LH pulse was detected within 30 min of the first injection in 0, 1, 2, 4 and 5 calves in the NAL group; 0, 0, 0, 1 and 0 SULP treated animals; 0, 1, 0, 1 and 2 heifers in the combined NAL+SULP group; and 0, 0, 0, 0 and 2 control calves, at 4, 14, 24, 36 and 48 weeks of age, respectively. Therefore, as the heifers grew older, more heifers in the NAL group had an LH pulse in response to the first NAL injection ($P < 0.05$) and at 36 and 48 weeks of age, more calves in the NAL group responded than in the other groups ($P < 0.05$). Over the full 10 h treatment period, LH pulse frequency was higher at 4 and 14 weeks of age in the NAL only treatment, compared with the combined NAL+SULP-treated calves ($P < 0.06$), resulting in increased mean LH concentrations at 14 weeks of age in NAL-treated calves ($P < 0.02$). In conclusion, endogenous opioids weakly suppress LH release in the early post-natal period, but this effect is stronger as heifer calves approach puberty. As SULP negated the effects of NAL on LH release, the suppressive effects of opioids are in part exerted through the inhibition of a dopaminergic neuronal system.

Plasma inhibin concentrations during the preovulatory period in the ewe. J. E. Wheaton, J. E. Romano, R. L. Meyer, M. T. Bailey and S. A. Christman, *Department of Animal Science, 495 Animal Science/Veterinary Medicine Building, 1988 Fitch Avenue, University of Minnesota, St. Paul, MN 55108, USA*

The objective of the experiment was to gain an insight into the possible differential functions of oestradiol and inhibin during the preovulatory period. The approach was to compare temporal relationships among oestradiol, inhibin, gonadotrophins and follicular development. During the mid-breeding season, eight mature Polypay ewes (75 ± 3 kg) were hemi-ovariectomized to facilitate transvaginal ultrasonography for monitoring follicular development. Ewes were administered a luteolytic dosage of $PGF_{2\alpha}$ and a progesterone-releasing pessary (CIDR-G), which was left in place for 12 days to synchronize oestrus. Upon CIDR-G removal, blood samples were drawn via a jugular vein cannula at 6 h intervals for 84 h. Ultrasonography was performed at 12 h intervals. All ewes showed oestrus (onset = 26 ± 2 h), had a preovulatory gonadotrophin surge (onset = 26 ± 2 h; peak LH values = 95 ± 7 ng ml^{-1} at 32 ± 2 h) and ovulated (52 ± 2 h; ovulation rate = 2 ± 0). During the preovulatory period (from −30 to −12 h preceding the time of peak LH surge values), plasma LH concentrations tended to increase, and FSH concentrations decreased ($P < 0.01$). Plasma oestradiol concentrations increased ($P < 0.01$) steadily from −30 h (2.2 ± 0.3 pg ml^{-1}) to −6 h (5.0 ± 0.6 pg ml^{-1}) and fell to low concentrations (< 1 pg ml^{-1}) by 6 h. Plasma concentrations of dimeric inhibin, measured using a two-site immunoradiometricassay, increased ($P < 0.05$) from −30 h (0.6 ± 0.1 ng ml^{-1}; Genentech Inc, Lot 13140-90B) and reached highest concentrations at −18 h (0.9 ± 0.1 ng ml^{-1}), after which concentrations decreased to 0.6 ± 0.1 ng ml^{-1} by −6 h. After the preovulatory gonadotrophin surge, concentrations of LH and oestradiol were low. Plasma FSH concentrations increased from 12 to 24 h and then decreased. Inhibin concentrations increased during the preovulatory surge and decreased at or near the time of the FSH increase. Overall, FSH and inhibin showed generalized reciprocal cyclic patterns. Clearly, different oestradiol and inhibin profiles indicate independent regulation. In contrast to oestradiol, inhibin concentrations remained nearly as high after the preovulatory gonadotrophin surge as before the surge. Evidently small- to medium-sized follicles provide a prominent source of inhibin, although an association was not apparent between inhibin and the number and size of non-ovulatory follicles. Temporal hormonal profiles are consistent with roles for oestradiol in signalling maturation of preovulatory follicles, and for inhibin in the feedback regulation of FSH for control of earlier stages of folliculogenesis.

Secretion of FSH and ovarian follicular development in female cattle of different ages and reproductive states. J. Koch, M. Mussard, L. Ehnis, H. Jimenez-Severiano, E. Zanella, V. Vega-Murillo and J. Kinder, *Department of Animal Science, University of Nebraska, Lincoln, NE 68583-0908, USA*

Increased FSH concentrations in blood precedes and is responsible for stimulation of development of a cohort of ovarian follicles (wave of follicular development) in female cattle pre- and post-puberty. The objective was to compare in female cattle of different ages and during different reproductive states concentrations of FSH in blood serum preceding and numbers and sizes of ovarian follicles during waves of ovarian follicular development. Prepubertal heifers (PRE; $n = 5$), post-pubertal, oestrous heifers 7 days post-oestrus (PPE; $n = 6$), post-pubertal, anoestrous heifers (PPA; $n = 6$), adult cows 7 days post-oestrus (AD7; $n = 5$) and adult cows 15 days post-oestrus (AD15; $n = 6$) were used. Concentrations of FSH and the numbers and sizes of ovarian follicles were compared among groups after ablation of ovarian follicles ≥ 5 mm in diameter. Blood samples were collected to evaluate FSH concentrations and ultrasonography was used to evaluate the ovarian follicles. FSH concentrations were greater ($P < 0.0001$) in females in the PPE group than in the PRE group (2.26 and 1.72 ng ml^{-1}, respectively), the PPE group compared with the AD7 group (2.26 and 1.83 ng ml^{-1}, respectively) and the PPA group compared with the PRE group (1.95 and 1.72 ng ml^{-1}, respectively). The number of follicles < 5 mm in diameter was greater ($P \leq 0.04$) in females in the PRE group compared with the PPE group from day 5 to day 8 (12.3 and 6.7, respectively), the PRE group compared with the PPA group on days 7 and 8 (14.5 and 8.6, respectively) and the AD7 group compared with the PPE group from day 1 to day 8 (11.3 and 6.2, respectively), after follicular ablation. The number of follicles ≥ 5 mm in diameter was greater ($P \leq 0.05$) in females in the PPA group compared with the PRE group on days 6 and 7 (5.1 and 1.3, respectively), in the AD7 group compared with the PRE group from day 2 to day 7 (6.8 and 3.1, respectively), in the AD15 group compared with the PRE group from day 5 to day 7 (5.3 and 2.0, respectively) and in the AD7 group compared with the PPE group from day 2 to day 6 (7.1 and 3.4, respectively), after follicular ablation. The diameter of the dominant follicle (mm) was greater ($P \leq 0.03$) in females in the AD15 group compared with the PRE group from day 5 to day 8 (14.1 and 10.1, respectively), and with the AD7 group on day 8 (15.1 and 12.0, respectively), after follicular ablation. In conclusion, age is an important factor involved in regulation of release of FSH and in the development of follicles during waves of follicular development in female cattle.

Secretion of LH in bull calves treated with analogues of GnRH. H. Jimenez-Severiano[1], M. Mussard[1], L. Ehnis[1], J. Koch[1], E. Zanella[1], B. Lindsey[1], W. Enright[2], M. D'Occhio[3] and J. Kinder[1], [1]*Department of Animal Science, University of Nebraska, Lincoln, NE 68583-0908, USA;* [2]*Intervet International, 5830 AA Boxmeer, The Netherlands; and* [3]*Animal Sciences and Production Group, Central Queensland University, North Rockhampton, Queensland 4701, Australia*

Infusion of GnRH for extended periods inhibits pulsatile LH release, but enhances testicular function of bulls. The reason long-term infusion of GnRH agonist does not suppress the pituitary–testicular axis of bulls in a similar way to what occurs in most species has not been delineated. Therefore, the primary objective of this study was to compare LH secretion in control bulls with those treated with GnRH analogues. Prepubertal bulls ($n = 24$; 5 months of age) were allocated to four groups: receiving a small (1 mg kg^{-1} body weight each day; N-small) or large (3 μg kg^{-1} body weight each day; N-large) dose of GnRH agonist azagly-nafarelin, GnRH antagonist SB-75 (5 μg kg^{-1} body weight each day; SB75), or vehicle (control). Treatments were administered via Alzet mini osmotic pumps for 27 days. Blood samples were collected at 20 min intervals for 24 h at days 1, 12 and 24 of treatment. Serum samples were assayed for LH. Mean LH concentrations were greater ($P < 0.05$) in both groups of bulls treated with GnRH agonist (2.36 and 2.76 ng ml^{-1}, N-small and N-large, respectively), compared with those treated with SB75 (1.46 ng ml^{-1}), or the control group (1.67 ng ml^{-1}) on day 1. Basal concentrations of LH were greater ($P < 0.05$) in agonist-treated bulls than in SB75-treated or control bulls during all three periods (mean concentration of the three periods was 1.77, 1.74, 0.74 and 0.79 ng ml^{-1}, for N-small, N-large, SB75 and control groups, respectively; SEM = 0.1). Agonist treatment decreased ($P < 0.05$) LH pulse frequency during days 12 and 24 to fewer than 1 pulse in 24 h compared with SB75 (3.8 pulses in 24 h) and control (2.9 pulses in 24 h) treatments. Control bulls had the greatest ($P < 0.001$) LH pulse amplitude during all three periods (average = 9.3 ng ml^{-1}) compared with SB75 (6.4 ng ml^{-1}) and the two agonist groups (approximately 1 ng ml^{-1}). In summary, prolonged infusion of bulls with GnRH agonist decreased the number of LH pulses and reduced their amplitude, and increased basal LH concentrations, whereas treatment with SB-75 only reduced the amplitude of LH pulses compared with controls. In conclusion, the greater testosterone secretion previously reported with prolonged treatment of bulls with GnRH agonist may result from enhanced basal secretion of LH.

Effect of reducing LH pulse frequency and amplitude on ovarian oestradiol production in the ewe. H. Dobson[1], B. K. Campbell[2], R. J. Scaramuzzi[3] and D. T. Baird[2], [1]*Department of Veterinary Clinical Science, University of Liverpool, Liverpool L64 7TE, UK;* [2]*Department of Obstetrics and Gynaecology, University of Edinburgh, Edinburgh EH3 9EW, UK; and* [3]*Royal Veterinary College, London NW1 0TU, UK*

At the end of the follicular phase, oestradiol production is maintained by pulses of LH with an abrupt decrease in LH pulsatility leading to a delay in the onset of the LH surge. To determine whether the latter effect is due to a reduction in follicular oestradiol secretion, 12 anoestrous ewes with ovarian transplants were administered progesterone for 10 days and follicular growth was initiated by injections of 2.5 μg ovine LH (S26) administered at 3 h intervals increasing hourly by 24 h. The dose of LH was then halved in all ewes. After a further 8 h, the six control ewes continued to receive 1.25 μg LH every hour, whereas LH was reduced to 0.625 μg every 2 h in the other six ewes. This latter treatment period continued for 12 h, after which all ewes were maintained on 1.25 μg LH per hour for the remaining 40 h. Throughout, ovarian vein blood was collected every 4 h with 10 min sampling for the last 4 h of the treatment period. Ten minutes after 1.25 μg LH injections, maximum plasma values were 1.1 ± 0.3 ng LH (NIAMDD-21) ml^{-1} (similar to spontaneous LH pulses during the late follicular phase in the breeding season) versus 0.4 ± 0.1 ng ml^{-1} after 0.625 μg LH. Exposure to LH injections at increasing frequency for 32 h enhanced oestradiol concentrations from 0.2 ± 0.03 ng ml^{-1} to 2.6 ± 0.8 ng ml^{-1}. Eight hours after the reduction in LH dose and frequency, there was a marked decline in oestradiol values to 0.4 ± 0.05 ng ml^{-1} ($P < 0.01$), whereas maximum values were maintained for a further 20 h in the control ewes. Interestingly, only one of the control ewes had a spontaneous LH surge (> 10 ng ml^{-1}), initiated 60 h after the start of the LH injections. In conclusion, halving LH pulse frequency and amplitude in the late follicular phase caused a dramatic reduction in oestradiol secretion.

This work was supported in part by MRC programme grant G8929853.

Cerebroventricular injection of neuropeptide Y differentially influences pituitary secretion of luteinizing hormone and growth hormone in cows. M. G. Thomas[1], O. S. Gazal[2], G. L. Williams[3], D. H. Keisler[4] and R. L. Stanko[5], [1]*Department of Animal and Range Sciences, New Mexico State University, Las Cruces, NM, USA;,* [2]*Animal Reproduction Laboratory, Fort Valley State University Agricultural Research Station, Fort Valley, GA, USA;* [3]*Animal Reproduction Laboratory, Texas A and M University Agricultural Research Station, Beeville, TX, USA;* [4]*Department of Animal Science, University of Missouri-Columbia, USA; and* [5]*Texas A and M University, Kingsville, TX, USA*

Previously we reported that elevated central concentrations of neuropeptide Y are associated with decreased serum concentrations of LH during chronic undernutrition in cattle. In addition, central injection of neuropeptide Y dramatically suppressed pituitary secretion of LH in well-nourished ovariectomized cows (*Biology of Reproduction* **54** Supplement **1** 153, 1996). Neuropeptide Y is also known to influence the neurones that regulate the growth hormone (GH) axis in rodents. These neurones, as well as the neurones that regulate the LH axis, are in an oestrogen-sensitive region of the hypothalamus. Thus, we hypothesized that neuropeptide Y would enhance pituitary secretion of GH as typically observed in undernourished cattle, and that the effects of neuropeptide Y would be altered in an oestrogen-influenced model. Two experiments were conducted to test these hypotheses using cows with surgically placed cerebroventricular cannula. In Expt 1, four well-nourished ovariectomized cows received central injections of either 50 or 500 µg of neuropeptide Y in a crossover design. Blood was collected via jugular cannulae at 10 min intervals from −4 h to +4 h relative to treatment. Neuropeptide Y suppressed tonic secretion of LH irrespective of dose and stimulated an abrupt increase in serum concentrations of GH. Pretreatment mean GH concentrations and amplitudes of pulses were less than post-treatment means (concentrations: $5.4 \pm 9 < 18.2 \pm 7.2$ ng ml^{-1}, respectively; $P = 0.07$; pulse amplitude: 8.8 ± 2 versus 93.1 ± 38 ng ml^{-1}, respectively; $P = 0.03$). In Expt 2, six cows that were well-nourished, ovariectomized and oestrogen-implanted (s.c. implant providing 2–6 pg oestradiol ml^{-1}) received central injections of 0, 50 or 500 µg neuropeptide Y in a cross-over design. Blood samples were collected before and at 10 min intervals after the injection for 240 min. The 500 µg injection of neuropeptide Y suppressed ($P < 0.05$) tonic secretion of LH relative to the 0 µg dose (mean concentration $2.7 \pm 0.4 < 4.7 \pm 1$ ng ml^{-1}; number of pulses $0.8 \pm 0.4 < 1.6 \pm 2.0$; pulse amplitude $2.1 \pm 0.9 < 7.7 \pm 2.4$ ng ml^{-1}). The 50 µg injection of neuropeptide Y tended ($P \leq 1.0$) to produce the same responses. These data provide evidence to suggest that neuropeptide Y acts as a chemical mediator of nutrient status to the LH and GH axes during periods of chronic undernutrition in cattle, and that these responses may be influenced by oestrogen.

The effect of active immunization against recombinant human leptin on reproductive and metabolic parameters in the ewe. C. G. Gutierrez[1], G. Baxter[2], B. K. Campbell[3], R. Webb[4] and D. G. Armstrong[2], [1]*Facultad de Medicina Veterinaria, Universidad Nacional Autonoma de Mexico, Mexico DF CP 04510;,* [2]*Division of Development and Reproduction, Roslin Institute (Edinburgh), Roslin, Midlothian EH25 9PS, UK;* [3]*University of Edinburgh, Department of Obstetrics and Gynaecology, Centre of Reproductive Biology, Chalmers Street, Edinburgh EH3 9EW, UK; and* [4]*University of Nottingham, School of Biological Sciences, Sutton-Bonington Campus, Loughborough LE12 5RD, UK*

Leptin has been postulated to have a permissive role for reproductive function in mice. Leptin administration corrects hypogonadotropism and restores fertility in *ob/ob* mice, prevents the delay in ovulation in nutritionally deprived mice and advances puberty in prepubertal mice. However, the relationship between leptin and reproductive function has not been established in non-rodent species. In this study, we report the effect of active immunization against leptin on reproductive and metabolic parameters in sheep. Eighteen ewes were immunized during the anoestrous season beginning on the 30 June with 100 µg of recombinant human leptin in Freund's complete adjuvant (FCA) (L; $n = 9$) or FCA with saline (S; $n = 9$), and boosted monthly for 4 months. Leptin immunization caused an increase in leptin antibody titres above 1:5000 in all cases, whilst antibody titres in control sheep did not differ from non-specific binding. Leptin immunization did not prevent the development of oestrous behaviour during the onset of the breeding season. Similarly, serum concentrations of GH (7.85 ± 1.7 versus 8.7 ± 2.5 ng ml^{-1}, L and S groups, respectively; $P > 0.05$), insulin (0.12 ± 0.01 versus 0.12 ± 0.02 ng ml^{-1}, L and S groups, respectively; $P > 0.05$) and body condition score (2.2 ± 0.3 versus 2.4 ± 0.3, L and S groups, respectively; $P > 0.05$) did not change after leptin immunization. After the third boost, oestrus was synchronized using $PGF_{2\alpha}$, and the ovulation rate was evaluated and found not to vary between groups (1.66 ± 0.16 versus 1.44 ± 0.17 ovulations, L and S groups, respectively; $P > 0.05$). We conclude that despite the high antibody titres achieved, active immunization against leptin had no effect on the metabolic or reproductive parameters examined in this study. The role of leptin in larger animal species remains to be clarified.

This work was supported by a BBSRC core strategic grant.

Pituitary expression of genes encoding steroidogenic factor 1 (SF-1) and gonadotrophin subunits during the ovine oestrous cycle. T. M. Nett[1], M. Baratta[2] and A. M. Turzillo[1], [1]*Animal Reproduction and Biotechnology Laboratory, Colorado State University, Fort Collins, CO, USA; and* [2]*Istituto di Fisiologia Veterinaria, Università di Parma, Italy*

Steroidogenic factor 1 (SF-1) is a transcription factor involved in the regulation of steroidogenic enzymes. Recent evidence indicates that SF-1 influences transcriptional activity of several genes expressed by gonadotropes. The purpose of this experiment was to measure pituitary gene expression for SF-1 during the ovine oestrous cycle and to determine if it is correlated with gonadotrophin subunit gene expression. Ewes were observed for oestrous behaviour at 4 h intervals using a vasectomized ram, and anterior pituitary glands were collected at five different time points ($n = 5$ ewes/time): (i) within 4 h of the first observation of oestrus (day 0); (ii) 24 h after the onset of oestrus (24 h); (iii) 48 h after the onset of oestrus (48 h); (iv) day 10; and (v) day 16. Polyadenylated RNA was isolated from pituitary tissues and steady-state levels of mRNA encoding SF-1, FSHβ subunit and LHβ subunit were measured by slot blot analysis using ovine SF-1 and bovine FSHβ and LHβ subunits as cDNA probes. Amounts of mRNA encoding SF-1 and FSHβ subunit were lowest on day 0, and increased 6.7-fold and 7.7-fold, respectively, by 24 h ($P < 0.001$). At 48 h, on day 10 and day 16, the amounts of FSHβ subunit mRNA were lower ($P < 0.05$) than those at 24 h, but higher ($P < 0.05$) than the nadir on day 0. Amounts of SF-1 mRNA and FSHβ subunit mRNA were correlated ($r = 0.65$, $P < 0.001$). Compared with the dynamic patterns of gene expression for SF-1 and FSHβ subunit, changes in LHβ subunit were less marked. Amounts of LHβ subunit mRNA were 1.8-fold higher ($P < 0.05$) on day 10 than on day 0, but were similar at all other time points. From these data we conclude that marked changes in SF-1 gene expression occur during the peri-ovulatory period in the ewe, with a significant decrease in amounts of SF-1 mRNA at oestrus. Coordinate expression of mRNAs encoding SF-1 and FSHβ subunit raises the possibility that SF-1 may be involved in transcriptional regulation of the FSHβ subunit gene in the ovine anterior pituitary gland.

Influence of milk yield on fertility and serum concentrations of thyroid hormones and glucose in Carora cows of Venezuela. H. Leyva-Ocariz, B. Reyes, D. Zambrano and M. Arteaga, *Universidad Centroccidental Lisandro Alvarado "UCLA," Decanato de Ciencias Veterinarias, Apartado 846, 7 Barquisimeto, Venezuela*

The effects of high (HY), medium (MY) and low (LY) milk yield on fertility, thyroid activity and glycaemia relationships were evaluated in Carora cows, a dairy cattle of Venezuela raised in tropical conditions. Cows (HY, $n = 9$; MY, $n = 14$; and LY $n = 11$) producing 29, 19 and 11 kg milk day^{-1}, respectively, were inseminated for the first time at 52 ± 10 days postpartum. A split-plot model with repeated measures over the first 5 weeks after first service or over 8 months postpartum was used to analyse effects of milk yield, pregnancy status and their interactions, on weekly or monthly serum concentrations of triiodothyronine (T_3), thyroxine (T_4) and glucose in six treatments: HY pregnant ($n = 8$), HY repeat breeder ($n = 1$), MY pregnant ($n = 6$), MY repeat breeder ($n = 8$), LY pregnant ($n = 7$) and LY repeat breeder ($n = 4$) cows. Mean serum concentrations of T_3 (116 ± 13 ng dl^{-1}) were greater ($P < 0.01$) during months 1–4 of lactation in HY compared with MY and LY cows. The mean serum concentration of glucose (57 ± 1.7 mg dl^{-1}) was greater ($P < 0.05$) during the first 4 months than during the last 4 months of lactation in LY cows. Significant ($P < 0.01$) interactions between milk yield group and pregnancy status were detected in serum concentrations of T_3, T_4 and glucose during the first 5 weeks post-insemination. A significant ($P < 0.05$) positive correlation was found between serum concentrations of T_3 and glucose. Breeding HY cows had decreased days in service ($P < 0.01$), number of services per conception ($P < 0.05$) and percentage of repeat breeder cows ($P < 0.05$). These results indicate that decreased concentrations of thyroid hormones associated with MY and LY may mediate the negative effect on fertility, contrasting with the results found in HY Carora cows.

Ultrasonographic examination of the postpartum bovine uterus. I. Situmbeko and L. Robertson, *Veterinary Reproductive Research Group, Division of Veterinary Anatomy, Department of Veterinary Preclinical Studies, Glasgow Veterinary School, Glasgow G61 6QH, UK*

The objective of this study was to characterize postpartum involution of the uterus ultrasonographically in dairy cows experiencing a normal or abnormal puerperium. Twenty-one normal cows and six that had experienced periparturient reproductive problems were examined once during each of the periods 2–9, 12–17, 23–29, 43–49 and 54–61 days postpartum (termed examinations 1–5, respectively), using a portable ultrasound scanner equipped with a 7.5 MHz linear-array rectal transducer. The uterus and cervix were scanned, noting whether a cervical lumen, uterine fluid and caruncles were visible. The diameters of the cervix and uterine horns at the level of the intercornual ligaments were measured. In normal cows, the cervical lumen was visible only during examination 1. Caruncles were visualized as oval echogenic structures during examination 1 but not thereafter. Echogenic uterine fluid was visible during examinations 1 and 2. By examination 3, fluid was visible only when associated with oestrus. During examination 1, the mean diameter of the uterine horns exceeded 60 mm. The size of the previously gravid horn reduced markedly from examinations 1 to 5, with a mean diameter of 45.1 mm during examination 2 and 29.7 mm by examination 5. Involution of the cervix was slower than that of the horns; mean diameter exceeded 60 mm until examination 3, when a mean diameter of 52 mm was recorded. By examination 5, the mean cervical diameter was 44.7 mm. After examination 3, the principal new features detected ultrasonically were associated with oestrus. Involution was delayed in the abnormal cows and was characterized ultrasonographically by the persistence of a visible cervical lumen and caruncles beyond examination 1, a uterine horn exceeding 60 mm or the presence of echogenic fluid after 23 days postpartum, and a cervical diameter exceeding 60 mm beyond 54 days postpartum. In conclusion, this study showed that it was possible to characterize the process of involution ultrasonographically and to recognize certain features by ultrasound scanning that were indicative of delayed involution or uterine pathology.

Disposition of intrauterine administered ampicillin in genital tissues of acyclic buffaloes. I. Singh and U. Singh, *Department of Gynaecology and Obstetrics, College of Veterinary Sciences, CCS Haryana Agricultural University, Hisar - 125 004, India*

Uteri of pluriparous non-lactating (> 300 days postpartum) and non-cycling buffaloes ($n = 4$) were transcervically infused with 2.0 g ampicillin in 40 ml distilled water. Peripheral plasma samples were collected at 0 (infusion), 15, 30, 45 and 60 min and then hourly until 6 h post-infusion. Two buffaloes were killed at 6 h and the other two at 24 h post-infusion. Immediately after death, tissues from different genital organs were collected, weighed (range 0.385–1.253 g) and triturated in phosphate buffer (pH 7.9, 10 ml g^{-1} tissue). After equilibration for 12 h at 4°C, supernatants were assayed in a large plate microbial assay system using Antibiotic Assay Medium no. 11 and *Sarcina lutea* (ATCC 9341) as test organism. Ampicillin standards (8.0–0.125 mcg ml^{-1}) were prepared fresh on the day of slaughter in phosphate buffer (pH 7.9) or ampicillin-free plasma, for tissue and plasma assays, respectively. After intrauterine infusion, ampicillin was detected in peripheral plasma at 15 min (average 0.62 mcg ml^{-1}), achieved peak values at 2 h (1.34 mcg ml^{-1}) before declining by 6 h post-infusion (0.52 mcg ml^{-1}). Average drug concentrations in different tissues collected at 6 and 24 h, resepectively, were 8.51 and 7.06 mcg g^{-1} of ovarian tissue, 44.96 and 21.94 mcg g^{-1} in fallopian tube, 13.32 and 7.02 mcg g^{-1} in uterine tissue, 44.62 and 18.91 mcg g^{-1} in endometrium and 14.94 and 4.95 mcg g^{-1} in myometrium. Except ovarian tissue, where ampicillin declined only marginally, concentrations in tubular genitalia tissues were reduced by half to one-third by 24 h as compared to 6 h values. These results indicate achievement of very high therapeutic concentrations in all genital tissues, including ovaries, following intrauterine infusion of 2.0 g ampicillin. With high ampicillin concentrations observed beyond 24 h post-infusion, one or two infusions may suffice for the treatment of genital infections in acyclic buffaloes, particularly important during post-partum period when the incidence of genital infections is high and cyclic ovarian activity is absent.

Vulnerability of early embryonic stages to ACTH-induced adrenal hyperactivity in ewes. I. Singh*, *Veterinary Field Station, University of Liverpool, Leahurst, Neston, UK*

Vulnerability of early embryonic stages to ACTH-induced adrenal hyperactivity was assessed in progestogen + $PFG_{2\alpha}$ synchronized and mated Welsh Mountain ewes. In Expt 1, ACTH (0.5 mg i.m. twice a day) was administered from day 1 to 11 post-mating ($n = 7$, saline-treated controls $n = 7$). On ultrasound examination at day 45, all treated ewes were non-pregnant and basal progesterone concentrations from day 15 to 19 in five of them suggested early pregnancy losses. Four control ewes maintained pregnancy to term while three returned to oestrus after a normal cycle. In Expt 2, following ACTH treatment (day 1–11, $n = 7$), uteri were flushed on day 15 after slaughter and the quality of recovered embryos was compared with those recovered from five saline-treated control ewes. Embryos were recovered from all five control ewes and five ACTH-treated ewes. In the control and treatment groups, respectively, the number of ovulations were seven and eight, embryos recovered seven and five, and good quality embryos were six and two.

Shorter exposure (day 1 to 4 post-mating, $n = 11$; day 5–8 $n = 12$; or day 9–12 $n = 10$) to exogenous ACTH in Expt 3, however, failed to compromise embryonic survival or pregnancy rates in comparison to controls ($n = 16$). In these experiments, peripheral cortisol concentrations remained consistently elevated during ACTH treatment compared with controls ($P < 0.001$), with maximum values at 1 and 2 h after each injection. After pooling data from all treatment groups, the non-pregnant ewes had higher basal and post-injection cortisol values ($P < 0.05$). These results suggest that early embryonic stages are vulnerable to chronic adrenal hyperactivity in sheep with maximum impact when the entire embryonic period prior to maternal recognition of pregnancy is involved.

**Present address* : Department of Gynaecology and Obstetrics, College of Veterinary Sciences, CCS Haryana Agricultural University, Hisar - 125 004 (Haryana), India

Ovarian activity postpartum in dual purpose cows evaluated through diverse management systems under the conditions of the humid Mexican tropics. F. Montiel[1], C. S. Galina[1] and C. Lamothe[2], [1]*Departments of Reproduction, Faculty of Veterinary Medicine, University of Mexico; and* [2]*Faculty of Veterinary Medicine, University of Veracruz, Mexico*

The objective of this study was to determine the effect of oestrous synchronization and/or supplementation on reproduction efficiency of grazing dual purpose cows after calving during the dry and wet seasons in the humid tropics of Mexico. A total of 327 crossbred cows (*Bos indicus* × *Bos taurus*) were distributed among 42 ranches located in the central–north region of Veracruz state, Mexico. Body condition score (BCS) was evaluated from 1 to 5 with good BCS categorized from 2.5 to 4.0 and poor BCS < 2.5. Oestrous responses to SMB from 36 to 48 h after implant removal in the dry and wet seasons were 96.4% (108/112) and 92.7% (115/124), respectively. Oestrous responses were higher ($P < 0.05$) than respective controls at 48 h (dry, 8.3% [4/48]; wet, 4.6% [2/43] seasons) and for the first 21 days (dry, 44% [21/48]; wet, 28% [12/43] seasons). Fertility following AI to a synchronized oestrus of cows in good body condition was higher ($P < 0.05$) in the dry season (62.5%, 35/56) than the wet season (40%, 26/65). A dry versus wet season effect was not observed in cows with poor body condition (42.3% [25/59] versus 37.5% [21/56], respectively). Pregnancy rates for 60 days for all cows (SMB-synchronized, SMB-unsynchronized and control) in the dry season were 53.75% [43/80] and 31.3% [25/80] for good and poor BCS statuses versus 33.3% [28/84] and 32.5% [27/83] for good and poor BCS statuses in the wet season (season by BCS interaction, $P < 0.05$). Short-term concentrate supplementation at 1.0% of body weight (2.8 Mcal DE kg^{-1} dry matter; 16% crude protein), for a 45 day period beginning at 30 days prior to synchronization, had no effect on fertility in either the dry or wet seasons regardless of body condition.

Reproductive strategies in arctic ungulates. J. E. Rowell, M. Sousa, J. E. Blake and R. G. White, *Large Animal Research Station, Institute of Arctic Biology, University of Alaska, Fairbanks, AK 99709– 7000, USA*

Only two ruminants are adapted to the arctic environment. Muskoxen (*Ovibos moschatus*) are anatomically very similar to sheep and goats, while caribou/reindeer (*Rangifer tarandus*) exhibit characteristics typical of temperate region deer. Because both of these species share many features in common with their southern counterparts, interpretation of reproductive events relies heavily on extrapolation from domestic animal studies. At the Large Animal Research Station (LARS), ongoing research on colonies of both species have verified many previously held assumptions and also revealed differences. The following focuses on two unusual features currently under investigation at LARS. In muskoxen we present evidence that pregnancy exerts control over lactation. Pregnant muskoxen exhibit a significant progesterone rise at weeks 10–12 and a decline in weeks 20–22 during a 33 week gestation. The progesterone rise follows regression of the corpus luteum of pregnancy so the putative source of this progesterone is placental. Muskoxen normally wean their calves from the previous year in mid-pregnancy (mid-January–early February), coincident with peak progesterone. We hypothesize that the sustained rise in progesterone during mid-pregnancy is a signal for the initiation of weaning. Although a preliminary survey of three pregnant muskoxen found no temporal association between the timing of weaning and the progesterone peak, in non-pregnant, lactating muskoxen, lactation is extended from 8 to 18 months. In *Rangifer*, the association of female antlers with the reproductive cycle presents a unique challenge for hormonal control. Antler cleaning and ossification is dependent on rising oestradiol at the beginning of the breeding season. However, in preliminary studies antler cleaning was closely associated with the onset of cyclic activity in only six of 14 animals. Antler casting in most pregnant females occurs around parturition. Our data indicate that early casting among healthy, pregnant cows is associated with breeding late in the season. We hypothesize that pregnant cows cast their antlers in response to photoperiodic cues in March, unless the female experiences a rise in oestradiol, such as occurs in late pregnancy.

Evaluation of the GnRH/PGF$_{2\alpha}$ protocols for synchronization of oestrus/ovulation in beef cows. T. W. Geary[1], J. C. Whittier[1], D. G. LeFever[1], G. D. Niswender[2], T. M. Nett[2] and D. M. Hallford[3], [1]*Department of Animal Science, Colorado State University, Fort Collins, CO 80523, USA;* [2]*Department of Physiology, Colorado State University, Fort Collins, CO 80523, USA;* [3]*Department of Animal and Range Sciences, New Mexico State University, Las Cruces, NM 88003, USA*

Beef cows ($n = 1939$) from seven locations were used to evaluate variations of the Ovsynch protocol for synchronization of oestrus and (or) ovulation. Treatment groups varied by location, but included direct comparisons of the Ovsynch protocol ($n = 220$) versus Syncro-Mate B ($n = 216$), Ovsynch protocol ($n = 627$) versus the CO-Synch protocol ($n = 594$), and Select Synch ($n = 116$) versus 2× PGF$_{2\alpha}$ ($n = 109$). The Ovsynch, CO-Synch and Select Synch protocols include the following injection schedules: GnRH – 7 days PGF$_{2\alpha}$ – 2 days GnRH – 1 day – AI; GnRH – 7 days PGF$_{2\alpha}$ – 2 days – GnRH + AI; and GnRH – 7 days PGF$_{2\alpha}$ + 5days AI 12 h after detection of oestrus, respectively. Calf removal (48 h) initiated at the time of the PGF$_{2\alpha}$ injection was evaluated in a subset of cows receiving the Ovsynch ($n = 235$) or CO-Synch ($n = 234$) protocol. Daily ultrasound and electronic oestrous detection were used to evaluate oestrus and ovarian response to the Select Synch protocol ($n = 57$). Progesterone analysis of two blood samples collected 10 days apart before treatment indicated that each of the GnRH/PGF$_{2\alpha}$ protocols was capable of inducing ovulation/oestrus in previously anoestrous cows. Pregnancy rates were greater ($P < 0.025$) for cows that received the Ovsynch protocol (54%) than in cows that received Syncro-Mate B (42%). Pregnancy rates were not different ($P > 0.1$) among cows that received the Ovsynch protocol (58%) or CO-Synch protocol (54%). However, 48 h calf removal improved ($P < 0.05$) pregnancy rates across both treatments by 8%. Synchronization, conception and pregnancy rates were not different among cows that received the Select Synch protocol (77%, 61% and 47%, respectively) or the 2× PGF$_{2\alpha}$ protocol (81%, 60%, and 49%, respectively). Ovarian response of cows that were between day 15 and 17 of their oestrous cycles at the time of the first GnRH injection was poor and these cows exhibited oestrus 11 ± 19 h before the $PGF_{2\alpha}$ injection . The GnRH/$PGF_{2\alpha}$ protocols can result in high pregnancy rates to timed inseminations or breeding by oestrus in beef cows.

The expression of insulin-like growth factor binding proteins 1–3 mRNA in the bovine uterus during early pregnancy. R. S. Robinson[1], G. E. Mann[2], G. E. Lamming[2] and D. C. Wathes[1], *[1]Department of Veterinary Basic Sciences Royal Veterinary College, Hawkshead Road, Potters Bar, Herts, EN6 1NB, UK; and [2]Division of Animal Physiology, School of Biological Sciences, University of Nottingham, Sutton Bonnington, Loughborough, Leics, LE12 5RD, UK*

Insulin-like growth factor (IGF) -I and -II are expressed in the uterus during the preimplantation period and are likely to play an important role in development of the embryo and uterus. IGFs actions are modulated by IGF-binding proteins (IGFBP), which generally influence IGF activity by modulating interaction with the IGF type 1 receptor. Uterine horn cross-sections were collected on day 16 from 15 pregnant cows (PREG), five inseminated cows with no embryo present (INP) and seven cyclic cows (CONT). The localization of IGFBP-1, -2 and -3 mRNA was determined by *in situ* hybridization and results were quantified by measuring the optical density (OD) units from autoradiographs. IGFBP-1 mRNA was localized to the luminal epithelium (LE), with the PREG group expressing higher concentrations of IGFBP-1 mRNA than cyclic cows, with intermediate values in the INP group (ODs: PREG 0.22 ± 0.03, CONT 0.12 ± 0.02; [$P < 0.05$; ANOVA], INP 0.18 ± 0.03). Low concentrations of IGFBP-2 mRNA were observed in the subepithelial stroma underlying the LE in four of five INP cows compared with only three of 15 PREG ($P < 0.025$; χ^2 test). Expression of IGFBP-3 mRNA was strongest in the caruncular stroma, the area of embryo attachment, but was also localized to the myometrium and stroma. IGFBP-3 mRNA expression in the stroma was lower in the PREG group compared with the cyclic cows but was not different from the INP group (ODs: PREG 0.04 ± 0.01, CONT 0.11 ± 0.02 [$P < 0.05$; ANOVA], INP 0.05 ± 0.01). In conclusion, pregnancy increased expression of IGFBP-1 mRNA, while decreasing IGFBP-2 mRNA expression, thereby probably increasing the bioavailability of the IGF for uterine and embryonic development. IGF-II mRNA expression was high in the caruncular stroma; therefore, IGFBP-3 is probably regulating IGF-II action in preparation for embryonic attachment.

Synchronization of ovulation in Gir cows with GnRH–PGF–GnRH treatment. C. M. Barros, A. L. G. Gambini, M. B. P. Moreira and C. Castilho, *Instituto de Biociências - UNESP, 18618.000 Botucatu - SP, Brazil*

Gir (*Bos indicus*) is a common dairy breed in Brazil because it is tolerant to heat stress. This study investigates the follicular dynamics in Gir cows and their response to a GnRH–PGF–GnRH protocol for synchronizing ovulation. In seven cows follicular development was monitored by daily ultrasonography. Most animals (71.4%) presented three follicular waves. Some parameters observed in these cows were: maximum diameter of CL (MD, 21.5 ± 1.4 mm), dominant follicle MD during the first (12.2 ± 0.8 mm), second (9.5 ± 0.7 mm) and third (13.5 ± 0.5 mm) follicular waves, duration of the first (14.4 ± 1.0 days), second (10.8 ± 0.9 days) and third (7.4 ± 0.7 days) follicular waves. In another experiment, 15 Gir cows were separated in two groups. Group 1 (G1, $n = 9$) was treated with a GnRH agonist (GA, Buserelin, 8 µg i.m.), 7 days later $PGF_{2\alpha}$ (25 mg) and 24 h later another injection of GA (8 µg). The control group (G2, $n = 6$) did not receive the second injection of GA. Ovarian morphology was monitored daily or every 6 h by ultrasonography. The first injection of GA caused ovulation in 7 of 9 (G1) and 4 of 6 (G2). The estimated interval between $PGF_{2\alpha}$ and ovulation was 51.2 ± 0.2 h (G1) and 83.3 ± 8.8 h (G2). The second injection of GA (G1) advanced ovulation approximately 22 h compared with the control group (G2) and synchronized it to occur within a 6 h period. Ovulation rates were 88.8% (G1) and 83.8% (G2) and most ovulatory follicles were from a new follicular wave. It is concluded that the follicular dynamics in Gir cows were similar to those observed in other Zebu and European breeds. Additionally, in the present protocol the first injection of GA was effective in promoting ovulation and $PGF_{2\alpha}$ followed by the second GA injection synchronized the ovulation of a new follicle.

Supported by FAPESP.

Effect of early social environment on sexual behaviour of rams. S. A. Wright, R. A. Dailey, E. C. Townsend, P. E. Lewis and E. K. Inskeep, *Division of Animal and Veterinary Sciences, West Virginia University, Morgantown, 26506-6108, USA*

A study tested the concept that early social environment affects sexual behaviour of rams as adults, and determined if sire or environment influenced peripheral concentrations of luteinizing hormone (LH), testosterone, and oestradiol. Twenty-four March-born ram lambs by three sires were weaned at 4 days of age and reared either in isolation (I, one per pen) or in groups (G, four per pen). During rearing, sexual behaviour of G rams was observed a total of 56 times for 30–60 min each time. Six of the 11 G rams mounted more than twice as often as they were mounted. To assess sexual behaviour, yearling rams were observed with individual oestrous ewes and in a mate-choice situation (ram versus ewe). Jugular samples were taken at 15 min intervals for 6 h in October (before testing) and March (after testing) and assayed for LH, testosterone and oestradiol. Data were examined using Principal Component (PC) analysis; 4 PCS accounted for significant portions of variance in behavioural traits and were compared for effects of sire, treatment, replicate and their interactions. G rams had greater ewe success (PC1) than I rams, and I rams were more inept (PC4) than G rams ($P < 0.01$), but increased in ewe success and decreased in ineptness over four test rounds ($P < 0.01$). Rams in both treatments decreased in ram interest (PC2) and ewe investigation (PC3) over four test rounds ($P < 0.001$). I rams had greater LH ($P < 0.05$) and offspring of sire A had lower LH in March ($P < 0.05$). I rams had greater testosterone than G rams ($P < 0.05$), but oestradiol did not differ. Rearing rams in isolation reduced sexual success but the effect declined with continued exposure. However, results did not confirm the hypothesis of Zenchak *et al.*, 1981(*Applied Animal Ethology* **7** 157) that rams reared in all-male groups were more likely to be homosexual than those reared in individual pens. At 18 months of age, rams were tested for breeding capacity by exposure to ten oestrous ewes for 30 min. G rams mounted and had intromission with more ewes ($P < 0.01$) more often ($P < 0.01$) and had more total ejaculations ($P < 0.05$) than I rams. Offspring of sire C mounted more frequently ($P < 0.01$) and had more intromissions ($P < 0.05$), but this effect was mainly due to one high-performing ram.

Regulation of prostaglandin $F_{2\alpha}$ ($PGF_{2\alpha}$) secretion from endometrium of cyclic and pregnant cattle. D. R. Arnold, M. Binelli, C. J. Wilcox and W. W. Thatcher, *Department of Dairy and Poultry Sciences, University of Florida, PO Box 110920, Gainesville, FL 32611–0920, USA*

The objective of this experiment was to compare *in vitro* secretion of $PGF_{2\alpha}$ from endometrial explants of cyclic and early pregnant cows in response to extracellular and intracellular regulators of $PGF_{2\alpha}$ secretion. Intercaruncular explants, from the uterine horn ipsilateral to the CL, were collected at day 17 (day 0 = oestrus) of the oestrous cycle from cyclic ($n = 9$) and pregnant ($n = 5$) cows. Explants (50–80 mg) were preincubated in 0.5 ml of Krebs-Hensliet buffer (KHB, control) at 37°C for 60 min. Fresh KHB (0.5 ml) was added for an additional 60 min. Explants were assigned randomly to receive 1.0 ml of KHB medium alone or containing the following treatments: phorbol 12,13 dibutyrate (PE, 10^{-6} mol l^{-1}), Ca^{2+} ionophore, A23187 (10^{-5} mol l^{-1}), melittin (10^{-4} mol l^{-1}), A23187+ melittin, or oxytocin (10^{-6} mol l^{-1}). Samples of media were removed at 0, 20, 40 and 60 min after addition of treatments and analysed for $PGF_{2\alpha}$ concentrations. $PGF_{2\alpha}$ secretion rates were analysed by homogeneity of regression. $PGF_{2\alpha}$ secretion rate was greater for cyclic status (0.407 versus 0.026 ng g^{-1} min^{-1}, $P < 0.01$). A status × time treatment interaction was detected for $PGF_{2\alpha}$ secretion rates ($P < 0.01$). Treatment effects on $PGF_{2\alpha}$ secretion rates (ng g^{-1} min^{-1}) differed in cyclic cows: KHB, 0.13; PE, 0.65; A23187, 0.20; melittin, 0.68; A23187+melittin, 0.50; and oxytocin, 0.28. Stimulatory effects of PE ($P < 0.025$) and melittin ($P < 0.01$) on $PGF_{2\alpha}$ secretion rates for explants from cyclic cows were abolished in endometrial explants of pregnant cows. In conclusion, regulators of PKC and PLA_2 stimulate $PGF_{2\alpha}$ secretion in explants from cyclic cows. Factors produced by the conceptus attenuate responses to intracellular regulators of $PGF_{2\alpha}$ secretion during the period of pregnancy recognition.

Utilization of glucose, pyruvate and glutamine during maturation of oocytes from pre-pubertal calves and adult cows. T. E. Steeves[1] and D. K. Gardner[1*], [1]*Centre for Early Human Development, Institute of Reproduction and Development, Monash University, Clayton, Victoria, 3168, Australia; *Present address: Colorado Center for Reproductive Medicine, Englewood, Colorado 80110, USA*

In vitro production of viable embryos from prepubertal heifers has the potential to increase the rate of genetic gain by significantly reducing the generation interval in elite breeding stock. While oocytes retrieved from ovaries of prepubertal animals can be matured and fertilized *in vitro*, subsequent developmental competence is often poor. Impaired embryo development *in vitro* is often due to perturbations in energy metabolism. The aim of the present study was therefore to determine whether poor embryo development in material from prepubertal animals was associated with impairment in energy metabolism during oocyte maturation. Ovaries were obtained from 2–4 week old crossbred beef calves and adult cows of unknown breed and age. At 0 h, 12 h and 24 h maturation in TCM 199, the activity of the Embden Meyerhof pathway (EMP: [5-^{3}H]glucose utilization) and the activity of two parts of the tricarboxylic acid cycle (TCA: [2-^{14}C]pyruvate utilization; [G-^{3}H]glutamine utilization), were measured for oocytes from both prepubertal calves (PCO) and adult cows (ACO). For metabolic measurements, denuded oocytes were incubated with radiolabelled substrates for 3 h in a modified synthetic oviduct fluid. Data were from three replicates. EMP activity increased during maturation of PCO and ACO. By 12 h, however, glucose utilization was significantly lower in PCO (PCO, 0.39 ± 0.04 pmol per oocyte per h; ACO, 0.63 ± 0.08 pmol per oocyte per h, $P < 0.05$). By 24 h maturation, glucose utilization in both groups was equivalent. During maturation of ACO, pyruvate utilization peaked at 12 h (0.95 ± 0.04 pmol per oocyte per h) and then fell to initial values by 24 h. A similar pattern was seen in PCO; however, pyruvate utilization was significantly lower at 12 h (0.82 ± 0.03 pmol per oocyte per h, $P < 0.05$). Glutamine utilization increased steadily during maturation of oocytes, peaking at 24 h. Utilization was significantly lower in PCO (1.08±0.03 pmol per oocyte per h) than in ACO (1.20 ± 0.05 pmol per oocyte per h, $P < 0.05$), after 24 h maturation. Furthermore, PCO were significantly smaller than ACO (0 h, 113.3 ± 0.80 μm versus 117.6 ± 0.93 μm, respectively, $P < 0.05$; 24 h, 112.2 ± 0.88 μm versus 117.0 ± 0.83 μm, respectively, $P < 0.05$). This study indicates that the activities of the energy generating pathways (EMP and TCA) differ during maturation of PCO and ACO. The observed differences in oocyte size and energy metabolism may account for subsequent compromised embryo development.

Supported by the Dairy Research and Development Corporation, Meat Research Corporation and Genetics Australia Coop. Ltd.

Identification of an acidic 58 kDa membrane-associated tyrosine phosphoprotein that interacts with the EGF receptor. S. Brûlé[1], F. Rabahi[1], R. Faure[2], J-F. Beckers[3], D. W. Silversides[1] and J. G. Lussier[1], [1]*CRRA, Université de Montréal, St-Hyacinthe, Québec, Canada, J2S 7C6;* [2]*CHUL, Université Laval, Ste-Foy, Québec, Canada; and* [3]*Université de Liège, Liège, Belgique*

The pattern of protein expression in granulosal cells is modified after the preovulatory LH surge. From the major proteins that are modified, a 58 kDa protein (58P) was characterized. The 58P is encoded by a 2100 bp cDNA, including an open reading frame of 1599 bp encoding a 533 amino acid protein (60.1 kDa, pI 4.2). We have shown that 58P mRNA is found in all tissues studied. Its pattern of expression is cell specific in the ovary: oocyte, granulosal, thecal, epithelial and luteal cells; and uterus: endometrial epithelial and glandular cells. Subcellular fractionation studies showed that p58 is associated with membrane and purified endosome fractions. Western zooblot analysis with hepatic protein extracts demonstrated 58P expression in all superior eukaryotes. P58 is tyrosine phosphorylated and structural analysis reveals characteristics of a protein involved in protein–protein interaction: an acidic domain flanked by three potential SH3 minimal ligand binding motifs (PXXP). Amino acid sequence comparison showed an overall 87% identity with a 59.4 kDa human protein named 80K-H (Sakai *et al.*, 1989). However, comparison of the protein–protein putative docking domain shows only a 63% identity. This analysis supports the hypothesis that these proteins are not species specific but may be members of a new protein family with molecular adapter functions. Our search for a biological function for P58 showed that P58 interacts specifically with the EGF receptor and also to unknown intracellular proteins of 64 kDa and 150 kDa in endometrial epithelial cells, and to 64 kDa and 70 kDa proteins in luteal cells. Our current hypothesis is that b58P is a molecular adapter involved in signalling and/or shuttling functions during receptor-mediated endocytosis.

Behaviour and endocrine correlates related to exposure of heterosexual, low-performing and male-oriented domestic rams to rams and ewes in oestrus. B. M. Alexander[1], J. N. Stellflug[4], J. D. Rose[2], J. A. Fitzgerald[4] and G. E. Moss[3], [1]*Reproductive Biology Program,* [2]*Department of Psychology, and* [3]*Department of Animal Science, University of Wyoming, Laramie 82071; and* [4]*US Department of Agriculture, Dubois, ID 83423, USA*

Heterosexual ($n = 10$), low-performing ($n = 8$) and male-oriented ($n = 9$) rams were used to investigate the neuroendocrine correlates of sexual interest and discrimination. Treatment consisted of visual and olfactory contact with stimulus animals (ewes in oestrus or other rams) for 4 h on each of 3 consecutive days. Before exposure to stimulus animals on day 1 and during the final hour of exposure on day 2, blood samples were collected every 15 min for 1 h to determine concentrations of luteinizing hormone (LH). Stimulus animals were rotated three times during each exposure period. During exposure to stimulus animals, rams were continuously observed and investigatory sniffs, vocalizations, flehmens and foreleg kicks were recorded. Owing to the limited number of vocalizations, flehmens and foreleg kicks exhibited over the 3 day test period, these behaviours were combined for statistical analysis. There were no day, day by group or day by treatment interactions for investigatory sniffs or other behaviours. Therefore, day tallies of recorded behaviours were combined for final analysis. Behaviours displayed by rams differed by group ($P < 0.05$), but not by exposure to oestrous ewes or rams ($P > 0.05$). Heterosexual rams exhibited more investigatory sniffs ($P < 0.01$) and other behaviours ($P < 0.05$) towards all stimulus animals than low-performing or male-orientated rams. Heterosexual rams showed an increase ($P < 0.05$) in serum concentrations of LH when exposed to oestrous ewes. However, no change ($P > 0.05$) in concentrations of LH was detected when heterosexual rams were exposed to other rams. Concentrations of LH in low-performing and male-orientated rams were unchanged ($P > 0.05$), regardless of exposure to rams or oestrous ewes. In conclusion, sexually active rams exhibit a high degree of investigatory behaviours towards oestrous ewes and towards other rams. They appear to discriminate the sex of stimulus animals and exhibit a neuroendocrine response (i.e. increased secretion of LH) only when exposed to ewes in oestrus. Low-performing and male-oriented rams appear to lack the ability to respond differentially to ewes in oestrus and to other males; this may be due, at least in part, to their low level of investigatory behaviour.

Associations among the insulin-like growth factor system, body composition and reproductive function. A. J. Roberts[1], R. N. Funston[2] and G. E. Moss[2], [1]*USDA, ARS, US Meat Animal Research Center, Clay Center, Nebraska 68933, USA; and* [2]*University of Wyoming Dept of Animal Sciences, Laramie, Wyoming 82071, USA*

The present study evaluated associations of systemic and tissue levels of IGF and IGFBPs with body composition and reproductive function in mature beef cows individually fed one of four levels of feed intake over a 5 year study. After weaning of their last calf, cows were maintained on their respective levels of feed intake until body weight equilibrated. Cows were then slaughtered and body composition determined. Percentage of empty body fat (range from 1.9 to 19.6%) was correlated with concentrations of IGF-I in serum at the time of slaughter ($n = 49$, $P < 0.02$, $r = 0.34$) and in the stalk median eminence (SME; $n = 26$, $P < 0.06$, $r = 0.38$), but not ($P > 0.3$) with concentrations of IGF-I in anterior pituitary or liver. A step-down regression was used to determine associations of percentage body fat and the IGF system with pituitary function. Independent variables in the initial model were percentage body fat, age of corpus luteum at slaughter (estimated from visual appearance and circulating concentration of progesterone), concentration of oestradiol in the dominant follicle (FFE_2), concentrations of IGF-I in liver, serum, SME and pituitary, and relative levels of pituitary IGFBPs. Concentrations of LH and FSH in serum from a single sample taken at slaughter and in anterior pituitary tissue were fit as dependent variables. The equation obtained for prediction ($R^2 = 0.18$, $P < 0.23$) of concentrations of LH in serum included effects of FFE2, liver IGF-I and pituitary IGFBP-2. Pituitary LH was predicted ($R^2 = 0.46$, $P < 0.01$) by liver IGF-I, pituitary IGF-I and pituitary levels of IGFBP-2 and -5. Serum FSH was predicted ($R^2 = 0.55$, $P < 0.01$) by FFE2, liver IGF-I, serum IGF-I, SME IGF-I and pituitary IGFBP-2. Pituitary concentrations of FSH were predicted ($R^2 = 0.56$, $P < 0.01$) by percentage body fat, serum IGF-I, SME IGF-I and pituitary levels of IGFBP-2, -3 and -5. These results provide evidence that alterations in components of the IGF system within the circulation and hepatic, hypothalamic and pituitary tissues are associated with gonadotrope function.

Immunolocalization of tissue inhibitor of metalloproteinase-1 (TIMP-1) in ovine corpora lutea during $PGF_{2\alpha}$ induced luteolysis. W. A. Ricke, F. N. Kojima, and M. F. Smith, *Department of Animal Science, University of Missouri, Columbia MO 65211, USA*

Luteolysis is characterized by decreased progesterone production, loss of cell adhesion to extracellular matrix (ECM), and apoptosis. Matrix metalloproteinases (MMPs) are a family of enzymes that cleave specific components of the ECM and are specifically inhibited by tissue inhibitor of metalloproteinase 1 (TIMP-1). Ovine large luteal cells (LLCs) contain receptors for $PGF_{2\alpha}$ and are the primary luteal source of TIMP-1. If ECM degradation is an important aspect of $PGF_{2\alpha}$-induced luteolysis, vesting LLCs with both maintenance of luteal ECM integrity and responsiveness to the uterine luteolysin ($PGF_{2\alpha}$) appears logical. Importantly, overexpression of stromelysin-1 (MMP-3) increased apoptosis in cultured mammary epithelial cells. In a previous study, luteal concentrations of ovine TIMP-1 were decreased by 6 h after $PGF_{2\alpha}$-induced luteolysis. Therefore, we hypothesize that $PGF_{2\alpha}$ decreases TIMP-1 content in ovine LLCs. A decrease in TIMP-1 may increase the MMP:TIMP ratio and result in apoptosis and structural luteolysis. The objective of this study was to immunolocalize TIMP-1 in corpora lutea following $PGF_{2\alpha}$ administration. After oestrous synchronization, corpora lutea were collected ($n = 5$ per group) at 0 h, 6 h, 12 h, 24 h and 36 h after injection of $PGF_{2\alpha}$ on day 10 of the oestrous cycle, fixed in 10% formalin and embedded in paraffin wax. Immunohistochemistry was performed as described previously using a modified avidin–biotin peroxidase procedure. Three independent evaluators examined the intensity of TIMP-1 immunolocalization within LLCs at the preceding time points. A ranking of 1, 2, 3 or 4 was used to determine intensity of immunolocalization (1 = low intensity through 4 = high intensity) and the data were analysed by ranked ANOVA procedure. As expected TIMP-1 was primarily localized within the cytoplasm of LLCs. No differences in intensity of TIMP-1 immunostaining were detected among 0 h, 12 h, 24 h and 36 h; however, at 6 h, intensity of TIMP-1 immunostaining was significantly lower ($P < 0.01$). Furthermore, by 6 h after $PGF_{2\alpha}$,TIMP-1 immunostaining within LLCs was punctate instead of uniformly distributed throughout the cytoplasm as observed at 0 h. These data support previous results indicating that concentration of TIMP-1 within ovine luteal tissue decreased by 6 h after administration of $PGF_{2\alpha}$. The apparent decrease in TIMP may increase ECM degradation during luteolysis.

IGF binding protein-1 is increased during prostaglandin $F_{2\alpha}$-induced regression of bovine corpora lutea. R. Taft, B. L Sayre, J. Killefer and E. K. Inskeep, *Division of Animal and Veterinary Sciences, West Virginia University, Morgantown, 26506–6108, USA*

In postpartum cows, regressing corpora lutea produce or stimulate secretion of factor(s) that are detrimental to embryonic survival during day 4 through day 8 after oestrus. Differential display reverse transcription–polymerase chain reaction (DD–PCR) was used to identify differences in gene expression from control corpora lutea during the first 72 h of luteal regression in animals treated with $PGF_{2\alpha}$ on day 4 to day 7 after oestrus. On day 4 after synchronized oestrus, cows ($n = 15$) received either $PGF_{2\alpha}$ (15 mg in 3 ml of saline) or saline (3 ml) intramuscularly every 8 h until CL removal at 24 ($n = 2$) or 48 ($n = 2$) h after initiation of saline, or 24 ($n = 4$), 48 ($n = 4$), or 72 ($n = 3$) h after initiation of $PGF_{2\alpha}$. RNA was harvested from CL, and DD-PCR was performed to determine differential expression of PCR fragments between non-regressing and regressing CL. IGF binding protein 1 (IGFBP-1) was identified by DD-PCR as a differentially expressed fragment that increased during regression. From the harvested RNA pool, cDNA was synthesized using an oligo $dT_{(16)}$ primer, and PCR was performed using primers specific for IGFBP-1. The bovine CL expressed IGFBP-1, and expression increased after treatment with $PGF_{2\alpha}$ ($P < 0.001$). Expression of IGFBP-1 was increased at 24 and 48 h after beginning treatment with $PGF_{2\alpha}$, then returned to control values by 72 h. In contrast, β-actin expression did not change during regression. Expression of mRNA for $PGF_{2\alpha}$ receptor increased at 24 h after the initiation of $PGF_{2\alpha}$ treatment on day 4 to day 7 (Sayre *et al.*, 1998 *Biology of Reproduction* **58** (Supplement 1)). $PGF_{2\alpha}$ is known to increase protein kinase C (PKC) activity, and phorbol esters are known to increase production of IGFBP-1. Thus, $PGF_{2\alpha}$ could stimulate production of IGFBP-1, through the PKC pathway. $PGF_{2\alpha}$ -induced luteal production of IGFBP-1 may interfere with either luteotropic and/or embryotropic effects of insulin-like growth factors.

Endothelin-1 and angiotensin II are luteolytic mediators in the bovine corpus luteum at days 8–12, but not at day 4 of the oestrous cycle. A. Miyamoto, K. Hayashi, S. Kobayashi and M. Ohtani, *Obihiro University of Agriculture and Veterinary Medicine, Obihiro 080, Japan*

Recent observations by ourselves and others suggest that luteal endothelin 1 (ET-1) interacts with prostaglandin $F_{2\alpha}$ ($PGF_{2\alpha}$) at luteolysis in the regressing bovine CL. The local role of ET-1 appears to be a direct inhibition of progesterone secretion as well as vasoconstriction of luteal arterioles. In this study, angiotensin II (ANG-II) was also found to be a similar luteolytic factor. We have examined the possible interaction of ET-1 and ANG-II with $PGF_{2\alpha}$ in the rapid suppression of progesterone release from the corpus luteum, by using an *in vitro* microdialysis system (MDS) with pieces of mid-cycle corpus luteum (days 8–12). Infusions with ANG-II (10^{-7}–10^{-5} mol l^{-1}) for two consecutive 2 h intervals induced a dose-dependent decrease in progesterone release (35–60% of baseline). When the corpus luteum explants were pre-perfused with $PGF_{2\alpha}$ (10^{-6} mol l^{-1}) for 2 h, the two consecutive perfusions of ANG-II (10^{-6} mol l^{-1}) also inhibited progesterone release (by 50%). Simultaneous infusion of ANG-II (10^{-6} mol l^{-1}), ET-1 (10^{-7} mol l^{-1}), or ANG-II+ET-1 with $PGF_{2\alpha}$ (10^{-6} mol l^{-1}) was more effective in suppressing production of progesterone (40% of baseline). The inhibitory effect of ANG-II combined with $PGF_{2\alpha}$ was blocked by infusion with an ANG-II antagonist. Infusion of $PGF_{2\alpha}$ for 2 h increased ANG-II release (200% of baseline). When young corpora lutea collected on day 4 were studied, a 30 min infusion of $PGF_{2\alpha}$ (10^{-5} mol l^{-1}), ET- 1 (10^{-7} mol l^{-1}), or ANG-II (10^{-6} or 10^{-5} mol l^{-1}) stimulated progesterone release (120–200% of baseline). When the corpora lutea explants were pre-perfused with $PGF_{2\alpha}$ for 30 min, a consecutive infusion of ET-1 or ANG-II (10^{-6} mol l^{-1}) for 30 min slightly increased progesterone release, but explants treated with ANG-II (10^{-5} mol l^{-1}) exhibited a dramatic increase in progesterone release during the experimental period (3 h thereafter). ANG-II was always detected in the MDS perfusates (10 pg ml^{-1} in young corpora lutea, 4 pg ml^{-1} in mid-cycle corpora lutea). These results suggest that luteal ANG-II and ET-1 interact with $PGF_{2\alpha}$ as local luteolytic mediators and that such local mechanisms are not functional in the young corpora lutea.

Expression of insulin-like growth factor (IGF) binding proteins -2, -3 and -4, and the type 1 IGF receptor mRNA in the bovine corpus luteum. K. J. Woad, D. G. Armstrong, G. Baxter, C. O. Hogg, T. A. Bramley[2] and R. Webb[1], *[1]Roslin Institute (Edinburgh), Roslin, Midlothian, EH25 9PS, UK; [2]Dept Obstetrics and Gynaecology, University of Edinburgh, EH3 9EW, UK; and [3]Dept Agriculture and Horticulture, Sutton Bonington Campus, University of Nottingham, Leicestershire, LE12 5RD, UK*

The insulin-like growth factors, IGF-I and IGF-II, have important effects on luteal function, including stimulation of the steroidogenic response to gonadotrophins. IGF action is mediated via interaction with specific cell surface receptors and is further modulated by the association of the IGFs with members of a family of IGF-binding proteins (IGFBPs). Since the IGFBPs affect the bioactivity of the IGFs and have been shown to attenuate their actions in all ovarian culture systems examined so far, we were interested in determining patterns of luteal expression of mRNA encoding the IGF-binding proteins and receptors. Expression of the type-1 IGF receptor and IGFBP-2, -3 and -4 mRNA was examined throughout the lifespan of the bovine corpus luteum using ribonuclease (RNase) protection assays and *in situ* hybridization. Bovine corpora lutea were obtained on days 5, 10, 15 and 19 of the oestrous cycle following synchronized oestrus. Corpora lutea were bisected and snap-frozen in liquid nitrogen prior to cryostat sectioning (14 µm) or total RNA extraction. Hybridization *in situ* and in solution utilised species-specific radiolabelled antisense and sense (control) RNA probes. RNase protection assays readily detected mRNA expression encoding the type-1 IGF receptor and IGFBP-3 and -4 in luteal RNA at all time points. Low levels of IGFBP-2 mRNA were also expressed. Expression of mRNA for the type-1 receptor and IGFBP-2, -3 and -4 showed limited temporal variation. *In situ* hybridization studies to determine the spatial pattern of expression for the binding proteins confirmed the presence of mRNA encoding IGFBP-2, -3 and -4 in the bovine corpus luteum. IGFBP-2 and -4 mRNA was detected at low levels throughout the CL, whilst mRNA encoding IGFBP-3 was associated predominantly with the luteal vasculature. In addition, a subset of large vessels in the periphery of the corpus luteum and stroma showed moderate to intense hybridization for IGFBP-2 mRNA. In summary, this study demonstrates bovine luteal expression of mRNAs for the type-1 IGF receptor and IGFBP-2, -3 and -4. Since local production of IGFs has been demonstrated previously these results support the hypothesis that the IGF system may have autocrine or paracrine actions in regulating the function of the corpus luteum in the cow.

Effect of dose of $PGF_{2\alpha}$ on steroidogenic components and oligonucleosomes in ovine luteal tissue. G. D. Niswender, J. L. Juengel, J. D. Haworth, P. J. Silva, M. K. Rollyson and E. McIntush, *Animal Reproduction and Biotechnology Laboratory, Colorado State University, Fort Collins, CO 80523, USA*

An experiment was conducted to test the hypothesis that a low dose of prostaglandin (PG) $F_{2\alpha}$ would have an antisteroidogenic effect in ovine luteal tissue, but would not induce oligonucleosome formation or loss of luteal tissue. On the basis of the results of preliminary experiments, ewes on day 9 or 10 of the oestrous cycle were administered 0, 3, 10 or 30 mg $PGF_{2\alpha}$ per 60 kg body weight by intrajugular injection. Concentrations of progesterone in sera, luteal weights, presence of oligonucleosomes and concentrations of mRNA encoding steroidogenic proteins were determined in luteal tissue collected 9 and 24 h after injection ($n = 5$ per dose per time). All doses of $PGF_{2\alpha}$ decreased ($P < 0.05$) concentrations of progesterone in sera compared with controls by 9 h after injection; however, ewes treated with 3 mg $PGF_{2\alpha}$ had concentrations of progesterone that were returning to control values at 24 h, and were higher ($P < 0.05$) than those in the 10 or 30 mg groups. Concentrations of progesterone in sera were correlated highly with luteal concentrations of mRNA encoding steroid acute regulatory protein (StAR) ($r = 0.74$; $P < 0.001$), cytochrome P450 side chain cleavage ($r = 0.40$; $P < 0.02$) and 3β-hydroxysteroid dehydrogenase ($r = 0.46$; $P < 0.01$). Corpora lutea collected from ewes treated with the 10 and 30 mg doses, but not the 3 mg dose of $PGF_{2\alpha}$ weighed less ($P < 0.05$) than those collected from saline-treated ewes. Oligonucleosomes, as determined by ethidium bromide staining of DNA following gel electrophoresis, were not present in luteal tissues collected from saline-treated ewes. Surprisingly, all doses of $PGF_{2\alpha}$ induced oligonucleosomes in a majority of animals at 9 h after treatment. At 24 h after injection, luteal tissue from the majority of ewes treated with 10 and 30 mg of $PGF_{2\alpha}$ contained oligonucleosomes; however, only one ewe treated with 3 mg $PGF_{2\alpha}$ had evidence of oligonucleosomes at 24 h. In conclusion, low doses of $PGF_{2\alpha}$ decreased secretion of progesterone, but did not induce luteolysis, whereas high doses of $PGF_{2\alpha}$ decreased secretion of progesterone and induced luteolysis. The induction of oligonucleosome formation was not a good indicator of tissue death in the corpus luteum. Concentrations of progesterone in sera were highly correlated with concentrations of mRNA encoding StAR; thus, downregulation of mRNA encoding StAR may be crucial for $PGF_{2\alpha}$-induced decreases in secretion of progesterone.

AUTHOR INDEX

SUBJECT INDEX